Pre
ALGEBRA

SHERRI MESSERSMITH
College of DuPage

NATHALIE M. VEGA-RHODES
Lone Star College—Kingwood

ROBERT S. FELDMAN
University of Massachusetts Amherst

With contributions from William C. Mulford, *McGraw-Hill Education*

Mc
Graw
Hill
Education

PREALGEBRA WITH P.O.W.E.R. LEARNING, SECOND EDITION

Published by McGraw-Hill Education, 2 Penn Plaza, New York, NY 10121. Copyright © 2018 by McGraw-Hill Education. All rights reserved. Printed in the United States of America. Previous editions © 2014. No part of this publication may be reproduced or distributed in any form or by any means, or stored in a database or retrieval system, without the prior written consent of McGraw-Hill Education, including, but not limited to, in any network or other electronic storage or transmission, or broadcast for distance learning.

Some ancillaries, including electronic and print components, may not be available to customers outside the United States.

This book is printed on acid-free paper.

1 2 3 4 5 6 7 8 9 LMN 21 20 19 18

ISBN 978–1–259–61029–5 (Student Bound Edition)
MHID 1–259–61029–2

ISBN: 978–1–260–37424–7 (Annotated Instructor's Edition)
MHID: 1–260–37424–6

ISBN: 978–1–260–37402–5 (Loose Leaf Edition)
MHID: 1–260–37402–5

Product Developer: *Luke Whalen*
Marketing Manager: *Noah Evans*
Content Project Manager: *Peggy Selle*
Buyer: *Sandy Ludovissy*
Design: *David W. Hash*
Content Licensing Specialist: *Lorraine Buczek*
Cover Image: *©percom/Shutterstock.com; wallpaper texture: ©leogri/Shutterstock.com*
Compositor: *SPi Global*

All credits appearing on page or at the end of the book are considered to be an extension of the copyright page.

Library of Congress Cataloging-in-Publication Data

Names: Messersmith, Sherri, author. | Vega-Rhodes, Nathalie, author. |
  Feldman, Robert S. (Robert Stephen), 1947- author. | Mulford, William C.,
  author.
Title: Prealgebra with P.O.W.E.R. learning / Sherri Messersmith (College of
  DuPage), Nathalie Vega-Rhodes (Lone Star College—Kingwood), Robert S.
  Feldman (University of Massachusetts Amherst); with contributions from
  William C. Mulford (McGraw-Hill Education).
Description: Second edition. | New York, NY : McGraw-Hill Education, [2018] |
  P.O.W.E.R. learning: Prepare, Organize, Work, Evaluate, and Rethink. |
  Includes index.
Identifiers: LCCN 2017052244 | ISBN 9781259610295 (alk. paper)
Subjects: LCSH: Mathematics—Textbooks. | Algebra—Textbooks.
Classification: LCC QA39.3 .M475 2018 | DDC 512.9—dc23
LC record available at https://lccn.loc.gov/2017052244

The Internet addresses listed in the text were accurate at the time of publication. The inclusion of a website does not indicate an endorsement by the authors or McGraw-Hill Education, and McGraw-Hill Education does not guarantee the accuracy of the information presented at these sites.

mheducation.com/highered

About the Authors

Sherri Messersmith
Professor of Mathematics, College of DuPage

©Phil Messersmith

Sherri Messersmith began teaching at the College of DuPage in Glen Ellyn, Illinois, in 1994 and has over 25 years of experience teaching many different courses from developmental mathematics through calculus. She earned a Bachelor of Science degree in the Teaching of Mathematics at the University of Illinois at Urbana-Champaign and taught high school for two years. Sherri returned to UIUC and earned a Master of Science degree in Applied Mathematics and stayed on at the university to teach and coordinate large sections of undergraduate math courses as well as teach in the Summer Bridge program for at-risk students. She is the author of many McGraw-Hill Education texts.

Sherri and her husband, Phil, are empty-nesters who recently relocated to the East Bay when her husband took at position at the University of California—Berkeley as a professor of Materials Science and Bioengineering. Sherri loves to cook, hang out with her dogs, study French, and travel; the manuscripts for this and her other books have accompanied her on her travels all over the globe.

Nathalie M. Vega-Rhodes
Professor of Mathematics, Lone Star College—Kingwood

©Nathalie M. Vega-Rhodes

Nathalie Vega-Rhodes' career in higher education began seventeen years ago and has encompassed a number of student-focused positions. For nearly a decade, she has taught mathematics ranging from developmental courses to calculus, as well as student success courses. She holds a Bachelor of Arts in Mathematics from the University of Houston and a Master of Science in Mathematics from the University of Houston—Clear Lake. In addition to teaching, as Mathematics Technology Coordinator at Lone Star College—Kingwood, she assists math faculty with technology-related pedagogical and implementation strategies. Her earliest work in higher education focused on academic support, first as a tutor and supplemental instruction (SI) leader, and then as a coordinator for a math tutoring and SI program. In her free time, Nathalie enjoys scuba diving and traveling with her husband, hanging out with her dog, and reading.

Robert S. Feldman
Deputy Chancellor and Professor of Psychological and Brain Sciences, University of Massachusetts—Amherst

©Robert S. Fedlman

Bob Feldman still remembers those moments of being overwhelmed when he started college at Wesleyan University. "I wondered whether I was up to the challenges that faced me," he recalls, "and although I never would have admitted it then, I really had no idea what it took to be successful at college."

That experience, along with his encounters with many students during his own teaching career, led to a life-long interest in helping students navigate the critical transition that they face at the start of their own college careers. Bob, who went on to graduate with High Honors from Wesleyan and receive a Doctorate in Psychology from the University of Wisconsin—Madison, teaches at the University of Massachusetts—Amherst, where he is the Deputy Chancellor and Professor of Psychological and Brain Sciences. He is founding director of the first-year experience course for incoming students at UMass and is Senior Fellow of the Center for Student Success Research.

Bob is a Fellow of the American Psychological Association, the American Association for the Advancement of Science, and the Association for Psychological Science. He has written and edited more than 250 scientific articles, book chapters, and books, including *P.O.W.E.R. Learning: Strategies for Success in College and Life,* 7e; *Understanding Psychology,* 14e; and *The First Year of College: Research, Theory, and Practice on Improving the Student Experience and Increasing Retention.* He is past president of the FABBS Foundation, an umbrella group of societies promoting the behavioral and brain sciences, and he is on the Board of Directors of New England Public Radio. Bob loves travel, music, and cooking. He and his wife live near the Holyoke mountain range in western Massachusetts.

The Story of the P.O.W.E.R. Textbooks

In the classroom, I viewed the textbook as merely a *guide* and did many other things on my own to better meet the needs of my students. For example, I taught in bite-sized pieces because developmental students in particular learn better when material is presented in more manageable chunks. As students' basic skills deteriorated, I created several Basic Skills Worksheets to help improve those skills. The students did them in class, and it only took up two or three minutes of precious class time. When McGraw-Hill Education saw some of the materials that I had made for my own classroom, they asked if I would write a textbook. So in 2004 I began writing my first book, *Beginning & Intermediate Algebra,* now in its 4th edition. The material for the book was written to align with how students learn best, and these strategies and activities were included in both the book and its accompanying supplements. Twenty-five years of teaching and writing that first book revealed that *developmental students don't want to fail; they just don't know how to succeed.*

Over the years, the books have evolved to include everything possible to give students *and* instructors the tools they need for success. Explicit student success skills were not included as part of the first several books. But it became more and more clear over time that students needed direct instruction in skills like how to *effectively* do homework, how to read a math textbook, and how to manage their time. I had been addressing all of this in my own classroom, and after meeting Bob Feldman, I began doing it more formally by using his research-based P.O.W.E.R. Learning framework. I asked myself, *"Why not incorporate P.O.W.E.R. into the textbooks to both teach study skills and organize the material according to the way research says students learn best?"* Happily, Bob agreed to come on as a coauthor to give us the math textbooks with P.O.W.E.R. Learning that we have today.

Even with this major evolution in our approach, *we made sure not to compromise the math whatsoever*! The math content and its level of rigor remains the same as it was in the books written before we added the P.O.W.E.R. framework—so the fact that our books contain student success skills does *not* mean they are lacking in rigor. These books are light-years ahead of the earlier ones because we have now addressed a huge weakness of many college students: *knowing how to learn.*

With these 2nd edition Math with P.O.W.E.R. Learning books, we have evolved even more. The power (no pun intended) of the Internet is that it allows us to enhance the learning and teaching experience for students and instructors. We can do things that we couldn't dream of even 10 years ago! That's where Nathalie Vega-Rhodes comes in. A long-time P.O.W.E.R. math textbook user and outstanding instructor, Nathalie has a lot of experience using digital tools. We share the same philosophy of teaching, love for students, and belief in addressing *all* of our students' needs in the classroom, so why not bring her on board as a coauthor with digital expertise? So that's what we did. An immediate benefit was the creation of the Integrated Video & Study Guides, or IVSGs, new to this edition. It was a perfect collaboration; I wrote the IVSGs, and Nathalie made the new online videos.

Our team is complete, but this isn't the end; our story is not finished. The books continue to evolve and improve because of engaged faculty like you. We are so happy that you have chosen our textbook and want to continue the conversation. We would love to hear from you. Tell us your stories, and share your suggestions. Be a part of our story and part of the evolution of the P.O.W.E.R. math textbooks.

Sherri Messersmith (sherri.messersmith@gmail.com)
Nathalie M. Vega-Rhodes (nvegarhodes@gmail.com)
Bob Feldman (feldman@chancellor.umass.edu)

Why should you use *Prealgebra with P.O.W.E.R. Learning?*

- **Solid, time-tested math content** with the amount of rigor needed to succeed in college-level courses
- **Written with friendly, conversational, non-intimidating language,** making it easier to read than most books while using all of the necessary mathematical language our students need
- **Written in bite-sized pieces** to make it easier to learn "complicated" material
- **Engaging applications** written with students and their interests in mind
- **Rewritten, easy-to-use, research-based student success materials** in every chapter *in the book* that instructors can use at their discretion

- **Research-based pedagogical features throughout the books**
  - **"Get Ready" exercises:** Located at the beginning of targeted exercise sets, these problems give students the chance to practice prerequisite skills before diving into the exercises on the learning objectives in the section. This helps students make connections between what they have already learned and what is new.
  - **24-hr problems:** Marked in each exercise set, these are indicated by an icon. Even if students do not complete their *entire* homework assignment immediately, they are encouraged to do the 24-hr problems within 24 hours of leaving class so that they better retain what they have learned.
  - **"You Try" exercises:** After almost every example in the book, students can do a You Try problem to work out a problem that is similar to what was presented.
  - **P.O.W.E.R. framework:** Since the 1st editions of our math with P.O.W.E.R. Learning textbooks, this research-based framework has informed the organization of chapters and sections to match the way that research shows students learn best.
  - **Rethink questions:** This is a *crucial* step in the learning process, and the "R" in P.O.W.E.R., yet it is the one that students overlook the most (or don't even realize exists). Every exercise set is followed by a set of Rethink questions that require students to reflect upon what they have just done. These questions are also assignable in ConnectMath Hosted by ALEKS.
  - **Student success skills, emPOWERme, and Study Strategies:** Every chapter has a study skills theme with an emPOWERme survey and a Study Strategies page *in the book* so that instructors can address students' weaknesses in this area. They are based upon Bob Feldman's research, and can be done in any order. If instructors do not want to include them in their courses, they can be skipped easily.
- *Time-saving supplements, especially for adjunct instructors:* Many adjuncts teach at more than one school, and they don't always have time to make materials specific to their classes and their students' needs. We offer many author-created supplements with every textbook.
  - **Basic Skills Worksheets**—Help students improve their basic skills in class while using only 2 or 3 minutes of class time. Their confidence improves, too, when they see their basic skills improve. They start to believe that they *can* learn math!
  - **Section Worksheets**
  - **Worksheets to Tie Multiple Concepts Together**
  - **Guided Student Notes**
  - *NEW!* **Integrated Video & Study Guides (IVSGs)**—Great for flipped classrooms or any class, these require students to actively engage with the videos not only for doing examples but also for filling in information for elements such as definitions and procedures. Students *must* pay attention to complete these guides. The IVSGs are written for every *objective* in every section and designed so that most videos are 3 or 4 minutes long. The videos, and exercises from the video guides aligned to them, are assignable in ConnectMath Hosted by ALEKS.
  - **PowerPoints**
- **Enhanced exercise sets** with more conceptual questions
- **Videos for selected homework exercises**
- **Putting It All Together** sections in chapters where it is appropriate
- **Group Activities** in every chapter
- **Instructor Resource Manual**
- **ConnectMath Hosted by ALEKS and ALEKS**

New to the 2nd Edition P.O.W.E.R. Series Books

- Approximately 1,000 new algorithmic exercises have been developed for the series within ConnectMath Hosted by ALEKS across the Prealgebra, Introductory Algebra, and Intermediate Algebra courses.
- Nearly 800 new online lecture videos for the series have been made for the eBooks in both ConnectMath Hosted by ALEKS and ALEKS, all of which are created and narrated by the authors themselves. These videos cover all Learning Objectives in the main print books and eBooks.
- A brand new workbook resource, the Integrated Video & Study Guide Workbook, has been created to be used in conjunction with the new online lecture videos, with a separate video guide provided for each new video. Practice exercises drawn from these video guides are assignable in ConnectMath Hosted by ALEKS.

Table of Contents

Instructor **P O W E R** Tool Kit

Find these resources in your P.O.W.E.R. Tool Kit

In the textbook

- Work Hints*
- In-Class Examples*
- You Trys*
- Get Ready Exercises*
- 24-hr Problems
- Enhanced Exercise Sets
- Putting It All Together sections*
- Group Activities
- Study Strategies and emPOWERme in each chapter*

Supplements

- Basic Skills Worksheets*
- Section Worksheets*
- Worksheets to Tie Multiple Concepts Together*
- Guided Student Notes*
- Comprehensive video package, including Integrated Video & Study Guides and videos for selected exercises*
- Power Points
- Instructor Solution Manual
- Computerized Test Bank and files
- Instructor Resource Manual**
- ConnectMath Hosted by ALEKS and ALEKS*

*Descriptions of these resources are included in the following pages.
**Additional information about how to use the resources can be found in the Instructor Resource Manual.

WORK HINTS highlight important steps in working out a problem, point out places of common student errors, or give a study tip for learning. Pulling these tips out of the main text makes them more noticeable to students.

W Hint
Try using graph paper to line up the numbers correctly.

$$\begin{array}{r} 6.2 \\ +\ 5.8\ 3 \\ \uparrow \end{array}$$
Line up the
decimal points.

$$\begin{array}{r} {\scriptstyle 1} \\ 6.2\ 0 \\ +\ 5.8\ 3 \\ \hline 1\ 2.0\ 3 \\ \uparrow \end{array}$$ ← Insert 0 in the hundredths place.
Line up the decimal point in the answer with the decimal points in the problem.

The **IN-CLASS EXAMPLES** exactly mirror the examples in the book, giving instructors additional problems to use while teaching in class. They are available only in the Annotated Instructor Edition, and they align with the examples in the **Guided Student Notes**.

EXAMPLE 4

Use an equation to solve each problem.

a)   7 is 10% of what number?   b) $3\frac{1}{2}\%$ of what number is 56?

Solution

a)   Let x represent the unknown value.

$$x = \text{the number}$$

YOU TRY problems follow almost every example in the book and exactly mirror the examples. After working through problems in class, the instructor can have the students do a **You Try** to practice on their own to reinforce the lesson. The **Examples**, **In-Class Examples**, and **You Try** problems are the same *types* of problems containing different numbers, providing consistency for students *and* instructors.

YOU TRY 9

Identify each as an expression or an equation. If it is an expression, find the sum. If it is an equation, solve it.

a) $\dfrac{2c}{7} + \dfrac{10}{21}$ b) $\dfrac{2c}{7} + \dfrac{10}{21} = \dfrac{c}{3}$

The **PUTTING IT ALL TOGETHER** sections tie multiple concepts together and take students through the thought processes necessary for problem recognition. For example, after learning and gaining practice using geometry vocabulary, properties, and formulas *individually,* the *Putting It All Together* section presents students with explanations and exercises with all of the concepts together.

Putting It All Together

P Prepare   **O Organize**

What are your objectives?	How can you accomplish each objective?
1 Review the Concepts of Sections 4.1–4.4	• Know the vocabulary associated with lines, segments, rays, and angles. • Be able to apply the properties and formulas associated with rectangles, squares, parallelograms, trapezoids, and triangles. • Be able to find the volume and surface area of the figures in Section 4.4. • Complete the given examples on your own. • Complete You Trys 1–3.

Video Package: Integrated Video and Study Guides as well as Exercise Videos

Would you like to flip your classroom or offer students a structured learning environment outside the classroom? The **Integrated Video and Study Guides,** IVSGs, allow instructors to do just that. New to this edition of the P.O.W.E.R. textbooks, the authors have created one **Integrated Video and Study Guide** for *each objective* in the book. Students are required to watch the video and fill out the **IVSG** as they go along, stopping along the way to do the similar *Your Turn* problems and *Think About It* conclusion questions designed for deeper conceptual thinking. Each companion video follows the procedures and examples, with detailed explanations aligning perfectly with the textbook. In addition to IVSG companion videos, hundreds of 3–5 minute **Exercise Video** clips show students how to solve various exercises from the textbook. All videos, as well as study strategies, are assignable in ConnectMath Hosted by ALEKS. Instructors also have the option to include the P.O.W.E.R. framework in assignments.

▶ *Play.* **Write out the examples as you watch the video.**

So far, we have learned how to solve equations with only one variable term. How do we solve an equation with *more than one* variable term? We can generalize the steps for solving linear equations with the following procedure.

> **Procedure How to Solve a Linear Equation**
>
> *Step 1:* Clear _____ and _____ like terms on each side of the equation.
>
> *Step 2:* Get the _____ on one side of the equal sign and the _____ on the other side of the equal sign (isolate the _____) using the addition or subtraction property of equality.
>
> *Step 3:* _____ for the variable using the multiplication or division property of equality.
>
> *Step 4:* _____ the solution in the original equation.

Example 1: Solve $15 - 7u - 6 + 2u = -1$.

⏸ *Pause and do Your Turn* 1: Solve $5 - 8k - 9 + 5k = 32$.

Procedure: How to Solve a Linear Equation

Step 1: Clear **parentheses** and **combine** like terms on each side of the equation.

Step 2: Get the **variable** on one side of the equal sign and the **constant** on the other side of the equal sign (isolate the **variable**) using the addition or subtraction property of equality.

Step 3: **Solve** for the variable using the multiplication or division property of equality.

Step 4: **Check** the solution in the original equation.

> division property of equality. Step 4: Check the solution in the original equation. Example 1: Solve 15-7u-6+2u=-1. Step 1: Because there

Example 1: Solve $15 - 7u - 6 + 2u = -1$.

$$15 - 7u - 6 + 2u = -1$$

$$-5u + 9 = -1 \qquad \text{Combine like terms.}$$

$$-5u + 9 - 9 = -1 - 9 \qquad \text{Subtract 9 from each side.}$$

$$-5u = -10 \qquad \text{Combine like terms.}$$

$$\frac{-5u}{-5} = \frac{-10}{-5} \qquad \text{Divide each side by } -5.$$

$$u = 2 \qquad \text{Simplify.}$$

> the division property of equality. Step 4: Check. The solution set is {2}. Now it's your turn. Pause the video and solve. Example 2:

Student Success Skills in Every Chapter

Start the study skills discussion by having your students do the emPOWERme activity. Found in *every chapter* before the Chapter Summary, most **emPOWERme** activities take the form of a survey so that students can learn something about themselves with respect to the student success skill of the chapter. Follow this up with the **Study Strategies** found at the beginning of the chapter to help your students use the P.O.W.E.R. framework to acquire skills such as learning about their school, reading a math textbook, doing their homework effectively, or managing their time.

em POWER me My School

Every school—whether it's a high school, community college, college, or university—operates under its own set of rules and procedures. Understanding how your school works and where to go for help are essential parts of being successful in college. It's important to understand how your school works so that, for example, you know where and when to turn in your financial aid application and you know where to get help if you have questions about choosing the classes you need for graduation. Take this survey to learn how well you know your school. Check all boxes that apply.

- ☐ I know the address of my school's website.
- ☐ I can navigate the school's website to find most information that I need.
- ☐ I am aware of whether my school has a handbook containing useful information.
- ☐ I have signed up to receive emergency campus messages by email, text, or automated phone call.
- ☐ I am aware of important dates such as when to register for classes, when tuition is due, and when financial aid forms are due.
- ☐ I know where to register for classes on campus.
- ☐ On campus, I know where to ask questions about financial aid.
- ☐ I can locate the bookstore.
- ☐ I know the location of the library.
- ☐ I know the difference between an adviser and a counselor.
- ☐ I know the location of the campus health center.
- ☐ I know the location of student services offices that might be of interest to me. Some examples are veterans' support services, the office to help students with disabilities, and child care.
- ☐ I know the locations of all of my instructors' offices as well as their office hours.
- ☐ I know the location of the tutoring center/math lab, and I know their procedures for getting help when I need it.
- ☐ I can locate the Testing Center and know its rules and hours of operation.
- ☐ I am aware of clubs, organizations, and activities on campus, and I know where to go to become involved in those that interest me.
- ☐ I know the location of the office where I can go if I have questions about or want help finding a job.

Think about the items that you have, and have *not*, checked in this survey. Which apply to you and might contribute to your success in college? In the Study Strategies at the beginning of this chapter, you will learn how to get to know your school.

Every Chapter Has a Student Success Theme

Find the **STUDY STRATEGIES** on the second page of every chapter. After doing the **emPOWERme** activity found before the Chapter Summary, continue the discussion of the chapter's student success skill by assigning or discussing the **Study Strategies.** Based on Bob Feldman's research, the **Study Strategies** explain, in easy-to-understand language, how to use each step of the P.O.W.E.R. framework to acquire student success skills such as getting to know your school, how to read a math textbook, and how to be a good time manager. The student success topic of each chapter is listed in the Table of Contents, and the topics can be done in any order.

Supplements Include a Suite of Ready-Made Worksheets

Want materials to use in class but don't have a lot of time to make them? Our package includes three types of author-created worksheets: **Basic Skills Worksheets, Section Worksheets,** and **Worksheets to Tie Multiple Concepts Together.** All are available in the student version (without answers) and instructor version (with answers), and *all worksheets are available as PDF or Word files so that instructors can download them and edit them as if they were their own Word documents.*

Worksheet 1A
MVF – PreAlgebra   Name: _____

1)	8·3	_____		16)	5·8	_____
2)	4·9	_____		17)	12·7	_____
3)	12·5	_____		18)	6·4	_____
4)	7·3	_____		19)	3·8	_____
5)	4·4	_____		20)	8·9	_____
6)	10·6	_____		21)	12·2	_____
7)	11·5	_____		22)	7·1	_____
8)	6·9	_____		23)	3·11	_____
9)	4·7	_____		24)	2·4	_____
10)	8·12	_____		25)	5·4	_____
11)	5·9	_____		26)	9·7	_____
12)	9·2	_____		27)	11·11	_____
13)	7·8	_____		28)	6·7	_____
14)	3·4	_____		29)	12·6	_____
15)	8·0	_____		30)	9·9	_____

BASIC SKILLS WORKSHEETS

The **Basic Skills Worksheets** help students improve their basic skills while using only a few minutes, sometimes only 2 or 3, of class time. Their confidence improves, too, when they see their basic skills improve. Use these *before* reaching the topic where the basic skill is needed so that students can strengthen their weaknesses and be better prepared to learn new content. Each type of **Basic Skills Worksheet** comes in six different versions, incorporating more difficult concepts as they move from version A to version F.

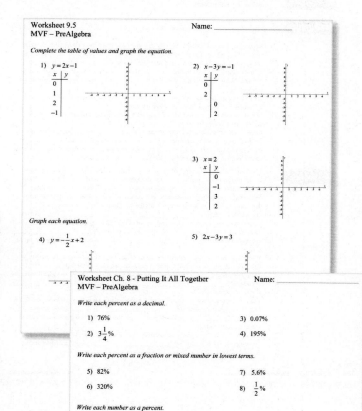

These author-created, ready-made **Section Worksheets** and **Worksheets to Tie Multiple Concepts Together** can be used for extra practice in class, with students working individually or in a group, or can be used for students to take home. They come in student versions (without answers) and instructor versions (with answers).

SECTION WORKSHEETS

Every section of *every book* comes with at least one **Section Worksheet** to help instructors teach new content or to give students extra practice problems. The **Section Worksheets** *exactly match* the material in the section and are a great way to standardize instruction across a department.

WORKSHEETS THAT TIE MULTIPLE CONCEPTS TOGETHER

When appropriate, there are **Worksheets That Tie Multiple Concepts Together** to help students with problem recognition and differentiation. These worksheets allow students to practice multiple concepts together after having learned them individually.

Help Students Learn How to Take Good Notes

GUIDED STUDENT NOTES

Use the **Guided Student Notes** to help your students become better note-takers. Every section of the book has a ready-made corresponding **Guided Student Note** that mirrors the material in the section. These are "skeleton note outlines" that help students learn how to structure their notes. Because the structure is already given to students, they don't have to write down *everything* that the instructors write in class so that students can concentrate better on what they are learning.

The **Guided Student Notes** save instructors time and help standardize the material that is taught across a math department. They come in both a P.O.W.E.R. format and a standard format and include answer keys. *The Guided Student Notes are available as PDF or Word files so that instructors can download them and edit them as if they were their own Word documents.*

Guided Student Notes
MVF - PreAlgebra

Definition of Like Fractions   **Definition of Unlike Fractions**

Steps for Adding and Subtracting Like Fractions

Add or Subtract.

2) $\dfrac{6}{13} + \dfrac{2}{13}$ 5) $\dfrac{5}{16} - \dfrac{13}{16}$

3) $\dfrac{7}{5} - \dfrac{4}{5}$ 6) $\dfrac{2}{15n} + \dfrac{8}{15n}$

4) $-\dfrac{1}{12} + \dfrac{11}{12}$

Steps for Writing a Fraction with a Different Denominator

7) Write $\dfrac{2}{5}$ as an equivalent fraction with a denominator of 10.

Guided Student Notes   Name:_____
MVF – PreAlgebra

3.4 Adding and Subtracting Like Fractions and Finding a Least Common Denominator

Prepare   What are my goals for this section?
_____

Organize   What am I going to do to accomplish these goals?
_____

Work

Steps for Drawing a Figure to Represent a Fraction

1) Add $\dfrac{3}{5} + \dfrac{1}{5}$:

a) by drawing a figure to represent each fraction.

b) using a number line.

McGraw Hill Education

connect MATH
HOSTED BY ALEKS

Looking for a consistent voice between text and digital? Problem solved!

McGraw-Hill Connect® Math Hosted by ALEKS® offers everything students and instructors need in one intuitive platform. ConnectMath is an online homework engine where the problems and solutions are consistent with the textbook authors' approach. It also offers integration with SmartBook, an assignable, adaptive eBook and study tool that directs students to the content they don't know and helps them study more efficiently. With ConnectMath, you get the tools you need to be the teacher you want to be.

©Steve Debenport/Getty Images

> "I like that ConnectMath reaches students with different learning styles . . . our students are engaged, attend class, and ask excellent questions."
> – Kelly Davis, South Texas College

Trusted Service and Support

A dedicated team of specialists and faculty consultants ensure that your ConnectMath implementation is seamless and painless . . . from start to finish.

ConnectMath Service and Support Offers:
- LMS integration that provides single sign-on capability and gradebook synchronization
- Industry-leading technical support and 99.97% uptime
- Resources for implementation, first day of class orientation, how-to videos and more
- Onsite seminars/worshops and webinars with McGraw-Hill and faculty consultants

How can ConnectMath help solve your students' challenges?

I like to learn by _____.

Whether it's reading, watching, discovering, or doing, ConnectMath has something for everyone. Instructors can create assignments that accommodate different learning styles, and students aren't stuck with boring multiple-choice problems. Instead they have a myriad of motivational learning and media resources at their fingertips. SmartBook delivers an interactive reading and learning experience that provides personalized guidance and just-in-time remediation. This helps students to focus on what they need, right when they need it.

I still don't get it. Can you do that problem again?

Because the content in ConnectMath is author-developed and goes through a rigorous quality control process, students hear one voice, one style, and don't get lost moving from text to digital. The high-quality, author-developed videos provide students ample opportunities to master concepts and practice skills that they need extra help with . . . all of which are integrated in the ConnectMath platform and the eBook.

How can ConnectMath help solve your classroom challenges?

I need meaningful data to measure student success!

From helping the student in the back row to tracking learning trends for your entire course, ConnectMath delivers the data you need to make an impactful, meaningful learning experience for students. With easy-to-interpret, downloadable reports, you can analyze learning rates for each assignment, monitor time on task, and learn where students' strengths and weaknesses are in each course area.

We're going with the _____ (flipped classroom, corequisite model, etc.) implementation.

ConnectMath can be used in any course setup. Each course in ConnectMath comes complete with its own set of text-specific assignments, author-developed videos and learning resources, and an integrated eBook that cater to the needs of your specific course. The easy-to-navigate home page keeps the learning curve at a minimum, but we still offer an abundance of tutorials and videos to help get you and your colleagues started.

©McGraw-Hill Education

Looking to motivate and engage students? Problem solved!

ALEKS® uses artificial intelligence to precisely map what each student knows, doesn't know, and is most ready to learn in a given course area. The system interacts with each student like a skilled human tutor, delivering a cycle of highly individualized learning and assessment that ensures mastery. Students are shown an optimal path to success, and instructors have the analytics they need to deliver a data-informed, impactful learning experience.

> "ALEKS has helped to create the best classroom experience I've had in 36 years. I can reach each student with ALEKS."
> — *Tommy Thompson, Cedar Valley College, TX*

How can ALEKS help solve your students' challenges?

I did all my homework, so why am I failing my exams?

The purpose of homework is to ensure mastery and prepare students for exams. ALEKS is the only adaptive learning system that ensures mastery through periodic assessments and delivers just-in-time remediation to efficiently prepare students. Because of how ALEKS presents lessons and practice, students learn by understanding the core principle of a concept rather than just memorizing a process.

I'm too far behind to catch up. - OR - I've already done this, I'm bored.

No two students are alike. So why start everyone on the same page? ALEKS diagnoses what each student knows and doesn't know, and prescribes an optimized learning path through your curriculum. Students only work on topics they are ready to learn, and they have a proven learning success rate of 93% or higher. As students watch their progress in the ALEKS Pie grow, their confidence grows with it.

How can ALEKS help solve your classroom challenges?

I need something that solves the problem of cost, time to completion, and student preparedness.

ALEKS is the perfect solution to these common problems. It provides an efficient path to mastery through its individualized cycle of learning and assessment. Students move through the course content more efficiently and are better prepared for subsequent courses. This saves both the institution and the student money. Increased student success means more students graduate.

My administration and department measure success differently. How can we compare notes?

ALEKS offers the most comprehensive and detailed data analytics on the market. From helping the student in the back row to monitoring pass rates across the department and institution, ALEKS delivers the data needed at all levels.

The customizable and intuitive reporting features allow you and your colleagues to easily gather, interpret, and share the data you need, when you need it.

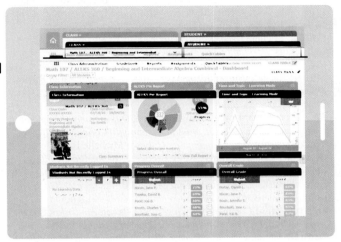

> **The ALEKS Instructor Module offers a modern, intuitive interface to manage courses and track student progress.**

Trusted Service and Support

A unique solution requires unique support. A dedicated team of specialists and faculty consultants ensure that your ALEKS implementation is seamless and painless . . . from start to finish.

ALEKS Service and Support Offers:
- LMS integration that provides single sign-on capability and gradebook synchronization
- Industry-leading technical support and 99.97% uptime
- Flexible courses that can align with any textbock and/or resources, for any classroom model
- Resources for implementation, first day of class orientation, how-to videos and more
- Onsite seminars/worshops and webinars with McGraw-Hill and faculty consultants

Acknowledgments

Manuscript Reviewers and Focus Group Participants

Thank you to all of the dedicated instructors who reviewed manuscript, participated in focus groups, and provided thoughtful feedback throughout the development of the *P.O.W.E.R.* series.

Darla Aguilar, *Pima Community College;* Scott Albert, *College of DuPage;* Bhagirathi Anand, *Long Beach City College;* Raul Arana, *Lee College;* Jan Archibald, *Ventura College;* Morgan Arnold, *Central Georgia Technical College;* Christy Babu, *Laredo Community College;* Michele Bach, *Kansas City Kansas Community College;* Kelly Bails, *Parkland College;* Vince Bander, *Pierce College, Pullallup;* Kim Banks, *Florence Darlington Technical College;* Michael Bartlett, *University of Wisconsin—Marinette;* Sarah Baxter, *Gloucester County College;* Michelle Beard, *Ventura College;* Annette Benbow, *Tarrant County College, Northwest;* Abraham Biggs, *Broward College;* Leslie Bolinger Horton, *Quinsigamond Community College;* Jessica Bosworth, *Nassau Community College;* Joseph Brenkert, *Front Range Community College;* Michelle Briles, *Gloucester County College;* Kelly Brooks, *Daytona State College (and Pierce);* Connie Buller, *Metropolitan Community College;* Rebecca Burkala, *Rose State College;* Gail Burkett, *Palm Beach State College;* Gale Burtch, *Ivy Tech Community College;* Jennifer Caldwell, *Mesa Community College;* Edie Carter, *Amarillo College;* Allison Cath, *Ivy Tech Community College of Indiana, Indianapolis;* Dawn Chapman, *Columbus Tech College;* Chris Chappa, *Tyler Junior College;* Charles Choo, *University of Pittsburgh at Titusville;* Patricia Clark, *Sinclair Community College;* Judy Kim Clark, *Wayne Community College;* Karen Cliffe, *Southwestern College;* Sherry Clune, *Front Range Community College;* Ela Coronado, *Front Range Community College;* Heather Cotharp, *West Kentucky Community & Tech College;* Danny Cowan, *Tarrant County College, Northwest;* Susanna Crawford, *Solano College;* George Daugavietis, *Solano Community College;* Joseph De Guzman, *Norco College;* Michaelle Downey, *Ivy Tech Community College;* Dale Duke, *Oklahoma City Community College;* Rhonda Duncan, *Midlands Technical College;* Marcial Echenique, *Broward College;* Sarah Ellis, *Dona Ana Community College;* Onunwor Enyinda, *Stark State College;* Chana Epstein, *Sullivan County Community College;* Karen Ernst, *Hawkeye Community College;* Stephen Ester, *St. Petersburg College;* Rosemary Farrar, *Southern West Virginia Community & Technical College;* John Fay, *Chaffey College;* Stephanie Fernandes, *Lewis and Clark Community College;* James Fiebiger, *Front Range Community College;* Angela Fipps, *Durham Technical Community College;* Jennifer Fisher, *Caldwell Community College & Technical Institute;* Elaine Fitt, *Bucks County Community College;* Carol Fletcher, *Hinds Community College;* Claude Fortune, *Atlantic Cape Community College;* Marilyn Frydrych, *Pikes Peak Community College;* Robert Fusco, *Broward College;* Jennifer Ganowsky, *Southern Utah University;* Jared Ganson, *Nassau Community College;* Kristine Glasener, *Cuyahoga Community College;* Ernest Gobert, *Oklahoma City Community College;* Linda Golovin, *Caldwell College;* Suzette Goss, *Lone Star College Kingwood;* Sharon Graber, *Lee College;* Susan Grody, *Broward College;* Leonard Groeneveld, *Springfield Tech Community College;* Joseph Guiciardi, *Community College of Allegheny County;* Susanna Gunther, *Solano College;* Lucy Gurrola, *Dona Ana Community College;* Frederick Hageman, *Delaware Technical & Community College;* Tamela Hanebrink, *Southeast Missouri State University;* Deborah Hanus, *Brookhaven College;* John Hargraves, *St. John's River State College;* Suzanne Harris-Smith,

Central New Mexico Community College; Michael Helinger, *Clinton Community College;* Mary Hill, *College of DuPage;* Jody Hinson, *Cape Fear Community College;* Kayana Hoagland, *South Puget Sound Community College;* Tracey Hollister, *Casper College;* Wendy Houston, *Everett Community College;* Mary Howard, *Thomas Nelson Community College;* Lisa Hugdahl, *Milwaukee Area Tech College—Milwaukee;* Larry Huntzinger, *Western Oklahoma State College;* Manoj Illickal, *Nassau Community College;* Sharon Jackson, *Brookhaven College;* Lisa Jackson, *Black River Technical College;* Christina Jacobs, *Washington State University;* Gretta Johnson, *Amarillo College;* Lisa Juliano, *El Paso Community College, Northwest Campus;* Elias M. Jureidini, *Lamar State College/Orange;* Ismail Karahouni, *Lamar University;* Cliffe Karen, *Southwestern College;* David Kater, *San Diego City College;* Joe Kemble, *Lamar University;* Joanne Kendall, *Lone Star College—CyFair;* Esmarie Kennedy, *San Antonio College;* Ahmed Khago, *Lamar University;* Michael Kirby, *Tidewater Community College VA Beach Campus;* Corrine Kirkbride, *Solano Community College;* Mary Ann Klicka, *Bucks County Community College;* Alex Kolesnik, *Ventura College;* Tatyana Kravchuk, *Northern Virginia Community College;* Randa Kress, *Idaho State University;* Julianne Labbiento, *Lehigh Carbon Community College;* Robert Leifson, *Pierce College;* Greg Liano, *Brookdale Community College;* Charyl Link, *Kansas City Kansas Community College;* Cassondra Lochard, *California State University—Dominguez Hills;* Wanda Long, *Saint Charles County Community College;* Lorraine Lopez, *San Antonio College;* Luke Mannion, *St. John's University;* Shakir Manshad, *New Mexico State University;* Robert Marinelli, *Durham Technical Community College;* Lydia Matthews-Morales, *Ventura College;* Melvin Mays, *Metropolitan Community College (Omaha NE);* Carrie McCammon, *Ivy Tech Community College;* Milisa Mcilwain, *Meridian Community College;* Valerie Melvin, *Cape Fear Community College;* Christopher Merlo, *Nassau Community College;* Leslie Meyer, *Ivy Tech Community College/Central Indiana;* Beverly Meyers, *Jefferson College;* Laura Middaugh, *McHenry County College;* Karen Mifflin, *Palomar College;* Kris Mudunuri, *Long Beach City College;* Sharon Muehlbacher, *California State Polytechnic University—Pomona;* Donald Munsey, *Louisiana Delta Community College;* Randall Nichols, *Delta College;* Joshua Niemczyk, *Saint Charles County Community College;* David Stumpf, *Lakeland Community College;* Katherine Ocker Stone, *Tusculum College;* Karen Orr, *Roane State;* Staci Osborn, *Cuyahoga Community College;* Steven Ottmann, *Southeast Community College, Lincoln Nebraska;* William Parker, *Greenville Technical College;* Joanne Peeples, *El Paso Community College;* Paul Peery, *Lee College;* Betty Peterson, *Mercer County Community College;* Carol Ann Poore, *Hinds Community College;* Hadley Pridgen, *Gulf Coast State College;* William Radulovich, *Florida State College @ Jacksonville;* Lakshminarayan Rajaram, *St. Petersburg College;* Kumars Ranjbaran, *Mountain View College;* Darian Ransom, *Southeastern Community College;* Nimisha Raval, *Central Georgia Technical College;* Amy Riipinen, *Hibbing Community College;* Janet Roads, *Moberly Area Community College;* Marianne Roarty, *Metropolitan Community College;* Jennifer Robb, *Scott Community*

College; Marie Robison, *McHenry County College;* Daphne Anne Rossiter, *Mesa Community College;* Anna Roth, *Gloucester County College;* Daria Santerre, *Norwalk Community College;* Kala Sathappan, *College of Southern Nevada;* Patricia Schubert, *Saddleback College;* William H. Shaw, *Coppin State University;* Azzam Shihabi, *Long Beach City College;* Jed Soifer, *Atlantic Cape Community College;* Lee Ann Spahr, *Durham Technical Community College;* Marie St. James, *Saint Clair County Community College;* Mike Stack, *College of DuPage;* Ann Starkey, *Stark State College of Technology;* Thomas Steinmann, *Lewis and Clark Community College;* Claudia Stewart, *Casper College;* Kim Taylor, *Florence Darlington Technical College;* Laura Taylor, *Cape Fear Community College;* Janet Teeguarden, *Ivy Tech Community College;* Janine Termine, *Bucks County Community College;* Yan Tian, *Palomar College;* Lisa Tolliver, *Brown Mackie South Bend;* David Usinski, *Erie Community College;* Hien Van Eaton, *Liberty University;* Diane Veneziale, *Rowan College at Burlington County;* Theresa Vecchiarelli, *Nassau Community College;* Val Villegas, *Southwestern College;* David Walker, *Hinds Community College;* Ursula Walsh, *Minneapolis Community & Tech College;* Dottie Walton, *Cuyahoga Community College;* LuAnn Walton, *San Juan College;* Thomas Wells, *Delta College;* Kathryn Wetzel, *Amarillo College;* Marjorie Whitmore, *North West Arkansas Community College;* Sandra Wildfeuer, *University of Alaska—Fairbanks;* Ross Wiliams, *Stark State College of Technology;* Gerald Williams, *San Juan College;* Michelle Wolcott, *Pierce College, Puyallup;* Mary Young, *Brookdale Community College;* Loris Zucca, *Lone Star College, Kingwood;* Michael Zwilling, *University of Mount Union*

Digital Contributors

Special thanks go to the faculty members who contributed their time and expertise to the digital offerings with *P.O.W.E.R.*

Jennifer Caldwell, *Mesa Community College*
Chris Chappa, *Tyler Junior College*
Tim Chappell, *MCC Penn Valley Community College*
Kim Cozean, *Saddleback College*
Katy Cryer
Cindy Cummins, *Ozarks Technical Community College*
Rob Fusco, *Bergen Community College*
Vicki Garringer, *College of DuPage*
Amy Hoherz, *Lone Star College*
Brian Huyvaert, *University of Oregon*
Sharon Jackson, *Brookhaven College*

Kelly Jackson, *Camden County College*
Theresa Killebrew, *Mesa Community College*
Corrine Kirkbride, *Solano Community College*
Brianna Ashley, *Daytona State College*
Jamie Manche, *Southwestern Illinois College*
Amy Naughten
Christy Peterson, *College of DuPage*
Melissa Rossi, *Southwestern Illinois College*
Lisa Rombes, *Washtenaw Community College*
Janine Termine, *Bucks County Community College*
Linda Schott, *Ozarks Technical Community College*

From the Authors

The authors would like to thank everyone at McGraw-Hill Education who helped make the publication of this book a reality.

In addition, words cannot adequately express our appreciation to Vicki Garringer and Amy Jo Hoherz. You are an integral part of our team, and your ability to turn work around quickly and accurately is nothing short of marvelous. You are truly rock stars! Special thanks also go out to Tim Chappell and Lisa Rombes for the patience and attention to detail you each demonstrated while helping us develop the digital side of our books. Working with both of you is always a pleasure.

From Sherri Messersmith: Thank you to my husband, Phil, and daughters, Alex and Cailen, for their support and for providing inspiration for applications throughout the books over the years. Shout out to the baristas at Philz on Shattuck, my new hometown coffee shop, for your excellent coffee-making skills and for letting me sit and work in the groovy Berkeley atmosphere for hours on end. Bob and Nathalie, I am extremely grateful that you agreed to become my coauthors; your expertise is essential to what these books have become. Bill Mulford (best student ever), thank you for all that you have done over the years and for introducing Bob and me in the first place. Working with our team of four has been a joy, and I look forward to continuing our commitment of creating the best possible resources for students and instructors in the future.

From Nathalie Vega-Rhodes: First, to my husband, Rob—there is no way I can adequately express in a few sentences how grateful I am for you. Thank you for your unconditional love and support over the years and, more recently, for making sure that I regularly eat (delicious food), sleep, and work out. I love you. And second, to Sherri—thank you. I've seen the impact these books have had on my students each semester. Being a part of helping students and instructors on a larger scale means the world to me. Thank you for our collaboration and also for the many talks we've had about teaching and life.

From Bob Feldman: I am grateful to my children, Jonathan, Joshua, and Sarah; my daughters-in-law Leigh and Julie and son-in-law Jeffrey; my extraordinarily smart, cute, and talented grandchildren Alex, Miles, Naomi, and Lilia; and most of all to my wife, Katherine (who no longer can claim to be the sole mathematician in our family). I thank them all, with great love.

Operations with Integers

©Blend Images/Ariel Skelley/Getty Images

Math at Work:

Physician's Assistant

Delia Cruz has always had an instinct to help people. Growing up, she was a volunteer tutor to younger students in her school, and she spent her summers interning at a local health clinic. So when it came time to choose a profession, Delia knew she wanted to enter a field that would allow her to make a positive impact on people's lives.

"Medicine was a perfect fit for me," Delia explains. "As a physician's assistant, I have the opportunity to be there for people when it matters the most."

Delia realized quickly that a key to such efforts was one she might not have expected: math. Her use of math on the job ranges from simple arithmetic to track changes in weight or blood pressure to more sophisticated calculations involving prescription dosage and the analysis of lab results.

"Math is part of everything I do to treat my patients," Delia says.

While helping others came naturally to Delia, math did not. But she understood that it was an essential component to the career—and the life—she wanted to build for herself. So she worked hard at it and used smart study skills. Eventually, Delia was able to turn math from a relative weakness into an area of strength—one that benefits her and the patients she treats every day.

In this chapter, we look at some basic concepts in math: operations with integers along with place value. We introduce the P.O.W.E.R. framework, which will help you master the skills of math or of any subject in the classroom and on the job.

Research shows that successful goal-achievers, and successful learners, do five things to achieve their goals: **P**repare, **O**rganize, **W**ork, **E**valuate, and **R**ethink. If we take the first five letters of those words, we get **P.O.W.E.R.**, as in the P.O.W.E.R. Learning Framework.

A great thing about this framework is that we can use it to do any task or achieve any goal whether it is cleaning out the basement, forming a band with friends, completing a project at work, or learning math! Let's learn what each step of P.O.W.E.R. means and apply it to an example to help you understand it.

- *Prepare means to explicitly state a goal.* Be very specific.
- Let's do an example: "*We will clean the basement by the end of the weekend.*"
- Keep in mind that there are long-term goals and short-term goals. "We will clean the basement by the end of the weekend" is the long-term goal, but we can set short-term goals along the way to help us achieve our ultimate goal of cleaning the basement. A short-term goal might be to buy boxes by Friday.

- *Organize means to **organize** the physical and mental tools you need to achieve your goal.*
- A *mental tool* for cleaning the basement is thinking ahead of time: How much time do we think it will take, and how can we fit this into our schedule? Where and how will we organize all of the stuff? We will have an area for stuff to save, another to donate, and another to throw away. Do we need storage boxes for stuff that we save?
- Some *physical tools* we might need are boxes, old clothes to wear while we clean, other people to help us, and a broom or vacuum cleaner.

- *Work means to **do the work** that needs to be done to achieve the goal.*
- In our example, that means go down and clean the basement! Look through old boxes and piles on the floor. Organize what we find according to what we will throw away, what we will keep, and what we will donate. Sweep or vacuum the floor.

- *Evaluate means you should **evaluate** what you have done.* Did you achieve your goal or not?
- For our example, we look at the basement and ask ourselves, "Did we get the basement cleaned by the end of the weekend?"

- **Rethink** *means to **rethink** and reflect upon your goal.* If you did *not* achieve your goal, ask yourself, "*Why not?*" If you did achieve your goal, ask yourself what you did *right*.
- If we did *not* get the basement cleaned, we should ask ourselves, "*Why not?*" "Did we procrastinate and run out of time? Did we run out of boxes? Did we need more help? What would we do differently next time?" Or, "Was it an unrealistic goal?"
- If we *did* achieve our goal of cleaning the basement, it is still important to understand the things we did right so that if we have to do something similar again we can apply similar strategies. We should ask ourselves questions like, "What went well in this process? Did we plan well ahead of time? Did we buy enough boxes? What would we do again that worked very well? Could we have done anything differently to make the job easier?"

You can apply these same steps to help you be a successful student! Throughout the book we will use the P.O.W.E.R. Learning Framework to learn math *and* to acquire study skills to help us succeed in *any* course. And, of course, you can use P.O.W.E.R. outside of school to help you achieve *any* goal!

Now we know that **P.O.W.E.R.** stands for **P**repare, **O**rganize, **W**ork, **E**valuate, and **R**ethink. We have learned how to apply the steps to a real-life situation, *cleaning the basement*, but how do we apply the steps to learning math? **Let's apply the P.O.W.E.R. framework to help a fictional student, Gabriela, succeed in this course.**

- *Prepare means to explicitly state a goal.* Remember, be specific.
- Gabriela's goal is **"I will make at least a B in this course."**

- *Organize means to **organize** the physical and mental tools you need to achieve your goal.*
- Some *physical tools* Gabriela will need are a backpack, a book (a print book or an e-book), a notebook, regular pencils, colored pencils (for taking good notes), and a quiet place to study.
- Some *mental tools* she will need are a positive attitude, a commitment to go to class every day and do her homework, good basic skills such as knowing the multiplication facts from 1 to 12 and basic geometry concepts, good time management so that she has enough time to work, go to class, and do her homework, and a knowledge of the locations of her instructor's office and the tutoring center.

- *Work means to **do the work** that needs to be done to achieve the goal.*
- Gabriela now has to **do the work** so that she can make at least a B in the course! That work includes going to class every day, putting away her cell phone, reviewing geometry concepts that she has trouble remembering, taking good notes, asking questions in class, studying in a place without distractions, rereading her notes when she gets home, reading the book and *writing out the examples as she reads them*, finishing her homework on time, and going to the tutoring center or to her instructor's office when she needs extra help.

- *Evaluate means you **evaluate** what you have done.* Did you achieve your goal or not?
- Gabriela should **evaluate** how she is doing *as the term goes along* so that she can make adjustments, if necessary. She should ask herself, *"Am I making the kinds of grades that I need to make on the quizzes and tests?"*
- She should **evaluate** how she has done at the *end of the term*. She should ask herself, *"Did I make at least a B in the course?"*

- **Rethink** *means to **rethink** and reflect upon your goal.* If you did *not* achieve your goal, ask yourself, *"Why not?"* If you did achieve your goal, ask yourself what you did *right*.
- First, Gabriela should **rethink** *throughout the term*. If she is doing well on her quizzes and tests, that's great! She should ask herself, *"What have I been doing right so far?"* so that she continues to do those things. If she is not making a B at this point, she should ask herself, *"Why not?"* as well as questions such as *"Did I miss too many classes? Did I put off doing my homework until the last minute? Should I go in for help more often? What can I do differently to improve before the end of the term?"* Then she can make adjustments so that she still has time to learn the material and improve her grade.
- Of course, Gabriela should **rethink** *at the end of the term*. If she made an A or a B in the course, then she has achieved her goal! *Still*, she should stop, think, and ask herself what she did to be successful so that she can apply similar strategies in the future. If she did *not* make at least a B in the course, she should ask herself, *"Why not?"* as well as ask herself the questions in the previous bullet. This way, she can do things differently in the future.

This is just one example of how you can apply the **P.O.W.E.R.** framework to learning math (or anything)! In the following chapters we will learn more *specific* skills to help you learn math and acquire study skills that will help you in any of your courses and in life.

Chapter 1 **POWER** Plan

P Prepare

O Organize

What are your goals for Chapter 1?	How can you accomplish each goal?
1 Be prepared before and during class.	• Don't stay out late the night before class, and be sure to set your alarm! • Bring a pencil, notebook paper, and textbook to class. • Avoid distractions by turning off your cell phone during class. • Pay attention, take good notes, and ask questions. • Complete your homework on time, and ask questions on problems you do not understand.
2 Understand the homework to the point where you could do it without needing any help or hints.	• Read the directions, and show all of your steps. • Go to the instructor's office for help. • Rework homework and quiz problems, and find similar problems for practice.
3 Learn and understand what each letter in P.O.W.E.R. represents so that you can apply it to learning math and acquiring study skills, and use it in your everyday life.	• Complete the emPOWERme that appears before the Chapter Summary. • Read the Study Strategies page that explains how to use the P.O.W.E.R. framework. Explain what each letter in P.O.W.E.R. stands for *and* what each step means. • Read the Study Strategies page that explains how to use P.O.W.E.R. to succeed in your math course.
4 Write your own goal. <u>Do the homework to</u> <u>help me succeed</u>	• <u>Good time management between</u> <u>classes, ask questions</u>

What are your objectives for Chapter 1?	How can you accomplish each objective?
1 Learn the terminology involved with whole numbers and place value.	• Take notes, and learn the definitions and procedures in Section 1.1. • Take good notes in class.
2 Learn the process of addition and subtraction.	• Master each objective of Sections 1.2–1.4 in order because they build upon each other. • Understand how to use a number line to perform operations. • Know the definition of the *commutative properties* and *associative properties of addition*.
3 Know how to round numbers.	• Take notes, and write the procedure in your own words. • Complete all of the exercises, and ask for help when you need it.
4 Learn how to add and subtract integers.	• Learn the rules for adding integers and the properties of addition. • Be able to find the additive inverse of an integer. • Understand how to rewrite a subtraction problem as an addition problem.
5 Estimate a sum or difference of integers.	• Understand how to estimate a sum or difference of integers by first rounding each number to an indicated place. • Learn front-end rounding. • Use front-end rounding to estimate a sum or difference.
6 Learn how to multiply and divide integers.	• Know the rules for multiplying or dividing two integers. • Know the properties of multiplication and division. • Use rounding to estimate a product or quotient.

7 Use exponents and square roots.	• Memorize the common powers of whole numbers. • Understand how to use exponents with negative numbers. • Be able to find a square root.
8 Use the order of operations.	• Understand how to use the order of operations to simplify an expression containing more than one operation. • Use **P**lease **E**xcuse **M**y **D**ear **A**unt Sally to help you remember the order of operations.
9 Write your own goal. _____ _____	• _____ _____

	W Work	Read Sections 1.1–1.8, and complete the exercises.	
E Evaluate Complete the Chapter Review and Chapter Test. How did you do?		**R Rethink**	• How did you perform on the goals for the chapter? Which steps could be improved for next time? If you had the chance to do this chapter over, what would you do differently? • Do you think you have all of the basic skills you need to be successful in this course, or do you need to review? Which, if any, basic skills do you need to improve? Appendix A contains prerequisite topics that may help you. • How has the P.O.W.E.R. framework helped you master the objectives of this chapter? Where else could you use this framework? Make it a point to use P.O.W.E.R. to complete another task this week.

1.1 Place Value and Rounding

P Prepare

What are your objectives for Section 1.1?

1 Identify Digits and Place Value

2 Write a Number in Words or Digits

3 Round Numbers

O Organize

How can you accomplish each objective?

- Understand the definitions of a *digit* and *whole numbers.*
- Draw the charts that help identify *place value* and *periods* in your notes.
- Complete the given example on your own.
- Complete You Try 1.

- Write the procedure for **Writing a Number in Words** in your own words.
- Complete the given examples on your own.
- Complete You Trys 2 and 3.

- Write the procedure for **Rounding Numbers** in your own words.
- Complete the given examples on your own.
- Complete You Trys 4 and 5.

W Work

Read the explanations, follow the examples, take notes, and complete the You Trys.

It is important to understand the language used in mathematics. So, let's start with some definitions.

1 Identify Digits and Place Value

A **digit** is a single character in a numbering system. We use the digits 0, 1, 2, 3, 4, 5, 6, 7, 8, and 9 to write numbers in the **decimal system.** If we put two or more digits together in a certain order, like 58, we get a *number*. (Single digits, like 5, are both digits and numbers.)

The first group of numbers we will study is the *whole numbers*. The **whole numbers** are

$$0, 1, 2, 3, 4, 5, 6, 7, 8, 9, 10, 11, 12, 13, 14, 15, \ldots$$

where the three dots mean that the list continues forever in this same way.

Here are some examples of numbers that *are* and that are *not* whole numbers:

Whole Numbers	*Not* Whole Numbers
0 4 59 203 571,684	$\frac{1}{2}$ 0.8 $2\frac{3}{8}$ -10

Each digit in a whole number represents a **place value** that is determined by where it appears in the whole number. Additionally, we use commas to separate each **group** of three digits in a number, starting from the right. These groups of numbers are also called **periods.**

Here is a place-value chart that shows the place value of each digit in the number 618,504,279,136,045 as well as the periods.

Place-Value and Period Chart for Whole Numbers

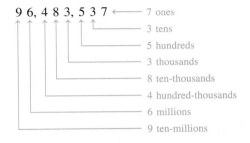

Note

Often, the comma is not written in four-digit numbers. For example, the number 2,758 can also be written as 2758.

EXAMPLE 1

Using the number 96,483,537,

a) identify the place value of each digit in the number.

9 6, 4 8 3, 5 3 7 ←——— 7 ones
————— 3 tens
————— 5 hundreds
————— 3 thousands
————— 8 ten-thousands
————— 4 hundred-thousands
————— 6 millions
————— 9 ten-millions

b) identify the digits in each period of the number.

9 6, 4 8 3, 5 3 7

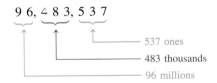

————— 537 ones
————— 483 thousands
————— 96 millions

[YOU TRY 1]

Using the number 815,742,459,

a) identify the place value of each digit in the number.

b) identify the digits in each period of the number.

---

2 Write a Number in Words or Digits

To read a number containing more than three digits, we use the following procedure.

> **Procedure** Writing a Number in Words
>
> Start at the left. Write or say the number in each period (group) followed by the name of the period (group) *except* for "ones." Leave off the plural "s" when reading the period. For example, 123,000 is "123 thousand," not "123 thousands." We do not use the word "ones" when writing or reading a number. When writing a number from 21 to 99, use a hyphen.

EXAMPLE 2

Write each number in words.

a) 64,189   b) 570,343,016   c) 3560

Solution

a) 64,189

$\underbrace{\text{sixty-four}}_{\substack{\text{number in} \\ \text{the period}}} \underbrace{\text{thousand,}}_{\substack{\text{name of} \\ \text{period}}} \underbrace{\text{one hundred eighty-nine}}_{\substack{\text{number in the period} \\ (\text{Do not write "ones.")}}}$

b) 570,343,016

$\underbrace{\text{five hundred seventy}}_{\substack{\text{number in the} \\ \text{period}}} \underbrace{\text{million,}}_{\substack{\text{name of} \\ \text{period}}} \underbrace{\text{three hundred forty-three}}_{\substack{\text{number in the} \\ \text{period}}} \underbrace{\text{thousand,}}_{\substack{\text{name of} \\ \text{period}}} \underbrace{\text{sixteen}}_{\substack{\text{number} \\ \text{in the} \\ \text{period}}}$

 Hint

Look back at the chart in Objective 1 for help if needed.

c) Remember that a four-digit number can be written with or without a comma. You should get used to seeing it both ways. Let's insert the comma to help us see the periods more clearly: 3,560.

$\underbrace{\text{three}}_{\substack{\text{number} \\ \text{in the} \\ \text{period}}} \underbrace{\text{thousand,}}_{\substack{\text{name of} \\ \text{period}}} \underbrace{\text{five hundred sixty}}_{\substack{\text{number in} \\ \text{the period}}}$

[**YOU TRY 2**]

Write each number in words.

a) 172,314   b) 38,206,975   c) 7409

BE CAREFUL

We do **not** use the word *and* when writing or reading whole numbers. The number 405, for example, is read as *four hundred five*, **not** as *four hundred and five*.

Next, let's take a number that is written in words and write it using digits.

EXAMPLE 3

Write each number using digits.

a) five thousand, two hundred ninety-eight

$$5,298 \quad \text{or} \quad 5298 \text{ with no comma}$$

b) seven hundred twenty-six thousand, fifty-three

726,053
└── There are no hundreds, so use a zero.

c) eighteen billion, three hundred seventy-one million, four hundred twenty

18,371,000,420
└── There are no thousands, so use zeros.

[YOU TRY 3] Write each number using digits.

a) eleven thousand, nine hundred seventy-four

b) twenty-one million, five hundred thirty-three thousand, eight

c) one billion, eight hundred fourteen thousand, six hundred twelve

---

3 Round Numbers

To **round** a number means to find another number close to the original number. Think about a number line. If we had to round the number 43 to the nearest ten, we could think about a number line on which each tick mark represents 10 units.

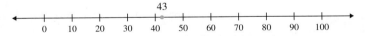

Is 43 closer to 40 or 50 on the number line? *It is closer to 40.* Therefore, we say that 43 rounded to the nearest ten is 40.

We can round numbers to make them easier to work with. Let's list the steps for rounding numbers.

Procedure Rounding Numbers

Step 1: Find the place to which we are asked to round. Underline the digit in that place.

Step 2: Look at the digit to the right of the underlined digit.
   a) If the digit to the right is **less than 5**, leave the underlined digit as it is.
   b) If the digit to the right is **5 or more**, increase the underlined digit by 1.

Step 3: Change all the digits to the right of the underlined digit to zeros.

EXAMPLE 4

Round each number to the indicated place.

a) 24,185 to the nearest hundred

b) 421,306 to the nearest ten-thousand

c) 70,522,917 to the nearest million

Solution

a) *Step 1:* Underline the number in the hundreds place: 24,185

Step 2: Look at the digit to the right of the underlined digit. Since 8 is greater than 5, we increase the underlined digit by 1 to make it 2.

Digit to right is 5 or more.

2 4, 1 8 5

Increase from 1 to 2.

Step 3: Change the digits to the right of the underlined digit to zeros: 24,200

24,185 rounded to the nearest hundred is 24,200. Therefore, 24,185 is closer to 24,200 than to 24,100.

b) *Step 1:* Underline the number in the ten-thousands place: 421,306

Step 2: Look at the digit to the right of the underlined digit. Since 1 is less than 5, we keep the underlined digit a 2.

Digit to right is less than 5.

421,306

Keep this a 2.

Step 3: Change the digits to the right of the 2 to zeros: 420,000

421,306 rounded to the nearest ten-thousand is 420,000.

c) *Step 1:* Underline the number in the millions place: 70,522,917

Step 2: Look at the digit to the right of the underlined digit. Since 5 is 5 or more, we increase the underlined digit by 1 to make it 1.

Digit to right is 5 or more.

70, 522,917

Increase from 0 to 1.

Step 3: Change the digits to the right of the 1 to zeros: 71,000,000

70,522,917 rounded to the nearest million is 71,000,000.

W Hint

Why is it helpful to underline the digit to which you are rounding?

[YOU TRY 4]

Round each number to the indicated place.

a) 94,270 to the nearest hundred

b) 578,413 to the nearest thousand

c) 63,503,881 to the nearest million

Sometimes, we have to use regrouping to round numbers.

EXAMPLE 5

Round 4,982 to the nearest hundred.

Solution

Step 1: Underline the digit in the hundreds place: 4,982

Step 2: Look at the digit to the right of the 9. Since 8 is greater than 5, we must increase the underlined digit by 1.

Because 9 hundreds + 1 hundred = 10 hundreds, we have to regroup the 10 hundreds as 1 thousand and add that to the 4 in the thousands place.

> **W Hint**
> This example involves rounding and regrouping.

┌─ Digit to right is 5 or more.
4,982
└─ Increase from 9 hundreds to 10 hundreds. Change to 0.
└─ Regroup 10 hundreds as 1 thousand, and increase from 4 to 5.

Step 3: Change the digits to the right of the underlined digit to zeros: 5,000
4,982 rounded to the nearest hundred is 5,000.

[YOU TRY 5] Round 39,801,042 to the nearest million.

ANSWERS TO [YOU TRY] EXERCISES

1) a) 8 hundred-millions, 1 ten-millions, 5 millions, 7 hundred-thousands, 4 ten-thousands, 2 thousands, 4 hundreds, 5 tens, 9 ones   b) 815 millions, 742 thousands, 459 ones
2) a) one hundred seventy-two thousand, three hundred fourteen   b) thirty-eight million, two hundred six thousand, nine hundred seventy-five   c) seven thousand, four hundred nine
3) a) 11,974   b) 21,533,008   c) 1,000,814,612   4) a) 94,300   b) 578,000   c) 64,000,000
5) 40,000,000

E Evaluate **1.1** Exercises   Do the exercises, and check your work.

Objective 1: Identify Digits and Place Value

1) What is the difference between a digit and a number?

2) Write down two different digits. Then, form a two-digit number with those two digits.

Identify the place value of the digit 5 in each whole number.

3) 7,591

4) 257,031

Identify the place value of the digit 2 in each whole number.

5) 327,055,149,619

6) 19,025,411,608

Identify the digit with the given place value in each whole number.

7) 83,962,519
   a) ten-millions
   b) millions
   c) ten-thousands
   d) hundreds

8) 94,257,108,637
   a) billions
   b) ten-millions
   c) hundred-thousands
   d) ones

9) 320,963
   a) thousands
   b) ones
   c) hundred-thousands

10) 423,574
   a) ten-thousands
   b) hundreds
   c) tens

11) 965,352,654

   a) hundred-millions

   b) ten-thousands

   c) hundreds

12) 807,004,219,505

   a) hundred-billions

   b) ten-millions

   c) thousands

Identify the digits in each period of the number.

13) 72,544

14) 956,128

 15) 803,001,216

16) 49,223,007,950

17) 41,906,553,213

18) 6,002,791

Objective 2: Write a Number in Words or Digits
Write each number in words.

19) 601

20) 84,915

21) 5,000,449

22) 30,008,723,010

23) 450,629,875

24) 1,201,402

Write each number using digits.

25) seven thousand, two hundred eighty-three

26) two hundred eleven thousand, three hundred ninety-five

27) forty-eight million, nine hundred two thousand, twenty

28) sixteen billion, five hundred fifty-eight million, four

29) eleven billion, two million, four

30) five hundred seventy-seven million, forty-two thousand

Rewrite each number in the problem using digits.

31) As of March 31, 2017, Facebook had approximately one billion, nine hundred forty million active monthly users. (newsroom.fb.com)

32) As of January 2017, there were ninety-five million photos shared on Instagram each day. (www.instagram.com)

33) During the 2016–2017 season, the Chicago Blackhawks' home attendance was eight hundred ninety-one thousand, eight hundred twenty-seven. (espn.go.com)

34) The distance from Earth to the sun is approximately ninety-three million, five hundred eighteen miles.

35) The population of California was approximately thirty-nine million, two hundred fifty thousand, seventeen in 2016. (quickfacts.census.gov)

36) The total area of Alaska is six hundred sixty-four thousand, nine hundred eighty-eight square miles.

Objective 3: Round Numbers

37) Explain, in your own words, how to round a number to the nearest hundred.

38) Explain, in your own words, how to round a number to the nearest thousand.

Round each number to the indicated place.

39) 38 to the nearest ten

40) 12 to the nearest ten

41) 157 to the nearest hundred

42) 253 to the nearest hundred

 43) 76,501 to the nearest thousand

44) 16,409 to the nearest thousand

45) 4,629 to the nearest hundred

46) 9,256 to the nearest hundred

47) 145,528 to the nearest thousand

48) 235,001 to the nearest thousand

49) 47 to the nearest hundred

50) 85 to the nearest hundred

51) 39,762 to the nearest thousand

52) 709,574 to the nearest thousand

53) 549,321 to the nearest hundred-thousand

54) 751,423 to the nearest hundred-thousand

55) 3,541,672 to the nearest hundred-thousand

56) 2,499,381 to the nearest hundred-thousand

 57) 399,129,721 to the nearest ten-million

58) 199,752,000 to the nearest ten-million

59) 299,318 to the nearest ten-thousand

60) 699,422 to the nearest ten-thousand

61) 999 to the nearest thousand

62) 35,999 to the nearest hundred-thousand

Round each number to the nearest ten, nearest hundred, and nearest thousand.

	Ten	Hundred	Thousand
63) 84	_____	_____	_____
64) 39	_____	_____	_____
65) 1,397	_____	_____	_____
66) 2,495	_____	_____	_____
67) 619,755	_____	_____	_____
68) 734,974	_____	_____	_____

69) In 1907, 1,004,756 immigrants passed through Ellis Island, more than in any other year. Round the number of immigrants to the nearest million.
(www.ellisisland.org)

70) The distance from Earth to the moon is 238,855 miles. Round this number to the nearest ten-thousand.
(solarsystem.nasa.gov)

Write a number that, when rounded to the nearest thousand, results in the given number.

71) 7,000

72) 34,000

73) 1,000

74) 0

R Rethink

R1) Write down an example of when you have rounded a number in the last week.

R2) Were there any problems you could not do? If so, write them down or circle them and ask your instructor for help.

1.2 Introduction to Integers

P Prepare

O Organize

What are your objectives for Section 1.2?	How can you accomplish each objective?
1 Understand Signed Numbers	• Write the definitions of *positive number* and *negative number* in your own words using the symbols < and >. • Summarize the procedure for **Writing Signed Numbers.** • Know the definition of an *integer*. • Complete the given example on your own. • Complete You Try 1.
2 Compare Integers	• Follow Example 2, and create a procedure you can follow to compare integers. • Complete You Try 2.
3 Evaluate Expressions Involving Absolute Value	• Write the definition and property of the *absolute value of a number* in your own words. • Complete the given example on your own. • Complete You Try 3.
4 Find the Opposite of a Number	• Write the definition of *opposites* in your own words. • Complete the given examples on your own. • Complete You Trys 4 and 5.

W Work

Read the explanations, follow the examples, take notes, and complete the You Trys.

1 Understand Signed Numbers

Until this point, we have worked with *positive numbers* and zero.

Definition

A **positive number** is a number greater than zero.

On a number line, positive numbers are to the *right* of 0. Here is how we graph the numbers 2, 5, and 9 on a number line.

Some numbers, however, are *less than* zero. These are called *negative numbers*.

Definition

A **negative number** is a number less than zero. To indicate that a number is negative, we put a negative sign, −, in front of it.

    Example: −16 is read as "negative sixteen."

Here are some examples of how we use negative numbers.

Statement	Use a Negative Number
9 degrees below zero	−9°
A golf score of 4 under par	−4
A loss of $281	−$281

Because numbers get smaller as we move to the *left* on a number line, negative numbers are to the *left* of zero. (Zero separates the negative numbers from the positive numbers, and zero is neither positive nor negative.)

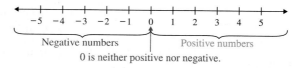

Positive numbers, negative numbers, and zero are also called **signed numbers.**

Procedure Writing Signed Numbers

1) A *negative* number is written with a negative sign in front of it. For example, −8 is "negative eight."

2) A *positive* number can be written with a "+" sign in front of it. For example, +13 is "positive thirteen."

3) If a number does *not* have a sign in front of it, the number is assumed to be positive. For example, 25 is "positive twenty-five" or just "twenty-five."

In this chapter, we begin our work with *integers*.

Definition

Integers consist of the positive and negative counting numbers as well as 0. Integers have no fractional or decimal parts. We list the integers like this:

$$..., -5, -4, -3, -2, -1, 0, 1, 2, 3, 4, 5,...$$ *Ellipsis Cont. forever*

The dots mean that the list continues, in this way, forever.

Note

We will learn about signed fractions and decimals in Chapters 3 and 5.

Next, let's graph some integers on a number line. Remember that negative numbers are to the *left* of 0.

EXAMPLE 1   Graph each pair of integers on a number line.

a) 5, −3   b) −4, 2

Solution

a)

−3 is 3 units to   The number 5 is 5
the *left* of 0.   units to the *right* of 0.

b)

−4 is 4 units to the   2 is 2 units to the
left of 0.   *right* of 0.

YOU TRY 1   Graph each pair of integers on a number line.

a) 3, −1   b) −5, 4

---

2 Compare Integers

Remember that numbers get *smaller* as we move to the *left* on a number line and that numbers get *larger* as we move to the *right* on a number line. We can use the *less than* symbol, <, and the *greater than* symbol, >, to compare integers.

For example, we can compare the numbers 1 and 4 in two ways.

1 < 4 because 1 is to the *left* of 4.   OR   4 > 1 because 4 is to the *right* of 1.

1 is less 4   4 is greater 1
  than   than

EXAMPLE 2

Fill in the blank with < or > to compare each pair of integers. Look at the number line to help you, if necessary.

a) 7 $>$ 3   b) −6 $<$ −1   c) −5 $>$ −8

d) 2 $>$ −4   e) −3 $<$ 0

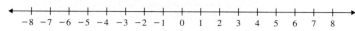

Solution

a) 7 > 3 because 7 is to the *right* of 3 on the number line.

b) −6 < −1 because −6 is to the *left* of −1 on the number line.

c) −5 > −8 because −5 is to the *right* of −8.

d) 2 > −4 because 2 is to the *right* of −4.

e) −3 < 0 because −3 is to the *left* of 0.

[YOU TRY 2] Fill in the blank with < or > to compare each pair of integers. Look at the number line in Example 2 to help you, if necessary.

a) −1 $>$ −4   b) 5 $<$ 8   c) −6 $<$ 2

d) 0 $<$ −7   e) −8 $<$ −3

3 Evaluate Expressions Involving Absolute Value

The concept of *absolute value* is one that is used often in mathematics. What *is* absolute value?

Definition

The **absolute value** of a number is the distance of the number from 0.

The absolute value of a number tells us the *distance* between that number and 0 on a number line, *not* which side of 0 the number is on. Therefore, the absolute value of a number is never negative.

Property Absolute Value of a Number

1) The absolute value of a number is the *distance* between that number and 0 on a number line, and distance is *not* negative. Therefore, *the absolute value of a number is never negative.*

2) The absolute value of a number is denoted by two vertical bars. For example, we read |4| as "the absolute value of 4."

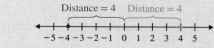

4 is a distance of 4 units from 0. Therefore, |4| = 4.
−4 is a distance of 4 units from 0. Therefore, |−4| = 4.

| EXAMPLE 3 | Evaluate each absolute value expression. |

a) $|7|$ 7   b) $|-3|$ 3   c) $|0|$ 0   d) $-|-9|$ -9

Solution

a) To evaluate $|7|$, ask yourself, *"What is the distance between 0 and 7?"*

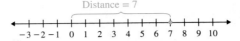

Distance = 7

The distance between 0 and 7 is 7, so $|7| = 7$.

| **W Hint**
The absolute value of a number is never negative. | b) To evaluate $|-3|$, ask yourself, *"What is the distance between 0 and -3?"* |

Distance = 3

The distance between 0 and -3 is 3, so $|-3| = 3$.

c) $|0| = 0$ because 0 is *zero* units from 0.

d) When evaluating an expression like $-|-9|$, evaluate the absolute value first then apply the negative outside the absolute values.

$$-|-9| = -(9) = -9$$ First evaluate $|-9|$, then apply the negative sign on the outside of the absolute value bars.

[**YOU TRY 3**]   Evaluate each absolute value expression.

a) $|13|$ 13   b) $|-5|$ 5   c) $-|8|$ -8   d) $-|-14|$ -14

4 Find the Opposite of a Number

Earlier we saw that, on a number line, 4 and -4 are the same distance from 0 but they are on *opposite* sides of 0.

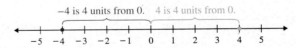

-4 is 4 units from 0.   4 is 4 units from 0.

We say that 4 and -4 are *opposites* of each other.

Definition

Two numbers are **opposites** of each other if they are the same distance from 0 on a number line but are on opposite sides of 0.

To find the opposite of a number, we write a negative sign in front of it.

Note

When we put a negative sign in front of a number to find its opposite, we also put the original number in parentheses. This will help us avoid confusion in the future.

EXAMPLE 4

Find the opposite of each number.

a) 2 _-2_   b) 7 _-7_   c) 0 _0_

Solution

a) The opposite of 2 is $-(2) = -2$.
   └─ Put a negative in front of the 2.

b) The opposite of 7 is $-(7) = -7$.

c) The opposite of 0 is $-(0) = 0$. Remember, 0 is neither positive nor negative.

[YOU TRY 4]

Find the opposite of each number.

a) 19 _-19_   b) 6 _-6_   c) 10 _-10_

W Hint

Why does this make sense? Draw a number line to visualize it.

Sometimes, a number will contain two negative signs. For example,

$$\overset{\nearrow}{-(-5)} \qquad \text{means} \qquad \text{the opposite of } -5.$$
The opposite of −5

Therefore, $-(-5) = 5$, and the opposite of −5 is 5.

Note

The opposite of a negative number is a positive number.

Example: The opposite of −8 is $-(-8) = 8$.
                                    ↑
                       This means "the opposite of."

EXAMPLE 5

Find the opposite of each number.

a) −15 _15_   b) −2 _2_

Solution

W Hint

Notice the use of parentheses to separate the two negative signs.

Number	Opposite
a) −15	$-(-15) = 15$
b) −2	$-(-2) = 2$

W Hint
Check your answers by hand first.

ANSWERS TO [YOU TRY] EXERCISES

1) a) 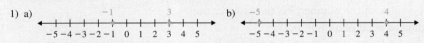   b)

2) a) > b) < c) < d) > e) <   3) a) 13 b) 5 c) −8 d) −14
4) a) −19 b) −6 c) −10   5) a) 38 b) 4

E Evaluate **1.2** Exercises   Do the exercises, and check your work.

Objective 1: Understand Signed Numbers

1) What is a negative number? *A number stating you owe something*

2) Is zero a positive number, negative number, or neither? *Neither*

Represent each statement with a signed number.

3) Justine owes $23. *+23*

4) Hector's checking account was overdrawn by $124.56. *−124.56*

5) Gary lost 3 lb this past week on a reduced-calorie diet. *−3*

6) The value of a stock decreased by $2.81/share. *−2.81*

7) Mount Whitney is the tallest mountain in the continental United States and rises 14,494 feet above sea level. (www.nps.gov/seki) *+14,494*

©robert cicchetti/Shutterstock

8) The deepest part of Lake Erie is 210 feet below the surface of the water. (www.great-lakes.net)

Graph each set of integers on the number line.

9) 0, 2, −1, −5, −3   10) −1, 0, −4, 4, −5

11) −4, 3, −2, 5, 1   12) 5, −3, 2, 1, −2

Objective 2: Compare Integers

Fill in the blank with *smaller* or *larger*.

13) As you move left on the number line, the numbers get *smaller*

14) As you move right on the number line, the numbers get *Bigger*

Fill in the blank with < or > to compare each pair of integers.

15) −13 _<_ 7   16) 11 _>_ −11

17) −4 _<_ −3   18) −11 _>_ −12

19) −3 _<_ 0   20) 0 _>_ −12

21) 8 _<_ 15   22) 11 _<_ 26

23) 9 _>_ −9   24) −14 _<_ −10

25) −107 _>_ −123   26) −182 _<_ −165

27) −2,000 _<_ −1,000   28) −5,000 _>_ −8,000

Use the number line for Exercises 29–34. Fill in the blank with < or > to make the statement true.

29) b _<_ c   30) 0 _>_ a

31) d _>_ b   32) a _<_ 0

33) 2 _>_ b   34) −3 _<_ c

Objective 3: Evaluate Expressions Involving Absolute Value

35) Your friend asks you to explain to him the meaning of the absolute value of a number. What would you tell him? *It's the closest thing to 0 and a Positive #*

36) The absolute value of the number 3 is the same as the absolute value of what other number?

Evaluate each absolute value expression.

37) $|8|$ *8*

38) $|1|$ *1*

39) $|19|$ *19*

40) $|87|$ *87*

41) $|-17|$ *17*

42) $|-13|$ *13*

43) $|0|$ *0*

44) $-|0|$ *0*

45) $-|35|$ *-35*

46) $-|15|$ *-15*

47) $-|-24|$ *-24*

48) $-|-56|$ *-56*

Objective 4: Find the Opposite of a Number

49) How are 5 and the opposite of 5 related on the number line? *Opposite of 5 is -5*

50) What number is its own opposite?
The negative number

Find the opposite of each number.

51) 1

52) 10

53) 47

54) 54

55) -23

56) -16

57) -147

58) -288

For Exercises 59–64, answer *true* or *false*.

59) $|-7| < 0$

60) $|-18| < |-23|$

61) $-(-6) > 0$

62) $-10 > -(-10)$

63) $-|-8| > -|-12|$

64) $-|-0| < -(-1)$

Use the number line for Exercises 65–68. Fill in the blank with < or > to make the statement true.

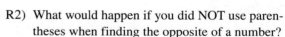

65) $-a$ ___<___ 0

66) $-d$ ___<___ c

67) $-d$ ___<___ $-c$

68) $|a|$ ___<___ b

R Rethink

R1) Explain how visualizing signed numbers on a number line helped you to understand absolute value and opposites.

It helped me understand by seening how it is numbered on the line.

R2) What would happen if you did NOT use parentheses when finding the opposite of a number?

It wouldn't be seperated

1.3 Adding Integers

What are your objectives for Section 1.3?	How can you accomplish each objective?
1 Add Integers Using a Number Line	• Review Appendix A.1 if you need more review on adding two whole numbers using a number line. • When adding a negative number, move left on the number line. • Complete the given example on your own. • Complete You Try 1.
2 Add Two Negative Integers	• Know that the sum of two negative numbers is always negative. • Write the procedure for **Adding Two Negative Numbers** in your own words. • Complete the given example on your own. • Complete You Try 2.
3 Add Integers with Different Signs	• Write the procedure for **Adding Two Numbers with Different Signs** in your own words. • Complete the given examples on your own by following the procedure you wrote down. • Complete You Trys 3 and 4.
4 Use Properties of Addition	• Write the **Identity, Commutative,** and **Associative Properties of Addition** in your own words. • Complete the given examples on your own. • Complete You Trys 5–11.

W Work

Read the explanations, follow the examples, take notes, and complete the You Trys.

1 Add Integers Using a Number Line

In an addition problem, the numbers being added together are called the **addends,** and the answer is called the **sum.** To add two *positive* numbers, like 2 + 4, we start at 0 and move 2 spaces to the *right* to reach 2. Then, we move 4 more spaces to the *right* to reach 6. The addends are 2 and 4, and the sum is 6.

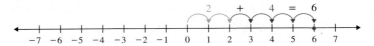

To add *negative* numbers, we move to the *left.*

EXAMPLE 1

Add using a number line.

a) $5 + (-3)$　　　b) $-7 + 4$　　　c) $-4 + (-1)$

Solution

a) Start at 0 and move 5 spaces to the right to reach 5. Then, to add -3, move 3 spaces to the left. We finish at 2.

b) To add $-7 + 4$, move 7 spaces to the *left* to reach -7. Then, to add 4, move 4 spaces to the right. We finish at -3.

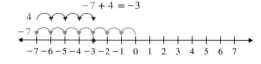

c) To add $-4 + (-1)$, first move 4 spaces to the left to reach -4. Then, move 1 more space to the left to add -1. The sum is -5.

[YOU TRY 1] Add using a number line.

a) $7 + (-9)$　　　b) $-3 + 8$　　　c) $-2 + (-4)$

Note

When a negative number follows an operation symbol like $+, -, \times, \div$, we usually put the number in parentheses.

Example: $-4 + (-1) = -5$

This negative number is *after* the +, so put it in parentheses.

2 Add Two Negative Integers

When we add two positive numbers, the sum is positive. When we add two negative numbers, the sum is *always* negative.

W Hint

Look at Example 1. Does the procedure make sense?

Procedure Adding Two Negative Numbers

Step 1: Find the absolute value of each number.

Step 2: Add the absolute values.

Step 3: Put a negative sign in front of the sum.

Note

The sum of two negative numbers is *always* negative.

EXAMPLE 2

Add.

a) $-6 + (-13)$ b) $-47 + (-29)$

Solution

a) **Step 1:** Find the absolute value of each number: $|-6| = 6, |-13| = 13$

Step 2: Add the absolute values: $6 + 13 = 19$

Step 3: Put a negative sign in front of the sum: $-6 + (-13) = -19$

b) **Step 1:** Find the absolute value of each number: $|-47| = 47, |-29| = 29$

Step 2: Add the absolute values: $47 + 29 = 76$

Step 3: Put a negative sign in front of the sum: $-47 + (-29) = -76$

YOU TRY 2

Add.

a) $-18 + (-9)$ b) $-52 + (-93)$

---

3 Add Integers with Different Signs

When we add a positive number and a negative number, sometimes the sum is positive (as in Example 1a) and sometimes the sum is negative (as in Example 1b).

> **Procedure** Adding Two Numbers with Different Signs
>
> **Step 1:** Find the absolute value of each number.
> **Step 2:** Subtract the smaller absolute value from the larger absolute value.
> **Step 3:** The sign of the *sum* will be the same as the sign of the number with the *greater* absolute value. Write the sum with this sign.

EXAMPLE 3

Find each sum, then verify the answer using a number line.

a) $-6 + 9$ b) $4 + (-10)$

Solution

a) **Step 1:** Find the absolute value of each number: $|-6| = 6, |9| = 9$

Step 2: Subtract the smaller absolute value from the larger absolute value.

$$9 - 6 = 3$$

Larger absolute value   Smaller absolute value

Step 3: The sign of the *sum* will be the same as the sign of the number with the *greater* absolute value.

Positive 9 has a greater absolute value than *negative* 6, so **the sum will be positive.**

$$-6 + 9 = 3$$
↑
The sum is positive.

We can see why the answer makes sense on a number line.

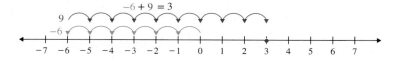

b) ***Step 1:*** Find the absolute value of each number: $|4| = 4$, $|-10| = 10$

Step 2: Subtract the smaller absolute value from the larger absolute value.

$$10 - 4 = 6$$
Larger absolute   Smaller absolute
value   value

Step 3: The sign of the *sum* will be the same as the sign of the number with the *greater* absolute value.

Negative 10 has a greater absolute value than *positive* 4, so **the sum will be negative.**

$$4 + (-10) = -6$$
↑
The sum is negative.

Let's look at this on a number line.

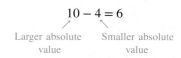

[YOU TRY 3]   Find each sum, then verify each answer using a number line.

a) $-5 + 1$ b) $7 + (-3)$

EXAMPLE 4   Add.

a) $38 + (-21)$ b) $-562 + 301$

Solution

a) ***Step 1:*** Find the absolute value of each number: $|38| = 38$, $|-21| = 21$

Step 2: Subtract the smaller absolute value from the larger absolute value.

$$38 - 21 = 17$$
Larger absolute   Smaller absolute
value   value

W Hint
Writing out the example as you read it is a good strategy for reading a math book!

Step 3: The sign of the sum will be *positive* because 38 has a greater absolute value than -21.

$$38 + (-21) = 17$$ The sum is positive.

b) **Step 1:** Find the absolute value of each number: $|-562| = 562$, $|301| = 301$

Step 2: Subtract the smaller absolute value from the larger absolute value.

$$562 - 301 = 261$$

Larger absolute   Smaller absolute
value   value

Step 3: Will the answer be positive or negative? It will be *negative* because -562 has a larger absolute value than 301.

$$-562 + 301 = -261 \quad \text{The sum is negative.}$$

[YOU TRY 4]   Add.

a) $-78 + 49$ b) $516 + (-497)$

---

4 Use Properties of Addition

What happens when we add 0 to a number? For example,

$$8 + 0 = 8 \qquad 0 + 41 = 41 \qquad -56 + 0 = -56$$

Whenever we add 0 to a number, the result is the number. This is called the *identity property of addition*.

> **Property** The Identity Property of Addition
>
> The **identity property of addition** says that if we add 0 to any number, the result is that number. (The number 0 is called the **identity element** for addition.)
>
> *Example:* $-9 + 0 = -9$

If we change the order in which we add numbers, does it change the sum?

EXAMPLE 5   Use a number line to add $5 + 2$ and $2 + 5$.

Solution

To find $5 + 2$, start at 0 and move 5 spaces to the right to reach 5. To add 2, move 2 more spaces to the right. We finish at 7.

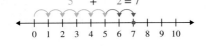

To find $2 + 5$, start at 0 and move 2 spaces to the right. To add 5, move 5 more spaces to the right. We finish at 7.

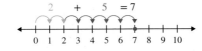

[YOU TRY 5]   Use a number line to add $3 + 7$ and $7 + 3$.

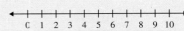

We can add numbers in any order and the result, or sum, will be the same. The *commutative property of addition* tells us this is true.

W Hint

You can add left to right or right to left.

Property The Commutative Property of Addition

The **commutative property of addition** says that changing the order in which we add numbers does not change the sum.

Example: $5 + 2 = 7$ and $2 + 5 = 7$

Note

It may be helpful to remember the commutative property this way: To *commute* to work means that each day we travel from home to our place of business and then back home again. Therefore, commuting refers to changing location. When we *commute* numbers, we are changing the locations of those numbers.

EXAMPLE 6 Add $8 + 1$, then rewrite the addition problem using the commutative property.

Solution

$8 + 1 = 9$. Using the commutative property, we get $1 + 8 = 9$.

YOU TRY 6 Add $4 + 7$, then rewrite the addition problem using the commutative property.

Let's add three numbers on a number line. Then we will learn another property of addition.

EXAMPLE 7 Use a number line to add $3 + 6 + 4$.

Solution

We extend the number line to 15.

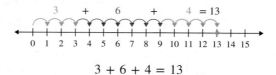

$$3 + 6 + 4 = 13$$

YOU TRY 7 Use a number line to add $5 + 2 + 8$.

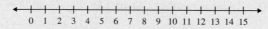

We can also use parentheses to add numbers. In addition to separating negative signs from operation symbols, like 5 + (−2), **parentheses** are grouping symbols that tell us the order in which we should perform arithmetic operations. Usually, we do the operations in parentheses before other operations in a problem.

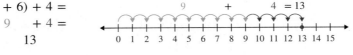

EXAMPLE 8   Use a number line to add (3 + 6) + 4.

Solution

Let's add the numbers in parentheses first. Then we will add on the number line.

$$(3 + 6) + 4 =$$
$$9 \quad + 4 =$$
$$13$$

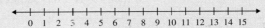

$$(3 + 6) + 4 = 13$$

> **W Hint**
>
> Notice that the same numbers are being added in Examples 7, 8, and 9.

[YOU TRY 8]   Use a number line to add (5 + 2) + 8.

When we add several numbers, can we change the placement of the parentheses without changing the sum?

EXAMPLE 9   Use a number line to add 3 + (6 − 4).

Solution

Add the numbers in parentheses first.

$$3 + (6 + 4) =$$
$$3 + \quad 10 \quad =$$
$$13$$

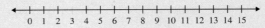

$$3 + (6 + 4) = 13$$

This is the same as the result in Example 8 when the parentheses were in a different place: (3 + 6) + 4 = 13.

[YOU TRY 9]   Use a number line to add 5 + (2 + 8).

Property The Associative Property of Addition

The **associative property of addition** says that we can change the position of grouping symbols when adding numbers and the sum remains the same.

Example: $(3 + 6) + 4 = 13$ and $3 + (6 + 4) = 13$

EXAMPLE 10

Add $9 + (3 + 8)$, then rewrite the addition problem using the associative property.

Solution

Original Sum	Using the Associative Property
$9 + (3 + 8) =$	$(9 + 3) + 8 =$
$9 + \quad 11 \quad = 20$	$12 \quad + 8 = 20$

Notice that we changed the position of the parentheses. We did not change the position of the numbers.

YOU TRY 10

Add $(5 + 2) + 8$, then rewrite the addition problem using the associative property.

Sometimes, using the commutative and associative properties together makes it easier to find the sum of numbers.

EXAMPLE 11

Add $-6 + 7 + 6 + 3$.

Solution

The commutative property says that we can add numbers in any order and the sum does not change. Let's rearrange the numbers.

$$-6 + 7 + 6 + 3 = -6 + 6 + 7 + 3 \qquad \text{Commutative property}$$

The associative property says that we can *group* the addends in any way.

$$= (-6 + 6) + (7 + 3)$$
$$= \quad 0 \quad + \quad 10 \quad = 10$$

> **W Hint**
> If possible, rearrange the numbers so that groups are easy to add.

Notice that -6 and 6 are opposites. When we add -6 and 6, the sum is 0. When possible, group numbers so they are easy to add.

YOU TRY 11

Add $-7 + 9 + 7 + 8$.

Note

The sum of a number and its opposite is 0.

Example: $-6 + 6 = 0$

ANSWERS TO [YOU TRY] EXERCISES

1) a) −2;

b) 5;

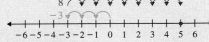

c) −6;

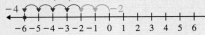

2) a) −27   b) −145

3) a) −4;

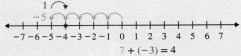

b) 4;

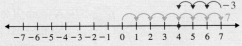

4) a) −29   b) 19

5)

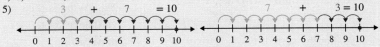

6) 4 + 7 = 11; 7 + 4 = 11

7)

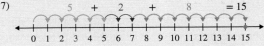

8)

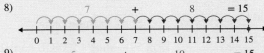

9)

10) 15; 5 + (2 + 8)   11) 17

E Evaluate   **1.3**   Exercises   Do the exercises, and check your work.

Get Ready

Add the whole numbers. Recall that writing the addition problems on graph paper will help you line up the numbers correctly.

1) 758 + 291

2) 6,024 + 1,397

3) 5,402 + 426,837 + 915

4) 89,433 + 59 + 562,178

Objective 1: Add Integers Using a Number Line

Add using a number line.

5) 8 + (−5)

6) 10 + (−3)

7) 5 + (−9)

8) 2 + (−3)

9) 1 + (−3)

10) 4 + (−7)

11) −1 + (−2)

12) −2 + (−5)

13) −1 + (−5)

14) −3 + (−3)

15) Explain, in your own words, how to add a positive number and a negative number on a number line.

16) Explain, in your own words, how to add two negative numbers on a number line.

Objective 2: Add Two Negative Integers

17) Is the sum of two negative numbers *always, sometimes,* or *never* negative?

18) Explain, in your own words, how to add two negative numbers.

Add.

19) $-8 + (-3)$

20) $-5 + (-7)$

21) $-12 + (-12)$

22) $-3 + (-24)$

23) $-53 + (-45)$

24) $-47 + (-32)$

25) $-376 + (-214)$

26) $-618 + (-152)$

27) $-4 + (-13) + (-15)$

28) $-11 + (-10) + (-8)$

29) $-22 + (-67) + (-36)$

30) $-49 + (-33) + (-51)$

31) Write an expression so that the sum of two negative numbers is -45. $-23 + -22 = -45$

32) Write an expression so that the sum of two negative numbers is -68. -68 $-34 + 34 = -68$

Objective 3: Add Integers with Different Signs

33) Explain, in your own words, how to add two numbers with different signs.

34) Is the sum of a negative number and a positive number *always, sometimes,* or *never* negative?

Add.

35) $-9 + 6$ -3

36) $-6 + 1$

37) $-3 + 11$ 8

38) $-7 + 9$

39) $10 + (-4)$ 6

40) $12 + (-8)$

41) $2 + (-13)$ -11

42) $5 + (-18)$

43) $-62 + 47$ -15

44) $-87 + 61$

45) $-135 + 328$

46) $-267 + 351$

47) $248 + (-156)$

48) $308 + (-172)$

49) $546 + (-795)$

50) $457 + (-603)$

51) Write an expression so that the sum of two integers with different signs is -28.

52) Write an expression so that the sum of two integers with different signs is -93.

53) Write an expression so that the sum of two integers with different signs is 51.

54) Write an expression so that the sum of two integers with different signs is 62.

Represent each statement with an addition problem, and solve the problem.

55) Yesterday, LaShonte received her paycheck in the amount of $736, and today she spent $258 at the shopping mall. How much of her paycheck remains?

56) Paulius received an email from the bank notifying him that his checking account was overdrawn by $132. He immediately went online and transferred $150 to his checking account from his savings account. Find the new balance of his checking account.

57) A football team lost 12 yd on the first play and gained 9 yd on the second play. What is their net yardage after these two plays?

©Thinkstock/Getty Images

58) While James had the flu, he lost 8 lb. After 2 weeks, he gained 5 lb back. What is his net weight gain?

Solve each problem.

59) Alma's 500-GB hard drive had 435 GB of data stored on it. After deleting 283 GB of data, Alma uploads 189 GB of family pictures and videos to the drive. How many gigabytes of storage are available on Alma's hard drive?

60) Molika made a purchase at a department store in the amount of $136. Later that day, she received an email stating that her debit account was overdrawn by $21 and was charged an $18 overdraft fee. What is the balance in Molika's account if she deposits $125 into her account?

61) Alexandria deposited her $750 scholarship check into her debit account, which had a balance of $247. Later that week, Alexandria used her debit card to pay her $540 tuition fee and withdrew $300 cash to pay for books and school supplies. What is Alexandria's new debit account balance?

62) Kapila had $589 in his checking account when he deposited his paycheck in the amount of $1782. To pay his bills, he wrote checks for $278, $78, $156, and $750. After the checks clear, what is the balance in Kapila's checking account?

Objective 4: Use Properties of Addition

63) In your own words, explain the commutative property of addition.

64) In your own words, explain the identity property of addition.

Add, then rewrite the problem using the commutative property.

65) $8 + 7$

66) $2 + 11$

67) $-16 + 9$

68) $4 + (-19)$

69) $7 + 0$

70) $0 + (-5)$

71) $2 + 4 + 8$

72) $9 + 7 + 1$

73) In your own words, explain the associative property of addition.

74) Answer true or false. $7 + (2 + 3) = (7 + 2) + 3$ Which property helped you get the answer?

Add, then rewrite the problem using the associative property.

75) $3 + (7 + 5)$

76) $(6 + 3) + 2$

77) $4 + (7 + 3)$

78) $2 + (4 + 6)$

79) $(-11 + 5) + 2$

80) $(-18 + 1) + 3$

81) $-7 + (-1 + 8)$

82) $-10 + (-2 + 12)$

Add. Use the properties to help you.

83) $-9 + 4 + 7 + 9$

84) $8 + (-5) + 11 + 5$

85) $12 + 9 + 8 + 1$

86) $7 + 2 + 13 + 18$

87) $-15 + (-2) + 2 + 15$

88) $6 + 19 + (-6) + (-19)$

Mixed Exercises: Objectives 2 and 3
Add.

89) $13 + (-4)$

90) $376 + (-508)$

91) $-58 + 21$

92) $-15 + 29$

93) $-408 + (-521)$

94) $-957 + (-604)$

95) $-351 + 186$

96) $209 + (-183) + (-417)$

Solve each problem.

97) A submarine dives to a depth of 675 ft and then rises 350 ft. Use an integer to represent the depth of the submarine.

©StockTrek/SuperStock

98) Rory plays two rounds of golf on the same day, posting scores of −4 and −3. What is his total score for the day?

99) On Monday morning, the temperature was −12°F. The next morning, the temperature was 17°F higher. Use an integer to represent Tuesday morning's temperature.

100) On Thursday, the value of a share of stock falls $3. On Friday, the value of the same stock decreases by $1. Use an integer to represent the total change in the value of the stock on those two days.

R Rethink

R1) Can you think of another way to add integers without using absolute value or a number line? Explain.

R2) Where have you encountered the addition of integers this past week? Write a problem explaining the situation, and solve it.

1.4 Subtracting Integers

P Prepare

O Organize

What are your objectives for Section 1.4?	How can you accomplish each objective?
1 Find the Additive Inverse of a Number	• Write the definition of *additive inverse* in your own words, and give an example. • Complete the given example on your own. • Complete You Try 1.
2 Subtract Integers	• Look at Appendix A.2 if you need to review how to subtract whole numbers. • Write the procedure for **Subtracting Two Numbers** in your own words, and write an example. • Complete the given example on your own. • Complete You Try 2.
3 Combine Adding and Subtracting of Integers	• Change subtraction to addition of the additive inverse, then add from left to right. • Complete the given example on your own, and write a procedure for solving problems involving addition and subtraction. • Complete You Try 3.

W Work   Read the explanations, follow the examples, take notes, and complete the You Trys.

1 Find the Additive Inverse of a Number

In Section 1.2, we learned that two numbers are opposites of each other if they are the same distance from 0 on a number line but are on opposite sides of 0.

For example, 3 and −3 are opposites.

The opposite of a number is also called its **additive inverse.** So, the additive inverse of 3 is −3, and the additive inverse of −3 is 3.

What do we get if we add 3 + (−3)?

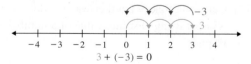

In fact, the sum of any number and its additive inverse is 0.

Definition

The opposite of a number is its **additive inverse.** The sum of a number and its additive inverse is 0.

Example: $3 + (-3) = 0$

EXAMPLE 1

Find the additive inverse of each number, then add the numbers.

a) 4   b) −7   c) 0

Solution

a) The additive inverse of 4 is −4. $4 + (−4) = 0$

⎿———Write a negative sign in
front of the number to
find its additive inverse.

b) To find the additive inverse of −7, put a negative sign in front of the number and simplify: $−(−7) = 7$

The additive inverse of −7 is 7. $−7 + 7 = 0$

c) The additive inverse of 0 is 0. $0 + 0 = 0$

[YOU TRY 1]   Find the additive inverse of each number, then add the numbers.

a) −16   b) 1

We can subtract integers using an additive inverse.

2 Subtract Integers

In a subtraction problem like $8 − 2$, we can subtract as we always have. Or, we can change the subtraction problem to an addition problem. Instead of subtracting the second number from the first, we *add the opposite,* or add the *additive inverse,* of the second number to the first number. We get the same answer.

$$8 − 2 = 6 \qquad 27 − 7 = 20$$
$$8 + (−2) = 6 \qquad 27 + (−7) = 20$$

Procedure Subtracting Two Numbers

To subtract two numbers, $a − b$,

1) Change subtraction to addition.
2) Find the additive inverse of b.
3) Add a and the additive inverse of b.

Notice that we keep the first number, a, the same.

Example: $9 − 4 = 5$ can be written as $9 + (−4) = 5$

Change subtraction to addition of the
additive inverse.

We can also state the procedure using the following formula:

W Hint

Can you write this formula using the letters *a* and *b*?

Procedure Subtracting Two Numbers

| First number | − | Second number | = | First number | + | Additive inverse of the second number |

Again, notice that the first number stays the same.

EXAMPLE 2

Subtract.

a) $3 - 9$ ➞ 6

b) $-14 - 11$ = 25

c) $-12 - (-25)$ 13

d) $-10 - (-3)$

e) $2,081 - (-1,476)$ -2,484+

Solution

a) Use the formula   First number − Second number = First number + Additive inverse of the second number

$$3 \quad - \quad 9 \quad = \quad 3 \quad + \quad (-9) \quad = -6$$

So, $3 - 9 = -6$.

W Hint

Notice that we use parentheses to separate a minus or a plus sign from a negative sign.

b) $-14 - 11 = -14 + (-11) = -25$
Change subtraction to addition of the additive inverse of 11.

c) $-12 - (-25) = -12 + (25) = 13$
Change subtraction to addition of the additive inverse of −25.

d) $-10 - (-3) = -10 + (3) = -7$
Change subtraction to addition of the additive inverse of −3.

e) Don't let larger numbers confuse you. We apply the rules the same way.

$$2,081 - (-1,476) = 2,081 + (1,476) = 3,557$$
Change subtraction to addition of the additive inverse of −1,476.

[YOU TRY 2]

Subtract.

a) $5 - 9$ ‒4

b) $-15 - 13$ ‒2

c) $10 - (-8)$

d) $-24 - (-47)$ 13

e) $-4,875 - (-2,391)$

BE CAREFUL Make sure you can tell the difference between a subtraction sign and a negative sign.

$$18 - (-4)$$ Read this as "18 minus negative 4."

Subtraction sign   Negative sign

3 Combine Adding and Subtracting of Integers

To add and subtract more than two integers, change subtraction to addition of the additive inverse. Then add from left to right.

| **EXAMPLE 3** | Perform the operations to simplify $-12 - 8 - (-15) + 11$. |

Solution

$$-12 - 8 - (-15) + 11 = -12 + (-8) + 15 + 11 \qquad \text{Change all subtraction to addition of the additive inverse.}$$

$$= \quad -20 \quad + 15 + 11 \qquad \text{Add from left to right. So, add } -12 + (-8) \text{ first.}$$

$$= \quad -5 \quad + 11 \qquad \text{Add } -20 + 15.$$

$$= 6 \qquad \text{Add } -5 + 11.$$

> **W Hint**
> Add two numbers at a time, moving from left to right until only one remains.

| **YOU TRY 3** | Perform the operations to simplify $-31 - (-16) + 10 - 17$. |

ANSWERS TO [YOU TRY] EXERCISES

1) a) 16; 0   b) −1; 0   2) a) −4   b) −28   c) 18   d) 23   e) −2,484   3) −22

E Evaluate   **1.4** Exercises   Do the exercises, and check your work.

Get Ready

Subtract the whole numbers. Recall that writing the subtraction problems on graph paper will help you line up the numbers correctly.

1) $835 - 204$ 631
2) $798 - 542$ 256
3) $5,617 - 895$ 4,722
4) $3,243 - 271$
5) $7,026 - 1,583$ 5,448
6) $9,250 - 4,269$

Objective 1: Find the Additive Inverse of a Number

7) How are a number and its additive inverse related on a number line? What is their sum?

8) Is the additive inverse of a number *always, sometimes,* or *never* positive?

Find the additive inverse of each number.

9) 6
10) 11
11) 0
12) 12
13) −8
14) −14
15) −23
16) −7

Objective 2: Subtract Integers

Rewrite the subtraction problem as an addition problem, then perform the operation using the number line.

17) $6 - 5$
18) $10 - 3$
19) $3 - 7$
20) $2 - 5$
21) $-2 - 3$
22) $-1 - 4$
23) $-2 - (-2)$
24) $-5 - (-9)$

25) Explain, in your own words, why $2 - (-3)$ is equivalent to $2 + 3$.

26) Explain, in your own words, how to tell a subtraction sign from a negative sign.

Rewrite each subtraction problem as an addition problem. Then, simplify.

27) $15 - 6$
28) $13 - 5$
29) $5 - 14$
30) $8 - 19$
31) $125 - 183$
32) $234 - 851$

33) −8 − 3

34) −7 − 5

35) −7 − 12

36) −6 − 15

37) −134 − 925

38) −117 − 893

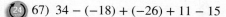 39) −10 − (−18)

40) −4 − (−9)

41) −23 − (−27)

42) −21 − (−28)

43) −29 − (−15)

44) −43 − (−12)

45) −791 − 683

46) −584 − 937

47) 3,508 − (−2,917)

48) 8,106 − (−5,876)

49) −54 − 54

50) −37 − 37

51) −21 − (−21)

52) −49 − (−49)

53) Write an expression so that the difference of two positive numbers is −13.

54) Write an expression so that the difference of two positive numbers is −18.

55) Write an expression so that the difference of two negative numbers is 9.

56) Write an expression so that the difference of two negative numbers is 4.

The summit of Mount Everest is 29,035 ft (8,850 m) above sea level and is the highest point on Earth. Mount McKinley is the highest mountain on the North American continent, and its summit is at 20,320 ft (6,194 m) above sea level. The table below notes several of the lowest places on Earth.

©Imageshop/Alamy

Lowest Places on Earth
Dead Sea (Jordan/Israel) −1,360 ft (−414 m)
Lake Assal (Djibouti, Africa) −509 ft (−155 m)
Turpan Pendi (China) −505 ft (−154 m)
Qattara Depression (Egypt) −435 ft (−133 m)
Denakil (Ethiopia) −410 ft (−125 m)
Laguna del Carbón (Argentina) −344 ft (−105 m)
Death Valley (United States) −282 ft (−86 m)
Salton Sea (California) −227 ft (−69 m)
Salinas Chicas (Argentina) −131 ft (−40 m)
Caspian Sea (Central Asia) −92 ft (−28 m)

(www.nps.gov)

Use this information for Exercises 57–62 to find the difference in elevation between the given geographical locations.

57) Mount McKinley and the Salton Sea; give your answer in feet.

58) Mount Everest and the Dead Sea; give your answer in feet.

59) The Caspian Sea and Lake Assal; give your answer in meters.

60) Salinas Chicas and Laguna del Carbón; give your answer in meters.

61) Turpan Pendi and Death Valley; give your answer in feet.

62) The Qattara Depression and Denakil; give your answer in meters.

Objective 3: Combine Adding and Subtracting of Integers

Perform the operations to simplify each expression.

63) −3 + (−14) − (−5)

64) −7 − (−8) + (−9)

65) 3 − (−26) + (−7) − 14

66) 6 − (−19) + (−2) − 9

67) 34 − (−18) + (−26) + 11 − 15

68) −59 − (−42) + (−10) + 23 − 7

69) 156 − 438 + (−257)

70) 106 − 357 + (−118)

71) |−37| − 82 + |−14| + (−25)

72) |−26| − 61 + |−11| + (−39)

Evaluate each expression. Perform the operations in the absolute values first.

73) |−8 + 3| + (−4) + 9

74) −4 + (−13) + |−7 − 1|

75) −571 − |−145 − 200| + (−87)

76) |−42 + 61| − 234 + 507

Mixed Exercises: Objectives 1–3

Find the additive inverse of each number.

77) −53

78) 71

Perform the operations.

79) −9 + 7 + (−15) − (−6)

80) 154 − 273

81) 81 − 159

82) $-4 - (-9) + 3 - 15$

83) $|-97| + (-158) - (-112)$

84) $102 + |-43| - |151| - (-62)$

85) $51 - |16 - 58| + |30| + (-100)$

86) Tiger Woods' highest 18-hole score in a professional tournament was +13, or 13 over par. His lowest score was −11, or 11 under par. What is the difference in these scores? (www.espn.com)

R Rethink

R1) Were any exercises especially difficult for you? If so, which ones? Why do you think they were difficult?

R2) Think of three different scenarios that would involve addition and subtraction. Represent what that might look like on a number line.

1.5 Estimating a Sum or Difference

P Prepare

O Organize

What are your objectives for Section 1.5?	How can you accomplish each objective?
1 Round to an Indicated Place to Estimate a Sum or Difference	• Review rounding in Section 1.1. • Round numbers to estimate a sum or difference. • Complete the given examples on your own. • Complete You Trys 1–3.
2 Use Front-End Rounding to Estimate a Sum or Difference	• Understand what front-end rounding is. • Complete the given examples on your own. • Complete You Trys 4 and 5.

W Work

Read the explanations, follow the examples, take notes, and complete the You Trys.

1 Round to an Indicated Place to Estimate a Sum or Difference

Have you ever been in a store and wanted to figure out approximately how much your items would cost? For example, if you wanted to buy a shirt that cost $27 and a pair of jeans that cost $44, approximately how much would you spend?

We find the exact total cost of the items by adding $27 + $44 = $71. Or, we can *estimate* the total cost of these two items by rounding each number to the nearest ten and adding them.

Exact Price		**Approximate Price**
$27	rounds to	$30
+ $44	rounds to	+ $40
Exact total cost → $71		$70 ← Estimated total cost

Estimating a sum or difference can be much easier than finding the exact value (especially if we are adding several numbers) and may be all we need if we want just an *approximate* cost of the items we are buying. That is why estimating sums and differences is a very useful skill for us to have.

> ## Procedure Estimating a Sum or Difference
>
> To estimate a sum or difference, round the numbers to the indicated place. Then, add or subtract.

EXAMPLE 1

Estimate the sum 217 + 556 + 134 by first rounding each number to the nearest

a) ten   b) hundred

Solution

a) Round each number to the nearest ten, then find the sum.

Round to nearest ten.

$$
\begin{array}{rcl}
217 & \longrightarrow & 220 \\
556 & \longrightarrow & 560 \\
+134 & \longrightarrow & +130 \\
\hline
& & 910 \quad \text{Estimated sum}
\end{array}
$$

Hint
Write out the example as you read it.

Rounded to the nearest ten, 217 + 556 + 134 ≈ 910. (Recall that the symbol ≈ means "is approximately equal to.")

b) Round each number to the nearest hundred, then find the sum.

Round to nearest hundred.

$$
\begin{array}{rcl}
217 & \longrightarrow & 200 \\
556 & \longrightarrow & 600 \\
+134 & \longrightarrow & +100 \\
\hline
& & 900 \quad \text{Estimated sum}
\end{array}
$$

Rounded to the nearest hundred, 217 + 556 + 134 ≈ 900.

> **Note**
> The exact sum of the numbers is 217 + 556 + 134 = 907. Our estimations in parts a) and b) are very close.

[YOU TRY 1]

Estimate the sum 385 + 182 + 633 by first rounding each number to the nearest
a) ten   b) hundred

EXAMPLE 2

Estimate the difference 5,893 − 2,152 by first rounding each number to the nearest

a) thousand   b) hundred

Solution

a) Round each number to the nearest thousand, then subtract.

Round to nearest thousand.

$$
\begin{array}{r}
5{,}893 \longrightarrow 6{,}000 \\
-\,2{,}152 \longrightarrow -\,2{,}000 \\
\hline
4{,}000 \quad \text{Estimated difference}
\end{array}
$$

b) Round each number to the nearest hundred, then subtract.

Round to nearest hundred.

$$
\begin{array}{r}
5{,}893 \longrightarrow 5{,}900 \\
-\,2{,}152 \longrightarrow -\,2{,}200 \\
\hline
3{,}700 \quad \text{Estimated difference}
\end{array}
$$

The exact difference is $5{,}893 - 2{,}152 = 3{,}741$.

[YOU TRY 2]

Estimate the difference $9{,}285 - 3{,}701$ by first rounding each number to the nearest

a) thousand   b) hundred

We can estimate a sum involving negative numbers as well.

EXAMPLE 3

Estimate the sum $-2{,}316 + 4{,}972$ by first rounding each number to the nearest thousand. Then, find the exact sum.

Solution

Notice on a number line that $-2{,}316$ is closer to $-2{,}000$ than to $-3{,}000$.

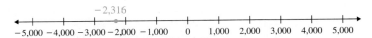

We use the same rules to round negative numbers that we use to round positive numbers. Since we are rounding $-2{,}316$ to the thousands place, underline the 2 in the thousands place and look at the digit to the right.

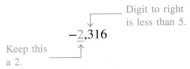

Change the digits to the right of the underlined digit to zeros: $-2{,}000$

To estimate the sum, round each number to the nearest thousand, then add.

Round to the nearest thousand.

$$
\begin{array}{r}
-2{,}316 \longrightarrow -2{,}000 \\
+\,4{,}972 \longrightarrow +\,5{,}000 \\
\hline
3{,}000
\end{array}
$$

Rounded to the nearest thousand: $-2{,}316 + 4{,}972 \approx 3{,}000$.

The exact sum is $-2{,}316 + 4{,}972 = 2{,}656$.

Ⓦ Hint

Picturing a number on a number line might help you round numbers more easily.

[YOU TRY 3] Estimate the sum −5,703 + 1,189 by first rounding each number to the nearest thousand. Then, find the exact sum.

2 Use Front-End Rounding to Estimate a Sum or Difference

Example 3, as well as Examples 1b) and 2a), are examples of *front-end rounding*. In **front-end rounding,** a number is rounded to the place furthest to the left. (In Example 3, we rounded −2,316 to the thousands place to get −2,000, and we rounded 4,972 to the thousands place to get 5,000.) Using front-end rounding is usually the easiest way to estimate an addition or subtraction problem. It is also a good way to check whether an answer is reasonable.

EXAMPLE 4

Use front-end rounding to estimate the difference 18,651 − 9,104. Then, find the exact answer.

Solution

To use front-end rounding, we round each number to the place furthest to the left:

Round 18,651 to the *ten-thousands* place:	20,000
Round 9,104 to the *thousands* place:	− 9,000
Estimate of the difference:	11,000

Hint
Do your work on graph paper to help keep your numbers lined up correctly.

The exact difference is

```
   O  18   4  11
    1  8, 6  5  1
 −     9, 1  0  4
       9, 5  4  7
```

It is quicker and easier to estimate the difference, and getting an estimate that is close to the exact answer helps to verify that we have computed the exact answer correctly.

[YOU TRY 4] Use front-end rounding to estimate the difference 7,365 − 959. Then, find the exact answer.

EXAMPLE 5

Tamika is remodeling her kitchen and has to buy new appliances. She chooses a stove that costs $489, a dishwasher that costs $416, and a refrigerator that costs $1,199. Use front-end rounding to estimate the total cost of the appliances. Then, find the exact answer.

Solution

Round each number to the place furthest to the left:

Round $489 to the *hundreds* place:	$500
Round $416 to the *hundreds* place:	$400
Round $1,199 to the *thousands* place:	+ $1,000
Estimate of the total cost:	$1,900

The exact cost is $489 + $416 + $1,199 = $2,104.

[YOU TRY 5] Joe puts the following items in his shopping cart at a sporting goods store: a football for $29, a helmet for $139, cleats for $109, and shoulder pads for $95. Use front-end rounding to estimate the total cost of the football gear. Then, find the exact answer.

ANSWERS TO [YOU TRY] EXERCISES

1) a) 1,200 b) 1,200   2) a) 5,000 b) 5,600   3) estimate: −5,000; exact: −4,514
4) estimate: 6,000; exact: 6,406   5) estimate: $330; exact: $372

E Evaluate **1.5** Exercises   Do the exercises, and check your work.

Objective 1: Round to an Indicated Place to Estimate a Sum or Difference

Estimate the answer to each problem by first rounding each number to the nearest ten. Then, find the exact answer.

1)   56
   + 23

2)   72
   + 14

3)   56
     64
   + 23

4)   72
     26
   + 55

5)   65
   − 42

6)   98
   − 52

7)   183
   − 122

8)   144
   − 118

9)   −327
   + 155

10) −568
   + 304

Estimate the answer to each problem by first rounding each number to the nearest hundred. Then, find the exact answer.

11)   672
    + 231

12)   318
    + 754

13)   183
      579
    + 617

14)   850
      238
    + 519

15)   726
    − 388

16)   316
    − 172

17)   4,063
    − 2,419

18)   9,251
    − 8,216

19)   −781
    + (−503)

20)   −454
    + (−427)

21)   −1,632
    + 8,265

22)   −2,180
    + 7,020

Solve each problem.

23) A flight from Sydney, Australia, to Chicago, Illinois, is listed in two legs. The flying distance from Sydney to Los Angeles is 7,438 mi, and the flying distance from Los Angeles to Chicago is 1,745 mi. Round each number to the nearest hundred to estimate the flying distance from Sydney to Chicago. Then, find the exact distance. (www.united.com)

©Taras Vyshnya/Shutterstock

24) In 2016, the population of Texas was 27,862,596 while the population of Florida was 20,612,439. Round each number to the nearest million to estimate how many more people lived in Texas than in Florida in 2016. Then, find the exact number.
(quickfacts.census.gov)

25) In 2010, there were approximately 62,600,000 smart phone users in the United States. By 2022, that number is expected to rise to 270,660,000. Round each number to the nearest million to estimate how many more smart phone users there will be in 2022 than in 2010. Then, find the exact number. (www.statista.com)

26) A community college has two campuses. One campus serves 8,743 students, and the other campus serves 13,266 students. Round each number to the nearest thousand to estimate the total number of students on both campuses. Then, find the exact number of students.

Objective 2: Use Front-End Rounding to Estimate a Sum or Difference

27) What is front-end rounding?

28) Junji rounds 573 to 570. Has he used front-end rounding? Explain.

Use front-end rounding to estimate the sum or difference. Then, find the exact answer.

29)   53
    + 68

30)   47
    + 72

31)   783
    + 154

32)   619
    + 457

33) 22,961 + 5,504

34) 7,338 + 35,096

35) −36 + (−42)

36) −53 + (−49)

37) −182 + 395

38) −271 + 695

39)   84
    − 27

40)   63
    − 18

41) 592 − 419

42) 874 − 332

43) 72,516 − 38,223

44) 51,080 − 16,411

45) −12 − 33

46) −44 − 61

47) 9,762 − (−999)

48) 9,845 − (−976)

49) 18 + 13 + 42 + 21

50) 56 + 11 + 34 + 27

51) −39 + (−25) + (−17)

52) −65 + (−41) + (−96)

Use front-end rounding to estimate the answer to each problem. Then, find the exact answer.

53) Last year, Tatiana sold 78 boxes of Girl Scout cookies. This year, she sold 94 boxes. How many more boxes did she sell this year?

54) Carlos bought a television for $409. Brett bought the same TV on sale for $325. How much more did Carlos pay for his TV?

©Digital Vision/Getty Images

55) Here are Ankur's expenses for one semester of college: tuition and fees, $1,762; books, $528; bus fare, $192; lunch, $235. Find Ankur's total expenses.

56) Danielle and Allison just got an apartment together. They had the following expenses during the first month: rent, $970; food, $362; utilities, $214; and cable and Internet, $99. Find their total expenses.

57) Paula's best golf score this year was −11 and her worst golf score was +16. What is the difference in these scores?

©Halfdark/Getty Images

58) Juergen likes to go mountain climbing and scuba diving. His highest climb was to a height of 17,804 ft, and his deepest dive was to a depth of 114 ft. What is the difference between his highest climb and his deepest dive?

59) In 2016, the three movies with the highest gross revenue on opening weekend in the United States were *Captain America: Civil War* with $179,139,142, *Batman v Superman: Dawn of Justice* with $166,007,347, and *Rogue One: A Star Wars Story*, which earned $155,081,681. Find the total gross revenue for these three movies during their opening weekends. (www.boxofficemojo.com)

60) In 2014, New York had 105,390 nursing home residents, Pennsylvania had 79,598, and New Jersey had 45,788. Find the total number of people living in nursing homes in these three states in 2014. (www.census.gov)

R Rethink

R1) How have you used estimation or rounding in the last week?

R2) Do you prefer rounding to an indicated place or front-end rounding? Why?

1.6 Multiplying Integers and Estimation

What are your objectives for Section 1.6?	How can you accomplish each objective?
1 Multiply Two Integers with Different Signs	• Look at Appendix A.3 if you need to review how to multiply whole numbers. • Write the procedure for **Multiplying Two Numbers with Different Signs** in your own words. • Complete the given example on your own. • Complete You Try 1.
2 Multiply Two Integers with the Same Sign	• Write the procedure for **Multiplying Two Numbers with the Same Sign** in your own words. • Complete the given example on your own. • Complete You Try 2.
3 Multiply More Than Two Integers	• Write the procedure for **Multiplying More Than Two Numbers** in your own words. • Complete the given example on your own. • Complete You Try 3.
4 Use Properties of Multiplication	• Write the multiplication properties in your own words. • Understand the **Distributive Property.** • Complete the given examples on your own. • Complete You Trys 4–7.
5 Use Front-End Rounding to Estimate a Product	• Write a procedure for multiplying numbers ending in zeros. • Understand how to solve a real-world problem. • Complete the given examples on your own. • Complete You Trys 8 and 9.

W Work   **Read the explanations, follow the examples, take notes, and complete the You Trys.**

1 Multiply Two Integers with Different Signs

Recall that multiplication represents repeated addition. For example, we can write $3 + 3 + 3 + 3 + 3 = 15$ as $3 \times 5 = 15$. We can use other notations for multiplication as well. We can write $3 \times 5 = 15$ as $3 \cdot 5 = 15$ and $3(5) = 15$ with no operation symbol between the 3 and the parenthesis.

How do we multiply a negative number and a positive number? Let's first add $-3 + (-3) + (-3) + (-3) + (-3)$ on a number line to see how we can write this sum as a multiplication problem.

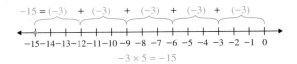

When we add −3 *five times,* we can say that −3 *times* 5 *equals* −15, which can be written as $-3 \times 5 = -15$ or $-3 \cdot 5 = -15$ or $-3(5) = -15$. Recall that the numbers being multiplied together are called the **factors,** and the answer to a multiplication problem is called the **product.**

$$-3 \times 5 = -15$$

−3 is a factor. 5 is a factor. The product is −15.

So, what is the rule for multiplying numbers with different signs?

> **Procedure** Multiplying Two Numbers with Different Signs
>
> The product of a *positive number* and a *negative number* is *negative.*
>
> *Example:* $-3 \times 5 = -15$

We can also illustrate this rule with a table.

As these numbers decrease by 1,	$3 \times 5 = 15$	the products decrease by 5.
	$2 \times 5 = 10$	
	$1 \times 5 = 5$	
	$0 \times 5 = 0$	
	$-1 \times 5 = -5$	
	$-2 \times 5 = -10$	
	$-3 \times 5 = -15$	

EXAMPLE 1

Multiply.

a) -2×9 b) $8 \cdot (-7)$

Solution

a) $-2 \times 9 = -18$ The product of a positive and a negative number is negative.

b) $8 \cdot (-7) = -56$ The product of a positive and a negative number is negative.

[YOU TRY 1] Multiply.

a) $9(-6)$ −42   b) -5×10 . −50

Note

Notice that, just like with addition and subtraction, we use parentheses to separate the multiplication symbol in Example 1b) from the negative sign in $8 \cdot (-7)$.

2 Multiply Two Integers with the Same Sign

We know that when we multiply two positive numbers, the product is positive. For example, $3 \times 5 = 15$. What is the sign of the product of two negative numbers? Again, let's make a table.

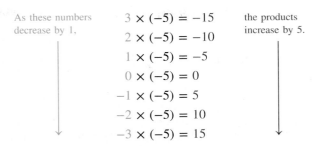

As these numbers decrease by 1,

$$3 \times (-5) = -15$$
$$2 \times (-5) = -10$$
$$1 \times (-5) = -5$$
$$0 \times (-5) = 0$$
$$-1 \times (-5) = 5$$
$$-2 \times (-5) = 10$$
$$-3 \times (-5) = 15$$

the products increase by 5.

The table illustrates that the product of two negative numbers is a positive number.

Procedure Multiplying Two Numbers with the Same Sign

The product of two numbers with the *same sign* is *positive*.

Example: $-2 \times (-5) = 10$

EXAMPLE 2

Multiply.

a) $-6 \cdot (-3)$ b) $-10(-9)$ c) 4×12

Solution

a) $-6 \cdot (-3) = 18$ The product of two negative numbers is positive.

b) $-10(-9) = 90$ The product of two negative numbers is positive.

c) $4 \times 12 = 48$ The product of two positive numbers is positive.

[YOU TRY 2] Multiply.

a) $-5(-10)$ b) $-4 \times (-3)$ c) $8 \cdot 7$

---

3 Multiply More Than Two Integers

Often, we have to multiply more than two integers. Use this procedure.

Procedure Multiplying More Than Two Numbers

If two factors are being multiplied inside parentheses, perform that multiplication first. If there is no product inside parentheses, multiply from left to right.

EXAMPLE 3

Multiply.

a) $-5 \cdot (-3 \cdot 4)$ b) $-2 \times 5 \times (-4) \times (-1)$

Solution

a) $-5 \cdot (-3 \cdot 4)$ contains a product inside parentheses, so perform that multiplication first.

$$-5 \cdot (-3 \cdot 4) \qquad \text{Multiply inside parentheses first.}$$
$$-5 \cdot \ (-12) \qquad \text{The product of two negative numbers is positive.}$$
$$60$$

b) $-2 \times 5 \times (-4) \times (-1)$ does *not* contain a product inside parentheses. Multiply from left to right.

$$-2 \times 5 \times (-4) \times (-1) \qquad \text{Multiply the first two factors on the left.}$$
$$-10 \times (-4) \times (-1) \qquad \text{Multiply } -10 \text{ and } -4.$$
$$40 \qquad \times (-1) \qquad \text{Multiply the last two factors.}$$
$$-40$$

$\begin{bmatrix} \textbf{YOU TRY 3} \end{bmatrix}$

Multiply.

a) $-2 \cdot (-2 \cdot 6) \cdot (-3)$ -72 b) $-1 \times 3 \times 2 \times (-7)$

Note

Notice that, in Example 3a), there are two negative factors, and the product is positive. In Example 3b), there are three negative factors, and the product is negative. These are examples of the following properties:

1)   If a multiplication problem has an **even** number of negative factors, the product will be **positive.**

2)   If a multiplication problem has an **odd** number of negative factors, the product will be **negative.**

4   Use Properties of Multiplication

In Section 1.3, we learned some properties of addition. Now we will learn some properties of multiplication.

What happens when we multiply a number by 0? For example,

$$0 \times 4 = 0 \qquad\qquad -8 \cdot 0 = 0 \qquad\qquad 0(51) = 0$$

Whenever we multiply a number and 0, the product is 0. This is called the *multiplication property of* 0.

Property The Multiplication Property of 0

The **multiplication property of 0** says that if we multiply a number by 0, the product is 0.

Examples: $7 \times 0 = 0$, $0(-2) = 0$, $-5 \cdot 3 \cdot 0 = 0$

What happens when we multiply a number by 1?

$$1 \cdot 6 = 6 \qquad -9 \times 1 = -9 \qquad 762(1) = 762$$

When we multiply a number by 1, the result is that number.

Property The Identity Property of Multiplication

The **identity property of multiplication** says that if we multiply any number by 1, the result is that number. (The number 1 is called the **identity element** for multiplication.)

Examples: $1 \times 8 = 8$, $-4(1) = -4$, $1 \cdot (-317) = -317$

EXAMPLE 4

Multiply. Then, name the property of multiplication that you used.

a) $-5 \cdot 1$ b) 0×12

Solution

a) $-5 \cdot 1 = -5$ Identity property of multiplication

b) $0 \times 12 = 0$ Multiplication property of 0

YOU TRY 4

Multiply. Then, name the property of multiplication that you used.

a) $0 \cdot 18$ b) -22×1

If we multiply numbers in any order, does the product remain the same? *Yes!* For example, $4 \cdot 9 = 36$ and $9 \cdot 4 = 36$. These are examples of the *commutative property of multiplication*.

Property The Commutative Property of Multiplication

The **commutative property of multiplication** says that changing the order in which we multiply numbers does not change the product.

Example: $7 \times 3 = 21$ and $3 \times 7 = 21$

EXAMPLE 5

Multiply $-6 \cdot 2$, then rewrite the problem using the commutative property.

Solution

$-6 \cdot 2 = -12$ Using the commutative property, we get $2 \cdot (-6) = -12$.

[**YOU TRY 5**]

Multiply $-10 \cdot (-11)$, then rewrite the problem using the commutative property.

Like addition, multiplication is also associative.

W Hint

How are the commutative and associative properties the same for addition and multiplication?

Property The Associative Property of Multiplication

The **associative property of multiplication** says that changing the position of grouping symbols will not change the product.

Example: $(2 \cdot 4) \cdot 5 = 40$ and $2 \cdot (4 \cdot 5) = 40$

EXAMPLE 6

Evaluate $(-3 \times 5) \times 2$, then rewrite the problem using the associative property.

Solution

Original Product	**Using the Associative Property**
$(-3 \times 5) \times 2$	$-3 \times (5 \times 2)$
$-15 \quad \times 2 = -30$	$-3 \times \quad 10 = -30$

[**YOU TRY 6**]

Evaluate $-2 \times (8 \times 3)$, then rewrite the problem using the associative property.

The last property we will discuss is the *distributive property*. It involves both multiplication and addition or multiplication and subtraction. Problems involving the distributive property look like this:

$$5(2 + 7) \qquad \text{and} \qquad 5(8 - 1)$$

This means   This means
$5 \cdot (2 + 7)$. $5 \cdot (8 - 1)$.

Because there is no operation symbol between the 5 and the parenthesis, the operation is multiplication.

Property The Distributive Property

The **distributive property** says that multiplication *distributes* over addition and subtraction.

Examples: $5(2 + 7) = 5 \cdot 2 + 5 \cdot 7$ and $5(8 - 1) = 5 \cdot 8 - 5 \cdot 1$

EXAMPLE 7

Evaluate each expression by first performing the operations in parentheses. Then, use the distributive property to evaluate each expression.

a) $4(8 + 3)$ b) $7(10 - 6)$ c) $-2(1 + 4)$ d) $-3(9 - 5)$

Solution

a) **Evaluate**

$4(8 + 3)$

$4 \cdot 11 = 44$

Notice that the results are the same.

Using the Distributive Property

$4(8 + 3) = 4 \cdot 8 + 4 \cdot 3$

$= 32 + 12 = 44$

b) **Evaluate**

$7(10 - 6)$

$7 \cdot 4 = 28$

Using the Distributive Property

$7(10 - 6) = 7 \cdot 10 - 7 \cdot 6$

$= 70 - 42 = 28$

c) **Evaluate**

$-2(1 + 4)$

$-2 \cdot 5 = -10$

Using the Distributive Property

$-2(1 + 4) = -2 \cdot 1 + (-2) \cdot 4$

$= -2 + (-8) = -10$

d) **Evaluate**

$-3(9 - 5)$

$-3 \cdot 4 = -12$

Using the Distributive Property

$-3(9 - 5) = -3 \cdot 9 - (-3) \cdot 5$

$= -27 - (-15)$

$= -27 + 15$ Change *minus* -15 to *plus* 15.

$= -12$

[YOU TRY 7]

Evaluate each expression by first performing the operations in parentheses. Then, use the distributive property to evaluate each expression.

a) $6(1 + 10)$ b) $8(11 - 4)$ c) $-5(7 + 3)$ d) $-4(6 - 8)$

---

5 Use Front-End Rounding to Estimate a Product

Recall that a quick way to multiply numbers with several zeros at the end is to multiply the digits at the beginning of the number and then count the total number of zeros in the factors to get the number of zeros in the product. For example,

$$8 \times 12 = 96$$

$$8{,}000 \times 120{,}000 = 960{,}000{,}000$$

3 zeros +   4 zeros =   7 zeros in the product

We can use front-end rounding to estimate a product in this way.

EXAMPLE 8

Estimate the product $-7{,}214 \cdot 386$ using front-end rounding, then find the exact product.

Solution

To use front-end rounding, we round each number to the place furthest to the left.

Round $-7{,}214$ to the *thousands* place: $-7{,}000$

Round 386 to the *hundreds* place: 400

$$-7 \cdot 4 = -28$$

The estimate of the product is $-7{,}000 \cdot 400 = -2{,}800{,}000$

3 zeros + 2 zeros = 5 zeros in the product

W Hint

Use graph paper to line up your numbers correctly.

The exact product is

```
    - 7, 2 1 4
  ×     3 8 6
    4 3 2 8 4
  5 7 7 1 2
2 1 6 4 2
- 2,7 8 4,6 0 4
```

YOU TRY 8

Estimate the product $-8{,}748 \cdot 235$ using front-end rounding, then find the exact product. $1{,}800{,}000$

We can use this method of estimating a product to help us solve real-world problems.

EXAMPLE 9

A car manufacturer recalls $80{,}644$ cars to fix their airbags at a cost of $\$292$ each. Use front-end rounding to estimate the amount of money the company loses on this recall. Then, find the exact answer. Because the company will be losing money, use a negative number for the cost.

Solution

Let's estimate the answer first. Round each number to the place furthest to the left:

Round $80{,}644$ to the *ten-thousands* place: $80{,}000$

Round $-\$292$ to the *hundreds* place: $-\$300$

$$8 \cdot (-3) = -24$$

The estimate of the product is $80{,}000 \cdot (-\$300) = -\$24{,}000{,}000$

4 zeros +   2 zeros =   6 zeros in the product

W Hint

How can estimating an answer help you check the exact answer?

The company will lose approximately $\$24{,}000{,}000$.

The exact product is $80{,}644 \cdot (-\$292) = -\$23{,}548{,}048$.

The company will lose exactly $\$23{,}548{,}048$.

[YOU TRY 9] Stephanie just signed the documents for a mortgage. She will pay $1,857 each month for 360 months. Use front-end rounding to estimate the total amount she will make in mortgage payments. Then, find the exact answer.

ANSWERS TO [YOU TRY] EXERCISES

1) a) −54 b) −50   2) a) 50 b) 12 c) 56   3) a) −72 b) 42
4) a) 0; multiplication property of 0   b) −22; identity property of multiplication
5) 110; −11 · (−10) = 110   6) −48; (−2 × 8) × 3 = −48
7) a) 6(1 + 10) = 6 · 11 = 66; 6(1 + 10) = 6 · 1 + 6 · 10 = 6 + 60 = 66
b) 8(11 − 4) = 8 · 7 = 56; 8(11 − 4) = 8 · 11 − 8 · 4 = 88 − 32 = 56
c) −5(7 + 3) = −5 · 10 = −50; −5(7 + 3) = −5 · 7 + (−5) · 3 = −35 + (−15) = −50
d) −4(6 − 8) = −4 · (−2) = 8; −4(6 − 8) = −4 · 6 − (−4) · 8 = −24 − (−32) = 8
8) estimate: −1,800,000; exact: −2,055,780   9) estimate: $800,000; exact: $668,520

E Evaluate **1.6** Exercises   Do the exercises, and check your work.

Get Ready

Multiply the whole numbers. Recall that writing the multiplication problems on graph paper will help you line up the numbers correctly.

1) 529 × 62 *32,798*   2) (473)(58) *27,434*

3) 18,044 · 705 *12,791,020* 4) 26,508 × 305

5) (6,970)(4,052)   6) 8,290 · 5,074

28,242,440

Objective 1: Multiply Two Integers with Different Signs

7) The product of a negative number and a *pos.* number is negative.

8) The product of a positive number and a negative number is *neg.*

Multiply.

9) −5 · 3   10) −10 · 4

11) −7 × 8   12) −12 × 6

13) 4 × (−8)   14) 6 × (−3)

15) 8(−12)   16) 11(−11)

17) (0)(−17)   18) (−30)(0)

19) (−1)(38)   20) (−19)(1)

21) 27(−34)   22) 39(−56)

Fill in the blank.

23) 7 · *−9* = −63   24) −4 × *5* = −20

25) *7* · −12 = −84   26) *−3* · 5 = −15

Objective 2: Multiply Two Integers with the Same Sign

27) The product of a negative number and a *neg.* number is positive.

28) The product of a *pos.* number and a positive number is positive.

Multiply.

29) −3(−2)   30) −7(−11)

31) −5 × (−9)   32) −8 × (−3)

33) −123 · (−8)   34) −108 · (−11)

Fill in the blank.

35) −10 · *−10* = 80   36) _____ × (−6) = 54

37) *−1* × (−9) = 9   38) −7 · *0* = 0

Objective 3: Multiply More Than Two Integers

Multiply.

39) −4 · 3 · (−2)   40) 3 · (−5) · (−2)

41) 5(−8)(7)   42) −4(6)(9)

43) 6 × (−2) × (−7) × (−1)

44) −8 × 3 × (−1) × (−3)

45) −5(−1)(−3)(2)(−1)   46) 2(−3)(−1)(−7)(−1)

47) −10 · (3 · 4)   48) −6 · (2 · 6)

49) −7(−8 × 9)   50) −4(−7 × 5)

51) 2 · (−5 · 3) · (−1)   52) 3 · (−4 · 2) · (−1)

Fill in the blank.

53) 3 · *7* · (−2) = 42

54) −4 × *5* × 3 = −60

Find the product. Simplify the expression in the absolute values first.

55) $-3 \times |-10| \times (-7)$

56) $-2 \cdot |-6| \cdot (-9)$

57) $8 \cdot |7 - 18| \cdot (-1)$

58) $5 \times |6 - 13| \, (-1)$

Objective 4: Use Properties of Multiplication

Which property of multiplication is illustrated by each statement? Choose from the multiplication property of 0, the identity property, the commutative property, the associative property, and the distributive property.

59) $-8 \cdot 3 = 3 \cdot (-8)$

60) $0 \times 9 = 0$

61) $4(1 + 9) = 4 \cdot 1 + 4 \cdot 9$

62) $-6 \cdot (2 \cdot 5) = (-6 \cdot 2) \cdot 5$

63) $1(-17) = -17$

64) $-4(-12) = -12(-4)$

Find the product, then rewrite the expression using the indicated property and show that the result is the same.

 65) $(5 \cdot 2) \cdot 7$; associative property

66) $3(2 - 8)$; distributive property

67) -9×8; commutative property

68) $-2 \times (-6 \times 4)$; associative property

 69) $-4(5 + 3)$; distributive property

70) $7(12)$; commutative property

71) $-3(9 - 1)$; distributive property

72) $-8(10 - 6)$; distributive property

Objective 5: Use Front-End Rounding to Estimate a Product

Use front-end rounding to estimate each product. Then, find the exact product.

73) 927×48

74) $591 \cdot 62$

75) $-34{,}615 \cdot 518$

76) $-68{,}239 \times 355$

77) $(-9{,}607)(-7{,}592)$

78) $(-4{,}356)(-9{,}918)$

Use front-end rounding to estimate each answer. Then, find the exact answer. Express the answer as a positive or negative number accordingly.

 79) A cable television company estimates it will lose 68 customers every week to a satellite television service. Find the change in the number of customers over a one-year period.

80) Tarun ate a package of 8 chocolate chip cookies having 26 calories per cookie. How many calories did he eat?

©ermingut/Getty Images

81) The tuition at a community college is $96 per unit. How much will Nascha pay if she enrolls for 12 units?

82) A small plane ascended 275 ft each minute during a 23-minute takeoff. How many feet did the plane ascend during takeoff?

83) A department store finds that it loses $1,183 of profit every month due to shoplifting. Find the change in profit due to shoplifting in one year.

©McGraw-Hill Education/
Andrew Resek, photographer

84) A high-protein diet promises that a person will lose 12 lb every month. If Jay goes on the diet, find his change in weight over a 9-month period.

Mixed Exercises: Objectives 1–3

Perform the indicated operations.

85) -384×7

86) $-62(-51)$

87) $-15{,}307 \cdot (-41)$

88) $-2{,}803 \times 84$

89) $-5 \cdot 3 \cdot (-2) \cdot (-1)$

90) $-4 \cdot (-2) \cdot 7 \cdot (-1)$

91) $(-8)(-4)(3 \cdot 6)$

92) $(-12)(-6)(5 \cdot 2)$

93) $-7 \cdot |-2 \cdot 3| \cdot (-5)$

94) $-5 \cdot |-4 \cdot 1| \cdot (-2)$

Solve each problem.

95) On a 100-point PreAlgebra test, Natalia missed five two-point questions and two three-point questions. She got the four-point extra credit question correct. What was Natalia's test score?

96) Ray has $492 in his checking account. He deposits four $30 checks that he received for mowing his neighbors' yards and $279 from his part-time job. Later that day, he withdraws $140 at the ATM machine. Find the balance in Ray's checking account.

R1) Demonstrate the equation $-2 \times 4 = -8$ on a number line. Why does this make sense?

R2) Write two statements that outline a *Procedure for Multiplying Signed Numbers*.

R3) Were there any problems you could not do? If so, write them down or circle them and ask your instructor for help.

1.7 Dividing Integers and Estimation

P Prepare

O Organize

What are your objectives for Section 1.7?	How can you accomplish each objective?
1 Divide Integers	• Refer to Appendices A.4 and A.5 if you need to review division of whole numbers. • Write the procedure for **Dividing Numbers** in your own words. • Understand the properties of division. • Complete the given example on your own. • Complete You Try 1.
2 Combine Multiplying and Dividing of Integers	• Perform operations in parentheses first. Then, perform operations from left to right. • Complete the given example on your own. • Complete You Try 2.
3 Use Front-End Rounding to Estimate a Quotient	• Write a procedure for dividing numbers ending in zeros. • Complete the given example on your own. • Complete You Try 3.
4 Solve Applications Involving Division	• Understand how to use front-end rounding to estimate the answer to a real-world problem. Then, find the exact answer. • Understand the meaning of a remainder in an application problem. • Complete the given examples on your own. • Complete You Trys 4 and 5.

W Work   Read the explanations, follow the examples, take notes, and complete the You Trys.

1 Divide Integers

There are several different ways to write a division problem. For example, $20 \div 4 = 5$ can be written as

$$
\text{Divisor} \to 4\overset{\overset{\text{Quotient}}{\downarrow}}{\overline{)20}} \qquad \text{or} \qquad \overset{\overset{\text{Dividend}}{\downarrow} \quad \overset{\text{Quotient}}{\downarrow}}{20 \div 4 = 5} \qquad \text{or} \qquad \overset{\overset{\text{Dividend}}{\downarrow}}{\dfrac{20}{4}} = 5 \leftarrow \text{Quotient}
$$

The answer to a division problem is called the **quotient.**

Division and multiplication are opposite operations and are related in the following ways:

$$\text{If } 20 \div 4 = 5, \text{ then } 5 \cdot 4 = 20.$$
$$\text{If } 5 \cdot 4 = 20, \text{ then } 20 \div 4 = 5.$$

Therefore, we can check every division problem using multiplication, and we can check every multiplication problem using division.

The rules for dividing integers are the same as the rules for multiplying integers.

Procedure Dividing Numbers

1) The quotient of two numbers with *different signs* is *negative.*

2) The quotient of two numbers with the *same sign* is *positive.*

EXAMPLE 1

Divide.

a) $-18 \div 3$ b) $-45 \div (-9)$ c) $\dfrac{28}{-7}$ d) $\dfrac{-72}{6}$

e) $\dfrac{-11}{-11}$ f) $3 \div 1$ g) $\dfrac{0}{-2}$ h) $\dfrac{-4}{0}$

Solution

a) $-18 \div 3 = -6$ The quotient of two numbers with *different signs* is *negative.*

b) $-45 \div (-9) = 5$ The quotient of two numbers with the *same sign* is *positive.*

W Hint

How can knowing the 1–12 multiplication facts help you with division?

c) $\dfrac{28}{-7} = -4$ The quotient of two numbers with *different signs* is *negative.*

d) $\dfrac{-72}{6} = -12$

e) $\dfrac{-11}{-11} = 1$ Remember that any nonzero number divided by itself is 1.

f) $3 \div 1 = 3$ Any nonzero number divided by 1 equals the number.

g) $\dfrac{0}{-2} = 0$ 0 divided by a nonzero number is 0.

h) $\dfrac{-4}{0}$ is undefined.   Division by 0 is undefined.

Divide.

a) $60 \div (-5)$ b) $-49 \div (-7)$ c) $\dfrac{-36}{-9}$ d) $\dfrac{8}{-2}$

e) $\dfrac{-54}{1}$ f) $\dfrac{-7}{0}$ g) $\dfrac{0}{-10}$ h) $\dfrac{-17}{-17}$

Examples 1e), f), g), and h) illustrate special properties of division. We summarize them here.

Properties Properties of Division

1) Any nonzero number divided by itself equals 1.   Example 1e): $\dfrac{-11}{-11} = 1$

2) Any nonzero number divided by 1 equals the number.   Example 1f): $3 \div 1 = 3$

3) Zero divided by any nonzero number equals zero.   Example 1g): $\dfrac{0}{-2} = 0$

4) Any number divided by zero is undefined.   Example 1h):
$\dfrac{-4}{0}$ is undefined.

(This means that there is no answer when we try to divide a number by 0.)

W Hint

Be sure you learn these properties. They are *very* important.

The easiest way to see that division by 0 is undefined is to try to write a related multiplication problem: If $\dfrac{-4}{0} = $ *some number,* then $0 \cdot $ *some number* $= -4$. But is there a number you can multiply by 0 to get -4? **No!** Division by 0 is undefined.

2 Combine Multiplying and Dividing of Integers

Sometimes, an expression will contain multiplication and division. If there is multiplication or division inside parentheses, perform those operations first. Then, perform the operations from left to right. When there are two sets of grouping symbols, we often use brackets, [ ]. (This is true when we use parentheses to separate an operation symbol from a negative sign, and then we use brackets to tell us which operations to perform first.)

EXAMPLE 2

Simplify.

a) $(-10 \cdot 6) \div (4 \cdot 5)$ b) $-3 \cdot [56 \div (-8)]$

c) $-20 \div 10(-18)(-2) \div (-72)$

Solution

a) $(-10 \cdot 6) \div (4 \cdot 5)$ contains products inside two sets of parentheses. Perform the multiplication in each set of parentheses first.

$(-10 \cdot 6) \div (4 \cdot 5)$ Multiply inside each set of parentheses first.

$-60 \div 20$

-3 The quotient of a negative number and a positive number is negative.

b) $-3 \cdot [56 \div (-8)]$ contains parentheses and brackets. The parentheses are used to just separate the division symbol, $\div$, from the negative sign on -8. Inside the brackets is a division problem. Perform the division first.

$-3 \cdot [56 \div (-8)]$ Divide inside the brackets first.

$-3 \cdot \quad -7$

21 The product of two negative numbers is positive.

c) $-20 \div 10(-18)(-2) \div (-72)$ does *not* contain any operations inside parentheses. Here, the parentheses separate the numbers from each other to indicate multiplication. Perform the operations from left to right, two numbers at a time.

$-20 \div 10(-18)(-2) \div (-72)$ Divide the first two numbers on the left.

$-2(-18)(-2) \div (-72)$ Multiply -2 and -18.

$36(-2) \div (-72)$ Multiply 36 and -2.

$-72 \div (-72)$ Divide.

1

[YOU TRY 2]   Simplify.

a) $(-54 \div 9) \cdot (-21 \div 7)$ b) $-4[-50 \div (-2)] \div 20$

c) $2(-36)(-1) \div (-9) \div 8$

---

3 Use Front-End Rounding to Estimate a Quotient

Recall that a quick way to divide numbers ending in zeros is to "divide out" zeros to create a simpler division problem. We can do this because it is like dividing by 10, 100, 1,000, or some other power of 10. For example,

$-15,0\cancel{0}\cancel{0} \div 5\cancel{0}\cancel{0} = -150 \div 5 = -30$
Dividing out two zeros is the same as dividing by 100.

$\dfrac{-8\cancel{0}\cancel{0}\cancel{0}}{-2\cancel{0}\cancel{0}\cancel{0}} = \dfrac{-8}{-2} = 4$
Dividing out three zeros is the same as dividing by 1,000.

We can use front-end rounding to estimate a division problem in this way.

EXAMPLE 3

Estimate $-9,843 \div 17$ using front-end rounding, then find the exact answer.

Solution

To use front-end rounding, we round each number to the place furthest to the left.

Round $-9,843$ to the *thousands* place: $-10,000$

Round 17 to the *tens* place: 20

The estimate is $-10,000 \div 20 = -1,000 \div 2 = -500.$

Divide out one zero
in each number.

Hint

Review long division, if necessary, and use graph paper to do the division.

To find the exact answer, do long division. We will not include the negative sign in the division process. We will put it in the final answer.

```
          5 7 9
   1 7 ) 9, 8 4 3
       - 8 5
         1 3 4
       - 1 1 9
           1 5 3
         - 1 5 3
               0
```

The exact answer is $-9,843 \div 17 = -579.$

[YOU TRY 3] Estimate $-19,008 \div (-396)$ using front-end rounding, then find the exact answer.

4 Solve Applications Involving Division

Many real-world problems involve division. Let's use rounding to estimate the answer to the first problem.

EXAMPLE 4

In the month of March (which has 31 days), Nasser spent $279 on gas for his car. Use front-end rounding to estimate the average amount he spent on gas per day. Then, find the exact answer.

Solution

To find the average amount spent on gas per day, we divide the total amount spent by the number of days:

$$\frac{\text{Total amount spent on gas}}{\text{Number of days}} = \text{Average amount spent per day}$$

To estimate the average amount Nasser spent per day,

Round $279 to the *hundreds* place: $300

Round 31 to the *tens* place: 30

The estimate of the average is $\dfrac{\$300}{30} = \$10$ per day.

To find the exact answer, we divide $\dfrac{\$279}{31}$. Use long division.

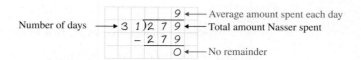

Notice that there is no remainder. The exact answer is that Nasser spent an average of $9 per day on gas during the month of March.

[**YOU TRY 4**] Phuong spent $120 on party favors for her daughter's birthday party, and the party favors for each child cost $8. How many children attended the party? Use front-end rounding to estimate the answer. Then, find the exact answer.

What does it mean if a division problem has a remainder?

EXAMPLE 5 A demolition company must haul away the waste from a building it will knock down. The foreman has to plan ahead so that there are enough dump trucks to move the waste from the site. He calculates that there will be 205 cubic yards of waste, and each truck can hold 18 cubic yards of material. How many trucks will be needed to carry away all of the waste?

Solution

W Hint

Are you writing out the example as you are reading it?

We can solve this problem using division.

$$(\text{Amount of waste}) \div \left(\begin{array}{c}\text{Amount each truck}\\ \text{will hold}\end{array}\right) = (\text{Number of trucks needed})$$

$$205 \text{ cubic yd} \quad \div \quad 18 \text{ cubic yd} \quad = \quad \text{Number of trucks needed}$$

Do long division.

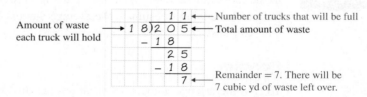

How do we interpret the division problem with a remainder? The foreman will need 11 trucks that will each contain 18 cubic yards of waste, and then there will be 7 cubic yards (the remainder) of waste left over. Therefore, another truck is needed to haul away the *remainder* of the waste.

The total number of dump trucks needed is 11 + 1 = 12.

Number of full trucks ↑ ↑ 1 truck to hold the remaining 7 cubic yd of waste

Total number of trucks

ANSWERS TO [YOU TRY] EXERCISES

1) a) −12 b) 7 c) 4 d) −4 e) −54 f) undefined g) 0 h) 1
2) a) 18 b) −5 c) −1   3) estimate: 50; exact: 48
4) estimate: 10; exact: 15   5) 4 vans

E Evaluate 1.7 Exercises   Do the exercises, and check your work.

Get Ready

Divide the whole numbers. Recall that writing the division problems on graph paper will help you line up the numbers correctly.

1) $5,622 \div 3$

2) $\dfrac{9,916}{4}$

3) $\dfrac{684}{9}$

4) $776 \div 8$

5) $\dfrac{43,176}{21}$

6) $\dfrac{68,442}{34}$

7) $54,000,000 \div 600$

8) $35,000,000 \div 700$

9) $815,300 \div 12$

10) $561,700 \div 15$

Objective 1: Divide Integers

11) The quotient of a negative number and a _____ number is positive.

12) The quotient of a positive number and a _____ number is positive.

Divide.

13) $-16 \div (-2)$

14) $-32 \div (-8)$

15) $-35 \div 7$

16) $-12 \div 3$

17) $\dfrac{48}{-6}$

18) $\dfrac{45}{-5}$

19) $\dfrac{-72}{0}$

20) $\dfrac{-50}{0}$

21) $63 \div (-1)$

22) $27 \div (-1)$

23) $\dfrac{14}{1}$

24) $\dfrac{38}{1}$

25) $\dfrac{-20}{-4}$

26) $\dfrac{-18}{-6}$

27) $0 \div (-5)$

28) $0 \div (-9)$

29) $\dfrac{-45}{-45}$

30) $\dfrac{-36}{-36}$

31) $\dfrac{-28}{28}$

32) $\dfrac{-12}{12}$

33) $-174 \div (-6)$

34) $-848 \div (-8)$

Fill in the blank.

35) $-55 \div$ _____ $= 5$

36) $36 \div$ _____ $= -3$

37) _____ $\div 6 = -7$

38) _____ $\div (-4) = 10$

39) Write an expression so that the quotient of two integers is −8.

40) Write an expression so that the quotient of two integers is −5.

Objective 2: Combine Multiplying and Dividing of Integers

Simplify.

41) $(9 \cdot 8) \div (-3 \cdot 4)$

42) $(-6 \cdot 10) \div (5 \cdot 3)$

43) $-2[140 \div (-7)]$

44) $-9[84 \div (-12)]$

45) $(-45 \div 9) \cdot [-6 \div (-2)]$

46) $[-54 \div (-6)] \cdot (-11 \div 11)$

47) $18(-5) \div (-3)(-2) \div (-15)$

48) $10(-14) \div (-7)(-4) \div (-5)$

Simplify each expression by evaluating the absolute value expression first.

49) $600 \div |-20| \cdot (-1)$

50) $150 \div |-15| \cdot (-1)$

51) $-8|-13 + 7| \div (-12)$

52) $-12|-9 + 4| \div (-4)$

Objective 3: Use Front-End Rounding to Estimate a Quotient

Divide.

53) $\dfrac{18,000}{300}$

54) $\dfrac{250,000}{500}$

55) $\dfrac{4,000,000}{-2,000}$

56) $\dfrac{-9,000,000}{30,000}$

Use front-end rounding to estimate each quotient. Then, find the exact answer.

57) $782 \div 23$

58) $672 \div 24$

59) $26,738 \div (-58)$

60) $-19,317 \div 47$

61) $-320,913 \div (-543)$

62) $-230,082 \div (-186)$

Objective 4: Solve Applications Involving Division

Solve each problem. Express the answer as a positive or negative number.

63) The number of employees at a factory has decreased by 252 workers over 4 years. What was the average change in the number of employees each year?

64) A survey showed that the number of homes with landline phones in a certain suburb decreased by 588 from 2011 to 2017. What was the average change in the number of landline phones each year?

65) From the year 2010 to 2015, the population of North Dakota increased by about 84,000. What was the average change in the population each year? (quickfacts.census.gov)

66) The number of students in public, U.S. high schools increased by about 3,600,000 from 1990 to 2015. What was the average change in the number of students per year? (www.census.gov)

Solve each problem. Use front-end rounding to estimate the answer. Then, find the exact answer.

67) Over its last 18 games, a basketball team scored a total of 1,710 points. Find the average number of points scored per game.

68) A concert sold out in 47 min. If 13,395 tickets were sold, find the number of tickets sold per minute.

69) During a 24-hour period, a certain smart-phone app was downloaded 32,880 times. Find the average number of downloads per hour.

70) Over the last 12 months, Neha saved $816. Find the average amount she saved each month.

Solve each problem. Be sure to think about how to interpret any remainders.

71) State regulations say that there must be at least one teacher for every 8 toddlers in a day-care center. If a day-care center has 53 toddlers, what is the minimum number of teachers the center must have?

72) A shuttle bus from a parking lot to a football stadium can hold 48 people. How many trips would it take for 288 people to get to the stadium?

73) A roller coaster can seat 24 people. How many trips would it take for 400 people to ride the roller coaster?

74) A student group held a coat drive to collect winter coats for people who need them. They received 376 coats, and each box will contain 6 coats. How many boxes are needed?

75) A cruise ship takes people to shore in smaller boats called *tenders*. If each tender can hold 48 passengers, how many are needed to carry 1,238 people to shore?

76) Once on shore, 438 people signed up for an excursion to a volcano. If each bus can hold 13 people, how many buses are needed?

R Rethink

R1) Do you understand how to interpret the meaning of a remainder in an application problem? Look at Exercise 73 and write out an explanation, in words, of how you solved this problem.

R2) Did you try using graph paper for long division? If so, did it help to keep the numbers lined up correctly? If not, do you think it could help you?

R3) How have you used division in the last week?

Putting It All Together

What are your objectives?	How can you accomplish each objective?
1 Review the Concepts of Sections 1.1–1.7	• Be sure that you can apply the objectives you have learned in the previous sections. • If you are not confident on a certain example, go back to the section that gives more explanation. • Complete the given example on your own. • Complete You Try 1.

W Work

Read the explanations, follow the examples, take notes, and complete the You Try.

1 Review the Concepts of Sections 1.1–1.7

Let's review all of the operations together.

EXAMPLE 1

Simplify each expression.

a) $-9(7)$ b) $-629 - (-158)$ c) $-74 + (-92)$

d) $\dfrac{-84}{-12}$ e) $\dfrac{-3}{0}$ f) $-120 \div (-6 \cdot 2)(-5)$

g) $-5(2)(-3)(-4)(-1)$

Solution

a) The product of a negative number and a positive number is negative: $-9(7) = -63$

b) Begin by changing the subtraction problem to addition.

$$-629 - (-158) = -629 + 158 \quad \text{Change subtraction to addition of the additive inverse of } -158.$$

W Hint

Be sure you can recognize when to use the rules for adding, subtracting, multiplying, and dividing integers.

Find the absolute value of each number: $|-629| = 629$, $|158| = 158$

Subtract the smaller absolute value from the larger absolute value.

$$629 - 158 = 471$$

Larger absolute value   Smaller absolute value

The sign of the final answer will be *negative* because -629 has a larger absolute value than 158.

$$-629 - (-158) = -471$$

c) To find $-74 + (-92)$, remember that the sum of two negative numbers is always negative.

Find the absolute value of each number: $|-74| = 74$, $|-92| = 92$

Add the absolute values: $74 + 92 = 166$

Put a negative sign in front of the sum: -166

$$-74 + (-92) = -166$$

d) The quotient of two negative numbers is positive: $\dfrac{-84}{-12} = 7$

e) A nonzero number divided by 0 is undefined: $\dfrac{-3}{0}$ is undefined.

f) To evaluate $-120 \div (-6 \cdot 2)(-5)$, do the multiplication in the parentheses first. Then perform the operations from left to right, two numbers at a time.

$$-120 \div (-6 \cdot 2)(-5) \qquad \text{Multiply inside the parentheses first.}$$

$$-120 \div \ \ (-12)(-5) \qquad \text{Perform the operations from left to right. Divide.}$$

$$10(-5) \qquad \text{Multiply.}$$

$$-50$$

g) Recall that if a multiplication problem has an **even** number of negative factors, the product will be **positive.** If a multiplication problem has an **odd** number of negative factors, the product will be **negative.** The expression $-5(2)(-3)(-4)(-1)$ has 4 negative factors, so the product will be positive. Because there are no operations within parentheses, we will evaluate this expression by multiplying from left to right.

$$-5(2)(-3)(-4)(-1) \qquad \text{Multiply } -5 \text{ and } 2.$$

$$-10(-3)(-4)(-1) \qquad \text{Multiply } -10 \text{ and } -3.$$

$$30(-4)(-1) \qquad \text{Multiply } 30 \text{ and } -4.$$

$$-120(-1) \qquad \text{Multiply.}$$

$$120$$

As expected, the product is positive.

[YOU TRY 1] Simplify each expression.

a) $\dfrac{0}{-19}$ b) $-4(-3)(1)(-5)(2)$ c) $107 - 694$

d) $-37 \times (-41)$ e) $-5{,}011 + (-6{,}292)$ f) $129 \div (-3)$

g) $(-144 \div 12) \div (2 \cdot 3) \cdot (-7)$

ANSWERS TO [YOU TRY] EXERCISES

1) a) 0   b) −120   c) −587   d) 1,517   e) −11,303   f) −43   g) 14

Putting It All Together Exercises

 **E Evaluate** Do the exercises, and check your work.

Objective 1: Review the Concepts of Sections 1.1–1.7

Simplify each expression.

1) $-341 - (-186)$ 2) $6 \times (-9)$

3) $-72 + (-109)$ 4) $0 \div 16$

5) $-114 \div (-6)$ 6) $5{,}217 - 9{,}608$

7) $-3(-2)(4)(-5)$ 8) $\dfrac{-305}{-5}$

9) $\dfrac{8}{0}$ 10) $7 \cdot (-60 \div 12)$

11) $-280 \div (-7 \cdot 4)(-3)$ 12) $-12(5)(0)(-2)$

13) $-7(-8)$ 14) $3{,}451 \div (-17)$

15) $-9[48 \div (-4)]$

16) $[8 \cdot (-9)] \div [-2 \cdot 6]$

17) $11(-7)(2)(0)$

18) $59 - (-93)$

19) $\dfrac{-1{,}404}{13}$

20) $-6(7)(-1)(-2)$

21) $\dfrac{0}{-28}$

22) $-3|1 - 16| \div (-9)$

23) $[1{,}000 \div (-25)] \cdot [6 \div (-6)6]$

24) $-216 + 143 - 198 + 74$

25) $2|17 - 29| \div (-3)$

26) $350 \div (-5 \cdot 2)(-4)$

27) $772 - 8{,}180 + 5{,}006 + (-49)$

28) $\dfrac{-6}{0}$

Fill in the blank.

29) _____ $\times 7 = -84$

30) $-213 +$ _____ $= 62$

31) $-109 -$ _____ $= -456$

32) _____ $\div (-1) = 8$

Round to the indicated place.

33) 2,748,190; hundred-thousands

34) 526,043; thousands

35) −16,981; hundreds

36) −83,527,664; millions

Estimate the sum or difference by first rounding each number to the nearest thousand. Then, find the exact answer.

37) $705{,}419 - 426{,}803$

38) $-61{,}156 + (-8{,}740)$

Use front-end rounding to estimate each sum, difference, product, or quotient. Then, find the exact answer.

39) $-5{,}271 \times 29$

40) $3{,}284 - 17{,}605$

41) $892 + 6{,}137$

42) $-52{,}125 \div (-125)$

Evaluate the expression, then rewrite the expression using the indicated property and show that the result is the same.

43) $9(4 - 6)$; distributive property

44) $-3(-10)$; commutative property

45) $-5 + (15 + 7)$; associative property

46) $-6(9 + 2)$; distributive property

Solve each problem. Find the exact answer. Use a signed number in the answer.

47) A football team gained 8 yd on its first play, gained 9 yd on its second play, lost 15 yd on the third play, and lost 7 yd on the fourth play. What was the team's net yardage on these four plays?

48) Deepak has $42 in his checking account. He is charged a $15 fee because his balance fell below $100. Then, he deposits $119, withdraws $80 from an ATM, and writes a check for $78. What is the balance in Deepak's account?

Solve each problem. Use front-end rounding to estimate each answer. Then, find the exact answer.

49) A baby stroller is recalled by its manufacturer. If 5,708 strollers were sold at a cost of $129 each, how much money will the company lose if each person returns the stroller?

50) A group of 394 students and chaperones will tour Washington D.C. in buses that each hold 36 people. Find the fewest number of buses needed.

R Rethink

R1) Could you remember all the rules for performing operations when the problems were mixed up? If not, can you think of a way to help you remember the rules?

R2) Is there a certain topic that was harder than others? If so, go back to that section and work the examples again.

1.8 Exponents, Roots, and Order of Operations

P Prepare

O Organize

What are your objectives for Section 1.8?	How can you accomplish each objective?
1 Use Whole Numbers with Exponents	• Understand why exponents are used, and write a sentence describing exponential expressions using the words *base* and *power*. • Memorize the common powers of whole numbers listed in the table in this section. • Complete the given example on your own. • Complete You Try 1.
2 Use Integers with Exponents	• If a negative number is being raised to a power, it is *always* in parentheses. • Complete the given examples on your own. • Complete You Trys 2 and 3.
3 Find Square Roots	• Understand how to find a *square root* and note what makes a *perfect square*. • Complete the given example on your own. • Complete You Try 4.
4 Use the Order of Operations	• Write the procedure for **The Order of Operations** in your own words using **Please Excuse My Dear Aunt Sally**. • Complete the given examples on your own. • Complete You Trys 5–7.

W Work

Read the explanations, follow the examples, take notes, and complete the You Trys.

1 Use Whole Numbers with Exponents

Exponents are used to represent repeated multiplication. For example,

$$3 \cdot 3 = 3^2 \leftarrow \text{Exponent or Power}$$
$$\uparrow$$
$$\text{Base}$$

The **base** is 3, and the **exponent,** or **power,** is 2. The exponent tells us how many times to use the base as a factor in the multiplication problem. We read the exponential expression 3^2 as "3 squared" or "3 to the second power." When we find that $3^2 = 9$, we are *evaluating* 3^2.

EXAMPLE 1 Identify the base and the exponent, then evaluate each expression.

a) 8^2 b) 5^3 c) 2^5

Solution

a) 8^2 ← Exponent   The exponent 2 tells us to use 8 as a factor 2 times.
   ↑
   Base

$$8^2 = 8 \cdot 8 = 64$$

b) 5^3 ← Exponent   The exponent 3 tells us to use 5 as a factor 3 times.
   ↑
   Base

$$5^3 = 5 \cdot 5 \cdot 5 = 125$$

We read 5^3 as "5 to the third power" or "5 cubed."

c) 2^5 ← Exponent   The exponent 5 tells us to use 2 as a factor 5 times.
   ↑
   Base

$$2^5 = 2 \cdot 2 \cdot 2 \cdot 2 \cdot 2 = 32$$

We read 2^5 as "2 to the fifth power."

[YOU TRY 1] Identify the base and the exponent, then evaluate each expression.

a) 7^2 b) 4^3 c) 2^6

Certain powers of whole numbers are used often in mathematics. We list them here for you to memorize.

 Hint

Quiz yourself by making flash cards.

Powers to Memorize							
$2^1 = 2$	$3^1 = 3$	$4^1 = 4$	$5^1 = 5$	$6^1 = 6$	$8^1 = 8$	$10^1 = 10$	
$2^2 = 4$	$3^2 = 9$	$4^2 = 16$	$5^2 = 25$	$6^2 = 36$	$8^2 = 64$	$10^2 = 100$	
$2^3 = 8$	$3^3 = 27$	$4^3 = 64$	$5^3 = 125$			$10^3 = 1,000$	
$2^4 = 16$	$3^4 = 81$						
$2^5 = 32$				$7^1 = 7$	$9^1 = 9$	$11^1 = 11$	
$2^6 = 64$				$7^2 = 49$	$9^2 = 81$	$11^2 = 121$	
$12^1 = 12$	$13^1 = 13$	$14^1 = 14$	$15^1 = 15$				
$12^2 = 144$	$13^2 = 169$	$14^2 = 196$	$15^2 = 225$				

2 Use Integers with Exponents

If an exponent is used to represent repeated multiplication of a *negative number*, we *must* put the number in parentheses.

EXAMPLE 2

Write each multiplication problem using an exponent.

a) $-6 \cdot (-6) \cdot (-6)$ b) $-9 \cdot (-9) \cdot (-9) \cdot (-9)$

Solution

a) $\underbrace{-6 \cdot (-6) \cdot (-6)}_{3 \text{ factors of } -6} = (-6)^3$ The factor of -6 is multiplied 3 times.

b) $\underbrace{-9 \cdot (-9) \cdot (-9) \cdot (-9)}_{4 \text{ factors of } -9} = (-9)^4$ The factor of -9 is multiplied 4 times.

[YOU TRY 2]

Write each multiplication problem using an exponent.

a) $-7 \cdot (-7) \cdot (-7) \cdot (-7)$ b) $-2 \cdot (-2) \cdot (-2) \cdot (-2) \cdot (-2)$

EXAMPLE 3

Evaluate.

a) $(-5)^3$ b) $(-3)^2$

Solution

a) $(-5)^3 = -5 \cdot (-5) \cdot (-5)$ Multiply -5 by itself 3 times.

       $= 25 \cdot (-5)$

       $= -125$

b) $(-3)^2 = \underbrace{-3 \cdot (-3)}_{2 \text{ factors of } -3} = 9$

> **W Hint**
>
> How can an objective from Section 1.6 help you to evaluate these problems more quickly?

[YOU TRY 3]

Evaluate.

a) $(-9)^2$ b) $(-2)^3$

---

3 Find Square Roots

> **W Hint**
>
> Finding a square root is the opposite of squaring a number.

At the beginning of this section, we found that $3^2 = 9$. The opposite of squaring a number is finding a *square root*. The **square root** of a number, like 9, is a number that, when squared, results in the given number. We use the symbol $\sqrt{}$ to represent a square root. Therefore, we can write $\sqrt{9} = 3$ because $3^2 = 9$. We say that 9 is a **perfect square** because it is the square of a whole number.

EXAMPLE 4

Evaluate.

a) $\sqrt{49}$ b) $\sqrt{144}$ c) $\sqrt{1}$ d) $\sqrt{0}$

Solution

a) $\sqrt{49} = 7$ since $7^2 = 49$.   b) $\sqrt{144} = 12$ since $12^2 = 144$.

c) $\sqrt{1} = 1$ since $1^2 = 1$.   d) $\sqrt{0} = 0$ since $0^2 = 0$.

[YOU TRY 4]

Evaluate.

a) $\sqrt{4}$ b) $\sqrt{25}$ c) $\sqrt{121}$ d) $\sqrt{100}$

Using the numbers we will study, negative numbers do *not* have square roots. We can see why if we try to relate it to the square of a number.

Finding $\sqrt{-16}$, for example, means that the *square* of some number equals -16. But,

$$4^2 = 4 \cdot 4 = 16 \quad \text{and} \quad (-4)^2 = -4 \cdot (-4) = 16$$

There is no number we can square to get -16, so we cannot find $\sqrt{-16}$. (In later math courses, you may learn about a special set of numbers called *imaginary numbers*. They can be used to find square roots of negative numbers. An electrician is someone who would work with imaginary numbers.)

4 Use the Order of Operations

If we are asked to evaluate an expression that contains different operation symbols, like $6 \times 7 - 5^2 + 12$, we need rules to tell us the order in which to perform the operations. We call these rules the **order of operations.**

Procedure The Order of Operations

Simplify expressions in the following order:

1) If **parentheses** or **other grouping symbols** appear in an expression, simplify what is inside these grouping symbols first.
2) Simplify expressions with **exponents** or **square roots.**
3) **Multiply** or **divide** moving from left to right.
4) **Add** or **subtract** moving from left to right.

Note

It may be helpful to remember the order of operations by remembering this sentence: Please Excuse My Dear Aunt Sally. The first letter of each word tells us the order in which we perform operations: Parentheses, Exponents, Multiplication or Division (from left to right), and Addition or Subtraction (from left to right). Remember that multiplication and division are at the same "level" in the process of performing operations, and that addition and subtraction are at the same "level."

EXAMPLE 5

Simplify each expression using the order of operations.

a) $19 - 10 + 3$ b) $-30 \div 3 \cdot 2$

c) $24 + 48 \div (-8)$ d) $6 \times 7 - 5^2 + 12$

Solution

a) When an expression contains only addition and subtraction, perform the operations from left to right.

$$19 - 10 + 3$$
$$\underbrace{9} + 3 = 12$$

 Hint

Remember "Please Excuse My Dear Aunt Sally."

b) When an expression contains only multiplication and division, perform the operations from left to right.

$$-30 \div 3 \cdot 2$$
$$\underbrace{-10} \cdot 2 = -20$$

c) Perform division before addition.

$$24 + 48 \div (-8)$$
$$24 + \underbrace{(-6)} = 18$$

d)
$$6 \times 7 - 5^2 + 12$$
$$6 \times 7 - 25 + 12 \qquad \text{Evaluate exponents before multiplication, addition, and subtraction.}$$
$$42 - 25 + 12 \qquad \text{Multiply before adding or subtracting.}$$
$$17 + 12 \qquad \text{Perform operations from left to right.}$$
$$29 \qquad \text{Add.}$$

YOU TRY 5

Simplify each expression using the order of operations.

a) $23 - 8 + 7$ b) $-84 \div 4 \times 3$

c) $10 + 35 \div (-5)$ d) $3 \cdot 12 - 4^2 + 9$

We must be careful when working with exponents. For example, are $(-3)^2$ and -3^2 the same or different? Let's see whether $(-3)^2$ and -3^2 mean the same thing.

EXAMPLE 6

Evaluate $(-3)^2$ and -3^2.

Solution

This is what $(-3)^2$ means: $(-3)^2 = \underbrace{-3 \cdot (-3)}_{\text{2 factors of } -3} = 9$ where the base is -3.

To evaluate -3^2, we use the order of operations. Because -3^2 does not contain parentheses, the base is 3 **not** -3.

$$-3^2 = -1 \cdot 3^2 \qquad -3^2 \text{ means } -1 \cdot 3^2.$$
$$= -1 \cdot 9 \qquad \text{Evaluate exponents before multiplying.}$$
$$= -9 \qquad \text{Multiply.}$$

Therefore, $(-3)^2 = 9$ but $-3^2 = -9$.

Evaluate $(-7)^2$ and -7^2.

 BE CAREFUL
When simplifying an exponential expression containing a signed number, remember the difference between having parentheses in the expression and **not** having parentheses.

Example: $(-3)^2 = \underbrace{-3 \cdot (-3)}_{\text{2 factors of } -3} = 9$ The base is -3. Multiply -3 by itself 2 times.

$-3^2 = -1 \cdot 3^2 = -1 \cdot 9 = -9$ The base is 3. Evaluate the exponent before multiplying.

W Hint
Don't forget to review this box!

Let's evaluate other expressions using the order of operations. When an expression contains parentheses or other grouping symbols, simplify what is inside these first.

EXAMPLE 7   Simplify each expression using the order of operations.

a) $\sqrt{100} \div (11 - 6)$ b) $-6^2 + 12\sqrt{4} - (3 - 10)^2$

c) $(-2)^3 \cdot 3^2 - (7 - 8)^3$

Solution

a) $\sqrt{100} \div (11 - 6)$

$\sqrt{100} \div \quad 5$ First, perform the operation in parentheses.

$10 \div \quad 5$ Evaluate the square root.

2 Divide.

b) $-6^2 + 12\sqrt{4} - (3 - 10)^2$

$\underline{-6^2} + 12\underline{\sqrt{4}} - \quad \underline{(-7)^2}$ First, perform the operation in parentheses.

$\downarrow \qquad \downarrow \qquad \quad \downarrow$

$-36 + 12 \cdot 2 - \quad\quad 49$ Evaluate exponents and the square root. $-6^2 = -1 \cdot 6^2 = -36$; $12\sqrt{4} = 12 \cdot \sqrt{4} = 12 \cdot 2$; $(-7)^2 = -7 \cdot 7 = 49$

$-36 + \quad 24 - 49$ Multiply.

$-12 - 49$ Add.

$-12 + (-49)$ Change subtraction to addition of the additive inverse of 49.

-61 Add.

c) Be careful working with all of the exponents.

$(-2)^3 \cdot 3^2 - (7 - 8)^3$

$\underline{(-2)^3} \cdot \underline{3^2} - \underline{(-1)^3}$ Perform the operation inside the parentheses first.

$\downarrow \qquad \downarrow \qquad \downarrow$

$-8 \quad \cdot \quad 9 \ - (-1)$ $(-2)^3 = -2 \cdot (-2) \cdot (-2) = -8$; $3^2 = 9$; $(-1)^3 = -1 \cdot (-1) \cdot (-1) = -1$

$-72 \quad - (-1)$ Multiply before subtracting.

$-72 + 1$ Change subtraction to addition of the additive inverse of -1.

-71 Add.

$$\boxed{\text{YOU TRY 7}}$$ Simplify each expression using the order of operations.

a) $\sqrt{36} \div 2$ b) $-9^2 + 8\sqrt{1} - (5 - 16)^2$

c) $100 - 4^3 \cdot 2^2 \div (1 - 5)^2$

ANSWERS TO $\boxed{\text{YOU TRY}}$ EXERCISES

1) a) base: 7; exponent: 2; 49 b) base: 4; exponent: 3; 64 c) base: 2; exponent: 6; 64
2) a) $(-7)^4$ b) $(-2)^5$ 3) a) 81 b) -8 4) a) 2 b) 5 c) 11 d) 10
5) a) 22 b) -63 c) 3 d) 29   6) $(-7)^2 = 49$; $-7^2 = -49$
7) a) 3 b) -194 c) 84

E Evaluate **1.8** Exercises Do the exercises, and check your work.

Objective 1: Use Whole Numbers with Exponents

Identify the base and the exponent, then evaluate each expression.

1) 10^2 2) 8^2

3) 5^2 4) 6^2

5) 0^2 6) 1^2

7) 2^3 8) 3^3

 9) 3^4 10) 2^6

11) 20^2 12) 30^2

Rewrite each product as an exponential expression. Do *not* evaluate.

13) $9 \cdot 9 \cdot 9 \cdot 9$ 14) $5 \cdot 5 \cdot 5 \cdot 5 \cdot 5 \cdot 5 \cdot 5$

15) $4 \cdot 4 \cdot 4 \cdot 4 \cdot 4 \cdot 4$ 16) $7 \cdot 7 \cdot 7$

Objective 2: Use Integers with Exponents

Fill in the blank with *always, sometimes,* or *never* to make the statement true.

17) Whenever the exponent is even and the base is nonzero, the result will _____ be positive.

18) Whenever the exponent is odd and the base is nonzero, the result will _____ be positive.

Write each multiplication problem using an exponent.

19) $-10 \cdot (-10) \cdot (-10)$ 20) $-4 \cdot (-4) \cdot (-4)$

21) $-13 \cdot (-13)$ 22) $-18 \cdot (-18)$

Evaluate.

23) $(-7)^2$ 24) $(-12)^2$

25) $(-2)^6$ 26) $(-10)^4$

27) $(-2)^5$ 28) $(-3)^5$

29) $(-10)^3$ 30) $(-2)^3$

31) $(-1)^4$ 32) $(-1)^6$

Objective 3: Find Square Roots

33) Explain how to find the square root of a whole number.

34) How do we know that $\sqrt{25} = 5$?

Evaluate.

35) $\sqrt{16}$ 36) $\sqrt{36}$

37) $\sqrt{81}$ 38) $\sqrt{100}$

39) $\sqrt{1}$ 40) $\sqrt{0}$

41) $\sqrt{49}$ 42) $\sqrt{4}$

43) $\sqrt{9}$ 44) $\sqrt{64}$

45) $\sqrt{400}$ 46) $\sqrt{900}$

Objective 4: Use the Order of Operations

47) Explain, in your own words, the order of operations.

48) Prof. Spahr asks her students to simplify this expression: $21 + 15 \div 3$. Laurel's answer is 12. Is this the correct answer? Why or why not?

Simplify each expression using the order of operations.

49) $8 - 2 + 3$

50) $9 - 5 + 2$

51) $-48 \div 12 \times 2$

52) $-90 \div 2 \times 3$

53) $2 \cdot 3 + (-4)$

54) $5 \cdot 4 + (-1)$

55) $15 - 9 \div 3 + 6$

56) $20 - 12 \div 4 + 2$

57) $-2 \times 7 + 4^2 - 19$

58) $-8 \times 3 + 6^2 - 27$

59) $6\sqrt{49} - 5\sqrt{36}$

60) $8\sqrt{16} - 6\sqrt{9}$

61) Evaluate.

    a) $(-9)^2$ b) -9^2

62) Evaluate.

    a) $(-8)^2$ b) -8^2

63) Evaluate.

    a) $(-5)^2$ b) -5^2

64) Evaluate.

    a) $(-7)^2$ b) -7^2

65) Identify the base of -7^2 and of $(-7)^2$ then evaluate each.

66) Explain, in your own words, the difference between evaluating -10^2 and $(-10)^2$

Evaluate.

67) $(-2)^4$

68) $(-3)^4$

69) -2^4

70) -3^4

71) $(-10)^3$

72) $(-5)^3$

73) $(-12)^2$

74) $(-13)^2$

75) -1^4

76) -1^6

Simplify each expression using the order of operations.

77) $4^2 + 7^2$

78) $3^2 + 5^2$

79) $8 - 8^2$

80) $7 - 7^2$

81) $-6(9 - 14) \div (-10)$

82) $-9(5 - 17) \div (-6)$

83) $6 + 8|2 - 9|$

84) $7 + 5|1 - 3|$

85) $-9(11 - 3) + 6\sqrt{144}$

86) $-7(9 - 6) + 3\sqrt{49}$

87) $(-2)^3 + (-4)^2 - 6$

88) $(-3)^3 + (-6)^2 - 12$

89) $(-8^2 + 2^2) \div \sqrt{25}$

90) $(-4^2 + 1^2) \div \sqrt{9}$

91) $4^2 \cdot 2^3 - (14 - 6) \div 2$

92) $3^2 \cdot 2^3 - (10 - 2) \div 4$

93) $-\sqrt{16}\,|12 - 18| - 5\sqrt{4}$

94) $-\sqrt{25}\,|16 - 25| - 3\sqrt{100}$

95) $63 \div (-9) - 4(7 - 18)^2$

96) $56 \div (-7) - 3(8 - 10)^2$

Each problem is done incorrectly. Find the error and correct it.

97) $48 - 4 \cdot (-2)$

    Work: $48 - 4 \cdot (-2) = 44 \cdot (-2) = -88$

98) $33 - 18 \div (-3)$

    Work: $33 - 18 \div (-3) = 15 \div (-3) = -5$

99) $-10^2 \div |12 - 37|$

    Work: $-10^2 \div |12 - 37| =$
           $100 \div |12 - 37| =$
           $100 \div \quad 25 \quad = 4$

100) $\sqrt{49} + (8 - 2)^2$

    Work: $\sqrt{49} + (8 - 2)^2 =$
           $7 + 64 - 4 =$
            $71 - 4 = 67$

R Rethink

R1) Write an explanation of the order of operations in your own words. Make up your own example and simplify it.

R2) What are some methods you could use to memorize the powers of whole numbers listed in this section? Do you know all of the powers yet?

Group Activity – Operations with Integers

Activity #1

There are several rules involved when we are performing operations with integers. Sometimes, it is helpful to first predict what the sign of the answer to a problem will be before we actually do the computations. For each problem below, predict whether the answer to the problem will be positive or negative. Fill in the blank space with the word *positive* or *negative*. You do not have to actually find the answer. Work with a partner, and then compare your results with the results of another group.

1) $-17 + 3$ The sum will be _____.

2) $23 - 36$ The difference will be _____.

3) $14 - (-3)$ The difference will be _____.

4) $-13 - 12$ The difference will be _____.

5) $19 + (-45)$ The sum will be _____.

6) $-1 - (-17)$ The difference will be _____.

7) $-25 - (-6)$ The difference will be _____.

8) $(-13)(4)(-11)$ The product will be _____.

9) $\dfrac{-15}{3}$ The quotient will be _____.

10) $\dfrac{-121}{-11}$ The quotient will be _____.

Activity #2: "Around the World"

- Work in groups of three students (students should work on the same piece of paper).
- One student performs the first step in simplifying one of the expressions below. That student then passes the paper to the next student, who performs the second step. Continue this pattern until the expression is simplified.
- Change the order of students for each new expression.

1) $-3(5 - 9)^2 + 6(10 - 12)^3$ 2) $\dfrac{8(-1) - (-4)(-8)}{2[-8 \div (-2 - 2)]}$

3) $-13(2 - 4) + (-4)^2 - (-2)^3$

One essential component to success—in the classroom and beyond—is taking responsibility for your results. Yes, there will always be things that happen that are beyond your control. But if you blame the alarm that didn't go off or the mosquito that wouldn't stop buzzing during the test for your poor results on a math test, you will never improve. You have the power, and the responsibility, to achieve your goals.

To get a sense of your ideas of why things happen to you, circle the statement from each of the pairs below that best describes your views.

1. A. In the long run, people get the respect they deserve in this world.
   B. Unfortunately, an individual's value often goes unrecognized no matter how hard he or she tries.

2. A. The idea that teachers are unfair to students is nonsense.
   B. Most students don't realize the extent to which their exam results are influenced by random events.

3. A. I have found that much of what happens will happen no matter what I do.
   B. Trusting fate has never turned out as well for me as making a decision to take a definite course of action.

4. A. For a well prepared student, there is rarely, if ever, such a thing as an unfair exam.
   B. Many times, exam questions are unrelated to coursework, and studying is often useless.

5. A. Becoming a success is a matter of hard work; luck has little or nothing to do with it.
   B. Getting a good job depends mainly on being in the right place at the right time.

6. A. It is not always wise to plan too far ahead because you can never predict what's going to happen to you.
   B. When I make plans, I am almost certain that I can make them work.

7. A. In my case, getting what I want has little or nothing to do with luck.
   B. I often feel like I might as well decide what to do by flipping a coin.

8. A. In general, I feel that I have little influence over the things that happen to me.
   B. It is impossible for me to believe that chance or luck plays an important role in my life.

9. A. What happens to me is my own doing.
   B. Sometimes I feel that I don't have enough control over the direction my life is taking.

10. A. Sometimes I can't understand how teachers arrive at the grades they give.
    B. There is a direct connection between how hard a person studies and the grades he or she gets.

(Continued on next page)

Scoring: Give yourself one point for each of the following answers and then add up your score:

1. A, 2. A, 3. B, 4. A, 5. A, 6. B, 7. A, 8. B, 9. A, 10. B

Your total score can range from 0 to 10. The higher your score, the more you believe that you have a strong influence over what happens to you and that you are in control of your life and your own behavior. The lower your score, the more you believe that your life is outside of your control and what happens to you is caused by luck or fate.

If you score below 5 on this questionnaire, consider how rethinking your views of the causes of what happens to you might lead to greater success.

———

Adapted from "Do you control what happens to you?" in Nathenson, M. (1985). *The Book of Tests*. New York: Penguin.

Source: Nathenson, M. (1985) *The Book of Tests*. New York: Penquin.

Chapter 1: Summary

Definition/Procedure	Example

1.1 Place Value and Rounding

A **digit** is a single character in a numbering system.

The **decimal system** consists of the digits 0, 1, 2, 3, 4, 5, 6, 7, 8, and 9.

The **whole numbers** are 0, 1, 2, 3, 4, 5, 6, 7, 8, 9, 10, 11, 12, ...

Each digit in a whole number represents a **place value** that is determined by where it appears in the number.

Write the number 78,250,361 in words and identify the place value of the digits 7 and 0.

In words, the number is *seventy-eight million, two hundred fifty thousand, three hundred sixty-one.*

The place value of 7 is 7 ten-millions, and the place value of the 0 is 0 thousands.

Rounding Numbers

To **round** a number means to find another number close to the original number. We can use these steps to round numbers.

Step 1: Find the place to which we are asked to round. Underline the digit in that place.

Step 2: Look at the digit to the right of the underlined digit.

    a) If the digit to the right is **less than 5,** leave the underlined digit as it is.

    b) If the digit to the right is **5 or more,** increase the underlined digit by 1.

Step 3: Change all the digits to the right of the underlined digit to zeros.

Round 2,378,541 to the nearest

a) million   b) thousand

                — Digit to right is less than 5.

a) 2,378,541

Since the digit to the right of the 2 is less than 5, keep the underlined digit 2. Make all digits to the right of the underlined digit zeros.

Rounding to the nearest million, we get 2,000,000.

                — Digit to right is 5 or more.

b) 2,378,541

Since the digit to the right of the 8 is 5 or more, increase the 8 by 1 to make it 9. Make all digits to the right of the underlined digit zeros.

Rounding to the nearest thousand, we get 2,379,000.

1.2 Introduction to Integers

Signed Numbers

A **positive number** is a number that is greater than 0. On a number line, positive numbers are to the *right* of 0.

A **negative number** is a number that is less than 0. To indicate that a number is negative, we put a negative sign, −, in front of it. On a number line, negative numbers are to the *left* of 0.

Zero is neither positive nor negative.

Graph the numbers on a number line.

$$2, -4, 0, -2, 5$$

Solution

Integers

Integers consist of the positive and negative counting numbers as well as 0. They have no fractional or decimal parts. We can list the integers like this:

    . . . , −5, −4, −3, −2, −1, 0, 1, 2, 3, 4, 5, . . .

The dots mean that the list continues, in this way, forever.

Which of the following numbers are integers?

$$9, -38, \frac{1}{2}, 0, -4.7, 296, 1.5, -2\frac{3}{5}$$

The integers are 9, −38, 0, and 296.

Definition/Procedure	Example
Comparing Integers Remember that numbers get *smaller* as we move to the *left* on a number line, and numbers get *larger* as we move to the *right* on a number line. We can use the < and > symbols to compare signed numbers.	Fill in the blank with < or > to compare the integers. a) 3 _____ -2 b) -5 _____ -1 **Solution** Let's look at a number line so that we can see how the numbers compare to each other. $$\begin{array}{c} \xleftrightarrow{\qquad\qquad\qquad\qquad} \\ -5\;-4\;-3\;-2\;-1\;\;0\;\;1\;\;2\;\;3\;\;4\;\;5 \end{array}$$ a) $3 > -2$ because 3 is to the *right* of -2 on the number line. b) $-5 < -1$ because -5 is to the *left* of -1 on the number line.
Absolute Value The **absolute value** of a number is the distance of the number from 0. Because distance is never negative, the absolute value of a number is never negative. The absolute value of a number is denoted by two vertical bars, $|\;\;|$.	Evaluate $|4|$ and $|-4|$. **Solution** Read $|4|$ as "the absolute value of 4." $|4| = 4$ Read $|-4|$ as "the absolute value of -4." $|-4| = 4$ Distance = 4, Distance = 4, so $|-4| = 4$ so $|4| = 4$ $$\xleftrightarrow{\qquad\qquad\qquad} \\ -5\;-4\;-3\;-2\;-1\;\;0\;\;1\;\;2\;\;3\;\;4\;\;5$$
The Opposite of a Number Two numbers are **opposites** of each other if they are the same distance from 0 on a number line but are on opposite sides of 0. To find the opposite of a number, we write a negative sign in front of it. The opposite of a positive number is a negative number. The opposite of a negative number is a positive number.	Find the opposite of each number. a) 8 b) -3 c) 0 **Solution** a) The opposite of 8 is -8. b) To find the opposite of -3, put a negative sign in front of the number: $-(-3)$. Now, evaluate: $-(-3) = 3$. c) The opposite of 0 is $-0 = 0$.

1.3 Adding Integers

Adding Two Negative Numbers *Step 1:* Find the absolute value of each number. *Step 2:* Add the absolute values. *Step 3:* Put a negative sign in front of the sum. The sum of two negative numbers is *always* negative. **(p. 21)**	Add $-16 + (-12)$. **Solution** *Step 1:* Find the absolute value of each number. $$|-16| = 16 \qquad |-12| = 12$$ *Step 2:* Add the absolute values: $16 + 12 = 28$ *Step 3:* Put a negative sign in front of the sum. $$-16 + (-12) = -28$$

Definition/Procedure	Example

Adding Two Numbers with Different Signs

Step 1: Find the absolute value of each number.

Step 2: Subtract the smaller absolute value from the larger absolute value.

Step 3: The sign of the *sum* will be the same as the sign of the number with the *greater* absolute value. Write the sum with this sign.

Add.

a) $-37 + 15$ b) $295 + (-141)$

Solution

a) *Step 1:* Find the absolute value of each number:
$$|-37| = 37, \quad |15| = 15$$

   Step 2: Subtract the smaller absolute value from the larger absolute value.
$$37 - 15 = 12$$
   Larger absolute value   Smaller absolute value

   Step 3: The sign of the *sum* will be the same as the sign of the number with the greater absolute value.

   Negative 37 has a greater absolute value than *positive* 15, so **the sum will be negative.**
$$-37 + 15 = -12$$
   The sum is negative.

b) *Step 1:* Find the absolute value of each number:
$$|295| = 295, \quad |-141| = 141$$

   Step 2: Subtract the smaller absolute value from the larger absolute value.
$$295 - 141 = 154$$
   Larger absolute value   Smaller absolute value

   Step 3: The sign of the *sum* will be the same as the sign of the number with the greater absolute value.

   Positive 295 has a greater absolute value than *negative* 141, so **the sum will be positive.**
$$295 + (-141) = 154$$
   The sum is positive.

Properties of Addition

1) The **identity property of addition** says that if we add 0 to any number, the result is that number. (The number 0 is called the **identity element** for addition.)

2) The **commutative property of addition** says that changing the order in which we add numbers does not change the sum.

3) The **associative property of addition** says that we can change the position of grouping symbols when adding numbers and the sum remains the same.

Here are some examples of the properties of addition:

1) Identity property: $-6 + 0 = -6, \; 0 + 91 = 91$

2) Commutative property: $7 + 4 = 4 + 7$

3) Associative property: $(1 + 5) + 2 = 1 + (5 + 2)$

Definition/Procedure	Example

1.4 Subtracting Integers

Additive Inverse

The opposite of a number is its **additive inverse.**

The sum of a number and its additive inverse is 0.

Find the additive inverse of each number.

a) 6   b) −17

Solution

a) The additive inverse of 6 is −6.   And, 6 + (−6) = 0.

b) The additive inverse of −17 is 17.   And, −17 + 17 = 0.

Subtracting Two Numbers

To subtract two numbers, $a − b$,

1) Change subtraction to addition.

2) Find the additive inverse of b.

3) Add a and the additive inverse of b.

Notice that we keep the first number, a, the same.

Subtract.

a) 4 − 10   b) −19 − 13   c) 25 − (−20)

Solution

a) $4 − 10 = 4 + (−10) = −6$

Change subtraction to addition
of the additive inverse of 10.

b) $−19 − 13 = −19 + (−13) = −32$

Change subtraction to addition
of the additive inverse of 13.

c) $25 − (−20) = 25 + 20 = 45$

Change subtraction to
addition of the additive
inverse of −20.

Combine Adding and Subtracting of Integers

To add and subtract more than two integers, change subtraction to addition of the additive inverse. Then, add from left to right.

Perform the operations to simplify
$−26 − 10 − (−19) + 4$.

Change to addition of the additive inverse of 10.

$$−26 − 10 − (−19) + 4 = −26 + (−10) + 19 + 4$$

Change to addition of the
additive inverse of −19.

$$= −36 \qquad + 19 + 4$$
$$= −17 \qquad + 4$$
$$= −13$$

1.5 Estimating a Sum or Difference

Round to an Indicated Place to Estimate a Sum or Difference

We can estimate a sum or difference by first

1) rounding each number to an indicated place

and then

2) adding or subtracting the rounded numbers.

The symbol $\approx$ means *is approximately equal to.*

Estimate the sum 4,581 + 1,922 + 705 by first rounding each number to the nearest hundred.

Round each number to the nearest hundred.

4,581	⟶	4,600
1,922	⟶	1,900
+ 705	⟶	+ 700
		7,200

$4,581 + 1,922 + 705 \approx 7,200$

The exact sum is $4,581 + 1,922 + 705 = 7,208$.

Definition/Procedure	Example

Use Front-End Rounding to Estimate a Sum or Difference

In **front-end rounding,** a number is rounded to the place furthest to the left.

To estimate a sum or difference, we can round each number using front-end rounding, then add or subtract.

Last year, José paid $885 per month in rent. This year he bought a house, and his monthly mortgage payment is $1,637. How much more is he paying each month for his mortgage than he paid in rent? Use front-end rounding to estimate the difference, then find the exact answer.

Solution

To estimate the answer, use front-end rounding first. Then, find the difference.

Round $1,637 to the *thousands* place: $2,000
Round $885 to the *hundreds* place: − $900
Estimate of the difference: $1,100

The exact difference is $1,637 − $885 = $752.

1.6 Multiplying Integers and Estimation

Multiplying Two Numbers with Different Signs

The product of a *positive number* and a *negative number* is a *negative number*.

Multiply.

a) -6×3 b) $52(-8)$

Solution

a) $-6 \times 3 = -18$ b) $52(-8) = -416$

The product of two numbers with different signs is negative.

Multiplying Two Numbers with the Same Sign

The product of two numbers with the *same sign* is *positive*.

Multiply.

a) $5 \cdot 9$ b) $-11 \times (-7)$

Solution

a) $5 \cdot 9 = 45$ b) $-11 \times (-7) = 77$

The product of two numbers with the same sign is positive.

Multiplying More Than Two Numbers

If two factors are being multiplied inside parentheses, perform that multiplication first. If there is no product inside parentheses, multiply from left to right.

Multiply $-8 \cdot (-5 \cdot 4)$.

$-8 \cdot (-5 \cdot 4)$ Multiply inside parentheses first.

$-8 \cdot (-20)$

160 The product of two negative numbers is positive.

Properties of Multiplication

1) The **multiplication property of 0** says that if we multiply a number by 0, the product is 0.

2) The **identity property of multiplication** says that if we multiply any number by 1, the result is that number. (The number 1 is called the **identity element** for multiplication.)

3) The **commutative property of multiplication** says that changing the order in which we multiply numbers does not change the product.

4) The **associative property of multiplication** says that changing the position of the grouping symbols will not change the product.

Here are some examples of the properties:

1) Multiplication Property of 0: $0 \times 14 = 0$, $-25(0) = 0$

2) Identity Property: $6 \cdot 1 = 6$, $1(-32) = -32$

3) Commutative Property: $-7 \times (-8) = -8 \times (-7)$

$56 \quad = \quad 56$

4) Associative Property: $-4 \cdot (2 \cdot 3) = (-4 \cdot 2) \cdot 3$

$-4 \cdot \quad 6 \quad = \quad -8 \quad \cdot 3$

$-24 \quad = \quad -24$

Definition/Procedure	Example

5) The **distributive property** says that multiplication *distributes* over addition and subtraction.

5) Distributive property:

a) $5(6 + 2) = 5 \cdot 6 + 5 \cdot 2 = 30 + 10 = 40$

b) $3(7 - 11) = 3 \cdot 7 - 3 \cdot 11 = 21 - 33 = -12$

We can use front-end rounding to estimate a product.

Estimate the product $-2{,}435 \times 78$ using front-end rounding, then find the exact answer.

Round $-2{,}435$ to the thousands place: $-2{,}000$

Round 78 to the tens place: 80

$$-2 \times 8 = -16$$

The estimate of the product is $-2{,}\underline{000} \times 80 = -16\underline{0{,}000}$

3 zeros + 1 zero = 4 zeros in the product

The exact product is

```
    -  2, 4  3  5
    ×         7  8
       1  9  4  8  0
    1  7  0  4  5
    - 1 8 9, 9 3 0
```

1.7 Dividing Integers and Estimation

There are several different ways to write a division problem. You should know the different parts of a division problem.

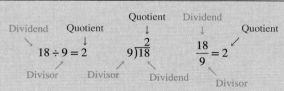

We read each of these as "18 divided by 9 equals 2."

Dividing Numbers

1) The quotient of two numbers with *different signs* is *negative*.

2) The quotient of two numbers with the *same sign* is *positive*.

Divide.

a) $-42 \div 6$ b) $\dfrac{-12}{-3}$ c) $\dfrac{40}{-8}$

Solution

a) $-42 \div 6 = -7$ The quotient of two numbers with *different signs* is *negative*.

b) $\dfrac{-12}{-3} = 4$ The quotient of two numbers with the *same sign* is *positive*.

c) $\dfrac{40}{-8} = -5$ The quotient of two numbers with *different signs* is *negative*.

Properties of Division

1) Any nonzero number divided by itself equals 1.

2) Any nonzero number divided by 1 equals the number.

Here are some examples of the properties:

1) $\dfrac{-4}{-4} = 1$

2) $13 \div 1 = 13$

Definition/Procedure	Example
3) Zero divided by any nonzero number equals zero.	3) $8\overline{)0}$ with 0 above
4) Any number divided by zero is undefined. (This means that there is no answer when we try to divide a number by 0.)	4) $\dfrac{-7}{0}$ is undefined.

Combining Multiplication and Division of Integers

If there is multiplication or division inside parentheses, perform those operations first. Then perform the operations from left to right.

When there are two sets of grouping symbols, we often use brackets, [ ].

Simplify $-4 \cdot [90 \div (-10)] \cdot (-2)$.

$-4 \cdot [90 \div (-10)] \cdot (-2)$

$-4 \cdot \quad (-9) \quad \cdot (-2)$ Divide inside the brackets first.

          $36 \quad\quad\quad \cdot (-2)$ Multiply the first two numbers on the left.

            -72 Multiply.

We can use front-end rounding to estimate a quotient.

Estimate $-874 \div 23$ using front-end rounding, then find the exact answer.

Solution

Round -874 to the *hundreds* place: -900

Round 23 to the *tens* place: 20

The estimate is $-900 \div 20 = -90 \div 2 = -45$.
          Dividing out one zero is the same as dividing by 10.

To find the exact answer, do long division. We will not include the negative sign in the division process. We will put it in the final answer.

$$
\begin{array}{r}
38 \\
23\overline{)874} \\
-69 \\
\hline
184 \\
-184 \\
\hline
0
\end{array}
$$

The exact answer is $-874 \div 23 = -38$.

Solving Applications Involving Division

Many real-world applications can be solved using division.

An airplane flew 3,696 miles in 8 hours. Use front-end rounding to estimate the plane's average speed, that is, the average miles the plane flew per hour. Then, find the exact answer.

Solution

$$\frac{\text{Total number of miles}}{\text{Number of hours}} = \text{Average miles per hour}$$

To estimate the average speed,

Round 3,696 miles: 4,000 miles

Round 8 hours: 10 hours

The estimate of the average is

$$\frac{4{,}000 \text{ miles}}{10 \text{ hours}} = 400 \text{ miles per hour}$$

The exact answer is

$$\frac{3{,}696 \text{ miles}}{8 \text{ hours}} = 462 \text{ miles per hour}$$

Definition/Procedure	Example

1.8 Exponents, Roots, and Order of Operations

An **exponent** represents repeated multiplication of the same number.

The **base** is the number that is being repeatedly multiplied. The **exponent,** or **power,** tells us the number of times to use the base as a factor in the multiplication problem.

Exponent or Power

$$5^3 = 5 \cdot 5 \cdot 5 = 125$$

Base

Use Integers with Exponents

We can use an exponent to represent repeated multiplication of a negative number. In this case, the negative number **must** be in parentheses.

Write $-8 \cdot (-8) \cdot (-8) \cdot (-8)$ using an exponent.

Solution

$$\underbrace{-8 \cdot (-8) \cdot (-8) \cdot (-8)}_{4 \text{ factors of } -8} = (-8)^4$$

Find Square Roots

The **square root** of a number is the number that, when squared, equals the original number.

$\sqrt{49} = 7$ because $7^2 = 49$

We read $\sqrt{49}$ as "the square root of 49."

Order of Operations

Simplify expressions in the following order:

1) If **parentheses** or **other grouping symbols** appear in an expression, simplify what is in these grouping symbols first.

2) Simplify expressions with **exponents** and **square roots.**

3) **Multiply** or **divide** moving from left to right.

4) **Add** or **subtract** moving from left to right.

Remember "**P**lease **E**xcuse **M**y **D**ear **A**unt **S**ally" to help you remember the order of operations.

Simplify $40 + 10^2 \div (23 - 3)$.

$40 + 10^2 \div (23 - 3)$	
$40 + 10^2 \div \quad 20$	First, perform the operation in parentheses.
$40 + 100 \div \quad 20$	Simplify the expression with the exponent.
$40 + \quad 5$	Do division before addition.
45	Add.

Be careful when you are simplifying expressions containing positive and negative numbers.

Simplify each expression.

a) $(-4)^2$ b) -4^2 c) $6 + 33 \div (-3) + (5 - 7)^3$

Solution

a) $(-4)^2 = -4 \cdot (-4) = 16$ The base is -4.

b) Notice that -4^2 does *not* contain parentheses. The order of operations tells us to evaluate exponents first.

$$-4^2 = -1 \cdot 4^2 = -1 \cdot 16 = -16$$

c) $6 + 33 \div (-3) + (5 - 7)^3$

$\quad 6 + 33 \div (-3) + (-2)^3$ Simplify inside parentheses first.

$\quad 6 + 33 \div (-3) + (-8)$ Evaluate the exponential expression.

$\quad 6 + (-11) + (-8)$ Divide.

$\quad -5 + (-8)$ Add and subtract from left to right.

$\quad -13$ Add.

Chapter 1: Review Exercises

(1.1) Identify the place value of the digit 7 in each whole number.

1) 7,015,388,602

2) 972,429

Identify the digits in each period.

3) 138,952,600

4) 56,033,421,007

Write each number in words.

5) 490,617,005,915

6) 98,468,040

Round each number as indicated.

7) 6,239 to the nearest hundred

8) 8,731 to the nearest thousand

9) 9,622,563 to the nearest million

10) 82,495,907 to the nearest ten-thousand

(1.2)

11) What is a negative number?

12) Which of these numbers are *not* integers?

$-9, 4, \frac{1}{2}, -7, 1.8, 0, 3$

Represent the number in each statement as a signed number.

13) The British luxury passenger liner *Titanic* was found approximately 13,000 feet below the ocean's surface. (www.britannica.com/titanic)

14) The Badwater area in Death Valley National Park is the lowest point in North America and is 282 feet below sea level. (www.nps.gov)

Graph each set of numbers on the number line.

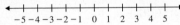

15) 1, 5, −1, −3, −2

16) −4, 2, 5, −2, 0

Fill in the blank with < or > to compare each pair of numbers.

17) 8 _____ −5

18) 2 _____ 11

19) 6 _____ 0

20) 0 _____ −9

21) −34 _____ −33

22) −45 _____ −44

Use the number line below for Exercises 23–26. Fill in the blank with < or > to make the statement true.

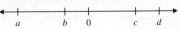

23) *c* _____ *a*

24) *b* _____ 0

25) *b* _____ *a*

26) 4 _____ *a*

27) In your own words, define the absolute value of a number.

28) The absolute value of 4 is the same as the absolute value of what other number?

Evaluate each expression.

29) |15| 30) |−6| 31) −|−26| 32) −|31|

Find the opposite of each number.

33) −35

34) 27

(1.3)

35) In your own words, explain how to add two integers.

36) The sum of a number and its opposite always equals what number?

Add.

37) −4 + (−1)

38) −10 + (−9)

39) −31 + 19

40) 7 + (−23)

41) −15 + 22

42) 54 + (−18)

43) 187 + (−94) + (−46) + 203

44) −219 + 166 + 78 + (−92)

Represent each statement with an addition problem, and solve the problem.

45) On his first run, Marcus gained 16 yd. On his second run, he lost 7 yd. What was his net yardage after these two plays?

46) Ellie's checking account was overdrawn by $57. She deposited $45 into the account. What is the balance of her account?

Solve each problem.

47) Svetlana deposited her $225 paycheck into her debit account, which had a balance of $37. Later that week, she used her debit card to purchase groceries costing $118 and withdrew $200 cash. What is Svetlana's new debit account balance?

48) Monte's 750-GB hard drive had 672 GB of data stored on it. After deleting 387 GB of data, Monte uploaded 157 GB of video and music files to the drive. How many gigabytes of storage are free on Monte's hard drive?

©Ingram Publishing/SuperStock

Add, then rewrite the problem using the commutative property.

49) 3 + 8

50) 8 + 7

51) −6 + 3 + 4

52) −4 + 1 + 6

Add, then rewrite the problem using the associative property.

53) 1 + (9 + 5)

54) (2 + 5) + 6

55) What is the additive inverse of a number?

56) What is the sum of a number and its additive inverse?

Find the additive inverse of each number.

57) 28   58) −94

For Exercises 59 and 60, fill in the blank with *always*, *sometimes*, or *never* to make the statement true.

59) A negative number subtracted from a positive number _____ equals a positive number.

60) A positive number subtracted from a positive number _____ equals a positive number.

Rewrite each subtraction problem as addition. Then, simplify.

61) $8 - 20$ 62) $11 - 19$

63) $-52 - (-89)$ 64) $-43 - (-91)$

65) $-6,142 - 1,087$ 66) $-3,519 - 2,114$

Perform the operations.

67) $4 + (-19) - (-15)$ 68) $2 + (-17) - (-5) - 23$

69) $|-38| - 72 + |-13| + (-60)$

70) $19 - |24| - (-41) + |-8 - 16|$

71) The highest temperature in North America was 134°F in Death Valley, California, in 1913. The coldest temperature in North America was approximately −81°F in the Yukon Territory of Canada in 1947. What is the difference in these two temperatures? (www.ncdc.noaa.gov)

(1.5)

72) What is the difference between rounding 4,726 to the hundreds place and rounding it using front-end rounding?

Estimate the answer to each problem by first rounding to the nearest hundred. Then, find the exact answer.

73) $5,322 - 1,849$ 74) $-7,851 + (-2,136)$

Use front-end rounding to estimate the sum or difference. Then, find the exact answer.

75) $-17,405 + 8,261$ 76) $337 - 192$

77) $79 - 104 + (-62) - (-14)$

Use front-end rounding to estimate the answer. Then, find the exact answer.

78) The number of customers at a coffee shop on five consecutive days were as follows: 342, 471, 489, 520, and 695. Find the total number of customers.

(1.6) For Exercises 79 and 80, fill in the blank with *positive* or *negative*.

79) The product of a negative number and a positive number is _____.

80) The product of a negative number and a _____ number is positive.

Multiply.

81) $-12 \cdot 5$ 82) $-1 \times (-11)$

83) $4(-2)(-3)(-1)(-2)$ 84) $-2|-5| \cdot (7) \cdot (-1)$

Find the product, then rewrite the expression using the indicated property and show that the result is the same.

85) $-3(7 + 2)$; distributive property

86) $4 \cdot (3 \cdot 5)$; associative property

87) $6(9)$; commutative property

88) $2(1 - 8)$; distributive property

Use front-end rounding to estimate each product. Then, find the exact answer.

89) $61 \cdot 473$ 90) $(-9,218)(285)$

91) A small plane descended 225 ft each minute during a 15-min landing. What was the change in the plane's altitude during that time? Express your answer as a signed number.

(1.7)

92) Explain how to divide two integers.

Divide.

93) $-44 \div (-11)$ 94) $\dfrac{54}{-6}$

95) $\dfrac{-405}{9}$ 96) $\dfrac{-265}{-5}$

97) $18 \div 0$ 98) $0 \div (-9)$

Simplify each expression.

99) $-5 \cdot 6 \div (-2)$

100) $|-12| \div (-6) \cdot |8|$

101) $[56 \div (-7)] \cdot (-9)(-1) \div 6$

Use front-end rounding to estimate each quotient. Then, find the exact answer.

102) $-8,944 \div (-43)$ 103) $1,552 \div (-16)$

Solve the application problem. Express your answer as a positive or negative number accordingly.

104) Krishna is downloading a large video file to his computer at a rate of 5 megabits per second. If the file size is 6,900 megabits, how many minutes will it take for Krishna to download the entire file?

105) An assembly line puts 16 packages of diapers in each box. Find the fewest number of boxes needed for 460 packages of diapers. Use front-end rounding to estimate the answer, then find the exact answer.

(1.8)

106) Explain the difference between evaluating $(-5)^2$ and -5^2, and find the value of each.

Write each multiplication problem using an exponent.

107) $8 \cdot 8 \cdot 8 \cdot 8 \cdot 8$

108) $-13 \cdot (-13) \cdot (-13) \cdot (-13)$

Evaluate.

109) 6^2

110) 9^2

111) 2^4

112) 3^4

113) $(-7)^2$

114) $(-10)^2$

115) -7^2

116) -10^2

117) $(-2)^3$

118) -4^3

119) $\sqrt{100}$

120) $\sqrt{64}$

Simplify each expression using the order of operations.

121) $3(8) - 7(5) + 4(6)$

122) $4(9) - 2(3) + 5(7)$

123) $-3^4 \div 9 - (-3)^2$

124) $-6^2 \div 2 - (-6)^3$

125) $-15 - 3\sqrt{16} \div (18 - 6)$

126) $-16 - 2\sqrt{81} \div (11 - 8)$

Each problem is done incorrectly. Find the error and correct it.

127) $\sqrt{81} + (7 - 4)^2$

Work: $\sqrt{81} + (7 - 4)^2 =$

$9 + 49 + 16 =$

$58 + 16 = 74$

128) $32 - 24 \div (-4)$

Work: $32 - 24 \div (-4) = 8 \div (-4) = -2$

Mixed Exercises

Perform the indicated operations.

129) -58×63

130) $308 - 527$

131) $-157 - (-421)$

132) $\dfrac{-648}{-18}$

133) $-|125| - (-308) + (-649)$

134) $33 - 3(-8) \div 2 - (-15)$

135) $-|-2| + (-16)(3) \div (7 - 9)^3 - 10$

Fill in the blanks.

136) _____ $\times 12 = -48$

137) $-42 \div$ _____ $= -7$

138) $-31 +$ _____ $= 9$

139) Write an expression so that the quotient of two integers is -9.

140) Write an expression so that the product of two integers is -48.

141) Write an expression so that the sum of a positive integer and a negative integer is 26.

142) Write an expression so that the sum of a positive integer and a negative integer is -75.

143) Write an expression so that the difference of two negative numbers is -37.

144) Write an expression so that the difference of two negative numbers is 56.

Solve each problem.

145) Marika is on a mountain at an altitude of 9,450 ft. At the same time, her friend is scuba diving at a depth of 68 ft. What is the difference between these two elevations?

146) An appliance store has seen its profits fall by $2,100 per month for the last year. Use a signed number to determine the store's change in profit over the last year.

Chapter 1:   Test

1) Identify the place value of the digit 6 in each number.

   a) 516,472

   b) 265,089,413

2) Write the number in part a) in words, and write the number in part b) using digits.

   a) 3,094

   b) eight million, one hundred fifteen thousand, six hundred twenty-two

3) Round 7,425,996 to the nearest

   a) ten-thousand.

   b) million.

   c) hundred.

4) Fill in the blank with < or > to compare the numbers.

   a) 19 _____ 8

   b) -6 _____ 0

   c) -5 _____ -2

   d) 4 _____ -7

5) Graph the numbers on the number line.
   $4, -2, -5, 0, 3$

6) Explain, in your own words, the definition of the absolute value of a number.

7) Evaluate each expression.

   a) $|-8|$

   b) $|32|$

   c) $|0|$

   d) $-|5|$

8) Find the opposite of each number.

   a) -13

   b) 6

9) Find the additive inverse of 10.

Perform the indicated operation.

10) $-7 \cdot (-4)$

11) $-11 - (-3)$

12) $-26 + (-35)$

13) $\dfrac{-24}{6}$

14) $197 - 832$

15) $\dfrac{-4}{0}$

16) $-85 - (-85)$

17) $\dfrac{0}{5}$

18) $-472 \div (-8)$

Evaluate each expression.

19) $-2 \cdot 5 \cdot (-4) \cdot (-1)$

20) $7 - 28 - (-19) + (-6)$

21) $(-8)^2$

22) -5^2

23) 2^5

24) $\sqrt{36}$

25) $\sqrt{81}$

26) Write $-7 \cdot (-7) \cdot (-7) \cdot (-7) \cdot (-7) \cdot (-7)$ using an exponent.

Simplify each expression using the order of operations.

27) $12 - 10 \cdot 3 + 13$

28) $-5(2 - 9) - 6^2 \div (-4)$

29) $(-9)^2 - 2\sqrt{144} + 15 \cdot (-2) - (4 - 1)^3$

30) Evaluate each expression. Then, rewrite each expression using the indicated property and show that the result is the same.

    a) $-8 + (12 + 3)$; associative property

    b) $4(5 - 7)$; distributive property

    c) $9 + 6$; commutative property

Use front-end rounding to estimate the answer. Then, find the exact answer.

31) $536 \cdot 77$

32) $-2,194 + (-382) + (-5,716)$

33) $\dfrac{-5,225}{19}$

34) Write an expression so that the product of two integers is -35.

35) Writen an expression so that the difference of two negative numbers is 72.

Solve each problem. Express the answer as a signed number.

36) A submarine that is cruising at a depth of 603 ft below sea level rises 149 ft. Find the current depth of the submarine.

37) The number of members at a health club has decreased by 318 people over 6 years. What is the average change in the number of members each year?

Solve each problem.

38) Janice bought four concert tickets for $78 each. She will also pay a $15 handling fee, $22 for parking, and an extra $26 for express delivery. Find the total cost.

39) A publishing company ships books to bookstores in boxes containing 12 books each. What is the fewest number of boxes needed to ship 740 books?

40) The *plus-minus* is a hockey statistic that measures a player's goal differential. With the exception of penalty shots and power-play goals, a player gets a "+1" if he is on the ice when his team scores a goal, but a player gets a

©Comstock/Getty Images

"−1" if he is on the ice when his team allows a goal by the other team. If a player was on the ice when the other team scored 3 goals, and he was on the ice when his own team scored 2 goals, what is his *plus-minus* for the game?

Expressions and Equations

2

©Kablonk/SuperStock

Math at Work:

Regional Sales Manager

Twenty-five copy machines were sold to a customer for $2000 each, and 52 copy machines were sold to a larger customer for $1800 dollars each. Two salespeople in a region generate $23,000 in sales per month, while one salesperson in a neighboring region generates $13,500 per month.

Andre Mathers went into sales because he was good with people and enjoyed the rush of closing a deal. But he found that the higher he rose in the office equipment supply company where he started as a salesperson, the more math came in handy.

"I always used math on the job," Andre says. "For example, customers would ask me how much it would cost for a certain number of units, and I had to be able to quickly do the multiplication in my head. Now, as a sales manager for an entire region, math is an even bigger part of my job." Whether Andre is dividing a sales bonus across his staff or determining the cost of giving a discounted price to a customer for 12 months, math in general—and solving problems in particular—is essential to his success.

"Sales is all about trust," Andre notes. "Your customers need to be able to trust you'll deliver what you tell them, and just as importantly, they need to know that any figures you give them are accurate."

In this chapter, we'll learn how to solve equations and application problems. We'll also introduce some strategies to help you read textbooks and master the skills you'll need to succeed in math class and on the job.

Some textbooks seem intimidating to read, especially math books! With some strategies, however, you can learn to read and use your textbook effectively.

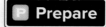

- *Prepare means to explicitly state a goal.* **I will learn how to read and use my math textbook.**

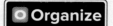

- *Organize means to* **organize** *the physical and mental tools you need to achieve your goal.*
- Read the preface and/or the introduction to learn about the features and structure of the book: the Table of Contents, the Chapter Summary and Test, the Answer Appendix in the back of the book, and the P.O.W.E.R. tables that introduce each chapter and section.
- Be aware of additional materials online such as review sections for prerequisite skills, an online homework system, videos, and other supplements.
- Gather the *physical tools* you may need: the book (print or e-book), a computer, pencils (including colored pencils for taking good notes), a notebook, a folder, and a highlighter. Locate a quiet place to study.

- *Work means to* **do the work** *that needs to be done to achieve the goal.*
- Reading a math textbook is different from reading, say, a history book. **The best way to read a math textbook is to** *write while you read*! Write out the examples in your notebook as you are reading them so that you are actually *doing* the problems.
- Use different colored pencils to write out the steps just like color is used in the book. This helps you to see the step performed as you go from one line to the next.
- Highlight and underline only the essential information in the book. No more than about 10% of what you read should be highlighted. Pay special attention to formulas, and write them in your notes.
- In this book, do the You Trys (and check the answers at the end of the section) after you have worked through the examples.
- When you read through definitions and procedures, jot down notes next to them explaining them in your own words.

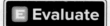

- *Evaluate means you should* **evaluate** *what you have done.* Did you achieve your goal?
- Did you understand what you read, or not? This can be evaluated in different ways. Could you explain major concepts, procedures, and definitions *in your own words* to a classmate? Could you do the You Trys on your own after reading the section? Could you do the homework exercises after reading?

- **Rethink** *means to* **rethink** *and reflect upon your goal.* If you did *not* achieve your goal, ask yourself, *"Why not?"* If you did achieve your goal, ask yourself what you did *right*.
- If you did not understand a section or could not do the You Trys or exercises ask yourself, "Why not?" Did you read in a place that was full of distractions? Did you write out the examples as you read them, or did you skip that part?
- If you did understand a section after reading it, think about what you did that led to that success. Did you read in a quiet place? Was writing out the examples using different colored pencils helpful to you? Be aware of what worked so that you can do it again.

©Andersen Ross/Blend Images LLC

Chapter 2 **P.O.W.E.R.** Plan

<table>
<tr>
<td colspan="2">P Prepare</td>
<td colspan="2">O Organize</td>
</tr>
<tr>
<td colspan="2">What are your goals for Chapter 2?</td>
<td colspan="2">How can you accomplish each goal?</td>
</tr>
<tr>
<td>1</td>
<td>Be prepared before and during class.</td>
<td colspan="2">
<ul>
<li>Don't stay out late the night before class, and be sure to set your alarm!</li>
<li>Bring a pencil, notebook paper, and textbook to class.</li>
<li>Avoid distractions by turning off your cell phone during class.</li>
<li>Pay attention, take good notes, and ask questions.</li>
<li>Complete your homework on time, and ask questions on problems you do not understand.</li>
</ul>
</td>
</tr>
<tr>
<td>2</td>
<td>Understand the homework to the point where you could do it without needing any help or hints.</td>
<td colspan="2">
<ul>
<li>Read the directions, and show all of your steps.</li>
<li>Go to the instructor's office for help.</li>
<li>Rework homework and quiz problems, and find similar problems for practice.</li>
</ul>
</td>
</tr>
<tr>
<td>3</td>
<td>Use the P.O.W.E.R. framework to learn how to read a math textbook.</td>
<td colspan="2">
<ul>
<li>Complete the emPOWERme that appears before the Chapter Summary.</li>
<li>Read the Study Strategies page that explains how to read a math textbook, and apply the steps of P.O.W.E.R. to become better at reading and using this book.</li>
</ul>
</td>
</tr>
<tr>
<td>4</td>
<td>Write your own goal.
_____
_____</td>
<td colspan="2">
<ul>
<li>_____
_____</li>
</ul>
</td>
</tr>
<tr>
<td colspan="2">What are your objectives for Chapter 2?</td>
<td colspan="2">How can you accomplish each objective?</td>
</tr>
<tr>
<td>1</td>
<td>Learn how to simplify expressions.</td>
<td colspan="2">
<ul>
<li>Understand the vocabulary used in algebra.</li>
<li>Be able to identify like terms.</li>
<li>Understand how to use the commutative, associative, and distributive properties to simplify an expression.</li>
<li>Evaluate an expression for given values of the variables.</li>
</ul>
</td>
</tr>
<tr>
<td>2</td>
<td>Learn to solve linear equations.</td>
<td colspan="2">
<ul>
<li>Know the difference between an equation and an expression.</li>
<li>Be able to use the addition, subtraction, and division properties of equality to solve an equation.</li>
<li>Apply the procedure for solving a linear equation to equations with variables on one side or both sides of the equation.</li>
</ul>
</td>
</tr>
<tr>
<td>3</td>
<td>Use the Five Steps for Solving Applied Problems to solve problems involving one or two unknowns.</td>
<td colspan="2">
<ul>
<li>Learn and understand how to use the basic steps for solving an applied problem.</li>
<li>Solve problems by writing an equation and by using the five-step process.</li>
</ul>
</td>
</tr>
<tr>
<td>4</td>
<td>Write your own goal.
_____
_____</td>
<td colspan="2">
<ul>
<li>_____
_____</li>
</ul>
</td>
</tr>
</table>

| | | Read Sections 2.1–2.7, and complete the exercises. |

E Evaluate	Complete the Chapter Review and Chapter Test. How did you do?	**R** Rethink	• Did you achieve your goals for the chapter? Which steps could be improved for next time? If you had the chance to do this chapter over, what would you do differently?
			• What were you able to learn about yourself by completing the emPOWERme at the end of the chapter?
			• Did you use the P.O.W.E.R. framework to help you read this textbook? Which strategies were most helpful? In the future, what would you do the same, and what would you do differently?

2.1 Introduction to Algebra

P Prepare   **O** Organize

What are your objectives for Section 2.1?	How can you accomplish each objective?
1 Evaluate Expressions	• Write the definitions of *variable* and *algebraic expression* in your own words. • Complete the given examples on your own. • Complete You Trys 1–4.
2 Use Exponents with Variables	• Understand that exponents are used the same way whether the base is a number or a variable. • Be able to identify the base. • Complete the given examples on your own. • Complete You Trys 5 and 6.
3 Identify Parts of an Expression	• Write the definitions of *term, constant,* and *coefficient* in your own words. • Complete the given examples on your own. • Complete You Trys 7 and 8.

W Work   **Read the explanations, follow the examples, take notes, and complete the You Trys.**

1 Evaluate Expressions

In arithmetic, we work with *specific* numbers and apply properties, show relationships, and perform operations. For example, $12 \cdot 3 = 36$. So why study algebra? If we use algebra, we can use symbols (usually letters) to represent unknown numbers and apply properties, show relationships, and perform operations on *general* quantities to draw *general* conclusions.

Definition

A **variable** is a symbol, usually a letter, used to represent an unknown number. (It is called a *variable* because its value can *vary*.)

Example: x can be used to represent an unknown number.

Definition

An **algebraic expression** is a collection of numbers, variables, and grouping symbols connected by operation symbols like $+$, $-$, $\times$, and $\div$.

For example, $x + 5$ is an expression, and the variable x represents a number. Let's see how we can take a specific arithmetic situation and represent it using algebra.

Marisol earns $12 per hour at her job. Find the total amount she earns if she works each of the following number of hours in a week.

Number of Hours Marisol Works	Hourly Wage $\downarrow$	Number of Hours Worked $\downarrow$	Marisol's Earnings $\downarrow$
8 hours:	12 $\cdot$	8	= $96
20 hours:	12 $\cdot$	20	= $240
33 hours:	12 $\cdot$	33	= $396
40 hours:	12 $\cdot$	40	= $480

Notice the pattern. Each time we calculate the total amount Marisol earns, we multiply the hourly wage and the number of hours worked. The hourly wage is always the same, $12, but the number of hours worked varies (changes). If we use algebra, we can write a *general* rule to fit each *specific* situation.

Let the letter h be the *variable* that represents the number of hours Marisol works. (We could have used any letter, but the letter h was chosen because it represents the number of *hours* worked.) Then, we can write an algebraic expression to represent Marisol's total earnings.

In English: Total earnings $=$ Hourly wage $\cdot$ Number of hours
Using Algebra: Total earnings $=$ 12 $\cdot$ h

The expression $12 \cdot h$, or $12h$, represents Marisol's total earnings. (In algebra, we usually do not write the multiplication symbol.)

Now, we can use the expression $12h$ to calculate Marisol's earnings for any number of hours she works. This is called **evaluating an expression.**

EXAMPLE 1

Let $h =$ the number of hours Marisol worked. The expression $12h$ represents Marisol's total earnings, in dollars. Evaluate $12h$ for each value of h, and explain the meaning of the answer.

a) $h = 5$

b) $h = 26$

Solution

a) To *evaluate* $12h$ when $h = 5$ means to find the value of the expression $12h$ when $h = 5$. To do this, substitute 5 for h and simplify. Use parentheses when substituting the value because there is no operation symbol between the 12 and the h.

Evaluate $12h$ for $h = 5$:

$$= 12(5) \quad \text{Substitute 5 for } h.$$
$$= 60 \quad \text{Multiply.}$$

When $h = 5$, $12h = 60$. When Marisol works 5 hours, she earns $60.

b) Evaluate $12h$ for $h = 26$:

$$= 12(26) \quad \text{Substitute 26 for } h.$$
$$= 312 \quad \text{Multiply.}$$

When $h = 26$, $12h = 312$. When Marisol works 26 hours, she earns $312.

[YOU TRY 1] See Example 1. Use the following values of h.

a) $h = 7$ b) $h = 31$

Let's evaluate other expressions.

EXAMPLE 2 Evaluate each expression when $x = 20$.

a) $x + 9$ b) $x - 7$ c) $\dfrac{x}{4}$ d) $3x$

Solution

a) Evaluate $x + 9$ when $x = 20$.

$$20 + 9 = 29 \quad \text{Substitute 20 for } x, \text{ then add.}$$

b) Evaluate $x - 7$ when $x = 20$.

$$20 - 7 = 13 \quad \text{Substitute 20 for } x, \text{ then subtract.}$$

c) Evaluate $\dfrac{x}{4}$ when $x = 20$.

$$\dfrac{20}{4} = 5 \quad \text{Substitute 20 for } x, \text{ then divide.}$$

d) The expression $3x$ means $3 \cdot x$. When substituting 20 for x, use parentheses to separate the 3 and the 20.

Evaluate $3x$ when $x = 20$.

$$3(20) = 60 \quad \text{Substitute 20 for } x, \text{ then multiply.}$$

[YOU TRY 2] Evaluate each expression when $m = 36$.

a) $m + 11$ b) $m - 15$ c) $\dfrac{m}{9}$ d) $2m$

So far, we have evaluated expressions containing only one operation. Let's look at some expressions containing more than one operation. We substitute the value for the variable, and use the order of operations to simplify.

EXAMPLE 3

Evaluate $2k + 9$ when $k = -7$.

Solution

$$2(-7) + 9 \qquad \text{Substitute } -7 \text{ for } k. \text{ Use parentheses.}$$
$$-14 + 9 \qquad \text{Use the order of operations. Multiply before adding.}$$
$$-5 \qquad \text{Add.}$$

[YOU TRY 3]

Evaluate $5r + 8$ when $r = -6$.

Some expressions contain more than one variable.

EXAMPLE 4

Evaluate $7a - 3b + 2$ when $a = 1$ and $b = -4$.

Solution

$$7a \;-\; 3b \;+\; 2$$
$$7(1) - 3(-4) + 2 \qquad \text{Substitute 1 for } a \text{ and } -4 \text{ for } b.$$
$$7 \;-\; (-12) + 2 \qquad \text{Multiply.}$$
$$7 \;+\; 12 \;+\; 2 = 21 \qquad \text{Change } -(-12) \text{ to } +12, \text{ then add.}$$

[YOU TRY 4]

Evaluate $-9m + 4n + 10$ when $m = -2$ and $n = -7$.

2 Use Exponents with Variables

In Section 1.8, we learned that exponents are used to represent repeated multiplication. For example,

$$\underbrace{2 \cdot 2 \cdot 2}_{3 \text{ factors of } 2} \text{ can be written as } 2^3 \qquad (\text{Read } 2^3 \text{ as } two \; cubed.)$$

We use exponents with variables in the same way. For example,

$$\underbrace{w \cdot w \cdot w}_{3 \text{ factors of } w} \text{ can be written as } w^3 \qquad (\text{Read } w^3 \text{ as } w \; cubed.)$$

Let's practice writing some products using exponents.

EXAMPLE 5

Write each product using exponents.

a) $h \cdot h$ b) $a \cdot a \cdot a \cdot b \cdot b \cdot b \cdot b$ c) $2 \cdot n \cdot n \cdot n \cdot n \cdot n$

d) $-5 \cdot x \cdot x \cdot y \cdot z \cdot z \cdot z$

Solution

a) $\underbrace{h \cdot h}_{\substack{2 \text{ factors} \\ \text{of } h}} = h^2$ Read this as h squared.

b) $\underbrace{a \cdot a \cdot a}_{\substack{3 \text{ factors} \\ \text{of } a}} \cdot \underbrace{b \cdot b \cdot b \cdot b}_{\substack{4 \text{ factors} \\ \text{of } b}} = a^3 \cdot b^4 = a^3 b^4$

c) $2 \cdot \underbrace{n \cdot n \cdot n \cdot n \cdot n \cdot n}_{6 \text{ factors of } n} = 2 \cdot n^6 = 2n^6$

d) $-5 \cdot \underbrace{x \cdot x}_{\substack{2 \text{ factors} \\ \text{of } x}} \cdot \overset{\overset{\textstyle 1 \text{ factor of } y}{\downarrow}}{y} \cdot \underbrace{z \cdot z \cdot z}_{\substack{3 \text{ factors} \\ \text{of } z}} = -5 \cdot x^2 \cdot y \cdot z^3 = -5x^2yz^3$

[YOU TRY 5] Write each product using exponents.

a) $v \cdot v \cdot v$ b) $c \cdot c \cdot c \cdot c \cdot d \cdot d$ c) $3 \cdot r \cdot r \cdot r \cdot r \cdot r$

d) $-7 \cdot x \cdot y \cdot y \cdot y \cdot y \cdot z \cdot z \cdot z$

Now we will evaluate some expressions containing exponents. Do you remember the list of powers that was given in Section 1.8?

Note
Get in the habit of always using parentheses when you substitute values into an expression containing exponents. It will help you avoid confusion with negative signs.

EXAMPLE 6

Evaluate each expression.

a) k^2 when $k = 5$ b) k^2 when $k = -5$ c) $-3m^4$ when $m = 2$

d) $7ab^3$ when $a = -4$ and $b = 3$

Solution

a) To evaluate k^2 when $k = 5$, substitute 5 for k and simplify. You can recall the powers of whole numbers listed in Section 1.8 or rewrite the expression as a multiplication problem.

$$(5)^2 = 25 \qquad \text{Recall the powers of whole numbers.}$$

We can also remember that $(5)^2 = \underbrace{5 \cdot 5}_{2 \text{ factors of } 5} = 25$. The result is the same.

b) To evaluate k^2 when $k = -5$, substitute -5 for k and simplify.

$$(-5)^2 = \underbrace{-5 \cdot (-5)}_{\text{2 factors of 5}} = 25$$

W Hint

Review the order of operations and powers of whole numbers in Chapter 1, if necessary.

c) To evaluate $-3m^4$ when $m = 2$, substitute 2 for m and simplify using the order of operations.

$$-3m^4 \text{ when } m = 2$$
$$\downarrow$$
$$-3(2)^4 = -3 \cdot (2)^4 \qquad \text{The exponent applies only to the 2.}$$
$$= -3 \cdot 16 \qquad \text{Evaluate the exponent before multiplying.}$$
$$= -48 \qquad \text{Multiply.}$$

d) To evaluate $7ab^3$ when $a = -4$ and $b = 3$, substitute -4 for a, 3 for b, then evaluate using the order of operations.

$$7ab^3 \text{ when } a = -4 \text{ and } b = 3$$
$$7(-4)(3)^3 = 7 \cdot (-4) \cdot (3)^3 \qquad \text{The exponent applies only to the 3.}$$
$$\downarrow$$
$$= 7 \cdot (-4) \cdot 27 \qquad \text{Evaluate the exponent before multiplying.}$$
$$= -28 \cdot 27 \qquad \text{Multiply from left to right.}$$
$$= -756 \qquad \text{Multiply.}$$

[YOU TRY 6] Evaluate each expression.

a) a^2 when $a = 9$ b) a^2 when $a = -9$

c) $-5t^3$ when $t = -2$ d) $8m^2n$ when $m = 6$ and $n = -1$

---

3 Identify Parts of an Expression

Here is an algebraic expression: $5x^4 - 8y^3 + z + 7$. Let's use it to define other words we need to know in algebra.

This expression contains three variables: x, y, and z. They are called *variables* because their values can vary. We call 7 the **constant** or **constant term** because its value remains *constant*; it does not change. The *terms* of this expression are $5x^4$, $-8y^3$, z, and 7. (Notice that the term $-8y^3$ takes the negative sign in front of it.) A **term** is a number or a variable or a product or quotient of numbers and variables. The **coefficient** of a term is the number that the variable is multiplied by. The coefficient of $5x^4$ is 5, the coefficient of $-8y^3$ is -8, and the coefficient of z is 1 because z can be written as $1z$.

W Hint

Be sure that you learn the vocabulary in this section!

Note

When no number is written in front of a variable, the coefficient is assumed to be 1.

EXAMPLE 7

List the terms and coefficients of the expression $4a^2 - b^5 + 6c - 2$. Also, identify the constant.

Solution

This expression contains four terms.

Term	Coefficient
$4a^2$	4
$-b^5$	-1
$6c$	6
-2	

$-b^5$ means $-1 \cdot b^5$, so the coefficient is -1.

-2 is the **constant.**

YOU TRY 7

List the terms and coefficients of the expression $-6m^4 + n^3 - 10p + 4$. Also, identify the constant.

We said earlier that a term is a number or a variable or a *product or quotient* of numbers and variables. It is important to remember this when we are identifying terms in an expression that contains more than one variable.

EXAMPLE 8

For the expression $-5x^2 + 4xy + 17$,

a) list the terms and coefficients of the expression, and identify the constant.

b) evaluate the expression for $x = 2$ and $y = -3$.

Solution

a) This expression contains three terms.

Term	Coefficient
$-5x^2$	-5
$4xy$	4
17	

$4xy$ is a term because it is a *product* of the number 4 and the variables x and y.
17 is the **constant.**

b) To evaluate, substitute the values for the variables. *Be sure to put the numbers in parentheses!*

$$-5x^2 + 4xy + 17 \text{ for } x = 2 \text{ and } y = -3$$

$$= -5(2)^2 + 4(2)(-3) + 17 \quad \text{Substitute the values for } x \text{ and } y.$$
$$= -5(4) + 4(2)(-3) + 17 \quad \text{Evaluate the exponent first: } 2^2 = 4.$$
$$= -20 + (-24) + 17 \quad \text{Multiply.}$$
$$= -27 \quad \text{Add.}$$

YOU TRY 8

For the expression $2c^3 - 7c^2d + 6$,

a) list the terms and coefficients of the expression, and identify the constant.

b) evaluate the expression for $c = -1$ and $d = 8$.

E Evaluate **2.1** Exercises   Do the exercises, and check your work.

Objective 1: Evaluate Expressions

Use the given situation and the value of the variable term to evaluate each expression.

1) Let h = the number of hours worked. Rebecca works in a bakery that pays $14 per hour. The expression $14h$ represents Rebecca's total earnings. Evaluate $14h$ for each value of h and explain the meaning of the answer.

   a) $h = 4$ b) $h = 18$

2) Let h = the number of hours worked. A department store pays a sales clerk $9 per hour. The expression $9h$ represents Nagar's total earnings. Evaluate $9h$ for each value of h and explain the meaning of the answer.

   a) $h = 6$ b) $h = 13$

 3) David works for a tree trimming service that pays him $85 for each day he works. The expression $85d$ represents David's total earnings. Evaluate $85d$ for each value of d and explain the meaning of the answer.

   a) $d = 5$ b) $d = 8$

4) Cante is paid $72 per day working for a landscaping service. The expression $72d$ represents Cante's total earnings. Evaluate $72d$ for each value of d and explain the meaning of the answer.

   a) $d = 4$ b) $d = 15$

 5) On a hiking trip, Julian can hike at an average speed of 3 miles per hour. The expression $3h$ represents Julian's total miles hiked. Evaluate $3h$ for each value of h and explain the meaning of the answer.

   a) $h = 3$ b) $h = 8$

6) The Diaz family decides to travel across the country in their family van. They travel at an average speed of 50 miles per hour. The expression $50h$ represents the family's total miles traveled. Evaluate $50h$ for each value of h and explain the meaning of the answer.

   a) $h = 5$ b) $h = 13$

7) Suppose that a fluid is dripping at a rate of 13 drops per minute. The expression $13m$ represents the number of drops that fall every 1 min. Evaluate $13m$ for each value of m and explain the meaning of the answer.

   a) $m = 15$ b) $m = 45$

8) A nurse measures Nacho's heart rate and finds it to be 72 beats per minute. The expression $72m$ represents the number of heartbeats every 1 minute. Evaluate $72m$ for each value of m and explain the meaning of the answer.

   a) $m = 30$ b) $m = 60$

Evaluate each expression for the given value of the variable.

9) $x = 25$

   a) $x + 15$ b) $x − 10$

   c) $\dfrac{x}{5}$ d) $5x$

10) $a = 16$

   a) $a + 8$ b) $a − 12$

   c) $\dfrac{a}{8}$ d) $3a$

11) $m = -12$

   a) $m + 17$ b) $m - 7$

   c) $\dfrac{m}{4}$ d) $4m$

12) $b = -21$

   a) $b + 29$ b) $b - 15$

   c) $\dfrac{b}{3}$ d) $5b$

13) $y = 4$

   a) $2y + 6$ b) $3y - 8$

   c) $\dfrac{7y}{-4}$

14) $c = 9$

   a) $4c + 8$ b) $8c - 15$

   c) $\dfrac{5c}{-9}$

15) $r = 7$

   a) $-5r + 16$ b) $-4r - 23$

   c) $\dfrac{14r}{5 - r}$

16) $q = 9$

   a) $-4q + 19$ b) $-6q - 45$

   c) $\dfrac{15q}{4 - q}$

17) By how much does the expression $2x - 5$ increase if x increases by 1?

18) By how much does the expression $-3y + 8$ decrease if y increases by 1?

Evaluate each expression for the given values of the variables.

19) $6x - 4y + 5$; $x = 2$; $y = -3$

20) $7s - 8t + 11$; $s = 5$; $t = -6$

21) $3p - 9q - 16$; $p = -7$; $q = 5$

22) $2j - 5k - 23$; $j = -6$; $k = 4$

23) $-8a + 7b - 15$; $a = -5$; $b = -8$

24) $-9f + 12g - 21$; $f = -2$; $g = -3$

25) $-5c - 10d - 16$; $c = -6$; $d = -12$

26) $-7w - 8z - 19$; $w = -9$; $z = -11$

Objective 2: Use Exponents with Variables

Write each of the products using exponents.

27) $5 \cdot 5$ 28) $7 \cdot 7 \cdot 7$

29) $x \cdot x$ 30) $z \cdot z \cdot z$

31) $r \cdot r \cdot r \cdot r \cdot r \cdot r \cdot r \cdot r$ 32) $u \cdot u \cdot u \cdot u \cdot u$

33) $a \cdot a \cdot a \cdot b \cdot b$ 34) $s \cdot s \cdot t \cdot t \cdot t$

35) $k \cdot k \cdot t \cdot t \cdot t \cdot t \cdot t$

36) $v \cdot v \cdot v \cdot w \cdot w \cdot w \cdot w \cdot w \cdot w$

37) $3 \cdot q \cdot q \cdot q \cdot q$

38) $7 \cdot m \cdot m \cdot m \cdot m \cdot m$

39) $-8 \cdot d \cdot d \cdot d \cdot f$

40) $-11 \cdot n \cdot n \cdot n \cdot n \cdot t$

41) $-15 \cdot q \cdot q \cdot r \cdot r \cdot r \cdot r \cdot s$

42) $-19 \cdot x \cdot y \cdot y \cdot y \cdot y \cdot y \cdot z$

43) $12 \cdot b \cdot b \cdot b \cdot c \cdot c \cdot d \cdot d \cdot d$

44) $5 \cdot h \cdot h \cdot r \cdot r \cdot r \cdot v \cdot v$

Evaluate each expression.

45) a) x^2 when $x = 4$ 46) a) z^2 when $z = 11$

    b) x^2 when $x = -4$ b) z^2 when $z = -11$

47) $2b^2$ when $b = 3$ 48) $4v^2$ when $v = 5$

49) $-3a^4$ when $b = -2$ 50) $-5c^4$ when $c = -2$

51) $6m^2n^3$ when $m = 3$ and $n = -1$

52) $5p^2q^3$ when $p = 1$ and $q = -2$

53) $3c^3d^2$ when $c = -2$ and $d = -3$

54) $2h^2j^3$ when $h = -4$ and $j = -3$

For Exercises 55–58, fill in the blank with *always*, *sometimes*, or *never* to make the statement true. Assume the variable does not equal 0.

55) The term $-x^4$ will _____ represent a positive value.

56) The term $-y^3$ will _____ represent a negative value.

57) The term $(-a)^5$ will _____ represent a negative value.

58) The term $(-z)^2$ will _____ represent a positive value.

Objective 3: Identify Parts of an Expression

59) Using your own words, explain the difference between a coefficient and a constant.

60) What is the minimum number of terms an expression must contain?

List the terms and coefficients of the given expression. Also, identify the constant.

 61) $3x^3 - y^2 + 5z - 6$

62) $4f^2 - g^4 + 7h^3 - 5$

63) $-7s^3 + 6t^2 - 12u + 16$

64) $-12d^5 - 2f^7 - 3g^9 + 15$

65) $-7a^2b + 36b^3c^2 - 51c + 32$

66) $-13x^2y^2 + 17y^2z^3 - 39z + 8$

For each expression, list the terms and coefficients of the expression, and identify the constant. Then, evaluate the expression for the given values of the variables.

67) $r^2 + 5r + 7; r = 6$

68) $u^2 + 4u + 9; r = 7$

 69) $3x^2 - 4xy + 7; x = 3$ and $y = -2$

70) $2a^2 - 3ab + 9; a = 2$ and $b = -4$

71) $-4g^2 + 8gh - 14; g = -5$ and $h = -3$

72) $-5s^2 + 6st - 22; s = -4$ and $t = -5$

73) $9m^3 + 2n^2 + mn + 3n; m = -2$ and $n = 5$

74) $2x^3 + 3y^2 + xy + 8y; x = -3$ and $y = 2$

75) $3d^2f - 4df^2 - 126; d = -1$ and $f = -6$

76) $2s^2t - 3st^2 - 65; s = -1$ and $t = -5$

R Rethink

R1) In your own words, explain the meaning of the terms *variable* and *algebraic expression* as if you are explaining them to someone who is learning algebra for the first time. Include examples for each.

R2) Were there any exercises you could not do or that you got wrong? If so, write them down on your paper or circle them and ask your instructor for help.

2.2 Simplifying Expressions

P Prepare

O Organize

What are your objectives for Section 2.2?	How can you accomplish each objective?
1 Use the Commutative and Associative Properties	• Learn the **commutative** and **associative properties** for addition and multiplication. • Complete the given examples on your own. • Complete You Trys 1 and 2.
2 Use the Distributive Property	• Learn the **distributive properties.** • Complete the given example on your own. • Complete You Try 3.
3 Combine Like Terms	• Understand what makes terms *like terms*. • Understand how to use the properties of numbers to combine like terms. • Complete the given examples on your own. • Complete You Trys 4–8.

1 Use the Commutative and Associative Properties

In Chapter 1, we learned several properties of addition and multiplication that can be used with numbers. Because variables represent numbers, we can use the same properties in algebra. Let's see how we can apply them to algebraic expressions.

The **commutative properties** say that we can add or multiply numbers in any order and the result will be the same.

Commutative Property of Addition	**Commutative Property of Multiplication**
$5 + 2 = 2 + 5$	$6 \cdot 9 = 9 \cdot 6$
$7 \ = \ 7$	$54 \ = \ 54$

The same is true if one or both of the terms is a variable.

Commutative Property of Addition	**Commutative Property of Multiplication**
$x + 2 = 2 + x$	$6 \cdot n = n \cdot 6$ or $6n = n6$

The **associative properties** say that the way in which we group numbers using parentheses in an addition or multiplication problem will not affect the result. (The *order* of the numbers does not change.)

Associative Property of Addition	**Associative Property of Multiplication**
$6 + (4 + 1) = (6 + 4) + 1$	$(3 \cdot 2) \cdot 5 = 3 \cdot (2 \cdot 5)$
$6 + 5 \ = \ 10 + 1$	$6 \ \cdot \ 5 = 3 \ \cdot \ 10$
$11 \ = \ 11$	$30 \ = \ 30$

This applies to variables, too.

Associative Property of Addition	**Associative Property of Multiplication**
$a + (b + 8) = (a + b) + 8$	$(7 \cdot x) \cdot y = 7 \cdot (x \cdot y)$ or $(7x)y = 7(xy)$

We can write general rules for the properties.

Property Commutative Properties

If a and b are numbers, then

1) $a + b = b + a$ Commutative property of addition

2) $ab = ba$ Commutative property of multiplication

EXAMPLE 1

Use the commutative property to rewrite each expression.

a) $9 + w$ b) $c \cdot 4$

Solution

a) $9 + w = w + 9$ b) $c \cdot 4 = 4 \cdot c$ or $4c$

[YOU TRY 1] Use the commutative property to rewrite each expression.

a) $p + 7$ b) $h \cdot 5$

Property Associative Properties

If a, b, and c are numbers, then

1) $(a + b) + c = a + (b + c)$ Associative property of addition

2) $(ab)c = a(bc)$ Associative property of multiplication

EXAMPLE 2 Rewrite each expression using the associative property and simplify, if possible.

a) $(m + 7) + 3$ b) $-5(9x)$

Solution

a) $(m + 7) + 3 = m + (7 + 3)$ Change the position of the parentheses.
$= m + 10$ Add.

b) $-5(9x) = (-5 \cdot 9)x$ Change the position of the parentheses.
$= -45x$ Multiply.

[YOU TRY 2] Rewrite each expression using the associative property and simplify, if possible.

a) $(a + 1) + 6$ b) $-4(-3t)$

---

2 Use the Distributive Property

In Section 1.6, we learned that the distributive property says that multiplication *distributes* over addition and subtraction. For example,

$$2(7 + 1) = 2 \cdot 7 + 2 \cdot 1 \qquad\qquad 5(6 - 9) = 5 \cdot 6 - 5 \cdot 9$$
$$= 14 + 2 \qquad\qquad\qquad\qquad = 30 - 45$$
$$= 16 \qquad\qquad\qquad\qquad\quad = -15$$

We can use the distributive property with expressions containing variables, too.

Property Distributive Properties

If a, b, and c represent numbers, then

1) $a(b + c) = ab + ac$

2) $a(b - c) = ab - ac$

EXAMPLE 3

Rewrite each expression using the distributive property.

a) $4(a + 2)$ b) $-3(5 + k)$ c) $11(7v - 2)$ d) $-(w - 8)$

Solution

a) $4(a + 2) = 4 \cdot a + 4 \cdot 2$ Apply the distributive property.

    $= 4a + 8$ Multiply.

b) $-3(5 + k) = -3 \cdot 5 + (-3) \cdot k$ Apply the distributive property.

    $= -15 + (-3k)$ Multiply.

    $= -15 - 3k$ Rewrite using the subtraction symbol.

Note

We can also write $-15 - 3k$ as $-3k - 15$:

$$-15 - 3k = -15 + (-3k) = -3k + (-15) = -3k - 15$$

Usually, we write the variable term first.

Hint

Explain the distributive property in your own words.

c) $11(7v - 2) = 11 \cdot 7v - 11 \cdot 2$ Apply the distributive property.

    $= (11 \cdot 7)v - 22$ Use the associative property. Multiply $11 \cdot 2$.

    $= 77v - 22$ Multiply.

d) A negative sign in front of parentheses is the same as multiplying by -1.

$$-(w - 8) = -1(w - 8)$$

$$= -1 \cdot w - (-1) \cdot 8$$ Apply the distributive property.

$$= -w + 8$$ Multiply.

$\left[\right.$ **YOU TRY 3** $\left.\right]$ Rewrite each expression using the distributive property.

a) $3(v - 11)$ b) $-5(6g + 7)$ c) $9(8y - 2)$ d) $-(4 + k)$

3 Combine Like Terms

Just as we can add numbers like $2 + 5$ to get 7, we can *combine like terms* to simplify expressions. **Simplifying an expression** means writing it in the simplest form possible.

In the expression $8a + 3a - 5a + 7a$, there are four **terms:** $8a, 3a, -5a, 7a$. In fact, they are *like terms*.

Definition

Like terms contain the same variables with the same exponents.

EXAMPLE 4

Determine whether the following groups of terms are like terms.

a) $2x$, $5x$, $-4x$ b) p^3, $-8p^3$, $6p^3$ c) $7a$, $2b$, 6 d) $3xy$, $4x^2y$

Solution

a)   Notice that there is no exponent written on the x. Therefore, the exponent is assumed to be 1. Because each term contains the same variable with the same exponent, $2x$, $5x$, and $-4x$ *are like terms*. (We say that they are x-terms.)

b)   The terms p^3, $-8p^3$, and $6p^3$ are like terms. They have the same variable with the same exponent.

c) $7a$, $2b$, and 6 are *not* like terms because two of the terms have different variables, and one term does not contain a variable.

d) $3xy$ and $4x^2y$ are *not* like terms. They contain the same variables, but the exponents of x are different.

[YOU TRY 4]

Determine whether the following groups of terms are like terms.

a) $9t$, $-8t$, $-3t$ b) $7r^2$, $-2r^2$, r^2 c) $6c$, d, 5 d) $4ab^2$, $-7a^2b$

To simplify an expression such as $8a + 3a - 5a + 7a$, we combine like terms using the distributive property "in reverse."

$$8a + 3a - 5a + 7a = (8 + 3 - 5 + 7)a \qquad \text{Distributive property}$$
$$= 13a \qquad \text{Perform the operations in parentheses.}$$

We can add and subtract only those terms that are like terms.

EXAMPLE 5

Simplify each expression by combining like terms.

a) $10h + 4h$ b) $w^2 + 9 - 6w^2 + 2$

Solution

a)   We can use the distributive property to combine like terms.

$$10h + 4h = (10 + 4)h = 14h$$

b) $w^2 + 9 - 6w^2 + 2 = w^2 - 6w^2 + 9 + 2$ Rewrite like terms together.

$= 1w^2 - 6w^2 + 9 + 2$ w^2 is the same as $1w^2$.

$= (1 - 6)w^2 + 11$ Use the distributive property. Add $9 + 2$.

$= -5w^2 + 11$ Combine like terms.

W Hint

We can combine terms only if they are *like* terms.

Note

In Example 5b, we could also write the answer as $11 - 5w^2$. However, we usually write the constant last.

Notice that using the distributive property to combine like terms is the same as combining the coefficients of the terms and leaving the variable and its exponent the same.

YOU TRY 5   Simplify each expression by combining like terms.

a) $14c + 9c$ b) $k^3 - 10 - 7k^3 + 8$

EXAMPLE 6

Simplify each expression.

a) $12a + 7b + 4a + 2b - 8$ b) $9c^2 + 2c - 1 - 5c^2 + c + 4$

c) $-10x + 6xy + y - 2xy + 7x$

Solution

a)   First, use the commutative property to write the like terms of $12a + 7b + 4a + 2b - 8$ together.

$$12a + 4a + 7b + 2b - 8 \qquad \text{Write the like terms together.}$$

$$16a \quad + \quad 9b \quad - 8 \qquad \text{Add like terms by adding the coefficients and keeping the variables the same.}$$

Ask yourself, "*Is $16a + 9b - 8$ completely simplified?*" Yes! None of the terms are like terms, so the final answer is $16a + 9b - 8$.

b) $9c^2 + 2c - 1 - 5c^2 + c + 4 = 9c^2 - 5c^2 + 2c + 1c - 1 + 4$ Write the like terms together; rewrite c as $1c$.

$$= \quad 4c^2 \quad + \quad 3c \quad + \quad 3 \qquad \text{Combine like terms.}$$

Ask yourself, "*Is $4c^2 + 3c + 3$ completely simplified?*" Yes! None of the terms are like terms, so the final answer is $4c^2 + 3c + 3$.

c) $-10x + 6xy + y - 2xy + 7x = -10x + 7x + 6xy - 2xy + y$ Write the like terms together.

$$= \quad -3x \quad + \quad 4xy \quad + y \qquad \text{Combine like terms.}$$

Ask yourself, "*Is $-3x + 4xy + y$ completely simplified?*" Yes! None of the terms are like terms, so the final answer is $-3x + 4xy + y$.

Note

When an expression contains more than one variable, as in Example 6a, the terms are usually written in alphabetical order from left to right: $16a + 9b - 8$.

When an expression contains the same variable with different exponents, as in Example 6b, it is usually written in descending powers; that is, write it from the highest power to the lowest power from left to right with the constant at the end: $4c^2 + 3c + 3$.

W Hint

Don't just *read* the examples. Write out each step on your paper as you are reading.

YOU TRY 6   Simplify each expression.

a) $5m + 2n + 3m + 9n - 10$ b) $7h^2 + 8h - 3 - 3h^2 + h + 12$

c) $-16a + 8ab + b - 7ab + 5a$

Before we simplify more complicated expressions, let's review the difference between using the associative property and the distributive property to simplify an expression.

EXAMPLE 7

Simplify each expression.

a) $4(9k)$

b) $4(9 + k)$

Solution

a) The expression $4(9k)$ does *not* have an addition or subtraction symbol inside the parentheses. So, to simplify $4(9k)$ we use the associative property.

$$4(9k) = (4 \cdot 9)k = 36k \qquad \text{Multiply.}$$

Change the position of
the parentheses using the
associative property.

W Hint

Be sure you notice the difference between the two expressions in Example 7.

b) The expression $4(9 + k)$ has an addition sign inside the parentheses between the 9 and k. *Use the distributive property* to simplify the expression.

$$4(9 + k) = 4 \cdot 9 + 4 \cdot k \qquad \text{Distribute the 4.}$$
$$= 36 + 4k \qquad \text{Multiply.}$$
$$= 4k + 36 \qquad \text{Rewrite the expression with the variable term first.}$$

[YOU TRY 7]

Simplify each expression.

a) $-8(7w)$

b) $-8(7 + w)$

Now we are ready to simplify expressions using more than one property.

EXAMPLE 8

Simplify each expression.

a) $7 + 2(p + 8)$

b) $-3(5d) + 4(2d + 1) + 6$

Solution

a) Be careful! To simplify $7 + 2(p + 8)$, we do *not* start by adding $7 + 2$. When we simplify expressions containing variables, we must **use the order of operations** just like when we simplify expressions containing only numbers. Multiplication comes before addition, so begin by using the distributive property to multiply.

$$7 + 2(p + 8) = 7 + 2 \cdot p + 2 \cdot 8 \qquad \text{Multiply before adding; distribute the 2.}$$
$$= 7 + 2p + 16 \qquad \text{Multiply before adding.}$$
$$= 2p + 7 + 16 \qquad \text{Write the like terms together.}$$
$$= 2p + 23 \qquad \text{Add.}$$

Ask yourself, "*Is* $2p + 23$ *completely simplified*?" Yes!

b)$\quad -3(5d) + 4(2d + 1) + 6 = \overbrace{(-3 \cdot 5)d + \overbrace{4 \cdot 2d + 4 \cdot 1}} + 6$

Use the associative property to move the parentheses. Use the distributive property.

$$= \quad -15d \quad + \quad 8d \quad + 4 \quad + 6 \qquad \text{Multiply.}$$
$$= \quad\quad\quad\quad -7d \quad\quad\quad + 10 \qquad \text{Add like terms.}$$

Ask yourself, "Is $-7d + 10$ *completely simplified*?" Yes!

[YOU TRY 8] Simplify each expression.

a) $5 + 6(y + 11)$ b) $-7(4z) + 8(3z - 5) + 1$

ANSWERS TO [YOU TRY] EXERCISES

1) a) $7 + p$ b) $5 \cdot h$ or $5h$ 2) a) $a + (1 + 6) = a + 7$ b) $(-4 \cdot (-3))t = 12t$
3) a) $3v - 33$ b) $-30g - 35$ c) $72y - 18$ d) $-4 - k$ or $-k - 4$
4) a) yes b) yes c) no d) no   5) a) $23c$ b) $-6k^3 - 2$
6) a) $8m + 11n - 10$ b) $4h^2 + 9h + 9$ c) $-11a + ab + b$
7) a) $-56w$ b) $-8w - 56$ 8) a) $6y + 71$ b) $-4z - 39$

E Evaluate ### 2.2 Exercises

Do the exercises, and check your work.

Get Ready

Use the commutative property to rewrite each expression and evaluate each to show that the result is the same.

1) $2 + 9$ 2) $6 + 1$

3) $-4 \cdot 7$ 4) $-5 \cdot 8$

Objective 1: Use the Commutative and Associative Properties

 5) In your own words, explain the commutative property of multiplication. Then, give an example.

6) In your own words, explain the associative property of addition. Then, give an example.

Use the commutative property to rewrite each expression.

 7) $3 + w$ 8) $4 + r$

 9) $c \cdot 2$ 10) $k \cdot 9$

11) $-7 + 8y$ 12) $-1 + 6t$

Use the associative property to rewrite each expression and evaluate each to show that the result is the same.

13) $(-9 + 4) + 3$ 14) $(-8 + 2) + 1$

15) $3 \cdot (2 \cdot 8)$ 16) $2 \cdot (5 \cdot 4)$

Use the associative property to rewrite each expression. Then, simplify.

17) $(h + 6) + 7$ _13 + h_   18) $(d + 5) + 4$ _9 + d_

19) $-10 + (2 + 5k)$ _-8 + 5k_   20) $-12 + (9 + 8t)$ _-3 + 8t_

21) $3(4u)$ 22) $7(9m)$

23) $-12(5p^2)$ 24) $-8(3g^2)$

25) $-6(-8ab)$ 26) $-5(-2xy)$ _10 xy_

27) $9(-n)$ 28) $4(-c)$ _4x - 4c_

Objective 2: Use the Distributive Property

Use the distributive property to evaluate each expression.

29) $7(2 + 9)$ 30) $4(6 + 3)$

31) $-(6 - 10)$

32) $-(7 - 2)$

Rewrite each expression using the distributive property.

33) $3(m + 5)$ 34) $6(k + 7)$

35) $2(c - 9)$ 36) $5(p - 11)$

37) $-4(t + 1)$ 38) $-9(z + 1)$

39) $8(2v - 3)$ 40) $7(2d - 5)$

 41) $-3(6 - 7r)$ 42) $-2(3 - 8y)$

43) $-(g + 4)$

44) $-(x + 10)$

45) $-(b - 23)$

46) $-(h - 18)$

Objective 3: Combine Like Terms

47) How do you know whether a group of terms are like terms?

48) Explain how to add like terms.

Determine whether each group of terms are like terms.

49) $8c, 2c$

50) $5n, 9n$

51) $3h^2, -7h^2, h^2$

52) $u^2, 4u^2, -2u^2$

53) $6x, 4y, 6$

54) $8c, 3d, 8$

55) $ab^2, 5ab^2, -3a^2b$

56) $-mn^2, 2mn^2, 7m^2n$

Simplify each expression by combining like terms.

57) $3g + 5g$

58) $4q + 2q$

59) $w + 9w$

60) $r + 8r$

61) $6p - 13p$

62) $2h - 10h$

63) $x - x$

64) $a - a$

65) $-7d + 2d + 3d$

66) $-13c + 3c + 6c$

67) $-2k + 6 + 11k + 9$

68) $-5z + 8 + 12z + 1$

69) $b^2 - 5 + 7b^2 + 3$

70) $t^2 - 9 + 4t^2 + 2$

71) $3xy + xy - 9xy$

72) $2ab + ab - 7ab$

73) $-mn + 4 + 8mn - 3 + mn$

74) $-cd + 10 + 4cd - 9 + cd$

75) $-8p + q - 7 + 2p + 9q + 7$

76) $-9r + t - 8 + 4r + 7t + 8$

77) $v^2 - 3v + 10 + 4v^2 - v + 9$

78) $g^2 - 5g + 12 + 6g^2 - g + 3$

79) $5a + 8ab - b - 11a + 2ab + 6b + 3a$

80) $6x + 7xy - y - 12x + 5xy + 4y + 2x$

81) $2c^2d + 5cd^2 + 4c^2d + 9cd^2 - 7cd$

82) $3m^2n + 3mn^2 + 2m^2n + 10mn^2 - 9mn$

For Exercises 83–86, determine whether you will use the associative or distributive property to simplify each expression. Then, use that property to simplify.

83) a) $5(6m)$
    b) $5(6 + m)$

84) a) $3(8z)$
    b) $3(8 + z)$

85) a) $9(-4 + c)$
    b) $9(-4c)$

86) a) $6(-7 + a)$
    b) $6(-7a)$

87) Professor Rossiter puts this problem on the board for her students to simplify: $4 + 2(k + 7)$. Here is how two of her students simplify it:

Kelly	Bashir
$4 + 2(k + 7) =$	$4 + 2(k + 7) =$
$6(k + 7) =$	$4 + 2 \cdot k + 2 \cdot 7 =$
$6 \cdot k + 6 \cdot 7 =$	$4 + 2k + 14 =$
$6k + 42$	$2k + 4 + 14 =$
	$2k + 18$

Who is right and why?

88) In your own words, explain why you would use the associative property to simplify $-2(10x)$ and why you would use the distributive property to simplify $-2(10 + x)$. Then, simplify each expression.

Simplify each expression.

89) $2 + 5(x + 9)$

90) $7 + 3(a + 4)$

91) $-7 + 3(c - 5) + 6c$

92) $-8 + 2(p - 11) + 5p$

93) $4(-6h + 1) + 3(h - 4) + 9$

94) $5(-2w + 1) + 4(w - 2) + 16$

95) $5(-3r - 2) - 9r + 7(6r) + 3(r + 5)$

96) $4(-5n - 3) - 7n + 8(4n) + 5(n + 6)$

97) $-2(8y) + 2(4y + 3) + 8y - 6$

98) $-6(3t) + 3(2t + 5) + 12t - 15$

99) Write an expression so that the difference of two terms is $-9p$.

100) Write an expression so that the sum of two terms is $16k$.

R Rethink

R1) How can the associative property make it easier to evaluate an expression like $(7 + 34) + 6$?

R2) When a cashier counts the money in the drawer at the end of the shift, how is that similar to combining like terms?

2.3 Solving Linear Equations Part I

 **Prepare**　　　　　　　　　**O Organize**

What are your objectives for Section 2.3?	How can you accomplish each objective?
1 Determine Whether a Number Is a Solution of an Equation	• Know the differences between expressions and equations and what solving an equation means. • Complete the given example on your own. • Complete You Try 1.
2 Use the Addition and Subtraction Properties of Equality	• Follow the explanation to understand and then learn the **Addition and Subtraction Properties of Equality.** • Complete the given examples on your own. • Complete You Trys 2–4.
3 Use the Division Property of Equality	• Learn the **Division Property of Equality.** • Complete the given examples on your own. • Complete You Trys 5 and 6.

W Work　　　**Read the explanations, follow the examples, take notes, and complete the You Trys.**

1 Determine Whether a Number Is a Solution of an Equation

What is an equation? It is a mathematical statement that two expressions are equal. For example, $5 + 3 = 8$ is an equation.

 Hint

Be sure you know the difference between an expression and an equation!

 Note

An equation contains an "=" sign and an expression does not.

$2x + 3 = 11$ is an *equation.*　　　$2x + 3x$ is an *expression.*

We can **solve** equations, and we can **simplify** expressions.

There are many different types of algebraic equations, and in Sections 2.3–2.5, we will learn how to solve *linear* equations.

To **solve an equation** means to find the value or values of the variable that make the equation true. For example, the **solution** of the equation $n - 3 = 2$ is $n = 5$ since substituting 5 for the variable makes the equation true.

$$n - 3 = 2$$
$$5 - 3 = 2 \quad \text{True}$$

We say that $n = 5$ is the *solution* of the equation $n - 3 = 2$. We also say that 5 *satisfies* the equation $n - 3 = 2$.

EXAMPLE 1

Determine whether 3 is a solution of each of the equations.

a) $18 + x = 21$ b) $13 = 5 + 2x$

Solution

a) $18 + x = 21$

$18 + 3 = 21$ Substitute 3 for x.

$21 = 21$ Add. The statement is true.

Because the statement $21 = 21$ is true, 3 **is** a solution of $18 + x = 21$.

W Hint

How is this similar to what you did in the previous section?

b) $13 = 5 + 2x$

$13 = 5 + 2(3)$ Substitute 3 for x.

$13 = 5 + 6$ Multiply.

$13 = 11$ Add. The statement is false.

Because the statement $13 = 11$ is false, 3 is **not** a solution of $13 = 5 + 2x$.

[**YOU TRY 1**]

Determine whether 5 is a solution of each of the equations.

a) $t + 9 = 14$ b) $20 = 4 + 3t$

---

2 Use the Addition and Subtraction Properties of Equality

Begin with the true statement $6 = 6$. What happens if we add the same number, say 1, to each side? Is the statement still true? Yes.

$$6 = 6$$
$$6 + 1 = 6 + 1$$
$$7 = 7 \qquad \text{True}$$

Will the statement remain true if we *subtract* the same number from each side? Let's begin with the true statement $9 = 9$ and subtract 4 from each side:

$$9 = 9$$
$$9 - 4 = 9 - 4$$
$$5 = 5 \qquad \text{True}$$

When we subtracted 4 from each side of the equation, the new statement was true.

$6 = 6$ and $6 + 1 = 6 + 1$ are *equivalent equations.*
$9 = 9$ and $9 - 4 = 9 - 4$ are *equivalent equations* as well.

Adding the same number to both sides of an equation or subtracting the same number from both sides of an equation will produce equivalent equations. We can use these principles to solve an algebraic equation because doing so will not change the equation's solution.

Property Addition and Subtraction Properties of Equality

Let a, b, and c be expressions representing numbers. Then,

1) If $a = b$, then $a + c = b + c$ Addition property of equality
2) If $a = b$, then $a - c = b - c$ Subtraction property of equality

---

EXAMPLE 2

Solve $x - 2 = 8$. Check the solution.

Solution

Remember, to solve the equation means to find the value of the variable that makes the statement true. To do this, we want to get the variable on a side by itself. We call this **isolating the variable.**

On the left side of the equal sign, the 2 is being **subtracted from** the x. To isolate x, we perform the "opposite" operation—that is, we **add 2** to each side.

$$x - 2 = 8$$
$$x - 2 + 2 = 8 + 2 \qquad \text{Add 2 to each side.}$$
$$x + 0 = 10$$
$$x = 10 \qquad \text{Simplify; } x + 0 = x.$$

Check: Substitute 10 for x in the original equation.

$$x - 2 = 8$$
$$10 - 2 = 8 \qquad \text{Substitute 10 for } x.$$
$$8 = 8 \checkmark$$

The solution is 10. We can also say that $x = 10$.

YOU TRY 2

Solve $b - 3 = 14$. Check the solution.

EXAMPLE 3

Solve $t + 5 = 12$. Check the solution.

Solution

Here, 5 is being added to t. To get the t by itself, subtract 5 from each side.

$$t + 5 = 12$$
$$t + 5 - 5 = 12 - 5 \qquad \text{Subtract 5 from each side.}$$
$$t + 0 = 7 \qquad \text{Subtract.}$$
$$t = 7 \qquad \text{Simplify; } t + 0 = t.$$

Check: Substitute 7 for t in the original equation.

$$t + 5 = 12$$
$$7 + 5 = 12 \qquad \text{Substitute 7 for } t.$$
$$12 = 12 \checkmark$$

The solution is 7.

YOU TRY 3   Solve $r + 8 = 3$. Check the solution.

Equations can contain variables on either side of the equal sign. Also remember that we can change subtraction to addition as in a problem like $-6 - 7 = -6 + (-7) = -13$.

EXAMPLE 4   Solve $-6 = m + 7$. Check the solution.

Solution

Notice that the 7 is being **added to** the variable, m. We will **subtract 7** from each side to isolate the variable.

$$-6 = m + 7$$
$$-6 - 7 = m + 7 - 7 \qquad \text{Subtract 7 from each side.}$$
$$-6 + (-7) = m + 0 \qquad \text{Change subtraction to addition of the additive inverse.}$$
$$-13 = m \qquad \text{Simplify.}$$

Check: Substitute -13 for m in the original equation.

$$-6 = m + 7$$
$$-6 = -13 + 7 \qquad \text{Substitute } -13 \text{ for } m.$$
$$-6 = -6 \checkmark$$

The solution is -13.

YOU TRY 4   Solve $-10 = y + 9$. Check the solution.

3   Use the Division Property of Equality

Let's start with the true statement $12 = 12$. What happens if we divide both sides by the same number, say 3? Is the equation still true?

$$12 = 12$$
$$\frac{12}{3} = \frac{12}{3} \qquad \text{Divide both sides by 3.}$$
$$4 = 4 \qquad \text{True}$$

Yes. If we divide both sides of an equation by the same nonzero number, then we will obtain an equivalent equation.

Property   Division Property of Equality

Let a, b, and c represent numbers where $c \neq 0$. Then,

     If $a = b$, then $\dfrac{a}{c} = \dfrac{b}{c}$ Division property of equality

     Example: Because $12 = 12$, it is also true that $\dfrac{12}{3} = \dfrac{12}{3}$.

Solve each equation. Check the solution.

a) $2c = -8$ b) $-45 = -5n$

Solution

a) Remember that $2c$ means $2 \cdot c$, so on the left-hand side of the equation, the c is being **multiplied** by 2. To solve for c, we will perform the "opposite" operation and **divide** each side by 2.

$$2c = -8$$

$$\frac{2c}{2} = \frac{-8}{2} \qquad \text{Divide each side by 2.}$$

$$1c = -4 \qquad \frac{2c}{2} = 2c \div 2 = 1c$$

$$c = -4 \qquad \text{Simplify; } 1c = c.$$

Check: Substitute -4 for c in the original equation.

$$2c = -8$$

$$2(-4) = -8 \qquad \text{Substitute } -4 \text{ for } c.$$

$$-8 = -8 \checkmark$$

The solution is -4. We can also say that $c = -4$.

b) Where is the variable in the equation $-45 = -5n$? It is on the *right* side of the equation, and the n is being **multiplied** by -5. To solve for n, we will perform the *opposite* operation: **divide** both sides by -5.

$$-45 = -5n$$

$$\frac{-45}{-5} = \frac{-5n}{-5} \qquad \text{Divide each side by } -5.$$

$$9 = 1n \qquad \text{Divide; } \frac{-5n}{-5} = -5n \div (-5) = 1n$$

$$9 = n \qquad \text{Simplify, } 1n = n.$$

Check: Substitute 9 for n in the original equation.

$$-45 = -5n$$

$$-45 = -5(9) \qquad \text{Substitute 9 for } n.$$

$$-45 = -45 \checkmark$$

The solution is 9.

[YOU TRY 5]

Solve each equation. Check the solution.

a) $3y = -21$ b) $-48 = -8r$

To solve the next equation, we must remember that an expression like $-x$ means $-1 \cdot x$.

| **EXAMPLE 6** | Solve $-a = 15$. Check the solution. |

Solution

The negative sign in front of the a tells us that the coefficient of a is -1. Since a is being **multiplied** by -1, we will **divide** each side by -1 to solve for a.

$$-a = 15$$

$$-1a = 15 \qquad \text{Rewrite } -a \text{ as } -1a.$$

$$\frac{-1a}{-1} = \frac{15}{-1} \qquad \text{Divide each side by } -1.$$

$$1a = -15 \qquad \frac{-1a}{-1} = -1a \div (-1) = 1a$$

$$a = -15 \qquad \text{Simplify; } 1a = a.$$

Check by substituting -15 for a.

$$-a = 15$$

$$-(-15) = 15 \qquad \text{Substitute } -15 \text{ for } a.$$

$$-1 \cdot (-15) = 15 \qquad -(-15) = -1 \cdot (-15)$$

$$15 = 15 \checkmark$$

W Hint
Remember that in a term like $-x$, the coefficient is -1.

[YOU TRY 6] Solve $-w = -4$. Check the solution.

ANSWERS TO [YOU TRY] EXERCISES

1) a) yes b) no   2) $b = 17$ 3) $r = -5$ 4) $y = -19$
5) a) $y = -7$ b) $r = 6$ 6) $w = 4$

E Evaluate **2.3** Exercises Do the exercises, and check your work.

Objective 1: Determine Whether a Number Is a Solution of an Equation

Identify each as an expression or an equation.

1) $8m + 7 - 2m$

2) $4a - 9 = 11$

3) $w + 3(2w + 1) = -8$

4) $6 + 5(k - 4)$

5) Can we solve $-5x + 9x$? Why or why not?

6) Can we solve $-5x + 9x = 16$? Why or why not?

7) In your own words, explain the difference between an expression and an equation.

8) Explain how to check the solution of an equation.

Determine whether the given value is a solution of the equation.

9) $y - 7 = -3; y = 2$

10) $7 = x + 15; x = -8$

11) $-4h = -12; h = 3$

12) $-3n = 15; n = -4$

13) $-11 = 2c - 9; c = -1$

14) $6t - 1 = 11; t = 2$

15) $-w = -8; w = -8$

16) $-a = 4; a = 4$

17) $5 = -p; p = -5$

18) $-7 = -y; y = 7$

Objective 2: Use the Addition and Subtraction Properties of Equality

Solve each equation and check the solution.

19) $b - 6 = 8$

20) $r + 9 = -3$

21) $y + 12 = 4$

22) $p - 1 = 6$

23) $m + 10 = 0$

24) $d - 8 = 0$

25) $13 = k - 8$

26) $22 = v + 16$

27) $-5 = -5 + x$

28) $-3 = -3 + a$

 29) $9 = t - 14$

30) $4 = q - 3$

 31) $-3 = w + 21$

32) $-7 = x + 9$

33) $-7 + y = 1$

34) $-5 + z = 13$

35) Write an equation that can be solved with the subtraction property of equality that has a solution of $w = -4$.

36) Write an equation that can be solved with the addition property of equality that has a solution of $x = 7$.

Objective 3: Use the Division Property of Equality

Solve each equation and check the solution.

 37) $2a = 10$

38) $3p = 27$

39) $-5u = 40$

40) $-8t = -56$

41) $16 = 8d$

42) $12 = 4m$

43) $-66 = -11r$

44) $-72 = 9w$

45) $28 = -7k$

46) $-60 = 5x$

47) $3b = 54$

48) $-2v = -78$

49) $-c = 1$

50) $-q = -6$

51) $-10 = -w$

52) $14 = -r$

53) Write an equation that can be solved with the division property of equality that has a solution of $n = 6$.

54) Write an equation that can be solved with the division property of equality that has a solution of $y = -1$.

Mixed Exercises: Objectives 2 and 3

For Exercises 55–58, determine which property should be used to solve each equation and explain why that property is used. Choose from the addition, subtraction, and division properties of equality. Then, solve each equation.

55) a) $4 + k = 20$ b) $4k = 20$

56) a) $6 + t = -54$ b) $6t = -54$

57) a) $18 = -3y$ b) $18 = y - 3$

58) a) $-36 = -12n$ b) $-36 = n - 12$

Solve and check each equation.

59) $-8c = 72$

60) $z + 16 = 4$

61) $23 = w + 14$

62) $66 = -6x$

63) $k - 37 = -84$

64) $-16 = -d$

65) $2 = -m$

66) $-78 = t - 25$

67) $x + 21 = 21$

68) $-5a = -5$

69) $6p = 6$

70) $-18 + r = -18$

R Rethink

R1) What is the difference between an expression and an equation?

R2) Which exercises in this section do you need to practice more?

R3) Which exercises were easiest for you? Why do you think they were easier than others?

2.4 Solving Linear Equations Part II

What are your objectives for Section 2.4?	How can you accomplish each objective?
1 Solve Equations Using the Properties of Division and Addition or Subtraction	<ul><li>Understand the properties learned in Section 2.3.</li><li>Follow Examples 1 and 2 to understand how and when to use each property.</li><li>Complete the examples on your own.</li><li>Complete You Trys 1 and 2.</li></ul>
2 Solve Equations by Combining Like Terms	<ul><li>Write the procedure for **How to Solve a Linear Equation** in your own words.</li><li>Follow the procedure for solving equations and notice the subtle differences between the examples.</li><li>Complete the given examples on your own.</li><li>Complete You Trys 3–5.</li></ul>

W Work   **Read the explanations, follow the examples, take notes, and complete the You Trys.**

1 Solve Equations Using the Properties of Division and Addition or Subtraction

So far, we have not combined the properties of addition, subtraction, and division to solve an equation. But that is exactly what we must do to solve equations like

$$3p + 8 = 20 \qquad \text{and} \qquad 6n + 9 - 8n + 2 = 17$$

Let's start by solving equations like $3p + 8 = 20$.

EXAMPLE 1   Solve $3p + 8 = 20$.

Solution

In this equation, there is a number, 8, being **added** to the term containing the variable, and the variable is being multiplied by a number, 3. **In general, we first eliminate the number being added to or subtracted from the variable term.** Then we eliminate the coefficient.

W Hint

To solve an equation like $3p + 8 = 20$, we use more than one property of equality. Which property or properties do we usually apply first?

$$3p + 8 = 20$$
$$3p + 8 - 8 = 20 - 8 \qquad \text{Subtract 8 from each side.}$$
$$3p = 12 \qquad \text{Combine like terms.}$$
$$\frac{3p}{3} = \frac{12}{3} \qquad \text{Divide by 3.}$$
$$p = 4 \qquad \text{Simplify.}$$

Check: $\quad 3p + 8 = 20$
$$3(4) + 8 = 20 \qquad \text{Substitute 4 for } p.$$
$$12 + 8 = 20 \qquad \text{Multiply.}$$
$$20 = 20 \checkmark$$

The solution is 4.

Solve $5m + 7 = 37$.

EXAMPLE 2

Solve each equation.

a) $-6c - 17 = 13$ b) $-19 = 2y - 5$

Solution

a) On the left-hand side, the c is being multiplied by -6, and 17 is being subtracted from the c-term. To solve the equation, **begin by eliminating the number being subtracted from the c-term.**

W Hint

Write out the example on your paper as you are reading it. Remember, use a colored pencil to indicate when you perform an operation in a problem.

$$6c - 17 = 13$$
$$-6c - 17 + 17 = 13 + 17 \quad \text{Add 17 to each side.}$$
$$-6c = 30 \quad \text{Combine like terms.}$$
$$\frac{-6c}{-6} = \frac{30}{-6} \quad \text{Divide each side by } -6.$$
$$c = -5 \quad \text{Simplify.}$$

The check is left to the student.

b) The variable is on the right-hand side of the equation, $-19 = 2y - 5$. First, we will add 5 to each side, then we will divide by 2.

$$-19 = 2y - 5$$
$$-19 + 5 = 2y - 5 + 5 \quad \text{Add 5 to each side.}$$
$$-14 = 2y \quad \text{Combine like terms.}$$
$$\frac{-14}{2} = \frac{2y}{2} \quad \text{Divide each side by 2.}$$
$$-7 = y \quad \text{Simplify.}$$

Verify that -7 is the solution.

YOU TRY 2

Solve each equation.

a) $-4a - 11 = 29$ b) $-32 = 6z - 8$

---

2 Solve Equations by Combining Like Terms

We have learned how to solve equations such as

$$x + 7 = 15 \qquad -3n = -12 \qquad 1 = 8c + 9$$

Each of these equations contains only one variable term. Now we will learn how to solve equations that require us to combine like terms as well as to use the distributive property to clear parentheses. We can use the following steps.

Procedure How to Solve a Linear Equation

Step 1: **Clear parentheses** and **combine like terms** on each side of the equation.

Step 2: **Get the variable on one side of the equal sign and the constant on the other side of the equal sign** (isolate the variable) using the addition or subtraction property of equality.

Step 3: **Solve for the variable** using the division property of equality.

Step 4: **Check the solution** in the original equation.

EXAMPLE 3

Solve $6n + 9 - 8n + 2 = 17$.

Solution

Step 1: Because there are two n-terms on the left side of the equal sign, begin by combining like terms.

$$6n + 9 - 8n + 2 = 17$$
$$6n + (-8n) + 9 + 2 = 17 \quad \text{Rewrite using the commutative property.}$$
$$-2n + 11 = 17 \quad \text{Combine like terms.}$$

Step 2: Isolate the variable. This means get the variable term on a side by itself.

$$-2n + 11 - 11 = 17 - 11 \quad \text{Subtract 11 from each side.}$$
$$-2n = 6 \quad \text{Combine like terms.}$$

W Hint

Do you know what *isolate the variable* means?

Step 3: Solve for n using the division property of equality.

$$\frac{-2n}{-2} = \frac{6}{-2} \quad \text{Divide each side by } -2.$$
$$n = -3 \quad \text{Simplify.}$$

Step 4: Check the solution in the original equation.

$$6n + 9 - 8n + 2 = 17$$
$$6(-3) + 9 - 8(-3) + 2 = 17 \quad \text{Substitute } -3 \text{ for each } n.$$
$$-18 + 9 + 24 + 2 = 17 \quad \text{Multiply.}$$
$$17 = 17 \checkmark$$

Therefore, $n = -3$.

[YOU TRY 3]

Solve $10 - 5h - 8 + 2h = -25$.

If an equation contains parentheses, the first step is to use the distributive property to clear the parentheses.

EXAMPLE 4

Solve $20 = 4(t + 5)$.

Solution

Step 1: Clear the parentheses and combine like terms, if possible.

$$20 = 4(t + 5)$$
$$20 = 4 \cdot t + 4 \cdot 5 \quad \text{Distribute the 4.}$$
$$20 = 4t + 20 \quad \text{Multiply.}$$

Now the equation looks like those we have already learned how to solve. Go to the next step.

Step 2: Isolate the variable. That is, get the variable on a side by itself.

$$20 - 20 = 4t + 20 - 20 \quad \text{Subtract 20 from each side.}$$
$$0 = 4t \quad \text{Simplify.}$$

Step 3: Solve for t using the division property of equality.

$$\frac{0}{4} = \frac{4t}{4} \quad \text{Divide each side by 4.}$$

$$0 = t \quad \text{Simplify. Remember, } \frac{0}{4} \text{ means } 0 \div 4 = 0.$$

Step 4: Check the solution in the *original* equation.

$$20 = 4(t + 5) \quad \text{This is the original equations.}$$
$$20 = 4(0 + 5) \quad \text{Substitute 0 for } t.$$
$$20 = 4(5) \quad \text{Use the order of operations; add inside the parentheses.}$$
$$20 = 20 \checkmark$$

The solution is $t = 0$.

YOU TRY 4

Solve $-32 = 2(3m - 4)$.

Finally, let's solve an equation that requires us to use the distributive property *and* to combine like terms.

EXAMPLE 5

Solve $-5(2x - 1) + 3(x - 7) = -23$.

Solution

Step 1: Clear the parentheses and combine like terms.

$$-5(2x - 1) + 3(x - 7) = -23$$
$$-10x + 5 + 3x - 21 = -23 \quad \text{Distribute.}$$
$$-10x + 3x + 5 - 21 = -23 \quad \text{Use the commutative property to write like terms together.}$$
$$-7x - 16 = -23 \quad \text{Combine like terms.}$$

> **W Hint**
>
> Combine like terms on each side of the equation *before* isolating the variable.

Step 2: Isolate the variable.

$$-7x - 16 + 16 = -23 + 16 \qquad \text{Add 16 to each side.}$$
$$-7x = -7 \qquad\qquad\qquad \text{Combine like terms.}$$

Step 3: Solve for x using the division property of equality.

$$\frac{-7x}{-7} = \frac{-7}{-7} \qquad \text{Divide each side by } -7.$$
$$x = 1 \qquad \text{Simplify.}$$

Step 4: The check is left to the student. The solution is $x = 1$.

$\left[\text{YOU TRY 5}\right]$ Solve $-6(w - 2) + 4(3w - 7) = -16$.

ANSWERS TO $\left[\text{YOU TRY}\right]$ **EXERCISES**

1) $m = 6$ 2) a) $a = -10$ b) $z = -4$ 3) $h = 9$ 4) $m = -4$ 5) $w = 0$

E Evaluate **2.4** Exercises   Do the exercises, and check your work.

Objective 1: Solve Equations Using the Properties of Division and Addition or Subtraction

Solve each equation, and check the solution.

1) $2k + 9 = 15$ 2) $3c + 4 = 22$

3) $6a - 1 = -25$ 4) $5p - 3 = -48$

5) $18 = 7m + 4$ 6) $37 = 4y + 9$

7) $-5r - 13 = 17$ 8) $-8g - 11 = 45$

9) $5 = -3k + 41$ 10) $6 = -7d + 27$

11) $-29 = 9w - 29$ 12) $-14 = 5h - 14$

13) $-11x + 17 = 6$ 14) $-6v + 19 = 13$

15) $6b - 54 = 0$ 16) $8k - 24 = 0$

17) $18 + 7n = -17$ 18) $23 + 6a = -13$

19) $-7 = 9 - 2z$ 20) $-8 = 4 - 3p$

21) $5 - x = 14$ 22) $6 - t = 17$

Objective 2: Solve Equations by Combining Like Terms

23) Explain, in your own words, the steps for solving a linear equation.

24) What is the first step for solving $9y + 4 + 2y - 7 = 8$? Do not solve the equation.

Solve each equation.

Fill It In

Fill in the blanks with either the missing mathematical step or reason for the given step.

25) $2x + 5 + 6x + 9 = 30$
     $8x + 14 = 30$
     _____   Subtraction property of equality
     $8x = 16$ _____
     _____   Division property of equality
     _____   Simplify.

26) $2(2k + 1) + 7 = 21$
     _____   Distribute.
             Combine like terms.
     $4k + 9 - 9 = 21 - 9$ _____
     $4k = 12$ _____
     $\dfrac{4k}{4} = \dfrac{12}{4}$ _____
     _____   Simplify.

Solve each equation. Be sure to check the solution.

27) $4h + 7 + 2h - 13 = 18$

28) $3y + 5 + 4y - 11 = 8$

29) $8d - 9 - 3d + 20 = 1$

30) $3w - 11 - 12w + 4 = 20$

31) $-14 = -10p + 1 + 8p + 5$

32) $-21 = -9n + 8 + 5n + 3$

33) $13 = 9 - 12x + 4 - 15 + 3x + 8x$

34) $-5 = 6 - 15k + 2 - 17 + 8k + 6k$

35) $11 + 9c + 5 - c + 8c = 32$

36) $14 + 7a + 3 + 3a - a = 26$

37) What is the first step for solving $3(z + 10) = 18$? Do *not* solve the equation.

38) What is the first step for solving $-65 = 5(2g - 3)$? Do *not* solve the equation.

Solve each equation. Be sure to check the solution.

39) $3(z + 10) = 18$ 40) $2(x + 9) = 14$

41) $4(3h - 5) = -8$ 42) $3(2t - 7) = -15$

43) $98 = -2(4w - 5)$ 44) $65 = -5(2g - 3)$

45) $-6(n + 1) - 29 = 7$ 46) $-8(y + 2) - 5 = 11$

47) $16 + 3(2k - 7) = -5$

48) $19 + 7(3b - 4) = -9$

49) $2(a + 6) + 5(a - 3) = 18$

50) $4(p + 5) + 3(p - 8) = 38$

51) $-11 = 3(2m - 7) + 4(m + 5)$

52) $3 = 2(3d - 8) + 5(d + 6)$

53) $-8(w - 2) + 7(w + 1) = 12$

54) $-9(z - 4) + 2(z + 3) = 7$

55) $1 + 2(x + 9) - 5(2x + 3) = -36$

56) $8 + 3(t + 7) - 2(5t + 1) = -29$

R Rethink

R1) Explain the steps for solving an equation to someone in your class.

R2) Circle or write down any problems you got wrong or did not know how to do, and ask for help.

2.5 Solving Linear Equations Part III

P Prepare

O Organize

What are your objectives for Section 2.5?	How can you accomplish each objective?
1 Solve Equations Containing Variables on Both Sides of the Equal Sign	• Use the procedure for solving equations given in Section 2.4 and notice the subtle differences between the examples. • Complete the given examples on your own. • Complete You Trys 1–3.
2 Write English Expressions as Algebraic Expressions	• Recognize the key words that imply addition, subtraction, multiplication, and division. • Understand the connection between expressions with only numbers and expressions with variables and numbers. • Complete the given examples on your own. • Complete You Trys 4–6.

W Work   **Read the explanations, follow the examples, take notes, and complete the You Trys.**

1 Solve Equations Containing Variables on Both Sides of the Equal Sign

We have one more type of equation to learn how to solve: equations that have variables on both sides of the equal sign. Follow the steps we learned in Section 2.4. After we have learned how to solve these equations, we will discuss how to write English expressions as algebraic expressions.

EXAMPLE 1

Solve $10c - 9 = 8c + 5$.

Solution

This equation has a variable term and a constant term on *each* side of the equal sign. **Our goal is to get *one* variable term on one side of the equal sign and *one* constant term on the other side.**

Step 1: There are no parentheses to clear and no like terms to combine on either side of the equal sign.

Step 2: Get the variable on one side of the equal sign and the constant on the other side.

First, let's subtract $8c$ from each side so that there will be one c-term on the left side of the equation.

$$10c - 8c - 9 = 8c - 8c + 5 \quad \text{Subtract } 8c \text{ from each side.}$$
$$2c - 9 = 5 \qquad\qquad\qquad \text{Combine like terms.}$$
$$2c - 9 + 9 = 5 + 9 \qquad\quad \text{Add 9 to both sides.}$$
$$2c = 14 \qquad\qquad\qquad\quad \text{Simplify.}$$

Step 3: Solve for c using the division property of equality.

$$\frac{2c}{2\cdot} = \frac{14}{2} \quad \text{Divide each side by 2.}$$
$$c = 7 \quad \text{Simplify.}$$

Step 4: Check in the *original* equation.

$$10c - 9 = 8c + 5 \qquad \text{This is the original equation.}$$
$$10(7) - 9 = 8(7) + 5 \quad \text{Substitute 7 for } c.$$
$$70 - 9 = 56 + 5 \qquad\quad \text{Multiply.}$$
$$61 = 61 \checkmark$$

The solution is $c = 7$.

> **W Hint**
>
> Are you using a colored pen or pencil as you are working through this problem?

[YOU TRY 1] Solve $13k - 15 = 4k + 3$.

We could have solved the equation in Example 1 by moving the variable to the right side of the equal sign like this:

$$10c - 9 = 8c + 5$$
$$10c - 10c - 9 = 8c - 10c + 5 \qquad \text{Subtract } 10c \text{ from each side.}$$
$$-9 = -2c + 5 \qquad \text{Combine like terms.}$$
$$-9 - 5 = -2c + 5 - 5 \qquad \text{Subtract 5 from each side.}$$
$$-14 = -2c \qquad \text{Combine like terms.}$$
$$\frac{-14}{-2} = \frac{-2c}{-2} \qquad \text{Divide each side by } -2.$$
$$7 = c \qquad \text{Simplify.}$$

The solution is the same. So remember that when we solve an equation, we can isolate the variable on either side of the equation.

EXAMPLE 2

Solve $-15y - 13 + 6y = 7 - 5y$.

Solution

Step 1: Combine like terms on the left side of the equal sign.

$$-15y - 13 + 6y = 7 - 5y$$
$$-9y - 13 = 7 - 5y \qquad \text{Combine like terms.}$$

Step 2: Isolate the variable. Combine like terms so that there is a single variable term on one side of the equation and a constant on the other side.

$$-9y + 5y - 13 = 7 - 5y + 5y \qquad \text{Add } 5y \text{ to each side.}$$
$$-4y - 13 = 7 \qquad \text{Combine like terms.}$$
$$-4y - 13 + 13 = 7 + 13 \qquad \text{Add 13 to each side.}$$
$$-4y = 20 \qquad \text{Combine like terms.}$$

Step 3: Solve for y using the division property of equality.

$$\frac{-4y}{-4} = \frac{20}{-4} \qquad \text{Divide each side by } -4.$$
$$y = -5 \qquad \text{Simplify.}$$

Step 4: The check is left to the student.

The solution is $y = -5$.

> **W Hint**
>
> Are you writing out the steps as you are reading the example?

YOU TRY 2

Solve $9w + 1 - 6w = -7w + 11$.

EXAMPLE 3

Solve $6(2t - 1) + 15 = 4t + 3(t + 3)$.

Solution

Step 1: Clear the parentheses and combine like terms.

$$6(2t - 1) + 15 = 4t + 3(t + 3)$$
$$12t - 6 + 15 = 4t + 3t + 9 \qquad \text{Distribute.}$$
$$12t + 9 = 7t + 9 \qquad \text{Combine like terms.}$$

Step 2: Isolate the variable.

$$12t - 7t + 9 = 7t - 7t + 9 \qquad \text{Subtract } 7t \text{ from each side.}$$
$$5t + 9 = 9 \qquad \text{Combine like terms.}$$
$$5t + 9 - 9 = 9 - 9 \qquad \text{Subtract 9 from each side.}$$
$$5t = 0 \qquad \text{Combine like terms.}$$

Step 3: Solve for t using the division property of equality.

$$\frac{5t}{5} = \frac{0}{5} \qquad \text{Divide each side by 5.}$$
$$t = 0 \qquad \text{Simplify.}$$

Step 4: Check:

$$6(2t - 1) + 15 = 4t + 3(t + 3)$$
$$6[2(0) - 1] + 15 = 4(0) + 3(0 + 3) \qquad \text{Substitute 0 for each } t.$$
$$6(0 - 1) + 15 = 0 + 3(3) \qquad \text{Multiply.}$$
$$6(-1) + 15 = 0 + 9$$
$$-6 + 15 = 9$$
$$9 = 9 \checkmark$$

The solution is $t = 0$.

[YOU TRY 3] Solve $12 + 5(r + 4) = 2(3r - 2) + 3r$.

---

2 Write English Expressions as Algebraic Expressions

Equations can be used to solve real-world problems. Before we learn how to do this, we need to learn how to recognize some key words and how to *translate* English expressions into algebraic expressions.

The tables in Examples 4 and 5 compare English phrases that can be written as *arithmetic* expressions with English phrases that can be written as *algebraic* expressions. **Notice that the way we translate the English into math is the same whether we are working with numbers or variables.**

Remember that, in algebra, a variable is used to represent a number. We can use *any* letter as the variable. In the next two examples, we will let x be the variable that represents the unknown number.

In Example 4, we learn key words that indicate addition and subtraction.

EXAMPLE 4

Write each English phrase as a mathematical expression. Let $x =$ the unknown number.

Addition		Subtraction	
English Expression	Math Expression	English Expression	Math Expression
2 plus 7	$2 + 7$ or $7 + 2$	8 minus 6	$8 - 6$
A number plus 7	$x + 7$ or $7 + x$	8 minus a number	$8 - x$
The sum of 4 and 3	$4 + 3$ or $3 + 4$	2 less than 9	$9 - 2$
The sum of a number and 3	$x + 3$ or $3 + x$	2 less than a number	$x - 2$
8 increased by 5	$8 + 5$ or $5 + 8$	10 decreased by 3	$10 - 3$
A number increased by 5	$x + 5$ or $5 + x$	A number decreased by 3	$x - 3$
2 more than −9	$-9 + 2$ or $2 + (-9)$	5 subtracted from 1	$1 - 5$
2 more than a number	$x + 2$ or $2 + x$	A number subtracted from 1	$1 - x$
6 added to 1	$1 + 6$ or $6 + 1$	4 subtracted from 7	$7 - 4$
A number added to 1	$1 + x$ or $x + 1$	4 subtracted from a number	$x - 4$

Because addition is commutative, ⌐ there are *two* ways to write addition expressions.

Because subtraction is *not* ⌐ commutative, there is only *one* way to write subtraction expressions.

Note

An expression like $x + 9$ can also be written as $9 + x$ because addition is commutative. But, a subtraction problem like $x - 6$ *cannot* be written as $6 - x$ because subtraction is *not* commutative. Think about it: Does $8 - 3 = 3 - 8$? No!

The word *difference* indicates subtraction. The **difference** is the answer we get when we subtract two numbers. For example, in the problem $9 - 5 = 4$, we say that the *difference* is 4.

[YOU TRY 4]

Write each English phrase as a mathematical expression. Let $x =$ the unknown number.

a) i) −5 increased by 2   ii) a number increased by 2

b) i) 4 subtracted from 11   ii) a number subtracted from 11

c) i) 8 less than −3   ii) 8 less than a number

d) i) the sum of 9 and 4   ii) the sum of 9 and a number

Example 5 introduces key words that indicate multiplication and division.

EXAMPLE 5

Write each English phrase as a mathematical expression. Let x = the unknown number.

Multiplication		Division	
English Expression	Math Expression	English Expression	Math Expression
4 times 5	$4 \cdot 5$	8 divided by 4	$\dfrac{8}{4}$
4 times a number	$4 \cdot x$ or $4x$	8 divided by a number	$\dfrac{8}{x}$
The product of 7 and 2	$7 \cdot 2$	-10 divided by 2	$\dfrac{-10}{2}$
The product of 7 and a number	$7 \cdot x$ or $7x$	A number divided by 2	$\dfrac{x}{2}$
Twice 9	$2 \cdot 9$	The quotient of 12 and 3	$\dfrac{12}{3}$
Twice a number	$2 \cdot x$ or $2x$	The quotient of 12 and a number	$\dfrac{12}{x}$
Double 6	$2 \cdot 6$	The quotient of 21 and -7	$\dfrac{21}{-7}$
Double a number	$2 \cdot x$ or $2x$	The quotient of a number and -7	$\dfrac{x}{-7}$
Triple 8	$3 \cdot 8$		
Triple a number	$3 \cdot x$ or $3x$		

The number that comes first in these phrases is the dividend (the number on top).

[YOU TRY 5] Write each English phrase as a mathematical expression. Let x = the unknown number.

a) i) 18 divided by -3 ii) 18 divided by a number

b) i) triple 10   ii) triple a number

c) i) the quotient of 36 and 4   ii) the quotient of a number and 4

d) i) the product of -8 and 5   ii) the product of -8 and a number

Let's keep all the key words in mind to write the following algebraic expressions.

EXAMPLE 6

Write an algebraic expression for each English phrase. Let x = the unknown number.

a)   4 less than a number   b)   the product of -6 and a number

c)   5 more than twice a number

Solution

a) First, **write down** what the variable represents. This is called *defining the variable*.

$$x = \text{the unknown number}$$

Slowly, break down the phrase. What does *less than* mean in the phrase "4 less than a number?" It means *subtract*.

The algebraic expression is $x - 4$.

b) **Define the variable.** That is, **write down** what the variable represents.

$$x = \text{the unknown number}$$

Slowly, break down the phrase. What does *product* mean in the phrase "the product of -6 and a number?" It means *multiply*.

The algebraic expression is $-6x$.

c) **Define the variable.** That is, **write down** what the variable represents.

$$x = \text{the unknown number}$$

Slowly, break down the phrase. Let's think about the phrase "5 more than twice a number" in parts:

1) How do you write an expression for **"5 more than"** something? _____ + 5

2) What does **"twice a number"** mean? It means two times the unknown number, or $2x$.

Put the information together:

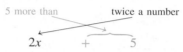

The algebraic expression is $2x + 5$. This can also be written as $5 + 2x$.

[YOU TRY 6]

Write an algebraic expression for each English phrase. Let $x =$ the unknown number.

a)   the quotient of -9 and a number   b)   a number increased by 7

c)   13 less than four times a number

ANSWERS TO [YOU TRY] EXERCISES

1) $k = 2$ 2) $w = 1$ 3) $r = 9$
4) a) i) $-5 + 2$ or $2 + (-5)$ ii) $x + 2$ or $2 + x$ b) i) $11 - 4$ ii) $11 - x$
    c) i) $-3 - 8$ ii) $x - 8$ d) i) $9 + 4$ or $4 + 9$ ii) $9 + x$ or $x + 9$
5) a) i) $\dfrac{18}{-3}$ ii) $\dfrac{18}{x}$ b) i) $3 \cdot 10$ ii) $3x$ c) i) $\dfrac{36}{4}$ ii) $\dfrac{x}{4}$ d) i) $-8 \cdot 5$ ii) $-8x$
6) a) $\dfrac{-9}{x}$ b) $x + 7$ or $7 + x$ c) $4x - 13$

E Evaluate **2.5** Exercises   Do the exercises, and check your work.

Objective 1: Solve Equations Containing Variables on Both Sides of the Equal Sign

1) In your own words, summarize the steps for solving an equation.

2) Does the variable always have to be isolated on the left side of the equal sign when you are solving an equation?

Solve each equation and check the solution.

3) $10x + 9 = 5x - 16$

4) $13n - 4 = 9n + 12$

5) $6a - 5 = 8a - 7$

6) $4w + 15 = 10w - 21$

7) $19k - 10 = 15k + 18$

8) $23c + 20 = 29c + 74$

9) $-7p - 8 + 5p = 4p + 34$

10) $-9y - 4 + 6y = 2y + 36$

11) $11 + 3g - 5 = 8g + 6 - g$

12) $9 + 7h - 1 = 4h + 8 - h$

13) $4r - 1 - 6r + 7 = 11r + 3 - 10r$

14) $18 - d + 5d - 11 = 9d + 19 - 3d$

15) $4 + 8z - 17 + 3z = 7z + 1 + 5z - 10$

16) $2b + 7 - 6b + 12 = 4b + 5 - 7b - 1$

17) $-15 + y = 2(4y - 5) + 30$

18) $-11 + m = 3(2m - 1) - 4m$

19) $3 + 4(w + 1) = 5w - 9$

20) $5 + 2(g + 3) = 3g - 8$

21) $10 + 2(v - 9) = 3(v + 1) - 6$

22) $1 + 5(4a - 7) = 4(7a - 3) - 30$

23) $13n + 6 - 5(2n - 3) = 1 + 4(n + 5)$

24) $2(1 - 8t) = 5 - 3(6t + 1)$

25) $2(3x - 4) - 6(x + 1) = -x + 4(x + 10)$

26) $-3(4c + 9) + 2(3c + 8) = -c + 2(c + 5)$

27) $2(6r + 5) = 14 - (7r - 4) + 11r$

28) $17 - (8k - 5) + 4k = 6(2k + 1)$

Objective 2: Write English Expressions as Algebraic Expressions

Determine whether the key words indicate addition, subtraction, multiplication, or division.

29) product

30) decreased by

31) sum

32) times

33) minus

34) more than

35) quotient

36) twice

37) Can the expression $x - 5$ also be written as $5 - x$? Explain your answer.

38) Can the expression $x + 3$ also be written as $3 + x$? Explain your answer.

Write each English phrase as a mathematical expression. Let $x =$ the unknown number.

39) a) 8 more than 1

   b) 8 more than a number

40) a) 2 increased by 7

   b) a number increased by 7

41) a) the product of 9 and 4

   b) the product of 9 and a number

42) a) -6 times 5

   b) -6 times a number

43) a) 3 subtracted from 10

   b) a number subtracted from 10

44) a) 8 subtracted from 19

   b) 8 subtracted from a number

45) a) the quotient of -42 and -7

   b) the quotient of -42 and a number

46) a) the quotient of 20 and -4

   b) the quotient of a number and -4

47) a) twice 15

   b) twice a number

48) a) triple 9

   b) triple a number

Write an algebraic expression for each English phrase. Let $x =$ the unknown number.

49) the sum of a number and 9

50) twice a number

51) a number decreased by 4

52) 12 more than a number

53) a number divided by -35

54) a number subtracted from 7

55) -8 times a number

56) the quotient of 6 and a number

57) double a number

58) a number increased by 1

59) 17 less than a number

60) triple a number

61) the product of 10 and a number

62) the product of a number and −5

63) the sum of 11 and twice a number

64) the sum of 8 and twice a number

65) 6 less than twice a number

66) 5 less than twice a number

67) 1 more than three times a number

68) 4 more than three times a number

69) 10 less than eight times a number

70) 7 less than six times a number

71) 3 subtracted from the quotient of a number and 5

72) 9 subtracted from the quotient of a number and 2

Write an English phrase to describe the math expression.
Let x = the number.

73) $\dfrac{x}{9}$

74) $x + 12$

75) $-x$

76) $2x$

77) $x - 15$

78) $-7x$

79) $2x + 11$

80) $4 - \dfrac{x}{3}$

R Rethink

R1) When we isolate the variable on one side of an equation, does it matter on which side we put the variable?

R2) Does $x = 5$ mean the same thing as $5 = x$?

R3) Write a paragraph that describes how to solve linear equations and include the properties learned so far.

2.6 Solve Applied Problems Involving One Unknown

P Prepare

O Organize

What are your objectives for Section 2.6?	How can you accomplish each objective?
1 Translate English Sentences into Algebraic Equations	• Learn the procedure **Five Steps for Solving Applied Problems** and write it in your own words. • Complete the given examples on your own. • Complete You Trys 1 and 2.
2 Solve Applied Problems	• Use **Five Steps for Solving Applied Problems.** • Complete the given examples on your own. • Complete You Trys 3 and 4.

 **W Work**   Read the explanations, follow the examples, take notes, and complete the You Trys.

1 Translate English Sentences into Algebraic Equations

In the previous section, we learned how to write English phrases as algebraic expressions. Now, we will learn how to write algebraic *equations* from information given to us in English. Remember, equations contain an equal sign.

Mathematical equations can be used to describe many situations in the real world. We must learn how to translate information presented in English into an algebraic equation. Although no single method will work for solving all applied problems, the following approach is suggested to help in the problem-solving process.

Procedure Five Steps for Solving Applied Problems

Step 1: **Read** the problem carefully, more than once if necessary, until you understand it. Draw a picture, if applicable. Identify what you are being asked to find.

Step 2: **Choose a variable** to represent an unknown quantity. If there are any other unknowns, define them in terms of the variable.

Step 3: **Translate** the problem from English into an equation using the chosen variable. Here are some suggestions for doing so:

- Restate the problem in your own words.
- Read and think of the problem in "small parts."
- Make a chart to separate these "small parts" of the problem to help you translate into mathematical terms.
- Write an equation in English, then translate it into an algebraic equation.

Step 4: **Solve** the equation.

Step 5: **Check** the answer in the original problem, and **interpret** the solution as it relates to the problem. Be sure your answer makes sense in the context of the problem.

EXAMPLE 1

Write the following statement as an equation, and find the number.
Three more than twice a number is seventeen. Find the number.

Solution

Step 1: **Read** the problem carefully. We must find an unknown number.

Step 2: **Choose a variable** to represent the unknown.

Let x = the number.

Step 3: **Translate** the information that appears in English into an algebraic equation by rereading the problem slowly and "in parts."

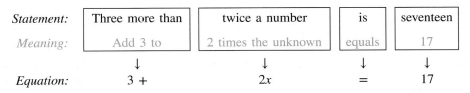

Statement:	Three more than	twice a number	is	seventeen
Meaning:	Add 3 to	2 times the unknown	equals	17
	↓	↓	↓	↓
Equation:	3 +	2x	=	17

The equation is $3 + 2x = 17$.

Step 4: **Solve** the equation.

$$3 + 2x = 17$$
$$3 - 3 + 2x = 17 - 3 \qquad \text{Subtract 3 from each side.}$$
$$2x = 14 \qquad \text{Combine like terms.}$$
$$x = 7 \qquad \text{Divide each side by 2.}$$

Step 5: **Check** the answer. Does the answer make sense? Three more than twice 7 is $3 + 2(7) = 17$. The answer is correct. The number is 7.

$\left[\text{YOU TRY 1}\right]$ Write the following statement as an equation, and find the number. *Nine more than twice a number is twenty-five.*

In Section 2.5, we were reminded that subtraction is *not* commutative. So when we write an expression involving subtraction, we must be sure the terms are in the correct order. For example, if we are asked to find *5 less than 9*, we write $9 - 5 = 4$ and *not* $5 - 9 = -4$. Keep this in mind as you read the next problem.

EXAMPLE 2   Write the following statement as an equation, and find the number.

Four less than three times a number is the same as the number increased by six. Find the number.

Solution

Step 1: **Read** the problem carefully. We must find an unknown number.

Step 2: **Choose a variable** to represent the unknown.

$$\text{Let } x = \text{the number.}$$

Step 3: **Translate** the information that appears in English into an algebraic equation by rereading the problem slowly and "in parts."

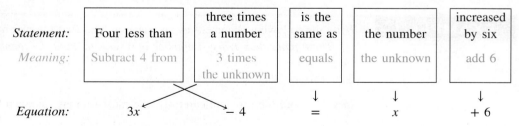

The equation is $3x - 4 = x + 6$.

Step 4: **Solve** the equation.

$$3x - 4 = x + 6$$
$$3x - x - 4 = x - x + 6 \qquad \text{Subtract } x \text{ from each side.}$$
$$2x - 4 = 6 \qquad \text{Combine like terms.}$$
$$2x - 4 + 4 = 6 + 4 \qquad \text{Add 4 to each side.}$$
$$2x = 10 \qquad \text{Combine like terms.}$$
$$x = 5 \qquad \text{Divide each side by 2.}$$

Step 5: **Check** the answer. Does it make sense? Four less than three times 5 is $3(5) - 4 = 11$. The number increased by six is $5 + 6 = 11$. The answer is correct. The number is 5.

[YOU TRY 2] Write the following statement as an equation, and find the number.

Two less than five times a number is the same as the number increased by ten. Find the number.

Let's apply this problem-solving technique to other types of applications.

2 Solve Applied Problems

EXAMPLE 3 Write an equation and solve.

Sonali went to the bank and deposited her $731 paycheck. Then, she went to the grocery store and paid $94 with her debit card. Her checking account balance is now $842. How much money was in the account before these transactions?

Solution

Step 1: **Read** the problem carefully, and identify what we are being asked to find.

We must find the amount of money in Sonali's account before she deposited her check and paid the grocery bill.

> **W Hint**
> Break down the statement in English and then write an algebraic equation as in Example 3.

Step 2: **Choose a variable** to represent the unknown.

$$x = \text{the original amount of money in Sonali's account}$$

Step 3: **Translate** the information that appears in English into an algebraic equation. Let's break up the problem slowly and restate it in our own words. We can think of the situation in this problem as:

The amount originally in the account *plus* the amount of the paycheck *minus* the amount paid at the store *equals* the new balance.

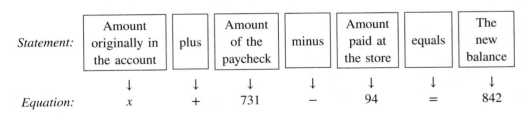

Statement:	Amount originally in the account	plus	Amount of the paycheck	minus	Amount paid at the store	equals	The new balance
	↓	↓	↓	↓	↓	↓	↓
Equation:	x	$+$	731	$-$	94	$=$	842

The equation is $x + 731 - 94 = 842$.

Step 4: **Solve** the equation.

$$
\begin{aligned}
x + 731 - 94 &= 842 \\
x + 637 &= 842 && \text{Subtract.} \\
x + 637 - 637 &= 842 - 637 && \text{Subtract 637 from each side.} \\
x &= 205 && \text{Subtract.}
\end{aligned}
$$

Step 5: **Check** the answer and **interpret** the meaning of the solution as it relates to the problem.

Sonali's bank account contained $205 before the transactions.

To check, start with the original balance of $205, add the deposit of $731, then subtract $94 for the groceries: $205 + $731 − $94 = $842. This is the correct account balance after the transactions.

EXAMPLE 4

Write an equation and solve.

Attendance at Comic-Con New York in 2015 was approximately 2000 more than five times the attendance when it was held in 2006. If 167,000 people attended Comic-Con in New York in 2015, how many attended in 2006? (www.forbes.com, www.ign.com)

Solution

Step 1: **Read** the problem carefully, and identify what we are being asked to find.

We must find how many people attended Comic-Con in New York in 2006.

Step 2: **Choose a variable** to represent an unknown.

x = the number of people who attended Comic-Con in 2006

W Hint
Always **write on your paper** what the variable represents.

Step 3: **Translate** the information that appears in English into an algebraic equation. One approach is to restate the problem in your own words.

167,000 people attended Comic-Con in 2015, which was 2000 more than five times the attendance in 2006.

Let's write this as an equation:

	167,000 attended in 2015	which was	2000 more than	five times the attendance in 2006
Statement:				
Meaning:	167,000	equals	add 2000 to	5 times the unknown
	↓	↓	↓	↓
Equation:	167,000	=	2000	+ 5x

The equation is $167{,}000 = 2000 + 5x$.

Step 4: **Solve** the equation

$$167,000 = 2000 + 5x.$$
$$167,000 - 2000 = 2000 - 2000 + 5x \qquad \text{Subtract 2000 from each side.}$$
$$165,000 = 5x \qquad \text{Combine like terms.}$$
$$\frac{165,000}{5} = \frac{5x}{5} \qquad \text{Divide each side by 5.}$$
$$33,000 = x \qquad \text{Simplify.}$$

Step 5: **Check** the answer and **interpret** the solution as it relates to the problem.

The attendance at Comic-Con New York in 2006 was approximately 33,000 people.

This makes sense because $167,000 = 5(33,000) + 2000$
$$= 165,000 + 2000 = 167,000 \checkmark$$

[YOU TRY 4] Write an equation and solve.

When Kim and Jason arrived at the airport for their honeymoon, Kim's suitcase weighed 27 lb less than twice the weight of Jason's suitcase. If Kim's suitcase weighed 63 lb, how much did Jason's weigh?

ANSWERS TO [YOU TRY] EXERCISES

1) $9 + 2x = 25$; 8   2) $5x - 2 = x + 10$; 3   3) $154   4) 45 lb

---

E Evaluate **2.6** Exercises   Do the exercises, and check your work.

Objective 1: Translate English Sentences into Algebraic Equations

1) What are the Five Steps for Solving Applied Problems?

2) If you are solving a problem in which you have to find the length of a picture frame, would -12 be a reasonable answer? Explain.

Write each statement as an equation, and find the number. Let x represent the number.

3) Nine more than a number is 24.

4) Twelve more than a number is 31.

5) Sixteen less than a number is eleven.

6) Fourteen less than a number is nineteen.

7) Eight more than twice a number is six.

8) Eleven more than twice a number is five.

9) Ten subtracted from three times a number is -4.

10) Thirteen subtracted from three times a number is -1.

11) The sum of -12 and a number equals the product of 5 and the number.

12) The sum of -18 and a number equals the product of -8 and the number.

13) Seven less than twice a number is the same as the number increased by 8.

14) Nine less than four times a number is the same as the number increased by 12.

15) Five times a number is the same as four more than six times the number.

16) Six times a number is the same as five more than seven times the number.

17) The sum of 6 and twice a number is the same as 8 less than four times the number.

18) The sum of 11 and triple a number is the same as 5 less than five times the number.

Write an English statement to describe each equation. Let $x =$ the number.

19) $x - 1 = 9$

20) $x + 6 = 15$

21) $4 + 3x = -11$

22) $2x - 5 = 23$

23) $4x = 7x - 18$

24) $-9x = 10 + x$

Objective 2: Solve Applied Problems

25) If you are asked to find the age of a man, why would -31 not be a reasonable answer?

26) If you are asked to find the balance of a bank account, what does an answer of $-\$56$ mean?

Write an equation, and solve each problem. Use the Five Steps for Solving Applied Problems.

27) Randa wrote a check for $149 for her car payment and then deposited her tax refund check of $517 into her checking account. The account balance is now $764. How much was originally in her account?

28) Vlad withdrew $160 from an ATM, then later he deposited his paycheck of $1023. His current account balance is $1318. How much was originally in his account?

29) Patrick earned $125 mowing lawns, and he deposited all of it into his checking account. Over the next two days, he used his debit card to pay $9 for lunch, $47 for gas, and $86 for clothes. If his account balance is now $216, how much was in his account before these transactions?

30) Alejandra deposited her babysitting money, $235, into her checking account. Over the next two days, she used her debit card to pay $74 for a textbook, $11 for a movie ticket, and $68 for a bus pass. Her current checking account balance is $374. How much was in her account before these transactions?

31) Shelby wrote a check to buy a train pass for $160 and then wrote a $420 check for day care. Later that day, she deposited a paycheck of $1371 and used her debit card to buy winter coats for her children for $87. What was the original balance of her account if her current balance is $645?

32) Maurice wrote a check for $361 to pay for car repairs and then wrote a check for $198 to register his twins for their baseball team. Later, he deposited his paycheck of $1249, then picked up dinner on his way home and paid $27 using his debit card. What was the original balance of his account if his current balance is $596?

33) Kenny's high score playing *Call of Duty: Black Ops* is 6180 points more than Kyle's high score. Find Kyle's high score if Kenny's is 33,615.

34) Pete Rose played in 576 more Major League Baseball games than Barry Bonds. If Rose played in 3562 games, in how many games did Barry Bonds play? (www.baseball-reference.com)

35) The distance from Chicago to St. Louis is approximately 118 mi less than the distance from Chicago to Minneapolis. If the distance from Chicago to St. Louis is 293 mi, what is the distance from Chicago to Minneapolis?

©Thinkstock

36) Disneyland Paris is approximately 25,558 acres smaller than Disney World in Orlando. If Disneyland Paris covers 4942 acres, how big is Disney World in Orlando? (www.frommers.com, disneybythenumbers.com)

37) In Game 1 of the 2016 NBA Finals, LeBron James scored 3 more than twice the number of points scored by his teammate, Tristan Thompson. If LeBron scored 23 points, how many did Thompson score? (www.nba.com)

38) In January, the number of résumés Justine sent out was five less than twice the number she sent out in December. If she sent 23 résumés in January, how many did Justine send in December?

39) In 2015, Taylor Swift was the highest-earning musician in the world while Rihanna came in at number 5. Taylor Swift's earnings were approximately $55 million less than three times Rihanna's earnings. If Taylor Swift earned $170 million, how much did Rihanna earn? (www.forbes.com)

40) In Professor Wainwright's Psychology class, the number of students who purchased an electronic textbook was seven more than twice the number who bought a traditional textbook. If 39 students purchased an electronic textbook, how many bought a traditional textbook?

41) Khadim's monthly cable bill is $121. This is $20 less than three times the amount of his bill five years ago. How much did he pay for cable five years ago?

42) Currently, 68 teams are chosen to compete in March Madness, the NCAA Men's Division I Basketball Tournament. This is 36 less than twice the number that participated in 1983. How many teams competed in March Madness in 1983? (www.espn.com)

43) There are 53 students in a high school Habitat for Humanity club. This is one more than four times the number in the Ultimate Frisbee club. How many students are in the Ultimate Frisbee club?

44) Thanh is a college student and spends 140 min each day doing homework. This is 5 min more than three times the number of minutes he spent on homework daily in high school. Determine the number of minutes he spent on homework in high school each day.

 45) The size of Yellowstone National Park is approximately 160 sq mi more than eight times the size of Rocky Mountain National Park. Find the size of Rocky Mountain National Park if Yellowstone contains 3472 sq mi of land. (www.nps.gov)

46) In Coretta's math class, the number of students with iPhones is four less than twice the number of students with a different type of phone. If 22 students own an iPhone, how many do not?

(24) 47) On the highway, the gas mileage of Jesse's motorcycle is twice that of his car. If his motorcycle gets 56 mpg on the highway, what is the gas mileage of his car on the highway?

48) A zoo has twice as many penguins as otters. If the zoo has 34 penguins, how many otters does it have?

Write an application problem that matches the given information.

49) v = the number of views of a kitten video
$v + 21{,}043$ = the number of views of a puppy video
$v + 21{,}043 = 93{,}688$

50) c = the number of Blake's concerts
$c - 3$ = the number of Luke's concerts
$c - 3 = 31$

R Rethink

R1) Look back at the applications in this exercise set. Did you *always* define the variable?

R2) Write your own application problem involving one unknown quantity. Then, solve it using the methods of this section.

2.7 Solve Applied Problems Involving Two Unknowns

P Prepare

O Organize

What are your objectives for Section 2.7?	How can you accomplish each objective?
1 Solve General Problems Involving Two Unknowns	• Use the **Five Steps for Solving Applied Problems** to complete the given examples on your own. • Complete You Trys 1 and 2.
2 Solve Applied Problems Involving Lengths	• Use the **Five Steps for Solving Applied Problems** to complete the given example on your own. • Complete You Try 3.

W Work   **Read the explanations, follow the examples, take notes, and complete the You Trys.**

So far, we have learned how to solve applications involving one unknown value. Now, we will learn how to solve problems that have two unknown values.

EXAMPLE 1

Write an equation and solve.

Together, *Desperate Housewives* and *Law and Order* were on television for 28 seasons. *Law and Order* was on for 12 more seasons than *Desperate Housewives*. For how many seasons was each show on television? (www.abc.com, www.nbc.com)

Solution

Step 1: **Read** the problem carefully, and identify what we are being asked to find.

We must find the number of seasons each show was on television.

©Francis Specker/Bloomberg via Getty Images

Step 2: **Choose a variable** to represent an unknown, and define the other unknown in terms of this variable.

In the statement "*Law and Order* was on for 12 more seasons than *Desperate Housewives*," the number of seasons of *Law and Order* is expressed *in terms of* the number of seasons of *Desperate Housewives*. Therefore, let

x = the number of seasons of *Desperate Housewives*

Define the other unknown (the number of seasons of *Law and Order*) in terms of x. The statement "*Law and Order* was on for 12 more seasons than *Desperate Housewives*" means

$$\boxed{\begin{array}{c}\text{Number of seasons of}\\ \textit{Desperate Housewives}\end{array}} + 12 = \boxed{\begin{array}{c}\text{Number of seasons of}\\ \textit{Law and Order}\end{array}}$$

$x + 12$ = number of seasons of *Law and Order*

Step 3: **Translate** the information that appears in English into an algebraic equation. One approach is to restate the problem in your own words.

Since the total number of seasons of these shows is 28, we can think of the situation in this problem as:

The number of seasons of *Desperate Housewives* **plus** the number of seasons of *Law and Order* is 28.

Let's write this as an equation.

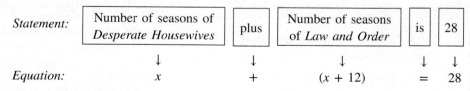

Statement:	Number of seasons of *Desperate Housewives*	plus	Number of seasons of *Law and Order*	is	28
	↓	↓	↓	↓	↓
Equation:	x	$+$	$(x + 12)$	$=$	28

The equation is $x + (x + 12) = 28$.

Step 4: **Solve** the equation.

$$x + (x + 12) = 28$$
$$2x + 12 = 28$$
$$2x + 12 - 12 = 28 - 12 \qquad \text{Subtract 12 from each side.}$$
$$2x = 16 \qquad \text{Combine like terms.}$$
$$\frac{2x}{2} = \frac{16}{2} \qquad \text{Divide each side by 2.}$$
$$x = 8 \qquad \text{Simplify.}$$

Step 5: **Check** the answer and **interpret** the meaning of the solution as it relates to the problem.

Since x represents the number of seasons of *Desperate Housewives,* this show was on for 8 seasons.

The expression $x + 12$ represents the number of seasons of *Law and Order,* so the number of seasons of *Law and Order* is $x + 12 = 8 + 12 = 20$ seasons.

The answer makes sense because the total number of seasons for these two shows is $8 + 20 = 28$. ✓

[**YOU TRY 1**] Write an equation and solve.

Rick flies his own plane, and on Saturday he flew 104 miles more than he flew on Sunday. If he flew a total of 848 miles over the weekend, how far did he fly each day?

EXAMPLE 2 Write an equation and solve.

During the 2016 Major League Soccer regular season, the Seattle Sounders FC had twice as many wins as the Chicago Fire. Together, they had 21 wins. How many wins did each soccer team have? (www.mlssoccer.com)

Solution

Step 1: **Read** the problem carefully, and identify what we are being asked to find.

We must find how many wins the Seattle Sounders and Chicago Fire had.

Step 2: **Choose a variable** to represent an unknown, and define the other unknown in terms of this variable.

W Hint

Write down the expressions for the two unknowns and say what they represent.

In the sentence "the Seattle Sounders FC had twice as many wins as the Chicago Fire," the number of Seattle Sounders wins is expressed *in terms of* the number of Chicago Fire wins. Therefore, let

$x =$ the number of Chicago Fire wins

Define the other unknown in terms of x.

$2x =$ the number of Seattle Sounders FC wins

Step 3: **Translate** the information that appears in English into an algebraic equation. One approach is to restate the problem in your own words.

Because the two teams have a total of 21 wins, we can think of the situation in this problem as:

The number of Chicago Fire wins **plus** the number of Seattle Sounders wins equals 21.

Let's write this as an equation.

Statement:	Number of Chicago Fire wins	plus	Number of Seattle Sounders wins	equals	21
	↓	↓	↓	↓	↓
Equation:	x	$+$	$2x$	$=$	21

The equation is $x + 2x = 21$.

Step 4: **Solve** the equation.

$$x + 2x = 21$$
$$3x = 21 \quad \text{Combine like terms.}$$
$$\frac{3x}{3} = \frac{21}{3} \quad \text{Divide each side by 3.}$$
$$x = 7 \quad \text{Simplify.}$$

Step 5: **Check** the answer and **interpret** the meaning of the solution as it relates to the problem.

The Chicago Fire had 7 wins.

The expression $2x$ represents the number of Seattle Sounders FC wins, so they had $2(7) = 14$ wins.

The answer makes sense because their total number of wins was $7 + 14 = 21$. ✓

[YOU TRY 2] Write an equation and solve.

Last month, Isaac worked twice as many hours as Mack. If they worked a total of 219 hr, how many hours did each of them work?

2 Solve Applied Problems Involving Lengths

Let's solve a problem involving length.

EXAMPLE 3

Write an equation and solve.

An electrician has a cable that is 30 ft long. He needs to cut it into two pieces so that one piece is 4 ft shorter than the other. How long will each piece be?

Solution

Step 1: **Read** the problem carefully, and identify what we are being asked to find.

We must find the length of each of the two pieces of cable.

Let's draw a picture.

30 ft

Step 2: **Choose a variable** to represent an unknown, and define the other unknown in terms of this variable.

One piece of cable must be 4 ft shorter than the other piece. Therefore, let

$$x = \text{the length of the longer piece}$$

Define the other unknown in terms of x.

$$x - 4 = \text{the length of the shorter piece}$$

©Ryan McVay/Getty Images

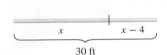

x $x - 4$

30 ft

Step 3: **Translate** the information that appears in English into an algebraic equation. Let's label the picture with the expressions representing the unknowns and then restate the problem in our own words.

From the picture at the left, we can see that

The length of one piece **plus** the length of the second piece equals 30 ft.

Let's write this as an equation.

Statement:	Length of one piece	plus	Length of second piece	equals	30 ft
	↓	↓	↓	↓	↓
Equation:	x	$+$	$(x - 4)$	$=$	30

The equation is $x + (x - 4) = 30$.

Step 4: **Solve** the equation.

$$x + (x - 4) = 30$$
$$2x - 4 = 30$$
$$2x - 4 + 4 = 30 + 4 \quad \text{Add 4 to each side.}$$
$$2x = 34 \quad \text{Combine like terms.}$$
$$\frac{2x}{2} = \frac{34}{2} \quad \text{Divide each side by 2.}$$
$$x = 17 \quad \text{Simplify.}$$

Step 5: **Check** the answer and **interpret** the solution as it relates to the problem.

One piece of cable is 17 ft long.

The expression $x - 4$ represents the length of the other piece of cable, so the length of the other piece is $x - 4 = 17 - 4 = 13$ ft.

The answer makes sense because the length of the original cable was
17 ft + 13 ft = 30 ft. ✓

[YOU TRY 3] Write an equation and solve.

A carpenter has a 36-in. board. He needs to cut it so that one piece is 6 in. shorter than the other. What will be the length of each piece?

ANSWERS TO [YOU TRY] EXERCISES

1) Saturday: 476 mi, Sunday: 372 mi   2) Mack: 73 hr, Isaac: 146 hr   3) 15 in., 21 in.

E Evaluate **2.7** Exercises   Do the exercises, and check your work.

Objective 1: Solve General Problems Involving Two Unknowns

1) The number of boys in a class is 4 less than the number of girls. If there are g girls in the class, write an expression for the number of boys.

2) The number of cars sold in May was 10 less than the number sold in June. If c cars were sold in June, write an expression for the number sold in May.

 3) Every day, Vivian drives 13 miles more (round-trip) to work than her husband. If her husband drives m miles every day, write an expression for the number of miles Vivian drives.

4) On her most recent math test, Latasha scored 9 points higher than on her first test. If Latasha's first test score was t, write an expression for her most recent test score.

5) Roberto's website had twice as many visitors in 2017 as in 2016. If the website had v visitors in 2016, write an expression for the number of visitors in 2017.

6) A rush-hour train carried three times as many passengers as a midday train. If the midday train had p passengers, write an expression for the number of passengers on the rush-hour train.

Solve each problem using the Five-Step Process.

 7) Keith's crab fishing boat carries 37 fewer pots than Sig's boat. Together, their boats hold 351 pots. How many pots are on each man's boat?

8) Elijah's farm has 58 more acres than Tom's farm, and they have a total of 432 acres. Find the size of each man's farm.

9) Bill has 29 more apps on his phone than Sherri, and they have a total of 99 apps. How many apps does each person have?

10) A street vendor sold 13 fewer pretzels than hot dogs. If he sold a total of 109 pretzels and hot dogs, how many of each did he sell?

©Lars A Niki

11) One cup of vanilla frozen yogurt has 60 fewer calories than one cup of vanilla ice cream. Together, they contain 460 calories. Determine the number of calories in each treat.

12) In 2008, there were 90 more deaths from tornadoes in the United States than in 2015. The total number of tornado deaths those two years was 162. How many people died in tornadoes each of these years? (www.spc.noaa.gov)

13) As of June 2016, Justin Bieber had about 22 million more Twitter followers than Rihanna. Together, they had approximately 144 million followers. How many Twitter followers did each performer have?

14) In Fall 2017, Eric paid $279 less in tuition than he did in Spring 2017. For both semesters, he paid a total of $3069 in tuition. How much was Eric's tuition each semester?

15) The number of Nicole's Facebook friends is 25 less than three times the number of her mom's. Together, they have 779 Facebook friends. How many Facebook friends does each of them have?

16) The length of a Cadillac Escalade EXT is approximately 8 in. less than twice the length of a Smart car. Together, their lengths total 310 in. Find the length of each car. (www.cadillac.com, www.smartusa.com)

17) The number of master's degrees awarded in the United States in 2013 was approximately 2000 more than twice the number awarded in 1993. Approximately 1,127,000 master's degrees were awarded these two years. How many were earned each year? (www.nced.ed.gov)

18) The number of people working as diagnostic medical sonographers in 2015 was approximately 3000 less than twice the number in the same profession in 2000. Approximately 93,000 people were working as medical sonographers during these two years. How many medical sonographers were employed in 2000 and in 2015? (www.bls.gov)

19) An adult-fare bus pass costs twice as much as a student-fare pass. Together, one student pass and one adult pass cost $129. How much does each pass cost?

20) One month, a car dealership sold twice as many cars as SUVs. If a total of 27 of these vehicles were sold, how many cars and SUVs were sold?

Objective 2: Solve Applied Problems Involving Lengths

21) A builder has a piece of window trim that is 36 in. long. It must be cut into two pieces so that one piece is 8 in. longer than the other. How long is each piece?

22) A 40-in.-long pipe must be cut into two pieces so that one piece is 6 in. shorter than the other. How long is each piece?

23) Eun Hee has a 33-in.-long piece of wire to make a necklace and a bracelet. She has to cut the wire so that the piece for the necklace is 9 in. longer than the piece for the bracelet. Find the length of each piece.

24) Anton ordered a 12-in.-long sub sandwich for him and his son Nikolai. Anton's piece was 2 in. longer than Nikolai's. Find the size of each of their sandwiches.

25) A 20-ft beam must be cut into two pieces so that one piece is 4 ft shorter than the other. Find the length of each piece.

26) A 25-ft pipe must be cut into two pieces so that one piece is 7 ft longer than the other. Find the length of each piece.

27) A 28-ft-long cable must be cut into two pieces so that one piece is three times the length of the other piece. Find the length of each piece.

28) A 27-ft-long log will be cut into two pieces so that one piece is twice as long as the other. Find the length of each piece.

29) A builder has a piece of siding that is 16 ft long. He must cut it into two pieces so that the length of one piece is 1 ft longer than twice the length of the other piece. How long is each piece?

30) Tree trimmers have a rope that is 35 ft long. They must cut it into two pieces so that the length of one piece is 1 ft shorter than twice the length of the other piece. How long is each piece?

R Rethink

R1) How does using the Five-Step Process help you solve application problems?

R2) Name two things you think you did well in this section.

R3) Name one thing you could do to improve learning the material in the next chapter.

Group Activity – Variables and Solving Equations

Finding an Unknown Value: Quiz Your Partner

- Work in groups of two.
- There are six values for x listed below. Each partner should take three of the values and create a translation problem in which the other partner has to solve for the given value of x.
- Once the partners have created their individual translation problems, each partner should write his/her problems on a separate sheet of paper.
- The partners should switch papers and try to solve each other's problems.
- Each partner already knows what the end result should be. The point of the activity is to translate the sentence into an equation and solve the equation to find the "unknown" value. See the example below.

Example:

Operations	Equations
Begin with	$x = -2$
Multiply both sides by 5:	$5x = -10$
Add 4 to both sides:	$5x + 4 = -6$
Write the translation problem:	**"Four more than the product of 5 and a number is −6. Find the number."**

1) Begin with $x = 8$

2) Begin with $x = -3$

3) Begin with $x = -12$

4) Begin with $x = 42$

> Each partner will write this final step on a separate sheet of paper. The partners should then switch papers.

Challenge: Include the distributive property and/or combining like terms in these last two examples.

5) Begin with $x = -4$

6) Begin with $x = 17$

You should know your reading attention span, the length of time you usually are able to sustain attention to a task, as you prepare for reading assignments. To get an idea of the length of your current attention span for reading, perform this exercise over the next few days.

1. Choose one of the textbooks that you've been assigned to read this semester.

2. Start reading a chapter, without any preparation, noting in the chart below the time that you start reading.

3. As soon as your mind begins to wander and think about other subjects, stop reading and note the time on the chart below.

4. Using the same textbook, repeat this process four more times over the course of a few days, entering the data on the chart below.

5. To find your reading attention span, calculate the average number of minutes across the five trials.
   Trial #1: Starting time: _____ Ending time: _____
   Number of minutes between start and end times: _____

   Trial #2: Starting time: _____ Ending time: _____
   Number of minutes between start and end times: _____

   Trial #3: Starting time: _____ Ending time: _____
   Number of minutes between start and end times: _____

   Trial #4: Starting time: _____ Ending time: _____
   Number of minutes between start and end times: _____

   Trial #5: Starting time: _____ Ending time: _____
   Number of minutes between start and end times: _____

Reading attention span (the average of the number of minutes in the last column, found by adding up the five numbers and dividing by 5) = _____ minutes.

Ask yourself these questions about your reading attention span:

1. Are you surprised by the length of your reading attention span? In what way?

2. Does any number in the set of trials stand out from the other numbers? For instance, is any number much higher or lower than the average? If so, can you account for this? For example, what time of day was it?

3. Do the numbers in your trials show any trend? For instance, did your attention span tend to increase slightly over the course of the trials, did it decrease, or did it stay about the same? Can you explain any trend you may have noted?

4. Do you think your attention span times would be very different if you had chosen a different textbook? Why or why not?

Chapter 2:   Summary

Definition/Procedure	Example

2.1 Introduction to Algebra

A **variable** is a symbol, usually a letter, used to represent an unknown number.

An **algebraic expression** is a collection of numbers, variables, and grouping symbols connected by operation symbols such as $+$, $-$, $\times$, and $\div$.

In the expression $9c + 2$, c is the variable.

Here is another example of an expression: $4y^2 - 7y + 6$

Evaluating Expressions
We can evaluate expressions for different values of the variables.

Evaluate $2x - 5y + 1$ when $x = -3$ and $y = 4$.

Substitute -3 for x and 4 for y and simplify.

$$
\begin{aligned}
2x - 5y + 1 &= 2(-3) - 5(4) + 1 \\
&= -6 - 20 + 1 \\
&= -6 + (-20) + 1 \\
&= -25
\end{aligned}
$$

Exponents are used to represent repeated multiplication.

$$\underbrace{k \cdot k \cdot k \cdot k}_{\text{4 factors of } k} = k^4$$

A **term** in an expression is a number or a variable or a product or quotient of numbers and variables.

The value of a **constant** does not change.

The **coefficient** of a variable is the number that the variable is multiplied by.

List the terms and coefficients of $7x^3 + x^2 - 9y + 8$.

Term	Coefficient
$7x^3$	7
x^2	1
$-9y$	-9
8	

8 is the constant.

2.2 Simplifying Expressions

If a, b, and c are numbers, then the following properties hold.

Commutative Properties:
$$a + b = b + a$$
$$ab = ba$$

Associative Properties:
$$(a + b) + c = a + (b + c)$$
$$(ab)c = a(bc)$$

Distributive Properties:
$$a(b + c) = ab + ac$$
$$a(b - c) = ab - ac$$

Examples:

Commutative Properties:
$$n + 3 = 3 + n$$
$$(-6)(h) = (h)(-6)$$

Associative Properties:
$$(x + y) + 2 = x + (y + 2)$$
$$(5 \cdot 2) \cdot c = 5 \cdot (2 \cdot c)$$

Distributive Properties:
$$
\begin{aligned}
6(w + 8) &= 6 \cdot w + 6 \cdot 8 \\
&= 6w + 48
\end{aligned}
$$

$$
\begin{aligned}
9(t - 2) &= 9t - 9 \cdot 2 \\
&= 9t - 18
\end{aligned}
$$

Like Terms
Like terms contain the same variables with the same exponents.

In the group of terms $5k^2$, $-8k$, $-4k^2$, $7k$, $5k^2$ and $-4k^2$ are like terms and $-8k$ and $7k$ are like terms.

Definition/Procedure	Example
Combining Like Terms We can simplify expressions by combining like terms.	Simplify the expression. $5p + 3(p + 2) + 1 = 5p + 3p + 6 + 1$ Distributive property $\qquad\qquad\qquad\qquad = 8p + 7$ Combine like terms.

2.3 Solving Linear Equations Part I

A number is a **solution** of an equation if it makes the equation true when the number is substituted for the variable.	Is 5 a solution of $4x - 9 = 11$? $4(5) - 9 = 11$ Substitute 5 for x. $20 - 9 = 11$ Multiply. $11 = 11$ Subtract. The statement is true. Yes, 5 is a solution of $4x - 9 = 11$.
The Addition and Subtraction Properties of Equality 1) If $a = b$, then $a + c = b + c$. 2) If $a = b$, then $a - c = b - c$.	Solve $3 + b = 20$ $3 - 3 + b = 20 - 3$ Subtract 3 from each side. $b = 17$ Check: $3 + b = 20$ $3 + 17 = 20$ Substitute 17 for b. $20 = 20$ Add. The statement is true. The solution is $b = 17$.
The Division Property of Equality If $a = b$, then $\dfrac{a}{c} = \dfrac{b}{c}$ $(c \neq 0)$.	Solve $2m = 16$ $\dfrac{2m}{2} = \dfrac{16}{2}$ Divide each side by 2. $1m = 8$ $\dfrac{2m}{2} = 1m$ $m = 8$ The solution is $m = 8$. The check is left to the student.

2.4 Solving Linear Equations Part II

Solve Equations Using the Properties of Division *and* Addition or Subtraction In general, first eliminate the number being added to or subtracted from the variable. Then, eliminate the coefficient.	Solve $4k - 7 = -27$. $4k - 7 = -27$ $4k - 7 + 7 = -27 + 7$ Add 7 to each side. $4k = -20$ Combine like terms. $\dfrac{4k}{4} = \dfrac{-20}{4}$ Divide each side by 4. $k = -5$ Simplify. The check is left to the student.

Definition/Procedure	Example

How to Solve a Linear Equation

Step 1: **Clear parentheses** and **combine like terms** on each side of the equation.

Step 2: **Get the variable on one side of the equal sign and the constant on the other side of the equal sign** (isolate the variable) using the addition or subtraction property of equality.

Step 3: **Solve for the variable** using the division property of equality.

Step 4: **Check the solution** in the original equation.

Solve $-5x + 2(x + 9) = 18$.

$$-5x + 2(x + 9) = 18$$
$$-5x + 2x + 18 = 18 \quad \text{Distribute.}$$
$$-3x + 18 = 18 \quad \text{Combine like terms.}$$
$$-3x + 18 - 18 = 18 - 18 \quad \text{Subtract 18 from each side.}$$
$$-3x = 0 \quad \text{Combine like terms.}$$
$$\frac{-3x}{-3} = \frac{0}{-3} \quad \text{Divide each side by } -3.$$
$$x = 0 \quad \text{Simplify.}$$

The check is left to the student.

2.5 Solving Linear Equations Part III

Solve Equations Containing Variables on Both Sides of the Equal Sign

Follow the steps we learned in Section 2.4.

Solve $8p - 2p + 15 = 4p + 1$.

$$8p - 2p + 15 = 4p + 1$$
$$6p + 15 = 4p + 1 \quad \text{Combine like terms.}$$
$$6p + 15 - 15 = 4p + 1 - 15 \quad \text{Subtract 15 from each side.}$$
$$6p = 4p - 14 \quad \text{Combine like terms.}$$
$$6p - 4p = 4p - 4p - 14 \quad \text{Subtract } 4p \text{ from each side.}$$
$$2p = -14 \quad \text{Combine like terms.}$$
$$\frac{2p}{2} = \frac{-14}{2} \quad \text{Divide each side by 2.}$$
$$p = -7 \quad \text{Simplify.}$$

The check is left to the student.

Recognizing key words can help us to write English expressions as algebraic expressions.

Notice that there are two ways to write addition expressions because addition is commutative.

Let's review the meaning of some key words. We will let x represent the number.

Addition	Subtraction
4 plus a number: $4 + x$ or $x + 4$	5 minus a number: $5 - x$
The sum of a number and 7: $x + 7$ or $7 + x$	8 less than a number: $x - 8$
A number increased by 2: $x + 2$ or $2 + x$	A number decreased by 6: $x - 6$
16 more than a number: $x + 16$ or $16 + x$	A number subtracted from -4: $-4 - x$
A number added to -9: $-9 + x$ or $x + (-9)$	1 subtracted from a number: $x - 1$

Multiplication	Division
9 times a number: $9 \cdot x$ or $9x$	10 divided by a number: $\dfrac{10}{x}$
The product of -5 and a number: $-5 \cdot x$ or $-5x$	A number divided by 3: $\dfrac{x}{3}$
Twice a number: $2 \cdot x$ or $2x$	
Double a number: $2 \cdot x$ or $2x$	The quotient of -8 and a number: $\dfrac{-8}{x}$
Triple a number: $3 \cdot x$ or $3x$	The quotient of a number and 7: $\dfrac{x}{7}$

Definition/Procedure	Example
Write English Expressions as Algebraic Expressions To write an English expression as an algebraic expression, 1) **Define the variable.** That is, *write down* what the variable represents. 2) **Slowly break down the phrase** and translate the English into math. Use the key words to help you.	Write the expression *seven more than twice a number* as an algebraic expression. Define the variable: Let $x =$ the number. 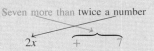 The expression is $2x + 7$. This can also be written as $7 + 2x$.

2.6 Solve Applied Problems Involving One Unknown

We can use these **Five Steps for Solving Applied Problems** to solve applications. *Step 1:* **Read** and reread the problem. Draw a picture, if applicable. Identify what we are being asked to find. *Step 2:* **Choose a variable** to represent an unknown. Define other unknown quantities in terms of the variable. *Step 3:* **Translate** from English into an equation. *Step 4:* **Solve** the equation. *Step 5:* **Check** the answer in the original problem, and interpret the solution as it relates to the problem.	Three less than twice a number is the same as the number plus eight. Find the number. *Step 1:* **Read** the problem carefully, then read it again. Identify what we are being asked to find. We must find the unknown number. *Step 2:* **Choose a variable** to represent the unknown. $$x = \text{the number}$$ *Step 3:* "Three less than twice a number is the same as the number plus eight" means $2x - 3 = x + 8$. *Step 4:* **Solve** the equation. $$\begin{aligned} 2x - 3 &= x + 8 \\ 2x - 3 + 3 &= x + 8 + 3 \qquad \text{Add 3 to each side.} \\ 2x &= x + 11 \qquad \text{Combine like terms.} \\ 2x - x &= x - x + 11 \qquad \text{Subtract } x \text{ from each side.} \\ x &= 11 \qquad \text{Simplify.} \end{aligned}$$ *Step 5:* The number is 11. The **check** is left to the student.

2.7 Solve Applied Problems Involving Two Unknowns

The **Five Steps for Solving Applied Problems** can be used to solve problems with two unknown values.	Write an equation and solve. A cable television installer has a wire that is 60 in. long. She must cut it into two pieces so that one piece is twice as long as the other. How long will each piece be? *Step 1:* **Read** the problem carefully, and identify what we are being asked to find. Let's draw a picture. 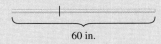 We must find the length of each of the two pieces of wire. *Step 2:* **Choose a variable** to represent an unknown, and define the other unknown in terms of this variable. One piece is *twice* as long as the other piece. Let $x =$ the length of the shorter piece. Then, $2x =$ the length of the longer piece.

Definition/Procedure	Example
	Step 3: Translate from English into an equation. Look at the picture. Let's write a statement in English and then write it as an algebraic equation. 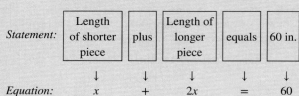 The equation is $x + 2x = 60$. **Step 4: Solve** the equation. $\begin{aligned} x + 2x &= 60 \\ 3x &= 60 \qquad \text{Combine like terms.} \\ \frac{3x}{3} &= \frac{60}{3} \qquad \text{Divide each side by 3.} \\ x &= 20 \qquad \text{Simplify.} \end{aligned}$ **Step 5: Check** the answer and **interpret** the solution as it relates to the problem. $x = 20$ means that the length of one piece is 20 in. The length of the other piece is $2x$, so the length of the other piece is $2(20) = 40$ in. The lengths of the two pieces are 20 in. and 40 in. The answer makes sense because the length of the original wire was 20 in. + 40 in. = 60 in.

Chapter 2:   Review Exercises

(2.1) Evaluate each expression for the given values.

1) $9p + 4$ for $p = -3$

2) $10k - 7$ for $k = 2$

3) $-3a + 2b + 1$ for $a = -4$ and $b = -9$

4) $m^2 - 2n^2 + 5m + 4$ for $m = 10$ and $n = -12$

List the terms and coefficients of the expression and identify the constant. Then, evaluate the expression for the given values of the variables.

5) $t^2 + 8t - 3$; $t = 6$

6) $-4x^3 + 3y^2 - 4xy + y$; $x = -2$ and $y = 5$

Use the commutative property to rewrite each expression.

7) $12 + k$

8) $w \cdot 5$

Use the associative property to rewrite each expression. Then, simplify.

9) $-8(7p)$

10) $(z + 6) + 1$

Rewrite each expression using the distributive property.

11) $7(k + 3)$

12) $-2(w + 9)$

13) $-(v - 6)$

14) $-5(-1 + d)$

(2.2)

15) Are the following terms *like terms*? Explain. $3x^2$, $-4x$, $8x^3$

Combine like terms.

16) $4a - 19 + 2a - 10$

17) $-6m + n - 3 - m - 3n + 4$

18) $x^2 + 5x - 9 + 4x^2 - x + 9$

19) $7h^2k - hk^2 + 8h + 9hk^2 + 2h^2k$

20) $-5 + 8(c + 2) + 3(-4c) - 1$

21) What is the difference between an expression and an equation?

22) Can we solve $2c + 7c$? Why or why not?

Determine whether the given value is a solution to the equation.

23) $n + 5 = 17; n = 12$

24) $7 = 8w - 1; w = -2$

Solve each equation, and check the solution.

25) $y - 9 = -2$

26) $d + 11 = 6$

27) $42 = 7w$

28) $15 = 3p$

29) $-x = -1$

30) $-m = 13$

31) $3z - 5 = 19$

32) $2d + 7 = -11$

33) $-29 = 16 - 5m$

34) $34 = -38 - 9x$

35) $15h + 4 - 8h + 7 = 32$

36) $-3r + 10 - 2r - 3 = 12$

37) $t - 2(3t - 1) + 8 = 10$

38) $4(2w - 5) + w + 11 = -9$

39) $4q + 9 = 6q + 17$

40) $7n - 15 = 13n - 21$

41) $-8x + 9 + 2x + 4 = -7 + x - 6x + 8$

42) $10 - 4p + 1 + p = 5p + 14 - 7p + 2$

43) $6(4a + 3) - 13a = 5(3a - 1) - 1$

44) $y - 2(3y - 5) = 4(7 - y) - 9$

Write a mathematical expression and simplify, if possible. Use x to represent the number.

45) 6 less than a number

46) 13 more than twice a number

47) The quotient of a number and 8

48) The product of -9 and a number

(2.6–2.7) Write each statement as an equation, and find the number. Use x to represent the number.

49) Eleven more than twice a number is five.

50) Six less than five times a number is the same as the number increased by ten.

51) What are the Five Steps for Solving Applied Problems?

52) If you are solving a problem in which you have to find the number of children at a birthday party, would -5 be a reasonable answer? Explain.

Write an equation, and solve each problem. Use the Five Steps for Solving Applied Problems.

53) The number of minutes it took Trina to do her homework was 10 min less than twice the amount of time it took Liz. If Trina spent 40 min doing her homework, how long did it take Liz?

54) On Monday, Andre had 64 new emails when he arrived at work. This was 7 more than three times the number he had on Tuesday. How many new emails did Andre have on Tuesday?

55) Lemi Berhanu Hayle won the 2016 Boston Marathon and his fellow Ethiopian, Lelisa Desisa, took second place. Hayle won twice as much as Desisa, and together their prize money totaled $225,000. How much did each man win? (www.bostonmarathonmediaguide.com)

56) Lakshmi is a senior in college while Dharani is a freshman. The number of credits Lakshmi has earned is three less than seven times the number that Dharani has earned. If Lakshmi has 102 credits, how many credits does Dharani have?

©Caia Image/Glow Images

57) The number of miles on the odometer of Maria's car is 11,000 less than three times the number of miles on her dad's car. Together, their cars have been driven 75,000 miles. Find the number of miles each car has been driven.

58) A 48-in.-long board must be cut into two pieces so that one piece is twice as long as the other. Find the length of each piece.

Mixed Exercises

For Exercises 59–66, look at each problem and determine whether it is an expression or an equation. If it is an expression, simplify it by combining like terms. If it is an equation, solve it.

59) $-8t + 7 = 15$

60) $3n - 48 + 29n + 12$

61) $9h + 2k - 11k + 3(h - 5)$

62) $-14 = w - 5(2w + 1) - 9$

63) $-54 = 9a$

64) $2p - 1 - p + 5$

65) $5 + 2(3c - 1) = c + 21 - 4c$

66) $-8x + 3 = 10 - 7x$

Fill in the blank with *always*, *sometimes*, or *never* to make the statement true.

67) If $x \neq 0$, then $-x^2$ is _____ negative.

68) If $x \neq 0$, then x^3 is _____ positive.

69) Write an expression so that the difference of two terms is $-5n$.

70) Write an expression so that the sum of two terms is $-8c$.

71) Write an equation that can be solved with the division property of equality that has a solution of $k = 3$.

72) Write an equation that can be solved with the subtraction property of equality that has a solution of $y = -7$.

Write an English statement to describe each equation. Let x = the number.

73) $x - 4 = 10$

74) $5x = 2x + 27$

Write an equation and solve. Use the Five Steps for Solving Applied Problems.

75) Susan has a 48-ft roll of lawn edging that needs to be cut into two pieces. One piece will be 8 ft shorter than the other. How long is each piece?

76) Hank has 19 lures in his tackle box. This is three less than twice the number of hooks. How many hooks are in Hank's tackle box?

©Sherri Messersmith

Chapter 2: Test

1) What is the difference between an expression and an equation?

Evaluate each expression for the given values.

2) $2a + 9$ for $a = 5$

3) $5h - k^2 + 6$ for $h = -2$ and $k = 3$

4) Write $-6 \cdot n \cdot n \cdot n \cdot p \cdot p$ using exponents.

5) List the terms and coefficients of $4x^2 + 7y^2 - 5x + y - 8$. Identify the constant.

Rewrite each expression using the distributive property.

6) $4(9 + n)$ 7) $-(y - 3)$

Simplify each expression by combining like terms.

8) $8c - 7d + 3 - 2c + d - 14$

9) $-2 + 5(k + 4) + 3(2k)$

10) Determine whether 3 is a solution to $7x - 5 = 16$.

Solve each equation. Check the solution.

11) $-5x = 45$ 12) $18 = w + 12$

13) $7p - 1 = 27$

14) $2y + 14 - 8y + 1 = 9$

15) $m - 3(2m - 5) + 2m - 1 = 14$

16) $4h - 5 = 9h + 5$

17) $5 + 3(k - 5) = 2(2k + 3) - 3$

Write each English phrase as an algebraic expression. Let x = the unknown number.

18) the product of 10 and a number

19) 7 less than a number

20) 15 increased by a number

21) the quotient of a number and -6

22) 8 more than twice a number

Write an equation to find the number. Use x to represent the number.

23) Five more than a number is sixteen.

24) Four less than three times a number is the same as eight less than the number.

Write an equation, and solve each problem. Use the Five Steps for Solving Applied Problems.

25) In 2016, the retail price of a pair of *Air Jordan Retro High OG* basketball shoes was $35 less than three times the retail price of the *Air Jordan 1* when it was released in 1985. If the retail price of the *Air Jordan Retro High OG* was $160, what was the price of the *Air Jordan 1*? (www.sneakerfiles.com)

26) Professor Chuca's 11 A.M. class has six more students than her 9 A.M. class. All together, the classes contain 62 students. How many students are in each class?

27) A 36-in.-long pipe must be cut into two pieces so that one piece is twice as long as the other. Find the length of each piece.

Fill in the blank with *always*, *sometimes*, or *never* to make the statement true.

28) If $x \neq 0$, then $x + 8$ is _____ negative.

29) If $x \neq 0$, then $(-x)^2$ is _____ positive.

30) Write an English phrase that describes the math expression $6x - 5$. Let x = the number.

31) Write an English statement to describe the equation $\frac{x}{4} = -3$. Let x = the number.

32) Write an expression so that the sum of two terms is $-19a$.

Chapter 2:   Cumulative Review for Chapters 1 and 2

1) Write 1873 in words.

2) Write *forty-five million, two thousand nine* in digits.

3) Round 9,981,462 to the nearest

   a) thousand.   b) million.

   c) hundred-thousand.

4) a) In your own words, explain the meaning of the absolute
      value of a number.

   b) Find $|15|$.   c) Find $|-8|$.

   d) Find $-|-3|$.

Perform the indicated operations.

5) $-156(43)$

6) $-23 + 40$

7) $161 - 507$

8) $\dfrac{712}{-4}$

9) $-19 - (-25)$

10) $-3 \cdot 2 \cdot (-4) \cdot (-1) \cdot (-5)$

11) Evaluate each expression.

   a) 8^2 b) -3^2

   c) 2^5 d) $\sqrt{49}$

12) In your own words, explain the order of operations.

Simplify each expression using the order of operations.

13) $-8 + 40 \div 4$

14) $\sqrt{16} + 5(3 - 9)$

15) $-5 \times 8 + 6^2 - 11$

16) $2^2 + 3^2 - |10 - 23|$

17) Austin has $253 in his checking account. He pays his $89
    cable bill, a $47 phone bill, then deposits a $450 paycheck.
    What is the balance of his account?

18) For $x^2 + 3y^2 - 7x + 2y - 4$,

   a) list the terms and coefficients of the expression and
      identify the constant.

   b) evaluate the expression for $x = 5$ and $y = -2$.

19) Rewrite $(w + 9) + 5$ using the associative property. Then,
    simplify the expression.

20) What are *like terms*?

21) Simplify $-3(2p) + 5(4p - 1) - 8$.

Solve each equation, and check the solution.

22) $-2h + 9 = 33$

23) $1 + 3(n - 4) + 5n = 4n + 9$

24) Let $x =$ the number. Write an algebraic expression for
    each English phrase.

   a) the quotient of a number and 8

   b) 5 less than a number

   c) the product of -4 and a number

   d) the sum of a number and 9

Write an equation to solve the problem.

25) Tara's age is two years more than three times her
    daughter's age. If Tara is 47, how old is her daughter?

3 Operations with Signed Fractions

©Hans Neleman/Getty Images

Math at Work:

Chef

For chef Jim Dion, the dream of opening his own restaurant has finally come true. Although Chez Dion Restaurant is a small place, with seating for just under 40 customers, he is thrilled.

Jim credits much of his success as a chef to the training he received in culinary arts school and to his creativity in combining unexpected flavors. However, he also credits his skill in math, which he uses on a daily basis. Cooking requires the regular use of fractions, as recipe ingredients are multiplied and divided according to the number of people being served. Jim routinely cuts ingredients into halves or quarters, or multiplies amounts in order to adjust the size of a recipe. In addition, he uses math to determine the amount of supplies that he needs to order depending on seasonal changes in the number of customers and to figure out whether or not he is making a profit.

Math is also key to another skill Jim applies on a daily basis: time management. "As the chef and owner of a restaurant, there are always a thousand things I need to do," Jim says. "So, I am very careful and organized about dividing the time I have during the day among all the tasks I need to accomplish."

In this chapter, we'll learn about working with fractions, and we'll also explore some strategies you can use to help improve your own time management.

 **Study Strategies** Time Management

Do you feel like you are always trying to juggle things such as school, work, activities, and family responsibilities? For college students in particular, it is essential to plan how you will use your time. Let's see how we can use the P.O.W.E.R. framework to manage our time wisely so that we can do everything we need to get done.

- *Prepare means to explicitly state a goal.* **I will get better at managing my time.**

- *Organize means to **organize** the physical and mental tools you need to achieve your goal.*
- Complete the emPOWERme survey that appears before the Chapter Summary to learn about how wisely you use your time.
- Create a *time log*: a record of how you actually spend your time, including interruptions, noting blocks of time in increments of as short as 15 minutes.
- Create a *weekly master calendar* that displays what you have to do during the week. Include items such as your class schedule, due dates for assignments and tests, work schedule, activities, and other responsibilities outside of school.
- Make a daily *to-do list* as well as one for the week.

- *Work means to **do the work** that needs to be done to achieve the goal.*
- Prioritize the items on your daily and weekly to-do lists. Think about which items are most important and which ones, such as assignments, are due first.
- Look at your time log and identify open blocks of time. Use this to help you decide when you have time to do your homework, study for a test, meet for a group project, and run errands. Write them on your master calendar.
- Look at your to-do lists and master calendar regularly. Check off items as you complete them, and add new ones as they arise.
- Be aware of distractions so that you don't waste time, and make a conscious effort to avoid them. For example, if you turn off your television while doing your homework, you will be able to concentrate better and will probably finish your homework faster!

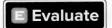

- *Evaluate means you should **evaluate** what you have done.* Did you achieve your goal?
- Look at your daily to-do list. Did you complete everything you needed to finish? Did you have enough time for school, work, and other activities, or did you run out of time?
- Consult your weekly to-do list and master calendar and ask yourself whether you are on track to complete the rest of the items on the list. At the end of the week, did you finish everything you needed to finish?

- **Rethink** *means to **rethink** and reflect upon your goal.*
- Reflect upon the day or week, and ask yourself how you felt. Did you feel rushed? Are you stressed? If necessary, adjust your master calendar. Reassess your priorities.
- If you completed all of your tasks, ask yourself which strategies helped you manage your time well so that you can do them again. Are there other things that you could try in the future to make your use of time even *more* efficient?
- If you did *not* finish what you needed to get done ask yourself, *"Why not?"* Should you have prioritized your tasks differently? Were there too many things on your schedule? Did you waste time by letting distractions get the best of you? What can you do differently in the future so that you can check *everything* off your to-do lists?

Not only will these strategies help you manage your time better for school, they will help you manage your time better for *everything* you need to get done in your life.

Chapter 3 **POWER** Plan

P Prepare

What are your goals for Chapter 3?	**O** Organize — How can you accomplish each goal?
1 Be prepared before and during class.	• Don't stay out late the night before class, and be sure to set your alarm! • Bring a pencil, notebook paper, and textbook to class. • Avoid distractions by turning off your cell phone during class. • Pay attention, take good notes, and ask questions. • Complete your homework on time, and ask questions on problems you do not understand.
2 Understand the homework to the point where you could do it without needing any help or hints.	• Read the directions and show all of your steps. • Go to the instructor's office for help. • Rework homework and quiz problems and find similar problems for practice.
3 Use the P.O.W.E.R. framework to learn how to manage your time.	• Complete the emPOWERme that appears before the Chapter Summary. • Read the Study Strategies. Think about the results of the emPOWERme exercise, and apply the steps of P.O.W.E.R. that can help you become better at managing your time.
4 Write your own goal. _____ _____	• _____ _____

What are your objectives for Chapter 3?	How can you accomplish each objective?
1 Learn how to identify different types of fractions.	• Take notes and learn the definitions and procedures in the chapter. • Take good notes in class.
2 Use the definitions of prime and composite numbers to find the prime factorization of a number and apply it to writing fractions in lowest terms.	• Master each objective in the sections, in order, because they build upon each other. • Understand the different methods to write fractions in lowest terms. • Watch the exercise videos for extra help if you get stuck on a certain problem.
3 Master the process of finding least common multiples, and understand how it helps you find a least common denominator.	• Learn the four different ways to find the LCM and which process is most efficient for you. • Practice and learn the procedure for **Writing a Fraction with a Different Denominator** in your own words. • Apply the process of finding the LCM to finding the LCD.
4 Learn how to multiply, divide, add, and subtract fractions.	• Take good notes on the different procedures you will need to use. • Understand when you need a common denominator and when you do not. • Read the book or reread your notes before doing the homework.
5 Learn how to solve applications involving fractions.	• Believe that you *can* do word problems. • Slow down when reading the word problem. Read it at least twice. • Underline key words in the problem.

6 Perform operations with mixed numbers.	• Know the procedures for adding, subtracting, multiplying, and dividing mixed numbers. • Understand how to regroup a mixed number for subtraction or addition. • Recognize and learn to add or subtract using improper fractions.
7 Compare fractions, and use the order of operations.	• Take notes, and write the procedures and definitions in your own words. • Complete all the exercises and ask for help when you need it.
8 Solve equations containing fractions.	• Understand how to use the properties we learned in Chapter 2 to solve equations containing fractions. • Understand how to eliminate fractions from an equation in order to solve it.
9 Write your own goal. _____ _____	• _____ _____

W **Work**	Read Sections 3.1–3.8, and complete the exercises.

| **E** **Evaluate** | Complete the Chapter Review and Chapter Test. How did you do? | **R** **Rethink** | • Are you confident that you can perform all of the operations with fractions and mixed numbers *without* looking at the book or your notes?
• Did you achieve your goals for the chapter? In terms of learning, what would you do the same and what would you do differently in the future?
• Do you think you budgeted your time wisely while studying this chapter?
• Have you tried using the steps of P.O.W.E.R. to help you manage your time? If so, what worked well? If not, why not? |

3.1 Introduction to Signed Fractions

What are your objectives for Section 3.1?	How can you accomplish each objective?
1 Understand What Fractions Represent	• Write your own definition of a *fraction,* and include the words *numerator* and *denominator.* • Complete the given examples on your own. • Complete You Trys 1 and 2.
2 Identify Proper and Improper Fractions	• Write the definition of a *proper fraction* and *improper fraction* in your own words, and be sure to include the words *numerator* and *denominator.* • Compare the relationship between the numerator and the denominator in the definitions. • Complete the given examples on your own. • Complete You Trys 3–4.
3 Understand Negative Fractions	• Explain, in your own words, the meaning of the absolute value of a fraction. • Explain how to graph a negative fraction on a number line. • Complete the given example on your own. • Complete You Try 5.

W Work   Read the explanations, follow the examples, take notes, and complete the You Trys.

Until now, we have been working with integers, sometimes known as the positive and negative counting numbers. We can list the integers like this:

$$\ldots, -5, -4, -3, -2, -1, 0, 1, 2, 3, 4, 5, \ldots$$

But what if we need a number to represent a *part* of a whole? We can use fractions.

1 Understand What Fractions Represent

What is a *fraction*? A **fraction** is a part of a whole. We will look at some figures and number lines to understand fractions. Let's begin with a circle divided into three equal parts.

Each of these parts is *one-third,* or $\frac{1}{3}$, of the circle.

The number $\frac{1}{3}$ is an example of a fraction.

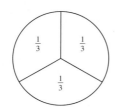

W Hint

Write your own definition of a fraction and include the words *numerator* and *denominator.*

Let's identify the parts of a fraction using the number $\frac{1}{3}$.

$$\text{Fraction bar} \rightarrow \frac{1}{3} \begin{array}{l} \leftarrow \text{Numerator} \\ \leftarrow \text{Denominator} \end{array}$$

The **denominator** is the number *below* the fraction bar. It represents the total number of equally-sized parts of a whole. The **numerator** is the number *above* the fraction bar. It represents the number of parts being considered.

Note

Remember from Chapter 1 that the fraction bar represents division.

Example: $\frac{1}{3} = 1 \div 3$

BE CAREFUL Because division by 0 is undefined, a fraction with a denominator of 0 is undefined.

Example: $8 \div 0$ is undefined, so $\frac{8}{0}$ is undefined.

Let's write fractions for shaded parts of the circle divided into thirds and also represent these numbers on a number line.

Each circle is divided into 3 equal parts, so *the denominator of each fraction is* 3.

Number of → $\frac{1}{3}$ of the circle is shaded. $\frac{2}{3}$ of the circle is shaded. $\frac{3}{3}$ of the circle is shaded.
shaded parts.

Where is each fraction on the number line? Let's look at the number line from 0 to 1, and divide it into 3 equal parts.

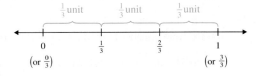

Just as each piece of the circle is $\frac{1}{3}$ of the circle, the space between each tick mark is $\frac{1}{3}$ of the distance from 0 to 1.

So, 0 can also be labeled as $\frac{0}{3}$, the next tick mark is $\frac{1}{3}$, the next is $\frac{2}{3}$, and 1 is equivalent to $\frac{3}{3}$.

EXAMPLE 1

Use a fraction to represent the shaded part of the rectangle, and represent the fraction on a number line.

Solution

The rectangle is divided into 4 equal parts: **The denominator is 4.**

Three parts are shaded: **The numerator is 3.**

Then, $\frac{3}{4}$, read as "three-fourths," of the rectangle is shaded. $\Big($ Notice that $\frac{1}{4}$, read as "one-fourth," of the rectangle is *not* shaded and that the shaded portion *plus* the unshaded portion is 1 whole. $\Big)$

To represent $\frac{3}{4}$ on the number line, first divide the number line from 0 to 1 into 4 equal parts. The space between consecutive tick marks is $\frac{1}{4}$ of a unit. Label the number line beginning at 0 with $\frac{0}{4}$.

Place the dot on $\frac{3}{4}$.

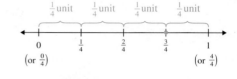

YOU TRY 1

Use a fraction to represent the shaded part of the rectangle, and represent the fraction on a number line.

EXAMPLE 2

Use a fraction to represent the *unshaded* part of the circle, and represent the fraction on a number line.

Solution

The circle is divided into 8 equal parts: **The denominator is 8.**

Five parts are *not* shaded: **The numerator is 5.**

Therefore, $\frac{5}{8}$ (five-eighths) of the circle is *not* shaded. $\Big($ Notice that $\frac{3}{8}$ of the circle *is* shaded and that the shaded portion *plus* the unshaded portion is 1 whole. $\Big)$

Divide the region on the number line from 0 to 1 into 8 equal parts. The space between consecutive tick marks is $\frac{1}{8}$ (one-eighth) of a unit. Label the number line beginning at 0 with $\frac{0}{8}$.

Place the dot on $\frac{5}{8}$.

YOU TRY 2

Use a fraction to represent the *unshaded* part of the circle, and represent the fraction on a number line.

2 Identify Proper and Improper Fractions

Let's learn more vocabulary associated with fractions.

Definition

If the numerator of a fraction is less than the denominator, then the fraction is a **proper fraction.** A proper fraction represents less than 1 whole.

Example: $\dfrac{5}{8}$ is a proper fraction.

W Hint

Compare the relationship between the numerator and the denominator in the definitions.

Most of the fractions we've seen so far have been proper fractions. However, in the beginning of this section we saw the following circle:

We said that $\dfrac{3}{3}$ of the circle is shaded, and that equals 1 whole circle. $\dfrac{3}{3}$ is an example of an *improper fraction*.

Definition

If the numerator of a fraction is greater than or equal to the denominator, then the fraction is an **improper fraction.** An improper fraction represents a quantity greater than or equal to 1 whole.

Examples:

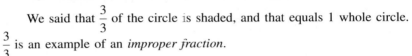

$\dfrac{3}{3}$ is an improper fraction.

Numerator equals denominator.

$\dfrac{9}{2}$ is an improper fraction.

Numerator is greater than denominator.

EXAMPLE 3   Identify each fraction as proper or improper.

a) $\dfrac{9}{8}$ b) $\dfrac{6}{11}$ c) $\dfrac{5}{5}$

Solution

a) $\dfrac{9}{8}$ is an *improper fraction* because the numerator is greater than the denominator.

b) $\dfrac{6}{11}$ is a *proper fraction* because the numerator is less than the denominator.

c) $\dfrac{5}{5}$ is an *improper fraction* because the numerator equals the denominator.

EXAMPLE 4

Use a fraction to represent the shaded part of the figure. Then, represent the fraction on a number line.

Solution

Notice that *more than* 1 *whole rectangle is shaded.* Therefore, the fraction will be improper.

Each rectangle is divided into 5 equal parts: The denominator is 5.

Seven parts are shaded: The numerator is 7.

$$\dfrac{7}{5} \textbf{ of the figure is shaded.}$$

Let's put $\dfrac{7}{5}$ on a number line. Notice that the number of *whole rectangles* shaded is greater than 1 and less than 2. Draw the number line from 0 to 2. Divide the region from 0 to 1 into 5 equal parts, and divide the region from 1 to 2 into 5 equal parts.

The space between consecutive tick marks is $\dfrac{1}{5}$ of a unit. Label each tick mark. Put a dot on $\dfrac{7}{5}$.

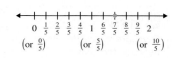

> **W Hint**
>
> Notice that 1 *whole rectangle* and then $\dfrac{2}{5}$ of the second rectangle are shaded. We can also represent the figure with the *mixed number* $1\dfrac{2}{5}$. We will learn about mixed numbers in Section 3.6.

[YOU TRY 4] Use a fraction to represent the shaded part of the figure, and represent the fraction on a number line.

3   Understand Negative Fractions

Just as negative numbers like -2 and -5 are to the left of 0 on the number line, negative fractions are to the left of 0 as well. For example, we can place $\dfrac{1}{2}$ and $-\dfrac{1}{2}$ on a number line like this:

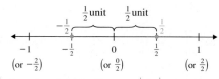

$-\dfrac{1}{2}$ is $\dfrac{1}{2}$ of a unit to the *left* of 0. $\dfrac{1}{2}$ is $\dfrac{1}{2}$ of a unit to the *right* of 0.

Remember that the absolute value of a number is its distance from 0. Because $\frac{1}{2}$ and $-\frac{1}{2}$ are each $\frac{1}{2}$ of a unit from 0, their absolute values are the same: $\left|\frac{1}{2}\right| = \frac{1}{2}$ and $\left|-\frac{1}{2}\right| = \frac{1}{2}$. We find the absolute value of a fraction the same way we find the absolute value of an integer.

Let's graph other fractions on a number line.

EXAMPLE 5

Represent each fraction on a number line.

a) $-\frac{3}{4}$ b) $-\frac{8}{3}$

Solution

a) The fraction $-\frac{3}{4}$ is $\frac{3}{4}$ of a unit to the *left* of 0 on the number line. Divide the number line from 0 to -1 into 4 equal parts. $\left(\text{The space between consecutive tick marks is } \frac{1}{4} \text{ of a unit.}\right)$ Label the number line beginning at 0 with $\frac{0}{4}$, and number the tick marks with $-\frac{1}{4}, -\frac{2}{4}, -\frac{3}{4}$, and $-\frac{4}{4}$ (or -1) as you move to the left. Place the dot on $-\frac{3}{4}$.

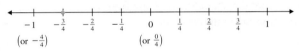

$\left(\text{Remember that numbers get smaller as you move to the left on the number line, so,} \right.$
for example, $\left. -\frac{3}{4} < -\frac{1}{4}.\right)$

b) The fraction $-\frac{8}{3}$ is $\frac{8}{3}$ units to the *left* of 0. It will be more than one unit to the left of 0 because $-\frac{8}{3}$ is an improper fraction. Let's divide the region from 0 to -1 into 3 equal parts, the region from -1 to -2 into 3 equal parts, and the region from -2 to -3 into 3 equal parts. Label the number line beginning at 0 with $\frac{0}{3}$, and number the tick marks with $-\frac{1}{3}, -\frac{2}{3}$, and so on, as you move to the left. Place the dot on $-\frac{8}{3}$.

[YOU TRY 5]

Represent each fraction on a number line.

a) $-\frac{1}{4}$ b) $-\frac{7}{5}$

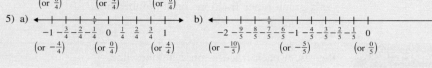
E Evaluate **3.1** Exercises   Do the exercises, and check your work.

Objective 1: Understand What Fractions Represent

1) Identify the numerator and denominator of $\frac{4}{7}$.

2) Does $\frac{0}{6} = \frac{6}{0}$? Explain your answer.

Use a fraction to represent the shaded part of the figure, and represent the fraction on a number line.

3)

4)

5)

6)

7)

8)

9)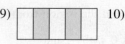

10)

11)

12)

13)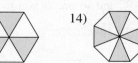

14)

15)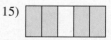

16)

Use a fraction to represent the *unshaded* part of the figure, and represent the fraction on a number line.

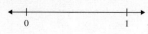

17)

18)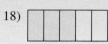

19)

20)

21)

22)

23)

24)

25) Draw a figure so that the shaded part represents $\frac{4}{6}$ of the whole figure.

26) Draw a figure so that the shaded part represents $\frac{3}{8}$ of the whole figure.

Exercises 27–32 show two circles with equally shaded areas. Each shaded area can be represented by a fraction, and the two fractions are equal. Write the two fractions that are equal to each other.

27)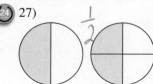

28)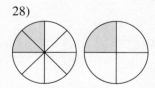

29)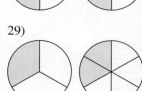

30)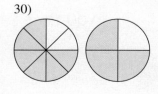

31)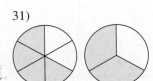

32)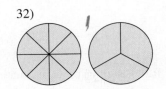

Objective 2: Identify Proper and Improper Fractions

Identify each fraction as proper or improper.

33) a) $\dfrac{13}{7}$ b) $\dfrac{5}{9}$ c) $\dfrac{12}{13}$

34) a) $\dfrac{7}{17}$ b) $\dfrac{13}{3}$ c) $\dfrac{9}{11}$

35) a) $\dfrac{21}{20}$ b) $\dfrac{14}{15}$ c) $\dfrac{23}{23}$

36) a) $\dfrac{21}{32}$ b) $\dfrac{52}{25}$ c) $\dfrac{11}{11}$

37) What is the difference between an improper fraction and a proper fraction?

The Denom is on the bottom

38) Is $\dfrac{2}{0}$ an improper fraction or a proper fraction?

39) Write an improper fraction with a denominator of 10.

40) Write a proper fraction with a numerator of 9.

Use a fraction to represent the shaded part of the group of figures, and represent the fraction on a number line.

41)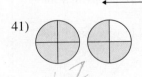

42)

43)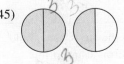

44)

45)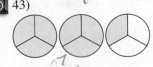

46)

47)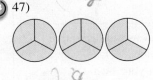

48)

49) Draw a group of figures that, all together, represents the fraction $\dfrac{11}{6}$.

50) Draw a group of figures that, all together, represents the fraction $\dfrac{9}{4}$.

Objective 3: Understand Negative Fractions

51) Where are negative fractions on a number line?

52) What does $\left|-\dfrac{2}{3}\right|$ mean?

Evaluate.

53) $\left|\dfrac{1}{3}\right|$ **54)** $\left|\dfrac{1}{5}\right|$ **55)** $\left|-\dfrac{1}{3}\right|$

56) $\left|-\dfrac{1}{5}\right|$ **57)** $\left|\dfrac{9}{4}\right|$ **58)** $\left|\dfrac{11}{6}\right|$

59) $\left|-\dfrac{9}{4}\right|$ **60)** $\left|-\dfrac{11}{6}\right|$

Represent each fraction on a number line.

61) $-\dfrac{1}{3}$ **62)** $-\dfrac{1}{5}$ **63)** $-\dfrac{2}{5}$

64) $-\dfrac{5}{6}$ **65)** $-\dfrac{7}{8}$ **66)** $-\dfrac{3}{8}$

67) $-\dfrac{3}{2}$ **68)** $-\dfrac{7}{4}$ **69)** $-\dfrac{13}{6}$

70) $-\dfrac{5}{2}$

Mixed Exercises: Objectives 1–3

Shade an appropriate amount of area on the figure according to the fraction represented on the number line.

71)

72)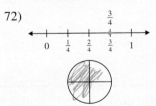

Represent each fraction on a number line.

73) $\dfrac{9}{4}$ **74)** $-\dfrac{3}{5}$

75) $-\dfrac{5}{8}$ **76)** $\dfrac{6}{6}$

R1) Write an explanation, in your own words, describing how to use a fraction to represent part of a whole.

R2) List three or more situations where you see or use fractions outside of school.

3.2 Writing Fractions in Lowest Terms

P Prepare

O Organize

What are your objectives for Section 3.2?	How can you accomplish each objective?
1 Identify Composite and Prime Numbers	• Review what you learned about factors in Section 1.6. • Learn and compare the definitions of a *composite number* and a *prime number*. • Learn the divisibility rules. • Complete the given examples on your own. • Complete You Trys 1 and 2.
2 Find the Prime Factorization of a Number	• How will the information you learned in Objective 1 help you in Objective 2? • Write the definition of *prime factorization* in your own words. • Note the different procedures you can use to find the prime factorization of a number. • Complete the given examples on your own. • Complete You Trys 3–6.
3 Understand Equivalent Fractions	• Write the definition of *equivalent fractions* in your own words. • Write the definition of *lowest terms* in your own words. • Complete the given example on your own. • Complete You Try 7.
4 Write Fractions in Lowest Terms Using Common Factors	• In your own words, take notes on the steps you would take to write a fraction in lowest terms using common factors. • Write the definition of a *greatest common factor* in your own words. • Complete the given example on your own. • Complete You Try 8.
5 Write Fractions in Lowest Terms Using Prime Factorization	• In your own words, take notes on the steps you would use to write a fraction in lowest terms using prime factorization. • Complete the given example on your own. • Complete You Try 9.
6 Determine Whether Two Fractions Are Equivalent	• Write down the steps you would use to determine whether two fractions are equivalent. • Complete the given example on your own. • Complete You Try 10.

Work | **Read the explanations, follow the examples, take notes, and complete the You Trys.**

Before we continue our study of fractions, let's review what we have learned about *factors* in Section 1.6 and add some new information.

1 Identify Composite and Prime Numbers

Factors are numbers that are multiplied together to get a product. (We consider only natural numbers as factors.) For example, 2 and 9 are factors of 18 because $2 \cdot 9 = 18$. Does 18 have any other factors?

EXAMPLE 1 | Find all factors of 18.

Solution

List the different pairs of numbers whose product is 18.

$$1 \cdot 18 = 18 \qquad 2 \cdot 9 = 18 \qquad 3 \cdot 6 = 18$$

All of these are ways to write 18 as the product of two numbers.
The factors of 18 are 1, 2, 3, 6, 9, and 18.

[YOU TRY 1] | Find all factors of 30.

 Hint
Learn and compare the definitions of a composite number and a prime number.

The number 18 is an example of a *composite number*.

> ## Definition
>
> A **composite number** is a number with factors other than 1 and itself.

The number 18 is composite because it has factors other than 1 and 18. Its other factors are 2, 3, 6, and 9.

> **Note**
> Another way to think of the factors of 18 is that 18 *divides evenly* by 1, 2, 3, 6, 9, and 18:
>
> $18 \div 1 = 18 \quad 18 \div 2 = 9 \quad 18 \div 3 = 6 \quad 18 \div 6 = 3 \quad 18 \div 9 = 2 \quad 18 \div 18 = 1$
>
> Because 18 *divides evenly* by 1, 2, 3, 6, 9, and 18, they are factors of 18.

What if the only two different factors of a number are 1 and itself? Then it is a *prime number*.

Definition

A **prime number** is a number whose only two different factors are 1 and itself.

For example, 5 is *prime* because $1 \cdot 5$ is the only way to write 5 as the product of two different numbers. So, the only factors of 5 are 1 and 5.

We can think of this in terms of division as well. Since 5 divides evenly *only* by 1 and 5, 5 is a prime number. ($5 \div 1 = 5$ and $5 \div 5 = 1$)

Note

1) The number 1 is not prime because it has only one factor: $1 \cdot 1 = 1$.

2) The numbers 0 and 1 are neither prime nor composite.

It is helpful to have rules to determine whether one number is divisible by another number. These are called the **divisibility rules.** We have not included a divisibility rule for 7 because it is difficult to use.

Divisibility Rules

A Number Is Divisible by	Example
…2 if it ends in 0, 2, 4, 6, or 8. If a number is divisible by 2, it is an **even number.**	7394 is divisible by 2 because it ends in 4. It is an even number.
…3 if the sum of its digits is divisible by 3.	837—Add its digits: $8 + 3 + 7 = 18$. Because 18 is divisible by 3, the number 837 is divisible by 3.
…4 if its last two digits form a number that is divisible by 4.	5932—The last two digits form the number 32. Since 32 is divisible by 4, the number 5932 is divisible by 4.
…5 if the number ends in 0 or 5.	645 is divisible by 5 since it ends in 5.
…6 if it is divisible by 2 and by 3.	1248—The number is divisible by 2 since it is an even number. The number is divisible by 3 since the sum of its digits is divisible by 3: $1 + 2 + 4 + 8 = 15$. Therefore, the number 1248 is divisible by 6.
…8 if its last three digits form a number that is divisible by 8.	5800—The last three digits form the number 800. Since 800 is divisible by 8, the number 5800 is divisible by 8.
…9 if the sum of its digits is divisible by 9.	79,542—Add its digits: $7 + 9 + 5 + 4 + 2 = 27$. Since 27 is divisible by 9, the number 79,542 is divisible by 9.
…10 if it ends in a zero.	490 is divisible by 10 because it ends in 0.

W Hint

The fastest way to memorize this table is to practice!

EXAMPLE 2 Identify each number as composite or prime.

a) 35   b) 19   c) 87   d) 2

Solution

a) Because 35 is divisible by 5, it is *not* prime: $35 \div 5 = 7$, which means that $5 \cdot 7 = 35$. So 35 has factors other than 1 and 35. Therefore, **35 is composite.**

b) Does 19 divide evenly by any numbers other than 1 and 19? *No.* **19 is prime.**

c) Think about 87 and the divisibility rules. Is it divisible by 2, 3, 5, or 10? *Yes,* it is divisible by 3. (87 is divisible by 3 since $8 + 7 = 15$ and 15 is divisible by 3.) Since 87 has factors other than 1 and itself, **87 is composite.**

d) The only way to write 2 as a product of two different numbers is $1 \cdot 2$. Therefore, **2 is a prime number.**

[YOU TRY 2] Identify each number as composite or prime.

a) 13 P   b) 4000   c) 171   d) 83 P

BE CAREFUL The only even number that is prime is 2. All of the other prime numbers are odd. But, not all odd numbers are prime! For example, 9 and 15 are odd but not prime.

Here is a list of the prime numbers from 1 to 100.

2	3	5	7	11	13	17	19	23
29	31	37	41	43	47	53	59	61
	67	71	73	79	83	89	97	

2 Find the Prime Factorization of a Number

To write fractions in lowest terms and to add and subtract fractions, it can be helpful to write a number as a product of its prime factors.

W Hint
Write the definition of prime factorization in your own words.

Definition

Finding the **prime factorization** of a number means writing the number as a *product* of prime factors.

We can find the prime factorization of a number using a *factor tree*.

EXAMPLE 3 Find the prime factorization of 18.

Solution

To make a factor tree, write 18 at the top. Then, think of *any* two numbers (except 1 and 18) that multiply to 18. When a factor is prime, circle it, and that part of the factor tree is complete.

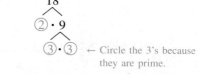

2 is prime, so → ②· 9 Think of *any* two numbers
circle it.   that multiply to 18.

Can 9 be written as the product of two numbers other than 1 and 9? *Yes.* 3 · 3 = 9. Continue with the tree.

18
②· 9
③·③ ← Circle the 3's because
        they are prime.

$$18 = 2 \cdot 3 \cdot 3 \quad \text{or} \quad 2 \cdot 3^2$$

> **W Hint**
> The factors in the prime factorization are written from smallest to largest. The prime factorization may be written with or without exponents.

When all of the factors at the end of the tree are primes, you have finished the tree. **The prime factorization is the product of all the circled primes.**

The prime factorization of 18 is $2 \cdot 3 \cdot 3$ or $2 \cdot 3^2$.

[YOU TRY 3] Find the prime factorization of 28.

BE CAREFUL In Example 1, we found that the *factors of* 18 are 1, 2, 3, 6, 9, and 18. In Example 3, we found that the *prime factorization of* 18 is $2 \cdot 3 \cdot 3$ or $2 \cdot 3^2$. Finding all the factors of a number and finding the prime factorization of a number are two different things.

EXAMPLE 4 Find the prime factorization of 462.

Solution

Can you think of two numbers that multiply to 462? Use the divisibility rules.

Is 462 divisible by 2? *Yes!* It is an even number.

Since 2 is a factor of 462, we can find another factor using long division.

$$\begin{array}{r} 2\ 3\ 1 \\ 2\overline{)4\ 6\ 2} \end{array}$$

Since $462 \div 2 = 231$, it follows that $231 \cdot 2 = 462$. Put these on the factor tree.

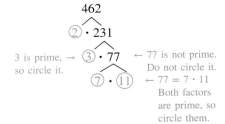

2 is prime, → so circle it.

Is 231 prime or composite? Let's use the divisibility rules again. 231 is not divisible by 2, 5, or 10. Is 231 divisible by 3? *Yes!*

Using long division, we get $231 \div 3 = 77$. It follows that $77 \cdot 3 = 231$. Put these on the factor tree.

3 is prime, → so circle it.

← 77 is not prime. Do not circle it.

← $77 = 7 \cdot 11$ Both factors are prime, so circle them.

Write the prime factorization of 462: $462 = 2 \cdot 3 \cdot 7 \cdot 11$.

[YOU TRY 4]   Find the prime factorization of each number.

a) 90   b) 396

We can also find the prime factorization of a number by modifying the long division procedure. When we use this method, *all* divisors must be prime.

EXAMPLE 5   Find the prime factorization of 12.

Solution

First ask yourself, *"What prime number divides evenly into 12?"* We could use either 2 or 3. Let's use 2.

$$\begin{array}{r} 6 \\ 2\overline{)12} \end{array}$$ Divide 12 by 2.

Now ask yourself, *"What prime number divides evenly into 6?"* We could use 2 or 3. We will use 2.

$$\begin{array}{r} 3 \\ 2\overline{)6} \end{array}$$ Divide 6 by 2.

Ask yourself, *"What prime number divides evenly into 3?"* That number is 3.

$$\begin{array}{r} 1 \\ 3\overline{)3} \end{array}$$ Divide 3 by 3.

When the quotient is 1, the division is finished. The prime factorization consists of the prime factors, in blue.

Write the prime factorization: $12 = 2 \cdot 2 \cdot 3$ or $12 = 2^2 \cdot 3$.

[YOU TRY 5]   Find the prime factorization of 63.

We can write the steps used in Example 5 as a continuous string of division problems.

EXAMPLE 6

Find the prime factorization of 180.

Solution

Remember, *all* divisors must be prime numbers. So, ask yourself, *"What prime number divides evenly into 180?"* Since 180 ends in 0, it is divisible by 2.

$$\begin{array}{r} 90 \\ 2\overline{)180} \end{array} \quad \text{Divide 180 by 2.}$$

$$\begin{array}{r} 45 \\ 2\overline{)90} \end{array} \quad \text{Divide 90 by 2.}$$

$$\begin{array}{r} 9 \\ 5\overline{)45} \end{array} \quad \text{Divide 45 by 5.}$$

$$\begin{array}{r} 3 \\ 3\overline{)9} \end{array} \quad \text{Divide 9 by 3.}$$

$$\begin{array}{r} 1 \\ 3\overline{)3} \end{array} \quad \text{Divide 3 by 3.}$$

 Hint

Write out the example as you are reading it!

When the quotient is 1, the division is finished. The prime factorization consists of the prime factors, in blue.

Write the prime factorization: $180 = 2 \cdot 2 \cdot 3 \cdot 3 \cdot 5$ or $180 = 2^2 \cdot 3^2 \cdot 5$.

[YOU TRY 6] Find the prime factorization of 364.

3 Understand Equivalent Fractions

Different fractions can describe the same quantity. For example, compare these two pizzas of the same size.

$\frac{1}{2}$ of the pizza is left. 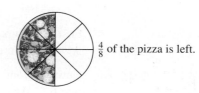 $\frac{4}{8}$ of the pizza is left.

Both of the fractions represent the same amount of pizza, so $\frac{1}{2} = \frac{4}{8}$. We say that $\frac{1}{2}$ and $\frac{4}{8}$ are *equivalent fractions*.

> ### Definition
>
> **Equivalent fractions** are different fractions that represent the same part of the whole.
>
> *Example:* $\frac{1}{2}$ and $\frac{4}{8}$ are equivalent fractions.

Although $\frac{1}{2}$ and $\frac{4}{8}$ represent the same amount of pizza, $\frac{1}{2}$ is in *lowest terms* and $\frac{4}{8}$ is not. $\frac{4}{8}$ is *not* in lowest terms because 4 and 8 share at least one *common factor:*

1, 2, and 4 are **common factors** of 4 and 8 because 1, 2, and 4 are factors of 4 and factors of 8. The number 4 is the *greatest common factor* of 4 and 8 because 4 is the largest of their common factors.

Definition

The **greatest common factor** of a group of numbers is the *largest* number that is a factor of each number in the group.

> *Example:* The greatest common factor of 4 and 8 is 4 because 4 is the largest number that is a factor of 4 and of 8.

The fraction $\frac{1}{2}$ *is* in lowest terms because the numerator and denominator do not have any common factor other than 1.

W Hint

Did you write these definitions in your own words?

Definition

A fraction is in **lowest terms** if the numerator and denominator have no common factors other than 1.

> *Example:* $\frac{1}{2}$ is in lowest terms.

Note

Writing a fraction in lowest terms is also called **simplifying** a fraction.

EXAMPLE 7

Determine whether each fraction is in lowest terms.

a) $\frac{5}{9}$ b) $\frac{14}{21}$

Solution

a) List the factors of 5 and 9.

　　　　Factors of 5: 1 and 5　　　　Factors of 9: 1, 3, and 9

$\frac{5}{9}$ *is* in lowest terms because 5 and 9 have no common factor other than 1.

b) Look at the fraction $\frac{14}{21}$.   ← 14 is divisible by 7.
                         ← 21 is divisible by 7.

14 and 21 have a common factor of 7. Therefore, $\frac{14}{21}$ is *not* in lowest terms.

| YOU TRY 7 | Determine whether each fraction is in lowest terms. |

a) $\dfrac{16}{40}$ b) $\dfrac{11}{18}$

4 Write Fractions in Lowest Terms Using Common Factors

Knowing how to write a fraction in lowest terms is a skill that we need in *many* areas of mathematics as well as in other subject areas like nursing, science, culinary arts, and more. There are two common methods for writing a fraction in lowest terms. The first method we will learn involves dividing the numerator and denominator by a common factor.

W Hint
Take notes on the steps you would take to write a fraction in lowest terms using common factors.

Procedure Writing a Fraction in Lowest Terms Using Common Factors

To write a fraction in lowest terms:

1) Ask yourself, *"What number divides evenly into both the numerator and denominator?"* (Use any number except 1.) Divide the numerator and denominator by that number.

2) Look at the result and ask yourself, *"Is the fraction in lowest terms?"* If the numerator and denominator still contain a common factor, repeat Steps 1 and 2.

Note

It is **very** important that you always look at the result and ask yourself, *"Is the fraction in lowest terms?"* If you divided by the greatest common factor, then it will be in lowest terms. But if you have *not* divided by the greatest common factor, the fraction will *not* be in lowest terms, and you will have to simplify it more.

EXAMPLE 8   Write each fraction in lowest terms.

a) $\dfrac{50}{80}$ b) $-\dfrac{56}{72}$ c) $\dfrac{198}{294}$ d) $\dfrac{24x}{40}$

Solution

a) Ask yourself, *"What number divides evenly into both 50 and 80?"* Let's use 10. (We could have used 2 or 5, but 10 is the **greatest common factor** of 50 and 80.) We will use the greatest common factor, 10, because that will simplify the fraction most quickly.

$$\frac{50}{80} = \frac{50 \div 10}{80 \div 10} = \frac{5}{8}$$

Ask yourself, *"Is $\frac{5}{8}$ in lowest terms?"* Yes it is, because 5 and 8 have no common factor other than 1.

Therefore, $\frac{50}{80} = \frac{5}{8}$, and $\frac{5}{8}$ is in lowest terms.

b) Keep the negative sign in front of the fraction and go through the steps to write it in lowest terms. Ask yourself, *"What number divides evenly into both 56 and 72?"* Let's use 8 because it is the greatest common factor of 56 and 72.

$$-\frac{56}{72} = -\frac{56 \div 8}{72 \div 8} = -\frac{7}{9}$$

Ask yourself, *"Is $-\frac{7}{9}$ in lowest terms?"* Yes. Therefore, $-\frac{56}{72} = -\frac{7}{9}$.

c) The numerator and denominator of $\frac{198}{294}$ are large numbers, so we probably will not divide by the *greatest* common factor the first time. But that is fine; we can divide by *any* common factor. Ask yourself, *"What number divides evenly into both 198 and 294?"* Both numbers divide by 2.

$$\frac{198}{294} = \frac{198 \div 2}{294 \div 2} = \frac{99}{147}$$

Ask yourself, *"Is $\frac{99}{147}$ in lowest terms?"* No! Think about the divisibility rules. Each number is divisible by 3.

$$\frac{99}{147} = \frac{99 \div 3}{147 \div 3} = \frac{33}{49}$$

"Is $\frac{33}{49}$ in lowest terms?" Yes. Therefore, in lowest terms, $\frac{198}{294} = \frac{33}{49}$.

d) This fraction, $\frac{24x}{40}$, contains one variable as a factor. We can write this fraction in lowest terms the same way we have written the other fractions in lowest terms.

Ask yourself, *"What number divides evenly into both 24 and 40?"* Let's divide both numbers by 8 because it is the greatest common factor of 24 and 40.

$$\frac{24x}{40} = \frac{24 \cdot x}{40} \qquad \text{Write the numerator as a product.}$$

$$= \frac{\overset{3}{24} \cdot x}{\underset{5}{40}} \qquad \text{Divide } 24 \div 8 = 3,\ 40 \div 8 = 5.$$

$$= \frac{3x}{5} \qquad \text{Multiply in the numerator.}$$

Ask yourself, *"Is $\frac{3x}{5}$ in lowest terms?"* Yes. Therefore, $\frac{24x}{40} = \frac{3x}{5}$.

W Hint

Example 8c) shows us why it is important to look at the result to determine whether the fraction can be simplified more.

Write each fraction in lowest terms.

a) $\dfrac{35}{60}$ b) $-\dfrac{48}{108}$ c) $\dfrac{1620}{2340}$ d) $\dfrac{14n}{63}$

5 Write Fractions in Lowest Terms Using Prime Factorization

We can also use the prime factorizations of the numerator and denominator to write a fraction in lowest terms.

> **W Hint**
>
> In your own words, take notes on the steps you would take to write a fraction in lowest terms using prime factorization.

Procedure Writing a Fraction in Lowest Terms Using Prime Factorization

To write a fraction in lowest terms:

1) Write the prime factorization of the numerator and denominator.

2) Divide out common factors. Write a 1 by each factor to indicate that you have performed the division.

3) Multiply the factors that are left in the numerator and in the denominator.

Note

Always look at the result and ask yourself, *"Is the fraction in lowest terms?"* in case you have not divided out all common factors. Divide out remaining common factors, if necessary.

EXAMPLE 9   Write each fraction in lowest terms using prime factorization.

a) $\dfrac{21}{126}$ b) $\dfrac{540}{45}$ c) $\dfrac{4a^2}{28a^3}$

Solution

a) Use one of the methods we learned in this section to write the prime factorizations of 21 and 126.

$$\frac{21}{126} = \frac{3 \cdot 7}{2 \cdot 3 \cdot 3 \cdot 7}$$ Write the prime factorizations of 21 and 126.

Divide out the common factors and write a 1 by each of these factors that have been divided.

$$\frac{21}{126} = \frac{\overset{1}{\cancel{3}} \cdot \overset{1}{\cancel{7}}}{2 \cdot \underset{1}{\cancel{3}} \cdot 3 \cdot \underset{1}{\cancel{7}}}$$

Multiply the factors that are left: $\dfrac{21}{126} = \dfrac{1 \cdot 1}{2 \cdot 1 \cdot 3 \cdot 1} = \dfrac{1}{6}$

Is $\dfrac{1}{6}$ in lowest terms? Yes. So, $\dfrac{21}{126} = \dfrac{1}{6}$, and $\dfrac{1}{6}$ is in lowest terms.

BE CAREFUL When all of the factors of the numerator were divided out, we were left with 1 so that the answer is $\dfrac{1}{6}$, *not* 6.

b) Write the prime factorization, and divide out the common factors.

$$\frac{540}{45} = \frac{2 \cdot 2 \cdot \overset{1}{\cancel{3}} \cdot \overset{1}{\cancel{3}} \cdot 3 \cdot \overset{1}{\cancel{5}}}{\underset{1}{\cancel{3}} \cdot \underset{1}{\cancel{3}} \cdot \underset{1}{\cancel{5}}} = \frac{2 \cdot 2 \cdot 1 \cdot 1 \cdot 3 \cdot 1}{1 \cdot 1 \cdot 1} = \frac{12}{1}$$

Ask yourself, *"Is $\dfrac{12}{1}$ in lowest terms?"* No! We can write $\dfrac{12}{1}$ as 12. In lowest terms, $\dfrac{540}{45} = 12$.

c) This fraction, $\dfrac{4a^2}{28a^3}$, contains variables in the numerator and denominator. Let's use prime factorization, and remember that exponents are a shorthand way to represent multiplication.

$$\frac{4a^2}{28a^3} = \frac{2 \cdot 2 \cdot a \cdot a}{2 \cdot 2 \cdot 7 \cdot a \cdot a \cdot a} \qquad \text{Write the prime factorization.}$$

$$= \frac{\overset{1}{\cancel{2}} \cdot \overset{1}{\cancel{2}} \cdot \overset{1}{\cancel{a}} \cdot \overset{1}{\cancel{a}}}{\underset{1}{\cancel{2}} \cdot \underset{1}{\cancel{2}} \cdot 7 \cdot \underset{1}{\cancel{a}} \cdot \underset{1}{\cancel{a}} \cdot a} \qquad \text{Divide out the common factors.}$$

$$= \frac{1}{7a} \qquad \text{Multiply.}$$

$\dfrac{1}{7a}$ is in lowest terms.

We could start this another way: Keep the coefficients, 4 and 28, as they are, write the prime factorizations of a^2 and a^3, then divide out common factors from the coefficients and from the variables.

$$\frac{4a^2}{28a^3} = \frac{4 \cdot a \cdot a}{28 \cdot a \cdot a \cdot a} \qquad \text{Write the prime factorizations of the variable factors.}$$

$$= \frac{\overset{1}{\cancel{4}} \cdot \overset{1}{\cancel{a}} \cdot \overset{1}{\cancel{a}}}{\underset{7}{\cancel{28}} \cdot \underset{1}{\cancel{a}} \cdot \underset{1}{\cancel{a}} \cdot a} \qquad \text{Divide 4 and 28 by 4. Divide out the common factors of } a.$$

$$= \frac{1}{7a} \qquad \text{Multiply.}$$

Write each fraction in lowest terms using prime factorization.

a) $\dfrac{18}{72}$ b) $\dfrac{378}{63}$ c) $\dfrac{5w^3}{60w^5}$

6 Determine Whether Two Fractions Are Equivalent

At the beginning of this section, a picture of two pizzas showed that $\dfrac{1}{2} = \dfrac{4}{8}$. That is, $\dfrac{1}{2}$ and $\dfrac{4}{8}$ are equivalent fractions. To determine whether two fractions are equivalent, write each of them in lowest terms.

EXAMPLE 10

Determine whether each pair of fractions is equivalent.

a) $\dfrac{6}{27}$ and $\dfrac{16}{72}$ b) $-\dfrac{22}{176}$ and $-\dfrac{28}{154}$

Solution

a) Write each fraction in lowest terms and see whether we get the same result. Let's simplify these fractions by dividing out the greatest common factor.

$$\dfrac{6}{27} = \dfrac{6 \div 3}{27 \div 3} = \dfrac{2}{9} \qquad \dfrac{16}{72} = \dfrac{16 \div 8}{72 \div 8} = \dfrac{2}{9}$$

W Hint

Write down the steps you would use to determine whether two fractions are equivalent.

Since $\dfrac{6}{27} = \dfrac{2}{9}$ and $\dfrac{16}{72} = \dfrac{2}{9}$, the fractions *are* equivalent.

b) Let's use prime factorization to write each fraction in lowest terms.

$$-\dfrac{22}{176} = -\dfrac{\overset{1}{2} \cdot \overset{1}{11}}{2 \cdot 2 \cdot 2 \cdot 2 \cdot 11} = -\dfrac{1 \cdot 1}{1 \cdot 2 \cdot 2 \cdot 2 \cdot 1} = -\dfrac{1}{8}$$

$$-\dfrac{28}{154} = -\dfrac{\overset{1}{2} \cdot 2 \cdot \overset{1}{7}}{2 \cdot 7 \cdot 11} = -\dfrac{1 \cdot 2 \cdot 1}{1 \cdot 1 \cdot 11} = -\dfrac{2}{11}$$

$-\dfrac{22}{176}$ and $-\dfrac{28}{154}$ are *not* equivalent.

[YOU TRY 10]

Determine whether each pair of fractions is equivalent.

a) $\dfrac{7}{42}$ and $\dfrac{12}{84}$ b) $-\dfrac{540}{1980}$ and $-\dfrac{45}{165}$

ANSWERS TO [YOU TRY] EXERCISES

1) 1, 2, 3, 5, 6, 10, 15, and 30   2) a) prime b) composite c) composite d) prime
3) $28 = 2 \cdot 2 \cdot 7$ or $28 = 2^2 \cdot 7$
4) a) $90 = 2 \cdot 3 \cdot 3 \cdot 5$ or $90 = 2 \cdot 3^2 \cdot 5$ b) $396 = 2 \cdot 2 \cdot 3 \cdot 3 \cdot 11$ or $396 = 2^2 \cdot 3^2 \cdot 11$
5) $63 = 3 \cdot 3 \cdot 7$ or $63 = 3^2 \cdot 7$ 6) $364 = 2 \cdot 2 \cdot 7 \cdot 13$ or $364 = 2^2 \cdot 7 \cdot 13$ 7) a) no b) yes
8) a) $\dfrac{7}{12}$ b) $-\dfrac{4}{9}$ c) $\dfrac{9}{13}$ d) $\dfrac{2n}{9}$ 9) a) $\dfrac{1}{4}$ b) 6 c) $\dfrac{1}{12w^2}$ 10) a) no b) yes

Get Ready

Use the divisibility rules to determine whether each number is divisible by 2, 3, 4, 5, and/or 10.

1) 75

2) 54

3) 320

4) 828

5) 47,925

6) 90,160

Objective 1: Identify Composite and Prime Numbers

Find all factors of each number.

7) 8

8) 6

9) 15

10) 26

11) 17

12) 11

13) 63

14) 42

15) In your own words, explain the difference between a prime number and a composite number.

16) Is 1 a prime number? Why or why not?

17) What is the smallest prime number?

18) Are there any prime numbers that are even? If so, give an example.

Identify each number as composite or prime.

19) 5

20) 9

21) 56

22) 17

23) 59

24) 101

25) 183

26) 1401

Objective 2: Find the Prime Factorization of a Number

27) What does it mean to write the prime factorization of a number?

28) Is 4 · 5 the prime factorization of 20? Explain your answer.

Find the prime factorization of each number using a factor tree.

29) 15

30) 35

31) 24

32) 50

33) 78

34) 98

35) 270

36) 210

37) 1300

38) 3600

Find the prime factorization of each number using the division process.

39) 8

40) 27

41) 99

42) 75

43) 330

44) 324

45) 495

46) 525

47) a) List all the factors of 24.

    b) Find the prime factorization of 24.

48) a) List all the factors of 70.

    b) Find the prime factorization of 70.

Objective 3: Understand Equivalent Fractions

49) How do you know whether a fraction is in lowest terms?

50) What is the greatest common factor of two numbers?

Determine whether each fraction is in lowest terms.

51) $\dfrac{2}{3}$

52) $\dfrac{2}{5}$

53) $\dfrac{11}{22}$

54) $\dfrac{7}{28}$

55) $\dfrac{63}{207}$

56) $\dfrac{36}{324}$

Objective 4: Write Fractions in Lowest Terms Using Common Factors

57) In your own words, explain how to write a fraction in lowest terms using common factors.

58) If you have divided the numerator and denominator by a common factor, will the result always be a fraction in lowest terms? Explain your answer.

Write each fraction in lowest terms by dividing out common factors.

59) $\dfrac{10}{16}$

60) $\dfrac{14}{21}$

61) $\dfrac{33}{77}$

62) $\dfrac{27}{36}$

63) $-\dfrac{4}{8}$

64) $-\dfrac{7}{42}$

65) $\dfrac{44}{28}$

66) $\dfrac{84}{30}$

67) $\dfrac{60}{12}$

68) $\dfrac{49}{7}$

69) $-\dfrac{225}{525}$

70) $-\dfrac{205}{615}$

71) $\dfrac{60m}{90}$

72) $\dfrac{15d}{35}$

73) $\dfrac{48}{68t^2}$

74) $\dfrac{63}{90r^3}$

Objective 5: Write Fractions in Lowest Terms Using Prime Factorization

Write each fraction in lowest terms using prime factorization.

75) $\dfrac{6}{15}$

76) $\dfrac{15}{35}$

77) $-\dfrac{36}{81}$

78) $-\dfrac{72}{84}$

79) $\dfrac{24}{288}$

80) $\dfrac{15}{135}$

81) $\dfrac{5t^3}{20t^4}$

82) $\dfrac{3b^4}{6b^5}$

83) $\dfrac{24x^6}{54x^2}$

84) $\dfrac{16a^5}{56a^2}$

85) $-\dfrac{51h^4}{39h^2}$

86) $-\dfrac{92z^6}{76z^3}$

Objective 6: Determine Whether Two Fractions Are Equivalent

Determine whether each pair of fractions is equivalent.

87) $\dfrac{12}{18}$ and $\dfrac{16}{24}$

88) $\dfrac{28}{35}$ and $\dfrac{16}{20}$

89) $\dfrac{72}{40}$ and $\dfrac{42}{18}$

90) $\dfrac{24}{16}$ and $\dfrac{36}{21}$

91) $\dfrac{45}{25}$ and $\dfrac{54}{30}$

92) $\dfrac{14}{12}$ and $\dfrac{63}{54}$

93) $-\dfrac{63}{168}$ and $-\dfrac{51}{136}$

94) $-\dfrac{112}{512}$ and $-\dfrac{49}{224}$

95) $-\dfrac{63}{144}$ and $-\dfrac{84}{182}$

96) $-\dfrac{126}{288}$ and $-\dfrac{168}{384}$

97) $-\dfrac{224}{512}$ and $-\dfrac{175}{400}$

98) $-\dfrac{104}{120}$ and $-\dfrac{960}{1200}$

Determine whether each statement is *true* or *false*.

99) Two equivalent fractions have the same location on a number line.

100) On the number line, $\dfrac{1}{10}$ is to the left of $\dfrac{1}{2}$.

101) All fractions are less than 1.

102) $\dfrac{10}{10,000} = \dfrac{1}{1000}$

103) $\dfrac{1}{1000}$ is greater than $\dfrac{1}{100}$.

104) $\dfrac{1}{3}$ is less than $\dfrac{1}{4}$.

R Rethink

R1) Do you prefer to find the prime factorization of a number using a factor tree or the division process? Why?

R2) Do you know the divisibility rules well enough to easily divide out common factors from the numerator and denominator of a fraction? If not, which rules do you need to learn better?

R3) Give some examples of where, outside of your math class, you would see fractions that could be written in lowest terms.

3.3 Multiplying and Dividing Signed Fractions

P Prepare | ## O Organize

What are your objectives for Section 3.3?	How can you accomplish each objective?
1 Multiply Fractions	• Write the procedure for **Multiplying Fractions** in your own words. • Complete the given examples on your own. • Complete You Trys 1 and 2.
2 Divide Out Common Factors Before Multiplying	• Add the step of dividing out a common factor before multiplying to the procedure you wrote for multiplying fractions. • Write the procedure for **Multiplying a Fraction and an Integer** in your own words. • Complete the given examples on your own. • Complete You Trys 3–6.
3 Find the Reciprocal of a Number	• Write the definition of a *reciprocal* in your own words. • Complete the given example on your own. • Complete You Try 7.
4 Divide Fractions	• Write the procedure for **Dividing Fractions** in your own words. • Complete the given examples on your own and notice the slight differences between them. • Complete You Trys 8 and 9.
5 Solve Applied Problems Involving Multiplication and Division of Fractions	• Make a list of key words that indicate multiplication and key words that indicate division. • Complete the given examples on your own, and notice when you multiply and when you divide. • Complete You Trys 10 and 11.

W Work

Read the explanations, follow the examples, take notes, and complete the You Trys.

1 Multiply Fractions

When we use the word *of* with fractions, it usually means multiplication. Why is that true? Let's consider this situation.

Rich is having a party and orders a party-size sub sandwich. He cuts the sandwich in half.

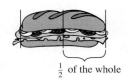

$\frac{1}{2}$ of the whole

He cuts each piece in half again and eats one piece. Rich eats $\frac{1}{2}$ *of* $\frac{1}{2}$ of the sandwich.

What fraction of the whole sandwich did Rich eat?

We can see from the figure that he ate $\frac{1}{4}$ of the whole sandwich. If we do not have a picture to look at, how can we find $\frac{1}{2}$ *of* $\frac{1}{2}$? We multiply.

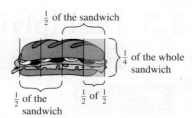

$\frac{1}{2}$ of the sandwich

$\frac{1}{4}$ of the whole sandwich

$\frac{1}{2}$ of the sandwich

$\frac{1}{2}$ of $\frac{1}{2}$

 Hint

Write the procedure for multiplying fractions in your own words.

Procedure Multiplying Fractions

To multiply fractions, multiply the numerators and multiply the denominators. Write the result in lowest terms.

Example: $\frac{1}{2}$ *of* $\frac{1}{2}$ means $\frac{1}{2} \cdot \frac{1}{2} = \frac{1 \cdot 1}{2 \cdot 2} = \frac{1}{4}$

EXAMPLE 1

Multiply. Write all answers in lowest terms.

a) $\frac{2}{5} \cdot \frac{4}{9}$ b) $\frac{7}{9} \cdot \frac{1}{4} \cdot \frac{5}{2}$ c) $-\frac{3}{8} \cdot \frac{5}{6}$

Solution

a) Multiply the numerators, and multiply the denominators.

$$\frac{2}{5} \cdot \frac{4}{9} = \frac{2 \cdot 4}{5 \cdot 9} = \frac{8}{45} \qquad \text{Multiply numerators, and multiply denominators.}$$

Is $\frac{8}{45}$ in lowest terms? *Yes.* So, the product is $\frac{8}{45}$.

b) $\frac{7}{9} \cdot \frac{1}{4} \cdot \frac{5}{2} = \frac{7 \cdot 1 \cdot 5}{9 \cdot 4 \cdot 2} = \frac{35}{72}$

Is $\frac{35}{72}$ in lowest terms? *Yes.* The answer is $\frac{35}{72}$.

 Hint

Remember: The product of a *positive* number and a *negative* number is *negative*.

c) $-\frac{3}{8} \cdot \frac{5}{6} = -\frac{3 \cdot 5}{8 \cdot 6} = -\frac{15}{48} \qquad \text{Recall that the product of a negative number and a positive number is negative.}$

Is $-\frac{15}{48}$ in lowest terms? *No!* We must write it in lowest terms.

$$-\frac{15}{48} = -\frac{15 \div 3}{48 \div 3} = -\frac{5}{16}$$

Is $-\frac{5}{16}$ in lowest terms? *Yes.* The answer is $-\frac{5}{16}$.

[YOU TRY 1] Multiply. Write all answers in lowest terms.

a) $\frac{8}{11} \cdot \frac{2}{3}$ b) $\frac{2}{9} \cdot \frac{7}{3} \cdot \frac{1}{5}$ c) $-\frac{5}{21} \cdot \frac{7}{2}$

Whenever you get a result that is a fraction, ask yourself, *"Is it in lowest terms?"* If it is not, write it in lowest terms.

Let's see how we multiply fractions containing variables.

EXAMPLE 2

Multiply. Write all answers in lowest terms.

a) $\dfrac{1}{8n} \cdot \dfrac{2n}{3}$ b) $\dfrac{a^4}{b} \cdot \dfrac{b^2}{a^2}$

Solution

a) We can begin by multiplying numerators and multiplying denominators. Then, we can write the answer in lowest terms.

$$\frac{1}{8n} \cdot \frac{2n}{3} = \frac{2n}{24n}$$

Ask yourself, *"Is $\dfrac{2n}{24n}$ in lowest terms?"* No. Divide out the common factors.

$$\frac{2n}{24n} = \frac{2 \cdot n}{24 \cdot n} = \frac{\overset{1}{\cancel{2}} \cdot \overset{1}{\cancel{n}}}{\underset{12}{\cancel{24}} \cdot \underset{1}{\cancel{n}}}$$ Divide 2 and 24 by 2. Divide out the common factor n.

$$= \frac{1}{12}$$ Multiply.

The fraction $\dfrac{1}{12}$ is in lowest terms, so $\dfrac{1}{8n} \cdot \dfrac{2n}{3} = \dfrac{2n}{24n} = \dfrac{1}{12}$.

b) We can also use prime factorization to multiply fractions and write the answer in lowest terms. So first, multiply numerators and multiply denominators. Next, write the prime factorization of each factor, divide out common factors, then multiply.

$$\frac{a^4}{b} \cdot \frac{b^2}{a^2} = \frac{a^4 \cdot b^2}{b \cdot a^2}$$ Multiply.

$$= \frac{\overset{1}{\cancel{a}} \cdot \overset{1}{\cancel{a}} \cdot a \cdot a \cdot \overset{1}{\cancel{b}} \cdot b}{\underset{1}{\cancel{b}} \cdot \underset{1}{\cancel{a}} \cdot \underset{1}{\cancel{a}}}$$ Write the prime factorization of each factor. Then, divide out common factors.

$$= \frac{a^2 b}{1}$$ Multiply.

$$= a^2 b$$ Write the answer in simplest form.

YOU TRY 2

Multiply. Write all answers in lowest terms.

a) $\dfrac{1}{15z} \cdot \dfrac{3z}{4}$ b) $\dfrac{r}{t} \cdot \dfrac{t^3}{r^5}$

---

2 Divide Out Common Factors Before Multiplying

In Example 1c, the product was *not* in lowest terms: $-\dfrac{3}{8} \cdot \dfrac{5}{6} = -\dfrac{15}{48}$. We had to perform

one more step to write the result in lowest terms: $-\dfrac{15}{48} = -\dfrac{15 \div 3}{48 \div 3} = -\dfrac{5}{16}$.

We could also find $-\dfrac{3}{8} \cdot \dfrac{5}{6}$ by dividing out common factors *before* finding the final product. We can write the prime factorization of 8 and 6 and divide out common factors.

$$-\frac{3}{8} \cdot \frac{5}{6} = -\frac{\overset{1}{\cancel{3}} \cdot 5}{2 \cdot 2 \cdot 2 \cdot \underset{1}{\cancel{3}}} = -\frac{1 \cdot 5}{2 \cdot 2 \cdot 2 \cdot 1} = -\frac{5}{16}$$

Because we have divided out all prime factors, the product, $-\dfrac{5}{16}$, is in lowest terms.

EXAMPLE 3

Multiply $\dfrac{4}{9} \cdot \dfrac{3}{10}$ by writing the prime factorization of each number and dividing out the common factors. Be sure the answer is in lowest terms.

Solution

$$\frac{4}{9} \cdot \frac{3}{10} = \frac{4 \cdot 3}{9 \cdot 10} = \frac{\overset{1}{\cancel{2}} \cdot 2 \cdot \overset{1}{\cancel{3}}}{\underset{1}{\cancel{3}} \cdot 3 \cdot 2 \cdot 5} = \frac{1 \cdot 2 \cdot 1}{1 \cdot 3 \cdot 1 \cdot 5} = \frac{2}{15}$$

The product is $\dfrac{2}{15}$, and it is in lowest terms.

[YOU TRY 3]

Multiply $\dfrac{15}{28} \cdot \dfrac{14}{25}$ by writing the prime factorization of each number and dividing out the common factors. Be sure the answer is in lowest terms.

Note

From this point onward, all fractional answers should be written in lowest terms unless stated otherwise.

There is another way to divide out common factors. We can divide them out *before* we multiply. Let's find the product in Example 3 another way.

EXAMPLE 4

Multiply $\dfrac{4}{9} \cdot \dfrac{3}{10}$ by dividing out common factors before multiplying.

Solution

First look at the 4 and 10 in $\dfrac{4}{9} \cdot \dfrac{3}{10}$. What is the greatest common factor of 4 and 10?

It is 2. Divide 4 and 10 by 2.

$$4 \div 2 = 2 \rightarrow \overset{2}{\cancel{4}} \cdot \dfrac{3}{\underset{5 \,\leftarrow\, 10 \div 2 = 5}{\cancel{10}}}$$

$$\dfrac{\overset{2}{\cancel{4}}}{9} \cdot \dfrac{3}{\cancel{10}}$$

W Hint

Add the step of dividing out a common factor before multiplying to the procedure you wrote for multiplying fractions.

Now, look at the 3 and 9. What is the greatest common factor of 3 and 9? It is 3.

$$\dfrac{\overset{2}{\cancel{4}}}{\underset{9 \div 3 = 3 \,\rightarrow\, 3}{\cancel{9}}} \cdot \dfrac{\overset{1 \,\leftarrow\, 3 \div 3 = 1}{\cancel{3}}}{\underset{5}{\cancel{10}}}$$

Multiply the numerators, and multiply the denominators: $\dfrac{2 \cdot 1}{3 \cdot 5} = \dfrac{2}{15}$.

This is the same as the result in Example 3.

[YOU TRY 4]

Multiply $\dfrac{15}{28} \cdot \dfrac{14}{25}$ by dividing out common factors before multiplying.

 BE CAREFUL

Be sure you divide the numerator and the denominator by the *same number. Always* look at the product to be sure it is in lowest terms!

EXAMPLE 5

Multiply by first dividing out common factors.

a) $-\dfrac{7}{40}\left(-\dfrac{16}{21}\right)$ b) $-\dfrac{36}{25} \cdot \dfrac{35}{12}$

Solution

a) In $-\dfrac{7}{40}\left(-\dfrac{16}{21}\right)$, we will divide 7 and 21 by 7, and divide 40 and 16 by 8. Then, multiply. Remember that the product of two negative numbers is positive.

$$-\dfrac{\overset{1}{\cancel{7}}}{\underset{5}{\cancel{40}}} \cdot \left(-\dfrac{\overset{2}{\cancel{16}}}{\underset{3}{\cancel{21}}}\right) = \dfrac{1 \cdot 2}{5 \cdot 3} = \dfrac{2}{15}$$

W Hint

Remember: The product of two numbers with the same sign is *positive*.

Ask yourself, "Is $\dfrac{2}{15}$ in lowest terms?" Yes.

b) Before multiplying $-\dfrac{36}{25} \cdot \dfrac{35}{12}$, divide 36 and 12 by 12, and divide 25 and 35 by 5.

$$-\dfrac{\overset{3}{\cancel{36}}}{\underset{5}{\cancel{25}}} \cdot \dfrac{\overset{7}{\cancel{35}}}{\underset{1}{\cancel{12}}} = -\dfrac{21}{5}$$ The product of a negative number and a positive number is negative.

$-\dfrac{21}{5}$ is in lowest terms.

[YOU TRY 5] Multiply by first dividing out common factors.

a) $-\dfrac{14}{15} \cdot \dfrac{5}{8}$ b) $-\dfrac{21}{10}\left(-\dfrac{25}{14}\right)$

Every integer can be written with a denominator of 1. For example,

$$4 = \dfrac{4}{1} \qquad 9 = \dfrac{9}{1} \qquad -20 = -\dfrac{20}{1}$$

We use this fact to multiply a fraction and an integer.

W Hint

Write the procedure for multiplying a fraction and an integer in your own words.

Procedure Multiplying a Fraction and an Integer

To multiply a fraction and an integer, rewrite the integer as a fraction with a denominator of 1. Then, multiply.

EXAMPLE 6

Multiply $\dfrac{2}{3} \cdot 12$.

Solution

$$\dfrac{2}{3} \cdot 12 = \dfrac{2}{3} \cdot \dfrac{12}{1}$$ Write 12 with a denominator of 1.

$$= \dfrac{2}{\underset{1}{\cancel{3}}} \cdot \dfrac{\overset{4}{\cancel{12}}}{1}$$ Divide 3 and 12 by 3.

$$= \dfrac{2 \cdot 4}{1 \cdot 1}$$ Multiply numerators, and multiply denominators.

$$= \dfrac{8}{1} = 8$$ Multiply, and write the answer in lowest terms.

Although this product is a whole number, the result of multiplying a fraction and a whole number is *not* necessarily a whole number.

[YOU TRY 6] Multiply.

a) $14 \cdot \dfrac{4}{7}$ b) $-32 \cdot \dfrac{9}{20}$

Let's use a number line to understand how to divide fractions. The division problem $12 \div 3$ means, *"How many threes does it take to make 12?"*

Likewise, the problem $2 \div \dfrac{1}{3}$ means, *"How many one-thirds does it take to make 2?"*

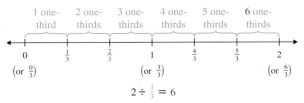

$$2 \div \tfrac{1}{3} = 6$$

It takes 6 *one-thirds* to make 2, so $2 \div \dfrac{1}{3} = 6$.

It is not a coincidence that $2 \div \dfrac{1}{3} = 6$ and $2 \cdot 3 = 6$. The numbers $\dfrac{1}{3}$ and 3 are *reciprocals*. We need to learn more about reciprocals before we learn more about dividing.

3 Find the Reciprocal of a Number

Definition

Two numbers are **reciprocals** if their product is 1. To find the reciprocal of a number, we interchange the numerator and denominator.

Example: The reciprocal of $\dfrac{2}{5}$ is $\dfrac{5}{2}$.

Check: $\dfrac{2}{5}$ and $\dfrac{5}{2}$ are reciprocals because $\dfrac{2}{5} \cdot \dfrac{5}{2} = \dfrac{10}{10} = 1$.

Note

We also say that we *invert* or *flip* a fraction to find its reciprocal.

EXAMPLE 7

Find the reciprocal of each number.

a) $\dfrac{3}{4}$ b) 2 c) $-\dfrac{1}{3}$

Solution

a) The reciprocal of $\dfrac{3}{4}$ is $\dfrac{4}{3}$. We can verify that $\dfrac{3}{4}$ and $\dfrac{4}{3}$ are reciprocals by multiplying.

$$\frac{3}{4} \cdot \frac{4}{3} = \frac{12}{12} = 1 \qquad \text{The product is 1, so they are reciprocals.}$$

b) How do we find the reciprocal of 2? First, write 2 as a fraction with a denominator of 1: $2 = \dfrac{2}{1}$

Then, flip $\dfrac{2}{1}$ to get its reciprocal, $\dfrac{1}{2}$. The reciprocal of 2 is $\dfrac{1}{2}$.

Check: $2 \cdot \dfrac{1}{2} = \dfrac{2}{1} \cdot \dfrac{1}{2} = \dfrac{2}{2} = 1$ ✓

c) The reciprocal of $-\dfrac{1}{3}$ is $-\dfrac{3}{1}$ or -3.

[YOU TRY 7]

Find the reciprocal of each number.

a) $\dfrac{7}{8}$ b) $-\dfrac{1}{5}$ c) 6

Does 0 have a reciprocal? *No.* The number 0 does *not* have a reciprocal because there is no number that can be multiplied by 0 to get 1. The product of any number and 0 is 0.

Note
The number 0 does *not* have a reciprocal.

4 Divide Fractions

Now that we understand reciprocals, we can learn how to divide fractions. At the beginning of this section, we found that

$$2 \div \dfrac{1}{3} = 6 \text{ because it takes 6 one-thirds to make 2.}$$

We also said that it is not a coincidence that $2 \div \dfrac{1}{3} = 6$ and $2 \cdot 3 = 6$. *To perform division involving fractions, multiply the first number by the reciprocal of the second number.*

W Hint
Write the procedure for dividing fractions in your own words.

Procedure Dividing Fractions

To perform division involving fractions, multiply the first number by the reciprocal of the second number.

Example: $\dfrac{1}{5} \div \dfrac{2}{3} = \dfrac{1}{5} \cdot \dfrac{3}{2} = \dfrac{3}{10}$

Change division to multiplication by the reciprocal.

Remember, write all answers in lowest terms.

EXAMPLE 8

Divide.

a) $\dfrac{5}{8} \div \dfrac{6}{7}$ b) $-\dfrac{9}{10} \div \left(-\dfrac{2}{5}\right)$ c) $\dfrac{3a}{8} \div 6$ d) $-4 \div \dfrac{1}{7}$

Solution

a) To divide fractions, *multiply by the reciprocal* of the second fraction.

$$\dfrac{5}{8} \div \dfrac{6}{7} = \dfrac{5}{8} \cdot \dfrac{7}{6} = \dfrac{35}{48}$$

Change division to multiplication
by the reciprocal.

Is $\dfrac{35}{48}$ in lowest terms? Yes.

 Hint

Do not divide out common
factors until *after*
changing the problem to
multiplication.

b) $-\dfrac{9}{10} \div \left(-\dfrac{2}{5}\right) = -\dfrac{9}{10} \cdot \left(-\dfrac{5}{2}\right) = -\dfrac{9}{\underset{2}{10}} \cdot \left(-\dfrac{\overset{1}{5}}{2}\right)$ Divide 5 and 10 by 5.

Change division to
multiplication by
the reciprocal.

$= \dfrac{9 \cdot 1}{2 \cdot 2}$ Multiply. The product of two negative
numbers is positive.

$= \dfrac{9}{4}$ Write the answer in lowest terms.

c) First, write 6 as $\dfrac{6}{1}$. Then, follow the procedure for dividing fractions.

$$\dfrac{3a}{8} \div 6 = \dfrac{3a}{8} \div \dfrac{6}{1}$$ Begin by writing 6 as $\dfrac{6}{1}$.

$$= \dfrac{3a}{8} \cdot \dfrac{1}{6}$$ Change division to multiplication by the reciprocal.

$$= \dfrac{\overset{1}{3a}}{8} \cdot \dfrac{1}{\underset{2}{6}}$$ Divide 3 and 6 by 3.

$$= \dfrac{a \cdot 1}{8 \cdot 2} = \dfrac{a}{16}$$ Multiply.

d) $-4 \div \dfrac{1}{7} = -\dfrac{4}{1} \div \dfrac{1}{7}$ Begin by writing -4 as $-\dfrac{4}{1}$.

$$= -\dfrac{4}{1} \cdot \dfrac{7}{1}$$ Change division to multiplication by the reciprocal.

$$= -\dfrac{28}{1} = -28$$ Multiply.

[YOU TRY 8]

Divide.

a) $\dfrac{1}{3} \div \dfrac{7}{8}$ b) $-\dfrac{13}{18} \div \left(-\dfrac{3}{10}\right)$ c) $-\dfrac{6k}{11} \div 16$ d) $7 \div \dfrac{1}{5}$

BE CAREFUL Remember that the denominator of a fraction cannot equal 0. So if you get an answer like $\frac{2}{3x}$, x cannot equal 0.

A division problem can also take the form of a *complex fraction*. A **complex fraction** contains a fraction in its numerator, its denominator, or both.

EXAMPLE 9

Divide $\dfrac{\frac{2}{9}}{\frac{5}{12}}$.

Solution

The fraction bar represents ⟶ division.

$$\frac{\frac{2}{9}}{\frac{5}{12}} = \frac{2}{9} \div \frac{5}{12} = \frac{2}{9} \cdot \frac{12}{5}$$ Change division to multiplication by the reciprocal.

$$= \frac{2}{\overset{}{9}_{3}} \cdot \frac{\overset{4}{12}}{5}$$ Divide 12 and 9 by 3.

$$= \frac{2 \cdot 4}{3 \cdot 5} = \frac{8}{15}$$ Multiply.

[YOU TRY 9]

Divide $\dfrac{\frac{3}{4}}{\frac{1}{2}}$.

---

5 Solve Applied Problems Involving Multiplication and Division of Fractions

We have learned that the following key words indicate multiplication: *product, times, double, twice, triple,* and *of.* Let's look at a problem involving multiplication of fractions.

EXAMPLE 10

Professor Hill teaches at a community college at night, and $\frac{5}{12}$ of her 144 students work full time. How many students work full time? How many do not?

©Don Hammond/Design Pics

Solution

The problem says that $\frac{5}{12}$ *of* 144 students work full time. The *of* indicates multiplication.

Multiply $\frac{5}{12}$ and 144 to find the number of students who work full time. Then, subtract that number from the total number of students to find how many do *not* work full time.

$$\frac{5}{12} \cdot 144 = \frac{5}{12} \cdot \frac{144}{1} = \frac{5}{\overset{12 \,\leftarrow\, 144 \div 12 = 12}{\underset{12 \div 12 = 1 \,\to\, 1}{12}}} \cdot \frac{144}{1} = \frac{60}{1} = 60$$

60 students work full time. $144 - 60 = 84$ students do not work full time.

To check the answer, notice that

Number who work full time	+	Number who do not work full time	=	Total number of students
60	+	84	=	144 students

[**YOU TRY 10**]

On a city bus, $\frac{3}{8}$ of the 56 passengers are using an electronic device such as an MP3 player, a cell phone, or an iPad. How many people are using an electronic device? How many are not?

Many applied problems are solved using division of fractions. Recall that some key words indicating division are *quotient, divided by, divided into, divided equally,* and *per.*

EXAMPLE 11

A jar contains 4 cups of spaghetti sauce. The label says that a serving size is $\frac{2}{3}$ of a cup. How many servings of sauce are in the jar?

©Purestock/SuperStock

Solution

We can think about the problem this way: How many $\frac{2}{3}$-cup servings does it take to make 4 cups? *Divide* 4 by $\frac{2}{3}$ to find the answer.

$$4 \div \frac{2}{3} = \frac{4}{1} \div \frac{2}{3} \qquad \text{Write 4 as } \frac{4}{1}.$$

$$= \frac{4}{1} \cdot \frac{3}{2} \qquad \text{Change division to multiplication by the reciprocal.}$$

$$= \frac{\overset{2}{4}}{1} \cdot \frac{3}{\underset{1}{2}} \qquad \text{Divide 4 and 2 by 2.}$$

$$= \frac{6}{1} = 6 \qquad \text{Multiply and simplify.}$$

The jar contains *six* $\frac{2}{3}$-cup servings of spaghetti sauce.

To check the answer, multiply the number of servings and the serving size to find the total amount of spaghetti sauce in the jar.

Size of each serving

$$\text{Number of servings} \rightarrow 6 \cdot \frac{2}{3} = \frac{6}{1} \cdot \frac{2}{3} = \frac{\overset{2}{\cancel{6}}}{1} \cdot \frac{2}{\underset{1}{\cancel{3}}} = \frac{4}{1} = 4 \quad \leftarrow \text{Total amount of sauce in the jar.}$$

The answer is correct.

[YOU TRY 11]

A 5-gallon office water cooler dispenses glasses of water in a serving size of $\frac{1}{20}$ of a gallon. How many servings of water are in the water cooler?

ANSWERS TO [YOU TRY] EXERCISES

1) a) $\frac{16}{33}$ b) $\frac{14}{135}$ c) $-\frac{5}{6}$ 2) a) $\frac{1}{20}$ b) $\frac{t^2}{r^4}$ 3) $\frac{3}{10}$ 4) $\frac{3}{10}$ 5) a) $-\frac{7}{12}$ b) $\frac{15}{4}$

6) a) 8 b) $-\frac{72}{5}$ 7) a) $\frac{8}{7}$ b) -5 c) $\frac{1}{6}$ 8) a) $\frac{8}{21}$ b) $-\frac{65}{27}$ c) $-\frac{3k}{88}$ d) 35 9) $\frac{3}{2}$

10) 21 people are using an electronic device; 35 people are not.

11) The water cooler contains 100 servings.

E Evaluate **3.3** Exercises   Do the exercises, and check your work.

Get Ready

1) The word *product* indicates _____.

2) The word *quotient* indicates _____.

3) Find the product of 12 and 6.

4) Find the product of 15 and 3.

5) Find the quotient of 20 and 5.

6) Find the quotient of 16 and 2.

Objective 1: Multiply Fractions
Multiply. Write all answers in lowest terms.

7) $\frac{3}{4} \cdot \frac{5}{8}$

8) $\frac{2}{7} \cdot \frac{5}{9}$

9) $-\frac{7}{8} \cdot \frac{4}{13}$

10) $-\frac{9}{8} \cdot \frac{5}{12}$

11) $\frac{1}{5} \cdot \frac{3}{4} \cdot \frac{2}{3}$

12) $\frac{1}{8} \cdot \frac{4}{3} \cdot \frac{1}{2}$

13) Find two fractions that have a product of $\frac{55}{48}$.

14) Find two fractions that have a product of $\frac{12}{35}$.

Objective 2: Divide Out Common Factors Before Multiplying
Multiply by first dividing out common factors.

15) $\frac{8}{9} \cdot \frac{7}{10}$

16) $\frac{3}{4} \cdot \frac{10}{13}$

17) $\left(-\frac{1}{6}\right)\left(-\frac{3}{4}\right)$

18) $\left(-\frac{2}{3}\right)\left(-\frac{1}{8}\right)$

19) $\frac{3}{20x} \cdot \frac{4x}{7}$

20) $\frac{7}{18d} \cdot \frac{6d}{5}$

21) $\left(\frac{55m^2}{144}\right)\left(-\frac{12}{77m^2}\right)$

22) $\left(\frac{27t^3}{7}\right)\left(-\frac{14}{63t^3}\right)$

23) $\frac{x}{y} \cdot \frac{y^2}{x^3}$

24) $\frac{a^4}{b^3} \cdot \frac{b}{a^2}$

25) $\frac{42v^5}{5u^2} \cdot \frac{5u^2}{7v}$

26) $\frac{32r^4}{3t} \cdot \frac{3t}{8r}$

27) $-\frac{45}{14} \cdot \frac{21}{5} \cdot \frac{1}{27}$

28) $-\frac{63}{62} \cdot \frac{31}{26} \cdot \frac{13}{36}$

29) $\frac{54}{60} \cdot \frac{84}{75} \cdot \frac{25}{14}$

30) $\frac{10}{21} \cdot \frac{11}{12} \cdot \frac{72}{15}$

31) Explain how to multiply a fraction and a whole number.

32) Is the product of a fraction and a whole number always a whole number?

Multiply.

33) $20 \cdot \dfrac{3}{5}$

34) $40 \cdot \dfrac{7}{8}$

35) $\dfrac{7}{3} \cdot (-21)$

36) $\dfrac{4}{9} \cdot (-72)$

37) $-\dfrac{5x}{14} \cdot 24$

38) $-\dfrac{5n}{12} \cdot 40$

39) $-5\left(-\dfrac{2}{15h}\right)$

40) $-2\left(-\dfrac{3}{8y}\right)$

Objective 3: Find the Reciprocal of a Number

41) If two numbers are reciprocals, then their product is _____.

42) Does every number have a reciprocal? Explain your answer.

43) In your own words, explain how to find the reciprocal of a fraction.

44) In your own words, explain how to find the reciprocal of a natural number like 7.

Find the reciprocal of each number.

45) $\dfrac{4}{5}$

46) $\dfrac{6}{13}$

47) $\dfrac{9k}{2}$

48) $\dfrac{12}{7a}$

49) $-\dfrac{1}{7}$

50) $-\dfrac{1}{4}$

51) 2

52) 3

Objective 4: Divide Fractions

53) In your own words, explain how to divide two fractions.

54) When dividing fractions, when can you divide out common factors?

Divide.

55) $\dfrac{3}{10} \div \dfrac{2}{5}$

56) $\dfrac{5}{8} \div \dfrac{7}{4}$

57) $\dfrac{15}{16} \div \left(-\dfrac{25}{21}\right)$

58) $\dfrac{7}{8} \div \left(-\dfrac{21}{10}\right)$

59) $-\dfrac{35y^3}{16} \div \left(-\dfrac{5y^2}{12}\right)$

60) $-\dfrac{28r^2}{11} \div \left(-\dfrac{32r}{33}\right)$

61) $\dfrac{9d}{7} \div 3$

62) $\dfrac{4c}{5} \div 2$

63) $-\dfrac{1}{2} \div 6$

64) $-\dfrac{1}{3} \div 15$

65) $6 \div \dfrac{1}{3x}$

66) $10 \div \dfrac{1}{5p}$

67) What is a complex fraction?

68) Is a complex fraction another way to write a division problem? Explain.

Divide.

69) $\dfrac{\dfrac{2}{7}}{\dfrac{1}{3}}$

70) $\dfrac{\dfrac{1}{8}}{\dfrac{3}{5}}$

71) $\dfrac{\dfrac{3}{5}}{\dfrac{6}{7}}$

72) $\dfrac{\dfrac{7}{24}}{\dfrac{9}{16}}$

73) $\dfrac{-\dfrac{121u^2}{15t^3}}{\dfrac{22u}{25t}}$

74) $\dfrac{-\dfrac{88b^3}{13a^4}}{\dfrac{12b^2}{5a}}$

75) $\dfrac{-\dfrac{1}{5}}{-20}$

76) $\dfrac{-18}{-\dfrac{2}{9}}$

Objective 5: Solve Applied Problems Involving Multiplication and Division of Fractions

Solve each problem.

77) Professor Velez's music class has 40 students. If $\dfrac{3}{5}$ of the class is female, how many students are female? How many students are male?

78) Professor Harding's math classroom at a community college has 45 students. If $\dfrac{2}{3}$ of the students are recent high school graduates, how many students are recent high school grads? How many are not?

79) In the 2015 Chicago Marathon, approximately $\frac{23}{50}$ of the runners who finished were female. Out of the approximate 37,000 runners who finished the race, how many were female? How many were male?
(www.running.competitor.com)

©Ingram Publishing

80) A hotel with 354 rooms allows smoking in $\frac{1}{6}$ of its rooms. How many rooms does the hotel offer to smokers? How many rooms are offered to nonsmokers?

81) A local 24-hour diner offers $\frac{1}{4}$ of its dinner plates as vegetarian meals. Of these vegetarian meals, $\frac{1}{3}$ are vegan. If the diner offers 36 different dinner plates, what fraction of the dinner plates are vegan? How many different vegan plates are there?

82) Concetta donates $\frac{1}{8}$ of her salary to a local women's group. The women's group uses $\frac{3}{4}$ of its collected donations to pay for its community services. If Concetta's salary is $2304, what fractional amount of her salary goes toward paying for community services? What amount is this?

The pie chart below shows how Kimi spends her monthly earnings of $2360. Use the information for Exercises 83–86.

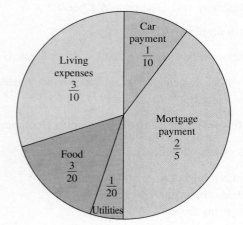

83) How much does Kimi spend on her food each month?

84) What is her monthly mortgage payment?

85) What is her combined budget for living expenses and utilities?

86) If Kimi decides to place $\frac{1}{3}$ of her living expenses budget into a savings account, how much will she deposit into the account?

87) Alexa makes 6 cups of vanilla frosting for her cupcakes. If each cupcake requires $\frac{1}{8}$ cup of frosting, how many cupcakes can Alexa frost?

©McGraw-Hill Education

88) Abhinav owns 12 acres of land and wants to subdivide the land into lots. If each lot is to be $\frac{1}{3}$ of an acre, how many lots can Abhinav make?

89) A pharmacist needs to divide $\frac{3}{4}$ of a quart of saline solution into 6 equal portions. What fraction of the saline solution is used for each portion?

90) A company needs to divide $\frac{5}{8}$ of its inventory into 10 equally sized shipments. What fraction of the total inventory is each shipment?

91) On a map, $\frac{3}{8}$ in. represents 1 mile. If two cities are 9 in. apart on the map, how many miles apart are the two cities?

92) Brandy needs to measure 2 cups of sugar for a recipe. She has only a $\frac{1}{8}$ cup measuring spoon. How many measuring spoons of sugar are required to get 2 cups of sugar?

93) If $\frac{5}{6}$ yd of decorative ribbon is required to decorate a single gift, how many gifts can be decorated with 30 yd of decorative ribbon?

94) A pest control worker uses $\frac{2}{7}$ of a gallon of insect spray to treat the exterior boundary of a home. How many homes can the worker treat with 10 gal of insect spray?

Mixed Exercises: Objectives 1–5

Perform the following operations.

95) $-\dfrac{11w}{12} \div \dfrac{3w^2}{8}$

96) $\left(-\dfrac{3}{10}\right)\left(-\dfrac{5}{21}\right)$

97) $a^5 \cdot \dfrac{b^3}{a^3}$

98) $\dfrac{\frac{21}{56}}{\frac{7}{18}}$

99) $\left(-\dfrac{6}{5}\right)\left(-\dfrac{8}{3}\right)\left(-\dfrac{1}{4}\right)$

100) $\dfrac{4}{5c^2} \cdot \left(-60c^2\right)$

101) $-\dfrac{5}{12} \div (-10)$

102) $2x \div \dfrac{4x}{9y}$

Solve each problem.

103) Stephan set up a savings account to pay for college. He deposited $300 every month for 1 year. At the end of the year, Stephan's parents deposited an additional $\frac{2}{3}$ of the amount he saved to the account.

How much money did Stephan's parents deposit into the account? What is the total amount in the savings account including the amount his parents deposited?

104) If $\frac{2}{3}$ of a gallon of cleaning solution needs to be divided into 4 equal portions, how much of a gallon is each portion?

105) Carlos has 1800 songs on his MP3 player, and he shares $\frac{2}{3}$ of them with his sister Mariana. She divides the songs equally into 8 playlists. How many songs are in each of Mariana's playlists?

106) A department store decides to have a $\frac{1}{2}$-off sale. Additionally, the first 50 customers get an additional $\frac{3}{5}$ off. If one of the first 50 customers purchases an item with an original sale price of $240, what fraction of the original sale price will she pay, and what is the sale price?

R Rethink

R1) Describe the last time you encountered a situation where you needed to multiply or divide by a fraction.

R2) Write an application problem similar to the problems you just solved, and have a friend solve it.

3.4 Adding and Subtracting Like Fractions and Finding a Least Common Denominator

P Prepare

O Organize

What are your objectives for Section 3.4?	How can you accomplish each objective?
1 Add and Subtract Like Fractions	<ul><li>Refer to Section 1.3 if you need to brush up on adding numbers using a number line.</li><li>Write the definitions of *like fractions* and *unlike fractions* in your notes.</li><li>Learn the procedure for **Adding and Subtracting Like Fractions.**</li><li>Complete the given examples on your own.</li><li>Complete You Trys 1 and 2.</li></ul>
2 Write Equivalent Fractions	<ul><li>Understand that this is the opposite process of writing a fraction in lowest terms.</li><li>Write the procedure for **Writing a Fraction with a Different Denominator** in your own words.</li><li>Complete the given examples on your own.</li><li>Complete You Trys 3 and 4.</li></ul>
3 Find the Least Common Multiple (LCM)	<ul><li>Write the definition of the *least common multiple* (LCM).</li><li>Describe the different ways to find the LCM.</li><li>Write the procedure for **Finding the LCM Using Prime Factorization** in your own words.</li><li>Complete the given examples on your own.</li><li>Complete You Trys 5 and 6.</li></ul>
4 Rewrite Fractions with the Least Common Denominator	<ul><li>Write the definition of the *least common denominator* (LCD) in your own words.</li><li>Complete the given examples on your own.</li><li>Complete You Trys 7 and 8.</li></ul>

W Work

Read the explanations, follow the examples, take notes, and complete the You Trys.

In this section, we will learn how to add and subtract like fractions. Let's begin by using figures and number lines to help us add fractions.

1 Add and Subtract Like Fractions

EXAMPLE 1

Add $\frac{3}{5} + \frac{1}{5}$

a) by drawing a figure to represent each fraction.

b) using a number line.

Solution

a) Let's shade $\frac{3}{5}$ of a rectangle and $\frac{1}{5}$ of a rectangle to find the sum $\frac{3}{5}+\frac{1}{5}$.

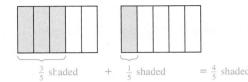

$\frac{3}{5}$ shaded $\quad+\quad$ $\frac{1}{5}$ shaded $\quad=\frac{4}{5}$ shaded

How many total fifths are shaded? 4 *fifths.* Therefore, $\frac{3}{5}+\frac{1}{5}=\frac{4}{5}$.

b) Now let's use a number line to add these fractions.

To add $\frac{3}{5}+\frac{1}{5}$, start at 0 and move $\frac{3}{5}$ unit to reach $\frac{3}{5}$. To add $\frac{1}{5}$ unit, move $\frac{1}{5}$ unit more to the right. We finish at $\frac{4}{5}$.

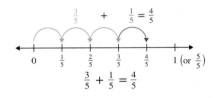

$\frac{3}{5}+\frac{1}{5}=\frac{4}{5}$

[YOU TRY 1]

Add $\frac{2}{7}+\frac{3}{7}$:

a) by drawing a figure to represent each fraction.

b) using a number line.

Notice in Example 1 that the denominators of $\frac{3}{5}$ and $\frac{1}{5}$ are the same. When fractions have the same denominator, we call them *like fractions*.

Definition

Fractions with the same denominator are called **like fractions.** Fractions with different denominators are called **unlike fractions.**

Example: **Like Fractions**   **Unlike Fractions**

$\frac{5}{8}$ and $\frac{1}{8}$ $\frac{4}{7}$ and $\frac{3}{4}$

$-\frac{2}{9}$ and $\frac{7}{9}$ $\frac{1}{2}$ and $-\frac{9}{10}$

$\frac{6}{x}$ and $\frac{4}{x}$ $\frac{8}{c}$ and $\frac{2}{5}$

To add and subtract like fractions, as in Example 1, we use these steps.

Procedure Adding and Subtracting Like Fractions

To add or subtract like fractions:

1) Add or subtract the numerators.

2) Use the denominator of the like fractions as the denominator of the result.

3) Write the answer in lowest terms.

We can also say that if a, b, and c represent numbers and if b does not equal zero, then

$$\frac{a}{b} + \frac{c}{b} = \frac{a+c}{b} \quad \text{and} \quad \frac{a}{b} - \frac{c}{b} = \frac{a-c}{b}$$

W Hint

Write a group of three like fractions.

Note

After you have added or subtracted fractions, always look at the result and ask yourself, *"Is the answer in lowest terms?"* If it is not, write the answer in lowest terms.

EXAMPLE 2   Add or subtract.

a) $\dfrac{4}{9} + \dfrac{1}{9}$ b) $\dfrac{5}{6} - \dfrac{4}{6}$ c) $-\dfrac{1}{15} + \dfrac{11}{15}$

d) $\dfrac{5}{8} - \dfrac{9}{8}$ e) $\dfrac{1}{9x} + \dfrac{5}{9x}$

Solution

a) Because the denominators are the same, we add the numerators and use the denominator 9.

$$\frac{4}{9} + \frac{1}{9} = \frac{4+1}{9} = \frac{5}{9} \qquad \text{Add the numerators, and keep the denominator the same.}$$

Ask yourself, *"Is $\dfrac{5}{9}$ in lowest terms?"* Yes. Therefore, $\dfrac{4}{9} + \dfrac{1}{9} = \dfrac{5}{9}$.

b) In the difference $\dfrac{5}{6} - \dfrac{4}{6}$, the denominators are the same. So, subtract the numerators and use the denominator 6.

$$\frac{5}{6} - \frac{4}{6} = \frac{5-4}{6} = \frac{1}{6} \qquad \text{Subtract the numerators, and keep the denominator the same.}$$

Ask yourself, *"Is $\dfrac{1}{6}$ in lowest terms?"*

Yes. Therefore, $\dfrac{5}{6} - \dfrac{4}{6} = \dfrac{1}{6}$. Let's verify this using a number line.

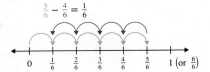

W Hint

Remember the rules for adding and subtracting integers.

c) $-\dfrac{1}{15} + \dfrac{11}{15} = \dfrac{-1+11}{15} = \dfrac{10}{15} \qquad \text{Add the numerators, and keep the denominator the same.}$

Ask yourself, *"Is $\dfrac{10}{15}$ in lowest terms?"* No! So, we must write it in lowest terms.

$$\frac{10}{15} = \frac{10 \div 5}{15 \div 5} = \frac{2}{3}$$

Therefore, $-\dfrac{1}{15} + \dfrac{11}{15} = \dfrac{2}{3}$.

d) $\dfrac{5}{8} - \dfrac{9}{8} = \dfrac{5-9}{8} = \dfrac{-4}{8} \text{ or } -\dfrac{4}{8} \qquad \text{Subtract the numerators, and keep the denominator the same.}$

Ask yourself, *"Is $-\dfrac{4}{8}$ in lowest terms?"* No! Let's write it in lowest terms.

$$-\frac{4}{8} = -\frac{4 \div 4}{8 \div 4} = -\frac{1}{2}$$

Therefore, $\dfrac{5}{8} - \dfrac{9}{8} = -\dfrac{1}{2}$.

e) If fractions contain variables, we add and subtract them the same way we add and subtract fractions that contain only numbers.

$$\frac{1}{9x} + \frac{5}{9x} = \frac{1+5}{9x} = \frac{6}{9x} \qquad \text{Add the numerators, and keep the denominator the same.}$$

Ask yourself, *"Is $\dfrac{6}{9x}$ in lowest terms?"* No! We can simplify fractions containing variables using prime factorization.

$$\frac{6}{9x} = \frac{2 \cdot \overset{1}{3}}{\underset{1}{3 \cdot 3} \cdot x} = \frac{2}{3x} \qquad \text{Divide out the common factor, 3, and multiply.}$$

Therefore, $\dfrac{1}{9x} + \dfrac{5}{9x} = \dfrac{2}{3x}$.

YOU TRY 2 Add.

a) $\dfrac{8}{11} + \dfrac{2}{11}$　　b) $\dfrac{10}{13} - \dfrac{4}{13}$　　c) $-\dfrac{3}{16} + \dfrac{9}{16}$

d) $\dfrac{2}{9} - \dfrac{5}{9}$　　e) $\dfrac{9}{20k} + \dfrac{7}{20k}$

BE CAREFUL

1) We can add and subtract fractions only if they are like fractions.

2) Always look at the sum or difference and ask yourself, *"Is it in lowest terms?"* If it is not, write it in lowest terms.

2 Write Equivalent Fractions

In Section 3.2, we learned how to write fractions in lowest terms. For example, we can write $\dfrac{18}{27}$ in lowest terms by dividing the numerator and denominator by 9.

$$\frac{18}{27} = \frac{18 \div 9}{27 \div 9} = \frac{2}{3}$$

In order to add and subtract *unlike* fractions, we need to know how to rewrite a fraction as an equivalent fraction with a different denominator. This process is the opposite of writing a fraction in lowest terms.

Let's look at the fractions $\dfrac{1}{2}$ and $\dfrac{3}{6}$ in terms of a figure and on a number line.

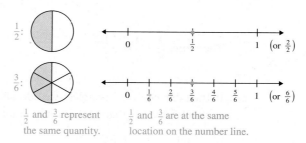

$\dfrac{1}{2}$ and $\dfrac{3}{6}$ represent the same quantity.　　$\dfrac{1}{2}$ and $\dfrac{3}{6}$ are at the same location on the number line.

The fractions $\dfrac{1}{2}$ and $\dfrac{3}{6}$ are *equivalent*. We can also show this by writing $\dfrac{3}{6}$ in lowest terms or by beginning with the fraction $\dfrac{1}{2}$ and rewriting it as $\dfrac{3}{6}$ like this:

$$\frac{1}{2} = \frac{1}{2} \cdot \frac{3}{3} = \frac{3}{6}$$

Note

When we multiply $\dfrac{1}{2}$ by $\dfrac{3}{3}$, we do not change the *value* of the fraction because $\dfrac{3}{3} = 1$. Multiplying $\dfrac{1}{2}$ by $\dfrac{3}{3}$ gives us the equivalent fraction $\dfrac{3}{6}$.

To add and subtract unlike fractions, we must know how to write a given fraction as an equivalent fraction with a different denominator. Let's practice that now.

EXAMPLE 3

Write $\dfrac{3}{4}$ as an equivalent fraction with a denominator of 20.

Solution

We must find a fraction that is equivalent to $\dfrac{3}{4}$ so that $\dfrac{3}{4} = \dfrac{?}{20}$.

Ask yourself, *"By what number do I multiply 4 to get 20?"* That number is 5: $4 \cdot 5 = 20$.

Multiply the numerator *and* denominator of $\dfrac{3}{4}$ by 5: $\dfrac{3}{4} \cdot \dfrac{5}{5} = \dfrac{15}{20}$

$\left(\text{Remember, } \dfrac{5}{5} = 1, \text{ so multiplying } \dfrac{3}{4} \text{ by } \dfrac{5}{5} \text{ does not change the } value \text{ of the fraction;}\right.$
$\left.\text{it gives us an } equivalent \text{ fraction.}\right)$

W Hint

Get in the habit of asking yourself the questions like they are asked in the examples!

YOU TRY 3

Write $\dfrac{7}{8}$ as an equivalent fraction with a denominator of 32.

Procedure Writing a Fraction with a Different Denominator

Step 1: Ask yourself, *"By what number (or variable) do I multiply the original denominator to get the new denominator?"*

Step 2: Multiply the numerator and denominator of the original fraction by that number (or variable) to get the equivalent fraction.

EXAMPLE 4

Write each fraction with the indicated denominator.

a) $\dfrac{4}{7} = \dfrac{?}{56}$ b) $-\dfrac{1}{12} = -\dfrac{?}{72}$ c) $\dfrac{3}{8} = \dfrac{?}{8m}$

Solution

a) **Step 1:** Ask yourself, *"By what number do I multiply 7 to get 56?"* That number is 8.

Step 2: Multiply the numerator and denominator of $\dfrac{4}{7}$ by 8: $\dfrac{4}{7} \cdot \dfrac{8}{8} = \dfrac{32}{56}$

Therefore, $\dfrac{4}{7} = \dfrac{32}{56}$.

b) **Step 1:** Ask yourself, *"By what number do I multiply 12 to get 72?"* That number is 6.

W Hint

Follow the procedure!

Step 2: Multiply the numerator and denominator of $-\dfrac{1}{12}$ by 6: $-\dfrac{1}{12} \cdot \dfrac{6}{6} = -\dfrac{6}{72}$

Therefore, $-\dfrac{1}{12} = -\dfrac{6}{72}$.

c) Notice that the second fraction contains a variable in the denominator while the first fraction does not.

Step 1: Ask yourself, *"By what variable do I multiply 8 to get 8m?"* That variable is m.

Step 2: Multiply the numerator and denominator of $\dfrac{3}{8}$ by m: $\dfrac{3}{8} \cdot \dfrac{m}{m} = \dfrac{3m}{8m}$

Therefore, $\dfrac{3}{8} = \dfrac{3m}{8m}$.

[YOU TRY 4] Write each fraction with the indicated denominator.

a) $\dfrac{6}{11} = \dfrac{?}{77}$ b) $-\dfrac{1}{15} = -\dfrac{?}{45}$ c) $\dfrac{2}{9} = \dfrac{?}{9z}$

---

3 Find the Least Common Multiple (LCM)

Next we will review what the *least common multiple* of a group of numbers is.

Definition

The **least common multiple,** or **LCM,** of a group of natural numbers is the smallest natural number divisible by each number in the group.

> *Example:* The LCM of 2 and 3 is 6 because 6 is the smallest number divisible by both 2 and 3.

There are different ways to find the least common multiple of a group of numbers. One way is to begin by listing the multiples of each number. The least common multiple (LCM) is the *smallest* number that appears on each list.

For example, the LCM of 4 and 6 is 12 because when we list the multiples of each number, the smallest number that is on each list is 12.

Multiples of 4: 4, 8, 12, 16, 20, 24, …
Multiples of 6: 6, 12, 18, 24, 30, 36, …

(Notice that 24 appears on each list. However, 24 is not the *least* common multiple.)

If the numbers are easy to work with, we should be able to find the LCM without making a list or writing anything on paper.

EXAMPLE 5

Find the least common multiple of 4 and 10 by inspection.

Solution

Ask yourself, *"What is the smallest number that is divisible by both 4 and 10?"* That number is 20. Therefore, the LCM of 4 and 10 is 20.

[YOU TRY 5] Find the least common multiple of 4 and 8 by inspection.

We can also use prime factorization to find the least common multiple of a group of numbers. This is a good method when the numbers are large.

Procedure Finding the Least Common Multiple Using Prime Factorization

Step 1: Write the prime factorization of each number.

Step 2: Identify the factors that will be in the least common multiple. The LCM will contain each different factor the *greatest* number of times it appears in any single factorization.

Step 3: The LCM is the *product* of the factors identified in Step 2.

EXAMPLE 6

Find the LCM of 6, 15, and 27 using prime factorization.

Solution

Step 1: Write the prime factorizations of 6, 15, and 27.

$$6 = 2 \cdot 3 \qquad 15 = 3 \cdot 5 \qquad \text{and} \qquad 27 = \underline{3 \cdot 3 \cdot 3}$$

1 factor of 2   1 factor of 5   3 factors of 3

W Hint

Write down the steps in the example on your own paper as you are reading them.

Step 2: The LCM will contain each different factor the *greatest* number of times it appears in any single factorization.

The LCM will contain 2, $\underline{3 \cdot 3 \cdot 3}$, and 5.   Use 1 factor of 2 and 5.

Use 3 factors of 3.

Step 3: The LCM is the *product* of the factors identified in Step 2.

The LCM of 6, 15, and 27 is $2 \cdot 3 \cdot 3 \cdot 3 \cdot 5 = 270$.

[YOU TRY 6]

Find the LCM of each group of numbers using prime factorization.

a) 15 and 21   b) 18, 54, and 60

Note

It is very important that you understand the different methods for finding the least common multiple. We will use these methods to find least common denominators of fractions with unlike denominators.

---

4 Rewrite Fractions with the Least Common Denominator

In order to add or subtract fractions with different denominators, we must first rewrite them with the same, or *common,* denominator. Although we can add or subtract fractions with *any* common denominator, we will rewrite them as equivalent fractions with a *least common denominator.*

Definition

The **least common denominator, or LCD,** of a group of fractions is the *least common multiple* of the denominators.

Example: The **least common denominator** of $\frac{2}{3}$ and $\frac{1}{6}$ is 6 because 6 is the *least common multiple* of 3 and 6.

EXAMPLE 7

For each group of fractions, identify the least common denominator, then write each fraction as an equivalent fraction with the LCD as its denominator.

a) $\frac{1}{4}$ and $\frac{5}{6}$ b) $\frac{c}{9}$ and $\frac{2}{3}$ c) $\frac{25}{36}$ and $\frac{17}{40}$

Solution

a) The least common denominator of $\frac{1}{4}$ and $\frac{5}{6}$ is the least common multiple of 4 and 6. Therefore, the **LCD = 12.**

Write each fraction with a denominator of 12. We want to find

$$\frac{1}{4} = \frac{?}{12} \quad \text{and} \quad \frac{5}{6} = \frac{?}{12}$$

$$\frac{1}{4} \cdot \frac{3}{3} = \frac{3}{12} \qquad \frac{5}{6} \cdot \frac{2}{2} = \frac{10}{12}$$

The LCD of $\frac{1}{4}$ and $\frac{5}{6}$ is 12. Then, $\frac{1}{4} = \frac{3}{12}$ and $\frac{5}{6} = \frac{10}{12}$.

b) The LCD of $\frac{c}{9}$ and $\frac{2}{3}$ is the least common multiple of 9 and 3: **LCD = 9**

$$\frac{c}{9} = \frac{?}{9} \quad \text{and} \quad \frac{2}{3} = \frac{?}{9}$$

$\frac{c}{9}$ is already written $\frac{2}{3} \cdot \frac{3}{3} = \frac{6}{9}$
with the LCD.

The LCD of $\frac{c}{9}$ and $\frac{2}{3}$ is 9, so $\frac{2}{3} = \frac{6}{9}$ and $\frac{c}{9}$ remains the same.

c) The denominators of $\frac{25}{36}$ and $\frac{17}{40}$ are large, so let's use the *prime factorizations* of 36 and 40 to find their least common multiple.

Prime factorizations: $36 = 2 \cdot 2 \cdot 3 \cdot 3$ $40 = 2 \cdot 2 \cdot 2 \cdot 5$

LCM of 36 and 40: $2 \cdot 2 \cdot 2 \cdot 3 \cdot 3 \cdot 5 = 360$

The LCD of $\frac{25}{36}$ and $\frac{17}{40}$ is 360.

Write each fraction with the LCD.

$$\frac{25}{36} = \frac{?}{360} \quad \text{and} \quad \frac{17}{40} = \frac{?}{360}$$

If you do not know, by inspection, what to multiply each fraction by to obtain the fraction with the LCD, go back to each prime factorization. Compare the prime factorizations of the denominators of $\frac{25}{36}$ and $\frac{17}{40}$ to the prime factorization of the LCD of 360 and ask yourself, *"What's missing?"*

Compare

$$36 = 2 \cdot 2 \cdot 3 \cdot 3$$

to

$$360 = 2 \cdot 2 \cdot 2 \cdot 3 \cdot 3 \cdot 5$$

These factors are *not* in 36.

What factors are in 360 that are missing from 36?   2 and 5

Because $2 \cdot 5 = 10$, multiply

$$\frac{25}{36} \cdot \frac{10}{10} = \frac{250}{360}$$

to obtain an equivalent fraction with a denominator of 360.

Compare

$$40 = 2 \cdot 2 \cdot 2 \cdot 5$$

to

$$360 = 2 \cdot 2 \cdot 2 \cdot 3 \cdot 3 \cdot 5$$

These factors are *not* in 40.

What factors are in 360 that are missing from 40? *Two* factors of 3

Because $3 \cdot 3 = 9$, multiply

$$\frac{17}{40} \cdot \frac{9}{9} = \frac{153}{360}$$

to obtain an equivalent fraction with a denominator of 360.

The LCD of $\frac{25}{36}$ and $\frac{17}{40}$ is 360, and $\frac{25}{36} = \frac{250}{360}$ and $\frac{17}{40} = \frac{153}{360}$.

[**YOU TRY 7**] For each group of fractions, identify the least common denominator, then write each fraction as an equivalent fraction with the LCD as its denominator.

a) $\frac{4}{5}$ and $\frac{2}{3}$ b) $\frac{w}{12}$ and $\frac{3}{4}$ c) $\frac{59}{84}$ and $\frac{41}{126}$

When a denominator contains a variable, use prime factorization to find the LCD.

EXAMPLE 8

Identify the least common denominator for $\frac{3}{8}$ and $\frac{1}{6r}$, then write each fraction as an equivalent fraction with the LCD as its denominator.

Solution

Begin by writing the prime factorizations of 8 and $6r$. Remember, the least common denominator will contain each different factor the *greatest* number of times it appears in any single factorization.

$$8 = 2 \cdot 2 \cdot 2 \quad \leftarrow \text{The LCD will contain three factors of 2.}$$

$$6r = 2 \cdot 3 \cdot r \quad \text{The LCD will contain one factor of 3 and one factor of } r.$$

This is only one factor of 2.

The LCD of $\dfrac{3}{8}$ and $\dfrac{1}{6r}$ is the product $2 \cdot 2 \cdot 2 \cdot 3 \cdot r = 24r$.

$$\text{LCD} = \mathbf{24r}$$

$$\frac{3}{8} = \frac{?}{24r} \qquad \text{and} \qquad \frac{1}{6r} = \frac{?}{24r}$$

$$\frac{3}{8} \cdot \frac{3r}{3r} = \frac{9r}{24r} \qquad\qquad \frac{1}{6r} \cdot \frac{4}{4} = \frac{4}{24r}$$

Therefore, $\dfrac{3}{8} = \dfrac{9r}{24r}$ and $\dfrac{1}{6r} = \dfrac{4}{24r}$.

[YOU TRY 8]

Identify the least common denominator for $\dfrac{1}{10}$ and $\dfrac{3}{4z}$, then write each fraction as an equivalent fraction with the LCD as its denominator.

Note

Remember that the denominator of a fraction cannot equal 0. Therefore, in Example 8, we assume that r does not equal 0.

ANSWERS TO [YOU TRY] EXERCISES

1) a)

$$\frac{2}{7} \qquad + \qquad \frac{3}{7} \qquad = \qquad \frac{5}{7}$$

b) $\dfrac{2}{7} + \dfrac{3}{7} = \dfrac{5}{7}$

2) a) $\dfrac{10}{11}$ b) $\dfrac{6}{13}$ c) $\dfrac{3}{8}$ d) $-\dfrac{1}{3}$ e) $\dfrac{4}{5k}$ 3) $\dfrac{28}{32}$

4) a) $\dfrac{6}{11} = \dfrac{42}{77}$ b) $-\dfrac{1}{15} = -\dfrac{3}{45}$ c) $\dfrac{2}{9} = \dfrac{2z}{9z}$ 5) 8   6) a) 105   b) 540

7) a) LCD = 15; $\dfrac{4}{5} = \dfrac{12}{15}, \dfrac{2}{3} = \dfrac{10}{15}$ b) LCD = 12; $\dfrac{w}{12}$ is already written with the LCD, $\dfrac{3}{4} = \dfrac{9}{12}$

c) LCD = 252; $\dfrac{59}{84} = \dfrac{177}{252}, \dfrac{41}{126} = \dfrac{82}{252}$ 8) LCD = 20z; $\dfrac{1}{10} = \dfrac{2z}{20z}, \dfrac{3}{4z} = \dfrac{15}{20z}$

E Evaluate   **3.4**   Exercises   Do the exercises, and check your work.

Objective 1: Add and Subtract Like Fractions

Add the like fractions. First, shade the figures appropriately to represent each fraction and the resulting sum. Then, use a number line to demonstrate the operation.

1) $\dfrac{3}{6} + \dfrac{2}{6}$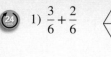

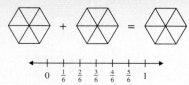

2) $\dfrac{1}{8} + \dfrac{6}{8}$

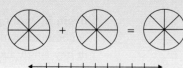

3) $\dfrac{2}{5} + \dfrac{2}{5}$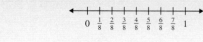

4) $\dfrac{3}{6}+\dfrac{3}{6}$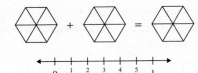

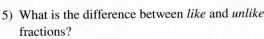

5) What is the difference between *like* and *unlike* fractions?

6) Explain, in your own words, how to add like fractions.

7) After adding fractions, you should look at your answer and ask yourself what question?

8) When Khinsi adds $\dfrac{1}{10}+\dfrac{3}{10}$, she writes down $\dfrac{4}{10}$ as her final answer. Is this correct? Why or why not?

Add or subtract.

9) $\dfrac{3}{11}+\dfrac{5}{11}$

10) $\dfrac{3}{7}+\dfrac{1}{7}$

11) $\dfrac{1}{4}+\dfrac{1}{4}$

12) $\dfrac{1}{10}+\dfrac{1}{10}$

13) $\dfrac{12}{13}-\dfrac{8}{13}$

14) $\dfrac{14}{15}-\dfrac{7}{15}$

15) $\dfrac{9}{10}-\dfrac{3}{10}$

16) $\dfrac{11}{14}-\dfrac{5}{14}$

17) $-\dfrac{8}{7}+\dfrac{3}{7}$

18) $-\dfrac{9}{5}+\dfrac{6}{5}$

19) $-\dfrac{12}{5}+\left(-\dfrac{3}{5}\right)$

20) $-\dfrac{20}{13}+\left(-\dfrac{6}{13}\right)$

21) $\dfrac{10}{9}-\dfrac{13}{9}$

22) $\dfrac{3}{8}-\dfrac{5}{8}$

23) $\dfrac{7}{20}+\dfrac{9}{20}+\dfrac{4}{20}$

24) $\dfrac{6}{28}+\dfrac{17}{28}+\dfrac{5}{28}$

25) $\dfrac{1}{9y}+\dfrac{4}{9y}$

26) $\dfrac{3}{11r}+\dfrac{7}{11r}$

27) $\dfrac{2}{5c}-\dfrac{1}{5c}$

28) $\dfrac{4}{7k}-\dfrac{2}{7k}$

29) $\dfrac{11}{12z}-\dfrac{5}{12z}$

30) $\dfrac{13}{18x}-\dfrac{7}{18x}$

31) $\dfrac{a}{4}+\dfrac{3}{4}$

32) $\dfrac{m}{5}+\dfrac{2}{5}$

Fill in the blank with a fraction to make both sides of the equation equal.

33) $\dfrac{3}{7}+$ _____ $=\dfrac{5}{7}$

34) $\dfrac{1}{9}+$ _____ $=\dfrac{5}{9}$

35) $\dfrac{7}{3}-$ _____ $=\dfrac{1}{3}$

36) $\dfrac{7}{4}-$ _____ $=\dfrac{3}{4}$

Objective 2: Write Equivalent Fractions
In Exercises 37–40, a fraction is given. Use the number lines below to identify its equivalent fraction(s).

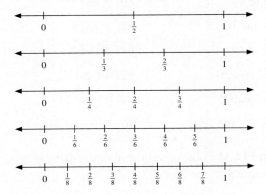

37) $\dfrac{1}{3}$

38) $\dfrac{3}{4}$

39) $\dfrac{4}{8}$

40) $\dfrac{1}{2}$

Write each fraction with the indicated denominator.

41) $\dfrac{1}{8}=\dfrac{?}{32}$

42) $\dfrac{5}{6}=\dfrac{?}{18}$

43) $\dfrac{7}{9}=\dfrac{?}{27}$

44) $\dfrac{7}{8}=\dfrac{?}{48}$

45) $-\dfrac{2}{3}=-\dfrac{?}{36}$

46) $-\dfrac{1}{12}=-\dfrac{?}{60}$

47) $-\dfrac{11}{15}=-\dfrac{?}{45}$

48) $-\dfrac{5}{12}=-\dfrac{?}{84}$

49) $\dfrac{15}{16}=\dfrac{?}{96}$

50) $\dfrac{13}{14}=\dfrac{?}{168}$

51) $\dfrac{4}{7}=\dfrac{?}{7p}$

52) $\dfrac{2}{3}=\dfrac{?}{3d}$

53) $-\dfrac{5}{2}=-\dfrac{?}{8k}$

54) $-\dfrac{9}{4}=-\dfrac{?}{20x}$

55) $\dfrac{c}{5}=\dfrac{?}{35}$

56) $\dfrac{n}{6}=\dfrac{?}{54}$

Objective 3: Find the Least Common Multiple (LCM)

57) In your own words, define the *least common multiple* of a group of numbers.

58) Is this statement true or false? Explain your answer. *The least common multiple of 6 and 12 is 24.*

Find the least common multiple of each group of numbers by inspection or by making a list of multiples of both numbers.

59) 3 and 12

60) 5 and 15

61) 4 and 10

62) 6 and 8

63) 3 and 7

64) 5 and 9

65) 12 and 20

66) 10 and 15

67) 15 and 25

68) 16 and 24

69) 3, 12, and 18

70) 4, 9, and 24

Find the LCM of each group of numbers using prime factorization.

71) 8 and 20

72) 9 and 15

73) 12 and 54

74) 16 and 28

75) 45 and 60

76) 45 and 54

77) 8, 15, and 36

78) 6, 16, and 18

79) 12, 21, and 56

80) 18, 24, and 40

Objective 4: Rewrite Fractions with the Least Common Denominator

81) What is the least common denominator of a group of fractions?

82) Dana says that the least common denominator of $\frac{5}{8}$ and $\frac{1}{6}$ is 48, but Inez says that it is 24. Who is right?

Identify the least common denominator of each group of fractions. Then, write each as an equivalent fraction with the LCD as its denominator.

83) $\frac{4}{7}$ and $\frac{3}{14}$

84) $\frac{3}{4}$ and $\frac{11}{12}$

85) $\frac{3}{8}$ and $\frac{5}{12}$

86) $\frac{5}{6}$ and $\frac{3}{10}$

87) $-\frac{7}{8}$ and $-\frac{5}{6}$

88) $-\frac{4}{9}$ and $-\frac{3}{4}$

89) $\frac{31}{54}$ and $\frac{2}{9}$

90) $\frac{13}{48}$ and $\frac{5}{12}$

91) $\frac{7}{12}$ and $\frac{3}{20}$

92) $\frac{3}{10}$ and $\frac{4}{15}$

93) $-\frac{19}{28}$ and $-\frac{7}{18}$

94) $-\frac{31}{42}$ and $-\frac{11}{24}$

95) $\frac{1}{52}$ and $\frac{42}{65}$

96) $\frac{1}{36}$ and $\frac{29}{66}$

97) $\frac{w}{10}$ and $\frac{3}{5}$

98) $\frac{a}{12}$ and $\frac{2}{3}$

99) $\frac{1}{6}$ and $\frac{x}{9}$

100) $\frac{1}{8}$ and $\frac{n}{10}$

101) $\frac{3}{8}$ and $\frac{5}{8c}$

102) $\frac{4}{7}$ and $\frac{2}{7b}$

103) $\frac{9}{10w}$ and $\frac{1}{6}$

104) $\frac{7}{12k}$ and $\frac{4}{9}$

105) $\frac{5}{12}, \frac{5}{6},$ and $\frac{4}{9}$

106) $\frac{1}{20}, \frac{4}{5},$ and $\frac{3}{8}$

107) $\frac{7}{10}, \frac{3}{4},$ and $\frac{1}{5}$

108) $\frac{5}{8}, \frac{8}{9},$ and $\frac{1}{6}$

109) Write two fractions, with different denominators, that have an LCD of 20.

110) Write two fractions, with different denominators, that have an LCD of 24.

R Rethink

R1) Why do you need common denominators to add and subtract fractions?

R2) How are finding a least common multiple and a least common denominator similar?

R3) In your own words, summarize how to find a least common denominator for two fractions.

3.5 Adding and Subtracting Unlike Fractions

P. Prepare

O Organize

What are your objectives for Section 3.5?	How can you accomplish each objective?
1 Add and Subtract Unlike Fractions	• Write the procedure for **Adding or Subtracting Unlike Fractions** in your own words, and add steps from previous sections as needed. • Be sure that the answer is in lowest terms! • Complete the given examples on your own. • Complete You Trys 1–3.
2 Add and Subtract Fractions with Large Denominators	• Use prime factorization to find the LCD if needed. • Complete the given example on your own. • Complete You Try 4.
3 Add and Subtract Integers and Fractions	• Follow Example 5 and outline a procedure for this objective. • Complete the given example on your own. • Complete You Try 5.
4 Solve Applied Problems Involving Adding and Subtracting Fractions	• Recognize when to add, subtract, multiply, or divide to solve a problem. • Complete the given examples on your own. • Complete You Trys 6 and 7.

W Work   **Read the explanations, follow the examples, take notes, and complete the You Trys.**

1 Add and Subtract Unlike Fractions

In this section, we will learn how to add and subtract unlike fractions. Let's think about the addition problem $\frac{1}{2} + \frac{1}{4}$.

As the fractions are written now, we cannot add them because their denominators are different. How *can* we add them? Let's look at some figures as well as some number lines.

We can represent $\frac{1}{2} + \frac{1}{4}$ like this:

We cannot add halves and fourths because they are not equally-sized pieces of the circles.

If we split the first circle in half again, it will be divided into four equal parts just like the second circle. Now, $\frac{2}{4}$ of the first circle is shaded, and we can add $\frac{2}{4} + \frac{1}{4}$ because fourths are equally-sized pieces of the circle.

$$\bigcirc + \bigcirc = \bigcirc$$

$$\frac{2}{4} \quad + \quad \frac{1}{4} \quad = \quad \frac{3}{4}$$

$\frac{2}{4}$ is equivalent to $\frac{1}{2}$.

Therefore, $\frac{1}{2} + \frac{1}{4} = \frac{2}{4} + \frac{1}{4} = \frac{3}{4}$.

Let's see how we can add $\frac{1}{2} + \frac{1}{4}$ using number lines.

In order to add $\frac{1}{2} + \frac{1}{4}$ using a number line, the denominators must be the same. Do you see on the number lines that $\frac{1}{2} = \frac{2}{4}$?

To add $\frac{1}{2} + \frac{1}{4}$ on a number line, write $\frac{1}{2}$ as $\frac{2}{4}$. Then, use the number line where the space between each tick mark is $\frac{1}{4}$ unit to add the fractions.

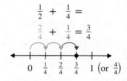

In order to add $\frac{1}{2} + \frac{1}{4}$ (or any fractions), they must be *like* fractions. In other words, the fractions **must** have a common denominator. The same is true for subtraction.

Note

In order to add or subtract fractions, they **must** have a common denominator.

W Hint

After finding the sum or difference, look at the result and ask yourself, *"Is the answer in lowest terms?"* If not, write it in lowest terms.

Procedure Adding or Subtracting Unlike Fractions

Step 1: Determine the least common denominator (LCD), and write it on your paper.

Step 2: Rewrite each fraction with the LCD.

Step 3: Add or subtract.

Step 4: Write the answer in lowest terms.

EXAMPLE 1

Add $\frac{2}{3} + \frac{1}{6}$.

Solution

Step 1: Write down the LCD of $\frac{2}{3}$ and $\frac{1}{6}$. **LCD = 6**

Step 2: Rewrite each fraction with the LCD.

$$\frac{2}{3} \cdot \frac{2}{2} = \frac{4}{6} \qquad \frac{1}{6} \text{ already has a denominator of 6.}$$

> **W Hint**
> Don't just think about the LCD "in your head." **Write it on your paper so you can *look* at it!**

Step 3: Add.

$$\frac{2}{3} + \frac{1}{6} = \frac{4}{6} + \frac{1}{6} = \frac{4+1}{6} = \frac{5}{6}$$

These fractions are equivalent.

Step 4: Ask yourself, *"Is $\frac{5}{6}$ in lowest terms?"* Yes. So, $\frac{5}{6}$ is the final answer.

[YOU TRY 1] Add $\frac{1}{2} + \frac{1}{12}$.

EXAMPLE 2

Add.

a) $\quad -\frac{5}{12} + \frac{2}{8}$ 　　 b) $\quad \frac{6}{7} + \frac{2}{9x}$ 　　 c) $\quad \frac{2}{15} + \frac{1}{6} + \frac{3}{10}$

Solution

a) **Step 1:** Write down the LCD of $-\frac{5}{12}$ and $\frac{2}{8}$. **LCD = 24**

Step 2: Rewrite each fraction with the LCD.

$$-\frac{5}{12} \cdot \frac{2}{2} = -\frac{10}{24} \qquad\qquad \frac{2}{8} \cdot \frac{3}{3} = \frac{6}{24}$$

> **W Hint**
> Remember the rules for adding positive and negative numbers.

Step 3: Add.

Equivalent fractions

$$-\frac{5}{12} + \frac{2}{8} = -\frac{10}{24} + \frac{6}{24} = \frac{-10}{24} + \frac{6}{24} = \frac{-10+6}{24} = \frac{-4}{24} \text{ or } -\frac{4}{24}$$

Equivalent fractions

Step 4: Ask yourself, *"Is $-\frac{4}{24}$ in lowest terms?"* No! Write it in lowest terms.

$$-\frac{4}{24} = -\frac{4 \div 4}{24 \div 4} = -\frac{1}{6}$$

The final answer is $-\frac{1}{6}$.

b) **Step 1:** Write down the LCD of $\dfrac{6}{7}$ and $\dfrac{2}{9x}$.   **LCD = 63x**

Step 2: Rewrite each fraction with the LCD.

$$\dfrac{6}{7} \cdot \dfrac{9x}{9x} = \dfrac{54x}{63x} \qquad \dfrac{2}{9x} \cdot \dfrac{7}{7} = \dfrac{14}{63x}$$

Step 3: Add.

Equivalent fractions

$$\dfrac{6}{7} + \dfrac{2}{9x} = \dfrac{54x}{63x} + \dfrac{14}{63x} = \dfrac{54x + 14}{63x}$$

Equivalent fractions

Step 4: Ask yourself, *"Is $\dfrac{54x + 14}{63x}$ in lowest terms?"* Yes. The final answer is $\dfrac{54x + 14}{63x}$.

c) **Step 1:** Write down the LCD of $\dfrac{2}{15}$, $\dfrac{1}{6}$, and $\dfrac{3}{10}$.   **LCD = 30**

Step 2: Rewrite each fraction with the LCD.

$$\dfrac{2}{15} \cdot \dfrac{2}{2} = \dfrac{4}{30} \qquad \dfrac{1}{6} \cdot \dfrac{5}{5} = \dfrac{5}{30} \qquad \dfrac{3}{10} \cdot \dfrac{3}{3} = \dfrac{9}{30}$$

Step 3: Add.

Equivalent fractions

$$\dfrac{2}{15} + \dfrac{1}{6} + \dfrac{3}{10} = \dfrac{4}{30} + \dfrac{5}{30} + \dfrac{9}{30} = \dfrac{4 + 5 + 9}{30} = \dfrac{18}{30}$$

Equivalent fractions   Equivalent fractions

Step 4: Ask yourself, *"Is $\dfrac{18}{30}$ in lowest terms?"* No! Write it in lowest terms.

$$\dfrac{18}{30} = \dfrac{18 \div 6}{30 \div 6} = \dfrac{3}{5}$$

The final answer is $\dfrac{3}{5}$.

[YOU TRY 2]   Add.

a) $-\dfrac{4}{9} + \dfrac{2}{6}$ b) $\dfrac{5}{9} + \dfrac{3}{8y}$ c) $\dfrac{1}{10} + \dfrac{7}{15} + \dfrac{1}{4}$

We must write fractions with their least common denominator before subtracting them.

EXAMPLE 3	Subtract.

a) $\dfrac{9}{10} - \dfrac{2}{5}$ b) $\dfrac{5}{12} - \dfrac{7}{9}$

Solution

a) **Step 1:** Write down the LCD of $\dfrac{9}{10}$ and $\dfrac{2}{5}$. **LCD = 10**

Step 2: Rewrite each fraction with the LCD.

$\dfrac{9}{10}$ already has the LCD. $\dfrac{2}{5} \cdot \dfrac{2}{2} = \dfrac{4}{10}$

Step 3: Subtract.

> **W Hint**
>
> Are you writing out the steps in the examples as you are reading them?

$$\dfrac{9}{10} - \dfrac{2}{5} = \dfrac{9}{10} - \dfrac{4}{10} = \dfrac{9-4}{10} = \dfrac{5}{10}$$

These fractions are equivalent.

Step 4: Ask yourself, *"Is $\dfrac{5}{10}$ in lowest terms?"* No! Write it in lowest terms.

$$\dfrac{5}{10} = \dfrac{5 \div 5}{10 \div 5} = \dfrac{1}{2}$$

The final answer is $\dfrac{1}{2}$.

b) **Step 1:** Write down the LCD of $\dfrac{5}{12}$ and $\dfrac{7}{9}$. **LCD = 36**

Step 2: Rewrite each fraction with the LCD.

$$\dfrac{5}{12} \cdot \dfrac{3}{3} = \dfrac{15}{36} \qquad \dfrac{7}{9} \cdot \dfrac{4}{4} = \dfrac{28}{36}$$

Step 3: Subtract.

Equivalent fractions

$$\dfrac{5}{12} - \dfrac{7}{9} = \dfrac{15}{36} - \dfrac{28}{36} = \dfrac{15-28}{36} = \dfrac{-13}{36} \text{ or } -\dfrac{13}{36}$$

Equivalent fractions

Step 4: Ask yourself, *"Is $-\dfrac{13}{36}$ in lowest terms?"* Yes.

[YOU TRY 3]	Subtract.

a) $\dfrac{11}{12} - \dfrac{3}{4}$ b) $\dfrac{2}{7} - \dfrac{5}{6}$

2 Add and Subtract Fractions with Large Denominators

If we are asked to add or subtract fractions with large denominators, remember that we can find the least common denominator using prime factorization.

EXAMPLE 4

Add $\dfrac{19}{60} + \dfrac{22}{45}$. Find the LCD using prime factorization.

Solution

Step 1: Find the LCD of $\dfrac{19}{60}$ and $\dfrac{22}{45}$ by first finding the prime factorizations of 60 and 45.

$$60 = 2 \cdot 2 \cdot 3 \cdot 5 \qquad 45 = 3 \cdot 3 \cdot 5$$

The LCD is $2 \cdot 2 \cdot 3 \cdot 3 \cdot 5 = 180$.

W Hint

When the denominators are large, making a list of multiples of the larger number is also a good way to find the LCD.

Step 2: Rewrite each fraction with the LCD.

$$\frac{19}{60} \cdot \frac{3}{3} = \frac{57}{180} \qquad \frac{22}{45} \cdot \frac{4}{4} = \frac{88}{180}$$

Step 3: Add.

Equivalent fractions

$$\frac{19}{60} + \frac{22}{45} = \frac{57}{180} + \frac{88}{180} = \frac{57 + 88}{180} = \frac{145}{180}$$

Equivalent fractions

Step 4: Ask yourself, *"Is $\dfrac{145}{180}$ in lowest terms?"* No! Write it in lowest terms.

$$\frac{145}{180} = \frac{145 \div 5}{180 \div 5} = \frac{29}{36}$$

The final answer is $\dfrac{29}{36}$.

[YOU TRY 4]

Add $\dfrac{5}{18} + \dfrac{3}{28}$. Find the LCD using prime factorization.

3 Add and Subtract Integers and Fractions

Sometimes, we must add or subtract a group of numbers that includes fractions as well as integers.

EXAMPLE 5

Add or subtract by first changing the integer to an improper fraction.

a) $1 - \dfrac{2}{5}$ b) $2 + \dfrac{n}{8}$

Solution

a) Write 1 as an improper fraction with a denominator of 1: $1 = \dfrac{1}{1}$

We can write the difference as $1 - \dfrac{2}{5} = \dfrac{1}{1} - \dfrac{2}{5}$.

Now, follow the steps for subtracting unlike fractions to find $\dfrac{1}{1} - \dfrac{2}{5}$.

Step 1: Write down the LCD of $\dfrac{1}{1}$ and $\dfrac{2}{5}$.   **LCD = 5**

Step 2: Rewrite each fraction with the LCD.

$$\dfrac{1}{1} \cdot \dfrac{5}{5} = \dfrac{5}{5} \qquad \dfrac{2}{5} \text{ is already written with the LCD.}$$

Step 3: Subtract.

Equivalent fractions

$$1 - \dfrac{2}{5} = \dfrac{1}{1} - \dfrac{2}{5} = \dfrac{5}{5} - \dfrac{2}{5} = \dfrac{3}{5}$$

Write 1 with a
denominator of 1.

Step 4: Ask yourself, "*Is $\dfrac{3}{5}$ in lowest terms?*" Yes.

b) Write 2 as an improper fraction with a denominator of 1: $2 = \dfrac{2}{1}$

We can write the sum as $2 + \dfrac{n}{8} = \dfrac{2}{1} + \dfrac{n}{8}$.

Write 2 as a fraction with a denominator of 1.

Now, follow the steps for adding unlike fractions to find $\dfrac{2}{1} + \dfrac{n}{8}$.

Step 1: Write down the LCD of $\dfrac{2}{1}$ and $\dfrac{n}{8}$.   **LCD = 8**

Step 2: Rewrite each fraction with the LCD.

$$\dfrac{2}{1} \cdot \dfrac{8}{8} = \dfrac{16}{8} \qquad \dfrac{n}{8} \text{ is already written with the LCD.}$$

Step 3: Add.

Equivalent fractions

$$2 + \dfrac{n}{8} = \dfrac{2}{1} + \dfrac{n}{8} = \dfrac{16}{8} + \dfrac{n}{8} = \dfrac{16 + n}{8}$$

Write 2 as $\dfrac{2}{1}$.

Step 4: Ask yourself, "*Is $\dfrac{16 + n}{8}$ in lowest terms?*" Yes. The final answer is $\dfrac{16 + n}{8}$.

W Hint

Write a procedure in your own words for this objective.

Note

$\dfrac{16 + n}{8}$ is in simplest form. Because the numerator is a *sum* of terms, we **cannot** divide out a common factor.

$$\dfrac{16 + n}{8} = \dfrac{\overset{2}{16} + n}{\underset{1}{8}} = 2 + n \quad \text{is \textbf{incorrect}.}$$

[YOU TRY 5] Add or subtract by first changing the integer to an improper fraction.

a) $1 - \dfrac{5}{8}$ b) $4 + \dfrac{a}{3}$

Many applications involve adding and subtracting fractions.

4 Solve Applied Problems Involving Adding and Subtracting Fractions

EXAMPLE 6

Mariah has a piece of craft wire that is $\dfrac{7}{8}$ ft long. She cuts off a piece that is $\dfrac{2}{3}$ ft long to make an ornament. How much wire remains?

Solution

Let's draw a picture to show what is happening in the problem.

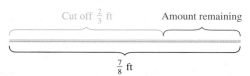

How do we find the amount remaining? Subtract $\dfrac{2}{3}$ from $\dfrac{7}{8}$.

$$\dfrac{7}{8} - \dfrac{2}{3} = \dfrac{21}{24} - \dfrac{16}{24} \qquad \text{Rewrite each fraction with the LCD.}$$

$$= \dfrac{5}{24} \qquad \text{Subtract.}$$

W Hint

Remember, drawing a picture can help you visualize what is happening in an application problem!

The amount of wire remaining is $\dfrac{5}{24}$ ft.

Check the answer.

Amount of wire cut off		Amount of wire remaining		= Total length of wire
$\dfrac{2}{3}$	$+$	$\dfrac{5}{24}$	$= \dfrac{16}{24} + \dfrac{5}{24} = \dfrac{21}{24} =$	$\dfrac{7}{8}$

The total length of the wire is $\dfrac{7}{8}$ ft. The answer is correct.

[YOU TRY 6]

A carpenter has a piece of window trim that is $\frac{11}{12}$ yd long. He cuts off a piece that is $\frac{3}{4}$ yd long. How much window trim remains?

We can read information from a chart to solve problems.

EXAMPLE 7

At an elementary school, many students come from homes where English is not the primary language. The pie chart shows the fractions of the 576 students who speak a given language at home.

a) What is the fraction of students who speak English or Chinese at home?

b) How many students speak English or Chinese at home?

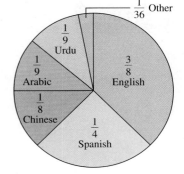

Language Spoken at Home as a Fraction of the Student Body

Solution

a) The $\frac{3}{8}$ in the English section means that $\frac{3}{8}$ of the students speak English at home. The $\frac{1}{8}$ in the Chinese section means that $\frac{1}{8}$ of the students speak Chinese at home. *Add the fractions* to determine the fraction of students who speak English or Chinese at home.

Fraction who speak English	+	Fraction who speak Chinese	=	Fraction who speak English or Chinese
$\frac{3}{8}$	$+$	$\frac{1}{8}$	$=$	$\frac{4}{8}$

W Hint

Recognize that all the pieces add up to 1 in this type of chart.

Write $\frac{4}{8}$ in lowest terms: $\frac{4}{8} = \frac{4 \div 4}{8 \div 4} = \frac{1}{2}$

$\frac{1}{2}$ of the students speak English or Chinese at home.

b) In part a), we determined that $\frac{1}{2}$ *of* the total number of students speak English or Chinese at home. The *of* indicates multiplication. To determine the *number* of students who speak English or Chinese at home, find $\frac{1}{2}$ *of* 576 students or $\frac{1}{2} \cdot 576$.

$$\frac{1}{2} \cdot 576 = \frac{1}{2} \cdot \frac{576}{1} = \frac{1}{2} \cdot \frac{\overset{288}{\cancel{576}}}{1} = \frac{288}{1} = 288$$

Write 576 with a denominator of 1.

288 students speak English or Chinese at home.

[YOU TRY 7]

Use the pie chart in Example 7 to answer these questions.

a) What is the fraction of students who speak English, Urdu, or "other" language at home?

b) How many students speak English, Urdu, or "other" language at home?

E Evaluate   **3.5**   Exercises   Do the exercises, and check your work.

Get Ready

Add or subtract. Write the answer in lowest terms.

1) $\dfrac{2}{7} + \dfrac{4}{7}$ 2) $\dfrac{9}{11} - \dfrac{5}{11}$

3) $\dfrac{14}{15} - \dfrac{8}{15}$ 4) $\dfrac{5}{18} + \dfrac{7}{18}$

5) $-\dfrac{21}{40} - \left(-\dfrac{13}{40}\right)$ 6) $-\dfrac{10}{27} + \left(-\dfrac{11}{27}\right)$

Objective 1: Add and Subtract Unlike Fractions

For Exercises 7–10, use the number lines below to identify the least common denominator, and add the fractions on the appropriate number line.

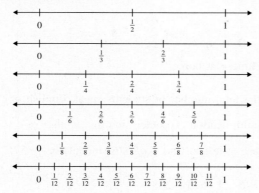

7) $\dfrac{1}{2}$ and $\dfrac{1}{3}$ 8) $\dfrac{1}{3}$ and $\dfrac{1}{4}$

9) $\dfrac{5}{12}$ and $\dfrac{1}{4}$ 10) $\dfrac{3}{4}$ and $\dfrac{1}{12}$

11) In your own words, explain how to add and subtract unlike fractions.

12) Diane adds these fractions this way: $\dfrac{1}{5} + \dfrac{3}{4} = \dfrac{4}{9}$.

   What did she do wrong? What is the right way to add these fractions?

Add.

13) $\dfrac{1}{3} + \dfrac{5}{12}$ 14) $\dfrac{1}{6} + \dfrac{7}{18}$

15) $\dfrac{1}{2} + \dfrac{2}{3}$ 16) $\dfrac{2}{3} + \dfrac{2}{5}$

17) $\dfrac{4}{9} + \dfrac{15}{27}$ 18) $\dfrac{3}{8} + \dfrac{35}{56}$

19) $-\dfrac{5}{12} + \dfrac{3}{8}$ 20) $-\dfrac{2}{9} + \dfrac{1}{6}$

21) $\dfrac{5}{6} + \left(-\dfrac{9}{20}\right)$ 22) $\dfrac{9}{10} + \left(-\dfrac{11}{15}\right)$

23) $\dfrac{1}{2} + \dfrac{1}{4} + \dfrac{1}{8}$ 24) $\dfrac{1}{3} + \dfrac{1}{6} + \dfrac{1}{9}$

25) $\dfrac{7}{x} + \dfrac{4}{5}$ 26) $\dfrac{8}{d} + \dfrac{3}{7}$

27) $\dfrac{11}{3} + \dfrac{1}{2k}$ 28) $\dfrac{7}{2} + \dfrac{1}{5n}$

29) $\dfrac{a}{12} + \dfrac{8}{9}$ 30) $\dfrac{w}{8} + \dfrac{5}{6}$

31) $\dfrac{6}{c^2} + \dfrac{7}{c}$ 32) $\dfrac{3}{y^2} + \dfrac{10}{y}$

Subtract.

33) $\dfrac{6}{7} - \dfrac{13}{21}$ 34) $\dfrac{3}{4} - \dfrac{3}{8}$

35) $\dfrac{5}{8} - \dfrac{7}{24}$ 36) $\dfrac{4}{5} - \dfrac{3}{10}$

37) $\dfrac{2}{5} - \dfrac{8}{9}$ 38) $\dfrac{1}{4} - \dfrac{10}{11}$

39) $\dfrac{11}{15} - \dfrac{3}{10}$ 40) $\dfrac{7}{8} - \dfrac{5}{12}$

41) $-\dfrac{5}{12} - \dfrac{3}{11}$ 42) $-\dfrac{7}{12} - \dfrac{2}{7}$

43) $\dfrac{9}{10} - \dfrac{1}{m}$

44) $\dfrac{11}{16} - \dfrac{1}{w}$

45) $\dfrac{h}{6} - \dfrac{5}{8}$

46) $\dfrac{x}{4} - \dfrac{9}{10}$

Objective 2: Add and Subtract Fractions with Large Denominators

47) When is prime factorization a good method for finding a least common denominator?

48) In addition to prime factorization, what is another good method for finding the LCD of fractions with large denominators?

Add or subtract as indicated. Find the LCD using prime factorization.

49) $\dfrac{14}{45} + \dfrac{17}{30}$

50) $\dfrac{7}{20} + \dfrac{12}{35}$

51) $-\dfrac{23}{42} + \dfrac{13}{24}$

52) $-\dfrac{31}{36} + \dfrac{19}{42}$

53) $\dfrac{11}{24} - \dfrac{9}{28}$

54) $\dfrac{31}{36} - \dfrac{3}{16}$

Objective 3: Add and Subtract Integers and Fractions

Add or subtract by first changing the integer to an improper fraction.

55) $5 + \dfrac{1}{4}$

56) $3 + \dfrac{1}{8}$

57) $4 + \dfrac{a}{7}$

58) $2 + \dfrac{r}{6}$

59) $1 - \dfrac{2}{9}$

60) $1 - \dfrac{3}{4}$

61) $\dfrac{z}{5} - 3$

62) $\dfrac{c}{9} - 8$

63) $-2 + \dfrac{w}{8}$

64) $-4 + \dfrac{x}{7}$

Objective 4: Solve Applied Problems Involving Adding and Subtracting Fractions

Solve the application problem.

65) An electrician has a wire that is $\dfrac{5}{6}$ yd long. He cuts off a piece that is $\dfrac{4}{9}$ yd long. How much wire is left?

66) A plumber has a piece of copper pipe that is $\dfrac{3}{4}$ ft long. She cuts off a piece that is $\dfrac{3}{8}$ ft long. How much of the pipe remains?

67) A surgeon had a 1-ft-long piece of surgical thread for closing an incision. She used $\dfrac{7}{12}$ ft during the surgery. How much of the thread was not used?

68) A computer technician had a 1-in.-long piece of wire, and he used $\dfrac{5}{16}$ in. while making repairs. How much of the wire was not used?

69) A lifeguard qualification event requires that prospective lifeguards be able to complete an open water swim of $\dfrac{1}{5}$ mi and a $\dfrac{1}{2}$-mi run on the sand within 25 min. What is the total distance traveled by the lifeguard applicants?

©Creatas Images/Getty Images

70) Kelsie wears a diamond ring that has a $\dfrac{1}{2}$-carat solitaire diamond with two $\dfrac{1}{8}$-carat diamonds mounted on each side. What is the total carat weight of the ring?

71) If a student spends $\dfrac{1}{3}$ of her scholarship check on tuition, $\dfrac{1}{4}$ on books, and $\dfrac{1}{8}$ on school supplies, what fraction of her scholarship check remains?

72) On Sunday just before finals week, Jason spends $\dfrac{1}{8}$ of his day on Facebook, $\dfrac{1}{6}$ watching TV, and $\dfrac{1}{4}$ playing video games. If Jason slept $\dfrac{1}{3}$ of the day (8 hours), what fraction of his day is left for studying?

73) Fiona drives an electric-powered car. When she sets off for work, the battery has $\dfrac{5}{6}$ of its capacity remaining before it begins using its gas generator. When she arrives at work, $\dfrac{3}{10}$ of the battery's electric capacity remains. What fraction of the battery's electric capacity was used on Fiona's drive to work?

©Martin Pickard/Getty Images

74) At the beginning of Lionel's school day, his laptop computer indicated that his battery was charged to $\frac{7}{8}$ capacity. At the end of the day, the battery was charged to $\frac{1}{3}$ capacity. What fraction of the laptop's battery capacity was used during the day?

A group of 620 college-bound seniors at Chavez High School was asked to indicate which of the five areas shown they might like to study. The pie chart shows the fraction of the 620 seniors that chose a specific area of study. Use this information for Exercises 75–82.

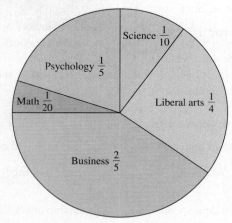

24) 75) What fraction of seniors chose math or science as an area of study?

76) What fraction of seniors chose psychology or liberal arts as an area of study?

24) 77) How many seniors chose math or science as an area of study?

78) How many seniors chose psychology or liberal arts as an area of study?

79) What fraction of seniors did not choose business as an area of study?

80) How many seniors did not choose business as an area of study?

81) Which two areas of study together equal the choice of exactly one-half the senior class?

R Rethink

R1) Were there any problems you could not do? If so, circle them or write them on your paper, and think about what made the problem difficult. Ask your questions in class.

82) Verify that the five fractional parts have a sum of 1.

Mixed Exercises: Objectives 1–4
Perform the indicated operations.

83) $-\frac{5}{12} + \left(-\frac{1}{3}\right)$

84) $1 - \frac{4}{11}$

85) $\frac{13}{15} - \frac{7}{24}$

86) $\frac{7}{12} + \frac{2}{9}$

87) $4 - \frac{7}{10}$

88) $-\frac{31}{42} + \frac{8}{35}$

89) $\frac{k}{7} + 3$

90) $\frac{1}{5} - \frac{6}{m}$

91) $\frac{1}{4} + \frac{7}{3a}$

92) $\frac{x}{8} - 2$

The pie chart represents the end-of-semester grade distribution for a 180-student biology lecture course. Use the pie chart for Exercises 93–96.

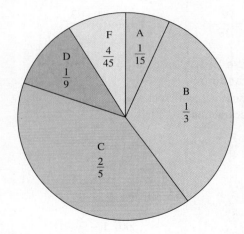

93) What fractional amount of students passed the course with a C or better?

94) What fractional amount of students earned a D or an F?

95) How many students passed the course with a C or better?

96) How many students earned a D or an F in the course?

R2) Create a word problem similar to those in Objective 4 from an experience you have had in the past month; then, solve it.

3.6 Operations with Mixed Numbers

P Prepare

O Organize

What are your objectives for Section 3.6?	How can you accomplish each objective?
1 Understand Mixed Numbers	• Write the definition of a *mixed number* in your own words. • Notice the relationship between improper fractions and mixed numbers. • Complete the given examples on your own. • Complete You Trys 1 and 2.
2 Convert Between Mixed Numbers and Improper Fractions	• In your own words, explain how to change a mixed number to an improper fraction. • In your own words, explain how to change an improper fraction to a mixed number. • Complete the given examples on your own. • Complete You Trys 3 and 4.
3 Multiply and Divide Mixed Numbers	• Write the procedure for **Multiplying Mixed Numbers** in your own words. • Write the procedure for **Dividing Mixed Numbers** in your own words. • Complete the given examples. • Complete You Trys 5 and 6.
4 Add and Subtract Mixed Numbers in Mixed-Number Form	• Write the procedures for **Adding or Subtracting Mixed Numbers** in your own words. • Complete the given example on your own. • Complete You Try 7.
5 Add and Subtract Mixed Numbers with Regrouping	• Write your own explanation of how to add and subtract mixed numbers when the answer needs regrouping. • Complete the given examples on your own. • Complete You Trys 8–12.
6 Add and Subtract Mixed Numbers Using Improper Fractions	• Write the procedure for **Adding and Subtracting Mixed Numbers Using Improper Fractions** in your own words. • Complete the given example on your own. • Complete You Try 13.
7 Solve Applied Problems Involving Mixed Numbers	• Review the notes you took for Section 2.6 on the **Five Steps for Solving Applied Problems,** and compare them to the examples for this objective. • Complete the given examples. • Complete You Trys 14 and 15.

W Work

Read the explanations, follow the examples, take notes, and complete the You Trys.

1 Understand Mixed Numbers

In this section, we will learn how improper fractions and *mixed numbers* are related. What *is* a mixed number?

> ## Definition
> A **mixed number** consists of a whole number and a fraction.
>
> *Example:* $2\frac{1}{2}$ is a mixed number. $2\frac{1}{2}$ means $2 + \frac{1}{2}$.
>
> A mixed number represents more than 1 whole.

In Example 4 of Section 3.1, we were given this figure:

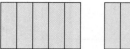

W Hint

Note the relationship between improper fractions and mixed numbers.

$\frac{7}{5}$ of the figure is shaded, and $\frac{7}{5}$ is an improper fraction because it represents a quantity greater than 1. Any fraction that is greater than 1 can be written as a mixed number. How can we write $\frac{7}{5}$ as a mixed number? Let's look at Example 1.

EXAMPLE 1

Use a mixed number to describe what portion of the figure is shaded, and represent the mixed number on a number line.

Solution

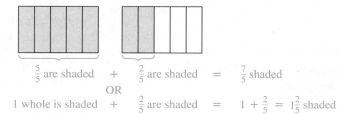

To represent $1\frac{2}{5}$ on a number line, first notice that this number is greater than 1 and less than 2. Draw the number line from 0 to 2. Because $1\frac{2}{5}$ means $1 + \frac{2}{5}$, start at 1 and then divide the region between 1 and 2 into 5 equal parts. Label the tick marks.

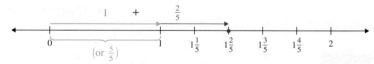

To add $1 + \frac{2}{5}$ start at 0 and move 1 space to the right to reach 1. Then, move $\frac{2}{5}$ of a unit to the right to reach $1\frac{2}{5}$. $\left(\text{This is the same as } \frac{7}{5}.\right)$

[YOU TRY 1] Use a mixed number to describe what portion of the figure is shaded, and represent the mixed number on a number line.

> **Note**
>
> If we are asked to round $1\frac{2}{5}$ to the nearest integer, it is rounded to 1. Notice on the number line in Example 1 that $1\frac{2}{5}$ is closer to 1 than to 2. This is because $\frac{2}{5}$ is *less than* $\frac{1}{2}$. If the fractional part of a positive mixed number is greater than or equal to $\frac{1}{2}$, we round up. For example, $5\frac{2}{3}$ rounds to 6.

Let's practice graphing more mixed numbers on a number line, including negative numbers. Remember that negative numbers are to the left of 0.

EXAMPLE 2 Graph each mixed number on a number line.

a) $4\frac{1}{2}$ b) $-1\frac{3}{4}$

Solution

a) Because $4\frac{1}{2}$ means $4 + \frac{1}{2}$, the mixed number is between 4 and 5. So, draw a number line from 0 to 5. Start at 0 and move to 4. Then, divide the space between 4 and 5 into two equal parts. Move another $\frac{1}{2}$ of a unit.

Move 4 units to the right, then move another $\frac{1}{2}$ of a unit.

Place the dot on $4\frac{1}{2}$.

b) The mixed number $-1\frac{3}{4}$ means $-\left(1\frac{3}{4}\right)$ or $-\left(1 + \frac{3}{4}\right)$ or $-1 + \left(-\frac{3}{4}\right)$ or $-1 - \frac{3}{4}$.

To graph $-1\frac{3}{4}$ on a number line, think of it as $-1 - \frac{3}{4}$. Start at 0 and move *to the left,* to -1. Then, move another $\frac{3}{4}$ of a unit *to the left* so that the number $-1\frac{3}{4}$ is between -1 and -2.

Move 1 unit to the left, then move another $\frac{3}{4}$ of a unit to the left.

Graph each mixed number on a number line.

a) $3\frac{7}{8}$ b) $-2\frac{1}{4}$ c) $-3\frac{1}{2}$

2 Convert Between Mixed Numbers and Improper Fractions

How can we change a mixed number to an improper fraction? The shaded part of this figure can be represented by the mixed number $4\frac{2}{3}$. We can also represent the shaded part with an improper fraction.

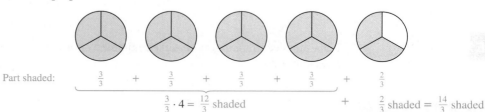

Part shaded: $\frac{3}{3}$ + $\frac{3}{3}$ + $\frac{3}{3}$ + $\frac{3}{3}$ + $\frac{2}{3}$

$\frac{3}{3} \cdot 4 = \frac{12}{3}$ shaded   + $\frac{2}{3}$ shaded $= \frac{14}{3}$ shaded

We can see that $4\frac{2}{3} = \frac{14}{3}$.

Here is another way to change a mixed number to an improper fraction.

W Hint

In your own words, explain the steps you would take to change a mixed number to an improper fraction.

Procedure Change a Mixed Number to an Improper Fraction

Step 1: Multiply the denominator and the whole number.

Step 2: Add the numerator to the result in Step 1.

Step 3: Write the improper fraction. The numerator is the result in Step 2 and the denominator is the original denominator.

EXAMPLE 3

Write each mixed number as an improper fraction.

a) $4\frac{2}{3}$ b) $7\frac{3}{5}$

Solution

a) $4\frac{2}{3}$	b) $7\frac{3}{5}$
Step 1: Multiply the denominator and whole number: $3 \cdot 4 = 12$	**Step 1:** Multiply the denominator and whole number: $5 \cdot 7 = 35$
Step 2: Add the numerator to the result in Step 1: $12 + 2 = 14$	**Step 2:** Add the numerator to the result in Step 1: $35 + 3 = 38$
Step 3: Write the improper fraction: $4\frac{2}{3} = \frac{14}{3}$ ← Numerator found in Step 2 ← Original denominator	**Step 3:** Write the improper fraction: $7\frac{3}{5} = \frac{38}{5}$ ← Numerator found in Step 2 ← Original denominator

[YOU TRY 3] Write each mixed number as an improper fraction.

a) $6\dfrac{1}{2}$ b) $2\dfrac{5}{8}$

To change an improper fraction to a mixed number, follow these steps.

Procedure Change an Improper Fraction to a Mixed Number

Step 1: Divide the numerator by the denominator.

Step 2: Write the mixed number. The *quotient* is the whole-number part. The *remainder* is the numerator of the fractional part, and the denominator is the same as the denominator of the improper fraction.

EXAMPLE 4 Write each improper fraction as a mixed number.

a) $\dfrac{9}{4}$ b) $\dfrac{52}{9}$ c) $\dfrac{48}{6}$

Solution

a) ***Step 1:*** Divide the numerator by the denominator.

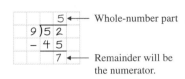

     Step 2: Write the mixed number. $\dfrac{9}{4} = 2\dfrac{1}{4}$

The denominator stays the same.

Let's look at a figure to see that these are the same.

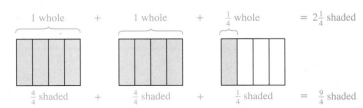

b) ***Step 1:*** Divide the numerator by the denominator.

     Step 2: Write the mixed number. $\dfrac{52}{9} = 5\dfrac{7}{9}$

The denominator stays the same.

c) **Step 1:** Divide the numerator by the denominator.

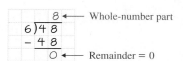

Since $\dfrac{48}{6} = 8$ (with *no* remainder), the improper fraction $\dfrac{48}{6}$ can be written as the whole number 8.

[YOU TRY 4] Write each improper fraction as a mixed number.

    a) $\dfrac{10}{7}$ b) $\dfrac{75}{8}$ c) $\dfrac{30}{5}$

Next, we will learn how to multiply and divide mixed numbers.

3   Multiply and Divide Mixed Numbers

Procedure Multiplying or Dividing Mixed Numbers

To multiply or divide mixed numbers,

1) Change each mixed number to an improper fraction.
2) Multiply or divide the fractions as indicated.
3) Write the answer in lowest terms. If it is an improper fraction, change it to a mixed number.

Note

In general, if we perform operations with mixed numbers and the result is an improper fraction, the final answer should be written as a mixed number.

EXAMPLE 5

Multiply.

    a) $1\dfrac{5}{6} \cdot 2\dfrac{2}{3}$ b) $5\dfrac{1}{3} \cdot 4\dfrac{1}{2}$ c) $-2\dfrac{5}{9} \cdot \dfrac{3}{10}$

Solution

a) First, change each mixed number to an improper fraction.

$$1\dfrac{5}{6} = \dfrac{11}{6} \qquad\qquad 2\dfrac{2}{3} = \dfrac{8}{3}$$

Multiply.

$$1\dfrac{5}{6} \cdot 2\dfrac{2}{3} = \dfrac{11}{6} \cdot \dfrac{8}{3} = \dfrac{11}{\cancel{6}} \cdot \dfrac{\overset{4}{\cancel{8}}}{3} \qquad \text{Divide 6 and 8 by 2.}$$

$$= \dfrac{11 \cdot 4}{3 \cdot 3} \qquad \text{Multiply.}$$

$$= \dfrac{44}{9} = 4\dfrac{8}{9} \qquad \text{Write the final answer as a mixed number.}$$

We can check to see whether the answer is reasonable by first rounding each mixed number to the nearest whole number and then multiplying to *estimate* the product.

$$1\frac{5}{6} \approx 2, \quad 2\frac{2}{3} \approx 3, \quad \text{and} \quad 2 \cdot 3 = 6$$

This is close to the exact answer, $4\frac{8}{9}$.

b) Change each mixed number to an improper fraction.

$$5\frac{1}{3} = \frac{16}{3} \qquad\qquad 4\frac{1}{2} = \frac{9}{2}$$

Multiply.

W Hint

Write the procedure for multiplying mixed numbers in your own words.

$$5\frac{1}{3} \cdot 4\frac{1}{2} = \frac{16}{3} \cdot \frac{9}{2} = \frac{\overset{8}{\cancel{16}}}{\cancel{3}} \cdot \frac{\overset{3}{\cancel{9}}}{\cancel{2}} = \frac{8 \cdot 3}{1 \cdot 1} = \frac{24}{1} = 24 \qquad \text{Simplify } \frac{24}{1} \text{ to } 24.$$

The product is 24. Is the answer reasonable? $5\frac{1}{3} \approx 5, \quad 4\frac{1}{2} \approx 5, \text{ and } 5 \cdot 5 = 25$

Yes. The estimate, 25, is close to the exact answer, 24.

c) To change $-2\frac{5}{9}$ to an improper fraction, remember that $-2\frac{5}{9}$ means $-\left(2\frac{5}{9}\right)$.

Then, we get

$$-2\frac{5}{9} = -\left(2\frac{5}{9}\right) = -\left(\frac{23}{9}\right) = -\frac{23}{9}$$

$$-2\frac{5}{9} = -\frac{23}{9}$$

Multiply: $\quad -2\frac{5}{9} \cdot \frac{3}{10} = -\frac{23}{9} \cdot \frac{3}{10} = -\frac{23}{\underset{3}{\cancel{9}}} \cdot \frac{\overset{1}{\cancel{3}}}{10} = -\frac{23}{30}$

As you can see here, sometimes the product is a *proper* fraction.

[YOU TRY 5] Multiply.

a) $\quad 5\frac{1}{8} \cdot 1\frac{3}{7}$ b) $\quad -3\frac{1}{5} \cdot 2\frac{3}{16}$ c) $\quad 3\frac{1}{18} \cdot \frac{4}{11}$

Now let's divide mixed numbers.

EXAMPLE 6

Divide.

a) $\quad 3\frac{3}{4} \div 2\frac{1}{8}$ b) $\quad 6\frac{1}{4} \div 15$ c) $\quad -12 \div \left(-4\frac{2}{7}\right)$

Solution

a) First, change each mixed number to an improper fraction.

$$3\frac{3}{4} = \frac{15}{4} \qquad 2\frac{1}{8} = \frac{17}{8}$$

Write the division problem using improper fractions.

$$3\frac{3}{4} \div 2\frac{1}{8} = \frac{15}{4} \div \frac{17}{8} \qquad \text{Change the mixed numbers to improper fractions.}$$

$$= \frac{15}{4} \cdot \frac{8}{17} \qquad \text{Change division to multiplication by the reciprocal.}$$

$$= \frac{15}{\overset{}{\underset{1}{\cancel{4}}}} \cdot \frac{\overset{2}{\cancel{8}}}{17} \qquad \text{Divide 4 and 8 by 4.}$$

$$= \frac{30}{17} = 1\frac{13}{17} \qquad \text{Multiply and write the answer as a mixed number.}$$

Therefore, $3\frac{3}{4} \div 2\frac{1}{8} = 1\frac{13}{17}$.

> **W Hint**
>
> Write the procedure for dividing mixed numbers in your own words. Include the word *reciprocal*.

b) Change $6\frac{1}{4}$ to an improper fraction, and write 15 as $\frac{15}{1}$.

$$6\frac{1}{4} = \frac{25}{4} \qquad 15 = \frac{15}{1}$$

Rewrite the division problem using the fractions.

$$6\frac{1}{4} \div 15 = \frac{25}{4} \div \frac{15}{1} = \frac{25}{4} \cdot \frac{1}{15} = \frac{\overset{5}{\cancel{25}}}{4} \cdot \frac{1}{\underset{3}{\cancel{15}}} = \frac{5}{12}$$

Change division to multiplication by the reciprocal.

c) Write -12 as $-\frac{12}{1}$, and change $-4\frac{2}{7}$ to an improper fraction.

$$-12 = -\frac{12}{1} \qquad -4\frac{2}{7} = -\frac{30}{7}$$

Multiply.

$$-12 \div \left(-4\frac{2}{7}\right) = -\frac{12}{1} \div \left(-\frac{30}{7}\right) = -\frac{12}{1} \cdot \left(-\frac{7}{30}\right) = -\frac{\overset{2}{\cancel{12}}}{1} \cdot \left(-\frac{7}{\underset{5}{\cancel{30}}}\right) = \frac{14}{5} = 2\frac{4}{5}$$

Change division to multiplication by the reciprocal.

[YOU TRY 6] Divide.

a) $7\frac{1}{5} \div 2\frac{7}{10}$ b) $11\frac{2}{3} \div 5$ c) $-4 \div \left(-5\frac{1}{5}\right)$

4 Add and Subtract Mixed Numbers in Mixed-Number Form

Now let's learn how to add and subtract mixed numbers.

> **Procedure** Adding or Subtracting Mixed Numbers
>
> To add mixed numbers, add the whole-number parts and add the fractional parts. Write the answer in lowest terms.
>
> To subtract mixed numbers, subtract the whole-number parts and subtract the fractional parts. Write the answer in lowest terms.

EXAMPLE 7

Add or subtract.

a) $2\dfrac{1}{5} + 7\dfrac{3}{10}$　　b) $9\dfrac{3}{4} - 5\dfrac{1}{6}$　　c) $-8\dfrac{11}{18} + 3\dfrac{2}{9}$

Solution

a) Because we will add the whole-number parts and add the fractional parts, begin by writing $\dfrac{1}{5}$ and $\dfrac{3}{10}$ as equivalent fractions with their least common denominator.

The LCD of $\dfrac{1}{5}$ and $\dfrac{3}{10}$ is 10.

$$\dfrac{1}{5}\cdot\dfrac{2}{2} = \dfrac{2}{10} \qquad \dfrac{3}{10} \text{ already has the LCD.}$$

$$2\dfrac{1}{5} + 7\dfrac{3}{10} = 2\dfrac{2}{10} + 7\dfrac{3}{10} \qquad \text{Write the fractions with the LCD.}$$

$$= 9\dfrac{5}{10} \qquad \text{Add whole-number parts, and add fractions.}$$

$$= 9\dfrac{1}{2} \qquad \text{Write } \dfrac{5}{10} \text{ in lowest terms.}$$

Is the answer, $9\dfrac{1}{2}$, reasonable? Let's estimate the answer:

$$2\dfrac{1}{5} + 7\dfrac{3}{10} \approx 2 + 7 = 9 \qquad \text{Estimate.}$$

Yes. The *exact* answer, $9\dfrac{1}{2}$, is close to the estimate, 9.

b) Begin by writing $\frac{3}{4}$ and $\frac{1}{6}$ as equivalent fractions with their LCD.

The LCD of $\frac{3}{4}$ and $\frac{1}{6}$ is 12.

$$\frac{3}{4} \cdot \frac{3}{3} = \frac{9}{12} \qquad \frac{1}{6} \cdot \frac{2}{2} = \frac{2}{12}$$

$$9\frac{3}{4} - 5\frac{1}{6} = 9\frac{9}{12} - 5\frac{2}{12} \qquad \text{Write the fractions with the LCD.}$$

$$= 4\frac{7}{12} \qquad \text{Subtract whole-number parts, and subtract fractions.}$$

Is $4\frac{7}{12}$ in lowest terms? Yes.

> **W Hint**
>
> The rules for adding and subtracting integers apply to fractions and mixed numbers also.

c) To add $-8\frac{11}{18} + 3\frac{2}{9}$, we find the absolute value of each number, then *subtract* the smaller absolute value from the larger absolute value. The final answer will have the same sign as the number with the larger absolute value: $-8\frac{11}{18}$ has the larger absolute value, so *the sum will be negative.*

$$8\frac{11}{18} - 3\frac{2}{9} = 8\frac{11}{18} - 3\frac{4}{18} = 5\frac{7}{18}$$

Get a common denominator.

$$\frac{2}{9} \cdot \frac{2}{2} = \frac{4}{18}$$

Therefore,

$$-8\frac{11}{18} + 3\frac{2}{9} = -5\frac{7}{18} \qquad \text{The final answer is negative because } -8\frac{11}{18} \text{ has the larger absolute value.}$$

> [**YOU TRY 7**] Add or subtract.
>
> a) $6\frac{3}{7} + 7\frac{2}{5}$ b) $8\frac{5}{6} - 3\frac{11}{15}$ c) $-9\frac{13}{16} + 2\frac{3}{4}$

5   Add and Subtract Mixed Numbers with Regrouping

Sometimes, adding mixed numbers involves *regrouping*. Let's learn what regrouping is.

EXAMPLE 8

Explain why $5\frac{7}{4}$ is not in simplest form. Then, write it in simplest form.

Solution

A mixed number is not in simplest form if its fractional part is an improper fraction.

$5\frac{7}{4}$ is not in simplest form because $\frac{7}{4}$ is an improper fraction.

To simplify $5\frac{7}{4}$, write $\frac{7}{4}$ as a mixed number to *regroup* the fractional part with the whole-number part.

$$5\frac{7}{4} = 5 + \frac{7}{4} \qquad \text{Meaning of a mixed number}$$

$$= 5 + 1\frac{3}{4} \qquad \text{Change } \frac{7}{4} \text{ to a mixed number to regroup the whole-number part.}$$

$$= 6\frac{3}{4} \qquad \text{Add.}$$

W Hint

Write a procedure for regrouping mixed numbers in this way.

Written in simplest form, $5\frac{7}{4} = 6\frac{3}{4}$. The fractional part is a *proper* fraction, and it is in lowest terms.

[YOU TRY 8]

Explain why $6\frac{8}{5}$ is not in simplest form. Then, write it in simplest form.

Now, let's add two mixed numbers using regrouping.

EXAMPLE 9

Add $3\frac{7}{9} + 1\frac{1}{2}$.

Solution

Write $\frac{7}{9}$ and $\frac{1}{2}$ as equivalent fractions with their least common denominator, 18.

W Hint

Remember: Write out the examples as you are reading them. That is the best way to read a math book!

$$\frac{7}{9} \cdot \frac{2}{2} = \frac{14}{18} \qquad \frac{1}{2} \cdot \frac{9}{9} = \frac{9}{18}$$

$$3\frac{7}{9} + 1\frac{1}{2} = 3\frac{14}{18} + 1\frac{9}{18} \qquad \text{Write the fractions with the LCD.}$$

$$= 4\frac{23}{18} \qquad \text{Add whole numbers, and add fractions.}$$

Is $4\frac{23}{18}$ in simplest form? *No!* $\frac{23}{18}$ is an improper fraction. Simplify $4\frac{23}{18}$ by regrouping.

$$4\frac{23}{18} = 4 + \frac{23}{18} = 4 + 1\frac{5}{18} = 5\frac{5}{18}$$

Change to a mixed number.

So, $3\frac{7}{9} + 1\frac{1}{2} = 5\frac{5}{18}$.

[YOU TRY 9]

Add $5\frac{3}{4} + 9\frac{5}{7}$.

Now let's learn about the type of regrouping, or *borrowing,* we need to use to subtract some mixed numbers.

EXAMPLE 10

Rewrite $7\dfrac{2}{3}$ as a mixed number with an improper fractional part.

Solution

To rewrite the fractional part as an improper fraction, we will regroup, or borrow, 1 from the whole-number part and add it to the fractional part.

Rewrite 7 as 6 + 1.

$$7\frac{2}{3} = 7 + \frac{2}{3} = 6 + 1 + \frac{2}{3} = 6 + \frac{3}{3} + \frac{2}{3} = 6\frac{5}{3}$$

Get a common denominator for 1 and $\frac{2}{3}$.

W Hint

Write a procedure for regrouping mixed numbers in this way.

Therefore, $7\dfrac{2}{3} = 6\dfrac{5}{3}$.

[YOU TRY 10]

Rewrite $5\dfrac{1}{6}$ as a mixed number with an improper fractional part.

Note

We use this type of regrouping in a subtraction problem when the fraction in the second mixed number is *larger than* the fraction in the first mixed number.

EXAMPLE 11

Subtract $11\dfrac{2}{5} - 4\dfrac{3}{4}$.

Solution

To find the difference $11\dfrac{2}{5} - 4\dfrac{3}{4}$, we must first write the fractional parts with their common denominator, 20.

W Hint

In your own words, explain when we have to regroup to subtract mixed numbers.

$$\frac{2}{5} \cdot \frac{4}{4} = \frac{8}{20} \qquad \frac{3}{4} \cdot \frac{5}{5} = \frac{15}{20}$$

$$11\frac{2}{5} - 4\frac{3}{4} = 11\frac{8}{20} - 4\frac{15}{20} \qquad \text{Write the fractions with the LCD.}$$

Can we perform the subtraction as it is written here? *No!* The second fraction is larger than the first.

Regroup (borrow) 1 from the whole-number part of the first mixed number.

Write 11 as 10 + 1.

$$11\frac{8}{20} = 11 + \frac{8}{20} = 10 + 1 + \frac{8}{20} = 10 + \frac{20}{20} + \frac{8}{20} = 10\frac{28}{20}$$

Get a common denominator for 1 and $\frac{8}{20}$.

Now we can subtract.

$$11\frac{2}{5} - 4\frac{3}{4} = 11\frac{8}{20} - 4\frac{15}{20} \qquad \text{Write the fractions with the LCD.}$$

$$= 10\frac{28}{20} - 4\frac{15}{20} \qquad \text{Rewrite } 11\frac{8}{20} \text{ as } 10\frac{28}{20}.$$

$$= 6\frac{13}{20} \qquad \text{Subtract.}$$

[YOU TRY 11] Subtract $12\frac{1}{5} - 11\frac{2}{3}$.

When we subtract a mixed number from a whole number, like $8 - 5\frac{2}{3}$, we will regroup, or borrow, 1 from the whole number.

EXAMPLE 12 Subtract $8 - 5\frac{2}{3}$.

Solution

Because 8 has no fractional part, we must rewrite it so that it does. The fractional part of $5\frac{2}{3}$ has a denominator of 3, so regroup 8 so that its fractional part has a denominator of 3.

Write 1 as $\frac{3}{3}$ so that it has the same denominator as $\frac{2}{3}$.

$$8 = 7 + 1 = 7 + \frac{3}{3} = 7\frac{3}{3}$$

Replace the 8 in $8 - 5\frac{2}{3}$ with $7\frac{3}{3}$.

$$8 - 5\frac{2}{3} = 7\frac{3}{3} - 5\frac{2}{3} = 2\frac{1}{3}$$

[YOU TRY 12] Subtract $9 - 5\frac{6}{11}$.

6 Add and Subtract Mixed Numbers Using Improper Fractions

Another way to add and subtract mixed numbers is to change them to improper fractions first. If you take an algebra class in the future, you will need to know this method.

> **Procedure** Adding and Subtracting Mixed Numbers Using Improper Fractions
>
> 1) Change each mixed number to an improper fraction.
> 2) Add or subtract the fractions.
> 3) Write the answer in lowest terms. If it is an improper fraction, change it to a mixed number.

EXAMPLE 13

Add or subtract by changing the mixed numbers to improper fractions.

a) $-1\frac{5}{6} + \left(-4\frac{2}{3}\right)$ b) $3\frac{1}{4} - 2\frac{6}{7}$

Solution

a) $-1\frac{5}{6} + \left(-4\frac{2}{3}\right) = -\frac{11}{6} + \left(-\frac{14}{3}\right)$ Write the mixed numbers as improper fractions.

$= -\frac{11}{6} + \left(-\frac{28}{6}\right)$ Multiply $-\frac{14}{3}$ by $\frac{2}{2}$ to get a common denominator.

$= -\frac{39}{6}$ Add. The sum of two negative numbers is negative.

$= -\frac{13}{2}$ Divide numerator and denominator by 3 to write in lowest terms.

$= -6\frac{1}{2}$ Write the result as a mixed number.

> **W Hint**
>
> Recall that the sum of two negative numbers is negative.

b) $3\frac{1}{4} - 2\frac{6}{7} = \frac{13}{4} - \frac{20}{7}$ Write the mixed numbers as improper fractions.

$= \frac{91}{28} - \frac{80}{28}$ Write each fraction with the LCD of 28.

$= \frac{11}{28}$ Subtract.

[YOU TRY 13]

Add or subtract by changing the mixed numbers to improper fractions.

a) $-3\frac{2}{9} + \left(-4\frac{1}{6}\right)$ b) $7\frac{1}{2} - 2\frac{3}{5}$

Note

If we use this method for adding and subtracting mixed numbers, we do not have to use regrouping.

7 Solve Applied Problems Involving Mixed Numbers

Many applications involve multiplying or dividing mixed numbers.

EXAMPLE 14

Amber sews dresses for Little Dresses for Africa, an organization that provides dresses for girls. Amber buys 14 yd of fabric, and each dress requires $1\frac{3}{4}$ yd. How many dresses can she make?

©SW Productions/Getty Images

Solution

We can think about the problem this way: How many $1\frac{3}{4}$-yd pieces of fabric are in 14 yd?

Divide 14 by $1\frac{3}{4}$ to find the number of dresses Amber can make.

$$14 \div 1\frac{3}{4} = \frac{14}{1} \div \frac{7}{4} \qquad \text{Write 14 as } \frac{14}{1}, \text{ and write } 1\frac{3}{4} \text{ as an improper fraction.}$$

$$= \frac{14}{1} \cdot \frac{4}{7} \qquad \text{Change division to multiplication by the reciprocal.}$$

$$= \frac{\overset{2}{\cancel{14}}}{1} \cdot \frac{4}{\underset{1}{\cancel{7}}} \qquad \text{Divide 14 and 7 by 7.}$$

$$= \frac{8}{1} = 8 \qquad \text{Multiply and simplify.}$$

Amber can make 8 dresses.

Check the answer. Multiply the 8 dresses she can make by the amount of fabric needed for each dress. The answer should be the total amount of fabric she bought.

$$8 \cdot 1\frac{3}{4} = \frac{8}{1} \cdot \frac{7}{4} = \frac{\overset{2}{\cancel{8}}}{1} \cdot \frac{7}{\cancel{4}} = \frac{14}{1} = 14 \leftarrow \begin{array}{l}\text{Total amount} \\ \text{of fabric}\end{array}$$

Number of dresses Amber can make ↗   ↑ Amount of fabric for each dress

The answer is correct.

YOU TRY 14

When the restaurant closes, Vijay must drain 9 gal of oil from the deep fryer and put it into $1\frac{1}{2}$-gal containers. How many containers will he need?

EXAMPLE 15

On Saturday, Reza worked $7\frac{1}{4}$ hours, and on Sunday he worked $6\frac{1}{2}$ hours. Find the total number of hours Reza worked over the weekend.

Solution

Let's think of the problem as

Statement:

Number of hours worked on Saturday	plus	Number of hours worked on Sunday	equals	Total number of hours worked

Equation: $7\dfrac{1}{4}$ + $6\dfrac{1}{2}$ =   Total number of hours worked

$$7\frac{1}{4} + 6\frac{1}{2} = 7\frac{1}{4} + 6\frac{2}{4} = 13\frac{3}{4} \qquad \text{Add the whole numbers, and add the fractions.}$$

$$\frac{1}{2} \cdot \frac{2}{2} = \frac{2}{4}$$

Reza worked $13\dfrac{3}{4}$ hours over the weekend.

Let's find an estimate for the answer to check to see whether our answer is reasonable: $7\dfrac{1}{4} + 6\dfrac{1}{2} \approx 7 + 7 = 14$ hours. This is very close to the exact answer of $13\dfrac{3}{4}$ hours.

[YOU TRY 15]

Tuesday night Liam played video games for $2\dfrac{2}{3}$ hours, and Wednesday night he played for $1\dfrac{1}{2}$ hours. Find the total number of hours Liam played video games those two nights.

ANSWERS TO [YOU TRY] EXERCISES

1) $1\dfrac{3}{4}$

2) a)

   b)

   c)

3) a) $\dfrac{13}{2}$ b) $\dfrac{21}{8}$ 4) a) $1\dfrac{3}{7}$ b) $9\dfrac{3}{8}$ c) 6

5) a) $7\dfrac{9}{28}$ b) -7 c) $1\dfrac{1}{9}$ 6) a) $2\dfrac{2}{3}$ b) $2\dfrac{1}{3}$ c) $\dfrac{10}{13}$

7) a) $13\dfrac{29}{35}$ b) $5\dfrac{1}{10}$ c) $-7\dfrac{1}{16}$

8) $6\dfrac{8}{5}$ is not in simplest form because $\dfrac{8}{5}$ is improper. $6\dfrac{8}{5} = 7\dfrac{3}{5}$

9) $15\dfrac{13}{28}$ 10) $4\dfrac{7}{6}$ 11) $\dfrac{8}{15}$ 12) $3\dfrac{5}{11}$ 13) a) $-7\dfrac{7}{18}$ b) $4\dfrac{9}{10}$

14) 6 containers   15) $4\dfrac{1}{6}$ hr

Objective 1: Understand Mixed Numbers

Use a mixed number to describe which portion of the figure is shaded, and represent the mixed number on a number line.

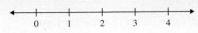

 1)

2)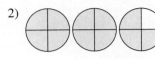

3)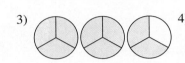

4)

Graph each group of numbers on a number line.

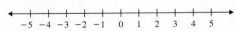

5) $4\frac{3}{4}$, $-2\frac{1}{2}$, $1\frac{4}{5}$, $-4\frac{7}{8}$, $-1\frac{1}{3}$

6) $4\frac{1}{3}$, $-1\frac{5}{6}$, $2\frac{1}{2}$, $-3\frac{1}{4}$, $-4\frac{3}{8}$

7) $-\frac{3}{4}$, $\frac{2}{3}$, $4\frac{2}{5}$, $-1\frac{1}{6}$, $3\frac{7}{10}$

8) $-\frac{4}{5}$, $\frac{3}{4}$, $3\frac{5}{8}$, $-2\frac{1}{6}$, $1\frac{3}{10}$

Objective 2: Convert Between Mixed Numbers and Improper Fractions

9) Explain, in your own words, how to change a mixed number to an improper fraction.

10) Explain, in your own words, how to change an improper fraction to a mixed number.

Write each mixed number as an improper fraction.

11) $1\frac{1}{3}$

12) $3\frac{1}{2}$

13) $2\frac{3}{5}$

14) $1\frac{1}{5}$

15) $6\frac{4}{9}$

16) $3\frac{7}{8}$

17) $10\frac{5}{6}$

18) $11\frac{5}{9}$

Fill in the blank with *proper* or *improper*.

19) A mixed number is the sum of a whole number and a(n) _____ fraction.

20) A(n) _____ fraction is a fraction that is less than one.

21) A(n) _____ fraction cannot be written as a mixed number.

22) A mixed number can always be represented by a(n) _____ fraction.

Write each improper fraction as a mixed number or whole number.

23) $\frac{5}{3}$

24) $\frac{11}{8}$

25) $\frac{9}{4}$

26) $\frac{24}{7}$

27) $\frac{58}{9}$

28) $\frac{37}{5}$

29) $\frac{64}{5}$

30) $\frac{107}{9}$

31) $\frac{36}{12}$

32) $\frac{42}{6}$

Objective 3: Multiply and Divide Mixed Numbers

33) In your own words, explain how to multiply mixed numbers.

34) When you multiply two mixed numbers, will the answer always be a mixed number?

Multiply.

35) $1\frac{2}{3} \cdot 2\frac{1}{7}$

36) $2\frac{1}{6} \cdot 4\frac{2}{7}$

37) $2\frac{4}{5} \cdot 3\frac{1}{2}$

38) $7\frac{1}{3} \cdot 1\frac{1}{6}$

39) $-6\frac{3}{7} \cdot 2\frac{1}{3}$

40) $-3\frac{2}{3} \cdot 6\frac{3}{11}$

41) $8 \cdot 2\frac{1}{6}$

42) $12 \cdot 1\frac{3}{8}$

43) $-11\frac{3}{5} \cdot \left(-\frac{2}{29}\right)$

44) $-3\frac{3}{4} \cdot \left(-\frac{7}{30}\right)$

45) In your own words, explain how to divide mixed numbers.

46) If you perform division involving one or more mixed numbers and the result is an improper fraction, how should you write the final answer?

Divide.

47) $7\frac{1}{5} \div 4\frac{1}{2}$

48) $4\frac{1}{6} \div 1\frac{1}{9}$

49) $-3\frac{3}{5} \div 5\frac{5}{8}$

50) $-1\frac{5}{9} \div 1\frac{5}{6}$

51) $-6\frac{3}{4} \div (-8)$

52) $7\frac{2}{3} \div 10$

53) $4\frac{7}{8} \div 3$

54) $-8\frac{2}{5} \div (-6)$

55) $15 \div 4\frac{1}{2}$

56) $6 \div 5\frac{1}{3}$

57) $8 \div \left(-6\frac{2}{5}\right)$

58) $10 \div \left(-4\frac{2}{3}\right)$

Objective 4: Add and Subtract Mixed Numbers in Mixed-Number Form

59) In your own words, explain how to add and how to subtract mixed numbers.

60) Is $4\frac{10}{18}$ in lowest terms? Why or why not?

Add or subtract, as indicated.

61) $8\frac{7}{9} - 6\frac{2}{9}$

62) $5\frac{8}{13} + 4\frac{3}{13}$

63) $1\frac{5}{14} + 3\frac{2}{7}$

64) $4\frac{7}{16} + 2\frac{1}{8}$

65) $5\frac{1}{3} + 6\frac{5}{12}$

66) $8\frac{1}{2} + 7\frac{3}{10}$

67) $9\frac{5}{6} - 7\frac{1}{18}$

68) $12\frac{5}{9} - 3\frac{2}{27}$

69) $14\frac{6}{7} - 9\frac{5}{21}$

70) $17\frac{4}{5} - 11\frac{3}{10}$

71) $-5\frac{5}{12} + 4\frac{1}{8}$

72) $-6\frac{3}{4} + 1\frac{1}{9}$

73) $10\frac{5}{6} + \left(-2\frac{4}{15}\right)$

74) $13\frac{7}{8} + \left(-5\frac{5}{6}\right)$

75) $-3\frac{3}{8} - 11\frac{2}{9}$

76) $-6\frac{1}{6} - 4\frac{3}{7}$

Objective 5: Add and Subtract Mixed Numbers with Regrouping

77) Is $4\frac{8}{7}$ in simplest form? Why or why not?

78) Is $6\frac{4}{9}$ in simplest form? Why or why not?

Write each mixed number in simplest form.

79) $3\frac{7}{5}$

80) $1\frac{11}{8}$

81) $2\frac{8}{7}$

82) $7\frac{19}{10}$

Add.

83) $4\frac{5}{6} + 2\frac{1}{2}$

84) $6\frac{2}{3} + 3\frac{7}{9}$

85) $5\frac{11}{24} + 7\frac{7}{8}$

86) $3\frac{9}{20} + 8\frac{4}{5}$

87) $11\frac{3}{8} + 16\frac{7}{10}$

88) $19\frac{11}{12} + 8\frac{3}{10}$

Regroup 1 from the whole-number part to rewrite the mixed number with an improper fractional part.

89) $8\frac{1}{2}$

90) $2\frac{1}{3}$

91) $9\frac{13}{16}$

92) $6\frac{17}{21}$

Subtract.

93) $9\frac{3}{8} - 5\frac{7}{8}$

94) $8\frac{1}{10} - 1\frac{9}{10}$

95) $16\frac{1}{5} - 8\frac{5}{6}$

96) $13\frac{3}{4} - 2\frac{4}{5}$

97) $14\frac{5}{12} - 13\frac{6}{7}$

98) $17\frac{5}{9} - 16\frac{10}{11}$

99) $11 - 10\frac{2}{5}$

100) $3 - 2\frac{5}{8}$

101) $6 - 3\frac{3}{7}$

102) $12 - 5\frac{4}{9}$

Objective 6: Add and Subtract Mixed Numbers Using Improper Fractions

Add or subtract by changing the mixed numbers to improper fractions.

103) $3\frac{1}{2} + 2\frac{1}{4}$

104) $5\frac{1}{3} + 2\frac{5}{6}$

105) $2\frac{2}{5} - 1\frac{2}{3}$ 106) $3\frac{3}{7} - 2\frac{3}{4}$

107) $-2\frac{5}{6} + \left(-1\frac{1}{4}\right)$ 108) $-4\frac{3}{5} + \left(-3\frac{1}{2}\right)$

109) $1\frac{5}{12} - 5\frac{3}{8}$ 110) $2\frac{7}{9} - 7\frac{1}{6}$

111) Do you prefer to add and subtract mixed numbers by leaving them as mixed numbers or by changing them to improper fractions? Why?

112) If we add and subtract mixed numbers by first changing them to improper fractions, will we have to do any regrouping?

Objective 7: Solve Applied Problems Involving Mixed Numbers

For Exercises 113 and 114, find the missing length.

 113)

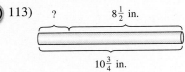

114)

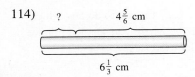

A healthcare professional monitors the weight of a patient who has been retaining fluids due to an illness. Use the table below for Exercises 115–118.

Day and Time	Weight in Pounds
Day 1, 8:00 A.M.	$141\frac{2}{3}$
Day 2, 9:00 A.M.	$142\frac{1}{2}$
Day 3, 8:00 A.M.	$143\frac{3}{4}$
Day 4, 10:00 A.M.	$144\frac{1}{2}$
Day 5, 8:00 A.M.	$142\frac{3}{4}$
Day 6, 10:00 A.M.	$141\frac{1}{4}$
Day 7, 9:00 A.M.	$139\frac{1}{3}$

115) How much weight did the patient gain from Day 1 to Day 4?

116) How much weight did the patient lose from Day 4 to Day 7?

117) From Day 3 to Day 4, did the patient gain or lose weight and how much?

118) Over the 7-day monitoring period, did the patient gain or lose weight and how much?

Solve each problem.

119) Sunny is going to crochet small squares for a throw blanket she wants to make. If each square requires $3\frac{3}{4}$ yd of yarn, how many squares can she make using 60 yd of yarn?

120) Colin needs to move 69 cubic feet of dirt to a truck using a wheelbarrow that can carry $5\frac{3}{4}$ cubic feet of dirt. How many times will Colin have to fill the wheelbarrow to move all the dirt?

 121) Spiro's study schedule requires that he spend $2\frac{1}{2}$ hr of study each week for every unit of class he is enrolled in. How many hours will he study per week if he is taking 15 units?

122) Fatima's commute to work is $1\frac{1}{4}$ hr each way every day. How many hours does she spend commuting to work and back each week if she works Monday through Friday only?

©Westend61/Getty Images

123) Ibada has a part-time job and worked $6\frac{1}{2}$ hr on Monday, $3\frac{3}{4}$ hr on Wednesday, and 8 hr on Friday.
   a) How many total hours did he work?
   b) How many more hours did he work on Monday than on Wednesday?

124) Araceli's sewing project needs three pieces of fabric with lengths of $\frac{3}{4}$, $2\frac{1}{2}$, and $4\frac{1}{3}$ yd.
   a) What is the total amount of fabric she needs?
   b) How much longer is the longest piece than the shortest piece?

125) Clarissa's chocolate chip cookie recipe requires $2\frac{1}{4}$ cups of all-purpose flour to make 5 dozen cookies. How much flour does she need to make $2\frac{1}{2}$ dozen cookies?

126) Heinrich's cheesecake recipe requires $1\frac{1}{2}$ cups of white sugar to make one cheesecake. If he needs to make one dozen cheesecakes for a conference event, how many cups of white sugar does he need?

127) If an MP3 song is encoded such that its file size is $1\frac{1}{5}$ megabytes per minute, how much memory is required to store an hour's length of music?

128) Romeo's phone takes digital pictures that are $2\frac{3}{4}$ megabytes each. How many mega-bytes are required to store 300 of Romeo's photos?

©Hero/Corbis/Glow Images

129) How many quarter-pound hamburger patties can be made from $8\frac{3}{4}$ lb of ground beef?

130) One gallon of water weighs approximately $8\frac{1}{3}$ lb. What is the approximate weight of a quart of water if a quart is $\frac{1}{4}$ gallon?

R Rethink

R1) Describe a situation that you have encountered recently that involved using mixed numbers.

R2) Which exercises were easy for you to do and which were difficult? Describe where you seemed to have trouble with the harder exercises.

Putting It All Together

P Prepare

O Organize

What are your objectives for Putting It All Together?	How can you accomplish the objective?
1 Review the Concepts of Sections 3.1–3.6	• Review the definitions and procedures that you have learned so far in this chapter. • Complete the given examples on your own and, if you struggled with specific concepts, go back to previous sections for more information on those concepts. • Complete You Trys 1 and 2.

W Work

Read the explanations, follow the examples, take notes, and complete the You Trys.

1 Review the Concepts of Sections 3.1–3.6

Let's review what we have learned about fractions so far.

A **fraction** is a part of a whole. In a **proper fraction**, like $\frac{2}{5}$, the numerator is less than the denominator. It represents a number less than one whole. In an **improper fraction**, like $\frac{11}{4}$, the numerator is greater than or equal to the denominator, and it represents a quantity greater than or equal to one whole. A **mixed number**, like $5\frac{1}{2}$, consists of a whole number and a fraction. It represents more than one whole.

When we perform operations with fractions and mixed numbers, we write the answers in lowest terms. To be sure a fraction is in lowest terms, you should ask yourself two questions: *"What number divides evenly into both the numerator and denominator?"* and, after dividing out the common factors, look at the result and ask, *"Is the fraction in lowest terms?"*

EXAMPLE 1

Determine whether each fraction is in lowest terms. If it is not, write it in lowest terms.

a) $\frac{12}{42}$ b) $\frac{10}{17}$ c) $\frac{350}{210}$

Solution

a) Look at $\frac{12}{42}$. Ask yourself, *"What number divides evenly into both 12 and 42?"* Let's use 6.

$$\frac{12}{42} = \frac{12 \div 6}{42 \div 6} = \frac{2}{7}$$

Ask yourself, *"Is $\frac{2}{7}$ in lowest terms?"* Yes. Therefore, $\frac{12}{42} = \frac{2}{7}$ and $\frac{2}{7}$ is in lowest terms.

b) Look at $\frac{10}{17}$ and ask yourself, *"What number divides evenly into both 10 and 17?"* Their only common factor is 1. Therefore, $\frac{10}{17}$ is in lowest terms.

c) Look at $\frac{350}{210}$ and ask yourself, *"What number divides evenly into both 350 and 210?"* Because both numbers end in 0, they are divisible by 10.

$$\frac{350}{210} = \frac{350 \div 10}{210 \div 10} = \frac{35}{21}$$

Ask yourself, *"Is $\frac{35}{21}$ in lowest terms?"* No! 35 and 21 are divisible by 7.

$$\frac{35}{21} = \frac{35 \div 7}{21 \div 7} = \frac{5}{3}$$

W Hint

Remember, sometimes you must divide out common factors more than once to write a fraction in lowest terms.

"Is $\frac{5}{3}$ in lowest terms?" Yes. In lowest terms, we can write $\frac{350}{210}$ as $\frac{5}{3}$ or as the mixed number $1\frac{2}{3}$.

[YOU TRY 1] Determine whether each fraction is in lowest terms. If it is not, write it in lowest terms.

a) $\dfrac{35}{48}$ b) $\dfrac{9}{72}$ c) $\dfrac{5280}{1620}$

Note

Recall that we can also write a fraction in lowest terms by writing the prime factorizations of the numerator and denominator. Then, divide out common factors.

W Hint

In your notes, summarize how to add, subtract, multiply, and divide fractions and mixed numbers.

Now let's put together everything we have learned about adding, subtracting, multiplying, and dividing fractions and mixed numbers. It is important to remember when a common denominator is needed and when it is not.

EXAMPLE 2   Perform the indicated operations.

a) $\dfrac{7}{12}+\dfrac{1}{6}+\dfrac{5}{8}$ b) $1\dfrac{9}{16}\cdot 2\dfrac{2}{15}$ c) $-\dfrac{21k}{8}\div\dfrac{14k}{3}$

d) $8\dfrac{9}{10}-3\dfrac{4}{5}$ e) $3\dfrac{5}{12}+9\dfrac{4}{5}$ f) $5-\dfrac{4}{7}$

Solution

a)   We have to find the sum $\dfrac{7}{12}+\dfrac{1}{6}+\dfrac{5}{8}$. Do we need a common denominator? *Yes!*
Write each fraction with the LCD, then add.

Step 1: Write down the LCD of $\dfrac{7}{12},\dfrac{1}{6}$, and $\dfrac{5}{8}$.   **LCD = 24**

Step 2: Rewrite each fraction with the LCD.

$$\dfrac{7}{12}\cdot\dfrac{2}{2}=\dfrac{14}{24} \qquad \dfrac{1}{6}\cdot\dfrac{4}{4}=\dfrac{4}{24} \qquad \dfrac{5}{8}\cdot\dfrac{3}{3}=\dfrac{15}{24}$$

Step 3: Add.

$$\dfrac{7}{12}+\dfrac{1}{6}+\dfrac{5}{8}=\dfrac{14}{24}+\dfrac{4}{24}+\dfrac{15}{24} \qquad \text{Write each fraction with the LCD.}$$

$$=\dfrac{33}{24} \qquad \text{Add.}$$

Ask yourself, "*Is $\dfrac{33}{24}$ in lowest terms?*" No! Write it in lowest terms.

$$\dfrac{33}{24}=\dfrac{33\div 3}{24\div 3}=\dfrac{11}{8} \text{ or } 1\dfrac{3}{8}$$

Therefore, $\dfrac{7}{12}+\dfrac{1}{6}+\dfrac{5}{8}=\dfrac{11}{8}$ or $1\dfrac{3}{8}$. Remember, we can write the answer as an improper fraction or as a mixed number as long as it is in lowest terms.

b) Do we need a common denominator to multiply mixed numbers? *No!* To multiply $1\frac{9}{16} \cdot 2\frac{2}{15}$, first change each mixed number to an improper fraction. Then multiply.

$$1\frac{9}{16} \cdot 2\frac{2}{15} = \frac{25}{16} \cdot \frac{32}{15}$$ Write each mixed number as an improper fraction.

$$= \frac{\overset{5}{\cancel{25}}}{\underset{1}{\cancel{16}}} \cdot \frac{\overset{2}{\cancel{32}}}{\underset{3}{\cancel{15}}}$$ Divide 25 and 15 by 5; divide 16 and 32 by 16.

$$= \frac{5 \cdot 2}{1 \cdot 3}$$ Multiply numerators, and multiply denominators.

$$= \frac{10}{3} \text{ or } 3\frac{1}{3}$$

Are $\frac{10}{3}$ and $3\frac{1}{3}$ in lowest terms? Yes.

c) Do we need a common denominator to find $-\frac{21k}{8} \div \frac{14k}{3}$? *No! We do not need a common denominator to divide, or multiply, fractions.* To divide fractions, multiply the first fraction by the reciprocal of the second.

$$-\frac{21k}{8} \div \frac{14k}{3} = -\frac{21k}{8} \cdot \frac{3}{14k}$$ Change division to multiplication by the reciprocal.

$$= -\frac{\overset{3}{\cancel{21k}}}{8} \cdot \frac{3}{\underset{2}{\cancel{14k}}}$$ Divide out common factors.

$$= -\frac{9}{16}$$ Multiply.

Ask yourself, *"Is $-\frac{9}{16}$ in lowest terms?"* Yes.

W Hint

We can also find $8\frac{9}{10} - 3\frac{4}{5}$ by changing each mixed number to an improper fraction.

d) What is the first step for finding $8\frac{9}{10} - 3\frac{4}{5}$? Write $\frac{9}{10}$ and $\frac{4}{5}$ as equivalent fractions with an LCD.

The LCD of $\frac{9}{10}$ and $\frac{4}{5}$ is 10.

$\frac{9}{10}$ already has the LCD. $\frac{4}{5} \cdot \frac{2}{2} = \frac{8}{10}$

$$8\frac{9}{10} - 3\frac{4}{5} = 8\frac{9}{10} - 3\frac{8}{10}$$ Write the fractions with the LCD.

$$= 5\frac{1}{10}$$ Subtract whole numbers, and subtract fractions.

Is $5\frac{1}{10}$ in lowest terms? Yes.

e) How do we find $3\frac{5}{12} + 9\frac{4}{5}$? Write $\frac{5}{12}$ and $\frac{4}{5}$ as equivalent fractions with their least common denominator. Then, add the whole-number parts, and add the fractional parts.

$$3\frac{5}{12} + 9\frac{4}{5} = 3\frac{25}{60} + 9\frac{48}{60} \qquad \text{Write the fractions with the LCD.}$$

$$= 12\frac{73}{60} \qquad \text{Add.}$$

Is $12\frac{73}{60}$ in simplest form? *No!* $\frac{73}{60}$ is an improper fraction. Simplify $12\frac{73}{60}$ by regrouping.

$$12\frac{73}{60} = 12 + \frac{73}{60} = 12 + 1\frac{13}{60} = 13\frac{13}{60}$$

So, $3\frac{5}{12} + 9\frac{4}{5} = 13\frac{13}{60}$. The answer is in lowest terms.

f) To find the difference $5 - \frac{4}{7}$, we first need a common denominator. Write 5 as an improper fraction with a denominator of 1: $5 = \frac{5}{1}$.

Now, subtract.

Rewrite with a denominator of 7.

$$5 - \frac{4}{7} = \frac{5}{1} - \frac{4}{7} = \frac{35}{7} - \frac{4}{7} = \frac{31}{7} \text{ or } 4\frac{3}{7}$$

Write 5 as a fraction.

BE CAREFUL

We need a common denominator *only* when we are adding or subtracting fractions and mixed numbers. When we are multiplying or dividing, a common denominator is *not* needed.

[YOU TRY 2] Perform the indicated operations.

a) $\frac{4}{15a} \div \frac{7}{6a}$

b) $-2\frac{1}{8} + 7\frac{3}{10}$

c) $9 - 5\frac{3}{5}$

d) $\frac{1}{24} \cdot \frac{18}{25} \cdot \frac{10}{63}$

e) $16 \div \left(-3\frac{1}{5}\right)$

f) $\frac{4}{7} + \frac{9}{14} + \frac{5}{8}$

Putting It All Together Exercises

 **Evaluate**   Do the exercises, and check your work.

Objective 1: Review the Concepts of Sections 3.1–3.6

Determine whether each statement is *true* or *false*. If it is false, explain why.

1) The denominator of a fraction can be any number.

2) The fraction $\frac{7}{1}$ is in lowest terms.

3) The number 35 is a prime number.

4) The prime factorization of 36 is $2^2 \cdot 3^2$.

Write each fraction in lowest terms.

5) $\frac{40}{75}$

6) $\frac{24}{96}$

7) $\frac{510t^2}{102t}$

8) $\frac{2520m^2}{9900m^3}$

Determine whether each pair of fractions is equivalent.

9) $\frac{28}{63}$ and $\frac{12}{27}$

10) $\frac{24}{40}$ and $\frac{48}{60}$

Change each number to an improper fraction.

11) $12\frac{6}{11}$

12) $3\frac{1}{8}$

Change each improper fraction to a mixed number.

13) $\frac{179}{12}$

14) $\frac{10}{7}$

Determine whether each statement is *true* or *false*. If it is false, explain why.

15) To multiply two fractions, we need a common denominator.

16) All answers should be written in lowest terms.

17) Before multiplying or dividing mixed numbers, we must change them to improper fractions.

18) The mixed number $3\frac{11}{6}$ is in simplest form.

19) $\frac{1}{2}$ is greater than $\frac{3}{2}$.

20) $\frac{3}{7}$ is less than 1.

Perform the indicated operations.

21) $\frac{7}{16} \cdot \left(-\frac{10}{21}\right)$

22) $\frac{14}{15} - \frac{8}{15}$

23) $3\frac{1}{4} + 6\frac{2}{7}$

24) $-\frac{20}{21} \div \frac{45}{77}$

25) $\frac{11}{18} - \frac{z}{6}$

26) $7\frac{3}{10} + 8\frac{4}{15}$

27) $\dfrac{\dfrac{8}{21}}{\dfrac{16}{7}}$

28) $-5\frac{1}{4} \cdot \left(-2\frac{3}{14}\right)$

29) $9\frac{11}{12} + \left(-8\frac{4}{9}\right)$

30) $1\frac{17}{20} \div 2\frac{3}{4}$

31) $-4 \cdot \left(-3\frac{1}{8}\right)$

32) $-\frac{7}{8} + \left(-\frac{5}{9}\right)$

33) $8 \div \frac{1}{10}$

34) $17 - 16\frac{5}{9}$

35) $12\frac{1}{2} - 5\frac{3}{4}$

36) $\frac{35}{54} \cdot \frac{8}{21} \cdot \frac{45}{32}$

37) $3\frac{9}{11} \cdot 1\frac{5}{7}$

38) $\frac{7}{16} + \frac{1}{20} + \frac{9}{14}$

39) $-\dfrac{5}{11}+\left(-\dfrac{1}{4}\right)$

40) $\dfrac{\dfrac{1}{4}}{\dfrac{2}{5}}$

41) $\dfrac{11}{30}+\dfrac{5}{16}+\dfrac{17}{40}$

42) $-6\dfrac{5}{7}+4\dfrac{1}{6}$

43) $-5\dfrac{1}{3}\div 1\dfrac{13}{15}$

44) $3\dfrac{7}{9}+8\dfrac{3}{5}$

45) $14-5\dfrac{7}{12}$

46) $-\dfrac{5d^4}{36}\cdot\left(-\dfrac{12}{13d^2}\right)$

47) $\dfrac{15}{28}\cdot\dfrac{16}{33}\cdot\dfrac{21}{40}$

48) $11\dfrac{3}{8}-2\dfrac{9}{16}$

49) $\dfrac{7}{10}+\dfrac{3}{8r}$

50) $-\dfrac{x^2}{y^3}\cdot\dfrac{y}{x^5}$

51) $\dfrac{1}{8}+2\dfrac{5}{6}+\dfrac{1}{2}+7\dfrac{3}{4}$

52) $-9\cdot 2\dfrac{2}{3}$

Fill in the blank.

53) a) $5\cdot\underline{\hspace{1cm}}=40$

b) $\dfrac{3}{4}\cdot\underline{\hspace{1cm}}=18$

54) a) $10+\underline{\hspace{1cm}}=23$

b) $\dfrac{2}{7}+\underline{\hspace{1cm}}=\dfrac{19}{42}$

55) a) $\underline{\hspace{1cm}}-37=19$

b) $\underline{\hspace{1cm}}-\dfrac{7}{12}=\dfrac{11}{36}$

56) a) $\underline{\hspace{1cm}}\div 2=6$

b) $\underline{\hspace{1cm}}\div\dfrac{4}{9}=\dfrac{3}{10}$

For Exercises 57–60, find two fractions with a

57) sum of $\dfrac{18}{29}$.

58) difference of $\dfrac{2}{7}$.

59) quotient of $\dfrac{21}{50}$.

60) product of $-\dfrac{10}{33}$.

61) Find two mixed numbers with a difference of $4\dfrac{5}{21}$.

62) Find two mixed numbers with a sum of $9\dfrac{11}{27}$.

Graph the numbers on a number line.

$\xleftarrow{\;\;}$ $\begin{array}{ccccccccccc}-5&-4&-3&-2&-1&0&1&2&3&4&5\end{array}$ $\xrightarrow{\;\;}$

63) $4\dfrac{2}{3},\dfrac{3}{4},-3\dfrac{1}{2},-\dfrac{5}{8},-1\dfrac{5}{6},2\dfrac{1}{5}$

64) $1\dfrac{4}{5},-\dfrac{2}{3},-4\dfrac{1}{2},\dfrac{1}{4},3\dfrac{3}{8},-2\dfrac{3}{4}$

Solve each problem.

65) Janae brewed 2 gallons of iced tea. Then, she put $\dfrac{2}{3}$-gallon portions into bottles. How many bottles did she use?

©National Cancer Institute/ Renee Comet, photographer

66) A worker has a 10-ft piece of sheet metal that must be cut into strips that are $\dfrac{5}{8}$ ft long. How many strips can be made?

67) On Saturday, Misaki ran $5\dfrac{3}{4}$ mi. On Sunday, she ran $7\dfrac{1}{2}$ mi. How much farther did Misaki run on Sunday than on Saturday?

68) On Wednesday, Maura practiced the piano for $1\dfrac{1}{2}$ hr, on Thursday she practiced for $1\dfrac{3}{4}$ hr, and on Friday she practiced for $\dfrac{2}{3}$ hr. Find the total number of hours Maura practiced.

69) Evan's history class is 50 min long, and he spent $\dfrac{2}{5}$ of the time on Facebook. How much time did he spend on Facebook during class?

70) During the 2015–2016 NBA regular season, Dwight Howard of the Houston Rockets made $\dfrac{31}{50}$ of his free throws. How many free throws did he make given that he had 600 attempts? (www.espn.go.com)

R Rethink

R1) Which types of exercises could you do without using the book or your notes? Which did you need to look up in order to do? Which types of problems do you need to practice more?

R2) How could you be sure that you are ready for a quiz or a test on these concepts right now?

3.7 Order Relations and Order of Operations

P Prepare | O Organize

What are your objectives for Section 3.7?	How can you accomplish each objective?
1 Compare Fractions	• Explain how to compare fractions in your own words. • Complete the given examples on your own. • Complete You Trys 1 and 2.
2 Use Exponents with Fractions	• Remember that exponents indicate repeated multiplication with integers *and* fractions. • Complete the given example on your own. • Complete You Try 3.
3 Use the Order of Operations with Fractions	• Remember the order of operations, and apply it to fractions. • Complete the given example on your own. • Complete You Try 4.

W Work

Read the explanations, follow the examples, take notes, and complete the You Trys.

1 Compare Fractions

In Chapter 1, we learned that the symbol > means *is greater than* and the symbol < means *is less than*. We have also learned that, on a number line, numbers get larger as you move to the right and numbers get smaller as you move to the left. For example,

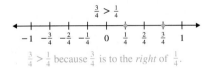

$\frac{3}{4} > \frac{1}{4}$ because $\frac{3}{4}$ is to the *right* of $\frac{1}{4}$.

$-\frac{4}{5} < -\frac{2}{5}$ because $-\frac{4}{5}$ is to the *left* of $-\frac{2}{5}$.

To compare fractions, we can think about where they are on a number line, or, if they have a common denominator, we can compare their numerators.

EXAMPLE 1

Fill in the blank with > or <.

a) $\frac{4}{7}$ ____ $\frac{6}{7}$ b) $-\frac{3}{8}$ ____ $-\frac{7}{8}$ c) $3\frac{8}{9}$ ____ $3\frac{4}{9}$

Solution

a) To compare these fractions, we can think about a shaded figure or the placement of $\frac{4}{7}$ and $\frac{6}{7}$ on a number line. Or, notice that the numerator of $\frac{4}{7}$ is *less than* the numerator of $\frac{6}{7}$.

$$\frac{4}{7} < \frac{6}{7}$$ The denominators are the same, and 4 *is less than* 6.

b) To fill in the blank for $-\dfrac{3}{8}$ _____ $-\dfrac{7}{8}$, first *rewrite each fraction with the negative sign in the numerator:*

$$\dfrac{-3}{8} \quad\underline{}\quad \dfrac{-7}{8}$$

Because $-3 > -7$, it follows that

$$\dfrac{-3}{8} > \dfrac{-7}{8} \qquad \text{or} \qquad -\dfrac{3}{8} > -\dfrac{7}{8}$$

We can see this on a number line:

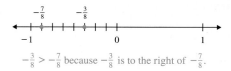

$-\dfrac{3}{8} > -\dfrac{7}{8}$ because $-\dfrac{3}{8}$ is to the right of $-\dfrac{7}{8}$.

c) $3\dfrac{8}{9} > 3\dfrac{4}{9}$ The whole-number part of each mixed number is the same, so compare the fractional parts. Because $\dfrac{8}{9}$ *is greater than* $\dfrac{4}{9}$, $3\dfrac{8}{9}$ *is greater than* $3\dfrac{4}{9}$.

[YOU TRY 1] Fill in the blank with > or <.

a) $\dfrac{9}{10}$ _____ $\dfrac{3}{10}$ b) $-\dfrac{19}{6}$ _____ $-\dfrac{7}{6}$ c) $5\dfrac{2}{11}$ _____ $5\dfrac{6}{11}$

How do we compare fractions with *unlike* denominators? We write them with the least common denominator and compare the numerators.

EXAMPLE 2 Fill in the blank with > or <.

a) $\dfrac{5}{8}$ _____ $\dfrac{9}{16}$ b) $-\dfrac{7}{4}$ _____ $-\dfrac{13}{9}$

Solution

a) First, write each fraction with the LCD: **LCD = 16**

$$\dfrac{5}{8} \cdot \dfrac{2}{2} = \dfrac{10}{16} \qquad \dfrac{9}{16} \text{ already has the LCD.}$$

Rewrite $\dfrac{5}{8}$ _____ $\dfrac{9}{16}$ as $\dfrac{10}{16}$ _____ $\dfrac{9}{16}$.

Because the numerator of $\dfrac{10}{16}$ *is greater than* the numerator of $\dfrac{9}{16}$, we write

$$\underbrace{\dfrac{10}{16} > \dfrac{9}{16} \qquad \text{or} \qquad \dfrac{5}{8} > \dfrac{9}{16}}_{\text{Equivalent fractions}}$$

W Hint
You can also draw a picture to visualize each fraction.

b) Write each fraction with a denominator of 36.

$$-\frac{7}{4}\cdot\frac{9}{9}=-\frac{63}{36} \text{ or } \frac{-63}{36} \qquad -\frac{13}{9}\cdot\frac{4}{4}=-\frac{52}{36} \text{ or } \frac{-52}{36}$$

Rewrite $-\frac{7}{4}$ —— $-\frac{13}{9}$ as $\frac{-63}{36}$ —— $\frac{-52}{36}$.

The numerator of $\frac{-63}{36}$ *is less than* the numerator of $\frac{-52}{36}$, so

Equivalent

$$-\frac{63}{36} < -\frac{52}{36} \qquad \text{or} \qquad -\frac{7}{4} < -\frac{13}{9}$$

Equivalent

[YOU TRY 2] Fill in the blank with > or <.

a) $\frac{5}{12}$ —— $\frac{11}{24}$ b) $-\frac{10}{7}$ —— $-\frac{9}{5}$

---

2 Use Exponents with Fractions

In Chapter 1, we learned that an exponent indicates repeated multiplication. For example,

$$5^2 = 5 \cdot 5 = 25 \qquad \text{and} \qquad 2^3 = 2 \cdot 2 \cdot 2 = 8$$

We can use exponents with fractions, too.

EXAMPLE 3   Evaluate.

a) $\left(\frac{7}{8}\right)^2$ b) $\left(\frac{1}{3}\right)^4$ c) $\left(-\frac{1}{2}\right)^3$

Solution

a) $\left(\frac{7}{8}\right)^2 = \frac{7}{8}\cdot\frac{7}{8} = \frac{49}{64}$ The exponent of 2 tells us to use $\frac{7}{8}$ as a factor 2 times.

b) $\left(\frac{1}{3}\right)^4 = \frac{1}{3}\cdot\frac{1}{3}\cdot\frac{1}{3}\cdot\frac{1}{3} = \frac{1}{81}$ The exponent of 4 tells us to use $\frac{1}{3}$ as a factor 4 times.

c) $\left(-\frac{1}{2}\right)^3 = -\frac{1}{2}\cdot\left(-\frac{1}{2}\right)\cdot\left(-\frac{1}{2}\right) = -\frac{1}{8}$ The exponent of 3 tells us to use $-\frac{1}{2}$ as a factor 3 times.

[YOU TRY 3] Evaluate.

a) $\left(\frac{2}{5}\right)^3$ b) $\left(\frac{1}{12}\right)^2$ c) $\left(-\frac{1}{4}\right)^3$

---

3 Use the Order of Operations with Fractions

We can use the order of operations to simplify expressions containing fractions. Let's review the order of operations that we first learned in Section 1.8.

> **Procedure** The Order of Operations
>
> Simplify expressions in the following order:
>
> 1) If **parentheses** or **other grouping symbols** appear in an expression, simplify what is in these grouping symbols first.
> 2) Simplify expressions with **exponents** and **square roots.**
> 3) **Multiply** or **divide,** moving from left to right.
> 4) **Add** or **subtract,** moving from left to right.

EXAMPLE 4

Simplify each expression using the order of operations.

a) $\left(-\dfrac{7}{8}\right)^2$ b) $-\left(\dfrac{7}{8}\right)^2$ c) $\dfrac{9}{10} - \dfrac{3}{10} \cdot \dfrac{5}{2}$

d) $-\dfrac{7}{20} \div \left(-\dfrac{1}{8} + \dfrac{4}{5}\right)$ e) $4 - 3\left(\dfrac{5}{6}\right)^2 + \dfrac{7}{12}$

Solution

a) In the expression $\left(-\dfrac{7}{8}\right)^2$, the base is $-\dfrac{7}{8}$.

$$\left(-\frac{7}{8}\right)^2 = -\frac{7}{8} \cdot \left(-\frac{7}{8}\right) = \frac{49}{64}$$

b) In the expression $-\left(\dfrac{7}{8}\right)^2$, the base is $\dfrac{7}{8}$. So, $-\left(\dfrac{7}{8}\right)^2$ means $-1 \cdot \left(\dfrac{7}{8}\right)^2$.

$$-\left(\frac{7}{8}\right)^2 = -1 \cdot \left(\frac{7}{8}\right)^2 = -1 \cdot \underbrace{\frac{7}{8} \cdot \frac{7}{8}}_{\text{2 factors of } \frac{7}{8}} = -\frac{49}{64}$$

Be very careful when you are working with negative signs and exponents!

Hint

Is this any different from using the order of operations in Section 1.8?

c) Be careful! Do not subtract first. Remember that we multiply before we subtract.

$$\frac{9}{10} - \frac{3}{10} \cdot \frac{5}{2} = \frac{9}{10} - \frac{3}{\overset{}{\underset{2}{10}}} \cdot \frac{\overset{1}{5}}{2} \qquad \text{Divide 10 and 5 by 5.}$$

$$= \frac{9}{10} - \frac{3}{4} \qquad \text{Multiply.}$$

$$= \frac{18}{20} - \frac{15}{20} \qquad \text{Write each fraction with the LCD, 20.}$$

$$= \frac{3}{20} \qquad \text{Subtract.}$$

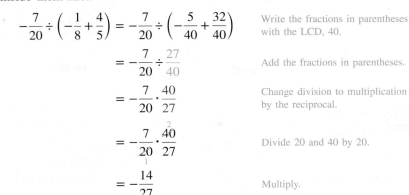

Hint

Are you writing out the examples as you are reading them?

d) When an expression contains parentheses or other grouping symbols, simplify what is inside them first.

$$-\frac{7}{20} \div \left(-\frac{1}{8} + \frac{4}{5}\right) = -\frac{7}{20} \div \left(-\frac{5}{40} + \frac{32}{40}\right)$$

Write the fractions in parentheses with the LCD, 40.

$$= -\frac{7}{20} \div \frac{27}{40}$$

Add the fractions in parentheses.

$$= -\frac{7}{20} \cdot \frac{40}{27}$$

Change division to multiplication by the reciprocal.

$$= -\frac{7}{20} \cdot \frac{\overset{2}{40}}{27}$$
$$\phantom{= -\frac{7}{20}}_{1}$$

Divide 20 and 40 by 20.

$$= -\frac{14}{27}$$

Multiply.

e) Look at the expression $4 - 3\left(\frac{5}{6}\right)^2 + \frac{7}{12}$. Notice that there is no operation symbol between the 3 and the expression in parentheses. Remember, this means the operation is multiplication.

What do we do first? Find $\left(\frac{5}{6}\right)^2$.

$$4 - 3\left(\frac{5}{6}\right)^2 + \frac{7}{12} = 4 - 3\left(\frac{25}{36}\right) + \frac{7}{12}$$

First, find $\left(\frac{5}{6}\right)^2$. Next, multiply.

$$= 4 - \frac{\overset{1}{3}}{1} \cdot \frac{25}{\underset{12}{36}} + \frac{7}{12}$$

Divide 3 and 36 by 3.

$$= 4 - \frac{25}{12} + \frac{7}{12}$$

Multiply.

When an expression contains addition and subtraction, perform the operations from left to right. Therefore, we must first find $4 - \frac{25}{12}$. Then, add $\frac{7}{12}$.

$$= \frac{4}{1} - \frac{25}{12} + \frac{7}{12}$$

Write 4 as $\frac{4}{1}$.

$$= \frac{48}{12} - \frac{25}{12} + \frac{7}{12}$$

Write $\frac{4}{1}$ with the LCD, 12.

$$= \frac{23}{12} + \frac{7}{12}$$

Subtract. Perform operations from left to right.

$$= \frac{30}{12}$$

Add.

$$= \frac{5}{2} \text{ or } 2\frac{1}{2}$$

Write the answer in lowest terms as an improper fraction or mixed number.

[YOU TRY 4] Simplify each expression using the order of operations.

a) $\left(-\frac{2}{9}\right)^2$

b) $-\left(\frac{2}{9}\right)^2$

c) $-\frac{1}{2} + \frac{2}{3} \div 8$

d) $9 - 7\left(\frac{3}{2} - \frac{5}{7}\right)$

e) $\frac{4}{11} + \frac{6}{11}\left(-\frac{1}{3}\right)^2 - \frac{14}{33}$

E Evaluate   **3.7**   Exercises   Do the exercises, and check your work.

Objective 1: Compare Fractions

For Exercises 1–6, use the number line to fill in the blank with > or <.

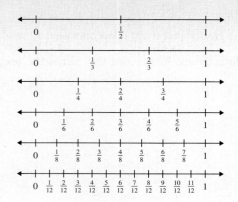

1) $\dfrac{1}{4}$ ___ $\dfrac{1}{3}$ 2) $\dfrac{1}{6}$ ___ $\dfrac{1}{4}$

3) $\dfrac{11}{12}$ ___ $\dfrac{7}{8}$ 4) $\dfrac{7}{8}$ ___ $\dfrac{5}{6}$

5) $\dfrac{5}{6}$ ___ 1   6) 1 ___ $\dfrac{2}{3}$

Fill in the blank with > or <.

7) $\dfrac{2}{3}$ ___ $\dfrac{1}{3}$ 8) $\dfrac{1}{8}$ ___ $\dfrac{3}{8}$

9) $-\dfrac{7}{12}$ ___ $-\dfrac{5}{12}$ 10) $-\dfrac{3}{11}$ ___ $-\dfrac{4}{11}$

11) $\dfrac{12}{5}$ ___ $\dfrac{13}{5}$ 12) $\dfrac{23}{14}$ ___ $\dfrac{19}{14}$

13) $3\dfrac{5}{7}$ ___ $3\dfrac{4}{7}$ 14) $2\dfrac{8}{9}$ ___ $2\dfrac{7}{9}$

15) $-5\dfrac{5}{6}$ ___ $-5\dfrac{1}{6}$ 16) $-4\dfrac{3}{8}$ ___ $-4\dfrac{5}{8}$

17) $\dfrac{4}{5}$ ___ 1   18) 1 ___ $\dfrac{5}{6}$

19) $\dfrac{63}{6}$ ___ 10   20) $\dfrac{53}{7}$ ___ 8

21) Explain, in your own words, how to compare fractions with different denominators.

22) Is this statement true or false? *As you move to the right on the number line, the numbers get larger.*

Fill in the blank with > or <.

23) $\dfrac{2}{3}$ ___ $\dfrac{5}{6}$ 24) $\dfrac{5}{8}$ ___ $\dfrac{11}{16}$

25) $\dfrac{5}{6}$ ___ $\dfrac{13}{18}$ 26) $\dfrac{3}{4}$ ___ $\dfrac{19}{24}$

27) $-\dfrac{5}{4}$ ___ $-\dfrac{4}{3}$ 28) $-\dfrac{13}{9}$ ___ $-\dfrac{11}{7}$

29) $\dfrac{5}{7}$ ___ $\dfrac{5}{6}$ 30) $\dfrac{3}{8}$ ___ $\dfrac{5}{12}$

31) $\dfrac{1}{2}$ ___ $\dfrac{1}{9}$ 32) $\dfrac{1}{3}$ ___ $\dfrac{1}{8}$

33) $1\dfrac{5}{12}$ ___ $1\dfrac{3}{8}$ 34) $2\dfrac{13}{15}$ ___ $2\dfrac{5}{6}$

35) $-4\dfrac{7}{15}$ ___ $-4\dfrac{3}{10}$ 36) $-3\dfrac{4}{5}$ ___ $-3\dfrac{11}{12}$

37) 0 ___ $\dfrac{1}{12}$ 38) $\dfrac{1}{4}$ ___ 0

Objective 2: Use Exponents with Fractions

Write each multiplication problem using an exponent.

39) $\dfrac{1}{5} \cdot \dfrac{1}{5} \cdot \dfrac{1}{5} \cdot \dfrac{1}{5}$ 40) $\dfrac{1}{9} \cdot \dfrac{1}{9} \cdot \dfrac{1}{9} \cdot \dfrac{1}{9} \cdot \dfrac{1}{9}$

41) $\left(-\dfrac{3}{4}\right)\left(-\dfrac{3}{4}\right)\left(-\dfrac{3}{4}\right)\left(-\dfrac{3}{4}\right)\left(-\dfrac{3}{4}\right)\left(-\dfrac{3}{4}\right)$

42) $\left(-\dfrac{8}{11}\right)\left(-\dfrac{8}{11}\right)\left(-\dfrac{8}{11}\right)$

Evaluate.

43) $\left(\dfrac{1}{8}\right)^2$ 44) $\left(\dfrac{1}{11}\right)^2$

45) $\left(\dfrac{5}{6}\right)^2$ 46) $\left(\dfrac{3}{7}\right)^2$

47) $\left(-\dfrac{10}{9}\right)^2$ 48) $\left(-\dfrac{12}{5}\right)^2$

49) $\left(\dfrac{2}{3}\right)^3$ 50) $\left(\dfrac{3}{4}\right)^3$

51) $\left(-\dfrac{1}{4}\right)^3$ 52) $\left(-\dfrac{1}{3}\right)^3$

53) $\left(-\dfrac{1}{3}\right)^4$ 54) $\left(-\dfrac{1}{2}\right)^4$

55) $\left(\dfrac{10}{3}\right)^3$ 56) $\left(\dfrac{5}{2}\right)^3$

57) $\left(\dfrac{2}{3}\right)^4$ 58) $\left(\dfrac{3}{10}\right)^4$

Fill in the blank with >, <, or =.

59) $\left(\dfrac{1}{3}\right)^2$ ____ $\dfrac{1}{3}$ 60) $\left(\dfrac{1}{4}\right)^2$ ____ $\dfrac{1}{4}$

61) $\dfrac{2}{9}$ ____ $\left(-\dfrac{2}{9}\right)^2$ 62) $\dfrac{3}{10}$ ____ $\left(-\dfrac{3}{10}\right)^2$

63) $\left(-\dfrac{3}{2}\right)^2$ ____ $\dfrac{9}{4}$ 64) $\left(-\dfrac{8}{5}\right)^2$ ____ $\dfrac{64}{25}$

Objective 3: Use the Order of Operations with Fractions

65) Explain, in your own words, the difference between evaluating $\left(-\dfrac{1}{2}\right)^2$ and $-\left(\dfrac{1}{2}\right)^2$.

66) Identify the base of $-\left(\dfrac{2}{3}\right)^2$ and $\left(-\dfrac{2}{3}\right)^2$, then evaluate each.

Evaluate.

67) a) $\left(-\dfrac{1}{8}\right)^2$ b) $-\left(\dfrac{1}{8}\right)^2$

68) a) $\left(-\dfrac{1}{6}\right)^2$ b) $-\left(\dfrac{1}{6}\right)^2$

69) a) $-\left(\dfrac{2}{3}\right)^3$ b) $\left(-\dfrac{2}{3}\right)^3$

70) a) $-\left(\dfrac{4}{5}\right)^3$ b) $\left(-\dfrac{4}{5}\right)^3$

71) a) $-\left(\dfrac{1}{2}\right)^4$ b) $\left(-\dfrac{1}{2}\right)^4$

72) a) $-\left(\dfrac{1}{3}\right)^4$ b) $\left(-\dfrac{1}{3}\right)^4$

Simplify each expression using the order of operations.

73) $\left(\dfrac{5}{3}\right)^2 \cdot \left(-\dfrac{6}{7}\right)^2$ 74) $\left(\dfrac{1}{6}\right)^2 \cdot \left(-\dfrac{3}{10}\right)^2$

75) $\dfrac{7}{12} - \dfrac{1}{5} \cdot \dfrac{1}{12}$ 76) $\dfrac{13}{6} - \dfrac{5}{6} \cdot \dfrac{2}{3}$

77) $\dfrac{40}{9} - \dfrac{25}{12} \div \dfrac{3}{5}$ 78) $\dfrac{7}{9} - \dfrac{1}{6} \div \dfrac{3}{8}$

79) $-\dfrac{13}{24} \div \left(\dfrac{5}{12} + \dfrac{1}{8}\right)$ 80) $-\dfrac{10}{27} \div \left(\dfrac{1}{6} + \dfrac{7}{9}\right)$

81) $-\dfrac{7}{8} + \dfrac{1}{3} - \dfrac{5}{6}$ 82) $-\dfrac{9}{10} + \dfrac{1}{5} - \dfrac{3}{4}$

83) $10 - 10\left(\dfrac{4}{5}\right)^2 - \dfrac{4}{15}$ 84) $1 - 3\left(\dfrac{2}{9}\right)^2 + \dfrac{5}{9}$

85) $\dfrac{3}{8}\left(-\dfrac{8}{9} + \dfrac{3}{4}\right)$ 86) $\dfrac{8}{5}\left(-\dfrac{8}{9} + \dfrac{3}{4}\right)$

87) $\left(\dfrac{2}{3}\right)^3 - \left(-\dfrac{2}{9}\right)^2$ 88) $\left(\dfrac{5}{4}\right)^3 - \left(-\dfrac{5}{8}\right)^2$

89) $\left(-\dfrac{3}{4} - \dfrac{1}{6}\right)^2$ 90) $\left(-\dfrac{2}{3} - \dfrac{1}{4}\right)^2$

91) $3\dfrac{3}{4} - 5 \div \dfrac{10}{3}$ 92) $3\dfrac{3}{5} - 8 \div \dfrac{16}{3}$

93) $\left(\dfrac{3}{4}\right)^2 + \left(-\dfrac{1}{2}\right)^3 \cdot \left(-\dfrac{1}{2} + \dfrac{3}{4}\right)$

94) $\left(-\dfrac{1}{3} + \dfrac{2}{15}\right) \div \dfrac{3}{5} \cdot \left(-\dfrac{2}{3}\right)^3$

R Rethink

R1) Why does it make sense to learn how to compare fractions before learning to use exponents and the order of operations?

R2) Based on Exercises 59 and 60, what happens when you square a proper fraction?

3.8 Solving Equations Containing Fractions

What are your objectives for Section 3.8?	How can you accomplish each objective?
1 Solve Equations Containing Fractions	• Review the **Addition, Subtraction, and Division Properties of Equality** from Chapter 2. • Learn the **Multiplication Property of Equality.** • Complete the given examples on your own. • Complete You Trys 1–6.
2 Solve Equations by First Eliminating the Fractions	• Write the procedure for **Eliminating Fractions from an Equation** in your own words, and add it to the procedure developed for solving linear equations in one variable. • Complete the given examples on your own. • Complete You Trys 7 and 8.
3 Distinguish Between Expressions and Equations	• Summarize the differences between expressions and equations. • Complete the given example on your own. • Complete You Try 9.

W Work   **Read the explanations, follow the examples, take notes, and complete the You Trys.**

1 Solve Equations Containing Fractions

In Section 2.3, we learned how to solve equations using the addition and subtraction properties of equality.

> **Property** Addition and Subtraction Properties of Equality
>
> Let a, b, and c be expressions representing numbers. Then,
>
> 1) If $a = b$, then $a + c = b + c$ Addition Property of Equality
> 2) If $a = b$, then $a - c = b - c$ Subtraction Property of Equality

We use these properties to solve equations like $x - 7 = 9$ and $y + 6 = 1$. These properties apply to equations containing fractions, too.

EXAMPLE 1   Solve $n + \dfrac{1}{2} = \dfrac{3}{4}$. Check the solution.

Solution

Here, $\frac{1}{2}$ is being **added** to n. To get the n by itself, **subtract** $\frac{1}{2}$ from each side.

$$n + \frac{1}{2} = \frac{3}{4}$$

$$n + \frac{1}{2} - \frac{1}{2} = \frac{3}{4} - \frac{1}{2} \qquad \text{Subtract } \frac{1}{2} \text{ from each side.}$$

$$n = \frac{3}{4} - \frac{2}{4} \qquad \text{Get an LCD: } \frac{1}{2} \cdot \frac{2}{2} = \frac{2}{4}.$$

$$n = \frac{1}{4} \qquad \text{Simplify.}$$

Check: Substitute $\frac{1}{4}$ for n in the original equation.

$$n + \frac{1}{2} = \frac{3}{4}$$

$$\frac{1}{4} + \frac{1}{2} = \frac{3}{4}$$

$$\frac{1}{4} + \frac{2}{4} = \frac{3}{4} \qquad \frac{1}{2} \cdot \frac{2}{2} = \frac{2}{4}$$

$$\frac{3}{4} = \frac{3}{4} \checkmark$$

Therefore, $n = \frac{1}{4}$ is the solution.

$\left[\text{YOU TRY 1}\right]$ Solve $t + \frac{3}{8} = \frac{5}{6}$. Check the solution.

The division property of equality says that we can *divide* both sides of an equation by a nonzero number and we get an equivalent equation. We can use it to solve an equation like $18 = 3k$. The multiplication property of equality says that we can *multiply* both sides of an equation by a nonzero number and we get an equivalent equation. We use it to solve equations like $\frac{w}{4} = -5$ and $\frac{3}{4}y = 12$. Let's state these two properties together.

W Hint

Did you notice that dividing both sides of an equation by c is the same as multiplying both sides by $1/c$?

Property Multiplication and Division Properties of Equality

Let a, b, and c be expressions representing numbers where $c \neq 0$. Then,

1) If $a = b$, then $ac = bc$.   Multiplication Property of Equality

2) If $a = b$, then $\dfrac{a}{c} = \dfrac{b}{c}$.   Division Property of Equality

EXAMPLE 2

Solve $\dfrac{w}{4} = -5$.

Solution

The w is being **divided** by 4. Therefore, we will **multiply** each side by 4 to get the w on a side by itself.

$$\dfrac{w}{4} = -5$$

$$4 \cdot \dfrac{w}{4} = 4 \cdot (-5) \qquad \text{Multiply each side by 4.}$$

$$1w = -20 \qquad \text{Simplify.}$$

$$w = -20$$

Check:

$$\dfrac{w}{4} = -5$$

$$\dfrac{-20}{4} = -5 \qquad \text{Substitute } -20 \text{ for } w.$$

$$-5 = -5 \checkmark$$

The solution of $\dfrac{w}{4} = -5$ is -20.

[**YOU TRY 2**]

Solve $\dfrac{z}{9} = -7$.

Sometimes we use the *reciprocal* of a number to solve an equation because when you multiply a number by its reciprocal, you get 1.

EXAMPLE 3

Solve each equation.

a) $\dfrac{3}{4}y = 12$ b) $-7 = \dfrac{1}{2}k$

Solution

a) On the left-hand side, the y is being **multiplied** by $\dfrac{3}{4}$. So, we could divide each side by $\dfrac{3}{4}$. However, recall that dividing a quantity by a fraction is the same as multiplying by the **reciprocal** of the fraction. Therefore, we will multiply each side by the reciprocal of $\dfrac{3}{4}$.

$$\dfrac{3}{4}y = 12$$

$$\dfrac{4}{3} \cdot \dfrac{3}{4}y = \dfrac{4}{3} \cdot 12 \qquad \text{The reciprocal of } \tfrac{3}{4} \text{ is } \tfrac{4}{3}. \text{ Multiply each side by } \tfrac{4}{3}.$$

$$1y = \dfrac{4}{3} \cdot \overset{4}{1\!\!\!/2} \qquad \text{Perform the multiplication.}$$

$$y = 16 \qquad \text{Simplify.}$$

Check:

$$\frac{3}{4}y = 12$$

$$\frac{3}{4}(16) = 12 \qquad \text{Substitute 16 for } y$$

$$\frac{3}{4}\left(\frac{\overset{4}{16}}{1}\right) = 12 \qquad \text{Write 16 as } \frac{16}{1}. \text{ Divide out the common factor 4.}$$

$$12 = 12 \checkmark \qquad \text{Multiply.}$$

The solution is 16.

b) On the right-hand side of the equation, the k is being multiplied by $\frac{1}{2}$. What is the reciprocal of $\frac{1}{2}$? It is $\frac{2}{1}$ or 2. Multiply both sides by 2 to solve for k.

W Hint

Notice that the variable is not always on the left-hand side of the equal sign.

$$-7 = \frac{1}{2}k$$

$$2(-7) = 2\left(\frac{1}{2}k\right) \qquad \text{Multiply both sides by 2.}$$

$$-14 = k$$

The solution of $-7 = \frac{1}{2}k$ is -14. The check is left to the student.

$\left[\text{YOU TRY 3}\right]$ Solve each equation.

a) $\frac{5}{9}z = 30$ \qquad b) $-3 = \frac{1}{8}r$

If there are fractions on both sides of the equation, make sure you notice which one is the coefficient of the variable so that you multiply by the reciprocal of the correct fraction.

EXAMPLE 4

Solve $-\frac{5}{6}a = -\frac{4}{9}$.

Solution

The variable is on the left-hand side of the equation, so we need to multiply by the reciprocal of $-\frac{5}{6}$.

W Hint

Are you working out the example on your paper as you read it?

$$-\frac{5}{6}a = -\frac{4}{9}$$

$$-\frac{6}{5}\left(-\frac{5}{6}a\right) = -\frac{6}{5}\left(-\frac{4}{9}\right) \qquad \text{The reciprocal of } -\frac{5}{6} \text{ is } -\frac{6}{5}. \text{ Multiply both sides by } -\frac{6}{5}.$$

$$a = -\frac{\overset{2}{6}}{5}\left(-\frac{4}{\underset{3}{9}}\right) \qquad \text{Divide out the common factor 3.}$$

$$a = \frac{8}{15} \qquad \text{Multiply.}$$

Therefore, $a = \frac{8}{15}$. The check is left to the student.

Solve $-\dfrac{7}{12}q = -\dfrac{3}{8}$.

Sometimes, it makes sense to multiply by the reciprocal of a coefficient even if the coefficient is an integer.

EXAMPLE 5

Solve $\dfrac{3}{8} = 2p$.

Solution

The coefficient of the variable is an integer, so we could solve the equation by dividing each side by 2. However, *because the other number is a fraction,* we will solve it by multiplying each side by the reciprocal of 2. Since $2 = \dfrac{2}{1}$, the reciprocal of 2 is $\dfrac{1}{2}$.

$$\dfrac{3}{8} = 2p$$

$$\dfrac{1}{2}\left(\dfrac{3}{8}\right) = \dfrac{1}{2}(2p) \qquad \text{The reciprocal of 2 is } \dfrac{1}{2}. \text{ Multiply both sides by } \dfrac{1}{2}.$$

$$\dfrac{3}{16} = p \qquad \text{Multiply.}$$

The check is left to the student. The solution is $\dfrac{3}{16}$.

YOU TRY 5

Solve $\dfrac{2}{3} = 5d$.

Remember that to solve an equation like $3n + 4 = 10$, we use more than one property of equality. The same may be true if an equation contains a fraction.

EXAMPLE 6

Solve $-\dfrac{2}{3}c - 5 = -1$.

Solution

On the left-hand side, the c is being multiplied by $-\frac{2}{3}$, and 5 is being subtracted from the c-term. To solve the equation, begin by eliminating the number being subtracted from the c-term.

$$-\frac{2}{3}c - 5 = -1$$

$$-\frac{2}{3}c - 5 + 5 = -1 + 5 \qquad \text{Add 5 to each side.}$$

$$-\frac{2}{3}c = 4 \qquad \text{Combine like terms.}$$

$$-\frac{3}{2} \cdot \left(-\frac{2}{3}c\right) = -\frac{3}{2} \cdot 4 \qquad \text{Multiply each side by the reciprocal of } -\frac{2}{3}.$$

$$1c = -\frac{3}{\underset{1}{2}} \cdot \overset{2}{4} \qquad \text{Divide out the common factor 2.}$$

$$c = -6 \qquad \text{Multiply.}$$

The check is left to the student. The solution is $c = -6$.

[YOU TRY 6]

Solve $-\frac{3}{4}u + 8 = -13$.

---

2 Solve Equations by First Eliminating the Fractions

In Section 2.4, we used these steps to solve a linear equation. (Notice that we have added the multiplication property of equality that we just learned.)

> **Procedure** How to Solve a Linear Equation
>
> **Step 1:** **Clear parentheses** and **combine like terms** on each side of the equation.
>
> **Step 2:** **Get the variable on one side of the equal sign and the constant on the other side of the equal sign** (isolate the variable) using the addition or subtraction property of equality.
>
> **Step 3:** **Solve for the variable** using the multiplication or division property of equality.
>
> **Step 4:** **Check the solution** in the original equation.

To solve $n + \frac{1}{2} = \frac{3}{4}$ in Example 1, we subtracted $\frac{1}{2}$ from each side, then performed the subtraction by getting the least common denominator. If an equation contains more than one fraction, however, it may be easier to solve it by first *eliminating* all of the fractions from the equation.

Procedure Eliminating Fractions from an Equation

To eliminate the fractions, determine the least common denominator (LCD) for all the fractions in the equation. Then multiply both sides of the equation by the LCD.

In the following equations, we will eliminate the fractions before applying the procedure in the box **How to Solve a Linear Equation.** Let's solve $n + \dfrac{1}{2} = \dfrac{3}{4}$ (the equation we solved in Example 1) by first eliminating the fractions.

EXAMPLE 7

Solve $n + \dfrac{1}{2} = \dfrac{3}{4}$ by first eliminating the fractions from the equation.

Solution

This equation contains more than one fraction. Before we apply the steps we have already learned for solving an equation, we will eliminate the fractions.

What is the least common denominator of $\dfrac{1}{2}$ and $\dfrac{3}{4}$? **The LCD is 4.** Multiply both sides of the equation by 4 to *eliminate* the fractions.

$$4\left(n + \frac{1}{2}\right) = 4\left(\frac{3}{4}\right)$$ Multiply both sides of the equation by 4, the LCD of $\frac{1}{2}$ and $\frac{3}{4}$.

> **W Hint**
> Look back at Example 1 to compare the different methods for solving this equation.

Now, apply the steps for solving a linear equation. Begin by distributing the 4 to clear the parentheses.

$$4n + 4\left(\frac{1}{2}\right) = 4\left(\frac{3}{4}\right) \qquad \text{Distribute.}$$

$$4n + 2 = 3 \qquad \text{Multiply.}$$

$$4n + 2 - 2 = 3 - 2 \qquad \text{Subtract 2 from each side.}$$

$$4n = 1 \qquad \text{Perform the subtraction.}$$

$$\frac{4n}{4} = \frac{1}{4} \qquad \text{Divide each side by 4.}$$

$$n = \frac{1}{4} \qquad \text{Simplify.}$$

The solution is $\dfrac{1}{4}$. This is the same as the result we obtained in Example 1.

[YOU TRY 7]

Solve $t + \dfrac{3}{8} = \dfrac{5}{6}$ by first eliminating the fractions from the equation.

EXAMPLE 8

Solve $\dfrac{1}{6}h - \dfrac{3}{8}h = 5$.

Solution

Because this equation contains more than one fraction, let's begin by eliminating them.

What is the LCD of $\dfrac{1}{6}$ and $\dfrac{3}{8}$? **The LCD is 24.** Multiply both sides of the equation by 24.

Even though 5 is not a fraction, we must still multiply it by 24! Then, distribute and solve.

$$24\left(\dfrac{1}{6}h - \dfrac{3}{8}h\right) = 24(5) \qquad \text{Multiply both sides of the equation by 24, the LCD of } \tfrac{1}{6} \text{ and } \tfrac{3}{8}.$$

$$24\left(\dfrac{1}{6}h\right) - 24\left(\dfrac{3}{8}h\right) = 120 \qquad \text{Distribute.}$$

$$4h - 9h = 120 \qquad \text{Multiply.}$$

$$-5h = 120 \qquad \text{Combine like terms.}$$

$$\dfrac{-5h}{-5} = \dfrac{120}{-5} \qquad \text{Divide each side by } -5.$$

$$h = -24 \qquad \text{Simplify.}$$

Therefore, $h = -24$. The check is left to the student.

[YOU TRY 8]

Solve $\dfrac{1}{6}m - \dfrac{2}{9}m = 2$.

---

3 Distinguish Between Expressions and Equations

We have learned how to perform operations with expressions, and we have learned how to solve equations. It is very important to be able to distinguish between the two, and this is often confusing for students. Let's compare equations and expressions.

Summary Equations and Expressions

1) An **equation** contains an equal (=) sign. We can *solve* equations. If an equation contains fractions, we can multiply the equation by the LCD of the fractions to *eliminate the denominators* and then solve the equation.

2) An **expression** does *not* contain an equal (=) sign. We *cannot* "solve" expressions. To add or subtract fractions in an expression, we must write each fraction with the LCD before adding or subtracting them. We *do not* eliminate the denominators.

EXAMPLE 9

Identify each as an expression or an equation. If it is an expression, find the sum. If it is an equation, solve it.

a) $\dfrac{3v}{2} + \dfrac{13}{10} = \dfrac{v}{5}$

b) $\dfrac{3v}{2} + \dfrac{13}{10}$

Solution

a) This is an **equation** because it contains an equal sign (=). Multiply both sides of the equation by the LCD to *eliminate* the fractions.

$$\frac{3v}{2} + \frac{13}{10} = \frac{v}{5}$$

$$10\left(\frac{3v}{2} + \frac{13}{10} = \frac{v}{5}\right) = 10\left(\frac{v}{5}\right) \qquad \text{Multiply both sides of the equation by 10, the LCD of all of the fractions..}$$

$$10\left(\frac{3v}{2}\right) + 10\left(\frac{13}{10}\right) = 2v \qquad \text{Distribute.}$$

$$15v + 13 = 2v \qquad \text{Multiply.}$$

$$15v - 15v + 13 = 2v - 15v \qquad \text{Subtract } 15v \text{ from each side.}$$

$$13 = -13v \qquad \text{Combine like terms.}$$

$$\frac{13}{-13} = \frac{-13v}{-13} \qquad \text{Divide each side by } -13.$$

$$-1 = v \qquad \text{Simplify.}$$

The check is left to the student. The solution is -1.

W Hint

Read these solutions carefully so that you know the difference between equations and expressions!

b) $\dfrac{3v}{2} + \dfrac{13}{10}$ is an **expression** because it is the sum of two fractions and does *not* contain an equal sign (=). We do *not* eliminate the fractions. We must write them with a common denominator so that we can add them.

$$\frac{3v}{2} + \frac{13}{10} = \frac{15v}{10} + \frac{13}{10} \qquad \text{Write } \frac{3v}{2} \text{ with the LCD of 10. } \frac{3v}{2} \cdot \frac{5}{5} = \frac{15v}{10}.$$

$$= \frac{15v + 13}{10} \qquad \text{Add.}$$

[YOU TRY 9] Identify each as an expression or an equation. If it is an expression, find the sum. If it is an equation, solve it.

a) $\dfrac{2c}{7} + \dfrac{10}{21}$ b) $\dfrac{2c}{7} + \dfrac{10}{21} = \dfrac{c}{3}$

Note

In Example 9, the fraction $\dfrac{3v}{2}$ can also be written as $\dfrac{3}{2}v$, and $\dfrac{v}{5}$ can also be written as $\dfrac{1}{5}v$. Be aware that these are equivalent ways to write algebraic fractions.

ANSWERS TO [YOU TRY] EXERCISES

1) $t = \dfrac{11}{24}$ 2) $z = -63$ 3) a) $z = 54$ b) $r = -24$ 4) $q = \dfrac{9}{14}$ 5) $d = \dfrac{2}{15}$

6) $u = 28$ 7) $t = \dfrac{11}{24}$ 8) $m = -36$ 9) a) expression; $\dfrac{6c + 10}{21}$ b) equation; $c = 10$

Get Ready

Solve each equation, and check the solution.

1) $n + 12 = 8$

2) $t - 9 = -2$

3) $15 = p - 3$

4) $1 = 7 + d$

5) $-6z = -54$

6) $4h = -4$

Objective 1: Solve Equations Containing Fractions

Solve each equation, and check the solution.

7) $w + \dfrac{7}{10} = \dfrac{4}{5}$

8) $k + \dfrac{3}{8} = \dfrac{19}{32}$

9) $\dfrac{5}{12} = b + \dfrac{2}{3}$

10) $\dfrac{11}{18} = x + \dfrac{5}{6}$

11) $a - \dfrac{4}{9} = \dfrac{1}{6}$

12) $n - \dfrac{1}{6} = \dfrac{3}{4}$

13) $-\dfrac{2}{7} = -\dfrac{3}{5} + d$

14) $-\dfrac{7}{15} = -\dfrac{1}{2} + y$

15) $m - 3 = -\dfrac{5}{8}$

16) $c - 4 = -\dfrac{6}{7}$

17) What is the reciprocal of $-\dfrac{2}{3}$?

18) What is the reciprocal of 5?

Solve each equation, and check the solution.

19) $2a = 10$

20) $3p = 27$

21) $\dfrac{d}{2} = 14$

22) $\dfrac{a}{3} = 12$

23) $9 = -\dfrac{x}{5}$

24) $12 = -\dfrac{k}{8}$

25) $\dfrac{h}{12} = \dfrac{4}{9}$

26) $\dfrac{m}{21} = \dfrac{11}{14}$

27) $\dfrac{1}{4}q = 6$

28) $\dfrac{1}{6}t = 7$

29) $-\dfrac{7}{18} = -\dfrac{1}{9}w$

30) $-\dfrac{3}{8} = -\dfrac{1}{10}b$

31) $\dfrac{1}{3} = -w$

32) $\dfrac{7}{2} = -r$

33) $\dfrac{5}{2}x = -35$

34) $-\dfrac{7}{6}k = -56$

35) $24 = \dfrac{3}{4}r$

36) $42 = \dfrac{6}{7}c$

37) $-\dfrac{7}{4}p = \dfrac{1}{6}$

38) $-\dfrac{13}{9}n = \dfrac{5}{12}$

39) $8a = \dfrac{4}{15}$

40) $6z = \dfrac{2}{7}$

41) $-\dfrac{11}{6} = -3d$

42) $-\dfrac{9}{10} = -2r$

Solve each equation.

43) $8g - 3 = 1$

44) $15f + 8 = 13$

45) $27 = 2 - 15x$

46) $-4 = 17 + 6y$

47) $\dfrac{1}{2}r + 12 = 7$

48) $\dfrac{1}{3}b + 11 = 4$

49) $\dfrac{4}{5}q - 5 = -3$

50) $\dfrac{12}{7}a - 3 = 5$

51) $6 = -\dfrac{11}{3}w + 4$

52) $10 = -\dfrac{9}{4}c + 8$

53) $2 - \dfrac{5}{6}z = -8$

54) $1 - \dfrac{3}{4}k = -5$

55) $\dfrac{1}{2} = \dfrac{3}{4} - \dfrac{1}{6}t$

56) $\dfrac{2}{5} = \dfrac{1}{2} - \dfrac{2}{3}h$

Objective 2: Solve Equations by First Eliminating the Fractions

57) In your own words, explain how to eliminate the fractions from an equation.

58) In your own words, summarize the steps for solving a linear equation. Include the step for eliminating the fractions.

Solve each equation by first eliminating the fractions.

59) $t + \dfrac{9}{10} = \dfrac{3}{4}$

60) $a + \dfrac{7}{8} = \dfrac{7}{12}$

61) $\dfrac{2}{3} + x = \dfrac{7}{6}$

62) $\dfrac{1}{2} + p = \dfrac{5}{6}$

63) $-\dfrac{4}{9} = \dfrac{2}{9}n - \dfrac{2}{5}$

64) $-\dfrac{9}{10} = \dfrac{3}{10}c - \dfrac{3}{4}$

65) $\dfrac{1}{2}y + \dfrac{1}{4}y = 3$

66) $\dfrac{1}{5}b + \dfrac{1}{3}b = 8$

67) $\dfrac{4k}{3} - 2 = \dfrac{3k}{2}$

68) $\dfrac{10z}{7} - 3 = \dfrac{3z}{2}$

69) $\dfrac{h}{2} + \dfrac{6}{5} = \dfrac{11h}{10} - \dfrac{4}{5}$

70) $\dfrac{4v}{9} + \dfrac{1}{3} = \dfrac{4v}{3} - \dfrac{7}{9}$

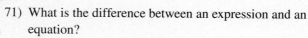
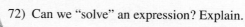

71) What is the difference between an expression and an equation?

72) Can we "solve" an expression? Explain.

Look at the pair in each exercise. Determine which part is an expression and which is an equation. If it is an equation, solve it. If it is an expression, simplify it by performing the indicated operation.

73) a) $\dfrac{r}{8} + \dfrac{1}{2} = \dfrac{3r}{16}$ b) $\dfrac{r}{8} + \dfrac{1}{2}$

74) a) $\dfrac{w}{21} + \dfrac{1}{2}$ b) $\dfrac{w}{21} + \dfrac{1}{2} = \dfrac{3w}{14}$

75) a) $\dfrac{2x}{3} - \dfrac{x}{4}$ b) $\dfrac{5}{6} = \dfrac{2x}{3} - \dfrac{x}{4}$

76) a) $\dfrac{4}{5} = \dfrac{a}{2} - \dfrac{3a}{10}$ b) $\dfrac{a}{2} - \dfrac{3a}{10}$

Mixed Exercises

The following exercises include all types of equations we have learned how to solve so far.

Solve each equation.

77) $9w + 4 = 7w + 16$ 78) $-18c = 12$

79) $-\dfrac{9}{10} = k - \dfrac{1}{4}$ 80) $6 = \dfrac{5}{8}t - 9$

81) $18 - \dfrac{3d}{5} = \dfrac{d}{5} - 30$

82) $2(4y - 5) - 6y = -15 + y$

83) $-\dfrac{9}{2}p + 11 = 5$ 84) $\dfrac{r}{6} + \dfrac{2r}{3} = 1$

Extension

Use what you learned about writing expressions in Chapter 2 to write an algebraic expression for each phrase. Use x for the unknown number.

85) half of a number

86) one-third of a number

87) the sum of 7 and two-thirds of a number

88) the sum of 5 and three-fourths of a number

89) four less than one-third of a number

90) one less than half of a number

Write an equation and solve it to find the missing number. Use x for the unknown number.

91) One more than half a number is eight.

92) Five more than half a number is thirteen.

93) Three-fourths of a number is -63.

94) Two-thirds of a number is -34.

Write an English statement to describe each equation. Let $x =$ the number.

95) $\dfrac{1}{3}x - 2 = 9$

96) $\dfrac{4}{5}x + 7 = 10$

Write an equation to solve each problem. Use the Five-Step Method you first learned in Chapter 2.

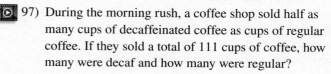

97) During the morning rush, a coffee shop sold half as many cups of decaffeinated coffee as cups of regular coffee. If they sold a total of 111 cups of coffee, how many were decaf and how many were regular?

98) In Jessica's history class, the number of students who bought an e-book was half the number who bought a hard-copy book. If a total of 39 students bought books, how many bought each type?

99) A 50-ft-long piece of crime-scene tape is cut into two pieces so that one piece is two-thirds as long as the other. How long is each piece?

100) A 28-ft-long cable must be cut into two pieces so that one piece is three-fourths as long as the other. Find the length of each piece.

R Rethink

R1) Do you prefer to eliminate fractions from an equation or do you prefer to work with fractions throughout the equation? Why?

R2) How do you recognize the difference between an expression and an equation?

R3) Were there any problems you could not do? If so, write them down or circle them, and ask for help.

Group Activity – Operations with Signed Fractions

- Students should work in one main group of 5 to 6 students. Each group should then break into two smaller groups (group #1 and group #2) of 2 or 3 students each.
- Each equation below contains fractions. Group #1 will solve equation set #1 and group #2 will solve equation set #2.
- Once the students in group #1 solve all the equations in set #1, they should order the solutions from smallest to largest, using the letter of each equation to represent the solution. The students in group #2 should do the same with the solutions from set #2.
- After each group has ordered the solutions, the students in group #1 should compare their ordered list with the ordered list of group #2. If all equations have been solved correctly, and if the solutions have been placed in the proper order, the list from group #1 should match the list from group #2. For example, if the ordered list from group #1 is B, E, D, A, C, then the ordered list from group #2 should also be B, E, D, A, C.
- If the groups find that the ordered lists do not match, then the groups should switch papers and try to identify the errors.

Equation Set #1

A: $\dfrac{1}{2}(x-5) = \dfrac{5}{16} - 4x$

B: $3x - \dfrac{9}{2} = -\dfrac{11}{2}$

C: $\dfrac{3}{8} - \dfrac{3}{2}x = 1\dfrac{1}{8}$

D: $\dfrac{3x}{10} - 2 = x - \dfrac{3}{5}$

E: $\dfrac{1}{4}(4 - 2x) = \dfrac{x}{6}$

Equation Set #2

A: $\dfrac{1}{2}x - \dfrac{14}{3} = \dfrac{5}{6}(2x - 7)$

B: $\dfrac{1}{4} - 3(x+1) = -\dfrac{11}{4}$

C: $\dfrac{104}{15} = \dfrac{x}{3} + 7$

D: $\dfrac{1}{5}(13x + 9) = x + \dfrac{7}{10}x$

E: $\dfrac{1}{12}(x - 7) = -\dfrac{1}{4}$

Do you procrastinate? To find out, circle the number that best applies for each question using the following scale:

1. I invent reasons and look for excuses for not acting on a problem.
   Strongly agree   4   3   2   1   Strongly disagree

2. It takes pressure to get me to work on difficult assignments.
   Strongly agree   4   3   2   1   Strongly disagree

3. I take half measures that will avoid or delay unpleasant or difficult tasks.
   Strongly agree   4   3   2   1   Strongly disagree

4. I face too many interruptions and crises that interfere with accomplishing my major goals.
   Strongly agree   4   3   2   1   Strongly disagree

5. I sometimes neglect to carry out important tasks.
   Strongly agree   4   3   2   1   Strongly disagree

6. I schedule big assignments too late to get them done as well as I know I could.
   Strongly agree   4   3   2   1   Strongly disagree

7. I'm sometimes too tired to do the work I need to do.
   Strongly agree   4   3   2   1   Strongly disagree

8. I start new tasks before I finish old ones.
   Strongly agree   4   3   2   1   Strongly disagree

9. When I work in groups, I try to get other people to finish what I don't.
   Strongly agree   4   3   2   1   Strongly disagree

10. I put off tasks that I really don't want to do but know that I must do.
    Strongly agree   4   3   2   1   Strongly disagree

Scoring: Total the numbers you have circled. If the score is below 15, you are not a chronic procrastinator and you probably have only an occasional problem. If your score is 16–25, you have a minor problem with procrastination. If your score is above 25, you procrastinate quite often and should work on breaking the habit.

Now, consider the following:

- If you do procrastinate often, why do you think you do it? What are some things you can do to complete tasks in a timely manner?
- Are there particular kinds of tasks that you are more likely to procrastinate on?
- Is there something that you are putting off doing right now? How might you get started on it?

Source: Ferner, J. D. (1980). *Successful Time Management.* NY: Wiley. p. 33.

Chapter 3: Summary

Definition/Procedure	Example

3.1 Introduction to Signed Fractions

A **fraction** is a part of a whole.

The **denominator** is the number *below* the fraction bar. It represents the total number of equally-sized parts of a whole.

The **numerator** is the number *above* the fraction bar. It represents the number of parts being considered.

$\frac{2}{3}$ of this circle is shaded.

The *numerator* of the fraction is 2, and the *denominator* is 3.

We can put the number $\frac{2}{3}$ on a number line.

1) $\frac{0}{\text{Number}} = 0$ if the denominator is not 0.

$\frac{0}{9} = 0$

2) $\frac{\text{Number}}{0}$ is undefined.

So, in a fraction like $\frac{5}{6n}$, *n* cannot equal 0.

$\frac{4}{0}$ is undefined.

If the numerator of a fraction is less than the denominator, then the fraction is a **proper fraction.** A proper fraction represents less than 1 whole.

$\frac{3}{5}$ is a proper fraction.

If the numerator of a fraction is greater than or equal to the denominator, then the fraction is an **improper fraction.** An improper fraction represents a quantity greater than or equal to 1 whole.

$\frac{11}{6}$ and $\frac{2}{2}$ are improper fractions.

On a number line, a **negative fraction** is to the left of 0.

Graph $-\frac{3}{4}$ on a number line.

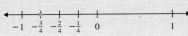

3.2 Writing Fractions in Lowest Terms

Factors are numbers that are multiplied together to get a product. (We consider only natural numbers as factors.)

The factors of 12 are 1, 2, 3, 4, 6, and 12.
Another way to think of the factors of 12 is that 12 *divides evenly* by 1, 2, 3, 4, 6, and 12.

Composite and Prime Numbers

A **composite number** is a number with factors other than 1 and itself.

Identify each number as composite or prime.

a) 11   b) 15

A **prime number** is a number whose only two different factors are 1 and itself.

a) 11 is *prime* because its only factors are 1 and 11.

The numbers 0 and 1 are neither prime nor composite.

b) 15 is *composite* because it has factors other than 1 and 15. For example, because 15 is divisible by 5, 5 is a *factor* of 15: $5 \cdot 3 = 15$.

Definition/Procedure	Example

Prime Factorization

Finding the **prime factorization** of a number means writing the number as a product of prime factors.

We can find the prime factorization of a number using a factor tree or by modifying the long division procedure.

Find the prime factorization of 18

a) using a factor tree.   b) using division.

$18 = 2 \cdot 3 \cdot 3$ or $2 \cdot 3^2$

Equivalent fractions are different fractions that represent the same part of the whole.

$\frac{1}{2}$ and $\frac{2}{4}$ are equivalent fractions.

$\frac{1}{2}$ of the figure is shaded. $\frac{2}{4}$ of the figure is shaded.

A fraction is in **lowest terms** if the numerator and denominator have no common factors other than 1.

$\frac{1}{2}$ is in lowest terms. $\frac{2}{4}$ is not in lowest terms because 2 and 4 have a common factor of 2.

Write a Fraction in Lowest Terms Using Common Factors

To write a fraction in lowest terms:

1) Ask yourself, *"What number divides evenly into both the numerator and denominator?"* (Use any number except 1.) Divide the numerator and denominator by that number.

2) Look at the result and ask yourself, *"Is the fraction in lowest terms?"* If the numerator and denominator still contain a common factor, repeat Steps 1 and 2.

Writing a fraction in lowest terms is also called **simplifying** a fraction.

Write $-\frac{36}{60}$ in lowest terms.

Ask yourself, *"What number divides evenly into both* 36 *and* 60?" Let's use 12. (We could have chosen other factors, but 12 is the **greatest common factor** of 36 and 60.)

$$-\frac{36}{60} = -\frac{36 \div 12}{60 \div 12} = -\frac{3}{5}$$

Ask yourself, *"Is* $-\frac{3}{5}$ *in lowest terms?"* Yes, because 3 and 5 have no common factor other than 1.

Therefore, $-\frac{36}{60} = -\frac{3}{5}$, and $-\frac{3}{5}$ is in lowest terms.

Write a Fraction in Lowest Terms Using Prime Factorization

To write a fraction in lowest terms:

1) Write the prime factorization of the numerator and denominator.

2) Divide out common factors. Write a 1 by each factor to indicate that you have performed the division.

3) Multiply the factors that are left in the numerator and denominator.

Write $\frac{90x^2}{15x}$ in lowest terms.

Write the prime factorization, and divide out the common factors.

$$\frac{90x^2}{15x} = \frac{2 \cdot \overset{1}{\cancel{3}} \cdot 3 \cdot \overset{1}{\cancel{5}} \cdot x \cdot x}{\underset{1}{\cancel{3}} \cdot \underset{1}{\cancel{5}} \cdot \underset{1}{x}} = \frac{6x}{1}$$

Ask yourself, *"Is* $\frac{6x}{1}$ *in lowest terms?"* No! We can write $\frac{6x}{1}$ as $6x$. In lowest terms, $\frac{90x^2}{15x} = 6x$.

Definition/Procedure	Example
Determine Whether Two Fractions Are Equivalent To determine whether two fractions are equivalent, write each of them in lowest terms.	Determine whether $\dfrac{8}{14}$ and $\dfrac{20}{35}$ are equivalent fractions. Begin by writing each fraction in lowest terms. $$\dfrac{8}{14} = \dfrac{8 \div 2}{14 \div 2} = \dfrac{4}{7} \qquad \dfrac{20}{35} = \dfrac{20 \div 5}{35 \div 5} = \dfrac{4}{7}$$ Each fraction equals $\dfrac{4}{7}$ in lowest terms. Therefore, $\dfrac{8}{14}$ and $\dfrac{20}{35}$ are equivalent fractions.

3.3 Multiplying and Dividing Signed Fractions

Multiply Fractions To multiply fractions, multiply the numerators and multiply the denominators. *Always write the result in lowest terms.* We can divide out common factors from numerators and denominators before multiplying.	Multiply. a) $\dfrac{2}{9} \cdot \dfrac{21}{16} = \dfrac{\overset{1}{\cancel{2}}}{\underset{3}{\cancel{9}}} \cdot \dfrac{\overset{7}{\cancel{21}}}{\underset{8}{\cancel{16}}}$ Divide 2 and 16 by 2; divide 9 and 21 by 3. $\qquad = \dfrac{1 \cdot 7}{3 \cdot 8}$ Multiply numerators; multiply denominators. $\qquad = \dfrac{7}{24}$ The answer is in lowest terms. b) $\dfrac{2a}{7a} \cdot \dfrac{3}{5} = \dfrac{2 \cdot \overset{1}{\cancel{a}} \cdot 3}{7 \cdot \underset{1}{\cancel{a}} \cdot 5} = \dfrac{6}{35}$
Multiply a Fraction and an Integer To multiply a fraction and an integer, rewrite the integer as a fraction with a denominator of 1. Then, multiply.	Multiply $-\dfrac{5}{8} \cdot 6$. $-\dfrac{5}{8} \cdot 6 = -\dfrac{5}{8} \cdot \dfrac{6}{1}$ Write 6 with a denominator of 1. $\qquad = -\dfrac{5}{\underset{4}{\cancel{8}}} \cdot \dfrac{\overset{3}{\cancel{6}}}{1}$ Divide 6 and 8 by 2. $\qquad = -\dfrac{5 \cdot 3}{4 \cdot 1}$ Multiply numerators; multiply denominators. $\qquad = -\dfrac{15}{4}$ The answer is in lowest terms.
Find the Reciprocal of a Number Two numbers are **reciprocals** if their product is 1. To find the reciprocal of a number, we interchange the numerator and denominator. We also say that we *invert* or *flip* a fraction to find its reciprocal. The number 0 does *not* have a reciprocal.	The reciprocal of $\dfrac{2}{11}$ is $\dfrac{11}{2}$. We can check by multiplying. $$\dfrac{2}{11} \cdot \dfrac{11}{2} = \dfrac{22}{22} = 1$$

Definition/Procedure	Example

Divide Fractions

To perform division involving fractions, multiply the first number by the reciprocal of the second number.

Divide.

a) $\dfrac{7}{10} \div \dfrac{42}{5}$

$$\dfrac{7}{10} \div \dfrac{42}{5} = \dfrac{7}{10} \cdot \dfrac{5}{42} = \dfrac{\overset{1}{\cancel{7}}}{\underset{2}{\cancel{10}}} \cdot \dfrac{\overset{1}{\cancel{5}}}{\underset{6}{\cancel{42}}} = \dfrac{1 \cdot 1}{2 \cdot 6} = \dfrac{1}{12}$$

Change division to multiplication by the reciprocal.

If a division problem contains a fraction and an integer, the first step is to write the integer as a fraction with a denominator of 1.

Do not divide out common factors until the problem has been changed to multiplication!

b) $\dfrac{9}{10} \div (-8)$

$$\dfrac{9}{10} \div (-8) = \dfrac{9}{10} \div \left(-\dfrac{8}{1}\right) \quad \text{Begin by writing } -8 \text{ as } -\dfrac{8}{1}.$$

$$= \dfrac{9}{10} \cdot \left(-\dfrac{1}{8}\right) \quad \text{Change division to multiplication by the reciprocal.}$$

$$= -\dfrac{9}{80} \quad \text{Multiply.}$$

Complex Fraction

A **complex fraction** contains a fraction in its numerator, its denominator, or both.

We simplify a complex fraction by treating it like a division problem.

Divide $\dfrac{\frac{3a^2}{8b}}{\frac{12a}{5b}}$.

$$\dfrac{\frac{3a^2}{8b}}{\frac{12a}{5b}} = \dfrac{3a^2}{8b} \div \dfrac{12a}{5b} = \dfrac{3a^2}{8b} \cdot \dfrac{5b}{12a} = \dfrac{\overset{1}{\cancel{3}} \cdot a \cdot a}{8b} \cdot \dfrac{5\overset{1}{\cancel{b}}}{\underset{4 \cdot 1}{\cancel{12a}}} = \dfrac{5a}{32}$$

Many applications involve multiplying or dividing with fractions.

Solve the problem.

In 2011, a survey of about 2400 adults revealed that $\dfrac{4}{25}$ of them have driven a car without auto insurance. How many of the people surveyed have driven without insurance? (www.motorwayamerica.com)

The problem states that $\dfrac{4}{25}$ of 2400 people surveyed have driven without auto insurance. The *of* in $\dfrac{4}{25}$ of 2400 means we should multiply.

$$\dfrac{4}{25} \cdot 2400 = \dfrac{4}{25} \cdot \dfrac{2400}{1} = \dfrac{4}{\underset{1}{\cancel{25}}} \cdot \dfrac{\overset{96}{\cancel{2400}}}{1} = \dfrac{4 \cdot 96}{1 \cdot 1} = \dfrac{384}{1} = 384$$

Write 2400 as $\dfrac{2400}{1}$.

384 of the 2400 people surveyed have driven without auto insurance.

Check the answer. Double-check your work to verify that the answer is correct.

Definition/Procedure	Example

3.4 Adding and Subtracting Like Fractions and Finding a Least Common Denominator

Adding and Subtracting Like Fractions

To add or subtract like fractions:

1) Add or subtract the numerators.

2) Use the denominator of the like fractions as the denominator of the sum.

3) Write the answer in lowest terms.

We can also say that if a, b, and c represent numbers and if b does not equal zero, then

$$\frac{a}{b} + \frac{c}{b} = \frac{a+c}{b} \quad \text{and} \quad \frac{a}{b} - \frac{c}{b} = \frac{a-c}{b}$$

Add or subtract.

a) $\dfrac{4}{11} + \dfrac{3}{11}$ b) $\dfrac{9}{10} - \dfrac{3}{10}$

a) $\dfrac{4}{11} + \dfrac{3}{11} = \dfrac{4+3}{11} = \dfrac{7}{11}$ The answer is in lowest terms.

b) $\dfrac{9}{10} - \dfrac{3}{10} = \dfrac{9-3}{10} = \dfrac{6}{10} = \dfrac{3}{5}$ Write the answer in lowest terms.

Writing Equivalent Fractions

Step 1: Ask yourself, *"By what number do I multiply the original denominator to get the new denominator?"*

Step 2: Multiply the numerator and denominator of the original fraction by that number to get the equivalent fraction.

When we multiply the numerator and denominator by the same number, we do *not* change the value of the fraction because we are multiplying the fraction by 1. We get an equivalent fraction.

Write $\dfrac{2}{7}$ as an equivalent fraction with a denominator of 21.

Step 1: Ask yourself, *"By what number do I multiply 7 to get 21?"* That number is 3.

Step 2: Multiply the numerator and denominator of $\dfrac{2}{7}$ by 3:

$$\frac{2}{7} \cdot \frac{3}{3} = \frac{6}{21}$$

Therefore, $\dfrac{2}{7} = \dfrac{6}{21}$.

The **least common multiple,** or **LCM,** of a group of natural numbers is the smallest natural number divisible by each number in the group. To find the LCM of a group of numbers, we can make a list of multiples of the numbers, find the LCM by inspection, or use prime factorization.

The LCM of 4 and 5 is 20 because 20 is the smallest number divisible by both 4 and 5.

Find the LCM Using Prime Factorization

Step 1: Write the prime factorization of each number.

Step 2: Identify the factors that will be in the least common multiple. The LCM will contain each different factor the *greatest* number of times it appears in any single factorization.

Step 3: The LCM is the *product* of the factors identified in Step 2.

Find the LCM of 9 and 15 using prime factorization.

Step 1: Write the prime factorization of each number.

1 factor of 3
↓
$$9 = 3 \cdot 3 \quad \text{and} \quad 15 = 3 \cdot 5$$
2 factors of 3   1 factor of 5

Step 2: The LCM will contain each different factor the *greatest* number of times it appears in any single factorization.

The LCM will contain $3 \cdot 3$ and 5.
2 factors of 3

Step 3: The LCM is the *product* of the factors identified in Step 2.

The LCM of 9 and 15 is $3 \cdot 3 \cdot 5 = 45$.

Definition/Procedure	Example

Least Common Denominator

The **least common denominator,** or **LCD,** of a group of fractions is the least common multiple of the denominators.

The least common denominator of $\frac{1}{4}$ and $\frac{5}{12}$ is 12 because 12 is the least common multiple of 4 and 12.

Writing a Group of Fractions with an LCD

To write a group of fractions with their least common denominator, identify the LCM of the denominators, then rewrite each fraction with the LCM as the least common denominator.

Write $\frac{9}{8}$ and $-\frac{4}{7}$ as equivalent fractions with the LCD as their denominators.

The LCM of 8 and 7 is 56. So, the LCD is 56.

Write each fraction with a denominator of 56. We want to find

$$\frac{9}{8} = \frac{?}{56} \qquad \text{and} \qquad -\frac{4}{7} = -\frac{?}{56}$$

$$\frac{9}{8} \cdot \frac{7}{7} = \frac{63}{56} \qquad\qquad -\frac{4}{7} \cdot \frac{8}{8} = -\frac{32}{56}$$

The LCD of $\frac{9}{8}$ and $-\frac{4}{7}$ is 56, and $\frac{9}{8} = \frac{63}{56}$ and $-\frac{4}{7} = -\frac{32}{56}$.

3.5 Adding and Subtracting Unlike Fractions

In order to add or subtract fractions, they **must** have a common denominator.

Adding or Subtracting Unlike Fractions

Step 1: Determine the least common denominator (LCD), and write it on your paper.

Step 2: Rewrite each fraction with the LCD.

Step 3: Add or subtract.

Step 4: Write the answer in lowest terms.

Add $\frac{4}{9} + \frac{5}{6}$.

Step 1: Write down the LCD of $\frac{4}{9}$ and $\frac{5}{6}$.   **LCD = 18**

Step 2: Rewrite each fraction with the LCD.

$$\frac{4}{9} \cdot \frac{2}{2} = \frac{8}{18} \qquad \frac{5}{6} \cdot \frac{3}{3} = \frac{15}{18}$$

Step 3: Add.

Equivalent fractions

$$\frac{4}{9} + \frac{5}{6} = \frac{8}{18} + \frac{15}{18} = \frac{8+15}{18} = \frac{23}{18} \text{ or } 1\frac{5}{18}$$

Equivalent fractions

Step 4: Ask yourself, *"Are $\frac{23}{18}$ and $1\frac{5}{18}$ in lowest terms?"*

Yes. We can write the answer as an improper fraction or mixed number.

Add and Subtract Integers and Fractions

We can add and subtract integers and fractions by changing the integer to an improper fraction first.

Add or subtract by first changing the integer to an improper fraction.

a) $3 + \frac{h}{8}$ b) $1 - \frac{3}{7}$

a)

Rewrite with a denominator of 8.

$$3 + \frac{h}{8} = \frac{3}{1} + \frac{h}{8} = \frac{24}{8} + \frac{h}{8} = \frac{24+h}{8}$$

Write 3 with a denominator of 1.

Definition/Procedure	Example

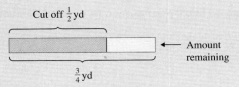

b) $\quad$ Rewrite with a denominator of 7.

$$1 - \frac{3}{7} = \frac{1}{1} - \frac{3}{7} = \frac{7}{7} - \frac{3}{7} = \frac{4}{7}$$

Write 1 as $\frac{1}{1}$.

Applications of Adding and Subtracting Fractions

Many applications involve adding or subtracting fractions.

A piece of trim for a kitchen countertop is $\frac{3}{4}$ yd long. A carpenter cuts off a piece that is $\frac{1}{2}$ yd long and installs it on the counter. How much of the trim remains?

Draw a picture to visualize the problem.

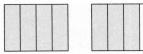

Cut off $\frac{1}{2}$ yd

Amount remaining

$\frac{3}{4}$ yd

To determine how much trim remains, we subtract.

$$\frac{3}{4} - \frac{1}{2} = \frac{3}{4} - \frac{2}{4} = \frac{1}{4}$$

Get a common denominator.

$\frac{1}{4}$ yd of the trim remains.

Check by adding:

$$\frac{1}{4} \quad + \quad \frac{1}{2} \quad = \frac{1}{4} + \frac{2}{4} = \frac{3}{4} \text{ yd}$$

$$\underset{\substack{\text{Amount} \\ \text{remaining}}}{\uparrow} + \underset{\substack{\text{Amount} \\ \text{used}}}{\uparrow} = \underset{\substack{\text{Original} \\ \text{amount}}}{\uparrow}$$

3.6 Operations with Mixed Numbers

A **mixed number** consists of a whole number and a fraction.	$1\frac{3}{4}$ is a mixed number.
A mixed number represents more than 1 whole.	Use a mixed number to represent the portion of the figure that is shaded. Then, represent the mixed number on a number line.

$1\frac{3}{4}$ of the figure is shaded. $\Big($ By counting each shaded region,

Mixed numbers can also be written as improper fractions.

we can also say that $\frac{7}{4}$ of the figure is shaded. $\Big)$

Let's put $1\frac{3}{4}$ on a number line.

$$\begin{array}{ccccc} & & & & \\ \hline 0 & & 1 & 1\frac{1}{4}\ 1\frac{2}{4}\ 1\frac{3}{4}\ 2 \end{array}$$

Definition/Procedure	Example
Change a Mixed Number to an Improper Fraction *Step 1:* Multiply the denominator and the whole number. *Step 2:* Add the numerator to the result in Step 1. *Step 3:* Write the improper fraction. The numerator is the result in Step 2 and the denominator is the original denominator.	Write $1\frac{3}{4}$ as an improper fraction. *Step 1:* Multiply the denominator and the whole number: $4 \cdot 1 = 4$ *Step 2:* Add the numerator to the result in Step 1: $4 + 3 = 7$ *Step 3:* Write the improper fraction: $$1\frac{3}{4} = \frac{7}{4}$$ This is the same as the result we obtained in the previous box when we determined the portion of the figure that was shaded.
Change an Improper Fraction to a Mixed Number *Step 1:* Divide the numerator by the denominator. *Step 2:* Write the mixed number. The *quotient* is the whole-number part. The *remainder* is the numerator of the fractional part, and the denominator is the same as the denominator of the improper fraction.	Write $\frac{17}{5}$ as a mixed number. *Step 1:* Divide the numerator by the denominator. $$\begin{array}{r} 3 \leftarrow \text{Whole-number part} \\ 5\overline{)1\ 7} \\ -\underline{1\ 5} \\ 2 \leftarrow \text{Remainder is the} \\ \text{numerator.} \end{array}$$ *Step 2:* Write the mixed number. $$\frac{17}{5} = 3\frac{2}{5}$$ Use the same denominator.
Multiply or Divide Mixed Numbers To multiply or divide mixed numbers, 1) Change each mixed number to an improper fraction. 2) Multiply or divide the fractions as indicated. 3) Write the answer in lowest terms. If it is an improper fraction, change it to a mixed number.	Multiply $2\frac{1}{2} \cdot 1\frac{3}{4}$. $2\frac{1}{2} \cdot 1\frac{3}{4} = \frac{5}{2} \cdot \frac{7}{4}$ Change each mixed number to an improper fraction. $= \frac{35}{8}$ Multiply. $= 4\frac{3}{8}$ Write the answer as a mixed number. Divide $3\frac{4}{7} \div 6\frac{1}{4}$. $3\frac{4}{7} \div 6\frac{1}{4} = \frac{25}{7} \div \frac{25}{4}$ Change each mixed number to an improper fraction. $= \frac{25}{7} \cdot \frac{4}{25}$ Change division to multiplication by the reciprocal. $= \frac{\overset{1}{\cancel{25}}}{7} \cdot \frac{4}{\underset{1}{\cancel{25}}}$ Divide by 25. $= \frac{4}{7}$ Multiply.

Definition/Procedure	Example

Adding Mixed Numbers

To add mixed numbers, add the whole-number parts and add the fractional parts. Write the answer in lowest terms.

Subtracting Mixed Numbers

To subtract mixed numbers, subtract the whole-number parts and subtract the fractional parts. Write the answer in lowest terms.

Add or subtract.

a) $4\frac{1}{12}+5\frac{4}{9}$ b) $8\frac{1}{2}-6\frac{3}{8}$

a) First, rewrite each fraction with the LCD, 36. Then, add the whole-number parts and add the fractional parts.

$4\frac{1}{12}+5\frac{4}{9}=4\frac{3}{36}+5\frac{16}{36}$ Write each fraction with the LCD.

$=9\frac{19}{36}$ Add the whole numbers, and add the fractions.

b) The LCD is 8, so rewrite $\frac{1}{2}$ with a denominator of 8.

Then, subtract the whole-number parts and subtract the fractional parts.

$8\frac{1}{2}-6\frac{3}{8}=8\frac{4}{8}-6\frac{3}{8}$ Write each fraction with the LCD.

$=2\frac{1}{8}$ Subtract the whole numbers, and subtract the fractions.

Add Mixed Numbers with Regrouping

Sometimes, the sum of mixed numbers must be simplified using regrouping.

Add $7\frac{3}{4}+1\frac{2}{3}$.

$7\frac{3}{4}+1\frac{2}{3}=7\frac{9}{12}+1\frac{8}{12}$ Write the fractions with the LCD.

$=8\frac{17}{12}$ Add.

Is $8\frac{17}{12}$ in simplest form? *No!* $\frac{17}{12}$ is an improper fraction.

Simplify $8\frac{17}{12}$ by regrouping.

$8\frac{17}{12}=8+\frac{17}{12}=8+1\frac{5}{12}=9\frac{5}{12}$

Write the improper fraction as a mixed number.

The final answer is $9\frac{5}{12}$.

Subtract Mixed Numbers with Regrouping

We regroup in a subtraction problem when the fraction in the second mixed number is *larger than* the fraction in the first mixed number.

Subtract $6\frac{2}{9}-3\frac{7}{9}$.

The fraction in the second mixed number is larger than the fraction in the first mixed number. Therefore, we must *regroup* or *borrow* 1 from the whole-number part of the *first* mixed number, $6\frac{2}{9}$.

Write 6 as 5 + 1.

$6\frac{2}{9}=6+\frac{2}{9}=5+1+\frac{2}{9}=5+\frac{9}{9}+\frac{2}{9}=5\frac{11}{9}$

Rewrite 1 as $\frac{9}{9}$.

Now, subtract.

$6\frac{2}{9}-3\frac{7}{9}=5\frac{11}{9}-3\frac{7}{9}=2\frac{4}{9}$

These are equivalent.

Definition/Procedure	Example

Add and Subtract Mixed Numbers Using Improper Fractions

1) Change each mixed number to an improper fraction.

2) Add or subtract the fractions.

3) Write the answer in lowest terms. If it is an improper fraction, change it to a mixed number.

Add $-2\frac{1}{10} + \left(-1\frac{4}{5}\right)$ by changing the mixed numbers to improper fractions.

$$-2\frac{1}{10} + \left(-1\frac{4}{5}\right) = -\frac{21}{10} + \left(-\frac{9}{5}\right) \quad \text{Write the mixed numbers as improper fractions.}$$

$$= -\frac{21}{10} + \left(-\frac{18}{10}\right) \quad \text{Multiply } -\frac{9}{5} \text{ by } \frac{2}{2} \text{ to get a common denominator.}$$

$$= -\frac{39}{10} \quad \text{Add.}$$

$$= -3\frac{9}{10} \quad \text{Write the result as an improper fraction.}$$

Because the numbers in the original problem were mixed numbers, we write the result as a mixed number, if possible.

3.7 Order Relations and Order of Operations

Comparing Fractions

As we move to the right on the number line, the numbers get larger.

Fill in the blank with > or <.

a) $\frac{3}{2}$ ____ $\frac{5}{2}$ b) $-\frac{6}{13}$ ____ $-\frac{10}{13}$

a) $\frac{3}{2} < \frac{5}{2}$ $\frac{3}{2}$ is less than $\frac{5}{2}$.

b) $-\frac{6}{13} > -\frac{10}{13}$ $-\frac{6}{13}$ is greater than $-\frac{10}{13}$.

Comparing Fractions with Unlike Denominators

Rewrite the fractions with the least common denominator.

Fill in the blank with > or <: $\frac{3}{4}$ ____ $\frac{5}{7}$

Write each fraction with the LCD, 28.

$$\frac{3}{4} \cdot \frac{7}{7} = \frac{21}{28} \qquad \frac{5}{7} \cdot \frac{4}{4} = \frac{20}{28}$$

Rewrite $\frac{3}{4}$ ____ $\frac{5}{7}$ as $\frac{21}{28}$ ____ $\frac{20}{28}$.

Therefore, $\frac{21}{28} > \frac{20}{28}$ or $\frac{3}{4} > \frac{5}{7}$.

We can use exponents with fractions. Be careful when working with negative signs.

Evaluate.   a) $\left(\frac{7}{11}\right)^2$ b) $\left(-\frac{7}{11}\right)^2$ c) $-\left(\frac{7}{11}\right)^2$

a) $\left(\frac{7}{11}\right)^2 = \frac{7}{11} \cdot \frac{7}{11} = \frac{49}{121}$

b) $\left(-\frac{7}{11}\right)^2 = -\frac{7}{11} \cdot \left(-\frac{7}{11}\right) = \frac{49}{121}$

   The base is $-\frac{7}{11}$.

c) $-\left(\frac{7}{11}\right)^2$ means $-1 \cdot \left(\frac{7}{11}\right)^2$.

$$-\left(\frac{7}{11}\right)^2 = -1 \cdot \left(\frac{7}{11}\right)^2 = -1 \cdot \frac{7}{11} \cdot \frac{7}{11} = -\frac{49}{121}$$

Definition/Procedure	Example

The Order of Operations

Simplify expressions in the following order:

1) If **parentheses** or **other grouping symbols** appear in an expression, simplify what is in these grouping symbols first.

2) Simplify expressions with **exponents** and **square roots.**

3) **Multiply** or **divide,** moving from left to right.

4) **Add** or **subtract,** moving from left to right.

Simplify $-\dfrac{1}{2} + \dfrac{8}{9}\left(\dfrac{3}{4}\right)^3$.

$$-\dfrac{1}{2} + \dfrac{8}{9}\left(\dfrac{3}{4}\right)^3 = -\dfrac{1}{2} + \dfrac{8}{9}\left(\dfrac{27}{64}\right)$$ Evaluate the exponent first.

$$= -\dfrac{1}{2} + \dfrac{\overset{1}{8}}{9} \cdot \dfrac{\overset{3}{27}}{\underset{8}{64}}$$ Multiply before adding; divide out common factors.

$$= -\dfrac{1}{2} + \dfrac{3}{8}$$ Multiply.

$$= -\dfrac{4}{8} + \dfrac{3}{8}$$ Get a common denominator.

$$= -\dfrac{1}{8}$$ Add.

3.8 Solving Equations Containing Fractions

The Properties of Equality

1) If $a = b$, then $a + c = b + c$.   Addition Property of Equality

2) If $a = b$, then $a - c = b - c$.   Subtraction Property of Equality

3) If $a = b$, then $ac = bc$.   Multiplication Property of Equality

4) If $a = b$, then $\dfrac{a}{c} = \dfrac{b}{c}$ $(c \neq 0)$.   Division Property of Equality

We can use the properties of equality with fractions the same way as with integers.

Solve $\dfrac{x}{2} = -7$.

$$\overset{1}{2} \cdot \dfrac{x}{\underset{1}{2}} = 2 \cdot (-7)$$ Multiply each side by 2.

$$x = -14$$

The solution is $x = -14$.

Check: $\dfrac{x}{2} = -7$

$$\dfrac{-14}{2} = -7$$

$$-7 = -7 \checkmark$$

If a variable has a fractional coefficient, we can multiply by the reciprocal of the coefficient in an equation like $\dfrac{3}{5}m = 6$.

Solve $\dfrac{3}{5}m = 6$.

$$\dfrac{5}{3} \cdot \dfrac{3}{5}m = \dfrac{5}{3} \cdot 6$$ Multiply each side by $\dfrac{5}{3}$.

$$m = 10$$

The solution is $m = 10$. The check is left to the student.

How to Solve a Linear Equation

Step 1: **Clear parentheses** and **combine like terms** on each side of the equation.

Step 2: **Get the variable on one side of the equal sign and the constant on the other side of the equal sign** (isolate the variable) using the addition or subtraction property of equality.

Step 3: **Solve for the variable** using the multiplication or division property of equality.

Step 4: **Check the solution** in the original equation.

Solve $2(c + 2) + 11 = 5c + 9$.

$2c + 4 + 11 = 5c + 9$	Distribute.
$2c + 15 = 5c + 9$	Combine like terms.
$2c - 5c + 15 = 5c - 5c + 9$	Get variable terms on one side.
$-3c + 15 = 9$	
$-3c = -6$	Get constants on one side.
$\dfrac{-3c}{-3} = \dfrac{-6}{-3}$	Division property of equality
$c = 2$	

The solution is $c = 2$. The check is left to the student.

Definition/Procedure	Example

Solve Equations Containing Fractions

To eliminate the fractions, determine the least common denominator (LCD) for all of the fractions in the equation. Then, multiply both sides of the equation by the LCD.

Use the steps we learned in Chapter 2 to solve the equation.

Solve $\frac{3}{4}y - 3 = \frac{1}{4}y - \frac{2}{3}$.　　　LCD = 12

$12\left(\frac{3}{4}y - 3\right) = 12\left(\frac{1}{4}y - \frac{2}{3}\right)$　Multiply each side of the equation by 12.

$12 \cdot \frac{3}{4}y - 12 \cdot 3 = 12 \cdot \frac{1}{4}y - 12 \cdot \frac{2}{3}$　Distribute.

$9y - 36 = 3y - 8$　Multiply.

$9y - 3y - 36 = 3y - 3y - 8$　Get the y-terms on one side.

$6y - 36 = -8$　Get the constants on the other side.

$6y - 36 + 36 = -8 + 36$

$6y = 28$

$\frac{6y}{6} = \frac{28}{6}$　Divide each side by 6.

$y = \frac{28}{6} = \frac{14}{3}$　Write the answer in lowest terms.

The check is left to the student.

Chapter 3:　Review Exercises

(3.1) Use a fraction to represent the shaded part of the figure, and represent the fraction on a number line.

1)

2)

3) Can the denominator of a fraction be any number? Explain.

4) What is the difference between an improper fraction and a proper fraction?

Identify each fraction as proper or improper.

5) $\frac{4}{7}$

6) $\frac{13}{9}$

Evaluate.

7) $\left|-\frac{8}{11}\right|$

8) $\left|\frac{2}{5}\right|$

Graph each group of fractions on a number line.

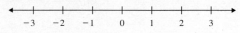

9) $-\frac{2}{3}, \frac{5}{8}, \frac{7}{4}, -\frac{5}{2}$

10) $-\frac{3}{4}, \frac{1}{2}, \frac{7}{3}, -\frac{11}{8}$

(3.2)

11) How do you know whether a number is divisible by 3? Write your own example of a four-digit number that is divisible by 3.

12) How do you know whether a number is divisible by 5? Write your own example of a four-digit number that is divisible by 5.

Identify each number as prime or composite.

13) 57

14) 41

Write the prime factorization of each number.

15) 42

16) 30

17) 1800

18) 792

Determine whether each fraction is in lowest terms.

19) $\frac{18}{27}$

20) $\frac{15}{28}$

Write each fraction in lowest terms.

21) $\frac{16}{40}$

22) $\frac{12}{45}$

23) $-\frac{550}{700}$

24) $-\frac{420}{560}$

25) $-\frac{63z}{99}$

26) $-\frac{16c}{36}$

27) $\frac{4a^2}{12a^5}$

28) $\frac{6n}{48n^3}$

Determine whether each pair of fractions is equivalent.

29) $-\frac{14}{21}$ and $-\frac{56}{88}$

30) $\frac{147}{462}$ and $\frac{126}{396}$

(3.3)

31) In your own words, explain how to divide fractions.

32) What question should you ask yourself after you find the product or quotient of fractions?

Perform the indicated operations.

33) $-\dfrac{3}{10} \div \dfrac{7}{15}$

34) $\left(-\dfrac{18}{5}\right)\left(-\dfrac{1}{12}\right)$

35) $\dfrac{13x}{15} \cdot 45$

36) $2 \div \dfrac{8}{3}$

37) $\dfrac{36}{11} \div 8$

38) $\dfrac{\frac{4t^2}{3}}{\frac{2t}{9}}$

39) $\left(\dfrac{a^2}{14b}\right)\left(\dfrac{7b^3}{a^4}\right)$

40) $-15 \cdot \dfrac{2w}{25}$

41) $\dfrac{\frac{3m}{8k^2}}{-\frac{7m^2}{18k^2}}$

42) $\dfrac{72}{13} \cdot \dfrac{1}{6} \cdot \dfrac{13}{8}$

43) $\dfrac{3}{34}\left(-\dfrac{8}{9}\right)\left(-\dfrac{17}{12}\right)$

44) $-\dfrac{28c}{19} \div 7c^3$

45) If $\dfrac{3}{4}$ of a liter of olive oil is divided into 6 equal portions, what is the size of each portion?

46) After a winter storm, a store owner uses $\dfrac{3}{4}$ of a 20-lb bag of rock salt to melt the ice in his parking lot. How much salt did he use?

©Ingram Publishing/SuperStock

(3.4)

47) Explain the difference between *like* and *unlike* fractions.

48) Given the fraction $\dfrac{2}{9}$,

   a) write a like fraction.

   b) write an unlike fraction.

Perform the indicated operations.

49) $\dfrac{1}{9} + \dfrac{7}{9}$

50) $\dfrac{6}{7} - \dfrac{2}{7}$

51) $\dfrac{13}{8} - \dfrac{5}{8}$

52) $\dfrac{1}{10} + \dfrac{3}{10}$

53) $-\dfrac{5}{6y} + \dfrac{1}{6y}$

54) $\dfrac{n}{11} + \dfrac{4}{11}$

Write each fraction with the indicated denominator.

55) $\dfrac{3}{10} = \dfrac{?}{50}$

56) $-\dfrac{5}{4} = -\dfrac{?}{36u}$

Find the least common multiple of each group of numbers.

57) 16 and 18

58) 9, 12, and 60

Identify the least common denominator of each group of fractions. Then, write each as an equivalent fraction with the LCD as its denominator.

59) $\dfrac{7}{10}$ and $\dfrac{5}{8}$

60) $\dfrac{3}{4p}, \dfrac{1}{6p}$ and $\dfrac{2}{3}$

(3.5)

61) In your own words, explain how to add unlike fractions.

62) Ted adds these fractions this way: $\dfrac{2}{3} + \dfrac{3}{5} = \dfrac{5}{8}$. Is this right or wrong? Explain.

Perform the indicated operations.

63) $\dfrac{1}{4} + \dfrac{1}{8}$

64) $\dfrac{1}{2} + \dfrac{1}{4}$

65) $\dfrac{8}{15} - \dfrac{7}{12}$

66) $\dfrac{3}{10} - \dfrac{5}{6}$

67) $\dfrac{h}{9} + \dfrac{4}{5}$

68) $\dfrac{9}{11} - \dfrac{2r}{3}$

69) $\dfrac{3}{4} - \dfrac{1}{10x}$

70) $\dfrac{8}{k} + \dfrac{5}{k^2}$

71) $-\dfrac{47}{56} + \dfrac{19}{40}$

72) $\dfrac{13}{21} + \left(-\dfrac{31}{35}\right)$

73) $1 - \dfrac{4}{11}$

74) $\dfrac{w}{7} - 4$

75) $-\dfrac{8}{15} + \dfrac{1}{2} + \left(-\dfrac{1}{30}\right)$

76) $\dfrac{5}{6} + \left(-\dfrac{7}{8}\right) + \dfrac{5}{12}$

The chart shows the recorded rainfall, measured in inches, for a Nevada desert area. Use the chart for Exercises 77 and 78.

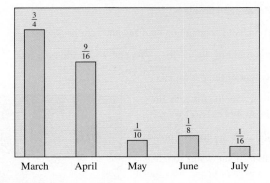

Recorded rainfall measured in inches

77) Find the total amount of rain during May, June, and July.

78) How much more rainfall did the area get in June than in May?

(3.6) Use a mixed number to describe which portion of the figure is shaded.

79)

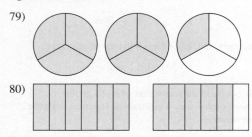

80)

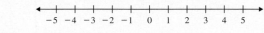

Graph each group of numbers on a number line.

$$\begin{array}{c}\longleftrightarrow\\ -5\ -4\ -3\ -2\ -1\quad 0\quad 1\quad 2\quad 3\quad 4\quad 5\end{array}$$

81) $3\frac{1}{4}, -4\frac{1}{2}, -\frac{2}{3}, \frac{5}{6}, 2\frac{3}{8}$ 82) $4\frac{3}{4}, -3\frac{1}{3}, -\frac{1}{2}, \frac{7}{8}, 3\frac{2}{5}$

Write each mixed number as an improper fraction.

83) $4\frac{2}{5}$ 84) $9\frac{5}{8}$

Write each improper fraction as a mixed number.

85) $\frac{26}{3}$ 86) $\frac{567}{11}$

87) For which operation(s) with mixed numbers do you need a common denominator? Choose from addition, subtraction, multiplication, and division.

88) Explain how to multiply mixed numbers.

Perform the indicated operations.

89) $5\frac{3}{8} \div 3\frac{1}{4}$ 90) $9\frac{1}{6} + 4\frac{1}{9}$

91) $-17\frac{4}{5} + 11\frac{3}{10}$ 92) $-3\frac{1}{7} \cdot \left(-8\frac{3}{4}\right)$

93) $15\frac{3}{8} - 5\frac{5}{6}$ 94) $6\frac{2}{3} + 7\frac{4}{5}$

95) $9\left(4\frac{1}{3}\right)$ 96) $2\frac{3}{10} - 5$

(3.7) Solve each problem.

97) Hyun sold some items on eBay and sent packages weighing $4\frac{3}{4}$ lb, $1\frac{5}{8}$ lb, $5\frac{1}{2}$ lb, and $2\frac{1}{4}$ lb. Find the total weight of the packages.

98) Toni's recipe for ciabatta bread makes two loaves and uses $2\frac{1}{3}$ cups of bread flour. If she wants to make 8 loaves of bread, how much flour will she need?

©Ingram Publishing

Fill in the blank with >, <, or =.

99) $-\frac{4}{9}$ _____ $-\frac{1}{9}$ 100) $\frac{12}{15}$ _____ $\frac{20}{25}$

101) $\frac{5}{8}$ _____ $\frac{7}{12}$ 102) $-2\frac{1}{6}$ _____ $-2\frac{3}{4}$

Evaluate.

103) $\left(-\frac{2}{3}\right)^2$ 104) $\left(\frac{1}{10}\right)^3$

105) $-\left(\frac{2}{3}\right)^2$ 106) $\left(-\frac{1}{10}\right)^3$

Simplify each expression using the order of operations.

107) $\frac{1}{10} - \frac{4}{5} \cdot \frac{1}{6}$ 108) $\left(\frac{3}{4} - \frac{5}{6}\right)^2$

109) $\left(\frac{3}{5}\right)^2 + \left(\frac{1}{2}\right)^2 \div \left(-\frac{3}{8} + \frac{3}{4}\right)$

110) $\frac{2}{15} + \frac{9}{10} \div 18 + \frac{11}{30}$

111) $-\frac{11}{12} + \frac{5}{8} - \frac{5}{6}$ 112) $12 - 10\left(\frac{3}{2}\right)^2 + 3\frac{1}{2}$

(3.8)

113) Solve $t + \frac{3}{4} = \frac{2}{5}$

a) by first subtracting $\frac{3}{4}$ from each side of the equation.

b) by first eliminating the fractions.

114) What is the difference between an expression and an equation?

Solve each equation.

115) $-\frac{6}{7}p = 18$ 116) $\frac{1}{3} = y + \frac{7}{8}$

117) $-2 = \frac{k}{4} - 7$ 118) $-\frac{3}{2}w = -\frac{1}{10}$

119) $\frac{6}{11} = 9x$ 120) $\frac{4}{5}c - 8 = -36$

121) $\frac{1}{3}n - 1 = \frac{1}{2}n$ 122) $\frac{h}{2} + \frac{2}{9} = \frac{13}{18} + \frac{2h}{3}$

123) Identify each as an expression or an equation. If it is an equation, solve it. If it is an expression, simplify it by performing the operation.

a) $\frac{d}{2} - \frac{2d}{5}$ b) $\frac{d}{2} - \frac{2d}{5} = \frac{1}{5}$

Write an equation and solve.

124) A 9-ft-long board is to be cut into two pieces so that one piece is half as long as the other. Find the length of each piece.

Mixed Exercises
Perform the indicated operations.

125) $\dfrac{7}{12} - \dfrac{3}{4}$

126) $\dfrac{1}{4}\left(2\dfrac{4}{7}\right)$

127) $\dfrac{8k^4}{9} \div (-2k)$

128) $-\dfrac{5}{8w} + \dfrac{2}{9w}$

129) $\dfrac{21}{22} \cdot \dfrac{25}{24} \cdot \dfrac{18}{35}$

130) $3\dfrac{7}{10} + 4\dfrac{5}{6}$

Chapter 3:   Test

For Exercises 1–7, fill in the blank with *always, sometimes,* or *never* to make the statement true.

1) A common denominator is _____ needed to multiply fractions.

2) A common denominator is _____ needed to subtract fractions.

3) An improper fraction is _____ greater than 1.

4) When multiplying or dividing mixed numbers, it is _____ necessary to begin by changing the mixed numbers to improper fractions.

5) A fraction is _____ undefined when the denominator equals 0.

6) The sum of a negative fraction and a positive fraction is _____ negative.

7) A negative fraction raised to the 6th power is _____ negative.

8) The least common denominator of two fractions is _____ the product of their denominators.

9) Use a fraction to represent the shaded part of the figure.

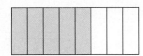

10) Use a mixed number to represent the shaded part of the figure.

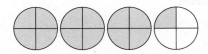

11) Graph the numbers on a number line.

$$-\dfrac{3}{4}, \ 3\dfrac{2}{3}, \ -\dfrac{9}{2}, \ \dfrac{1}{2}, \ 2\dfrac{3}{8}, \ -2\dfrac{1}{5}$$

<---+---+---+---+---+---+---+---+---+---+--->
$-5 \ -4 \ -3 \ -2 \ -1 \ \ 0 \ \ 1 \ \ 2 \ \ 3 \ \ 4 \ \ 5$

12) a) Find all factors of 24.

  b) Write the prime factorization of 24.

Write each fraction in lowest terms.

13) $-\dfrac{48}{72}$

14) $\dfrac{510w^4}{1050w}$

15) Are $\dfrac{9}{12}$ and $\dfrac{24}{32}$ equivalent fractions? Explain your answer.

16) Write each fraction with the indicated denominator.

  a) $\dfrac{4}{7} = \dfrac{?}{63}$

  b) $-\dfrac{5}{6} = -\dfrac{?}{18z}$

17) Is the mixed number $6\dfrac{5}{3}$ in lowest terms? Explain your answer.

18) a) When you are dividing fractions, do you need to get a common denominator?

  b) Explain how to divide fractions.

Perform the indicated operations.

19) $\dfrac{7}{12} - \dfrac{1}{4}$

20) $-\dfrac{9}{20} \cdot \left(-\dfrac{2}{3}\right)$

21) $5\dfrac{2}{7} + 2\dfrac{3}{8}$

22) $\dfrac{5}{18x} + \dfrac{7}{18x}$

23) $\dfrac{5c^3}{6d^3} \cdot \dfrac{2d}{9c^2}$

24) $\dfrac{2}{3} + \dfrac{3}{20} + \dfrac{1}{4}$

25) $\dfrac{7}{9} - 1$

26) $4\dfrac{7}{8} + 12\dfrac{2}{5}$

27) $\dfrac{-\dfrac{w}{8}}{\dfrac{9w}{4}}$

28) $-32 \cdot \dfrac{3r}{8}$

29) $\dfrac{15}{16} \div 10$

30) $-2\dfrac{4}{9} \div 1\dfrac{5}{6}$

31) $-\dfrac{19}{24} + \left(-\dfrac{23}{40}\right)$

32) $\dfrac{a}{6} - 2$

33) $\dfrac{4}{m} - \dfrac{3}{2m^2}$

34) $\dfrac{3}{5} + \dfrac{1}{8z}$

Fill in the blank with >, <, or = .

35) $-\dfrac{5}{11}$ _____ $-\dfrac{9}{11}$

36) $2\dfrac{5}{9}$ _____ $2\dfrac{7}{12}$

37) Evaluate.

  a) $\left(\dfrac{2}{5}\right)^3$

  b) $\left(-\dfrac{3}{8}\right)^2$

  c) $-\left(\dfrac{3}{8}\right)^2$

Simplify using the order of operations.

38) $\dfrac{1}{6} - \dfrac{1}{12} \div \dfrac{3}{16}$

39) $2 - 12\left(\dfrac{1}{3}\right)^2 + \left(-\dfrac{1}{4}\right)$

40) $\left(\dfrac{1}{3} - \dfrac{1}{2}\right)^3$

Solve each problem.

41) The label on a 20-oz bottle of root beer says that it contains $2\dfrac{1}{2}$ servings. How many ounces are in each serving?

42) Sig is a crab fisherman, and his crew hauled up a pot containing 320 opi crabs. Because they cannot keep females or crabs that are too small, the crew had to throw back $\dfrac{3}{8}$ of their catch. How many crabs did they keep?

©earl_of_omaha/iStock/
Getty Images

43) A recipe for Caribbean black beans uses $\dfrac{3}{4}$ tsp of ginger, $\dfrac{1}{3}$ tsp of cumin, and $\dfrac{1}{2}$ tsp of allspice.

a) Find the total amount of these ingredients in the recipe.

b) How much more ginger is used than cumin?

Solve each equation.

44) $\dfrac{3}{4}p = 18$

45) $n - \dfrac{3}{5} = -\dfrac{5}{6}$

46) $\dfrac{4x}{9} = \dfrac{2x}{3} + 2$

47) $\dfrac{14}{11} = 7y$

48) $6 - \dfrac{2}{5}h = 16$

49) $\dfrac{4}{9}c + \dfrac{1}{3} = \dfrac{1}{2}c - \dfrac{1}{9}$

50) Write an expression for "nine less than half a number." Use x for the unknown number.

Chapter 3:   Cumulative Review for Chapters 1–3

Add, subtract, multiply, or divide as indicated.

1) $458 + 3017 + (-839)$

2) $\begin{array}{r} 6435 \\ -\,4709 \\ \hline \end{array}$

3) $(628)(-509)$

4) $25\overline{)15{,}042}$

Round each number to the nearest ten, nearest hundred, and nearest thousand.

	Ten	Hundred	Thousand
5) 675	_____	_____	_____
6) 58,399	_____	_____	_____

7) A group of 9 lottery players share a winning ticket that is worth \$4,007,106 after taxes. If the money is divided up evenly, how much will each person receive?

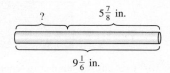

Simplify each expression using the order of operations.

8) $46 - 4\sqrt{79 - 15} \div 4 \cdot 3 - \sqrt{81}$

9) $4^2 \cdot 2^2 - (14 - 6) \div 2$

10) Write $\dfrac{1140}{3720}$ in lowest terms.

Perform the indicated operations.

11) $\dfrac{14}{15} \div \dfrac{8}{25}$

12) $\dfrac{7}{10} - \dfrac{1}{6x}$

13) $5\dfrac{7}{8} + 6\dfrac{3}{4}$

14) $-2c \cdot \dfrac{5}{24c^2}$

15) $-\dfrac{20}{27} \cdot \dfrac{14}{11} \cdot \dfrac{9}{28}$

16) $\dfrac{7}{8} - \dfrac{2}{5} \div \dfrac{8}{3}$

17) $\dfrac{k}{9} - \dfrac{5}{12}$

For Exercises 18–19, answer true or false.

18) $\dfrac{7}{6} > \dfrac{9}{8}$

19) $-\dfrac{8}{7} < -\dfrac{13}{12}$

20) Find the missing length.

? $\underbrace{\qquad}$ $5\dfrac{7}{8}$ in.
$9\dfrac{1}{6}$ in.

Solve each equation.

21) $2(y + 8) - 13 = 5y + 21$

22) $9 = 4 - \dfrac{10}{3}v$

23) $20w = 5$

24) $\dfrac{3a}{14} + \dfrac{2a}{7} = 2$

Write an equation and solve. Use the Five Steps for Solving Applied Problems.

25) After playing two games of *Angry Birds,* Amanda had a total of 44,000 points. She had 6000 fewer points in her first game than in the second. How many points did she get in each game?

Design elements: Yellow diamond road sign ©Nora Development; Texas Instruments TI-83 graphing calculator ©mbbirdy/Getty Images; Yellow pencil ©McGraw-Hill Education.

Basic Geometry Concepts and Algebra

©Corbis Premium RF/Alamy

Math at Work:

Interior Designer

Jamie Callum walks into a room and sees possibilities. She immediately starts imagining pictures on the walls, rugs across the floor, lights hanging from the ceiling. As an interior designer, her job, in her words, "is to make spaces comfortable, beautiful, and functional."

Jamie grew up reading fashion and architecture magazines. What she never expected, however, was how much math interior design would require. "I couldn't do my job without geometry," Jamie explains. "I have to account for the angles of walls, the perimeters of rooms, and which furniture should be parallel and perpendicular to the walls. Math helps me bring my creative ideas to life." Jamie's combination of mathematical skill and creative inspiration has enabled her to design the interiors of homes, restaurants, and offices.

Interior design, Jamie explains, is ultimately a series of decisions. "Every room is like a blank page," she describes. "You have to choose how to fill that page. But I have to listen to what my clients need, what they want, and determine what is possible. So it is very important for me to be a good listener and to take good notes so that when I go back to the office, I can design something my client will love."

In this chapter, we will explore geometry, the math of angles, lines, and shapes. We'll also discuss some ways to become better note-takers.

Taking good notes is not always easy, but it is essential for college success. What an instructor says in class usually reflects what he or she thinks are the most important points on any subject. How do you feel about *your* note-taking skills? Let's see how we can use the P.O.W.E.R. framework to take good notes.

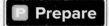

- *Prepare means to explicitly state a goal.* **I will take good notes.**

- *Organize means to **organize** the physical and mental tools you need to achieve your goal.*
- Complete the emPOWERme survey that appears before the Chapter Summary to determine whether you are an active listener. Consider the results of the survey so that you know what you do well and where you need to improve.
- Buy one notebook for each class. Bring your book, math notebook, and pencils or pens to class. Bring colored pencils or pens, too.
- Take a seat in the classroom where you can see and hear the instructor clearly.
- Turn off your cell phone and put it away to minimize distractions.

- *Work means to **do the work** that needs to be done to achieve the goal.*
- Be an *active* listener. Think about what your instructor is saying, and ask questions if you do not understand something. Remember, if *you* have a question chances are good that someone else has the same question too!
- Take notes in a notebook that is only for the class you are in. At the top of the paper, write down the date as well as the section number for the day's lesson. Write neatly!
- Don't write down *everything*, but be sure to write down the key definitions, procedures, formulas, and examples, especially those that your instructor writes on the board.
- When copying the examples, use colored pencils or pens when a step is performed, such as dividing both sides of an equation by 3, so that you can *easily* see the operation performed in each step when you look at your notes at a later time.
- While you are taking notes, include comments *in your own words* that may help you to better understand a procedure or example.
- Be sure to write down all information about homework due dates, quizzes, and tests.

- *Evaluate means you should **evaluate** what you have done.* Did you achieve your goal?
- Look over your notes as class is ending. Are they neat and easy to read? Do you understand everything you wrote? Is there anything that is incomplete or that you do not understand? If so, ask another student or your instructor about it before you leave class. If necessary, see your instructor after class, talk with another student, or go to the tutoring center for extra help.

- **Rethink** *means to **rethink** and reflect upon your goal.*
- Sometime after class, especially before doing your homework, read over your notes. Are they easy to understand? Could one of your classmates understand what you have written? If not, what could you do differently to take better notes?

©Hero Images/Getty Images

Chapter 4 **P.O.W.E.R** Plan

P Prepare

O Organize

What are your goals for Chapter 4?	How can you accomplish each goal?
1 Be prepared before and during class.	• Don't stay out late the night before class, and be sure to set your alarm! • Bring a pencil, notebook paper, and textbook to class. • Avoid distractions by turning off your cell phone during class. • Pay attention, take good notes, and ask questions. • Complete your homework on time, and ask questions on problems you do not understand.
2 Understand the homework to the point where you could do it without needing any help or hints.	• Read the directions, and show all of your steps. • Go to the instructor's office for help. • Rework homework and quiz problems, and find similar problems for practice.
3 Use the P.O.W.E.R. framework to learn how to take good notes in class.	• Complete the emPOWERme that appears before the Chapter Summary to learn about your listening style. • Read the Study Strategy to learn how to use the P.O.W.E.R. framework to take notes. Then try some of the strategies to improve your notetaking. • Exchange notes with a classmate, and see if you can understand each other's notes. Make suggestions to each other on how you can improve your notetaking.
4 Write your own goal. _____ _____	• _____ _____

What are your objectives for Chapter 4?	How can you accomplish each objective?
1 Understand the basic components of geometry.	• Know the definitions of lines, segments, rays, different types of angles, and parallel and perpendicular lines.
2 Be able to find perimeter and area as well as classify various types of geometric figures.	• Know the definitions of a rectangle, square, triangle, parallelogram, and trapezoid. • Find missing measures of different figures. • Identify different types of triangles. • Use the formulas to find the perimeter and area of a rectangle, square, triangle, parallelogram, and trapezoid.
3 Be able to compute the volume and surface area of different geometric figures.	• Use the formulas to find the volume of a rectangular solid and pyramid. • Be able to find the surface area of a rectangular solid.
4 Use geometry formulas and algebra to solve problems.	• Identify the geometry formula needed to solve the problem. • Be able to solve an equation using the techniques learned in previous chapters. • Use the **Five Steps for Solving Applied Problems.**
5 Write your own goal. _____ _____	• _____ _____

E Evaluate	Complete the Chapter Review and Chapter Test. How did you do?	R Rethink	• Can you use the definitions and formulas in this chapter without looking at your book or your notes? It is important to know them for future chapters.
			• Are you happy with the results of your quizzes and tests in this chapter? Why or why not? What would you do the same and what would you do differently in the next chapter?
			• Did you use P.O.W.E.R. to help you take notes? If so, what was helpful? If not, why not?

4.1 Introduction to Geometry

P Prepare　　　　　　　**O Organize**

What are your objectives for Section 4.1?	How can you accomplish each objective?
1 Identify Lines, Line Segments, and Rays	• Write the definitions of *space, plane, point, line, line segment,* and *ray* in your notes with an example next to each. • Complete the given example on your own. • Complete You Try 1.
2 Identify and Classify Angles	• Learn the different parts that make up an *angle* such as *rays* and *vertex,* and understand the different ways to name an angle using $\angle$. • Understand what a *degree* represents. • Learn the definitions of an *acute angle, right angle, obtuse angle,* and a *straight angle.* • Complete the given examples on your own. • Complete You Trys 2 and 3.
3 Identify Parallel and Perpendicular Lines	• Draw examples and write the definitions of *intersecting lines, parallel lines,* and *perpendicular lines* in your notes. • Complete the given example on your own. • Complete You Try 4.

W Work　　　　**Read the explanations, follow the examples, take notes, and complete the You Trys.**

Thousands of years ago, the Egyptians developed techniques to measure the amount of land a person owned. They used this information to collect taxes. Later, the Greeks formalized this process of measurement into a branch of mathematics we call **geometry.** The word *geometry* comes from the Greek words for "Earth measurement."

Today, geometry is used in many different ways. We use geometry when we do home improvement projects, buy carpeting, or work in the garden. And it is used by people in different careers like interior design, air traffic control, science, construction, and many more.

In this chapter, we will learn some basics of geometry beginning with some terms.

1 Identify Lines, Line Segments, and Rays

Space is an unlimited, three-dimensional expanse. The Earth is an example of an object in space. A **plane** is a flat surface that continues indefinitely. A floor is part of a plane, as is a wall or a piece of paper. The most basic concept in geometry is a *point*. A **point** is a location in space with no length, width, or height. A point is represented by a dot, and we usually name a point with a capital letter. For example, here is point P.

$$\bullet\, P$$
Point P

A **line** is a straight set of points that continues forever in two directions. When we draw a line through two points, we draw arrows at the ends to show that the line never ends. We name the line using any two points on the line and put the notation $\longleftrightarrow$ above the letters to indicate that it is a line. We can also name a line with a lowercase letter. For example, here is line PQ or $\overleftrightarrow{PQ}$ or line l.

line PQ or $\overleftrightarrow{PQ}$ or line l

A **line segment** is a piece of a line with two endpoints. We name a line segment using its two endpoints. For example, this line segment with endpoints A and B can be named $\overline{AB}$ or $\overline{BA}$.

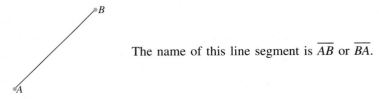

The name of this line segment is $\overline{AB}$ or $\overline{BA}$.

A **ray** is a part of a line. It has one endpoint and continues forever in the other direction. *We name a ray using the endpoint first and any other point on the ray.* Here is ray RT, which we write as $\overrightarrow{RT}$.

Ray RT or $\overrightarrow{RT}$

EXAMPLE 1

Identify each figure as a line, a line segment, or a ray. Then, name it using the correct notation.

a)

b)

c) X •————————————• Y

Solution

a) This figure has *one* endpoint and an arrowhead on the other end indicating that it goes on forever in that direction. **This is a ray.** Write its name as $\overrightarrow{MN}$. (Notice that the *M must* come first because it is the endpoint of the ray.)

b) This figure is straight and continues forever in both directions. **This is a line.** Write its name as $\overleftrightarrow{AB}$ or $\overleftrightarrow{BA}$.

c) This figure is straight with two endpoints. **This is a line segment.** Write its name as $\overline{XY}$ or $\overline{YX}$.

[**YOU TRY 1**] Identify each figure as a line, a line segment, or a ray. Then name it using the correct notation.

a)

b)

c)

2 Identify and Classify Angles

If we join two rays at their endpoints, we get an **angle.** The **vertex** of the angle is the common endpoint. This angle is formed by joining rays $\overrightarrow{BA}$ and $\overrightarrow{BC}$.

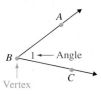

An angle is denoted by the symbol ∠. We can name this angle in different ways:

Put the vertex in the middle	Using only the vertex	Using the number label given to the angle
∠ABC or ∠CBA	∠B	∠1

Note

Sometimes two or more angles share the same vertex. In this case, we do **not** name the angle using only the vertex.

EXAMPLE 2   Name each of the numbered angles in this figure in two different ways.

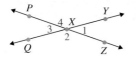

Solution

W Hint

How many different rays, line segments, and lines are represented in this example?

First, notice that X is the vertex of *all* the angles. So, **we cannot use just the vertex to name any of the angles in this figure.**

We can give each of the angles the following names:

$\angle 1$: $\angle YXZ$ or $\angle ZXY$ $\angle 2$: $\angle ZXQ$ or $\angle QXZ$

$\angle 3$: $\angle PXQ$ or $\angle QXP$ $\angle 4$: $\angle PXY$ or $\angle YXP$

[YOU TRY 2]   Name each of the numbered angles in this figure in two different ways.

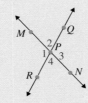

Angles can be different sizes. In Example 2, $\angle 1$ is smaller than $\angle 2$. We can measure the size of an angle using **degrees.** The symbol for degrees is °. For example, if the measure of an angle is 45 degrees, we write 45°. Let's look at a circle to understand the sizes of degree measures.

If we start at a ray and go all the way around the circle, we have moved 360°. We say that a circle contains 360°.

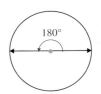

If we start at a ray and form an angle halfway around a circle, the angle measure is 180°. A **straight angle** is an angle whose measure is 180°.

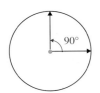

The angle formed by moving a quarter of the way around a circle has a measure of 90°. A **right angle** is an angle whose measure is 90°. (It may also be called a *square angle*.) A right angle is denoted with a small square at the vertex:

A right angle has a measure of 90°.

Definition

1) An **acute angle** is an angle whose measure is greater than 0° and less than 90°.

   Examples:

   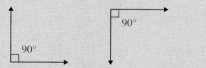

2) A **right angle** is an angle whose measure is 90°.

   Examples:

   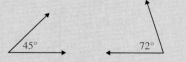

3) An **obtuse angle** is an angle whose measure is greater than 90° and less than 180°.

   Examples:

   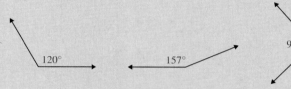

4) A **straight angle** is an angle whose measure is 180°.

   Examples:

   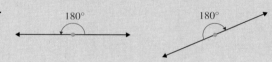

Note

Angles can also be measured in radians. This topic is studied in future math courses.

EXAMPLE 3   Classify each angle as acute, right, obtuse, or straight.

a)
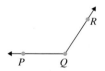

b)
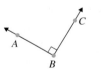

c)

d)
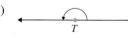

Solution

a) The measure of ∠*PQR* is greater than 90° and less than 180°. It is an **obtuse angle.**

b) The small square inside the angle tells us that its measure is *exactly* 90°. ∠*ABC* is a **right angle.**

c) The measure of ∠*YXZ* is less than 90°, so it is an **acute angle.**

d) The measure of ∠*T* is 180°, so it is a **straight angle.**

Classify each angle as acute, right, obtuse, or straight.

a) K b) R S c) Y Z d)

                            T A

Note

The measure of an angle is denoted by m∠. For example, in Example 3b, the measure of ∠ABC is 90°. We write this as m∠$ABC = 90°$.

3   Identify Parallel and Perpendicular Lines

At the beginning of this section, we said that a plane is a flat surface that continues indefinitely. A surface like a piece of paper is part of a plane.

Two lines that cross each other are called **intersecting lines.**

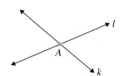

Lines l and k *intersect* at point A.

Parallel lines are lines in the same plane that do *not* intersect. The symbol ‖ means "is parallel to."

Lines x and y do *not* intersect. They are *parallel*. We can write $x \parallel y$.

Perpendicular lines intersect at right angles. The symbol ⊥ means "is perpendicular to."

Lines $\overleftrightarrow{PQ}$ and $\overleftrightarrow{RT}$ intersect at right angles. They are *perpendicular*. We can write $\overleftrightarrow{PQ} \perp \overleftrightarrow{RT}$.

EXAMPLE 4

Classify each pair of lines as parallel, perpendicular, or neither. If they are parallel or perpendicular, use the appropriate notation.

a)

b)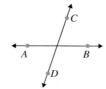

c)

Solution

a) Lines m and n meet at right angles, so they are perpendicular. Write $m \perp n$.

b) Lines $\overleftrightarrow{AB}$ and $\overleftrightarrow{CD}$ intersect but not at right angles. They are neither parallel nor perpendicular.

c) Lines $\overleftrightarrow{JK}$ and $\overleftrightarrow{WX}$ do *not* intersect, so they are parallel. Write $\overleftrightarrow{JK} \parallel \overleftrightarrow{WX}$.

[YOU TRY 4]

Classify each pair of lines as parallel, perpendicular, or neither. If they are parallel or perpendicular, use the appropriate notation.

a)

b)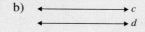

c)

Note

Even if two lines or rays *look like* they intersect at right angles, we cannot assume they are right angles unless the angle is denoted with the square.

is a right angle   may *not* be a right angle

ANSWERS TO [YOU TRY] EXERCISES

1) a) line segment; $\overline{CD}$ or $\overline{DC}$ b) ray; $\overrightarrow{HG}$ c) line; $\overleftrightarrow{MN}$ or $\overleftrightarrow{NM}$
2) $\angle 1$: $\angle RPM$ or $\angle MPR$; $\angle 2$: $\angle MPQ$ or $\angle QPM$; $\angle 3$: $\angle QPN$ or $\angle NPQ$; $\angle 4$: $\angle RPN$ or $\angle NPR$
3) a) right   b) acute   c) straight   d) obtuse
4) a) neither   b) $c \parallel d$ c) $\overleftrightarrow{AB} \perp \overleftrightarrow{FG}$

Objective 1: Identify Lines, Line Segments, and Rays

1) Explain the difference between a line segment and a ray.

2) Draw a picture of a vertical ray pointing downward. Name the ray.

Identify each figure as a line, line segment, or a ray. Then name it using the correct notation.

3)
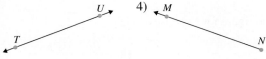

4)
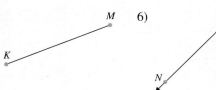

5)

6)

7) U   V
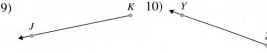

8)

9)
10)

11) Draw a line named $\overleftrightarrow{CT}$.

12) Draw a ray named $\overrightarrow{RD}$.

Objective 2: Identify and Classify Angles

13) What is an obtuse angle?

14) What is an acute angle?

15) What is a right angle, and how is it indicated when an angle is drawn?

16) What kind of angle has a measure of 180°?

Name each of the numbered angles in the figures in two different ways.

17)
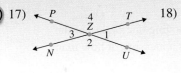

18)
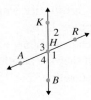

19)

20)
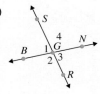

21)

22)

Classify each angle as acute, right, obtuse, or straight.

23)

24)

25)

26)

27) E   B   S
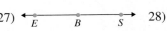

28)

29)

30)

31) Draw an obtuse angle named ∠*TNW*.

32) Draw a right angle named ∠*AYC*.

33) The angle formed by moving one-third of the way around a circle has a measure of how many degrees?

34) The angle formed by moving one-eighth of the way around a circle has a measure of how many degrees?

Objective 3: Identify Parallel and Perpendicular Lines

Classify each pair of lines as parallel, perpendicular, or neither. If they are parallel or perpendicular, use the appropriate notation.

35)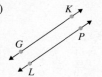

36)

37)

38)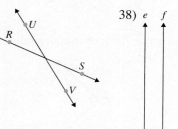

39)

40)

Mixed Exercises: Objectives 1–3

Classify each angle as acute, right, obtuse, or straight.

41)

42)

43)

44)

Use the figure for Exercises 45–54.

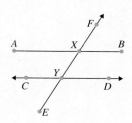

45) Identify the ray in the figure.

46) Identify the line segment in the figure.

Answer *true* or *false* for Exercises 47–54.

47) ∠*BXF* is an acute angle.

48) ∠*DYE* is an obtuse angle.

49) $\overline{AB} \parallel \overrightarrow{EF}$

50) $\overleftrightarrow{CD} \perp \overline{AB}$

51) m∠*EYD* is greater than m∠*CYE*.

52) m∠*AXF* is less than m∠*BXF*.

53) m∠*AXY* could be 116°.

54) m∠*XYD* could be 77°.

R Rethink

R1) Look at and think about this homework assignment. Which vocabulary and concepts do you understand well and which do you need to work on more?

R2) If a circle were divided into only 100 degrees instead of 360 degrees, how many degrees would represent a right angle and a straight angle?

4.2 Rectangles, Squares, Parallelograms, and Trapezoids

P Prepare

O Organize

What are your objectives for Section 4.2?	How can you accomplish each objective?
1 Find the Perimeter and Area of a Rectangle	<ul><li>Write the definition of a *rectangle* in your own words, draw an example, and note its main characteristics.</li><li>Write the formula for determining the **Perimeter of a Rectangle** in your notes.</li><li>Write the formula for determining the **Area of a Rectangle** in your notes.</li><li>Complete the given examples on your own.</li><li>Complete You Trys 1–3.</li></ul>
2 Find the Perimeter and Area of a Square	<ul><li>Write the definition of a *square* in your own words, draw an example, and note its main characteristics.</li><li>Write the formula for determining the **Perimeter of a Square** in your notes.</li><li>Write the formula for determining the **Area of a Square** in your notes.</li><li>Complete the given example on your own.</li><li>Complete You Try 4.</li></ul>
3 Find the Perimeter and Area of a Parallelogram	<ul><li>Write the definition of a *parallelogram* in your own words, draw an example, and note its main characteristics.</li><li>Know how to find the perimeter of a parallelogram.</li><li>Write the formula for determining the **Area of a Parallelogram** in your notes.</li><li>Complete the given example on your own.</li><li>Complete You Try 5.</li></ul>
4 Find the Perimeter and Area of a Trapezoid	<ul><li>Write the definition of a *trapezoid* in your own words, draw an example, and note its main characteristics.</li><li>Know how to find the perimeter of a trapezoid.</li><li>Write the formula for determining the **Area of a Trapezoid** in your notes.</li><li>Complete the given example on your own.</li><li>Complete You Try 6.</li></ul>
5 Find the Perimeter and Area of an Irregular Figure	<ul><li>After following the example, write your procedure for finding the perimeter and area of an irregular figure.</li><li>Complete the given example on your own.</li><li>Complete You Try 7.</li></ul>

 **Work**   **Read the explanations, follow the examples, take notes, and complete the You Trys.**

In this section, we will learn how to find the *perimeter* and *area* of different geometric figures. Here is the definition of perimeter.

Definition

The **perimeter** of a figure is the distance around the figure.

Let's begin by applying this definition to a *rectangle*.

1   Find the Perimeter and Area of a Rectangle

Definition

A **rectangle** is a four-sided figure containing four right angles.

Here are other important characteristics of a rectangle.

1) The opposite sides are parallel and have the same length.

2) The longer side of a rectangle is called the **length,** often abbreviated **l.** The shorter side is called the **width,** often abbreviated **w.**

3) We can use a small square to indicate that a four-sided figure has 90° angles. We can do this in a couple of ways.

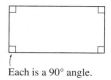

Each is a 90° angle.

If the sides are parallel and *one* 90° angle is indicated, then the other angles also measure 90°.

We can find the perimeter of a rectangle by adding the lengths of the sides. Or, if we are given the length *l* and width *w*, we can use the following formulas to find its perimeter. Perimeter is often denoted by *P*.

Formula   Perimeter of a Rectangle

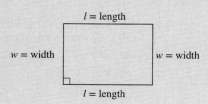

$l = $ length
$w = $ width
$w = $ width
$l = $ length

We can find the perimeter, *P*, of a rectangle using one of the following formulas:

1) $P = 2 \cdot length + 2 \cdot width$ or

2) $P = 2l + 2w$

EXAMPLE 1   Find the perimeter of the rectangle.

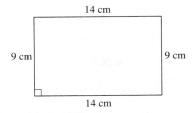

14 cm

9 cm   9 cm

14 cm

Solution

We can find the perimeter by adding the lengths of the sides: $P = 14 \text{ cm} + 9 \text{ cm} + 14 \text{ cm} + 9 \text{ cm} = 46 \text{ cm}$. Or, we can use the formula. Identify the length, l, and the width, w. Then, use the formula $P = 2l + 2w$.

$l = 14 \text{ cm}, w = 9 \text{ cm}$

$$P = 2l + 2w$$
$$P = 2(14 \text{ cm}) + 2(9 \text{ cm}) \qquad \text{Substitute the values for } l \text{ and } w.$$
$$P = 28 \text{ cm} + 18 \text{ cm} \qquad \text{Multiply.}$$
$$P = 46 \text{ cm} \qquad \text{Add.}$$

The perimeter of the rectangle is 46 cm.

[YOU TRY 1]   Find the perimeter of the rectangle.

5 mm

21 mm

What is the *area* of a figure?

Definition

The **area** of a figure is the size of the region enclosed in the figure.

For example, if we have a rectangular rug that is 8 ft long and 5 ft wide, the amount of surface that it covers is the *area* of the rug.

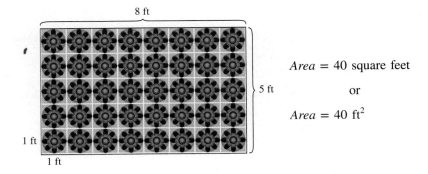

8 ft

5 ft

1 ft

1 ft

Area = 40 square feet

or

Area = 40 ft^2

Each square inside this rectangle has a length and width of 1 foot. The area of each *square* is 1 *square* foot. This can be abbreviated as 1 ft².

How many squares are inside this rectangular rug? There are 40 squares. Therefore, the area of the rectangular rug is 40 *square* feet or 40 ft².

We can also find the area of a rectangle by multiplying its length and width.

Formula Area of a Rectangle

We can find the area, *A*, of a rectangle using this formula:

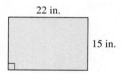

l = length

w = width

$Area = length \cdot width$ or

$A = lw$

The units for area are *square* units like square feet (ft²), square inches (in²), square meters (m²), etc.

We can apply the area formula to find the area of the rug: $A = lw = (8 \text{ ft})(5 \text{ ft}) = 40 \text{ ft}^2$.

EXAMPLE 2

Find the area of the rectangle.

22 in.

15 in.

Solution

Identify the length and width: $l = 22$ in., $w = 15$ in.
Use the formula $A = lw$.

$$A = lw$$
$$A = (22 \text{ in.})(15 \text{ in.}) \qquad \text{Substitute the values for } l \text{ and } w.$$
$$A = 330 \text{ in}^2 \qquad \text{Multiply.}$$

The area of the rectangle is 330 in².

[**YOU TRY 2**]   Find the area of the rectangle.

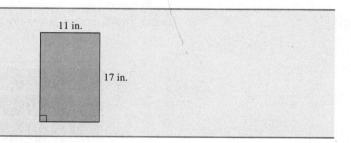

11 in.

17 in.

EXAMPLE 3

Dimitri buys a rectangular plot of land to build a house. The lot measures $23\frac{1}{2}$ m by 40 m. What is the area of the lot?

Solution

Draw and label a rectangle to represent the land.

Identify the length and width: $l = 40$ m,

$w = 23\frac{1}{2}$ m

Use the formula $A = lw$.

$A = lw$

$A = (40 \text{ m})\left(23\frac{1}{2} \text{ m}\right)$ Substitute the values for l and w.

$A = (40 \text{ m})\left(\frac{47}{2} \text{ m}\right)$ Change the mixed number to an improper fraction.

$A = 940 \text{ m}^2$ Multiply.

The area of the lot is 940 m².

[40 m across top, $23\frac{1}{2}$ m on right side of shaded rectangle]

[YOU TRY 3]

A rectangular truck bed liner is $1\frac{4}{5}$ m long and $1\frac{1}{10}$ m wide. Find the area of the liner.

---

2 Find the Perimeter and Area of a Square

Definition

A **square** is a rectangle with all four sides of equal length.

Example:

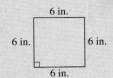

6 in.

To find the perimeter of the square in the definition box, we can add the lengths of all of the sides:

$$P = 6 \text{ in.} + 6 \text{ in.} + 6 \text{ in.} + 6 \text{ in.} = 24 \text{ in.}$$

Notice that we have added the length of a side *four times*. So, we can find the perimeter of a square by multiplying the length of a side by 4.

Formula Perimeter of a Square

The **perimeter,** *P*, of a square with side length *s* can be found using one of these formulas:

$$Perimeter = side + side + side + side \quad \text{or}$$
$$P = 4 \cdot side \quad \text{or}$$
$$P = 4s$$

We find the area of a square by multiplying length times width, but since those values are the same, we can use one of the following formulas.

Formula Area of a Square

The **area,** *A,* of a square with side length *s* can be found using one of these formulas:

$$Area = side \cdot side \quad \text{or}$$
$$A = s \cdot s \quad \text{or}$$
$$A = s^2$$

The units for area are *square* units.

EXAMPLE 4

For the square pictured here, find a) its perimeter and b) its area.

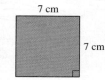

Solution

a) Identify the side length, *s*:   *s* = 7 cm

Use the formula *P* = 4*s* or add up all the side lengths.

Formula	*Add side lengths*
$P = 4s$	$P = side + side + side + side$
$P = 4(7 \text{ cm})$	$P = 7 \text{ cm} + 7 \text{ cm} + 7 \text{ cm} + 7 \text{ cm}$
$P = 28 \text{ cm}$	$P = 28 \text{ cm}$

The perimeter is 28 cm. We get the same result using either method.

b) The side length is *s* = 7 cm. Use one of the formulas to find the area.

$A = side \cdot side$	or	$A = s^2$
$A = (7 \text{ cm}) \cdot (7 \text{ cm})$		$A = (7 \text{ cm})^2$
$A = 49 \text{ cm}^2$		$A = 49 \text{ cm}^2$

The area is 49 cm², and we get the same answer using either formula.

[YOU TRY 4]

For the square pictured here, find a) its perimeter and b) its area.

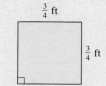

Next, let's learn about parallelograms.

3 Find the Perimeter and Area of a Parallelogram

Definition

A **parallelogram** is a four-sided figure whose opposite sides are parallel and the same length.

Here are two examples of parallelograms.

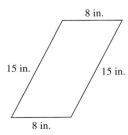

8 in.

15 in.   15 in.

8 in.

The opposite sides are the same length, and they are parallel.

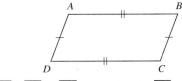

$\overline{AD} \parallel \overline{BC}$ ($\overline{AD}$ is parallel to $\overline{BC}$)

and

$\overline{AB} \parallel \overline{DC}$ ($\overline{AB}$ is parallel to $\overline{DC}$)

Opposite sides are the same length, as indicated by the hash marks "|" and "||".

Note Rectangles and squares are also parallelograms.

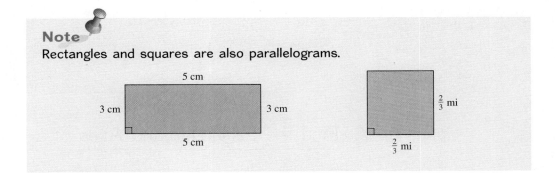

5 cm

3 cm   3 cm

5 cm

$\frac{2}{3}$ mi

$\frac{2}{3}$ mi

To find the *perimeter* of a parallelogram, we add the lengths of the four sides. To find the *area* of a parallelogram, we use the following formula.

Formula Area of a Parallelogram

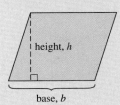

height, h

base, b

The **area**, A, of a parallelogram is

$$Area = base \cdot height \quad \text{or}$$
$$A = bh$$

where b = the length of the base and h = the height. Notice that the base is the side that forms a right angle with the height.

W Hint

The derivation of the formula of the area of a parallelogram can be found in Appendix B.1.

EXAMPLE 5   Find the a) perimeter and b) area of the parallelogram.

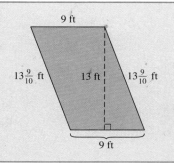

Solution

a)   To find the perimeter of the parallelogram,
     add the lengths of the sides.

$$P = 16 \text{ cm} + 10 \text{ cm} + 16 \text{ cm} + 10 \text{ cm} = 52 \text{ cm}$$

The perimeter is 52 cm. (Notice that we did not need the height, 6 cm.)

b)   *Area = base · height*, so identify the base and height. Remember, the base is the side
     that forms the right angle with the height.

$$b = 16 \text{ cm} \qquad h = 6 \text{ cm}$$

(We do not need the 10 cm to find the area.)

$$A = bh$$
$$A = (16 \text{ cm})(6 \text{ cm}) \qquad \text{Substitute the values.}$$
$$A = 96 \text{ cm}^2 \qquad \text{Multiply.}$$

The area of the parallelogram is 96 cm^2.

[YOU TRY 5]   Find the a) perimeter and b) area of the parallelogram.

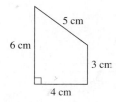

---

4 Find the Perimeter and Area of a Trapezoid

Definition

A **trapezoid** is a four-sided figure with exactly one pair of parallel sides.

Here are some examples of trapezoids.

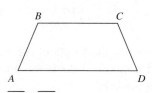

$\overline{BC} \parallel \overline{AD}$
($\overline{BC}$ is parallel to $\overline{AD}$)

The sides of length
6 cm and 3 cm
are parallel.

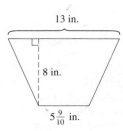

The sides of length
13 in. and $5\frac{9}{10}$ in.
are parallel.

> **W Hint**
>
> The opposite sides of a trapezoid do *not* have to be the same length.

To find the perimeter of a trapezoid, we add the lengths of the four sides. To find the *area* of a trapezoid, use this formula:

Formula Area of a Trapezoid

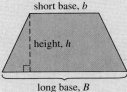

short base, *b*

height, *h*

long base, *B*

The **area**, *A*, of a trapezoid is

$$A = \frac{1}{2} \cdot height \cdot (short\ base + long\ base) \quad \text{or}$$

$$A = \frac{1}{2}h(b + B)$$

where *h* = the height, *b* = the length of the short base, and *B* = the length of the long base.

The bases are always the parallel sides.

EXAMPLE 6   Find the a) perimeter and b) area of the trapezoid.

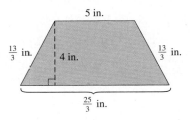

5 in.

$\frac{13}{3}$ in.   4 in. $\frac{13}{3}$ in.

$\frac{25}{3}$ in.

Solution

a) *Perimeter:* The perimeter is the distance around the figure, so add the lengths of the sides. (Do not add the 4 in. because it is the height of the trapezoid and not a side.)

$$P = 5 \text{ in.} + \frac{13}{3} \text{ in.} + \frac{25}{3} \text{ in.} + \frac{13}{3} \text{ in.}$$

$$P = \frac{15}{3} \text{ in.} + \frac{13}{3} \text{ in.} + \frac{25}{3} \text{ in.} + \frac{13}{3} \text{ in.} \qquad \text{Get a common denominator.}$$

$$P = \frac{66}{3} \text{ in. or 22 in.} \qquad \text{Add and simplify.}$$

The perimeter of the trapezoid is 22 in.

b) *Area:*   Use the formula:

$$A = \frac{1}{2}h(b + B)$$

where *h* = height, *b* = short base, and *B* = long base. Identify *h*, *b*, and *B*. *The bases are the parallel sides.*

$$h = 4 \text{ in.} \qquad b = 5 \text{ in.} \qquad B = \frac{25}{3} \text{ in.}$$

$$\left(\text{We do } not \text{ need either of the } \frac{13}{3} \text{ in. sides.} \right)$$

$$A = \frac{1}{2}h(b + B)$$

$$A = \frac{1}{2}(4 \text{ in.})\left(5 \text{ in.} + \frac{25}{3} \text{ in.}\right)$$ Substitute the values.

$$A = \frac{1}{2}(4 \text{ in.})\left(\frac{15}{3} \text{ in.} + \frac{25}{3} \text{ in.}\right)$$ Get a common denominator.

$$A = \frac{1}{2}(4 \text{ in.})\left(\frac{40}{3} \text{ in.}\right)$$ Add.

$$A = \frac{80}{3} \text{ in}^2 \text{ or } 26\frac{2}{3} \text{ in}^2$$ Multiply.

The area of the trapezoid is $\frac{80}{3}$ in² or $26\frac{2}{3}$ in².

Hint

Are the bases of a trapezoid always on the top and bottom?

[YOU TRY 6] Find the a) perimeter and b) area of the trapezoid.

5 cm

6 cm

3 cm

4 cm

5 Find the Perimeter and Area of an Irregular Figure

Sometimes, we are asked to find the perimeter and area of a figure that consists of other figures put together. In this case, we must try to break down the figure into shapes that we know.

Remember that the perimeter of a figure is the distance *around* the figure, while the area of a figure is the size of the region enclosed in the figure.

EXAMPLE 7   Luciano's garden has the shape pictured here.

a)   How much fencing will it take to enclose the garden?

b)   What is the area of the garden?

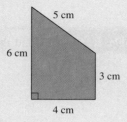

14 ft

11 ft

16 ft

9 ft

5 ft

5 ft

Solution

a)   If Luciano wants to enclose the garden with a fence, that means the fence will go *around* the garden. The **perimeter** is the distance around a figure, so add up the lengths of all the sides to determine how much fencing is needed.

$$P = 14 \text{ ft} + 11 \text{ ft} + 9 \text{ ft} + 5 \text{ ft} + 5 \text{ ft} + 16 \text{ ft}$$
$$P = 60 \text{ ft}$$

The perimeter is 60 ft, so Luciano needs 60 ft of fencing to enclose his garden.

b) The **area** is the size of the region enclosed in the garden. Let's break up this figure into two regions: a rectangle and a square.

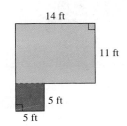
14 ft

11 ft

5 ft

5 ft

Total area = Area of rectangle + Area of square

$$= \quad lw \quad + \quad s^2$$
$$= (14 \text{ ft})(11 \text{ ft}) \quad + \quad (5 \text{ ft})^2$$
$$= 154 \text{ ft}^2 \quad + \quad 25 \text{ ft}^2$$
$$= 179 \text{ ft}^2$$

The area of the garden is 179 ft².

W Hint

Write explanations, in your own words, of how to find the perimeter and area of an irregularly-shaped figure.

YOU TRY 7

A construction site has the shape pictured here.

a) How much fencing is needed to enclose the site?

b) What is the area?

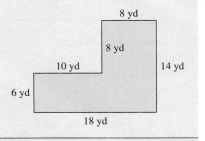

8 yd

8 yd

10 yd   14 yd

6 yd

18 yd

ANSWERS TO **YOU TRY** **EXERCISES**

1) 52 mm   2) 187 in²   3) $1\frac{49}{50}$ m²   4) a) 3 ft b) $\frac{9}{16}$ ft²   5) a) $P = 45\frac{4}{5}$ ft b) $A = 117$ ft²

6) a) $P = 18$ cm b) $A = 18$ cm²   7) a) 64 yd b) 172 yd²

E Evaluate **4.2** Exercises   Do the exercises, and check your work.

Get Ready

Perform the operations.

1) $23 \cdot 18$

2) $45 \cdot 127$

3) $2 \cdot \left(\frac{3}{4}\right) + 2 \cdot \left(\frac{5}{8}\right)$

4) $2 \cdot \left(\frac{2}{3}\right) + 2 \cdot \left(\frac{1}{4}\right)$

5) $1\frac{1}{2} \times 4\frac{7}{12}$

6) $3\frac{5}{6} \times 2\frac{1}{2}$

Objective 1: Find the Perimeter and Area of a Rectangle

7) In your own words, define *perimeter of a rectangle*. Then, state the formula for finding the perimeter.

8) In your own words, define *area of a rectangle*. Then, state the formula.

Find the perimeter and area of the given rectangle.

9)

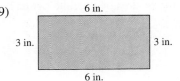
6 in.

3 in.   3 in.

6 in.

10)

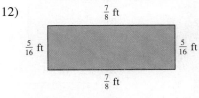
9 ft

7 ft

11)

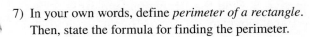
$\frac{3}{4}$ yd

$\frac{1}{8}$ yd

12)

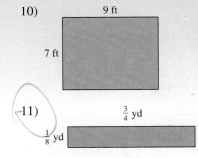
$\frac{7}{8}$ ft

$\frac{5}{16}$ ft $\frac{5}{16}$ ft

$\frac{7}{8}$ ft

13)

$6\frac{1}{2}$ in.

$3\frac{1}{2}$ in.

14)

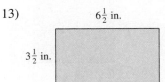
$6\frac{3}{4}$ m

$1\frac{1}{4}$ m

15) A regulation bowling lane is 60 ft long from the foul line to the head pin and $41\frac{1}{2}$ in. wide. Find the perimeter and area of this portion of the lane in units of inches. (United States Bowling Congress)

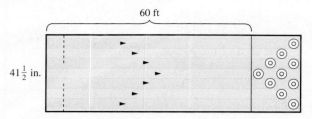
60 ft

$41\frac{1}{2}$ in.

16) Higinio wants to make eight cloth place mats for his dining table. Using the dimensions for one place mat shown in the figure at right, what is the combined area of all eight place mats?

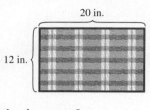
20 in.

12 in.

17) Suppose a rectangle has a perimeter of 20 ft. If the rectangle has a length of 8 ft, what is its width?

18) What is the sum of the measures of all the angles inside a rectangle?

Objective 2: Find the Perimeter and Area of a Square

19) Suppose a square has a perimeter of 10 ft. What is its side length?

20) Answer *true* or *false*. If the area of a square is 16 m², its side length is 8 m.

Find the perimeter and area of the given square.

21)
12 cm

12 cm

22)
17 in.

17 in.

23)
$\frac{3}{5}$ m

$\frac{3}{5}$ m

$\frac{3}{5}$ m

24)
$2\frac{1}{4}$ yd

$2\frac{1}{4}$ yd

$2\frac{1}{4}$ yd

$2\frac{1}{4}$ yd

25)
$1\frac{3}{8}$ m

$1\frac{3}{8}$ m

26)
$\frac{5}{6}$ yd

$\frac{5}{6}$ yd

27) A square has an area of 25 ft². What is the length of each side of the square?

28) The perimeter of a square is 34 mm. How long is each side of the square?

Mixed Exercises: Objectives 1 and 2
Solve each problem.

29) A regulation football field is 120 yd long and $53\frac{1}{3}$ yd wide. Find the perimeter and the area of the field. (www.nfl.com)

30) A standard sheet of paper measures $8\frac{1}{2}$ in. by 11 in. Find its perimeter and area.

31) The area of a baseball infield within the baselines is a square with a side length of 90 ft. Find the area within the baselines. How far must a baseball player travel when he hits a home run? (www.mlb.com)

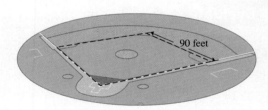
90 feet

32) Mackenzie has 22 ft of decorative fencing to line her rectangular flower garden. If her garden must have a width of 4 ft, what will the length of her garden have to be to use all 22 ft of fencing?

Objective 3: Find the Perimeter and Area of a Parallelogram

33) Make a drawing showing how to form a rectangle and a parallelogram using two identically sized right triangles.

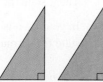

34) What is the formula for the area of a parallelogram?

Find the perimeter and area of each parallelogram.

35)
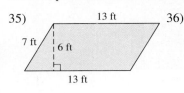
13 ft
7 ft | 6 ft
13 ft

36)
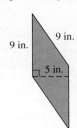
9 in. | 9 in.
5 in.

having a custom winter cover made at a cost of $12 per square meter. What is the total cost of the cover?

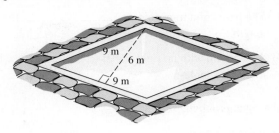
9 m
6 m
9 m

24 37)
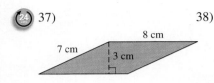
8 cm
7 cm
3 cm

38)
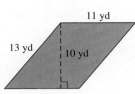
11 yd
13 yd | 10 yd

Objective 4: Find the Perimeter and Area of a Trapezoid

45) Write two forms of the formula for the area of a trapezoid.

46) How do you know which sides of the trapezoid are the bases?

Find the perimeter and area of each trapezoid.

39)
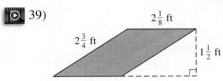
$2\frac{1}{8}$ ft
$2\frac{3}{4}$ ft
$1\frac{1}{2}$ ft

24 47)
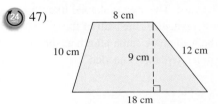
8 cm
10 cm | 9 cm | 12 cm
18 cm

40)
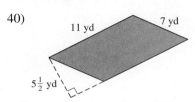
11 yd | 7 yd
$5\frac{1}{2}$ yd

48)
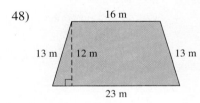
16 m
13 m | 12 m | 13 m
23 m

41) A parallelogram with no right angles *always, sometimes,* or *never* has two acute angles and two obtuse angles.

42) In your own words, explain why the rectangle and the parallelogram below have the same area.

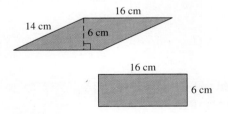
16 cm
14 cm | 6 cm

16 cm
6 cm

49)
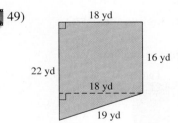
18 yd
22 yd
18 yd
16 yd
19 yd

50)
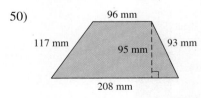
96 mm
117 mm | 95 mm | 93 mm
208 mm

Solve each problem.

43) Husna will sew together pieces of fabric shaped as parallelograms to make a comforter. Each piece will have the shape shown here. What is the total area of the comforter if she uses 160 pieces?

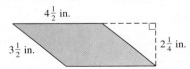
$4\frac{1}{2}$ in.
$3\frac{1}{2}$ in. | $2\frac{1}{4}$ in.

51)
$4\frac{3}{4}$ ft
7 ft | $5\frac{1}{2}$ ft | $2\frac{1}{2}$ ft
$5\frac{1}{4}$ ft

52)
$6\frac{1}{4}$ in.
$7\frac{1}{2}$ in. | $9\frac{1}{2}$ in.
$11\frac{3}{4}$ in.

44) Maarten has a swimming pool shaped like the parallelogram shown in the figure. For the winter months, he is

Fill in the blank with *always, sometimes,* or *never* to make the statement true.

53) The bases of a trapezoid are _____ the same length.

54) A trapezoid with no right angles _____ has four acute angles.

Solve each problem.

55) Jerome, who is 3 years old, drew a family portrait on a sheet of paper shaped like a trapezoid. His parents want to have a custom frame built for the picture. How much wood is needed for the frame?

56) Mrs. Saldivar wants to install a chain-link fence around her pet area in the backyard. The area is shaped like a trapezoid with one side up against the wall of the home. (The side by the wall is the same length as the opposite side.) How much fencing is needed?

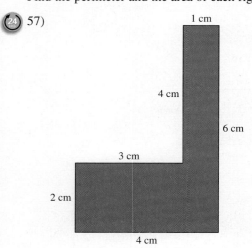

Objective 5: Find the Perimeter and Area of an Irregular Figure
Find the perimeter and the area of each figure.

57)

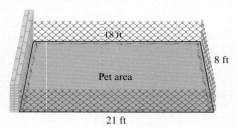

58)

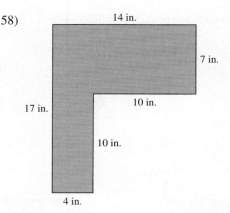

59)

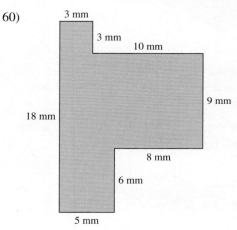

60)

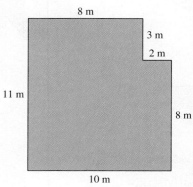

61) Mr. and Mrs. Szabo are having a new driveway installed at a cost of $40 per square meter. The driveway is represented by the shaded region in the figure. What is the total cost for the new driveway?

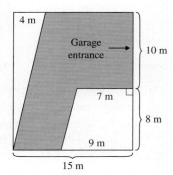

62) The entranceway to the Green family home will be paved with red brick at a cost of $12 per square foot including installation. The entranceway is represented by the shaded area in the figure. What is the total cost?

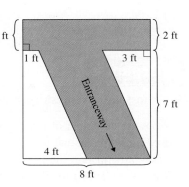

Find the perimeter and the area of the given figure. In Exercises 67 and 68, you must first find the missing length.

63)

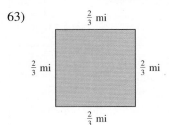

64)

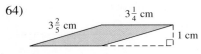

65)

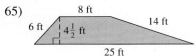

66)

 67)

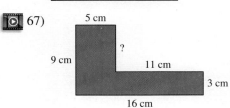

68)

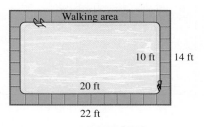

Solve each problem.

69) Cailen is an artist and wants to frame one of her paintings by herself. The frame she chooses costs $12 per foot. Find the cost of the frame if the painting is $1\frac{1}{2}$ ft by 3 ft.

70) Frank will replace the broken glass in his back door with a rectangular piece of glass that costs $4 per square foot. If he needs a piece of glass that is $1\frac{3}{4}$ ft ft by 3 ft, find the cost of the glass.

71) A walking area surrounding the pool in the diagram is going to be refinished with a nonslip surface. If it costs $17 per square foot to install the surface, what is the total cost? (Hint: Multiply the nonslip area by $17, the cost per square foot.)

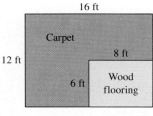

72) Mr. and Mrs. Anh want to install wood flooring and carpet in an area of their home as shown in the diagram. Installation costs are $6 per square foot for carpet and $7 per square foot for the wood flooring. What is the total cost for the installation of both? (Hint: The cost of installing the carpet equals the area of the carpet times $6, the cost per square foot. The cost of installing the wood flooring equals the area of the wood flooring times $7, the cost per square foot.)

R Rethink

R1) Look at Exercises 57 through 59 and use subtraction to find the total area if you did not do it that way the first time. Describe what you did to find the area.

R2) Describe how a good knowledge of common squares and multiplication has helped you with this section.

4.3 Triangles

P Prepare

O Organize

What are your objectives for Section 4.3?	How can you accomplish each objective?
1 Find the Perimeter and Area of a Triangle	• Write the definition of a *triangle* in your own words, draw an example, and note its main characteristics. • Know how to find the perimeter of a triangle. • Write the formula for finding the **Area of a Triangle** in your notes. • Complete the given examples on your own. • Complete You Trys 1–3.
2 Find Angle Measures in a Triangle	• Write the property for the **Sum of the Angle Measures in a Triangle** in your notes. • Complete the given example on your own. • Complete You Try 4.
3 Classify Triangles	• Learn how to classify triangles by their angles: *acute triangle*, *right triangle*, and *obtuse triangle*. • Learn how to classify triangles by their sides: *equilateral triangle*, *isosceles triangle*, and *scalene triangle*. • Complete the given examples on your own. • Complete You Trys 5 and 6.

W Work

Read the explanations, follow the examples, take notes, and complete the You Trys.

In this section, we will learn about triangles.

1 Find the Perimeter and Area of a Triangle

W Hint

Make and use flash cards to help you learn the terms and formulas in this section.

Definition

A **triangle** is a closed figure with exactly three sides.

Example:

As with other figures, the perimeter of a triangle is the distance around the triangle. To find the perimeter, we add the lengths of the three sides.

EXAMPLE 1 Find the perimeter of this triangle.

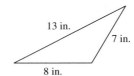
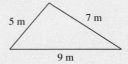

Solution

Add the lengths of the sides: $P = 13$ in. $+ 7$ in. $+ 8$ in. $= 28$ in.

[**YOU TRY 1**] Find the perimeter of this triangle.

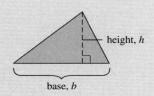

To find the *area* of a triangle, we use the following formula. (The derivation of the area formula is given in Appendix B.1.)

Formula Area of a Triangle

The **area,** A, of a triangle is

$$Area = \frac{1}{2} \cdot base \cdot height \qquad \text{or}$$

$$A = \frac{1}{2}bh$$

where $b =$ the length of the base and $h =$ the height.

height, h

base, b

BE CAREFUL The *base* of the triangle is always the side of the triangle that forms the right angle with the height. The height is not always labeled inside the triangle.

Examples:

height

base

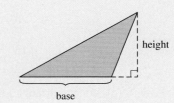

height

base

EXAMPLE 2 Find the area of each triangle.

a)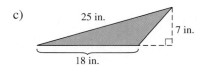

17 cm   8 cm   10 cm
21 cm

b)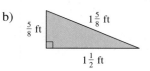

$\frac{5}{8}$ ft $1\frac{5}{8}$ ft
$1\frac{1}{2}$ ft

c)
25 in.
7 in.
18 in.

Solution

W Hint

Remember, the best way to read a math book is to write out the example as you are reading it.

a) Identify the base, b, and height, h. Remember, the base is the side that forms the right angle with the height.

$$b = 21 \text{ cm} \qquad h = 8 \text{ cm}$$

(We do not need the lengths 17 cm and 10 cm.)

$$A = \frac{1}{2}bh$$

$$A = \frac{1}{2}(21 \text{ cm})(\overset{4}{\underset{1}{\cancel{8}}} \text{ cm}) \qquad \text{Substitute the values, and divide out 2.}$$

$$A = 84 \text{ cm}^2 \qquad \text{Multiply.}$$

The area is 84 cm^2. Remember to use *square* units for area.

b) Identify the base, b, and height, h: $\quad b = 1\frac{1}{2}$ ft, $\qquad h = \frac{5}{8}$ ft

$$A = \frac{1}{2}bh$$

$$A = \frac{1}{2}\left(1\frac{1}{2} \text{ ft}\right)\left(\frac{5}{8} \text{ ft}\right) \qquad \text{Substitute the values.}$$

$$A = \frac{1}{2}\left(\frac{3}{2} \text{ ft}\right)\left(\frac{5}{8} \text{ ft}\right) \qquad \text{Change the mixed number to a fraction.}$$

$$A = \frac{15}{32} \text{ ft}^2 \qquad \text{Multiply.}$$

The area is $\frac{15}{32}$ ft^2.

c) In this triangle, the base must be extended with a dotted line to find the height. (The base is *still* the base of the actual triangle.)

Identify the base, b, and the height, h: $\quad b = 18$ in. and $h = 7$ in.

$$A = \frac{1}{2}bh$$

$$A = \frac{1}{2}(18 \text{ in.})(7 \text{ in.}) \qquad \text{Substitute the values.}$$

$$A = 63 \text{ in}^2 \qquad \text{Multiply.}$$

The area is 63 in^2.

[YOU TRY 2] Find the area of each triangle.

a)

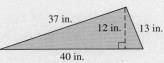

b)

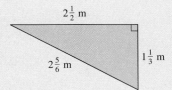

c)

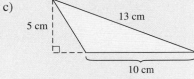

Let's put together what we know about different shapes to solve a problem.

EXAMPLE 3 Find the area of the shaded region.

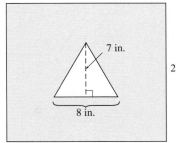

Solution

The area of the shaded region is the area of the square *minus* the area of the triangle.

Square

Length of a side: $s = 20$ in.

$A = s^2 = (20 \text{ in.})^2 = 400 \text{ in}^2$

Triangle

$b = 8$ in. and $h = 7$ in.

$A = \dfrac{1}{2}bh = \dfrac{1}{2}(8 \text{ in.})(7 \text{ in.}) = 28 \text{ in}^2$

$$\begin{array}{rl}
\text{Area of the shaded region} = & \text{Area of the square} - \text{Area of the triangle} \\
= & 400 \text{ in}^2 - 28 \text{ in}^2 \\
= & 372 \text{ in}^2
\end{array}$$

The area of the shaded region is 372 in^2.

[YOU TRY 3] Find the area of the shaded region.

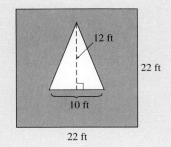

Next we will learn about angle measures in a triangle.

2 Find Angle Measures in a Triangle

Property The Sum of the Angle Measures in a Triangle

The measures of the angles in a triangle add up to 180°.

Example:

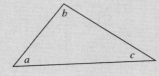

$$m\angle a + m\angle b + m\angle c = 180°$$

We can use this fact to find missing angle measures in a triangle.

EXAMPLE 4

Find m∠x in each triangle.

a)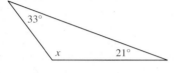

b)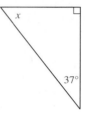

Solution

a) Since the measures of all the angles in a triangle add up to 180°, we find m∠x by subtracting:

$$m\angle x = 180° - 33° - 21° = 126°$$

W Hint

Find 90 − 37. How could that help you with triangles that have a right angle?

b) It may look like only one angle is labeled (37°), but remember that the angle marked with the small square is a *right* angle, so its measure is 90°.

$$m\angle x = 180° - 37° - 90° = 53°$$

     ↑   ↑

All of the   The measure of
angles add   the right angle
to 180°.

[YOU TRY 4]

Find m∠x in each triangle.

a)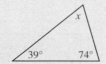

b)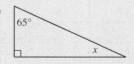

3 Classify Triangles

We can classify triangles by the measures of their angles and by the lengths of their sides.

Definition

Acute Triangle	**Right Triangle**	**Obtuse Triangle**
	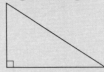	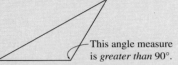 This angle measure is *greater than* 90°.
An **acute triangle** is one in which all three angles are acute. (All three angles measure *less than* 90°.)	A **right triangle** contains one *right*, or 90°, angle.	An **obtuse triangle** contains one *obtuse* angle. (The measure of an obtuse angle is *greater than* 90°.)

EXAMPLE 5

Classify each triangle as acute, right, or obtuse.

a)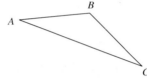

b)

c)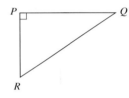

Solution

a) This is an *obtuse triangle* because ∠B is an obtuse angle. (m∠B is *greater than* 90°.)

W Hint

A right triangle will always have a small square indicating a right angle.

b) This is an *acute triangle* because each angle is acute. (The measure of each angle is *less than* 90°.)

c) This is a *right triangle* because ∠P is a right angle. (m∠P = 90°)

[YOU TRY 5]

Classify each triangle as acute, right, or obtuse.

a)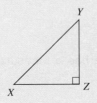

b)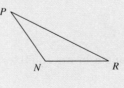

c)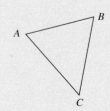

Definition

Equilateral Triangle   **Isosceles Triangle**   **Scalene Triangle**

An **equilateral triangle** has *three* sides of equal length.

An **isosceles triangle** has *two* sides of equal length.

A **scalene triangle** has *no* sides of equal length.

When the sides of triangles are the same length, we mark them with a hash mark like this: |

EXAMPLE 6   Classify each triangle as equilateral, isosceles, or scalene.

a)
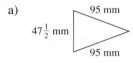

b)
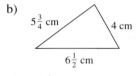

c)
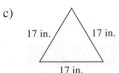

Solution

a)   This is an *isosceles triangle* because two of the sides are the same length.

b)   This is a *scalene triangle* because no sides are the same length.

c)   What do you notice about the lengths of the sides? *All sides are the same length.* This is an *equilateral triangle.*

[YOU TRY 6]   Classify each triangle as equilateral, isosceles, or scalene.

a)
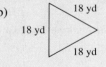

b)
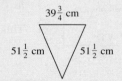

c)

$39\frac{3}{4}$ cm

$51\frac{1}{2}$ cm $51\frac{1}{2}$ cm

Note

Each angle in an equilateral triangle has a measure of 60°.

ANSWERS TO [YOU TRY] EXERCISES

1) 21 m   2) a) 240 in^2 b) $1\frac{2}{3}$ m^2 c) 25 cm^2 3) 424 ft^2 4) a) 67°   b) 25°

5) a) right   b) obtuse   c) acute   6) a) scalene   b) equilateral   c) isosceles

---

Objective 1: Find the Perimeter and Area of a Triangle

1) What is the formula for the area of a triangle?

2) How do you identify the base of a triangle?

Find the perimeter of the given triangle.

 3)

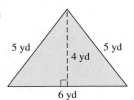

4)

5)

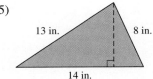

6)

Find the area of the given triangle.

 7)

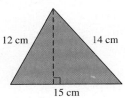

8)

9)

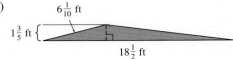

10)

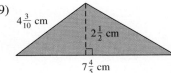

Find the area of the shaded region.

11)

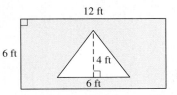

12)

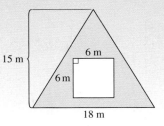

Find the area of the *unshaded* region in each figure.

13)

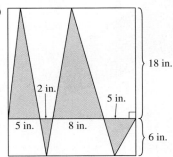

14)

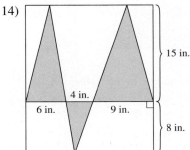

Find the perimeter and the area of each figure.

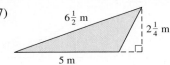 15)

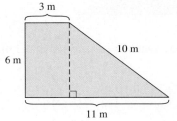

16)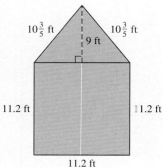

$10\frac{3}{5}$ ft $10\frac{3}{5}$ ft

9 ft

11.2 ft   11.2 ft

11.2 ft

17)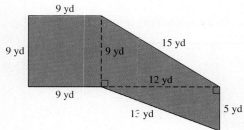

9 yd

9 yd   9 yd   15 yd

9 yd   12 yd

1? yd   5 yd

18)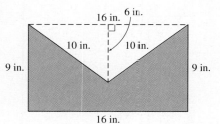

16 in.   6 in.

10 in.   10 in.

9 in.   9 in.

16 in.

19) Find the area of each triangle in the figure. Then, find the area of the rectangle that is formed by both triangles. What do you notice?

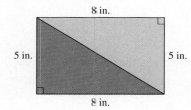

8 in.

5 in.   5 in.

8 in.

20) How many different triangles can you find in the figure below?

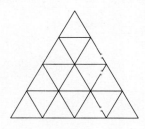

Objective 2: Find Angle Measures in a Triangle

21) The measures of all the angles of a triangle add up to what number?

22) Answer *true* or *false*. A triangle can have two right angles.

Find m∠x in each triangle.

23)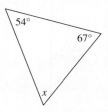

54°

67°

x

24)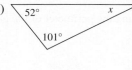

52°   x

101°

25)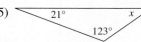

21°   x

123°

26)

32°

128°   x

27)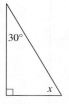

31°

136°   x

28)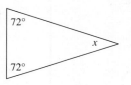

72°

x

72°

29)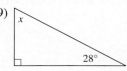

x

28°

30)

30°

x

Objective 3: Classify Triangles

31) What is an acute triangle?

32) What is an obtuse triangle?

33) What is a scalene triangle?

34) What is an equilateral triangle?

Classify each triangle as acute, right, or obtuse.

35)

36)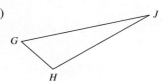

37)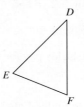

38)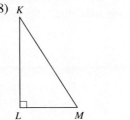

39)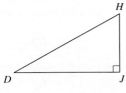

40)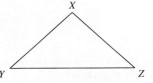

Classify each triangle as equilateral, isosceles, or scalene.

41)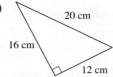

42)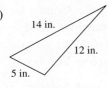

43)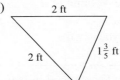

44)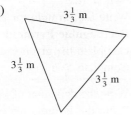

45)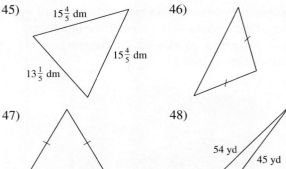

46)

47)

48)

For Exercises 49–52, answer *true* or *false*.

49) A triangle can have two obtuse angles.

50) Each of the angles in an equilateral triangle has a measure of 60°.

51) Every equilateral triangle has three acute angles.

52) A right triangle can have an obtuse angle.

Solve each problem.

53) A hexagonal gazebo rooftop is made using equally sized triangles. Use the picture to calculate the total area of the rooftop.

(Note: A hexagon is a six-sided figure.)

54) Find the area of the front side of the gingerbread house below.

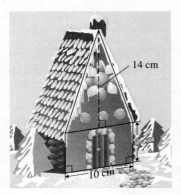

55) A community center wishes to have its courtyard area covered with natural stone pavers. If the installation cost is $12 per square foot, use the diagram below to determine the total cost.

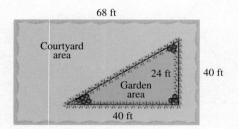

68 ft

Courtyard area

24 ft   40 ft

Garden area

40 ft

R Rethink

R1) Can a triangle have more than one obtuse or right angle? Why or why not? Draw some triangles to help you reach your conclusion.

56) Shirley buys a piece of remnant fabric that is triangular in shape with a base of length 2 yd and a height measuring $\frac{1}{2}$ yd. If the fabric sells for $4 per square yard, how much does Shirley pay for the remnant fabric?

Draw your own example of each type of triangle. Label the lengths of each side.
57) isosceles   58) equilateral

Draw your own example of each type of triangle. Label each angle measure.
59) scalene   60) right

R2) How do the lengths of the sides of a triangle compare with the sizes of the angles in the triangle?

4.4 Volume and Surface Area

P Prepare

O Organize

What are your objectives for Section 4.4?	How can you accomplish each objective?
1 Find the Volume of a Rectangular Solid	• Write the definition of *volume* in your own words, and draw a few examples. • Write the formula for finding the **Volume of a Rectangular Solid** in your notes. • Complete the given example on your own. • Complete You Try 1.
2 Find the Volume of a Rectangular Pyramid	• Write the definition of a *rectangular pyramid* in your own words. • Write the formula for finding the **Volume of a Rectangular Pyramid** in your notes. • Complete the given example on your own. • Complete You Try 2.
3 Find the Surface Area of a Rectangular Solid	• Write the definition of *surface area* in your own words. • Write the formula for finding the **Surface Area of a Rectangular Solid** in your notes. • Complete the given example on your own. • Complete You Try 3.

W Work   **Read the explanations, follow the examples, take notes, and complete the You Trys.**

Until now, we have worked with two-dimensional figures; that is, we have worked with figures in a *flat* plane, such as rectangles and triangles. A rectangle, for example, has the two dimensions of length and width.

1 Find the Volume of a Rectangular Solid

In this section, we will learn about three-dimensional (or solid) objects like a rectangular solid and a pyramid. (A rectangular solid, or a *box,* has the three dimensions of length, width, and height.)

We have found the *area* of two-dimensional figures, and now we will find the *volume* of some three-dimensional objects.

W Hint

Write a few sentences explaining the differences between area and volume. Include a statement about the units.

Definition

The **volume** of a three-dimensional object is a measure of the amount of space occupied by the object or the amount of space inside the object. Volume is measured in *cubic* units.

Here are two examples of cubic units used to measure volume.

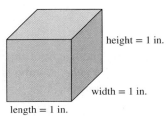

height = 1 in.

width = 1 in.

length = 1 in.

1 cubic inch (1 in³)

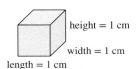

height = 1 cm

width = 1 cm

length = 1 cm

1 cubic centimeter (1 cm³)

If we say that the volume of this box is 18 cm³, it means that we can fit 18 of the 1-cm³ boxes inside this larger box.

Let's begin finding volumes of solid objects (or *solids*) with a familiar shape: a box. A **rectangular solid** is a box-like shape with dimensions of length, width, and height.

Volume = 18 cm³

W Hint

Make and use flashcards to help you learn the terms and formulas in this section.

Formula Volume of a Rectangular Solid

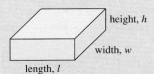

height, h

width, w

length, l

The **volume,** V, of a rectangular solid with length l, width w, and height h is

$$Volume = length \cdot width \cdot height \quad \text{or}$$
$$V = lwh$$

Use *cubic units* to measure volume.

EXAMPLE 1

Find the volume of each box.

a)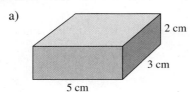

b)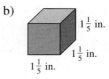

Solution

a) Identify length, width, and height.

$$l = 5 \text{ cm}, \ w = 3 \text{ cm}, \ h = 2 \text{ cm}$$

Use the volume formula.

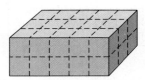

$V = lwh$

$V = (5 \text{ cm})(3 \text{ cm})(2 \text{ cm})$ Substitute the values.

$V = 30 \text{ cm}^3$ $cm \cdot cm \cdot cm = cm^3$

This box will hold 30 of the 1-cm³ boxes or 30 cm³.

b) Notice that the length, width, and height of this box are the same. This is a special type of rectangular solid; it is a *cube*. A **cube** is a rectangular solid in which the length, width, and height are the same.

$V = lwh$

$V = \left(1\frac{1}{5} \text{ in.} \right)\left(1\frac{1}{5} \text{ in.} \right)\left(1\frac{1}{5} \text{ in.} \right)$ Substitute the values.

$V = \left(\frac{6}{5} \text{ in.} \right)\left(\frac{6}{5} \text{ in.} \right)\left(\frac{6}{5} \text{ in.} \right)$ Change the mixed numbers to improper fractions.

$V = \frac{216}{125} \text{ in}^3$ Multiply.

The volume is $\frac{216}{125} \text{ in}^3$ or $1\frac{91}{125} \text{ in}^3$.

[YOU TRY 1]   Find the volume of each box.

a)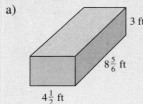

b)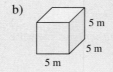

Since all sides of a **cube** are the same length, we can also label the length of each side as s and write the formula for its volume like this:

$$\text{Volume} = s \cdot s \cdot s \quad \text{or}$$
$$V = s^3$$

2 Find the Volume of a Rectangular Pyramid

A **rectangular pyramid** is a pyramid with a rectangular (or square) base.

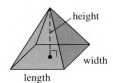

Rectangular pyramid

The volume of a rectangular pyramid equals one-third of the area of the base *times* the height of the pyramid.

Formula Volume of a Rectangular Pyramid

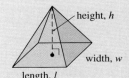

The **volume,** V, of a rectangular pyramid with height h, and a base of length l and width w is

$$\text{Volume} = \frac{1}{3} \cdot \text{length} \cdot \text{width} \cdot \text{height} \quad \text{or}$$

$$V = \frac{1}{3} lwh$$

Because lw equals the *area* of the base, we can also think of the volume formula as $V = \frac{1}{3} Ah$, where $A = lw$, the area of the base of the pyramid.

EXAMPLE 2

Find the volume of the pyramid with the given dimensions.

Solution

Identify the values.

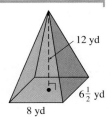

$$l = 8 \text{ yd} \qquad w = 6\frac{1}{2} \text{ yd} \qquad h = 12 \text{ yd}$$

Use the formula.

$$V = \frac{1}{3} lwh$$

$$V = \frac{1}{3}(8 \text{ yd})\left(6\frac{1}{2} \text{ yd}\right)(12 \text{ yd}) \qquad \text{Substitute the values.}$$

$$V = \frac{1}{3}\left(\frac{8}{1} \text{ yd}\right)\left(\frac{13}{2} \text{ yd}\right)\left(\frac{12}{1} \text{ yd}\right) \qquad \text{Write the numbers as fractions.}$$

$$V = 208 \text{ yd}^3 \qquad \text{Multiply and simplify.}$$

Find the volume of the pyramid with the given dimensions.

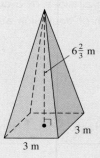

$6\frac{2}{3}$ m

3 m

3 m

Note

In future math courses, you may study pyramids with other kinds of bases, such as triangles or hexagons (a six-sided figure).

3 Find the Surface Area of a Rectangular Solid

W Hint
Be sure you understand the difference between surface area and volume.

Definition

The **surface area** of a three-dimensional object is the total area on the surface of the object.

We will learn how to find the surface area of a rectangular object.

Let's say that, in your job, you must determine how much cardboard is needed to make a box, with the dimensions shown here, for a laptop computer. The amount of cardboard needed equals the *surface area* of the box.

$h = 2$ in.

$w = 9$ in.

$l = 13$ in.

Unfold the box, then flatten it.

back

left side | top | right side | bottom

front

The surface area of the box is the *sum* of the areas of each of its sides. Notice that the *area of the top = area of the bottom*, the *area of the front = area of the back*, and the *area of the left side = area of the right side*.

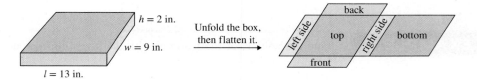

$$\begin{aligned}
\text{Surface Area} &= w\;\boxed{\text{top}} + \boxed{\text{bottom}}\,w + h\;\boxed{\text{front}} + \boxed{\text{back}}\,h + w\;\boxed{\text{left side}} + \boxed{\text{right side}}\,w \\
&\qquad\quad\; l \qquad\quad l \qquad\qquad\quad l \qquad\quad l \qquad\qquad\qquad\quad h \qquad\quad h \\
\text{Surface Area} &= \quad lw \;+\; lw \quad + \quad lh \;+\; lh \quad + \quad wh + wh \\
SA &= \qquad 2lw \qquad + \qquad 2lh \qquad + \qquad 2wh
\end{aligned}$$

Formula Surface Area of a Rectangular Solid

The **surface area,** *SA*, of a rectangular solid with length *l*, width *w*, and height *h* is

$$SA = 2lw + 2lh + 2wh$$

Because we are finding the total *area* on the surface of the object, surface area is measured in *square* units.

EXAMPLE 3

Find the surface area of the box of macaroni and cheese.

18 cm

3 cm

9 cm

Solution

Identify *l*, *w*, and *h*:   *l* = 9 cm, *w* = 3 cm, and *h* = 18 cm.

$SA = 2lw + 2lh + 2wh$

$SA = 2(9 \text{ cm})(3 \text{ cm}) + 2(9 \text{ cm})(18 \text{ cm}) + 2(3 \text{ cm})(18 \text{ cm})$ Substitute the values.

$SA = 54 \text{ cm}^2 + 324 \text{ cm}^2 + 108 \text{ cm}^2$ Multiply.

$SA = 486 \text{ cm}^2$ Add.

[YOU TRY 3] Find the surface area of the box.

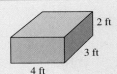

2 ft

3 ft

4 ft

ANSWERS TO [YOU TRY] EXERCISES

1) a) $\dfrac{477}{4}$ ft³ or $119\dfrac{1}{4}$ ft³   b) 125 m³   2) 20 m³   3) 52 ft²

E Evaluate **4.4** Exercises   Do the exercises, and check your work.

Get Ready

Perform the operations. Write the answers in lowest terms.

1) $7 \cdot 4 \cdot 3\dfrac{1}{2}$

2) $9\dfrac{2}{3} \cdot 6\dfrac{1}{2} \cdot 8$

3) $\dfrac{1}{3}(5)\left(2\dfrac{3}{4}\right)(10)$

4) $\dfrac{5}{6}\left(2\dfrac{2}{9}\right)(2)\left(1\dfrac{7}{8}\right)$

5) $3\dfrac{2}{5} \cdot 3\dfrac{2}{5} \cdot 3\dfrac{2}{5}$

6) $\dfrac{5}{6} \cdot \dfrac{5}{6} \cdot \dfrac{5}{6}$

Objective 1: Find the Volume of a Rectangular Solid

7) Write the formula that represents the volume of a rectangular solid with length *l*, width *w*, and height *h*.

8) When can you call a rectangular solid a cube?

Find the volume of each rectangular solid.

9)

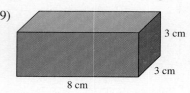

3 cm
3 cm
8 cm

10)

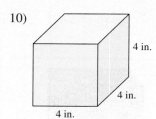

4 in.
4 in.
4 in.

11)

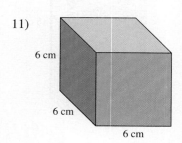

6 cm
6 cm
6 cm

12)

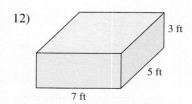

3 ft
5 ft
7 ft

13)

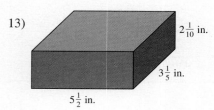

$2\frac{1}{10}$ in.
$3\frac{1}{5}$ in.
$5\frac{1}{2}$ in.

14)

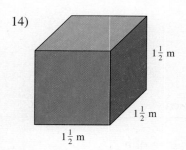

$1\frac{1}{2}$ m
$1\frac{1}{2}$ m
$1\frac{1}{2}$ m

Objective 2: Find the Volume of a Rectangular Pyramid

Find the volume of the pyramid with the given dimensions.

15)

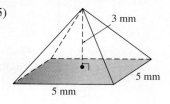

3 mm
5 mm
5 mm

16)

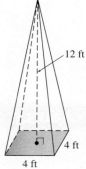

12 ft
4 ft
4 ft

17)

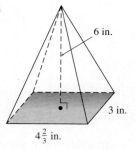

6 in.
3 in.
$4\frac{2}{3}$ in.

18)

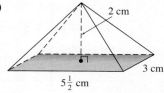

2 cm
3 cm
$5\frac{1}{2}$ cm

Mixed Exercises: Objectives 1 and 2

Find the volume of each solid.

19)

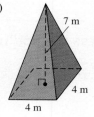

7 m
4 m
4 m

20)

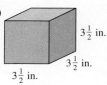

$3\frac{1}{2}$ in.
$3\frac{1}{2}$ in.
$3\frac{1}{2}$ in.

21)

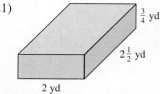

$\frac{3}{4}$ yd
$2\frac{1}{2}$ yd
2 yd

22)

14 m
20 m
30 m

Solve each problem.

23) The Great Pyramid of Giza is the world's largest pyramid. The base of the pyramid has four side lengths all approximately equal to 230 m. The height of the pyramid is approximately 147 m. Calculate the volume of the Great Pyramid of Giza using these dimensions.
(www.nationalgeographic.com)

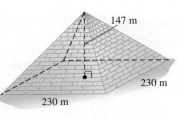

24) 24) A speaker system has five small speakers and a module. Each of the small speakers is $3\frac{1}{2}$ in. long, $3\frac{1}{2}$ in. wide, and 4 in. high. The module is $22\frac{1}{2}$ in. long, 8 in. wide, and $16\frac{1}{2}$ in. high. Find the combined volume of the speaker system.

25) A flat-screen television is 26 in. tall, 43 in. wide, and $1\frac{1}{4}$ in. deep. What is the volume of the TV?

26) A paperweight is in the shape of a rectangular pyramid. The base is 9 cm wide and 9 cm long, and the paperweight is 8 cm tall. What is its volume?

27) A cube has a volume of 27 cm³. What is the length of each side?

28) A cube has a volume of $\frac{64}{125}$ in³. What is the length of each side?

Objective 3: Find the Surface Area of a Rectangular Solid
Find the surface area of each figure.

 29)

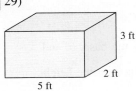

30)

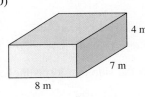

31)

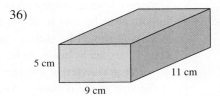

32)

33) How much cardboard is needed to make a gift box with a length of 10 in., width of 8 in., and height of $6\frac{1}{2}$ in.?

34) How much plastic is needed to make a storage container with a length of 15 ft, width of $12\frac{1}{3}$ ft, and height of 9 ft?

Find the volume and surface area of each figure.

35)

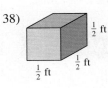

36)

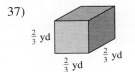

37)

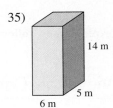

38)

R Rethink

R1) Explain the difference between volume and surface area.

R2) Do you find it more difficult to use the volume and surface area formulas when fractions are involved? Which topics in Chapter 3 could you review to help you?

Putting It All Together

What are your objectives?	How can you accomplish each objective?
1 Review the Concepts of Sections 4.1–4.4	• Know the vocabulary associated with lines, segments, rays, and angles. • Be able to apply the properties and formulas associated with rectangles, squares, parallelograms, trapezoids, and triangles. • Be able to find the volume and surface area of the figures in Section 4.4. • Complete the given examples on your own. • Complete You Trys 1–3.

W Work

Read the explanations, follow the examples, take notes, and complete the You Trys.

1   Review the Concepts of Sections 4.1–4.4

Concepts from geometry show up in different ways and in different situations, so it is important to be able to distinguish among different figures and formulas to solve problems. Also, you should be able to distinguish among a line, line segment, and ray, and be able to classify angles as acute, right, or obtuse.

> **W Hint**
>
> Make a chart of the perimeter and area formulas we have learned.

    The **perimeter** of a figure is the distance around the figure, while the **area** of a figure is the size of the region enclosed in the figure. Area is measured in *square* units like square inches (in^2) or square meters (m^2).

EXAMPLE 1

Find the area and perimeter of each figure.

a)

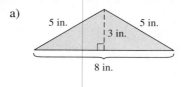

b)

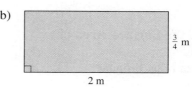

Solution

a)   *Area:* The formula for the area of a triangle is $A = \dfrac{1}{2}bh$. Identify the base, b, and height, h. Remember, the base is the side that forms the right angle with the height.

$$b = 8 \text{ in.} \qquad h = 3 \text{ in.}$$

$$A = \frac{1}{2}bh$$

$$A = \frac{1}{2}(8 \text{ in.})(3 \text{ in.}) \qquad \text{Substitute the values.}$$

$$A = \frac{1}{2}(\overset{4}{8} \text{ in.})(3 \text{ in.}) \qquad \text{Divide out 2.}$$

$$A = 12 \text{ in}^2 \qquad \text{Multiply.}$$

The area is 12 in^2. Remember to use *square* units for area.

Perimeter: The perimeter of the triangle is the distance around the triangle, so add the lengths of the sides: $P = 5$ in. $+ 5$ in. $+ 8$ in. $= 18$ in.

b) *Area:* The formula for the area of a rectangle is $A = lw$. Identify the length and width: $l = 2$ m, $w = \dfrac{3}{4}$ m.

Use the formula $A = lw$.

$$A = lw$$

$$A = (2 \text{ m})\left(\frac{3}{4}\text{ m}\right) \qquad \text{Substitute the values for } l \text{ and } w.$$

$$A = \frac{3}{2}\text{ m}^2 \qquad\qquad\quad \text{Multiply.}$$

We can write the answer as an improper fraction or mixed number. The area of the rectangle is $\dfrac{3}{2}$ m^2 or $1\dfrac{1}{2}$ m^2.

Perimeter: The perimeter of the rectangle is the distance around the rectangle, so add the lengths of the sides or use the formula $P = 2l + 2w$.

$$P = 2l + 2w$$

$$P = 2(2 \text{ m}) + 2\left(\frac{3}{4}\text{ m}\right) \qquad \text{Substitute the values for } l \text{ and } w.$$

$$P = 4 \text{ m} + \frac{3}{2}\text{ m} \qquad\qquad \text{Multiply.}$$

$$P = \frac{8}{2}\text{ m} + \frac{3}{2}\text{ m} \qquad\qquad \text{Get a common denominator.}$$

$$P = \frac{11}{2}\text{ m} \qquad\qquad\qquad \text{Add.}$$

The perimeter of the rectangle is $\dfrac{11}{2}$ m or $5\dfrac{1}{2}$ m.

[YOU TRY 1] Find the area and perimeter of each figure.

a)
$\frac{3}{5}$ ft

$\frac{3}{5}$ ft

b)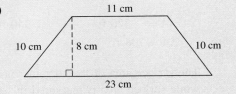
11 cm

10 cm   8 cm   10 cm

23 cm

W Hint

Make a chart of the volume formulas we have learned.

The **volume** of a three-dimensional object is a measure of the amount of space occupied by the object or the amount of space inside the object. Volume is measured in *cubic* units like cubic feet (ft^3) or cubic centimeters (cm^3).

The **surface area,** *SA,* of a three-dimensional object is the *total area* on the surface of the object. Therefore, surface area is measured in *square* units.

We can find the surface area of a rectangular solid with length *l*, width *w*, and height *h* using the formula $SA = 2lw + 2lh + 2wh$.

EXAMPLE 2

A container used to ship cargo overseas by ship has the given measurements. Find the volume and surface area of the container.

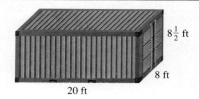

$8\frac{1}{2}$ ft

8 ft

20 ft

Solution

Let's find the volume first. We will use the formula $V = lwh$. Identify the length, width, and height.

$$l = 20 \text{ ft} \qquad w = 8 \text{ ft} \qquad h = 8\frac{1}{2} \text{ ft}$$

$$V = lwh$$

$V = (20 \text{ ft})(8 \text{ ft})\left(8\frac{1}{2} \text{ ft}\right)$ Substitute the values.

$V = (20 \text{ ft})(8 \text{ ft})\left(\dfrac{17}{2} \text{ ft}\right)$ Change to an improper fraction.

$V = (20 \text{ ft})(\overset{4}{8} \text{ ft})\left(\dfrac{17}{\underset{1}{2}} \text{ ft}\right)$ Divide out 2.

$V = 1360 \text{ ft}^3$ Multiply.

The container will hold 1360 ft³ of cargo.

To find the surface area, use the formula $SA = 2lw + 2lh + 2wh$.

$SA = \qquad 2lw \qquad + \qquad 2lh \qquad + \qquad 2wh$

$SA = 2(20 \text{ ft})(8 \text{ ft}) + 2(20 \text{ ft})\left(8\frac{1}{2} \text{ ft}\right) + 2(8 \text{ ft})\left(8\frac{1}{2} \text{ ft}\right)$ Substitute the values.

$SA = 2(20 \text{ ft})(8 \text{ ft}) + 2(20 \text{ ft})\left(\dfrac{17}{2} \text{ ft}\right) + 2(8 \text{ ft})\left(\dfrac{17}{2} \text{ ft}\right)$ Change the mixed numbers to improper fractions.

$SA = \qquad 320 \text{ ft}^2 \qquad + \qquad 340 \text{ ft}^2 \qquad + \qquad 136 \text{ ft}^2$ Multiply.

$SA = 796 \text{ ft}^2$ Add.

The surface area of the container is 796 ft². Notice that volume is measured in *cubic* feet while surface area is measured in *square* feet.

[YOU TRY 2]

American Airlines regulations state that the dimensions of the largest allowable carry-on baggage are 22 in. long, 14 in. wide, and 9 in. tall. Find the volume and surface area of a suitcase with those dimensions. (www.aa.com)

In addition to learning how to find the area and perimeter of a triangle, we learned how to classify triangles according to their angles as well as the lengths of their sides.

Acute Triangle	Right Triangle	Obtuse Triangle
An **acute triangle** is one in which all three angles are acute. (All angles measure *less than* 90°.)	A **right triangle** contains one right, or 90°, angle.	An **obtuse triangle** contains one obtuse angle. (The measure of an obtuse angle is *greater than* 90°.)

Note

The measures of the angles in a triangle add to 180°.

EXAMPLE 3

Classify each triangle as acute, right, or obtuse.

a)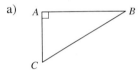

b)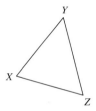

c)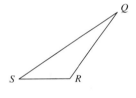

Solution

a)   This is a *right triangle* because ∠A is a right angle. (m∠A = 90°)

b)   This is an *acute triangle* because each angle is acute. (The measure of each angle is *less than* 90°.)

c)   This is an *obtuse triangle* because ∠R is an obtuse angle. (m∠R is *greater than* 90°.)

YOU TRY 3

Classify each triangle as acute, right, or obtuse.

a)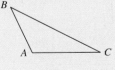

b)

c)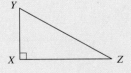

We can also classify triangles by the lengths of their sides. When the sides of triangles are the same length, we mark them with a hash mark like this: |

Note

An **equilateral triangle** has *three* sides of equal length.	An **isosceles triangle** has *two sides* of equal length.	A **scalene triangle** has *no* sides of equal length.

ANSWERS TO [YOU TRY] EXERCISES

1) a) area $= \dfrac{9}{25}$ ft^2; perimeter $= \dfrac{12}{5}$ ft or $2\dfrac{2}{5}$ ft   b) area $= 136$ cm^2; perimeter $= 54$ cm

2) $V = 2772$ in^3; $SA = 1264$ in^2 3) a) obtuse   b) acute   c) right

Putting It All Together Exercises

E Evaluate   Do the exercises, and check your work.

Objective 1: Review the Concepts of Sections 4.1–4.4

Identify each figure as a line, a line segment, or a ray. Then, name it using the correct notation.

1)

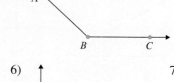

2)

$\overset{\bullet}{A} \underline{\hspace{3cm}} \overset{\bullet}{B}$

3)

4) Draw a ray named $\overrightarrow{CK}$.

Classify each angle as acute, right, obtuse, or straight, and name each angle.

5)

6)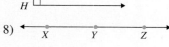

7)

8)

9) Draw lines $\overleftrightarrow{AB}$ and $\overleftrightarrow{CD}$ so that they are parallel.

10) Draw lines $\overleftrightarrow{MN}$ and $\overleftrightarrow{XY}$ so that they are perpendicular.

11) Write the formulas for the area, A, and perimeter, P, of a rectangle with length l and width w.

12) Write the formula for the area, A, of the triangle with height h and base of length b.

Find the area and perimeter of each figure.

13)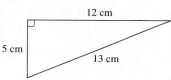

14)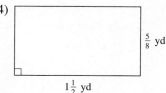

15)

16)

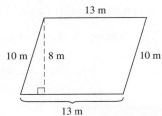

13 m · 10 m · 8 m · 10 m · 13 m

17)

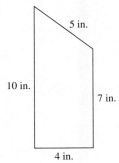

5 in. · 10 in. · 7 in. · 4 in.

18)

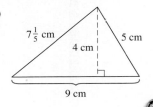

$7\frac{1}{5}$ cm · 4 cm · 5 cm · 9 cm

19)

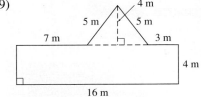

4 m · 5 m · 5 m · 7 m · 3 m · 4 m · 16 m

20)

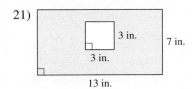

6 ft · 10 ft · 6 ft · 17 ft

Find the area of the shaded region.

21)

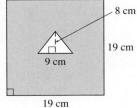

3 in. · 7 in. · 3 in. · 13 in.

22)

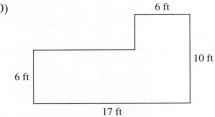

8 cm · 19 cm · 9 cm · 19 cm

23) What is the sum of the measures of the angles in a triangle?

24) What is an obtuse triangle?

25) What is an equilateral triangle?

26) What is an isosceles triangle?

Fill in the blank with *always*, *sometimes*, or *never* to make the statement true.

27) A right triangle can _____ contain an obtuse angle.

28) Parallel lines _____ intersect.

29) The bases of a trapezoid are _____ the parallel sides.

30) An acute angle _____ has a measure greater than 90°.

Find m∠x in each triangle.

31)

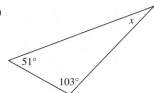

x · 51° · 103°

32)

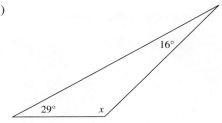

16° · 29° · x

33)

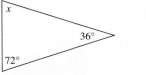

x · 36° · 72°

34)

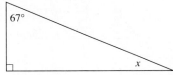

67° · x

Classify each triangle as acute, obtuse, or right.

35)

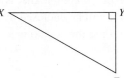

X · Y · Z

36)

B · A · C

37)

38)

48) A company is designing a tablet computer that is $7\frac{1}{2}$ in. wide and $9\frac{1}{2}$ in. long. The width of the border is $\frac{3}{4}$ in. What is the actual viewing area of the screen?

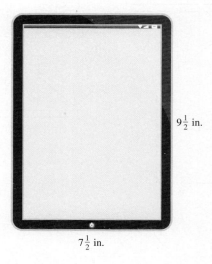

$9\frac{1}{2}$ in.

$7\frac{1}{2}$ in.

Classify each triangle as equilateral, isosceles, or scalene.

39)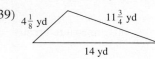
$4\frac{1}{8}$ yd $11\frac{3}{4}$ yd   14 yd

40)
$\frac{1}{2}$ m $\frac{1}{2}$ m $\frac{1}{2}$ m

41)
5 cm   6 cm   5 cm

42)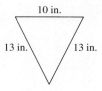
10 in.   13 in.   13 in.

Find the volume and surface area of each figure.

43)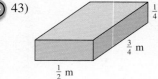
$\frac{1}{4}$ m $\frac{3}{4}$ m $\frac{1}{2}$ m

44)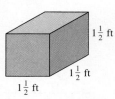
$1\frac{1}{2}$ ft $1\frac{1}{2}$ ft $1\frac{1}{2}$ ft

49) Shivendra plans to lay down new sod in his yard. The sod costs $2 per square yard, but he will also need to rent a rototiller for $75 and a lawn roller for $15, and then buy fertilizer for $140. Find the total amount Shivendra will spend to install the sod.

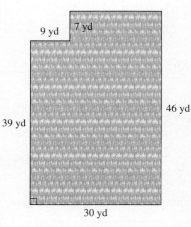
9 yd   7 yd   46 yd   39 yd   30 yd

Find the volume of each figure.

45)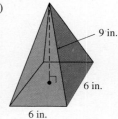
9 in.   6 in.   6 in.

46)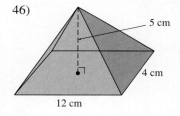
5 cm   4 cm   12 cm

50) A hybrid car battery is approximately 106 mm wide, 285 mm long, and 20 mm high. Find the volume of the battery.

Solve each problem.

47) A playground is in the shape of a rectangle that is 180 ft long and 120 ft wide. It will be enclosed by a fence that costs $13 per foot. Find the cost to fence in the playground.

R Rethink

R1) Could you do this exercise set without looking back at the book or your notes? If not, what could you do to review the topics you don't know very well?

R2) Are you comfortable working with fractions? Which sections in Chapter 3 could you review to get more practice with fractions?

R3) Choose a topic in this section and write out an explanation of it in your own words.

4.5 Solving Geometry Applications Using Algebra

W Work   **Read the explanations, follow the examples, take notes, and complete the You Trys.**

We can combine geometry formulas with what we know about solving equations to solve for an unknown value.

1 Solve for the Unknown Value

EXAMPLE 1

The formula for the area of a triangle is $A = \frac{1}{2}bh$. If $A = 42$ cm when $b = 12$ cm, find the height, h.

Solution

Substitute the known values for A and b. Then, solve for the height, h. (We will not include the units when we substitute the values. But, don't forget to write the correct units in the answer!)

W Hint
If necessary, review Chapter 2 on solving equations.

$$A = \frac{1}{2}bh$$

$$42 = \frac{1}{2}(12)h \qquad \text{Substitute the values for } A \text{ and } b.$$

$$42 = 6h \qquad \text{Multiply.}$$

$$\frac{42}{6} = \frac{6h}{6} \qquad \text{Divide each side by 6.}$$

$$7 = h \qquad \text{Simplify.}$$

The height, h, is 7 cm.

[YOU TRY 1]

The formula for the area of a triangle is $A = \frac{1}{2}bh$. If $A = 12$ ft when $b = 3$ ft, find the height, h.

EXAMPLE 2

The formula for the perimeter of a rectangle is $P = 2l + 2w$. Find the width, w, when $P = 26$ in. and $l = 9$ in.

Solution

$$P = 2l + 2w$$

$26 = 2(9) + 2w$	Substitute the values for P and l.
$26 = 18 + 2w$	Multiply.
$26 - 18 = 18 - 18 + 2w$	Subtract 18 from each side.
$8 = 2w$	Combine like terms.
$\dfrac{8}{2} = \dfrac{2w}{2}$	Divide each side by 2.
$4 = w$	Simplify.

The width, w, is 4 in.

[YOU TRY 2]

The formula for the perimeter of a rectangle is $P = 2l + 2w$. Find the length, l, when $P = 44$ m and $w = 8$ m.

---

2 Solve Applications Involving Geometry

Next, we will solve some applications that require us to use geometry formulas as well as solve an equation. We will use the same five-step process that we have used throughout this book.

EXAMPLE 3

The area of a tennis court is 2808 ft². Find the width of the court if the length is 78 ft.

Solution

Step 1: **Read** the problem carefully, and identify what we are being asked to find.

We must find the width of the tennis court.

A picture will be very helpful in this problem.

Step 2: **Choose a variable** to represent the unknown.

$w =$ the width of the tennis court

Label the picture with the length, 78 ft, and the width, w.

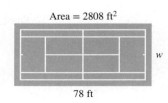

Area = 2808 ft²

w

78 ft

Step 3: Translate the information that appears in English into an algebraic equation. We will use a known geometry formula. How do we know which formula to use? List the information we are given and what we want to find:

The court is in the shape of a *rectangle;* its *area* = 2808 ft² and its *length* = 78 ft. We must find the *width.* Which formula involves the area, length, and width of a rectangle?

$$A = lw$$

W Hint

On a sheet of paper, make a list of all the geometric formulas that are used in this section.

Substitute the known values into the formula for the area of a rectangle, and solve for w.

$$A = lw$$
$$2808 = 78w \qquad \text{Substitute the known values.}$$

Step 4: Solve the equation.

$$2808 = 78w$$
$$\frac{2808}{78} = \frac{78w}{78} \qquad \text{Divide by 78.}$$
$$36 = w \qquad \text{Simplify.}$$

W Hint

Remember to include the correct units in your answer!

Step 5: Check the answer and **interpret** the solution as it relates to the problem.

If $w = 36$ ft, then $lw = (78 \text{ ft})(36 \text{ ft}) = 2808 \text{ ft}^2$. Therefore, the width of the tennis court is 36 ft.

[YOU TRY 3]

Write an equation and solve.
The area of a rectangular room is 232 ft². Find the length of the room if the width is $14\frac{1}{2}$ ft.

Sometimes, a problem will contain two unknown values.

EXAMPLE 4

Michelle built a new rectangular stall for her horse. She used 52 ft of fencing, and the length of the stall is 2 ft more than the width. Find the dimensions of the stall.

Solution

Step 1: Read the problem carefully, and identify what we are being asked to find.

We must find the length and width of the stall.

A picture will be very helpful.

Step 2: Choose a variable to represent an unknown, and define the other unknown in terms of this variable.

The length is 2 ft longer than the width. Therefore, let

$$w = \text{the width of the stall}$$

w

$w + 2$

Define the other unknown in terms of w.

$$w + 2 = \text{the length of the stall}$$

Label the picture with the expressions for the width and length.

Step 3: **Translate** the information that appears in English into an algebraic equation.

Use a known geometry formula. What does the 52 ft of fencing represent? *Because the fencing goes around the horse stall, the 52 ft represents the perimeter of the stall.* We need to use a formula that involves the length, width, and perimeter of a rectangle. The formula we will use is

$$P = 2l + 2w$$

Substitute the known values and expressions into the formula.

$$P = 2l + 2w$$
$$52 = 2(w + 2) + 2w \qquad \text{Substitute.}$$

W **Hint**

Write out the steps as you are reading the example!

Step 4: **Solve** the equation.

$$52 = 2(w + 2) + 2w$$
$$52 = 2w + 4 + 2w \qquad \text{Distribute.}$$
$$52 = 4w + 4 \qquad \text{Combine like terms.}$$
$$52 - 4 = 4w + 4 - 4 \qquad \text{Subtract 4 from each side.}$$
$$48 = 4w \qquad \text{Combine like terms.}$$
$$\frac{48}{4} = \frac{4w}{4} \qquad \text{Divide each side by 4.}$$
$$12 = w \qquad \text{Simplify.}$$

Step 5: **Check** the answer and **interpret** the meaning of the solution as it relates to the problem.

The width of the horse stall is 12 ft. The length is $w + 2 = 12 + 2 = 14$ ft.

The answer makes sense because the perimeter of the stall is
$2(14 \text{ ft}) + 2(12 \text{ ft}) = 28 \text{ ft} + 24 \text{ ft} = 52 \text{ ft}.$ ✓

[**YOU TRY 4**] Write an equation and solve.

Agnieszka wants to put Christmas lights around her rectangular picture window. The length is 1 ft more than the width, and it would take 18 ft of lights to go around the window. Find the dimensions of the window.

Recall from Section 4.3 that the sum of the angle measures in a triangle is 180°. We will use this fact in our next example.

EXAMPLE 5   Find the missing angle measures.

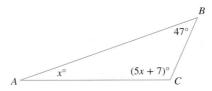

Solution

Step 1: **Read** the problem carefully, and identify what we are being asked to find.

Find the missing angle measures.

Step 2: The unknowns are already defined. We must find x, the measure of one angle, and then $5x + 7$, the measure of the other angle.

Step 3: **Translate** the information into an algebraic equation. Since the sum of the angles in a triangle is 180°, we can write

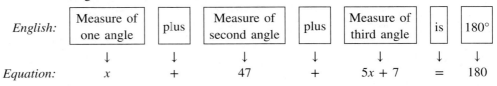

The equation is $x + 47 + (5x + 7) = 180$.

Step 4: **Solve** the equation.

$x + 47 + (5x + 7) = 180$	
$6x + 54 = 180$	Combine like terms.
$6x + 54 - 54 = 180 - 54$	Subtract 54 from each side.
$6x = 6$	Combine like terms.
$\dfrac{6x}{6} = \dfrac{126}{6}$	Divide each side by 6.
$x = 21$	Simplify.

Step 5: **Check** the answer and **interpret** the solution as it relates to the problem.

One angle, x, has a measure of 21°. The other unknown angle measure is $5x + 7 = 5(21) + 7 = 112°$.

The answer makes sense because the sum of the angle measures is $21° + 47° + 112° = 180°$.

$\left[\text{YOU TRY 5}\right]$ Find the missing angle measures.

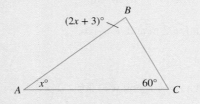

ANSWERS TO $\left[\text{YOU TRY}\right]$ **EXERCISES**

1) 8 ft   2) 14 m   3) 16 ft   4) 4 ft by 5 ft   5) $m\angle A = 39°$; $m\angle B = 81°$

Get Ready

Write down each of the following formulas.

1) The perimeter, P, of a rectangle with length l and width w

2) The area, A, of a rectangle with length l and width w

3) The area, A, of a triangle with base b and height h

4) The perimeter, P, of a square with side length s

5) The sum of the measures of the angles of a triangle if the angle measures are a, b, and c

6) The volume, V, of a rectangular solid with length l, width w, and height h

Objective 1: Solve for the Unknown Value

7) If the area of a rectangle is given in square feet and its length is given in feet, the width will be in which units?

8) If you substitute values into $A = bh$, the formula for the area of a parallelogram, would it be reasonable to solve for b and get $b = -4$? Explain.

Substitute the given values into the geometry formula, and solve for the remaining variable. Be sure to use the correct units in your answer.

9) $A = lw$; If $A = 36$ ft^2 when $w = 4$ ft, find l.

10) $A = bh$; If $A = 56$ mm^2 when $h = 7$ mm, find b.

11) $A = \frac{1}{2}bh$; If $A = 18$ cm^2 when $b = 3$ cm, find h.

12) $A = \frac{1}{2}bh$; If $A = 35$ yd^2 when $b = 5$ yd, find h.

13) $P = 2l + 2w$; If $P = 96$ in. when $l = 31$ in., find w.

14) $P = 2l + 2w$; If $P = 76$ m when $w = 15$ m, find l.

15) $V = lwh$; If $V = \frac{3}{4}$ m^3 when $l = 2$ m and $h = \frac{1}{2}$ m, find w.

16) $V = \frac{1}{3}lwh$; If $V = \frac{8}{3}$ ft^3 when $l = 4$ ft and $w = 3$ ft, find h.

17) $A = \frac{1}{2}h(b + B)$; If $A = 19\frac{1}{2}$ ft^2 when $b = 4$ ft and $B = 9$ ft, find h.

18) $A = \frac{1}{2}h(b + B)$; If $A = 17\frac{1}{2}$ cm^2 when $b = 2$ cm and $B = 3$ cm, find h.

19) $A = \frac{1}{2}h(b + B)$; If $A = 90$ mm^2 when $h = 10$ mm and $b = 7$ mm, find B.

20) $A = \frac{1}{2}h(b + B)$; If $A = 108$ in^2 when $h = 12$ in. and $B = 13$ in., find b.

Objective 2: Solve Applications Involving Geometry

21) Summarize the Five Steps for Solving Applied Problems.

22) If you are solving for the length of a plot of land, is -140 ft a reasonable answer? Explain.

Solve each problem. Use the Five Steps for Solving Applied Problems.

23) Find the height of a parallelogram if its base measures 15 cm and its area is 195 cm^2.

24) Find the width of a rectangle if its area is 102 in^2 and its length is 17 in.

25) The volume of a rectangular pyramid is $13\frac{1}{3}$ m^3. The length of its base is 5 m, and the width is 2 m. What is the height of the pyramid?

26) The volume of a box is 56 ft^3. It is $2\frac{1}{3}$ ft high and 4 ft wide. What is the length of the box?

27) A square has a perimeter of 14 in. How long is each side?

28) A square has a perimeter of 10 cm. How long is each side?

29) The *Mona Lisa* is one of the most famous paintings in the world. It is 77 cm tall and has an area of 4081 cm^2. Find its width.

30) A rectangular table has an area of $7\frac{1}{2}$ ft^2. Find the length if it is $2\frac{1}{2}$ ft wide.

©Shutterstock/Everett - Art

31) A rectangular fish tank holds 8640 in³ of water. Find the length of the tank if it is 15 in. wide and 16 in. high.

32) A rectangular compost bin holds 20,280 in³ of waste. Find the height of the bin if both the length and width of its base are 26 in.

33) A cell phone is $10\frac{1}{2}$ cm long and 1 cm thick. What is the width of the phone if its volume is 63 cm³?

34) A rectangular stack of sticky notes is 3 in. long and 3 in. wide. How tall is the stack if the volume is 18 in³?

35) The perimeter of a table tennis table is 28 ft, and it is 4 ft longer than it is wide. Find the length and width.

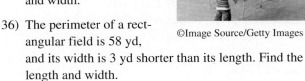
©Image Source/Getty Images

36) The perimeter of a rectangular field is 58 yd, and its width is 3 yd shorter than its length. Find the length and width.

37) Miyuki wants to put a rectangular safety fence around the backyard pool. She will need 92 ft of fencing, and the width will be 6 ft less than the length. What are the dimensions of the fence?

38) A shopping mall will build a rectangular children's play area that will be enclosed by 180 ft of fence. If it will be 14 ft longer than it is wide, find the dimensions of the play area.

39) A contractor used 510 ft of fencing to enclose a rectangular construction site. Find the dimensions of the site if the length is twice the width.

40) Raheem uses 156 in. of wood to build a rectangular frame. Find the dimensions of the frame if the width is half the length.

41) A college pennant is in the shape of a triangle and has a perimeter of 46 in. The two longer sides have the same length, while the third side is 8 in. shorter than a longer side. Find the lengths of the sides of the pennant.

42) A triangular piece of fabric has a perimeter of 80 cm. The length of one side is 29 cm. Of the two remaining sides, one is 11 cm longer than the other. How long is each unknown side?

43) The measures of the angles in a triangle add to what number?

44) If a triangle has an angle with a measure of 105°, can another of its angles have a measure of 98°? Explain.

Find the missing angle measures. Use the Five Steps for Solving Applied Problems.

45)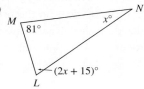

46)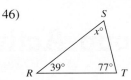

47)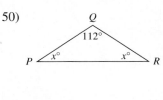

48)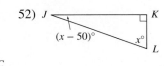

49) U x° x° V / 44° / W

50) Q 112° / P x° x° R

51) B x° / A (x − 40)° C

52) J K / (x − 50)° x° / L

53) G x° / F (1/3 x)° / (1/6 x)° H

54)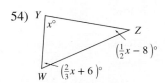

R1) Why is it important to be able to check your answers by hand? Did any of your solutions automatically seem unreasonable before checking it? Why?

R2) How could you use the procedure for solving applied problems in other classes or at your job?

Group Activity – Geometry

The Buyer family recently purchased a foreclosed home at a real estate auction. The Buyers got a great deal on the price of the home, but the home needs many repairs. The Buyers decide to install new floor molding (baseboards) in the living room. The rectangular living room is 24 ft long and 15 ft wide. The floor plan of the room is shown below.

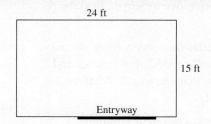

1) Mr. and Mrs. Buyer are going to a home improvement store to purchase the floor molding for the living room. They each estimated the amount of floor molding that is needed (ignoring the entryway for now). Mr. Buyer thinks that they will need 360 ft of molding, and Mrs. Buyer thinks they will need 78 ft of molding. Which estimate is correct? Give reasons to support your answer.

2) Describe in your own words how to determine the perimeter of a rectangle if you know the length and width.

3) Calculate the amount of floor molding required for all six of the rooms in the table below.

Room	Length	Width	Calculations: $2l + 2w = P$	Perimeter
Living	24 ft	15 ft	$2(24) + 2(15) = 48 + 30 = 78$	78 ft
Dining	16 ft	12 ft		
Family	30 ft	16 ft		
Bedroom 1	18 ft	15 ft		
Bedroom 2	14 ft	14 ft		
Bedroom 3	13 ft	13 ft		

4) In the rooms above, there is a total of 4 entryways and 3 doorways. Each entryway is 7 ft wide and each doorway is 3 ft wide. Accounting for the space in the entryways and doorways, how much molding is needed for all six rooms?

5) The type of floor molding that the Buyers chose is sold in lengths of 12 ft for $3 each. How many lengths will they have to buy? What will the total cost be?

em**POWER**me Active Listening

If you want to improve your math skills, just attending class is not enough. You need to do the necessary work outside of class, and when you are in class, you need to listen actively.

To assess whether you are an active listener, consider the following pairs of statements. Place a check next to the statement in each pair that more closely describes your classroom listening style.

☐ 1a. When I'm listening in class, I lean back and get as comfortable as possible.
☐ 1b. When I'm listening in class, I sit upright and even lean forward a little.

☐ 2a. I let the instructor's words wash over me, generally going with the flow of the lecture.
☐ 2b. I try to guess in advance what the instructor is going to say and what direction the lecture is taking.

☐ 3a. I regard each lecture as a separate event, not necessarily related to what the instructor has said before or will say the next time.
☐ 3b. As I listen, I regularly ask myself how this relates to what was said in previous classes.

☐ 4a. When I take notes, I try to reproduce the instructor's words as closely as possible.
☐ 4b. When I take notes, I try to interpret and summarize the ideas behind the instructor's words.

☐ 5a. I don't usually question the importance of what the instructor is saying or why it's the topic of a lecture or discussion.
☐ 5b. I often ask why the content of the lecture is important enough for the instructor to be speaking about it.

(Continued on next page)

☐ 6a. I rarely question the accuracy or logic of a presentation, assuming that the instructor knows the topic better than I do.

☐ 6b. I often ask myself how the instructor knows something and find myself wondering how it could be proved.

☐ 7a. I just about never make eye contact with the instructor.

☐ 7b. I often make eye contact with the instructor.

If you tended to prefer the "a" statements in most pairs, you have a more passive style of listening. If you preferred the "b" statements, you have a more active style of listening. Based on your responses, consider ways that you can become a more active listener.

Chapter 4:   Summary

Definition/Procedure	Example

4.1 Introduction to Geometry

Identify Lines, Line Segments, and Rays

Space is an unlimited, three-dimensional expanse.

A **plane** is a flat surface that continues indefinitely.

A **point** is a location in space with no length, width, or height.

A **line** is a straight set of points that continues forever in two directions.

A **line segment** is a piece of a line with two endpoints.

A **ray** is a part of a line that has one endpoint and that continues forever in the other direction.

Identify each figure as a line, a line segment, or a ray. Then, name it using correct notation.

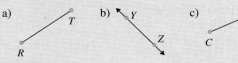

a)   b)   c)

Solution

a) This figure is straight with two endpoints. **This is a line segment.** Write its name as $\overline{RT}$ or $\overline{TR}$.

b) This figure is straight and continues forever in both directions. **This is a line.** Write its name as $\overleftrightarrow{YZ}$ or $\overleftrightarrow{ZY}$.

c) This figure has *one* endpoint and an arrowhead on the other end indicating it goes on forever in that direction. **This is a ray.** Write its name as $\overrightarrow{CA}$. (Notice that the C *must* come first because it is the endpoint of the ray.)

Identifying Angles

If we join two rays at their endpoints, we get an **angle.** The **vertex** of the angle is the common endpoint. An angle is denoted by the symbol ∠.

Name each of the angles in this figure in two different ways.

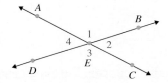

Solution

First, notice that E is the vertex of all the angles. So, we cannot use just the vertex to name any of the angles in this figure.

We can give each of the angles the following names:

∠1: ∠*AEB* or ∠*BEA*
∠2: ∠*BEC* or ∠*CEB*
∠3: ∠*CED* or ∠*DEC*
∠4: ∠*DEA* or ∠*AED*

Classifying Angles

We measure the size of an angle using **degrees.** The symbol for degrees is °.

An **acute angle** is an angle whose measure is greater than 0° and less than 90°.

A **right angle** is an angle whose measure is 90°.

An **obtuse angle** is an angle whose measure is greater than 90° and less than 180°.

A **straight angle** is an angle whose measure is 180°.

Classify each angle as acute, right, obtuse, or straight.

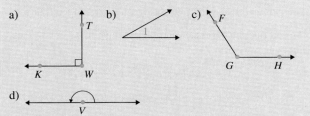

a)   b)   c)

d)

Solution

a) The small square inside the angle tells us that its measure is exactly 90°. ∠*KWT* is a *right* angle.

b) The measure of ∠1 is less than 90°, so it is an *acute* angle.

c) The measure of ∠*FGH* is greater than 90° and less than 180°. It is an *obtuse* angle.

d) The measure of ∠*V* is 180°, so it is a *straight* angle.

Definition/Procedure	Example

Identifying Parallel and Perpendicular Lines

Two lines that cross each other are called **intersecting lines.**

Parallel lines are lines in the same plane that do *not* intersect. The symbol ‖ means "is parallel to."

Perpendicular lines intersect at right angles. The symbol ⊥ means "is perpendicular to."

Classify each pair of lines as parallel, perpendicular, or neither. If they are parallel or perpendicular, use the appropriate notation.

a)

b)

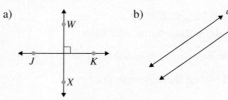

c)

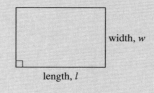

Solution

a) Lines $\overleftrightarrow{JK}$ and $\overleftrightarrow{WX}$ meet at right angles, so they are perpendicular. Write $\overleftrightarrow{JK} \perp \overleftrightarrow{WX}$.

b) Lines a and b do not intersect, so they are parallel. Write $a \parallel b$.

c) Lines $\overleftrightarrow{MN}$ and $\overleftrightarrow{OP}$ intersect but not at right angles. They are neither parallel nor perpendicular.

4.2 Rectangles, Squares, Parallelograms, and Trapezoids

Rectangles

A **rectangle** is a four-sided figure containing four right angles.

The opposite sides are parallel and congruent.

The longer side of a rectangle is called the length, l. The shorter side is called the width, w.

A small square indicates that the rectangle has four 90° angles.

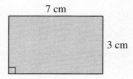

Perimeter of a Rectangle

The **perimeter** of a figure is the distance around the figure.

We can find the perimeter, P, of a rectangle using one of the following formulas:

    1) $P = 2 \cdot length + 2 \cdot width$ or

    2) $P = 2l + 2w$

Find the perimeter of the rectangle.

7 cm

3 cm

$P = 2l + 2w$

$P = 2(7 \text{ cm}) + 2(3 \text{ cm})$ Substitute the values.

$P = 14 \text{ cm} + 6 \text{ cm}$ Multiply.

$P = 20 \text{ cm}$ Add.

Definition/Procedure	Example
Area of a Rectangle The **area** of a rectangle is the size of the region enclosed in the figure. We can find the area, A, of a rectangle using this formula: $$Area = length \cdot width \quad \text{or}$$ $$A = lw$$ The units for area are *square* units.	Find the area of the rectangle in the previous box in the Summary. $$A = lw$$ $$A = (7 \text{ cm})(3 \text{ cm}) \quad \text{Substitute the values.}$$ $$A = 21 \text{ cm}^2 \quad \text{Multiply.}$$
Perimeter and Area of a Square A **square** is a rectangle with all four sides of equal length. The **perimeter,** P, of a square with side length s can be found using one of these formulas: 1) $Perimeter = side + side + side + side$ 2) $\quad P = 4 \cdot side$ 3) $\quad P = 4 \cdot s$ The **area,** A, of a square with side length s can be found using one of these formulas: 1) $Area = side \cdot side$ 2) $\quad A = s \cdot s$ 3) $\quad A = s^2$	Find a) the perimeter and b) the area of this square. $\frac{1}{2}$ ft $\frac{1}{2}$ ft a) Use $P = 4s$. $$P = 4\left(\frac{1}{2}\text{ ft}\right) \quad \text{Substitute the length of the side.}$$ $$P = 2 \text{ ft} \quad \text{Multiply.}$$ b) Use $A = s^2$. $$A = \left(\frac{1}{2}\text{ ft}\right)^2 \quad \text{Substitute the length of the side.}$$ $$A = \frac{1}{4}\text{ ft}^2$$
Area of a Parallelogram A **parallelogram** is a four-sided figure whose opposite sides are parallel and the same length. The **area,** A, of a parallelogram is $$Area = base \cdot height \quad \text{or}$$ $$A = bh$$ where $b =$ the length of the base and $h =$ the height. The base is the side that forms a right angle with the height.	Find the area of the parallelogram. 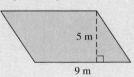 5 m 9 m $$A = bh$$ $$A = (9 \text{ m})(5 \text{ m}) \quad \text{Substitute the values.}$$ $$A = 45 \text{ m}^2 \quad \text{Multiply.}$$
Area of a Trapezoid A **trapezoid** is a four-sided figure with exactly one pair of parallel sides. The **area,** A, of a trapezoid is $$A = \frac{1}{2} \cdot height \cdot (short\ base + long\ base) \quad \text{or}$$ $$A = \frac{1}{2}h(b + B)$$ where $h =$ the height, $b =$ the length of the short base, and $B =$ the length of the long base. **The bases are always the parallel sides.**	Find the area of the trapezoid. 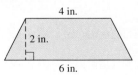 4 in. 2 in. 6 in. $$A = \frac{1}{2}h(b + B)$$ $$A = \frac{1}{2}(2 \text{ in.})(4 \text{ in.} + 6 \text{ in.}) \quad \text{Substitute the values.}$$ $$A = \frac{1}{2}(2 \text{ in.})(10 \text{ in.}) \quad \text{Add inside the parentheses.}$$ $$A = 10 \text{ in}^2 \quad \text{Multiply.}$$

Definition/Procedure	Example

4.3 Triangles

Perimeter and Area of a Triangle

A **triangle** is a closed figure with exactly three sides.

The **perimeter** of a triangle is the distance around the triangle.

The **area**, A, of a triangle is

$$1)\ Area = \frac{1}{2} \cdot base \cdot height \quad \text{or}$$

$$2)\quad A = \frac{1}{2}bh$$

where b = the length of the base and h = the height.

Find a) the perimeter and b) the area of this triangle.

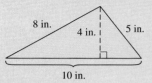

a) To find the perimeter, add the lengths of the sides of the triangle.

$$P = 8 \text{ in.} + 5 \text{ in.} + 10 \text{ in.} = 23 \text{ in.}$$

b) To find the area, we will use $A = \frac{1}{2}bh$.

$$A = \frac{1}{2}(10 \text{ in.})(4 \text{ in.}) \qquad \text{Substitute the values.}$$

$$A = 20 \text{ in}^2 \qquad \text{Multiply.}$$

The Sum of the Angle Measures in a Triangle

The measures of the angles in a triangle add up to 180°.

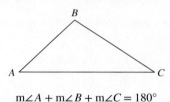

$$m\angle A + m\angle B + m\angle C = 180°$$

Classifying Triangles

An **acute triangle** is one in which all three angles are acute.

Acute triangle:

A **right triangle** contains one *right*, or 90°, angle.

Right triangle:

An **obtuse triangle** contains one *obtuse* angle.

Obtuse triangle:

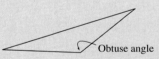

Obtuse angle

An **equilateral triangle** has *three* sides of equal length.

Equilateral triangle:

5 cm   5 cm

5 cm

An **isosceles triangle** has *two* sides of equal length.

Isosceles triangle:

A **scalene triangle** has *no* sides of equal length.

Scalene triangle:

Definition/Procedure	Example

4.4 Volume and Surface Area

The **volume** of a three-dimensional object is a measure of the amount of space occupied by the object or the amount of space inside the object.

Volume is measured in *cubic* units, the number of *cubic* units it takes to fill the object.

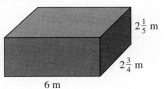
1 in.
4 in.
3 in.

If we say that the volume of this box is 12 in³, it means that we can fit *twelve* 1-in³ boxes inside this larger box.

Volume of a Rectangular Solid

A **rectangular solid** is a box-like shape with dimensions of length, width, and height.

The **volume**, V, of a rectangular solid (or box) with length l, width w, and height h is

$$Volume = length \cdot width \cdot height \quad \text{or}$$
$$V = lwh$$

A **cube** is a rectangular solid in which all sides are the same length. The volume of a cube with side length s is

$$Volume = s \cdot s \cdot s \quad \text{or}$$
$$V = s^3$$

Find the volume of this box.

$2\frac{1}{5}$ m

$2\frac{3}{4}$ m

6 m

$V = lwh$

$V = 6 \text{ m}\left(2\frac{3}{4}\text{ m}\right)\left(2\frac{1}{5}\text{ m}\right)$ Substitute the values.

$V = 6 \text{ m}\left(\frac{11}{4}\text{ m}\right)\left(\frac{11}{5}\text{ m}\right)$ Change to improper fractions.

$V = \frac{363}{10}$ m³ or $36\frac{3}{10}$ m³   Multiply.

We can express our answer in either of these ways.

Volume of a Rectangular Pyramid

The **volume**, V, of a rectangular pyramid with height h, and a base of length l and width w is

$$Volume = \frac{1}{3} \cdot length \cdot width \cdot height \quad \text{or}$$
$$V = \frac{1}{3}lwh$$

Because lw equals the *area* of the base, we can also think of the volume formula as $V = \frac{1}{3}Ah$, where $A = lw$, the area of the base of the pyramid.

Find the volume of the pyramid with the given dimensions.

2 yd
3 yd
1 yd

$V = \frac{1}{3}lwh$

$V = \frac{1}{3}(3 \text{ yd})(1 \text{ yd})(2 \text{ yd})$ Substitute the values.

$V = 2 \text{ yd}^3$ Multiply.

Definition/Procedure	Example
Surface Area of a Rectangular Solid The **surface area** of a three-dimensional object is the total area on the surface of the object. The **surface area**, *SA*, of a rectangular solid with length *l*, width *w*, and height *h* is $$SA = 2lw + 2lh + 2wh$$ Because we are finding the total *area* on the surface of the object, surface area is measured in *square* units.	Find the surface area of this box. 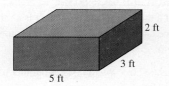 Identify the length, width, and height, then substitute them into the formula. $$l = 5 \text{ ft}, w = 3 \text{ ft}, h = 2 \text{ ft}$$ $SA = 2lw + 2lh + 2wh$ $SA = 2(5 \text{ ft})(3 \text{ ft}) + 2(5 \text{ ft})(2 \text{ ft}) + 2(3 \text{ ft})(2 \text{ ft})$ $SA = \quad 30 \text{ ft}^2 \quad + \quad 20 \text{ ft}^2 \quad + \quad 12 \text{ ft}^2$ $SA = 62 \text{ ft}^2$

4.5 Solving Geometry Applications Using Algebra

We can use geometry formulas to solve for unknown values.	The formula for the area of a triangle is $A = \dfrac{1}{2}bh$. If $A = 72$ cm when $b = 16$ cm, find the height, *h*. $A = \dfrac{1}{2}bh$ $72 = \dfrac{1}{2}(16)h$ Substitute the given values. $72 = 8h$ Multiply. $\dfrac{72}{8} = \dfrac{8h}{8}$ Divide both sides by 8. $9 = h$ Simplify.

To solve some applications, we can use the five-step process to write an equation and solve.	A rectangular piece of poster board is 6 in. longer than it is wide. If its perimeter is 132 in., find its dimensions.
	
	Step 1: **Read** the problem carefully, and identify what we are being asked to find.
	
	    We must find the length and width of the poster board.
	
	    Draw a picture.
	
	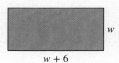
	
	Step 2: **Choose a variable** to represent the unknown, and define the other unknown in terms of this variable.
	
	    $w = $ the width of the poster board
	    $w + 6 = $ the length of the poster board
	
	    Label the picture with the expressions for the width and length.

Definition/Procedure	Example
	Step 3: **Translate** the information that appears in English into an algebraic equation. The formula for the perimeter of a rectangle is $$P = 2l + 2w$$ Substitute the known values and expressions into the formula. $P = 2l + 2w$ $132 = 2(w + 6) + 2w$ Substitute. **Step 4:** **Solve** the equation. $132 = 2(w + 6) + 2w$ $132 = 2w + 12 + 2w$ Distribute. $132 = 4w + 12$ Combine like terms. $132 - 12 = 4w + 12 - 12$ Subtract 12 from each side. $120 = 4w$ Combine like terms. $\dfrac{120}{4} = \dfrac{4w}{4}$ Divide each side by 4. $30 = w$ Simplify. **Step 5:** **Check** the answer and **interpret** the meaning of the solution as it relates to the problem. The width of the poster board is 30 in. The length is $w + 6 = 30 + 6 = 36$ in. The answer makes sense because the perimeter of the poster board is $2(36$ in.$) + 2(30$ in.$) = 72$ in. $+ 60$ in. $= 132$ in.

Chapter 4: Review Exercises

(4.1) Name each of the numbered angles in the figures in two different ways.

1)
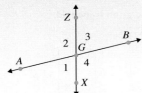

2)
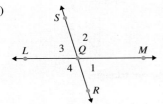

Classify each angle as acute, right, obtuse, or straight.

3)
J   F   U

4)
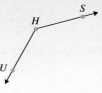

5)

6)
H   F
L

Classify each pair of lines as parallel, perpendicular, or neither. If they are parallel or perpendicular, use the appropriate notation.

7)
y
z

8)

9)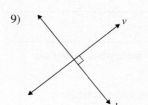

10)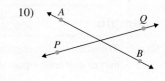

(4.2) Find the perimeter and area of each rectangle.

11)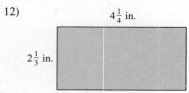

10 cm

8 cm   8 cm

10 cm

12)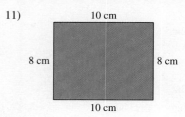

$4\frac{1}{4}$ in.

$2\frac{1}{3}$ in.

Find the perimeter and area of each square.

13)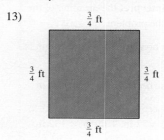

$\frac{3}{4}$ ft

$\frac{3}{4}$ ft $\frac{3}{4}$ ft

$\frac{3}{4}$ ft

14)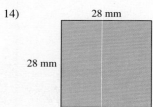

28 mm

28 mm

Find the perimeter and the area of the given figure. Note that you must first find the missing lengths.

15)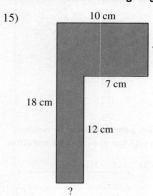

10 cm

?

7 cm

18 cm

12 cm

?

16)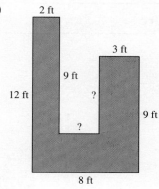

2 ft

3 ft

9 ft

12 ft   ?

?   9 ft

8 ft

Solve each problem.

17) Regulation soccer fields vary in size depending on age and skill level. What is the difference between the areas of the two soccer fields shown below?

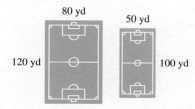

80 yd

50 yd

120 yd   100 yd

18) An archaeologist discovers some dinosaur fossils and measures off a rectangular area around the fossils that is 16 ft long and 12 ft wide. How much fencing is needed to enclose this area?

Find the perimeter and area of the given parallelogram.

19)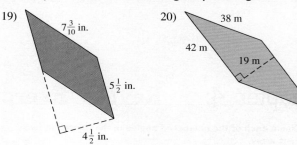

$7\frac{3}{10}$ in.

$5\frac{1}{2}$ in.

$4\frac{1}{2}$ in.

20) 38 m

42 m

19 m

Find the perimeter and area of the given trapezoid.

21)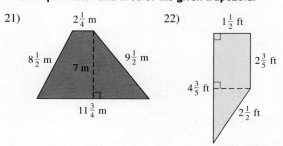

$2\frac{1}{4}$ m

$8\frac{1}{2}$ m   7 m $9\frac{1}{2}$ m

$11\frac{3}{4}$ m

22) $1\frac{1}{2}$ ft

$2\frac{3}{5}$ ft

$4\frac{3}{5}$ ft

$2\frac{1}{2}$ ft

(4.3) Find the perimeter of the given triangle.

23)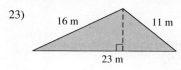

16 m   11 m

23 m

24)

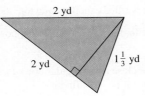
2 yd

2 yd
$1\frac{1}{3}$ yd

Find the area of the given triangle.

25)

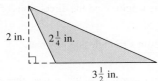
2 in.
$2\frac{1}{4}$ in.
$3\frac{1}{2}$ in.

26)

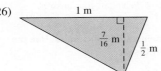
1 m
$\frac{7}{16}$ m
$\frac{1}{2}$ m

27) Find the area of the shaded region.

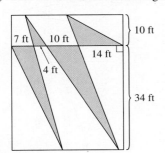
7 ft   10 ft
14 ft
4 ft
10 ft
34 ft

28) Find the area of the unshaded region.

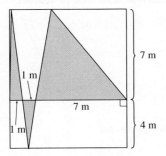
7 m
1 m
7 m
1 m
4 m

Solve each problem.

29) Mr. and Mrs. Nguyen have a reflecting pond in their backyard with an irregular shape shown in the figure. They hire a monthly service at a cost of $2 per square foot to maintain the pond. What is the cost of the Nguyens' monthly service?

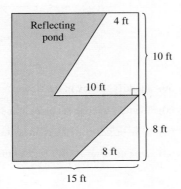
Reflecting pond
4 ft
10 ft
10 ft
8 ft
8 ft
15 ft

30) The entranceway to a newly renovated hotel is going to be paved with natural stone costing $126 per square meter for the installation. The entranceway is represented by the shaded area below. What is the total cost for the new entranceway?

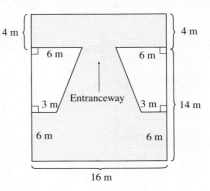
4 m   4 m
6 m   6 m
3 m   Entranceway   3 m   14 m
6 m   6 m
16 m

Find m∠x in each triangle.

31)

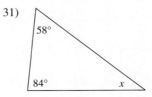
58°
84°   x

32)

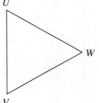
124°
37°   x

Classify each triangle as acute, right, or obtuse.

33) U

W
V

34)

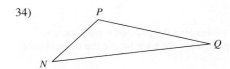
P
Q
N

Classify each triangle as equilateral, isosceles, or scalene.

35)

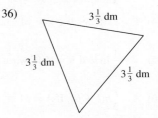
20 cm
16 cm
12 cm

36)

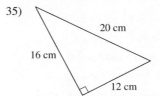
$3\frac{1}{3}$ dm
$3\frac{1}{3}$ dm
$3\frac{1}{3}$ dm

(4.4) Find the volume and surface area of each box.

37)

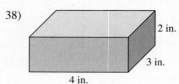
9 mm   4 mm   3 mm

38)

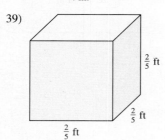
4 in.   3 in.   2 in.

39)

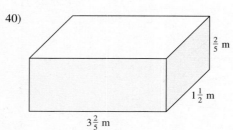
$\frac{2}{5}$ ft $\frac{2}{5}$ ft $\frac{2}{5}$ ft

40)

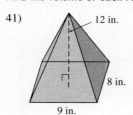
$3\frac{2}{5}$ m $1\frac{1}{2}$ m $\frac{2}{5}$ m

Find the volume of each rectangular pyramid.

41)

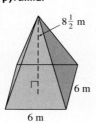
12 in.   8 in.   9 in.

42)

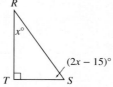
$8\frac{1}{2}$ m   6 m   6 m

(4.5) Substitute given values into the geometry formula, and solve for the remaining variable. Be sure to use the correct units in your answer.

43) $A = \frac{1}{2}bh$; If $A = 20$ cm^2 when $h = 8$ cm, find b.

44) $P = 2l + 2w$; If $P = 9$ yd when $l = 3\frac{1}{2}$ yd, find w.

45) $V = \frac{1}{3}lwh$; If $V = 15$ ft^3 when $l = 10$ ft and $h = \frac{1}{2}$ ft, find w.

46) $A = \frac{1}{2}h(b + B)$; If $A = 24$ m^2 when $h = 4$ m and $B = 7$ m, find b.

Write an equation, and solve each problem. Use the Five Steps for Solving Applied Problems.

47) A rectangular parking space is $5\frac{1}{2}$ ft wide and has an area of 55 ft^2. What is the length of the parking space?

48) If a rectangular baking dish is filled exactly to the top, its volume is 209 in^3. If it is 11 in. long and $9\frac{1}{2}$ in. wide, how deep is the baking dish?

49) It takes 62 ft of fencing to enclose a rectangular dog run. If the dog run is 9 ft longer than it is wide, find its dimensions.

50) Frank used 120 in. of trim around a rectangular window. The width is 12 in. less than the length. Find the dimensions of the window.

51) The measures of the angles in a triangle add to what number?

52) If a triangle has an angle with a measure of 95°, can another of its angles have a measure of 102°? Explain your answer.

Write an equation to find the missing angle measures. Use the Five Steps for Solving Applied Problems.

53)

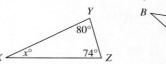
Y   80° $x°$ 74°   Z   X

54)

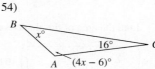
B $x°$ 16°   C   A $(4x - 6)°$

55)

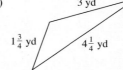
R $x°$ $(2x - 15)°$ T   S

56)

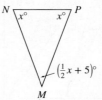
N $x°$ $x°$ P $\left(\frac{1}{2}x + 5\right)°$ M

Classify each triangle as equilateral, isosceles, or scalene.

57)

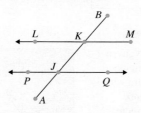
3 yd $1\frac{3}{4}$ yd $4\frac{1}{4}$ yd

58)

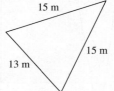
15 m   15 m   13 m

Use the figure below for Exercises 59–62.

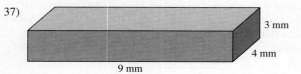
B   L   K   M   J   P   Q   A

59) Identify the ray in the figure.

60) Identify the line segment in the figure.

61) *True* or *false*: $\angle PJK$ is an acute angle.

62) *True* or *false*: $\angle MKJ$ is an obtuse angle.

Solve each problem.

63) Mr. and Mrs. Fraser want to install wood flooring on their rectangular living room floor that measures 20 ft by 14 ft. If the installation charge is $5 per square foot, what is the total installation charge?

64) Suppose a rectangle has a perimeter of 30 ft. If the rectangle has a length of 6 ft, what is its width?

65) The base of a pyramid is a square with a side length of 50 ft. If the height of the pyramid is 42 ft, find the volume of the pyramid.

66) The base of a triangular sign is $1\frac{1}{2}$ m long, and the height of the sign is 2 m. What is the area of the sign?

Find the area of the figure.

67)

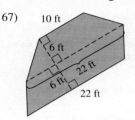

10 ft, 6 ft, 22 ft, 6 ft, 22 ft

Mixed Exercises

Fill in the blank with _always_, _sometimes_, or _never_ to make the statement true.

68) Parallel lines _____ intersect at 90° angles.

69) The measure of an acute angle is _____ greater than the measure of obtuse angle.

70) A triangle _____ contains three acute angles.

71) The area of a figure is _____ measured in _square_ units.

72) The bases of a trapezoid are _____ perpendicular.

73) The height of a triangle _____ forms a right angle with the base of the triangle.

74) Find the measure of each angle.

$(2x + 10)°$ at B, $(x + 2)°$ at A, $x°$ at C

75) A 3 × 3 Rubik's cube has a side length of approximately $2\frac{1}{4}$ in. Find the combined area of all six faces of the cube.

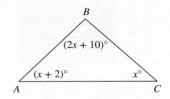

©McGraw-Hill Education

Write an equation and solve. Use the Five Steps for Solving Applied Problems.

76) Lorenzo is building a workbench for his garage. The top is in the shape of a rectangle, and it is 4 ft longer than it is wide. Find the dimensions of the top of the workbench if its perimeter is 20 ft.

Chapter 4:   Test

1) Identify each figure as a line, line segment, or ray, and name it using the correct notation.

a)
K, N

b)
F, G

c)
A, R

2) a) What does it mean if two lines are perpendicular?

   b) What does it mean if two lines are parallel?

3) What is the difference between an acute angle and an obtuse angle?

4) Classify each angle as acute, obtuse, right, or straight.

a)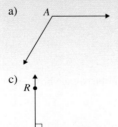
A

b) Z, Y, X

c)
R, S, T

For Exercises 5–8,

a) identify the figure.

b) find the area.

c) find the perimeter.

5)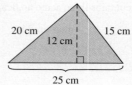
20 cm   15 cm
12 cm
25 cm

6)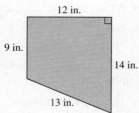
12 in.
9 in.
14 in.
13 in.

7)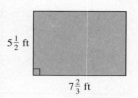
$5\frac{1}{2}$ ft
$7\frac{2}{3}$ ft

8)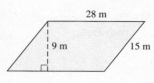
28 m
9 m
15 m

9) For this figure, find
   a) the area.
   b) the perimeter.

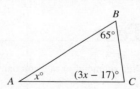
6 ft
7 ft
10 ft
5 ft
6 ft
4 ft
5 ft
9 ft

10) The angles in a triangle add up to how many degrees?

For Exercises 11 and 12, classify each triangle as

a) acute, obtuse, or right

b) equilateral, isosceles, or scalene

11)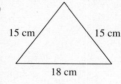
15 cm   15 cm
18 cm

12)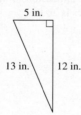
5 in.
13 in.   12 in.

13) Find the volume of the figure.

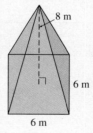
8 m
6 m
6 m

14) Find the volume and surface area of the storage box.

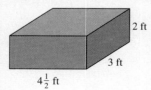
2 ft
3 ft
$4\frac{1}{2}$ ft

Solve each problem.

15) A town's park district will install a flower garden in the shape shown here. It will be enclosed by decorative fencing that costs \$18/ft. Find the cost of the fence.

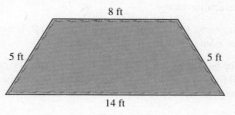
8 ft
5 ft   5 ft
14 ft

16) Supinda is printing out an essay on paper that measures $8\frac{1}{2}$ in. by 11 in. She sets the top and bottom margins at 1 in. and the left and right margins at $1\frac{1}{4}$ in. Find the area of the region on the paper available for print.

Substitute the given values into the geometry formula, and solve for the remaining variable. Be sure to use the correct units in your answer.

17) $P = 2l + 2w$; If $P = 34$ mm when $l = 11$ mm, find w.

18) $A = \frac{1}{2}h(b + B)$; If $A = 26$ in^2 when $b = 3$ in. and $B = 10$ in., find h.

Solve each problem. Use the Five Steps for Solving Applied Problems.

19) The area of a triangular window is 225 in^2, and the length of its base is 25 in. What is the height of the window?

20) Suzanne will sew a border around the rectangular quilt she is making. The width of the quilt is 2 ft less than the length. If Suzanne will need 26 ft of material for the border, what are the dimensions of the quilt?

21) Find the missing angle measures.

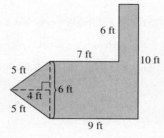
B
$65°$
A $x°$ $(3x - 17)°$ C

Fill in the blank with *always*, *sometimes*, or *never* to make the statement true.

22) The volume of a figure is _____ measured in *square* units.

23) The sides of an equilateral triangle are _____ the same length.

24) The opposite sides of a parallelogram are _____ the same length.

25) The measure of a right angle is _____ greater than the measure of an obtuse angle.

Chapter 4:   Cumulative Review for Chapters 1–4

Perform the indicated operations. Write the answers in lowest terms.

1) $(-517)(-38)$

2) $-2906 + 84 - (-531) - 1277$

3) $-\dfrac{20}{21} \cdot \dfrac{14}{45}$

4) $-3\dfrac{3}{4} \div \left(-5\dfrac{5}{8}\right)$

5) $\dfrac{5}{6} - \dfrac{7}{12} + \dfrac{3}{8}$

6) -7^2

7) $\dfrac{2}{9} \div (-4)$

8) $-9|7 - 12| + \sqrt{144}$

9) $48 - 6^2 \div 12$

10) Use front-end rounding to estimate $8892 \div 13$. Then, find the exact answer.

Simplify each expression.

11) $12x - 5(7x + 4) + 19 + 11x$

12) $-\dfrac{2}{3}n + \dfrac{3}{4} + \dfrac{7}{8}n - 1$

Solve each equation.

13) $-7p = 28$

14) $11 = \dfrac{3}{2}a - 10$

15) $-\dfrac{3}{5}k + \dfrac{1}{2} = \dfrac{1}{5}k + \dfrac{3}{10}$

16) $5y + 3(3y + 4) = 8y + 2$

17) Identify each figure as a line, line segment, or ray, and name it using the correct notation.

a)

b)

18) Find the area of the triangle.

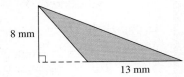

8 mm

13 mm

19) Find the area and perimeter of the rectangle.

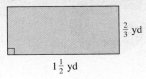

$\frac{2}{3}$ yd

$1\frac{1}{2}$ yd

20) Find the volume and surface area of the box.

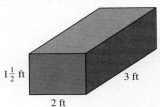

$1\frac{1}{2}$ ft   3 ft

2 ft

Write an equation to solve each problem. Use the Five Steps for Solving Applied Problems.

21) Seven less than four times a number is 65. Find the number.

22) Online spending during the holiday season in 2016 was about $9 billion more than in 2015. Together during both years, holiday shoppers spent about $175 billion online. How much was spent each year? (www.washingtonpost.com)

23) A carpenter has a board that is 51 in. long. He needs to cut it into two pieces so that one piece is half as long as the other. What will be the length of each piece?

24) Find the measure of each angle.

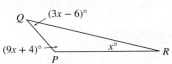

Q $(3x - 6)°$

$(9x + 4)°$ $x°$ R

P

25) A town has a rectangular, fenced-in storage area where it keeps its snowplows, backhoes, and other heavy equipment. Find the length of the storage area if it is 190 ft wide and covers 45,600 ft^2.

5 Signed Decimals

©Dragon Images/Shutterstock

Math at Work:

Bookkeeper

Tamika Jones has always loved math. From elementary school all through college, it was her favorite subject. "It's not that math ever seemed very easy to me," Tamika describes. "Like everyone else, I had to work at it. But there's just something about working with numbers that really appeals to me."

It's no surprise, then, that she entered a career as a bookkeeper, managing the financial records for an insurance agency of over 60 people. Her job allows her to apply her math skills on a daily basis: creating profit-and-loss statements, tracking income and expenditures, managing payroll, and so on. "Bookkeeping requires you to use math in a very precise way," Tamika says. "A single decimal out of place could potentially cost the company thousands of dollars."

Yet, while Tamika was glad to take on the math duties bookkeeping required, one thing she did not expect was how stressful her job became around tax time. "Every spring, there is a huge rush to prepare the documents we need to file our taxes," she says. In order to cope with the pressure and the tight deadlines, Tamika calls on strategies she used in a similar context in college: taking math tests. "What worked for me when I was doing math for an exam also works when I'm doing math for my job on April 14th," Tamika says.

We'll discuss the use of decimals in this chapter and also introduce some strategies to help you succeed on math tests. As Tamika's story suggests, these are skills that will serve you well in any high-intensity situation.

Study Strategies   Taking Math Tests

For many students, math tests cause a great deal of anxiety. However, if you have attended class regularly, taken good notes, and done all of the homework then you should be able to do well on the quizzes and exams. Maybe you need some new test-taking strategies! Here is how you can use the P.O.W.E.R. framework to help you perform your best on math tests.

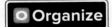

- **I will get a good grade on my math test.** (Your goal should include the grade you want.)

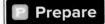

- Complete the emPOWERme survey that appears before the Chapter Summary to learn about your level of test anxiety.
- Days before the test, know exactly which sections, chapters, or material will be included.
- Attend a review session or organize a study group.
- Find a quiet, **distraction-free** place to study.
- Begin studying a few days before the test by reading your notes, making a list of key definitions and formulas, and working out some problems in every section.
- Make a list of questions about problems or topics you do not understand. Ask about them in class, in your instructor's office, or at a tutoring center.
- Do the review that your instructor has given the class or the Chapter Review in the book. Ask questions if you have any.
- When you think you are ready for the test, do the Chapter Test that appears at the end of the chapter in the book *without looking back at your notes or the book.* If you can do the problems, great! If not, then you are not ready for the test. Keep on studying and/or see your instructor for more help.
- Get a good night's sleep before the test so that you are well rested and can think clearly.
- The day of the test, warm up for it by *doing* (not just reading) a few problems before walking into the classroom. This way, you will already be in the groove of doing problems.
- Arrive early for the test so that you do not feel rushed and stressed. Be sure you have everything you need such as pencils, an eraser, and a calculator (if that is allowed).

- It's time for the test. Stay calm, focused, and confident. If you have prepared well, believe that you will do well!
- When you get the test, read all of the directions *carefully.*
- Look over the whole test to get an idea of how many questions it has.
- Do the easiest problems first. Circle the ones you do not know how to do and come back to them later.
- Write neatly and in an organized way, and show all of your work. You may receive partial credit even if you do not finish a problem.
- Ask the instructor a question if something is unclear.
- Double-check your work if you have extra time. Sometimes, it is better to redo a problem and see if you get the same answer than to just read over the work you have already done.
- Keep track of time so that you can pace yourself.

- Your instructor returns the tests. Did you get the grade you wanted?

- If you got the grade you wanted, then congratulations! Ask yourself and think about what you did that helped you do well on the test. Use the same approach next time but think of other strategies you could use, too.
- If you did not achieve your goal, ask yourself, *"Why not?"* (Be honest!) What could you do differently next time to improve the outcome?
- No matter how you did on the test, be sure you understand the mistakes you made in the problems you did wrong. Know (and write down) how to do the problems correctly. Math concepts build on each other, so what you did not understand on this test may make it difficult for you to learn the material in the next chapter. Plus, there's always the final exam to consider.

Chapter 5 **POWER** Plan

What are your goals for Chapter 5?	How can you accomplish each goal?
1 Be prepared before and during class.	• Don't stay out late the night before class, and be sure to set your alarm! • Bring a pencil, notebook paper, and textbook to class. • Avoid distractions by turning off your cell phone during class. • Pay attention, take good notes, and ask questions. • Complete your homework on time, and ask questions on problems you do not understand.
2 Understand the homework to the point where you could do it without needing any help or hints.	• Read the directions, and show all of your steps. • Go to the instructor's office for help. • Rework homework and quiz problems, and find similar problems for practice.
3 Use the P.O.W.E.R. framework to learn ways to improve the way you take math tests: *Is Anxiety the Hardest Problem on the Test?*	• Complete the emPOWERme that appears before the Chapter Summary to learn how much text anxiety affects your test taking. • Read the Study Strategies to learn how to use P.O.W.E.R. when studying for and taking tests. • Think about which strategies could help you the most, and use them when studying for and taking your next test.
4 Write your own goal. _____ _____	• _____ _____

What are your objectives for Chapter 5?	How can you accomplish each objective?
1 Learn to read, write, and round decimals.	• Use place value, number lines, and writing a decimal as a fraction or mixed number to help understand what a decimal represents. • The same rounding principles you learned previously still apply to decimals.
2 Learn how to perform basic operations on decimals.	• Write the procedures for **Adding, Subtracting, Multiplying, and Dividing with Signed Decimals** in your own words. • Know how to multiply or divide by a power of 10. • Be able to solve applied problems using decimals.
3 Learn to write a fraction as a decimal.	• Understand the two ways to write a fraction as a decimal. One way is to use division, and the other is to write an equivalent fraction with a denominator that is a power of 10. • Know how to compare a decimal and a fraction.
4 Understand how to use measures of central tendency.	• Be able to find a mean, weighted mean, median, and mode. • Know what the different measures represent.
5 Learn how to solve equations containing decimals.	• Understand that we use the same steps to solve equations containing decimals that we used to solve equations containing integers and fractions. • Know how to eliminate decimals to solve an equation. • Be able to solve applied problems by writing an equation involving decimals.

6 Use the Pythagorean theorem.	• Learn how to approximate square roots. • Learn the Pythagorean theorem, and use it to find a missing side of a triangle. • Know how to solve applied problems.
7 Find the area and circumference of a circle.	• Understand the relationship between the radius and the diameter. • Apply the area and circumference formulas.
8 Find the volume and surface area of three-dimensional objects.	• Learn the volume and surface area formulas and apply them. • Understand how to find an *exact* value and an *approximate* value.
9 Write your own goal.	• _____

	W Work	Read Sections 5.1–5.10, and complete the exercises.	
E Evaluate	Complete the Chapter Review and Chapter Test. How did you do?	**R Rethink**	• Many, if not all, concepts in math build upon each other. How was that true for this chapter? • Do you feel like you worked and studied effectively in this chapter? Why or why not? • Did you use P.O.W.E.R. to help you prepare for and take your quizzes and test? If so, what worked well and what did not? If you did not use P.O.W.E.R., why not?

5.1 Reading and Writing Decimals

What are your objectives for Section 5.1?	How can you accomplish each objective?
1 Understand What Decimals Represent	• Know the definition of a *decimal* and how it represents a fraction. • Be able to make the same visual representations of decimals as you did with fractions. • Complete the given examples on your own. • Complete You Trys 1 and 2.
2 Use Place Value	• Understand the **Place Value Chart** and how it applies to numbers containing decimals. • Be able to write fractions and mixed numbers as decimals. • Complete the given examples on your own. • Complete You Trys 3–5.
3 Read and Write Decimals in Words	• Use the procedure for **Reading a Decimal Number,** and write it in your own words. • Complete the given examples on your own. • Complete You Trys 6 and 7.
4 Write Decimals as Fractions or Mixed Numbers	• Write your own procedure for writing a decimal as a fraction or mixed number. • Complete the given examples on your own. • Complete You Trys 8 and 9.

W Work   **Read the explanations, follow the examples, take notes, and complete the You Trys.**

In Chapter 3, we learned about fractions. We learned how to represent the shaded part of a figure with a fraction, and we learned how to represent the value of a fraction on a number line. Let's do the same with *decimals*.

1 Understand What Decimals Represent

Definition

A **decimal** is a number, containing a *decimal point*, that is another way to represent a fraction with a denominator that is a power of 10.

$$\textit{Example:} \quad 0.1 = \frac{1}{10} \quad \text{Both of these are read as } \textit{one tenth.}$$

↑
Decimal point

Look at the figure.

One tenth of the figure is shaded. We can write this with a fraction or a decimal.

We can say that $\frac{1}{10}$ of the figure is shaded, or 0.1 of the figure is shaded.

To represent $\frac{1}{10}$ (one tenth) on a number line, divide the number line from 0 to 1 into 10 equal parts so that the space between each tick mark is $\frac{1}{10}$ (one tenth) of a unit. Place a dot on $\frac{1}{10}$.

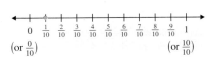

W Hint

Remember, a decimal represents a fraction with a denominator that is a *power of 10.*

To represent 0.1 (one tenth) on a number line, divide the number line from 0 to 1 into 10 equal parts so that the space between each tick mark is 0.1 (one tenth) of a unit. Label the number line using decimal notation.

Notice from their placement on the number lines that $\frac{1}{10}$ and 0.1 represent the same quantity!

EXAMPLE 1

Use a decimal to represent the shaded part of the rectangle, and represent the decimal on a number line.

Solution

The rectangle is divided into 10 equal parts. 7 parts are shaded. Therefore,

0.7 (seven tenths) of the rectangle is shaded.

(Notice that, as a fraction, we say $\frac{7}{10}$ is shaded. Both 0.7 and $\frac{7}{10}$ are read as *seven tenths.*) To represent 0.7 on a number line, divide the number line from 0 to 1 into 10 equal parts. The space between each tick mark is 0.1 (one tenth) of a unit. Label the number line using decimal notation.

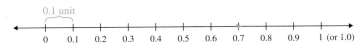

Place the dot on 0.7.

[YOU TRY 1] Use a decimal to represent the shaded part of the rectangle, and represent the decimal on a number line.

Decimals represent fractions with denominators that are powers of 10. So, let's look at a figure that is divided into 100 equal parts.

EXAMPLE 2 Use a decimal to represent the shaded part of the figure.

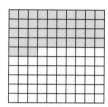

Solution

The square is divided into 100 equal parts. 43 parts are shaded. As a *fraction,* we say that $\frac{43}{100}$ (forty-three hundredths) is shaded. In *decimal* form, we say that

0.43 (forty-three hundredths) is shaded.

W Hint

Notice that there is no need to simplify a decimal.

Each small square is $\frac{1}{100}$ (one hundredth) or 0.01 (one hundredth) of the entire figure.

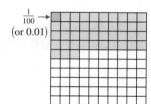

[YOU TRY 2] Use a decimal to represent the shaded part of the figure.

2 Use Place Value

Now, let's learn about place value. For example, the number 572,364 has no fractional parts, so we do not include a decimal point in the number. (If we *did* include a decimal point, it would go at the end of the number: 572,364**.**) But, what if we have a number like 643,081.29753? What are the place-value names for the digits to the *right* of the decimal point?

b) There are two methods we can use to write $\dfrac{139}{100}$ as a decimal.

Method 1: $\dfrac{139}{100}$ is read as "one hundred thirty-nine hundredths." The number farthest to the right in the numerator, 9, must be written in the hundredths place.

$$\dfrac{139}{100} = 1.39$$

hundredths place — 9
tenths
ones

Method 2: Write $\dfrac{139}{100}$ as the mixed number $1\dfrac{39}{100}$. Read the mixed number as "one *and* thirty-nine hundredths."

The *and* tells us where to put the decimal point.

$$\dfrac{139}{100} = 1\dfrac{39}{100} = 1.39$$

one and thirty-nine hundredths

[YOU TRY 4] Write each fraction or mixed number as a decimal.

a) $4\dfrac{87}{100}$ b) $\dfrac{61}{10}$

Note

Because an integer has no fractional part, it is usually written without a decimal point. We should know, however, that an integer *can* be written with a decimal point at the end of the number, after the ones place.

$$5 = 5.\qquad\qquad -294 = -294.$$

The decimal point goes at the end of the number.   The decimal point goes at the end of the number.

EXAMPLE 5   Identify the place value of each digit.

a) 0.0086   b) 372.04

Solution

a) 0 . 0 0 8 6

ones
tenths
hundredths
thousandths
ten-thousandths

b) 3 7 2 . 0 4

hundreds
tens
ones
tenths
hundredths

[YOU TRY 5] Identify the place value of each digit.

a) 0.09058   b) 71.403

3 Read and Write Decimals in Words

We have seen that reading a fraction helps us to write it as a decimal. To read and write a decimal, we must identify the place value of the digit farthest to the right. Let's start with decimals that have no whole-number part.

EXAMPLE 6

Write each decimal in words.

a) 0.5   b) 0.008   c) 0.263   d) 0.0701

Solution

a) There is no whole-number part, and the number ends in the *tenths* place. Read the number to the right of the decimal point like you would read the whole number, then follow it with *tenths*.

0.5 is read as "five *tenths*."   Think of 0.5 as $\frac{5}{10}$, read as "five *tenths*."

b) There is no whole-number part, and the number ends in the *thousandths* place. Read the number to the right of the decimal point like you would read the whole number, then follow it with *thousandths*.

0.008 is read as "eight *thousandths*."   Think of 0.008 as $\frac{8}{1000}$, read as "eight *thousandths*."

c) The number ends in the *thousandths* place. Read the number to the right of the decimal point like you would read the whole number, then follow it with *thousandths*.

0.263 is read as "two hundred sixty-three *thousandths*." $0.263 = \frac{263}{1000}$

d) The number ends in the *ten-thousandths* place. Read the number to the right of the decimal point like you would read the whole number, then follow it with *ten-thousandths*.

0.0701 is read as "seven hundred one *ten-thousandths*." $0.0701 = \frac{701}{10,000}$

[YOU TRY 6] Write each decimal in words.

a) 0.7   b) 0.002   c) 0.926   d) 0.0065

When the decimal contains a whole-number part, read that first. We read the decimal point as "and"; then we read the part to the right of the decimal point.

Hint

If the number does not have a whole-number part, we read the part to the right of the decimal point as in Example 6.

Procedure Reading a Decimal Number

1) Read the whole-number part first.
2) Read the decimal point as "and."
3) Read the fractional part, the digits to the *right* of the decimal point, last.

EXAMPLE 7

Write each decimal in words.

a) 34.9   b) 207.04   c) 1.625

Solution

a) Read the whole-number part first, read the decimal point as "and," then read the part to the right of the decimal point.

34.9

thirty-four and nine tenths

Read 34.9 as "thirty-four and nine tenths."

b) Read the whole-number part first, read the decimal point as "and," then read the part to the right of the decimal point.

207.04

two hundred seven and four hundredths

Read 207.04 as "two hundred seven and four hundredths."

c) Read the whole-number part first, read the decimal point as "and," then read the part to the right of the decimal point.

1.625

one and six hundred twenty-five thousandths

Read 1.625 as "one and six hundred twenty-five thousandths."

YOU TRY 7

Write each decimal in words.

a) 524.6   b) 901.006   c) 2.3808

---

4 Write Decimals as Fractions or Mixed Numbers

Let's learn more about writing decimals as fractions and change decimals to mixed numbers as well.

EXAMPLE 8

Write each decimal as a fraction or mixed number.

a) 0.93   b) 0.0149   c) 2.007

Solution

a) Read 0.93 as "ninety-three hundredths."
$$\underbrace{\text{ninety-three}}_{\text{numerator} = 93} \underbrace{\text{hundredths}}_{\text{denominator} = 100}$$

$$0.93 = \frac{93}{100}$$

Also notice that in 0.93, the digit farthest to the right is in the *hundredths* place. So, the *denominator* is 100, and the numerator is 93.

$$0.93 = \frac{93}{100}$$

3 is in the *hundredths* place, so the denominator = 100.

b) Read 0.0149 as "one hundred forty-nine ten-thousandths."
$$\underbrace{\text{one hundred forty-nine}}_{\text{numerator} = 149} \underbrace{\text{ten-thousandths}}_{\text{denominator} = 10{,}000}$$

$$0.0149 = \frac{149}{10{,}000}$$

9 is in the *ten-thousandths* place, so the denominator = 10,000.

Also notice that in 0.0149, the digit farthest to the right, 9, is in the *ten-thousandths* place. So, the *denominator* is 10,000, and the numerator is 149.

c) Because 2.007 has a whole-number part, it will be a mixed number. Read 2.007 as "two and seven *thousandths*."

whole-number part = 2   numerator = 7   denominator = 1000

$$2.007 = 2\frac{7}{1000}$$

7 is in the *thousandths* place, so the denominator = 1000.

[YOU TRY 8]

Write each decimal as a fraction or mixed number.

a) 0.41   b) 0.0903   c) 5.029

In Chapter 3, we said that all fractional answers must be in lowest terms. The same is true when we change a decimal to a fraction or mixed number.

Note

When we change a decimal to a fraction or mixed number, we *must* write the answer in lowest terms.

EXAMPLE 9

Write each decimal as a fraction or mixed number in lowest terms.

a) 0.6   b) 3.124   c) −8.75

Solution

a) $0.6 = \dfrac{6}{10}$

      6 is in the *tenths* place, so the denominator = 10.

     Ask yourself, *"Is $\dfrac{6}{10}$ in lowest terms?"* No! Write it in lowest terms.

$$0.6 = \frac{6}{10} = \frac{6 \div 2}{10 \div 2} = \frac{3}{5}$$

b) $3.124 = 3\dfrac{124}{1000}$ Write $\dfrac{124}{1000}$ in lowest terms.

      4 is in the *thousandths* place, so the denominator = 1000.

$$3.124 = 3\frac{124}{1000} = 3\frac{124 \div 4}{1000 \div 4} = 3\frac{31}{250}$$

c) $-8.75 = -8\dfrac{75}{100}$ Write $\dfrac{75}{100}$ in lowest terms.

      5 is in the *hundredths* place, so the denominator = 100.

$$-8.75 = -8\frac{75}{100} = -8\frac{75 \div 25}{100 \div 25} = -8\frac{3}{4}$$

[YOU TRY 9]

Write each decimal as a fraction or mixed number in lowest terms.

a) 0.4   b) 9.528   c) −17.25

ANSWERS TO [YOU TRY] EXERCISES

1) 0.3;   ← + + + + +⊹+ + + + + + + →   2) 0.61
         0 0.1 0.2 0.3 0.4 0.5 0.6 0.7 0.8 0.9 1

3) a) 0.07   b) 0.613   c) −0.0909   d) −0.0002

4) a) 4.87   b) 6.1

5) a) 0.—ones   b) 7—tens
     .0—tenths   1—ones
     9—hundredths   4—tenths
     0—thousandths   0—hundredths
     5—ten-thousandths   3—thousandths
     8—hundred-thousandths

6) a) seven tenths   b) two thousandths   c) nine hundred twenty-six thousandths
   d) sixty-five ten-thousandths   7) a) five hundred twenty-four and six tenths
   b) nine hundred one and six thousandths
   c) two and three thousand eight hundred eight ten-thousandths

8) a) $\dfrac{41}{100}$ b) $\dfrac{903}{10,000}$ c) $5\dfrac{29}{1000}$ 9) a) $\dfrac{2}{5}$ b) $9\dfrac{66}{125}$ c) $-17\dfrac{1}{4}$

Objective 1: Understand What Decimals Represent

1) How are decimals and fractions related?

 2) Explain how to make a number line, from 0 to 1, that is divided into tenths.

Use a fraction with a denominator of 10 to represent the shaded part of the rectangle and represent the fraction as a decimal on a number line.

3)

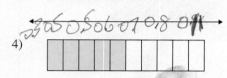

4)

 5)

6)

Use a fraction and a decimal to represent the shaded part of the figure. Write the fraction in lowest terms, if possible.

7) 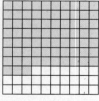   8)

9)    10)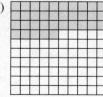

Approximate the location of the decimal value on the given number line.

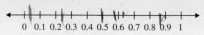

11) 0.58   12) 0.63   13) 0.06

14) 0.89   15) 0.25   16) 0.50

Objective 2: Use Place Value

Write each fraction as a decimal.

17) $\dfrac{3}{100}$ 18) $\dfrac{7}{100}$

19) $\dfrac{81}{100}$ 20) $\dfrac{29}{100}$

21) $-\dfrac{141}{1000}$ 22) $-\dfrac{719}{1000}$

23) $\dfrac{67}{1000}$ 24) $\dfrac{43}{1000}$

25) $\dfrac{893}{10,000}$ 26) $\dfrac{557}{10,000}$

27) $\dfrac{2051}{10,000}$ 28) $\dfrac{8049}{10,000}$

Write each fraction or mixed number as a decimal.

29) $6\dfrac{7}{10}$ 30) $4\dfrac{1}{10}$

31) $-38\dfrac{1}{1000}$ 32) $-22\dfrac{7}{1000}$

33) $10\dfrac{533}{1000}$ 34) $16\dfrac{829}{1000}$

35) $\dfrac{81}{10}$ 36) $\dfrac{57}{10}$

37) $\dfrac{409}{100}$ 38) $\dfrac{703}{100}$

39) $-\dfrac{2443}{1000}$ 40) $-\dfrac{5119}{1000}$

Identify the place value of each digit.

41) 0.3572   42) 0.1489   43) 237.804

44) 895.601   45) 40.16259   46) 90.51437

Objective 3: Read and Write Decimals in Words

Write each decimal in words.

47) 0.4

48) 0.8

 49) 0.36

50) 0.52

51) 0.007

52) 0.005

 53) 0.7415

54) 0.6213

55) 57.3

56) 24.1

57) −809.56

58) −302.97

59) 3.0576

60) 5.0804

61) −6.00017

62) −9.0056

Write each word statement as a decimal.

63) fifteen hundredths

64) twelve hundredths

65) ninety-six and seven tenths

66) forty-nine and three tenths

67) thirty-two thousandths

68) sixty-seven thousandths

69) negative eight and four ten-thousandths

70) negative one and nine ten-thousandths

71) five thousand five and five hundred-thousandths

72) two thousand two and two hundred-thousandths

Objective 4: Write Decimals as Fractions or Mixed Numbers

73) Explain how to write a decimal as a fraction.

74) Can a whole number be written with a decimal point? Explain your answer.

Write each decimal as a fraction or mixed number in lowest terms.

75) 0.73

76) 0.81

77) 0.0207

78) 0.0503

79) 4.9

80) 6.7

R Rethink

R1) When do you consistently use decimals? Think about when you buy something!

R2) What did you previously not know about decimals that you learned by completing this section? How will it help you every day?

81) −0.6

82) −0.4

83) −0.60

84) −0.40

85) 0.68

86) 0.96

87) 9.8

88) 7.6

89) −5.144

90) −1.238

91) 1.00015

92) 2.00035

93) Explain why 0.20 is equivalent to 0.2.

94) Explain why 1.3200 is equivalent to 1.32.

Mixed Exercises: Objectives 1–4

Write each word statement as a decimal.

95) eight and fourteen ten-thousandths

96) three hundred thousand thirty-nine and sixty-two hundredths

Write each decimal in words.

97) −204.8

98) 781.005

Use a simplified fraction and a decimal to represent the shaded part of the figure.

99)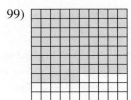

Identify the place value of each digit.

100) 574.01368

Write each fraction as a decimal.

101) $\dfrac{37}{1000}$

102) $-\dfrac{189}{10}$

Write each decimal as a fraction or mixed number in lowest terms.

103) −2.84

104) 0.0725

5.2 Rounding Decimals

P Prepare

O Organize

What are your objectives for Section 5.2?	How can you accomplish each objective?
1 Round Decimals	• Write the procedure for **Rounding Decimals** in your own words. • Understand how to regroup with decimals. • Complete the given examples on your own. • Complete You Trys 1–4.
2 Round Money Amounts to the Nearest Cent	• Use the same procedure for rounding decimals for this section. • Complete the given examples on your own. • Complete You Trys 5 and 6.
3 Round Money Amounts to the Nearest Dollar	• Use the same procedure for rounding decimals for this section, and know how to regroup $1 if needed. • Complete the given examples on your own. • Complete You Trys 7 and 8.

W Work

Read the explanations, follow the examples, take notes, and complete the You Trys.

In Section 1.1, we learned that to *round* a number means to find another number close to the original number. For example, 748 rounded to the nearest ten is 750, and 5391 rounded to the nearest thousand is 5000.

1 Round Decimals

How do we round decimals? Let's use a number line to round 0.67 to the nearest tenth. Put 0.67 on a number line in which each tick mark represents 0.1 unit.

Is 0.67 closer to 0.6 or 0.7 on the number line? It is closer to 0.7. Therefore, we can say that 0.67 rounded to the nearest tenth is 0.7.

Rounding decimals is especially useful when working with money. We will see examples using money later in this section.

Here are the rules for rounding decimals.

Procedure Rounding Decimals

Step 1: Find the place to which we are asked to round. Underline the digit in that place, and draw a vertical line after it.

Step 2: Look at the digit to the right of the vertical line.

a) If the digit to the right of the vertical line is **less than 5,** "drop off" the digits to the right of the vertical line and leave the underlined digit as it is.

b) If the digit to the right of the vertical line is **5 or more,** "drop off" the digits to the right of the vertical line and increase the underlined digit by 1.

W Hint

Once again, 5 is the "cutoff" for rounding!

EXAMPLE 1

Round 90.4637 to the nearest hundredth.

Solution

Step 1: Underline the digit in the hundredths place, and draw a vertical line after it.

$$90.4\underline{6}|37$$

Step 2: Look at the digit to the right of the vertical line. Because 3 is less than 5, we will "drop off" the digits to the right of the vertical line and keep the underlined digit the same.

Keep this digit the same.
3 is less than 5.

$$90.4\underline{6}|37$$

Drop these.

Round to 90.46.

Therefore, 90.4637 rounded to the nearest hundredth is 90.46. We can also say that 90.4637 ≈ 90.46. (Remember, ≈ means *is approximately equal to*.)

[YOU TRY 1]

Round 19.5439 to the nearest hundredth.

Note

90.4637 rounded to the nearest hundredth is *not* 90.4600. (We do *not* replace the dropped digits with zeros.) **The place to which we are rounding should be the last place in our rounded number.** That is why, rounded to the nearest hundredth, 90.4637 is 90.46.

EXAMPLE 2

Round 7.2814 to the nearest tenth.

Solution

Step 1: Underline the digit in the tenths place and draw a vertical line after it.

$$7.\underline{2}|814$$

Step 2: Look at the digit to the right of the vertical line. Because 8 is more than 5, "drop off" the digits to the right of the vertical line and increase the underlined digit by 1.

Increase by 1.
8 is more than 5.

7.2|814

Drop these.

Round to 7.3.

Rounded to the nearest tenth, 7.2814 is 7.3. (Because we are rounding to the *tenths* place, the last digit in the answer will be in the *tenths* place.)

[YOU TRY 2]

Round 8.1609 to the nearest tenth.

Note

In Example 2, we increased the digit in the *tenths* place by 1. This is the same as adding 1 *tenth* (or 0.1) to the number after dropping off the digits that follow the vertical line.

$$7.2|814 \longrightarrow \begin{array}{r} 7.2 \\ +\ 0.1 \\ \hline 7.3 \end{array}$$ Increase the tenths digit by 1.

Some rounding problems involve regrouping. We can regroup 10 *tenths* as 1 *whole*.

EXAMPLE 3

Round each number to the indicated place.

a)   12.973 to the nearest tenth   b)   0.8395 to the nearest thousandth

Solution

a)   **Step 1:** 12.9|73   Underline the digit in the tenths place, and draw a vertical line after it.

Increase by 1.
7 is more than 5.

Step 2: 12.9|73

Drop these.

$$\begin{array}{r} 12.9 \\ +\ 0.1 \\ \hline 13.0 \end{array}$$

Increasing the 9 by 1 is the same as adding 1 tenth to 9 tenths.
9 tenths + 1 tenth = 10 tenths
Regroup 10 tenths as 1 *whole* (or 1 *one*), and add.

12.973 rounded to the nearest tenth is 13.0.

W Hint

Be sure you understand why you must keep the zero in the tenths place!

The correct answer is 13.0 and *not* 13 because we are rounding to the nearest *tenth*. Therefore, we must keep the 0 in the *tenths* place.

b) **Step 1:** 0.839|5   Underline the digit in the thousandths place, and draw a
vertical line after it.

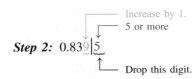

Step 2: 0.839|5

        Drop this digit.

           0.839   Increasing the 9 by 1 is the same as adding 1 thousandth to 9 thousandths.
      +0.001   9 thousandths + 1 thousandth = 10 thousandths
       0.840   Regroup this as 1 hundredth, and add.

0.8395 rounded to the nearest thousandth is 0.840. Remember to keep the 0 at the end because it is in the thousandths place.

[YOU TRY 3]

Round each number to the indicated place.

a)   45.991 to the nearest tenth   b)   0.7295 to the nearest thousandth

---

In some cases, we are asked to round to the *first* decimal place (the *tenths* place), the *second* decimal place (the *hundredths* place), to *three* decimal places (the *thousandths* place), etc. These are different ways to indicate rounding.

EXAMPLE 4

Round 6.71958 to

a)   the first decimal place.   b)   four decimal places.

Solution

a)   Round 6.71958 to the *first* decimal place means to round it to the *tenths* place.

    Step 1: 6.7|1958   Underline the digit in the first decimal place (tenths place),
                      and draw a vertical line after it.

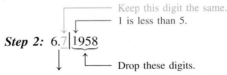

    Step 2: 6.7|1958

                      Drop these digits.

  Round to 6.7.

b)   Round 6.71958 to *four* decimal places means to round it to the *fourth* place after the decimal point. This is the *ten-thousandths* place.

    Step 1: 6.7195|8   Underline the digit in the fourth decimal place,
                      and draw a vertical line after it.

                  Increase by 1.
                  8 is more than 5.

    Step 2: 6.7195|8

                      Drop this digit.

  Round to 6.7196.

<table>
<tr>
<td>

[YOU TRY 4]

</td>
<td>

Round 9.04253 to

a)   the first decimal place.   b)   three decimal places.

</td>
</tr>
</table>

2 Round Money Amounts to the Nearest Cent

We use decimals every day, especially in terms of money. Let's look at the decimal places in a dollar amount.

$$\$36.28$$

— hundredths place
— tenths place

The value of a penny is 1¢ (1 cent). We can also write it in terms of a dollar as $0.01 or 0.01 dollar. $\left(\text{We can also say that it is } \dfrac{1}{100} \text{ dollar.}\right)$ Because it is the smallest denomination of money in the United States, most everyday money amounts are rounded *to the nearest cent*. This is the same as rounding *to the nearest hundredth of a dollar*.

Some *exact* calculations, like computing sales tax or computing the amount of a discount on an item, will actually give a dollar amount with digits to the right of the hundredths place (or number of cents). These are some examples of when money amounts would be rounded to the nearest cent.

<table>
<tr>
<td>

EXAMPLE 5

</td>
<td>

When Helena computes the amount of tax she will owe on the makeup she bought, she gets the exact amount of $1.5362. Round this amount to the nearest cent to determine the amount of tax she will actually pay.

</td>
<td>

©Getty Images

</td>
</tr>
</table>

Solution

Round $1.5362 *to the nearest cent* means to round it to the *hundredths place*. Underline the 3 in the hundredths place, and round.

— Increase by 1.
— 6 is more than 5.

$$\$1.53|62$$

— Drop these digits.

Round to $1.54.

$1.5362 rounded to the nearest cent is $1.54. This answer makes sense because $1.5362 is closer to $1.54 than $1.53.

<table>
<tr>
<td>

[YOU TRY 5]

</td>
<td>

Jignesh computes the amount of tax he will owe on the jeans he bought, and that exact amount is $2.7456. Round this to the nearest cent to determine the amount of tax he will actually pay.

</td>
</tr>
</table>

<table>
<tr>
<td>

EXAMPLE 6

</td>
<td>

Makoto computed the exact amount of the discount on the batteries he bought as $0.783. Round this to the nearest cent to determine how much money he will actually save on the batteries.

</td>
</tr>
</table>

376   CHAPTER 5   **Signed Decimals**   www.mhhe.com/messersmith

Solution

Underline the 8 in the hundredths (or cents) place, and round.

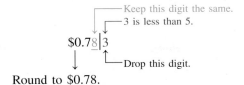

Round to $0.78.

Makoto will save $0.78. This can also be written as 78¢.

[YOU TRY 6] Oksana computed the exact amount of the discount on a mechanical pencil as $0.352. Round this to the nearest cent to determine how much money she will actually save.

Note

Some stores round all discount amounts *up* to the nearest cent. For example, some stores would round the $0.783 to $0.79 even though the rounding rules tell us to keep the 8 the same.

---

3 Round Money Amounts to the Nearest Dollar

Sometimes we round money amounts to the nearest dollar. We might do this if we want to get an estimate of how much something costs. Rounding to the nearest *dollar* means rounding to the *ones* place.

EXAMPLE 7 Round $27.39 to the nearest dollar.

Solution

We can solve this problem in two ways.

Method 1: Rounding $27.39 to the nearest dollar means rounding it to the ones place.

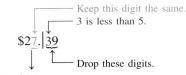

Round to $27.

Write the answer as $27, *not* $27.00. Writing $27.00 would indicate rounding to the nearest cent.

Method 2: $27.39 is between $27 and $28. Ask yourself, *"Is $27.39 closer to $27 or $28?"* It is closer to $27.

[YOU TRY 7] Round $85.42 to the nearest dollar.

BE CAREFUL When rounding a number like $27.39 to the nearest dollar, the answer is $27, *not* $27.00. The last place in the rounded number should be the ones place.

EXAMPLE 8   Round each number to the nearest dollar.

a) $406.81   b) $0.59   c) $199.79   d) $320.50

Solution

a) Underline the 6 in $406.81, and round.

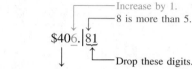

Round to $407.

Or ask yourself, *"Is $406.81 closer to $406 or $407?"* It is closer to $407.

b) Underline the 0 in $0.59, and round.

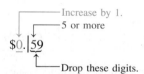

Round to $1.

Or ask yourself, *"Is $0.59 closer to $0 or $1?"* It is closer to $1.

c) Underline the digit in the *ones* place, and round.

$199.|79

$199
+ $1   Increasing $199 by $1 gives us $200.
$200

$199.79 is closer to $200 than to $199.

d) Underline the digit in the ones place, and round.

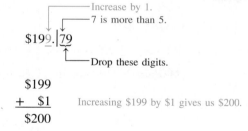

Round to $321.

Although $320.50 is *exactly* halfway between $320 and $321, the rounding rules tell us to round $320.50 to $321.

YOU TRY 8 Round each number to the nearest dollar.

    a) $601.73   b) $0.64   c) $599.89   d) $1270.50

Note

In some areas of science, there are other rules for rounding numbers that are exactly halfway between two numbers. Sometimes, these numbers are not rounded up.

ANSWERS TO [YOU TRY] **EXERCISES**

1) 19.54   2) 8.2   3) a) 46.0 b) 0.730   4) a) 9.0 b) 9.043
5) $2.75   6) $0.35 or 35¢   7) $85   8) a) $602 b) $1 c) $600 d) $1271

E Evaluate **5.2** Exercises   Do the exercises, and check your work.

Objective 1: Round Decimals

Round the given decimal to the nearest tenth. Then represent the rounded decimal on the number line.

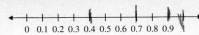

0 0.1 0.2 0.3 0.4 0.5 0.6 0.7 0.8 0.9 1

1) 0.39   2) 0.93

3) 0.02   4) 0.97

5) 0.653   6) 0.158

Round each number to the indicated place.

7) 0.683   8) 0.592

    a) tenth   a) tenth

    b) hundredth   b) hundredth

9) 94.349   10) 34.179

    a) hundredth   a) tenth

    b) tens   b) ones

11) 5628.49241   12) 4972.67516

    a) thousandth   a) thousandth

    b) thousand   b) hundred

13) 7620.183849   14) 1195.073264

    a) hundred   a) thousand

    b) ten-thousandth b) ten-thousandth

15) 0.0982   16) 0.0954

    a) ones   a) ones

    b) hundredth   b) hundredth

17) 5.271   18) 8.629

    a) the first decimal   a) the first decimal
       place   place

    b) two decimal places b) two decimal places

19) 43.917995   20) 56.618995

    a) five decimal   a) three decimal
       places   places

    b) four decimal   b) five decimal
       places   places

21) Explain, in your own words, how to round a decimal number.

22) If you are asked to round 28.4578 to the nearest hundredth, is the answer 28.4600? Explain your answer.

For Exercises 23 and 24, choose from thousandths, ten-thousandths, and hundred-thousandths.

23) Rounding to which place value would give you the most accurate estimate of the number 0.0050589? Write down the rounded number.

24) Rounding to which place value would give you the most accurate estimate of the number 1.00273? Write down the rounded number.

For Exercises 25–32, fill in the blank with >, <, or =.

25) 0.63 _____ 0.60

26) 0.40 _____ 0.43

27) 1.0299 _____ 1.0300

28) 8.4940 _____ 8.4900

29) 3.05 _____ 3.0500

30) 1.0800 _____ 1.08

31) 0.1 _____ 0.01

32) 0.0090 _____ 0.090

Objective 2: Round Money Amounts to the Nearest Cent

Solve each problem.

33) Lyndsey computed the amount of tax she will owe for her new spring outfit as $5.2275. Round to the nearest cent to determine the amount of tax she will actually pay.

34) When Sheena computes the amount of tax she will owe for her new bedroom set, she gets the exact amount of $85.405. Round to the nearest cent to determine the amount of tax she will actually pay.

35) When Cameron bought his used car, he computed the amount of tax to be exactly $722.1175. Round to the nearest cent to determine the amount of tax he will actually pay.

©Ingram Publishing

36) Juanita computed the amount of tax she will owe on a purchase of video editing software to be exactly $16.1875. Round to the nearest cent to determine the amount of tax she will actually pay.

37) Erin purchased a new pair of earrings for $275.50, and she computed the amount of tax she will owe to be exactly $22.72875. How much must Erin pay at the cash register, including the sales tax?

38) Michael buys a rear spoiler for his sports car. The purchase price is $142.50, and he computes the amount of tax he will owe to be exactly $9.61875. What is the total cost of the rear spoiler, including the sales tax?

In Exercises 39–42, round each number in bold to the nearest cent.

39) Three boxes of macaroni and cheese sell for $3.29, so **the cost of one box is $1.09667.**

40) Paper towels are on sale for two for $4.19, so **the cost of one roll is $2.095.**

41) Four cans of tomato paste cost $2.69, so **the cost of one can is $0.6725.**

42) Three containers of yogurt cost $2.29, so **the cost of one container is $0.76333.**

Objective 3: Round Money Amounts to the Nearest Dollar

Round each amount to the nearest dollar.

43) $18.48

44) $11.36

45) $39.05

46) $76.24

47) $42.61

48) $58.73

49) $681.57

50) $912.58

51) $1599.91

52) $2099.97

53) $45.50

54) $12.50

55) $0.48

56) $0.76

57) $0.92

58) $0.34

59) Rahim puts the following items into his cart at a home improvement store, and he wants to estimate how much he is spending. Round the price of each item to the nearest dollar, then add the rounded numbers to estimate the total cost of his items.

©Andrew Resek/McGraw-Hill Education

Faucet:	$87.26
Towel bar:	$24.99
Light fixture:	$129.95
Scale:	$19.35
Plunger:	$6.58

60) Sandy puts the following items into her cart at a toy store, and she wants to estimate how much she is spending. Round the price of each item to the nearest dollar, then add the rounded numbers to estimate the total cost of her items.

Doll stroller:	$9.49
Doll:	$15.99
Electronic math game:	$27.35
Jump rope:	$4.68
Booster seat:	$109.99

Mixed Exercises: Objectives 1–3

Round each number to the indicated place.

61) 1.85 to the nearest tenth

62) 46.0944 to the nearest thousandth

63) 0.00723 to the nearest ten-thousandth

64) 925.8996 to the nearest hundredth

65) 6999.586 to the nearest ones place

66) 74.53 to the nearest tens place

Round each amount a) to the nearest cent and b) to the nearest dollar.

67) $375.854

68) $71.9983

69) $0.359

70) $0.573

R Rethink

R1) Since you have learned about rounding in a previous section, explain how it was helpful for understanding how to round decimals.

R2) Give two examples of where you encounter rounding money to the nearest dollar outside of class.

5.3 Adding and Subtracting Signed Decimals

P Prepare

O Organize

What are your objectives for Section 5.3?	How can you accomplish each objective?
1 Add Positive Decimals	• Write the procedure for **Adding or Subtracting Decimals** in your own words. • Complete the given examples on your own. • Complete You Trys 1 and 2.
2 Subtract Positive Decimals	• Use the same procedure from Objective 1. • Complete the given examples on your own. • Complete You Trys 3–6.
3 Add and Subtract with Negative Decimals	• Write the procedure for **Adding Signed Numbers** in your own words. • Write a procedure for **Subtracting Signed Numbers.** Look back at Section 1.4, if necessary. • Complete the given examples on your own. • Complete You Trys 7–9.
4 Solve an Applied Problem	• Complete the given example on your own. • Complete You Try 10.

Read the explanations, follow the examples, take notes, and complete the You Trys.

In this section, we will learn how to add and subtract decimals.

1 Add Positive Decimals

To add or subtract *decimal* numbers, we **line up the decimal points**—that is, we line up the numbers in the tenths place, line up the numbers in the hundredths place, and so on. Then, add or subtract.

Procedure Adding or Subtracting Decimals

1) Write the numbers vertically so that the decimal points are lined up.

2) If any numbers are missing digits to the right of the decimal point, insert zeros. Then, add or subtract the same way we add or subtract whole numbers.

3) Place the decimal point in the answer *directly below* the decimal point in the problem.

Note
Using graph paper will help us line up the numbers correctly.

EXAMPLE 1

Add.

a) $9.7 + 2.8$ b) $14.223 + 7.501 + 0.884$

Solution

W Hint
Write out the steps as you are reading the example.

a)   Write the numbers vertically so that the decimal points are lined up.

```
    9.7
 +  2.8
```
Line up the
decimal points.

Now, add just like we add whole numbers.

```
    1
    9.7
 1  2.8
 1  2.5
```
Line up the decimal point
in the answer with the
decimal points in the problem.

So, $9.7 + 2.8 = 12.5$.

b)   Write the numbers vertically so that the decimal points are lined up.

```
  1 4.2 2 3
    7.5 0 1
 +  0.8 8 4
```
Line up the
decimal points.

Now, add just like we add whole numbers.

```
    1 1 1
  1 4.2 2 3
    7.5 0 1
 +  0.8 8 4
  2 2.6 0 8
```
Line up the decimal point in
the answer with the decimal
points in the problem.

So, $14.223 + 7.501 + 0.884 = 22.608$.

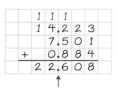

www.mhhe.com/messersmith

[YOU TRY 1] Add.

   a) 8.6 + 5.7   b) 27.958 + 9.042 + 0.737

In Example 1a, both numbers had just one decimal place. In Example 1b, all numbers had three decimal places. Sometimes, the numbers in a problem do *not* have the same number of decimal places. When this happens, we insert zeros as placeholders so that they *will* have the same number of decimal places.

EXAMPLE 2   Add.

   a) 6.2 + 5.83   b) 11.36 + 7.295 + 8

Solution

a) Write the numbers vertically so that the decimal points are lined up.

> 5.83 has a digit in the hundredths place, but 6.2 does not. **Insert a 0 in the hundredths place of 6.2.** Then, add.

W Hint

Try using graph paper to line up the numbers correctly.

Line up the
decimal points.

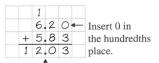

Line up the decimal point in
the answer with the decimal
points in the problem.

We can insert the 0 in the hundredths place of 6.2 because 6.2 is equivalent to 6.20.

Therefore, 6.2 + 5.83 = 12.03.

b) 11.36 + 7.295 + 8

Write the numbers vertically so that the decimal points are lined up. Remember that 8 can be written as 8.

> **Insert zeros in the thousandths place of 11.36 and in three places to the right of the decimal point in 8.** Then, add.

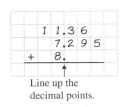

Line up the
decimal points.

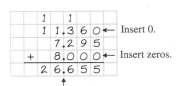

Line up the decimal point in
the answer with the decimal
points in the problem.

So, the sum is 26.655.

[YOU TRY 2] Add.

   a) 9.4 + 7.81   b) 25.7 + 7 + 8.643

2 Subtract Positive Decimals

The first step in subtracting decimals is to line up the decimal points. Follow the steps in the Procedure box at the beginning of this section.

EXAMPLE 3

Subtract 46.97 − 21.32.

Solution

Write the numbers vertically so that the decimal points are lined up.

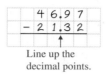

Line up the
decimal points.

Then subtract just like we subtract whole numbers.

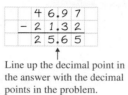

Line up the decimal point in
the answer with the decimal
points in the problem.

We can check the answer using addition. The answer is correct.

$$\begin{array}{r} 2\,5.6\,5 \\ +\ 2\,1.3\,2 \\ \hline 4\,6.9\,7 \end{array}$$

(If the sum did *not* equal 46.97, then we made a mistake and we need to work the problem again.)

YOU TRY 3

Subtract 85.37 − 55.16.

We regroup with decimal numbers just like we regroup with whole numbers.

EXAMPLE 4

Subtract 48.639 from 129.386.

Solution

Line up the decimal points, then subtract. We will need to regroup (or borrow).

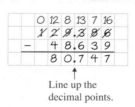

Line up the
decimal points.

To check, add 80.747 + 48.639. The sum is 129.386, so the answer is correct.

YOU TRY 4

Subtract 64.478 from 114.239.

Sometimes, we need to insert zeros as placeholders in subtraction problems.

EXAMPLE 5

Find each difference.

a) 29.8 minus 6.174

b) Subtract 36.2 from 89.06.

Solution

a) Line up the decimal points.

Insert zeros after the 8 in 29.8. Then, subtract.

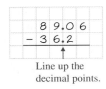

Line up the
decimal points.

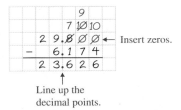

Line up the
decimal points.

W Hint

Be sure you are lining up the decimal points when you subtract.

Remember, we can add zeros at the end of 29.8 because 29.8 is equivalent to 29.800.

b) Line up the decimal points.

Insert 0 after the 2 in 36.2. Then, subtract.

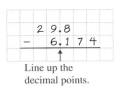

Line up the
decimal points.

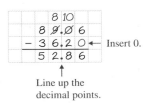

Line up the
decimal points.

YOU TRY 5

Find each difference.

a) 61.7 minus 4.329

b) Subtract 51.6 from 98.18.

When subtracting decimal numbers and whole numbers, we must remember that a whole number can be written with a decimal point after it. For example, 6 = 6.

EXAMPLE 6

Subtract 12 − 5.386.

Solution

First, rewrite 12 with a decimal point: 12 = 12.

Line up the decimal points.

Insert zeros and subtract.

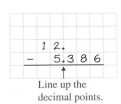

Line up the
decimal points.

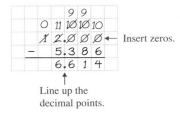

Line up the
decimal points.

Remember, we can check the answer with addition.

[YOU TRY 6] Subtract 31 − 14.601.

Note

Remember, we can add zeros at the end of decimal numbers to add or subtract because it does not change the *value* of the number.

$$12 = 12.000 \qquad 8.3 = 8.30$$

3 Add and Subtract with Negative Decimals

We add with negative decimals the same way that we add integers. Let's review the rules here.

Procedure Adding Signed Numbers

To **add two negative numbers,** find the absolute value of each number. Then, add the absolute values. The sum of two negative numbers is negative, so put a negative sign in front of the sum.

To **add two numbers with different signs,** find the absolute value of each number. Subtract the smaller absolute value from the larger absolute value. The sign of the sum will be the same as the sign of the number with the greater absolute value.

EXAMPLE 7   Add.

a)   −5.8 + (−12.371)   b)   −0.47 + 0.9

Solution

a)   Find the absolute value of each number: $|-5.8| = 5.8$, $|-12.37| = 12.371$. Add the absolute values:

W Hint

Do your work on graph paper.

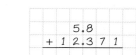

The sum of two negative numbers is negative, so the final answer will be negative:

$$-5.8 + (-12.371) = -18.171$$

b)   Find the absolute value of each number: $|-0.47| = 0.47$, $|0.9| = 0.9$. Subtract the smaller absolute value from the larger absolute value.

The sum will be *positive* because 0.9 has a larger absolute value than −0.47.

$$-0.47 + 0.9 = 0.43 \qquad \text{The sum is positive.}$$

Add.

a) $-57.29 + (-46.1)$ b) $0.035 + (-0.7)$

In Section 1.4, we subtracted integers by changing the subtraction to an addition problem like this:

$$8 - 13 = 8 + (-13) = -5$$

Change subtraction to addition of the additive inverse of 13.

We work with signed decimal numbers the same way.

EXAMPLE 8

Subtract.

a) $17.3 - 19.84$ b) $-9.2 - 3.5$

Solution

a) $17.3 - 19.84 = 17.3 + (-19.84)$ Change subtraction to addition of the additive inverse of 19.84.

Find the absolute value of 17.3 and -19.84: $|17.3| = 17.3$, $|-19.84| = 19.84$.

Subtract the absolute values:

W Hint

Don't forget to insert zeros when necessary.

```
  1 9.8 4
-   1 7.3
```

```
  1 9.8 4
- 1 7.3 0 ← Insert a zero.
  2.5 4
```

The final answer will be *negative* because -19.84 has a larger absolute value than 17.3.

Therefore, $17.3 - 19.84 = -2.54$.

b) $-9.2 - 3.5 = -9.2 + (-3.5)$ Change subtraction to addition of the additive inverse of 3.5.

Find the absolute value of each number: $|-9.2| = 9.2$, $|-3.5| = 3.5$.
Add the absolute values:

```
    9.2
+   3.5
  1 2.7
```

The sum of two negative numbers is negative, so the final answer is negative:

$$-9.2 + (-3.5) = -12.7$$

Subtract.

a) $-102.753 - 86.4$ b) $95.29 - 115.8$

The order of operations applies to decimal numbers just like it does to integers and fractions.

EXAMPLE 9

Simplify $64 - (-18.2 + 5.09)$.

Solution

Add inside the parentheses first.

$64 - (-18.2 + 5.09)$

$= 64 - (-13.11)$ Add the numbers in parentheses first.

$= 64 + 13.11$ Change subtraction to addition of the additive inverse of -13.11.

$= 77.11$ Add.

[YOU TRY 9] Simplify $-73.805 - (25.6 - 62.94)$.

4 Solve an Applied Problem

Decimals are used in many types of real-world problems.

[EXAMPLE 10]

Liam's bill at a fast-food restaurant is $6.83. If he pays with a $10 bill, how much change will he receive?

Solution

To determine the amount of change Liam should get, we subtract 6.83 from 10.

W Hint

You can also use the Five Steps for Solving Applied Problems to solve these problems.

Write 10 with a decimal point: 10.
Insert the zeros as placeholders, then subtract.

Liam's change will be $3.17.

We can check by adding: $3.17 + $6.83 = $10.00 ✓

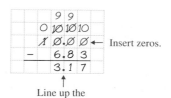

Line up the decimal points.

[YOU TRY 10] Tanya's grocery bill is $14.67. If she pays with a $20 bill, how much change will she receive?

ANSWERS TO [YOU TRY] EXERCISES

1) a) 14.3   b) 37.737   2) a) 17.21   b) 41.343   3) 30.21   4) 49.761   5) a) 57.371   b) 46.58
6) 16.399   7) a) -103.39 b) -0.665 8) a) -189.153 b) -19.51 9) -36.465 10) $5.33

E Evaluate 5.3 Exercises

Do the exercises, and check your work.

Objective 1: Add Positive Decimals

Add.

1) 8.7 + 5.8

2) 9.3 + 6.8

3) 0.42 + 0.53

4) 0.18 + 0.71

 5) 2.501 + 6.089 + 4.374

6) 6.435 + 3.436 + 1.607

7) 4.9 + 1.65

8) 1.3 + 2.88

9) 17.91 + 2.391 + 1

10) 13.2 + 6.482 + 9

11) 50.009 + 0.003 + 2.3

12) 30.005 + 0.0004 + 6.5

13) 8.234 + 419 + 26.076 + 7.39

14) 14.15 + 6.746 + 582 + 34.904

15) The sign shows the price of 1 gallon of regular gasoline, in cents, as a mixed number. Represent the mixed number as a sum. Write the exact price of 1 gallon of gas as a dollar amount in decimal form.

16) For which gasoline grade is the price displayed incorrectly on the sign? Why is it incorrect?

Objective 2: Subtract Positive Decimals

17) Explain, in your own words, how to add and subtract decimals.

18) Does 46 = 46.00? Explain your answer.

Subtract.

19) 9.6 − 4.3

20) 8.7 − 2.5

21) 53.08 − 31.92

22) 92.47 − 46.63

23) 963.381 − 732.482

24) 578.036 − 254.139

25) 85.14 − 23.8

26) 64.36 − 18.5

27) 76.5 − 51.34

28) 59.8 − 12.27

29) 120.4 − 38.661

30) 140.3 − 75.792

Find the difference.

31) 82.4 minus 43.983

32) 46.8 minus 27.909

33) Subtract 13.707 from 32.5.

34) Subtract 9.732 from 11.5.

35) Subtract 38 from 74.8.

36) Subtract 59 from 81.6.

Subtract.

37) 54.47 − 29

38) 32.25 − 8

39) 5 − 1.6

40) 9 − 4.3

41) 8 − 2.07

42) 6 − 3.04

43) 261 − 139.4952

44) 483 − 228.3794

Objective 3: Add and Subtract with Negative Decimals

Add.

 45) $-2.5 + (-5.91)$ 46) $-8.6 + (-9.43)$

47) $-0.162 + (-0.49)$ 48) $-0.741 + (-0.15)$

49) $24.6 + (-17.3)$ 50) $28.7 + (-19.6)$

51) $-152.8 + 84.095$ 52) $-146.3 + 59.088$

53) $-1 + 0.94$ 54) $-1 + 0.02$

Subtract.

55) $2.53 - 7.06$ 56) $4.12 - 9.08$

57) $-61.48 - (-99.5)$ 58) $-57.93 - (-87.4)$

59) $726 - (-156.179)$ 60) $495 - (-294.523)$

61) $-62.325 - 43.8$ 62) $-28.986 - 71.4$

63) $0.0054 - 6$ 64) $0.0072 - 8$

Simplify. Remember to use the order of operations.

65) $13.4 + (2.17 - 11.6)$

66) $19.3 + (4.82 - 15.7)$

67) $(-21 - 15.8) - (-72.59 + 38.77)$

68) $(-34 - 12.7) - (-90.29 + 45.64)$

69) $-96.07 - |-53 - (-41.52)| + 0.008$

70) $-89.24 - |-68 - (-50.37)| + 0.009$

Objective 4: Solve an Applied Problem

Solve each problem.

71) Namiko buys five pieces of Mochi ice cream for $5.67. If she pays with a $10 bill, how much change will she receive?

72) Luciano buys four pastries from a bakery for $6.58. If he pays with a $10 bill, how much change will he receive?

73) DeMarcus buys a small popcorn and a drink at the movie theater for $9.86. If he gives the cashier a $20 bill, how much change will he receive?

©Erica Simone Leeds

74) Caitlin and her friend each buy a milk tea drink with boba at a local cafe. Their total bill is $7.48. If Caitlin gives the cashier a $20 bill, how much change will she receive?

75) Ahdoja fills up the gas tank in her hybrid vehicle for $43.67, and she also buys a candy bar and soda for $2.94. If she pays with two $20 bills and a $10 bill, how much change will she receive?

76) Soren's bill for a taxi ride from the airport to his home is $28.64, and he gives the driver a $4 tip. If he pays the driver with two $20 bills, how much change will he receive?

77) At a county fair, Kathleen buys three corn dogs and three lemonades for her children. Her total bill is $16.25. If Kathleen gives the cashier a $20 bill, a $1 bill, and a quarter, how much change does she receive?

78) Ignacio's bill at a sandwich shop is $5.77. He gives the cashier a $10 bill and two pennies. How much change does he receive?

79) At a farmers' market, Morgan buys some fresh fruit costing $8.51. If Morgan gives the vendor two $5 bills and one penny, how much change does he receive?

©Amanda Mills/CDC

80) Keely's bill at a smoothie café is $6.33. She gives the cashier a $10 bill and eight pennies. How much change does she receive?

81) Jesse buys the following items and gives the cashier two $20 bills and a $5 bill. If the tax on the items is $2.17, how much change will he receive?

Sunscreen:	$7.49
Towel:	$10.75
T-shirt:	$9.98
Flip-flops:	$11.99

82) LaVonda buys the following items and gives the cashier a $20 bill and a $10 bill. If the tax on the items is $1.34, how much change will she receive?

Picture frame:	$6.99
Wastebasket:	$8.19
Vase:	$12.75

Mixed Exercises: Objectives 1–4

Perform the indicated operations.

83) $49.58 - 620.7$

84) $9.427 + 138.6 + 58 + 19.09$

85) $185.1 + 0.007$ 86) $51 - 0.014$

87) $23.75 - 12$ 88) $16.46 + (-28.45)$

89) $-17 + 4.827 + 0.083 - (-12.6)$

90) $5.978 + 37.201 + 7.846$

91) $46 - 12.597$

92) $-762.3 - 255$

93) Find the sum of 19.4 and 6.98.

94) Subtract 50.37 from 42.6.

Solve each problem.

95) Before she went away to college, Ileana's grandmother gave her a $150 gift card. Ileana used the gift card to buy the items listed here. If tax on the items was $7.54, how much money is left on her gift card?

Shelves:	$48.99
Shower caddy:	$6.35
Backpack:	$39.98
Desk lamp:	$15.77

96) Takahiro's bills for the month are listed here. If he had $1061.82 in his checking account when he sat down to pay the bills, how much remains after he pays them?

Rent:	$425
Electricity:	$37.14
Cable:	$41.93
Car payment:	$183.07

R Rethink

R1) How comfortable do you feel adding and subtracting with negative decimals?

R2) Which topics do you still need to master in this section?

R3) Where have you recently encountered the objectives of this section in your life? Write an applied problem similar to the ones you just completed and solve.

5.4 Multiplying Signed Decimals

P Prepare

O Organize

What are your objectives for Section 5.4?	How can you accomplish each objective?
1 Multiply Signed Decimals	• Write the procedure for **Multiplying Two Decimal Numbers** in your own words. • Complete the given examples on your own. • Complete You Trys 1 and 2.
2 Multiply a Number by a Power of 10	• Write the procedure for **Multiplying a Number by a Power of 10** in your own words. • Write the procedure for **Multiplying a Number by 0.1, 0.01, 0.001, etc.,** in your own words. • Complete the given examples on your own. • Complete You Trys 3 and 4.
3 Solve Applied Problems	• Complete the given example on your own. • Complete You Try 5.

1   Multiply Signed Decimals

When we add and subtract decimals, we must line up the decimal points. This is *not* true when we multiply decimals. We can understand why the procedure for multiplying decimals is different if we multiply decimals by changing them to fractions first.

$$0.2 \times 0.37 = \frac{2}{10} \times \frac{37}{100} = \frac{74}{1000} = 0.074$$

<table>
<tr><td>↑</td><td></td><td>↑</td><td></td><td></td><td>↑</td></tr>
<tr><td>1 decimal
place</td><td>+</td><td>2 decimal
places</td><td></td><td>=</td><td>3 decimal
places</td></tr>
</table>

The number of decimal places in the *product* is the *sum* of the numbers of decimal places in the factors. (Recall that the *factors* are the numbers being multiplied, and the *product* is the answer.)

Procedure Multiplying Two Decimal Numbers

1) **Multiply the numbers (factors) just like you would multiply whole numbers.** (Line up the numbers on the right; the decimal points do *not* have to be lined up.)

2) **Determine the total number of decimal places in the answer (the product).** This will be the *total* number of decimal places in the factors.

3) **Insert the decimal point in the answer.** Start at the right side of the product, and count the *total* number of places you determined in 2). Sometimes, you may need to insert zeros as placeholders on the left side of the answer.

4) **Determine the sign of the product.** If the two factors have the *same sign,* the product is *positive.* If the two factors have *different signs,* the product is *negative.*

BE CAREFUL When we *multiply* decimals, we do **not** have to line up the decimal points. When we add or subtract decimals, we **must** line up the decimal points.

EXAMPLE 1   Multiply -15.83×4.6.

Solution

Multiply the numbers just as if they were whole numbers. (Line up the numbers on the right.) Do **not** line up the decimal points.

W Hint
Use graph paper when you multiply decimals to line up the numbers correctly.

```
      1 5 . 8 3
    ×       4 . 6
      9 4 9 8
    6 3 3 2
    7 2 8 1 8
```

Determine the total number of decimal places in the answer. This will be the total number of decimal places in the factors.

```
      1 5 . 8 3 →   2 decimal places
    ×       4 . 6 → + 1 decimal place
      9 4 9 8       3 decimal places
    6 3 3 2         in the answer
    7 2 . 8 1 8
```

Start at the right side of the number and count 3 places to the left. Insert the decimal point.

The factors have *different signs,* so the product is *negative.*

$$-15.83 \times 4.6 = -72.818$$

Sometimes, we have to insert zeros on the left side of the product to have the correct number of decimal places in the answer.

EXAMPLE 2

Find each product.

a) $-0.029 \times (-0.05)$ b) $(0.3)^2$

Solution

a) **Multiply the numbers just as if they were whole numbers.** Line up the numbers on the right.

$$
\begin{array}{r}
0.0\,2\,9 \\
\times \quad 0.0\,5 \\
\hline
1\,4\,5
\end{array}
$$

Determine the total number of decimal places in the answer. This will be the total number of decimal places in the factors.

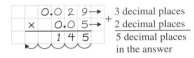

Start at the right side and count 5 places to the left. Insert zeros in the blank spaces.

$$
\begin{array}{r}
0.0\,2\,9 \\
\times \quad 0.0\,5 \\
\hline
0\;0\;1\;4\;5
\end{array}
$$

There are 5 places to the right of the decimal point.

The final answer will be positive because both factors are negative. Write the final answer with a zero to the left of the decimal point, in the ones place.

$$-0.029 \times (-0.05) = 0.00145$$

Put a zero in the ones place.

W Hint

Can you think of a faster way to square 0.3?

b) $(0.3)^2$ means 0.3×0.3.

Multiply the numbers just as if they were whole numbers, and determine the number of decimal places in the product.

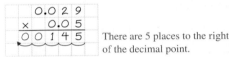

Start at the right, and count 2 places to the left. Insert a zero in the blank space.

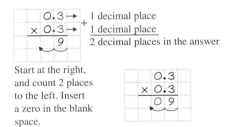

Write the final answer with a zero to the left of the decimal point, in the ones place.

$$(0.3)^2 = 0.09$$

Put a zero in the ones place.

[YOU TRY 2] Find each product.   a) $-0.019 \times (-0.07)$ b) $(0.02)^2$

2 Multiply a Number by a Power of 10

Multiplying a number by a power of 10 can be simple if we notice a pattern. Let's multiply 6.72 by 10, 100, and 1000, and see what happens.

$$6.72 \times 10 = 67.20$$
or 67.2

$$6.72 \times 100 = 672.00$$
or 672

$$6.72 \times 1000 = 6720.00$$
or 6720

Do you notice the pattern?

$$6.72 \times 10 = 67.2$$

$$6.72 \times 100 = 672.$$

$$6.72 \times 1000 = 6720.$$

Multiply by 10, move the decimal point *right* 1 place.

Multiply by 100, move the decimal point *right* 2 places.

Multiply by 1000, move the decimal point *right* 3 places.

The number of zeros in the power of 10 tells us how many places to move the decimal point to obtain the product.

Procedure Multiplying a Number by a Power of 10

To multiply a number by a power of 10,

1) Count the number of zeros in the power of 10.

2) Move the decimal point in the number *to the right* the same number of spaces as the number of zeros in the power of 10.

3) If necessary, add zeros as placeholders on the right.

EXAMPLE 3

Multiply.

a) 0.5941 · 100   b) −89 × 1000

Solution

a) $0.5941 \cdot 100 = 59.41$

       ↑   Move decimal point
   2 zeros 2 places to the right.

            3 zeros
            ↓

b) $-89 \times 1000 = -89. \times 1000 = -89.000 = -89{,}000$

        ↑
   The decimal point comes   Move the decimal point 3 places
   at the end of the number.   to the right and insert zeros.

We usually do *not* write a decimal point at the end of a number if there are no digits after it.

[**YOU TRY 3**] Multiply.

    a) 0.4459×1000 b) $-74 \cdot 100$

There is a similar procedure for multiplying a number by 0.1, 0.01, 0.001, and so on. Let's multiply 34.9 by 0.1, 0.01, and 0.001 and see whether we notice a pattern.

3 4 . 9
×   0 . 1
3 . 4 9

3 4 . 9
× 0 . 0 1
0 . 3 4 9

3 4 . 9
× 0 . 0 0 1
0 . 0 3 4 9

$34.9 \times 0.1 = 3.49$ $34.9 \times 0.01 = 0.349$ $34.9 \times 0.001 = 0.0349$

1 decimal place — Move decimal point 1 place to the left.

2 decimal places — Move decimal point 2 places to the left.

3 decimal places — Move decimal point 3 places to the left.

W Hint

Compare this with the procedure for multiplying by a power of 10.

Procedure Multiplying a Number by 0.1, 0.01, 0.001, etc.

To multiply a number by:

1) 0.1, move the decimal point in the number 1 place to the left.

2) 0.01, move the decimal point 2 places to the left.

3) 0.001, move the decimal point 3 places to the left.

and so on.

EXAMPLE 4

Multiply.

    a) 0.299×0.001 b) $-650{,}000 \cdot 0.0001$

Solution

    a) $0.299 \times 0.001 = 0.000299$ Move the decimal point 3 places to the left.

           3 decimal places

    b) $-650{,}000. \cdot 0.0001 = -65.0000 = -65$ Move the decimal point 4 places to the left.

           Insert the decimal point.   4 decimal places

[**YOU TRY 4**] Multiply.

    a) -0.038×0.01 b) $12{,}000{,}000 \cdot 0.0001$

3 Solve Applied Problems

Let's solve a problem using multiplication of decimals.

EXAMPLE 5

Shu Fang's car payment is $189.65 per month. Determine the total amount she pays in one year.

Solution

Each month, Shu Fang pays $189.65 for her car. *There are* 12 *months in one year,* so multiply $189.65 by 12 to determine the total amount of her car payments in one year.

$$
\begin{array}{r}
1\,8\,9.6\,5 \rightarrow \text{2 decimal places} \\
\times \qquad 1\,2 \rightarrow + \text{0 decimal places} \\
\hline
3\,7\,9\,3\,0 \qquad \text{2 decimal places in} \\
1\,8\,9\,6\,5 \qquad \text{the answer} \\
\hline
2\,2\,7\,5.8\,0
\end{array}
$$

Shu Fang's total car payment in one year is $2275.**80**.

We *must* leave the 0 on the end of the number because money amounts are written to the nearest hundredths place, or to the nearest cent.

Note
Remember that, usually, money amounts are written to the nearest cent or to the *hundredths* place. Sometimes, it will be necessary to round the final answer to the nearest cent.

[YOU TRY 5]

Gaurav pays $397.15 per month to repay his student loans. How much does he pay in one year?

ANSWERS TO [YOU TRY] EXERCISES

1) −401.076   2) a) 0.00133 b) 0.0004   3) a) 445.9 b) −7400
4) a) −0.00038 b) 1200   5) $4765.80

Using Technology

Putting gasoline in our cars is something we often do.

Suppose you pay $379\dfrac{9}{10}$ cents, or $3.799, for a gallon of gas and the pump indicates you purchased 13.527 gallons of gasoline. To find the amount you should pay in dollars, multiply 3.799 by 13.527. First perform the calculation by hand. Now enter 3 . 7 9 9 × 1 3 . 5 2 7 = into the calculator. The answer is 51.389073. Remember, however, to round the answer to the nearest hundredth since we are working with dollar amounts. The cost of the gas will be $51.39.

Objective 1: Multiply Signed Decimals

1) Explain, in your own words, how to multiply decimals.

2) Which operations with decimals require that we line up the decimal points?

Multiply.

3) $\begin{array}{r} 6.3 \\ \times\, 4.1 \end{array}$

4) $\begin{array}{r} 7.5 \\ \times\, 2.9 \end{array}$

5) $\begin{array}{r} -12.7 \\ \times\quad 5.4 \end{array}$

6) $\begin{array}{r} -17.4 \\ \times\quad 4.3 \end{array}$

7) $\begin{array}{r} 782.29 \\ \times\quad 0.4 \end{array}$

8) $\begin{array}{r} 913.63 \\ \times\quad 0.8 \end{array}$

9) 13.42×8.7

10) 11.56×9.3

11) $-27.4(-19.35)$

12) $-40.8(-22.85)$

13) $(0.0005)(-4018.6)$

14) $(0.0002)(-8307.5)$

15) $31.03(21)$

16) $16.09(14)$

17) $-0.024 \cdot 0.03$

18) $-0.032 \cdot 0.06$

19) $(-5.004)(-32,800)$

20) $(-7.006)(-29,400)$

21) 0.6×1005

22) 0.2×4005

23) Given that $\dfrac{1}{8} = 0.125$, write a product that is equivalent to the decimal representation of $\dfrac{5}{8}$. Calculate the decimal representation of $\dfrac{5}{8}$.

24) Given that $\dfrac{1}{16} = 0.0625$, write a product that is equivalent to the decimal representation of $\dfrac{9}{16}$. Calculate the decimal representation of $\dfrac{9}{16}$.

Evaluate each exponential expression.

25) $(0.4)^2$

26) $(0.7)^2$

27) $(-1.2)^2$

28) $(-1.4)^2$

29) $(-0.12)^2$

30) $(-0.14)^2$

31) $(0.05)^2$

32) $(0.08)^2$

33) $(0.007)^2$

34) $(0.006)^2$

Objective 2: Multiply a Number by a Power of 10

35) Explain how to multiply a number by 1000.

36) Explain how to multiply a number by 0.01.

Find each product.

37) 0.2587×100

38) 0.7134×100

39) $-3.66 \cdot 10$

40) $-1.09 \cdot 10$

41) $(-0.0000608)(-10,000)$

42) $(-0.0000407)(-10,000)$

43) $(1000)(43.67)$

44) $(1000)(82.18)$

45) -64×100

46) -73×100

Multiply.

47) $38.01 \cdot 0.1$

48) $70.08 \cdot 0.1$

49) $(-1925.8)(-0.01)$

50) $(-4173.4)(-0.01)$

51) $-0.702(0.01)$

52) $-0.559(0.01)$

53) $0.0001 \times 94,000$

54) $0.0001 \times 67,000$

55) $0.00001(-32,000,000)$

56) $0.00001(-56,000,000)$

Objective 3: Solve Applied Problems

Solve each problem. For all problems involving money, the final answer should be to the nearest cent.

57) Talog's cable television bill is $54.75 per month. Determine the total amount he pays in one year.

58) Molly treats her friends to a late-night snack at a taco stand. If Molly buys 16 tacos costing $1.35 each, what is her total bill?

59) Kathy is a real estate agent, and she will earn a commission rate of 0.04 for selling a house for $279,000. Find the amount of Kathy's commission. (Multiply the commission rate by the sale price of the house.)

60) Felipe buys six tiki torches for his backyard. How much did he spend if they cost $11.97 each?

©Stockdisc/PunchStock

61) A storage rack holds four canoes and each one weighs 78.6 lb. Find the total weight of the canoes on the rack.

62) A warehouse worker uses a forklift to stack five boxes on top of each other. If each box is 3.9 ft tall, what is the height of the stack?

©ColorBlind Images/Blend Images LLC

63) Every morning on her way to work, Nicole buys the same drink at her favorite coffee shop. If she spends $3.41 per day, five days each week, how much does she spend in one month?

64) Jadvyga takes the train to work every day. She buys a monthly pass for $102.25. How much does she spend for her train pass each year?

65) Justin downloads 13 songs from iTunes at $1.29 each and six songs at $0.99 each. How much did he spend for his music?

66) For his son's soccer team, Huang buys 16 bottles of sports drinks at $1.19 each and six boxes of granola bars at $3.49 each. How much did Huang pay for these snacks?

67) Selena's car gets 27.8 mi per gal. How far can she drive on 10.7 gallons of gas?

68) Noor drove 3.25 hr at an average speed of 67.8 mi per hr. How far did she go?

69) An electronic reader is 7.5 in. long and 4.8 in. wide. Find the area of its surface and the perimeter of the reader.

70) A whiteboard is 48.5 in. wide and 96.2 in. long. Find the area of its surface and the perimeter of the whiteboard.

71) Aiko buys 3.5 lb of basmati rice at $1.98 per lb and 2.8 lb of jasmine rice at $2.79 per lb. Find the total cost of the rice.

72) Bill buys 1.5 lb of a Costa Rican coffee at $6.79 per lb, 3.25 lb of a Kenyan coffee at $9.20 per lb, and 2 lb of Italian espresso beans at $6.29 per lb. How much did Bill pay for the coffee?

73) Salim is an inspector for his city's building department, and he earns $23.89 per hr. Here is a list of the hours he worked last week:

Monday	7:30 A.M.–3:30 P.M.
Tuesday	8:30 A.M.–4:00 P.M.
Wednesday	9:00 A.M.–5:00 P.M.
Thursday	7:30 A.M.–3:00 P.M.
Friday	7:45 P.M.–3:00 P.M.

a) How many hours did Salim work last week?

b) Find his gross pay. (Gross pay is the amount earned before deductions.)

c) Multiply his gross pay by 0.09 to determine the amount of money that will be deducted from his paycheck for taxes.

d) Find Salim's take-home pay if, in addition to the amount deducted for taxes, $29.36 will be deducted for insurance and $25.68 will be deducted for his retirement plan. (Take-home pay is the gross pay minus the amount of all the deductions.)

74) Joyce makes three sizes of wreaths that she sells at craft fairs. The small size uses 1.25 ft of wire, the medium size uses 2.50 ft of wire, and the large wreath uses 3.50 ft of wire.

a) If the wire costs $0.64 per ft, how much does it cost to make one of each size wreath?

©Dynamic Graphics Group/ Creatas/Alamy

b) For the next craft fair, Joyce plans to make 7 small wreaths, 15 medium wreaths, and 8 large wreaths. How much wire will she need?

c) Find the cost of the wire to make all the wreaths in part b).

Multiply.

75) $\begin{array}{r} 0.076 \\ \times\ 0.05 \end{array}$

76) $(-2.25)^2$

77) $-16{,}974 \times 0.0001$

78) $100{,}000(20.06)$

79) $(-3.8)(-28.04)$

80) $-0.054 \cdot 0.08$

81) $(1.41)^2$

82) 0.1×0.7062

83) $1000 \cdot 25.9$

84) $\begin{array}{r} -\ 503.06 \\ \times\quad\ 25 \end{array}$

85) $(-1.75)(-0.001)$

86) $(0.006)(400.8)$

87) $0.00003 \times 1{,}000{,}000$

88) $0.000705(-36{,}000)$

Solve each problem. The answer should be to the nearest cent.

89) Fermin works part-time and earns \$12.80 per hr. One week, he worked three days from 4:00 P.M. to 9:30 P.M. How much did Fermin earn?

90) Laycee purchases 12 yd of cloth costing \$5.85 per yd. What is the purchase price for the cloth?

R Rethink

R1) What is the only extra step in multiplying decimals compared with multiplying whole numbers?

R2) Where do you encounter the multiplication of decimal numbers on a daily basis?

R3) Which objective(s) do you still need help mastering?

5.5 Dividing Signed Decimals and Order of Operations

P Prepare

O Organize

What are your objectives for Section 5.5?	How can you accomplish each objective?
1 Divide a Decimal by a Whole Number	• Write the procedure for **Dividing a Decimal by a Whole Number** in your own words. • Add details to this procedure as you follow the examples. Include information about *adding extra zeros, nonrepeating decimals,* and *repeating decimals.* • Complete the given examples on your own. • Complete You Trys 1–4.
2 Divide a Number by a Decimal	• Write the procedure for **Dividing a Number by a Decimal** in your own words. • Understand why you are able to move the decimal point in both the dividend and the divisor. • Complete the given examples on your own. • Complete You Trys 5–7.
3 Use the Order of Operations with Decimals	• Use the same rules as before (PEMDAS) for the order of operations, and apply them to decimals. • Complete the given example on your own. • Complete You Try 8.

Read the explanations, follow the examples, take notes, and complete the You Trys.

Do you remember how to perform long division with whole numbers? For example, to find $45 \div 7$, we can divide as follows:

$$\text{Divisor} \rightarrow 7\overline{)45} \leftarrow \text{Dividend} \qquad 45 \div 7 = 6\text{ R}3$$
$$\underline{-\;4\;2}$$
$$3 \leftarrow \text{Remainder}$$

with Quotient marked above the 6 and Dividend marked to the 45.

In this section, we will learn how to divide decimals.

1 Divide a Decimal by a Whole Number

Let's start by learning how to divide a decimal by a whole number. That is, the divisor is a whole number.

Procedure Dividing a Decimal by a Whole Number

1) Write the problem in long division form.
2) Write the decimal point in the quotient directly above the decimal point in the dividend.
3) Perform the division as if the numbers were whole numbers.
4) Determine the sign of the quotient. If both numbers have the *same sign*, the quotient is *positive*. If the numbers have *different signs*, the quotient is *negative*.

Note

When performing division involving decimals, we do *not* write a remainder in an answer like we did when we divided whole numbers. Later in this section, we will learn ways to work with the remainder.

EXAMPLE 1

Divide.

a) $37.52 \div 4$ 　　 b) $\dfrac{0.00945}{-7}$

Solution

a) Write the problem in long division form, and write the decimal point in the quotient directly above the decimal point in the dividend.

W Hint

Remember, work out the examples on your paper as you are reading them.

Perform the division as if the numbers were whole numbers.

$$
\begin{array}{r}
9.38 \\
4\overline{)37.52} \\
-36 \\
\hline
15 \\
-12 \\
\hline
32 \\
-32 \\
\hline
0
\end{array}
$$

Check by multiplying:

$$
\begin{array}{r}
9.38 \\
\times 4 \\
\hline
37.52 \quad \checkmark
\end{array}
$$

The quotient is *positive* because both numbers have the *same sign*: $37.52 \div 4 = 9.38$.

---

b) Write the problem in long division form, and write the decimal point in the quotient directly above where it appears in the dividend.

Perform the division as if the numbers were whole numbers. Because $0 \div 7 = 0$, we must begin by putting zeros in the quotient.

$$
\begin{array}{r}
.00135 \\
7\overline{)0.00945} \\
\underline{-7} \\
24 \\
\underline{-21} \\
35 \\
\underline{-35} \\
0
\end{array}
$$

Check by multiplying:

$$
\begin{array}{r}
0.00135 \\
\times 7 \\
\hline
0.00945 \quad \checkmark
\end{array}
$$

$\dfrac{0.00945}{-7} = -0.00135.$ The quotient is *negative* because the numbers have different signs.

[YOU TRY 1] Divide.

a) $\dfrac{25.41}{3}$ b) $-0.00785 \div 5$

In Example 1, both problems had a remainder of zero. Sometimes, however, we reach the end of the dividend, the remainder is *not* zero, and there are no more digits to bring down. If this happens, write extra zeros on the *right* end of the dividend and keep dividing.

EXAMPLE 2 Divide. $8\overline{)3.94}$

Solution

Begin the division process like we did in Example 1.

$$
\begin{array}{r}
0.49 \\
8\overline{)3.94} \\
\underline{-32} \\
74 \\
\underline{-72} \\
2
\end{array}
$$

← No more digits to bring down

← Remainder ≠ 0

We have reached the end of the dividend, the remainder is *not* zero, and there are no more digits to bring down.

Write extra zeros on the *right* end of the dividend and keep on dividing.

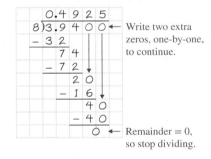

← Write two extra zeros, one-by-one, to continue.

← Remainder = 0, so stop dividing.

The answer is 0.4925.

Check by multiplying: $0.4925 \times 8 = 3.94$ ✓

[YOU TRY 2] Divide. $8\overline{)6.26}$

SECTION 5.5 **Dividing Signed Decimals and Order of Operations**

> **Note**
>
> We can add zeros at the right end of a number containing a decimal point because it does *not* change the value of the number.
>
> Example: 3.94 = 3.9400

Sometimes, the remainder will *never* equal 0. One way to write the answer to such a division problem is to round it.

EXAMPLE 3

Divide −8.6 by −7. Round the answer to the nearest thousandth.

Solution

Because we will be rounding to the *thousandths* place, we must continue the division until the quotient has a digit in the *ten-thousandths* place, one place to the right of where we must round.

W Hint

Are you using graph paper to divide decimals?

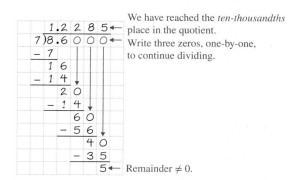

This division will never give a remainder of 0. So, our quotient is an approximation.

$$-8.6 \div (-7) \approx 1.229 \longleftarrow \text{Round the quotient.}$$

Because 1.229 is not the *exact* answer, when we check the answer, it will be different from (but should be very close to) −8.6.

Check: 1.229 × (−7) = −8.603. This is very close to −8.6.

[YOU TRY 3] Divide −12.8 by −7. Round the answer to the nearest thousandth.

Sometimes, a digit (or digits) in a decimal will repeat forever. For example, the fraction $\frac{5}{6}$ is equivalent to the decimal 0.8333..., where the 3 repeats forever. (We will learn how to write fractions as decimals in Section 5.6.) A decimal like 0.8333... is called a *repeating decimal*.

> **Definition**
>
> A **repeating decimal** is a decimal in which a digit or a group of digits repeats forever.
>
> Example: 0.8333...

A repeating decimal can be written in two ways:

<table>
<tr><td>1) Use the three dots at the end of the number.</td><td>or</td><td>2) Use a bar above the repeating digit or digits.</td></tr>
</table>

1) Use the three dots at the end of the number.

0.8333... means the 3 repeats forever.

or 2) Use a bar above the repeating digit or digits.

0.8333... can also be written as $0.8\overline{3}$.

1.4525252... can also be written as $1.4\overline{52}$.

Hint

Add your own details to the procedure for this objective.

Sometimes a quotient will be a repeating decimal.

EXAMPLE 4

Find $7.9 \div 3$. Give the exact answer and an approximation rounded to the nearest thousandth.

Solution

When we divide, notice that we must write extra zeros on the right end of the dividend, 7.9, because the remainder is not 0.

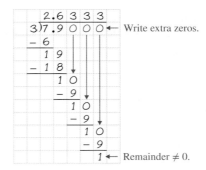

Also notice that the 3 in the quotient keeps repeating because the remainder of 1 keeps repeating. **This pattern will continue forever, so the remainder will never equal 0.**

We can stop dividing and write the exact answer as

$$7.9 \div 3 = 2.6333...\qquad \text{or} \qquad 7.9 \div 3 = 2.6\overline{3}$$

The dots mean the 3 will repeat forever.

The bar above the 3 means the 3 will repeat forever.

Rounded to the nearest thousandth, $7.9 \div 3 \approx 2.633$.

Check: $2.633 \times 3 = 7.899$. This is not *exactly* 7.9, but it is very close.

YOU TRY 4

Find $9.1 \div 6$. Give the exact answer and an approximation rounded to the nearest thousandth.

2 Divide a Number by a Decimal

Now we will learn how to divide a number by a decimal. That is, the *divisor* is a decimal.

Procedure Dividing a Number by a Decimal

1) Move the decimal point to the right end of the divisor. This makes the divisor a whole number. Count the number of decimal places you have moved.

2) Move the decimal point in the dividend the *same* number of places to the right. (If necessary, write in zeros.)

3) Write the decimal point in the quotient directly above the decimal point in the dividend. Then, divide.

4) Determine the sign of the quotient. If both numbers have the *same sign,* the quotient is *positive.* If the numbers have *different signs,* the quotient is *negative.*

EXAMPLE 5

Divide. $0.5\overline{)6.85}$

Solution

Start with the original problem and *move the decimal point one place to the right in the divisor* to change 0.5 to the whole number 5. Also, *move the decimal point in 6.85 one place to the right.*

W Hint

Are you writing out the example as you read it?

Divisor → $0.5\overline{)6.85}$ Make the divisor a whole number. → $05.\overline{)68.5}$ Move the decimal points one place to the right.

↑
Dividend

Divide. Rewrite the divisor as just 5.

$$
\begin{array}{r}
13.7 \\
5\overline{)68.5} \\
-5 \\
\hline
18 \\
-15 \\
\hline
35 \\
-35 \\
\hline
0
\end{array}
$$

Check:
$$
\begin{array}{r}
13.7 \\
\times\ 0.5 \\
\hline
6.85 \quad \checkmark
\end{array}
$$

Therefore, when we divide $0.5\overline{)6.85}$, the quotient is 13.7.

[YOU TRY 5]

Divide. $0.8\overline{)9.76}$

Why can we move the decimal points? Let's write $0.5\overline{)6.85}$ as a division problem in fraction form.

$$0.5\overline{)6.85} \quad \text{can be written as} \quad \frac{6.85}{0.5}$$

$$\frac{6.85}{0.5} = \frac{6.85}{0.5} \cdot \frac{10}{10} = \frac{68.5}{5}$$

Multiplying the numerator and denominator of a fraction by the same number gives us an equivalent fraction. Therefore,

$$\frac{6.85}{0.5} = \frac{68.5}{5} \quad \text{and} \quad \frac{68.5}{5} \quad \text{can be written as} \quad 5\overline{)68.5}$$

So, $0.5\overline{)6.85}$ is equivalent to $5\overline{)68.5}$. **Moving the decimal point one place to the right in the divisor and the dividend is the same as multiplying each number by 10.**

Note

Moving the decimal point to the right:

1) *one place* in the divisor and the dividend is like multiplying both of them by 10.

2) *two places* in the divisor and the dividend is like multiplying both of them by 100.

3) *three places* in the divisor and the dividend is like multiplying both of them by 1000.

Moving the decimal point the **same number of places** in the divisor and the dividend does *not* change the quotient.

EXAMPLE 6

Divide -116.4 by 0.012.

Solution

Set up the division problem, leaving out the negative, then *move the decimal point in the divisor,* 0.012, *three places to the right* so that it is at the end of the number. *Move the decimal point in the dividend three places to the right.* (This is like multiplying both numbers by 1000.) Write in zeros.

$$0.012\overline{)116.4} \quad \longrightarrow \quad 012.\overline{)116400.}$$

Move the decimal points three places to the right. Write the decimal point in the quotient directly above the decimal point in the dividend. Remember, the quotient of two numbers with different signs is negative.

Divide. $\quad \begin{array}{r} 9700. \\ 12\overline{)116400.} \end{array} \quad$ Check by multiplying.

The final answer is $-116.4 \div 0.012 = -9700$.

YOU TRY 6

Divide 124.6 by -0.014.

In the next example, we will divide a whole number by a decimal.

EXAMPLE 7   Divide. Round the answer to the nearest hundredth. $9 \div 2.6$

Solution

Set up the division problem. $2.6\overline{)9}$

Because the dividend, 9, does not contain a decimal point as it is written, begin by inserting the decimal point after the 9. Then, move the decimal point *one place to the right,* and divide.

 Hint

Are you using graph paper to do these division problems?

$$2.6\overline{)9.} \xrightarrow{\substack{\text{Move decimal} \\ \text{points one place} \\ \text{to the right.}}} 26\overline{)90.}$$

Insert the decimal point.

Write in one zero and put the decimal point in the quotient.

Divide.

```
            3. 4 6 1  ←
  2 6 ) 9 0 . 0 0 0
      − 7 8
        1 2 0
      − 1 0 4
          1 6 0
        − 1 5 6
              4 0
            − 2 6
              1 4
```

Because we are rounding to the *hundredths* place, carry out the division to the *thousandths* place. Round 3.461 to 3.46.

Therefore, $9 \div 2.6 \approx 3.46$.

[YOU TRY 7]   Divide. Round the answer to the nearest hundredth. $5 \div 2.3$

3 Use the Order of Operations with Decimals

We use the order of operations with decimals just like we did with whole numbers and fractions.

EXAMPLE 8   Simplify each expression using the order of operations.

a) $(-0.9)^2 - 4.2 \div (-7) + 5.3$ b) $2 + 10(1.78 - 8.6)$

Solution

a) $(-0.9)^2 - 4.2 \div (-7) + 5.3 = 0.81 - 4.2 \div (-7) + 5.3$ Evaluate exponents first.

$\qquad\qquad\qquad\qquad\qquad\quad = 0.81 - (-0.6) + 5.3$ Divide before adding and subtracting.

$\qquad\qquad\qquad\qquad\qquad\quad = 0.81 + 0.6 + 5.3$ Change subtraction to addition.

$\qquad\qquad\qquad\qquad\qquad\quad = 1.41 + 5.3$ Perform addition and subtraction from left to right.

$\qquad\qquad\qquad\qquad\qquad\quad = 6.71$ Add.

 Hint

Remember "Please Excuse My Dear Aunt Sally."

b) $2 + 10(1.78 - 8.6) = 2 + 10(-6.82)$ Perform operations in parentheses first.

$\qquad\qquad\qquad\qquad\quad = 2 + (-68.2)$ Multiply before adding.

$\qquad\qquad\qquad\qquad\quad = -66.2$ Add.

[YOU TRY 8] Simplify each expression using the order of operations.

a) $(-1.1)^2 + 5.4 \div (-6) - 1.29$　　　b) $9.6 + 100(4.38 - 7)$

ANSWERS TO [YOU TRY] EXERCISES

1) a) 8.47　b) −0.00157　　2) 0.7825　　3) 1.829　　4) exact: 1.51666... or $1.51\overline{6}$; approximation: 1.517
5) 12.2　　6) −8900　　7) 2.17　　8) a) −0.98　b) −252.4

Using Technology

Calculators can perform arithmetic operations, but they will give us a wrong answer if we tell the calculator to perform arithmetic operations that do not follow the order of operations.

First, calculate $4.4 \div 0.2 + 2 \times 0.75$ by hand, using the order of operations. To use a calculator, we must first perform the division operation by entering ④ . ④ ÷ ⓪ . ② ⼀. The display will likely show 22. Next, perform the multiplication by entering ② ⓧ ⓪ . ⑦ ⑤ ⼀ into the calculator. The display will likely show 1.5. Now, find the sum of these two results by entering ② ② ⼦ ① . ⑤ ⼀ into the calculator. Your final result should be 23.5.

Note: If your calculator has a parenthesis function, you could use the parenthesis buttons to tell the calculator how to make the calculation. In this case, you would enter ⼁ ④ . ④ ÷ ⓪ . ② ⼂ ⼦ ⼁ ② ⓧ ⓪ . ⑦ ⑤ ⼂ to get the correct result.

E Evaluate **5.5** Exercises　　Do the exercises, and check your work.

Objective 1: Divide a Decimal by a Whole Number

1) Explain, in your own words, how to divide a decimal by a whole number.

2) What do we do if we reach the end of the dividend and the remainder is not zero?

Divide.

3) $23.52 \div 8$　　　　　　4) $14.16 \div 6$

5) $9\overline{)46.08}$　　　　　　6) $5\overline{)21.85}$

7) $\dfrac{0.02112}{-4}$　　　　　8) $\dfrac{0.01932}{-7}$

9) $-147.2 \div 16$　　　　10) $-101.4 \div 13$

11) $\dfrac{-484.89}{-21}$　　　　12) $\dfrac{-806.52}{-26}$

13) $6\overline{)2.319}$　　　　　14) $8\overline{)4.836}$

15) $\dfrac{12.14}{4}$　　　　　16) $\dfrac{17.16}{8}$

17) $8\overline{)3.46}$　　　　　18) $4\overline{)2.07}$

19) Divide 7.5 by −6.　　　　20) Divide 49.4 by −5.

Divide. Round the answer to the nearest thousandth.

21) $8.1 \div 7$　　　　　　22) $23.5 \div 9$

23) Divide −9.8 by 3.　　　24) Divide −12.7 by 6.

25) $11\overline{)0.144}$　　　　26) $16\overline{)0.166}$

27) $\dfrac{2472.9}{-31}$　　　　28) $\dfrac{5358.2}{-78}$

Divide. Give the exact answer and an approximation rounded to the nearest thousandth.

29) $8.6 \div 6$　　　　　　30) $7.7 \div 3$

31) $\dfrac{-28.34}{3}$　　　　32) $\dfrac{-51.28}{9}$

33) $9\overline{)0.92}$　　　　　34) $6\overline{)0.56}$

35) $11\overline{)4.7}$　　　　　36) $11\overline{)8.1}$

Objective 2: Divide a Number by a Decimal

37) Explain, in your own words, how to divide a number by a decimal.

38) Is $0.29\overline{)7.283}$ equivalent to $29\overline{)728.3}$? Explain your answer.

Divide.

39) $0.4\overline{)4.48}$

40) $0.6\overline{)7.62}$

41) $0.8\overline{)5.48}$

42) $0.4\overline{)1.02}$

43) $0.645 \div (-0.43)$

44) $0.756 \div (-0.28)$

45) $\dfrac{-0.3552}{-9.6}$

46) $\dfrac{-0.4615}{-7.1}$

47) $-937.8 \div 0.18$

48) $-879.2 \div 0.14$

49) Divide 646.8 by 1.32.

50) Divide 486.2 by 1.43.

51) $2.5\overline{)38}$

52) $3.6\overline{)63}$

53) Divide -522 by (-0.04).

54) Divide -434 by (-0.02).

Divide. Round the answer to the nearest hundredth.

55) $-9 \div 3.2$

56) $-7 \div 1.5$

57) $5.7\overline{)3}$

58) $8.3\overline{)4}$

59) Divide -150.7 by -4.1.

60) Divide -287.4 by -3.3.

61) $\dfrac{0.26}{-0.15}$

62) $\dfrac{0.94}{-0.16}$

63) $0.008\overline{)7.309}$

64) $0.009\overline{)6.058}$

Objective 3: Use the Order of Operations with Decimals

Simplify each expression using the order of operations.

65) $(-1.5)^2 - 8.1 \div 9 + 2.7$

66) $(-1.3)^2 - 6.4 \div 4 + 8.3$

67) $5 + 10(1.73 - 2.6)$

68) $4 + 10(2.84 - 5.3)$

69) $9.7 - 0.8 \times 3.4 \div 2$

70) $1.5 - 0.3 \times 7.2 \div 0.6$

71) $3 - \dfrac{(2.4)^2}{-6}$

72) $7 - \dfrac{(2.1)^2}{-9}$

73) $0.1(1.7 - 0.94) \div (-0.002)$

74) $0.1(1.5 - 0.86) \div (-0.004)$

75) $160(0.025) + (-2.8) \div 0.7$

76) $200(0.075) + (-4.5) \div 0.3$

Solve each problem. For all answers involving money, give the final answer to the nearest cent.

77) Janine buys a package of diapers for $9.99. The package contains 31 diapers. Find the cost of each diaper.

78) Carlos is a waiter. He worked 7.5 hr and earned $137.25 in tips. How much did he earn per hour, in tips?

©DreamPictures/Blend Images LLC

79) During the 2016 regular season of the National Football League, Ezekiel Elliot of the Dallas Cowboys ran for 1631 yd in 15 games. Find the average number of yards he ran per game, rounded to the nearest tenth. (This is his total number of yards divided by the number of games.) (www.nfl.com)

80) A baseball player's batting average is the number of hits divided by the number of "at bats." In the Major League Baseball 2016 regular season, Anthony Rizzo of the Chicago Cubs had 170 hits in 583 at bats. What is his batting average rounded to the nearest thousandth? (www.mlb.com)

81) Razeena spends $36.86 for 14.4 gal of gas. What is the cost per gallon?

82) Three boxes of the same size are shipped to a restaurant. The total weight of the boxes is 37.5 lb. How much does each box weigh?

Mixed Exercises: Objectives 1–3

Divide.

83) $0.0037\overline{)4.81}$

84) $-23.76 \div 6$

85) $\dfrac{-47.79}{-8.1}$

86) $0.004\overline{)0.17}$

87) $840 \div (-4.2)$

88) $\dfrac{0.0065}{-1.3}$

Divide. Give the exact answer and an approximation to the nearest hundredth.

89) $5.23 \div 0.3$

90) $1.5\overline{)19}$

Divide. Round the answer to the nearest thousandth.

91) Divide 8 by -2.1.

92) $\dfrac{-2.5415}{-42.5}$

Simplify each expression using the order of operations.

93) $59.5 + (-2.8) \div (-0.7) - 63.5$

94) $-16.2(1.5) - (-0.09)^2$

Solve each problem. Give the final answer to the nearest cent.

95) At the end of the year, Trevor receives a bank statement that says he has paid a total of $17,247.36 for his mortgage over the last year. If he pays the same amount each month, what is his monthly mortgage payment?

96) Charise fills up her car's gas tank about once each week. Last month, she spent a total of $205.39 on gas. Find the average amount she spent per week. (The average per week is the total amount divided by the number of weeks.)

R Rethink

R1) Which example from this section would help you divide 1 by 3? Perform the division, and round the answer to the nearest thousandth.

R2) Which topics in this section were most difficult for you? Specifically, where did you have trouble while doing these exercises?

Putting It All Together

P Prepare

What are your objectives?

O Organize

How can you accomplish each objective?

What are your objectives?	How can you accomplish each objective?
1 Review the Concepts of Sections 5.1–5.5	• Understand what a decimal represents and how to round a decimal. • In your own words, summarize how to perform operations with decimals. • Complete the given examples on your own. • Complete You Trys 1–3.

W Work

Read the explanations, follow the examples, take notes, and complete the You Trys.

1 Review the Concepts of Sections 5.1–5.5

In this chapter, we have learned that a decimal is another way to represent a fraction with a denominator that is a power of 10.

$$0.7 = \frac{7}{10} \qquad 0.61 = \frac{61}{100} \qquad 0.837 = \frac{837}{1000}$$

It is important to understand the place values of the digits in the decimals so that we can understand how to read and write them as well as perform operations with decimals.

EXAMPLE 1

Write each decimal in words, then write it as a fraction or mixed number in lowest terms.

a) 0.43   b) 3.028

Solution

a) Read 0.43 as "forty-three hundredths." Therefore, $0.43 = \dfrac{43}{100}$.

hundredths place   numerator = 43   denominator = 100

b) Read 3.028 as "three and twenty-eight thousandths."

thousandths place   whole-number part   numerator = 28   denominator = 1000

> **W Hint**
>
> When we change a decimal to a fraction or a mixed number, we will write the answer in lowest terms.

As a mixed number, $3.028 = 3\dfrac{28}{1000}$. Is $\dfrac{28}{1000}$ in lowest terms? No!

$$\frac{28}{1000} = \frac{28 \div 4}{1000 \div 4} = \frac{7}{250}$$

Therefore, $3.028 = 3\dfrac{28}{1000} = 3\dfrac{7}{250}$.

[YOU TRY 1]

Write each decimal in words, then write it as a fraction or mixed number in lowest terms.

a) 0.91   b) 5.075

Often, we round decimals when working with money or when performing operations.

EXAMPLE 2

Round each number to the indicated place.

a) 7.3815 to the nearest thousandth   b) $19.6239 to the nearest cent

Solution

a) **Step 1:** $7.381|5$ Underline the digit in the *thousandths* place, and draw a vertical line after it.

Increase by 1.
5 or more

Step 2: $7.381|5$

Drop this digit.

7.3815 rounded to the nearest thousandth is 7.382.

b) When we are working with a money amount, like $19.6239, we round to the nearest *cent*, which is the nearest *hundredth*.

Step 1: $19.62|39$ Underline the digit in the *cent* place.

Keep this digit the same.
3 is less than 5.

Step 2: $19.62|39$

Drop these digits.

Rounded to the nearest cent, $19.6239 is $19.62. We can also say that $19.6239 \approx $19.62.

410   CHAPTER 5   **Signed Decimals**   www.mhhe.com/messersmith

Round each number to the indicated place.

a) 4.0683 to the nearest thousandth   b) $22.9657 to the nearest cent

Let's review adding, subtracting, multiplying, and dividing decimals.

EXAMPLE 3

Perform the indicated operations. If the remainder will never be 0 in a division problem, give the exact answer and an approximation of the answer rounded to the nearest hundredth.

a) -2.908×7.4 b) $0.147 \div 3$ c) $57.8 - 21.546$

d) $0.18\overline{)4.3}$ e) $(-6.7)^2 - 10(14.76 - 11.65)$

Solution

a) When we **multiply** decimals, we do *not* have to line up the decimal points.

$$\begin{array}{r} 2.908 \rightarrow \text{3 decimal places} \\ \times \quad 7.4 \rightarrow +\text{1 decimal place} \\ \hline 11632 \quad \text{4 decimal places in} \\ 20356 \quad \text{the answer} \\ \hline 21.5192 \end{array}$$

The factors have *different signs*, so the product is negative. $-2.908 \times 7.4 = -21.5192$

Hint

In your notes, summarize how to add, subtract, multiply, and divide decimals.

b) We are **dividing a decimal by a whole number:** $0.147 \div 3$. Write the decimal point in the quotient directly above the decimal point in the dividend. Then, divide.

$$\begin{array}{r} 0.049 \\ 3\overline{)0.147} \\ -12 \\ \hline 27 \\ -27 \\ \hline 0 \end{array}$$

So, $0.147 \div 3 = 0.049$.

Check by multiplying.

$$\begin{array}{r} 0.049 \\ \times \quad 3 \\ \hline 0.147 \end{array} ✓$$

c) To **subtract decimals,** write the numbers vertically so that the decimal points are lined up.

$$\begin{array}{r} 57.8 \\ -21.546 \end{array} \qquad \begin{array}{r} 9 \\ 7\,\cancel{10}\,10 \\ 57.8\cancel{0}\cancel{0} \leftarrow \text{Insert two zeros.} \\ -21.546 \\ \hline 36.254 \end{array}$$

The answer is 36.254.

To check, add $36.254 + 21.546$. The sum is 57.800 or 57.8. The answer is correct.

d) In this problem, $0.18\overline{)4.3}$, we are **dividing a number by a decimal.**

$$0.18\overline{)4.3} \xrightarrow[\substack{\text{Move the} \\ \text{decimal points} \\ \text{2 places to the} \\ \text{right.} \\ \text{This is the same} \\ \text{as multiplying} \\ \text{both numbers by} \\ \text{100.}}]{} 018\overline{)430.}$$

Write a zero in the dividend, and put the decimal point in the quotient.

Divide.

```
              2 3 . 8 8 8
        1 8 ) 4 3 0 . 0 0 0 ←── Write extra zeros to continue
            − 3 6                       dividing.
              7 0
            − 5 4
              1 6 0
            − 1 4 4
              1 6 0
            − 1 4 4
              1 6 0
            − 1 4 4
              1 6 ←── Remainder will never equal 0.
```

The exact answer is $23.\overline{8}$ Rounded to the nearest hundredth, $4.3 \div 0.18 \approx 23.89$.

Check: $23.89 \times 0.18 = 4.3002$. This is not *exactly* 4.3, but it is very close. (When we round a quotient, we will *not* get the exact dividend, but it should be close.)

e) $(-6.7)^2 - 10(14.76 - 11.65) = 44.89 - 10(14.76 - 11.65)$ Evaluate exponents first.

$$= 44.89 - 10(3.11)$$ Perform operations in parentheses.

$$= 44.89 - 31.1$$ Multiply before subtracting.

$$= 13.79$$ Subtract.

[YOU TRY 3] Perform the indicated operations. If the remainder will never be 0 in a division problem, give the exact answer and an approximation of the answer rounded to the nearest hundredth.

a) $42.1 + 138.976 + 2.05$

b) $100 \times (-0.03)^2 + 16 \times (-2.7)$

c) $6)\overline{0.231}$

d) $-91.45 \times (-4.1)$

e) $-2.29 \div 0.09$

ANSWERS TO [YOU TRY] EXERCISES

1) a) ninety-one hundredths; $\dfrac{91}{100}$ b) five and seventy-five thousandths; $5\dfrac{3}{40}$ 2) a) 4.068   b) $22.97

3) a) 183.126   b) −43.11   c) 0.0385   d) 374.945   e) exact: $-25.\overline{4}$; approximation: −25.44

Putting It All Together Exercises

 **E Evaluate**   Do the exercises, and check your work.

Objective 1: Review the Concepts of Sections 5.1–5.5

Write each fraction as a decimal.

1) $\dfrac{79}{10}$

2) $\dfrac{23}{1000}$

Write each decimal in words; then write it as a fraction or mixed number in lowest terms.

3) 0.31

4) 0.125

 5) 1.6

6) Write as a decimal: *two and eight hundredths.*

Round each number to the indicated place.

7) 831.562 to the nearest tenth

8) 74.0698 to the nearest thousandth

9) $33.575 to the nearest dollar

10) Explain, in your own words, how to add decimals.

Perform the indicated operations.

11) $527.92 - 82.38$

12) $0.001(6.7)$

13) $-0.0216 \div 0.8$

14) $\dfrac{135.45}{-15}$

15) $10,000(0.041)$

16) $-7.2 + 4(16.59 - 20.7)$

17) $-905.47 \times (-0.073)$

18) $216.58 + 97 + 36.9$

19) $4.216 + 387.5 + 29.96$

20) $(-1.4)^2$

21) $77 - 5.64(3.5)^2$

22) $2.5\overline{)160}$

23) $18\overline{)234.918}$

24) $53 - 28.09$

25) $10 - 10(0.092 + 0.37) + (-35) \div (-0.7)$

26) -81.6×7.42

Divide. Give an exact answer and an approximation to the nearest thousandth.

27) $\dfrac{43}{0.9}$

28) $7.7 \div 9$

29) $-0.08 \div 1.1$

30) $1.76\overline{)144}$

Divide. Round the answer to the nearest hundredth.

31) $31\overline{)7.2}$

32) $\dfrac{6.2}{-0.023}$

Solve each problem.

©Ingram Publishing

33) Nikos buys the following items to make Greek salad. How much did he spend if the tax on his purchase was $2.03?

2 cucumbers at $0.99 each

3 yellow peppers at $2.49 each

2.5 lb of tomatoes at $2.39 per lb

1.6 lb of red onions at $1.25 per lb

1.5 lb of feta cheese at $12.98 per lb

34) During the first three months of 2016, Facebook had approximately 1.090 billion daily active users. That number rose to 1.284 billion during the first three months of 2017. How many more daily active users were there in 2017? (www.facebook.com)

35) General admission tickets for a concert cost $38.50 each. The revenue from these tickets was $10,048.50. How many tickets were sold?

36) Laura earns $13.60 per hour. During a two-week period, she worked 38.5 hours. Determine the amount of her paycheck if $74.92 was deducted for taxes and $56.31 was deducted for health insurance.

37) Find the missing length.

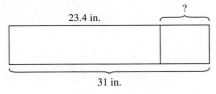

38) Find the perimeter of the triangle.

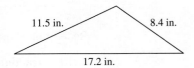

R Rethink

R1) Write an explanation, in your own words, of how to work with the decimal point when adding, subtracting, multiplying, and dividing numbers containing decimals. Include an example of each.

R2) Were you able to do these exercises without looking at the book or your notes? Which topics do you need to practice more?

5.6 Writing Fractions as Decimals

What are your objectives for Section 5.6?	How can you accomplish each objective?
1 Write a Fraction as a Decimal Using Division	• Write the procedure for **Writing a Fraction as a Decimal Using Division.** • Complete the given examples on your own. • Complete You Trys 1 and 2.
2 Write a Fraction as a Decimal Using Equivalent Fractions with a Denominator of 10, 100, etc.	• Write your own procedure for **Writing a Fraction as a Decimal Using an Equivalent Fraction with a Denominator of 10, 100, etc.** • Complete the given example on your own. • Complete You Try 3.
3 Compare Fractions and Decimals	• Know the three different ways to compare fractions and decimals: number line, written as fractions, or written as decimals. • Complete the given examples on your own. • Complete You Trys 4 and 5.

W Work

Read the explanations, follow the examples, take notes, and complete the You Trys.

We have learned that a decimal is another way to represent a fraction with a denominator that is a power of 10. For example,

$$\frac{3}{10} = 0.3 \qquad \frac{481}{100} = 4.81 \qquad -\frac{257}{1000} = -0.257$$

But how can we write other fractions, like $\frac{3}{4}$, as a decimal?

1 Write a Fraction as a Decimal Using Division

Recall that a fraction is one way to represent division. Therefore, $\frac{3}{4}$ means $3 \div 4$. **To write a fraction in decimal form, divide the numerator by the denominator.**

> **Procedure** Writing a Fraction as a Decimal Using Division
>
> To write a fraction as a decimal, divide the numerator by the denominator.

EXAMPLE 1

Write each fraction or mixed number as a decimal.

a) $\dfrac{3}{4}$ b) $-\dfrac{5}{8}$ c) $3\dfrac{4}{5}$

Solution

a) $\dfrac{3}{4}$ means $3 \div 4$. To perform this division, begin by putting a decimal point after the 3 and inserting a 0 so that we can divide 30 by 4. Then, complete the division process.

$$
\begin{array}{r}
0.75 \\
4\overline{)3.00} \\
-28 \\
\hline
20 \\
-20 \\
\hline
0
\end{array}
$$
← Write in zeros.

← Remainder = 0.

$\dfrac{3}{4} = 0.75$

> **W Hint**
>
> Remember that you can check your division using multiplication.

b) We will insert the negative sign in the final answer. $\dfrac{5}{8}$ means $5 \div 8$. Put a decimal point after the 5 and insert a 0 so that we can divide 50 by 8. Continue the division process.

$$
\begin{array}{r}
0.625 \\
8\overline{)5.000} \\
-48 \\
\hline
20 \\
-16 \\
\hline
40 \\
-40 \\
\hline
0
\end{array}
$$
← Write in zeros.

← Remainder = 0.

$-\dfrac{5}{8} = -0.625$

c) We can use two different methods to write $3\dfrac{4}{5}$ as a decimal.

Method 1: Remember, $3\dfrac{4}{5}$ means $3 + \dfrac{4}{5}$. Change $\dfrac{4}{5}$ to a decimal, then add 3.

$$
\begin{array}{r}
0.8 \\
5\overline{)4.0} \\
-40 \\
\hline
0
\end{array}
$$

$\dfrac{4}{5} = 0.8$, so $3\dfrac{4}{5} = 3 + \dfrac{4}{5} = 3 + 0.8 = 3.8$

Method 2: Write $3\dfrac{4}{5}$ as an improper fraction. Then, divide.

$$3\dfrac{4}{5} = \dfrac{19}{5}$$

$\dfrac{19}{5}$ means $19 \div 5$.

$$
\begin{array}{r}
3.8 \\
5\overline{)19.0} \\
-15 \\
\hline
40 \\
-40 \\
\hline
0
\end{array}
$$

$3\dfrac{4}{5} = 3.8$

[YOU TRY 1] Write each fraction or mixed number as a decimal.

a) $\dfrac{1}{4}$ b) $-\dfrac{3}{8}$ c) $2\dfrac{9}{25}$

Sometimes, we will never get a remainder of 0 when we divide.

EXAMPLE 2

Write $\frac{2}{3}$ as a decimal. Give the exact answer and an approximation rounded to the nearest thousandth.

Solution

$\frac{2}{3}$ means $2 \div 3$, so divide. In addition to giving the exact answer, we are asked to round the answer to the nearest *thousandth*. Therefore, carry out the division one more place, to the *ten-thousandths* place.

W Hint

This is similar to what you learned in Section 5.5.

$\frac{2}{3}$ means $2 \div 3$.

```
        0.6 6 6 6
    3)2.0 0 0 0  ← Write in zeros.
     - 1 8
         2 0
       - 1 8
           2 0
         - 1 8
             2 0
           - 1 8
               2  ← The remainder
                     will always be 2.
```

The *exact* answer is $\frac{2}{3} = 0.\overline{6}$.

The *approximation* to the nearest thousandth is $\frac{2}{3} \approx 0.667$.

[YOU TRY 2]

Write $\frac{5}{9}$ as a decimal. Give the exact answer and an approximation rounded to the nearest thousandth.

2 Write a Fraction as a Decimal Using Equivalent Fractions with a Denominator of 10, 100, etc.

If we can write a fraction as an equivalent fraction with a denominator of 10, 100, 1000, or another power of 10, then we can convert the fraction to a decimal.

EXAMPLE 3

Write each fraction as a decimal.

a) $\frac{2}{5}$ b) $-\frac{18}{25}$

Solution

a)   We can write $\frac{2}{5}$ with a denominator of 10: $\frac{2}{5} \cdot \frac{2}{2} = \frac{4}{10}$

Write $\frac{4}{10}$ as a decimal: 0.4.   Therefore, $\frac{2}{5} = 0.4$.

b) *Can we write* $-\dfrac{18}{25}$ *as a fraction with a denominator of* 10 *or* 100? Yes, we can write it with a denominator of 100.

$$-\frac{18}{25} \cdot \frac{4}{4} = -\frac{72}{100}$$

Write $-\dfrac{72}{100}$ as a decimal: -0.72. So, $-\dfrac{18}{25} = -0.72$.

[YOU TRY 3] Write each fraction as a decimal.

a) $\dfrac{4}{5}$ b) $-\dfrac{7}{20}$

Note

Not all fractions can be written as an equivalent fraction with a denominator that is a power of 10.

3 Compare Fractions and Decimals

We used number lines with fractions in Chapter 4, and we placed decimals on number lines in Sections 5.1 and 5.2. Now, let's compare decimals.

EXAMPLE 4

Use $=$, $>$, or $<$ to compare the numbers.

a) 0.8 _____ 0.2 b) -0.461 _____ -0.43

Solution

a)

0.8 is to the *right* of 0.2, so 0.8 *is greater than* 0.2: $0.8 > 0.2$

b) To compare -0.461 and -0.43, write -0.43 with the same number of decimal places as -0.461 by writing a zero at the end of -0.43: $-0.43 = -0.430$

Compare -0.461 and -0.430.

This is $-\dfrac{461}{1000}$.   ↑   ↑ This is $-\dfrac{430}{1000}$.

$-\dfrac{461}{1000}$ *is less than* $-\dfrac{430}{1000}$, so $-0.461 < -0.43$.

[YOU TRY 4] Use $=$, $>$, or $<$ to compare the numbers.

a) 0.5 _____ 0.6 b) -0.825 _____ -0.81

To compare fractions and decimals, we can think about where they appear on a number line. Or, we can write both numbers as fractions or both as decimals.

EXAMPLE 5 Use =, >, or < to compare the numbers.

a) -0.6 _____ $-\dfrac{1}{8}$ b) 0.75 _____ $\dfrac{3}{4}$ c) 1.4 _____ $1\dfrac{1}{5}$

Solution

a) Think about where -0.6 and $-\dfrac{1}{8}$ appear on a number line.

-0.6 is to the *left* of $-\dfrac{1}{8}$, so $-0.6 < -\dfrac{1}{8}$.

b) Let's change 0.75 to a fraction to compare it to $\dfrac{3}{4}$.

$$0.75 = \frac{75}{100} = \frac{75 \div 25}{100 \div 25} = \frac{3}{4}$$

$$0.75 = \frac{3}{4}$$

W Hint

Choose the most efficient way to compare numbers.

c) Let's change $1\dfrac{1}{5}$ to a decimal to compare it to 1.4.

$1\dfrac{1}{5}$ means $1 + \dfrac{1}{5}$. Change $\dfrac{1}{5}$ to a decimal: $\dfrac{1}{5} \cdot \dfrac{2}{2} = \dfrac{2}{10} = 0.2$

$1\dfrac{1}{5} = 1.2$. Compare 1.4 _____ 1.2.

     one and *four* tenths   one and *two* tenths

$1.4 > 1.2$, so $1.4 > 1\dfrac{1}{5}$.

[YOU TRY 5] Use =, >, or < to compare the numbers.

a) $-\dfrac{2}{3}$ _____ -0.3 b) $1\dfrac{9}{20}$ _____ 1.4 c) 0.24 _____ $\dfrac{6}{25}$

Note

In Example 5b, we could have changed $\dfrac{3}{4}$ to the decimal 0.75 to see that $0.75 = \dfrac{3}{4}$.

In Example 5c, we could have changed 1.4 to a mixed number to compare it to $1\dfrac{1}{5}$.

$$1.4 = 1\frac{4}{10} = 1\frac{2}{5} \qquad \text{Divide 4 and 10 by 2.}$$

Then, we can think of 1.4 _____ $1\dfrac{1}{5}$ as $1\dfrac{2}{5}$ _____ $1\dfrac{1}{5}$. We get $1\dfrac{2}{5} > 1\dfrac{1}{5}$ or $1.4 > 1\dfrac{1}{5}$.

Using Technology

We can use a calculator's division function to change a fraction to a decimal. In many cases, we must represent the displayed answer by indicating that there is a repeating decimal. In some cases, we may be asked to round our answer. For example,

convert $-\dfrac{10}{11}$ to a decimal using a calculator by entering $\boxed{+/-}\ \boxed{1}\ \boxed{0}\ \boxed{\div}\ \boxed{1}\ \boxed{1}\ \boxed{=}$
or $\boxed{1}\ \boxed{0}\ \boxed{+/-}\ \boxed{\div}\ \boxed{1}\ \boxed{1}\ \boxed{=}$ into the calculator. The display will likely show −0.909090909, indicating that there is a repeating decimal. To represent the displayed repeating decimal, we would write $-0.\overline{90}$ for our answer. However, if we are asked to round to the nearest hundredth, we would write −0.91 for the final answer.

E Evaluate **5.6** Exercises   Do the exercises, and check your work.

Objective 1: Write a Fraction as a Decimal Using Division

Use division to write each fraction or mixed number as a decimal.

1) $\dfrac{1}{2}$

2) $\dfrac{1}{4}$

3) $\dfrac{3}{5}$

4) $\dfrac{4}{5}$

5) $-\dfrac{1}{8}$

6) $-\dfrac{7}{8}$

7) $\dfrac{7}{16}$

8) $\dfrac{5}{16}$

9) $-2\dfrac{2}{5}$

10) $-1\dfrac{3}{4}$

11) $4\dfrac{1}{16}$

12) $9\dfrac{1}{8}$

13) When we use a 12-inch ruler to find the length of an object, the measurement often requires that we break up an inch into a fractional amount or a decimal equivalent. To help master this skill, first complete the following tables. Look for a pattern in both columns of the tables. Then using the ruler diagrams to the right, label the tick marks of one diagram with the appropriate fractions in lowest terms, and label the other with their decimal equivalents.

Fraction	Decimal Equivalent	Fraction	Decimal Equivalent
$\frac{1}{16}$	0.0625	$\frac{1}{8}$	0.125
$\frac{2}{16}=\frac{1}{8}$		$\frac{2}{8}=\frac{1}{4}$	
$\frac{3}{16}$		$\frac{3}{8}$	
$\frac{4}{16}=\frac{2}{8}=\frac{1}{4}$		$\frac{4}{8}=\frac{1}{2}$	
$\frac{5}{16}$		$\frac{5}{8}$	
$\frac{6}{16}=\frac{3}{8}$		$\frac{6}{8}=\frac{3}{4}$	
$\frac{7}{16}$		$\frac{7}{8}$	
$\frac{8}{16}=\frac{4}{8}=\frac{1}{2}$		$\frac{8}{8}=1$	
$\frac{9}{16}$			
$\frac{10}{16}=\frac{5}{8}$			
$\frac{11}{16}$			
$\frac{12}{16}=\frac{6}{8}=\frac{3}{4}$			
$\frac{13}{16}$			
$\frac{14}{16}=\frac{7}{8}$			
$\frac{15}{16}$			
$\frac{16}{16}=1$			

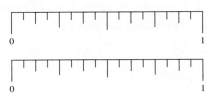

Use the results of Exercise 13 to write the decimal equivalent of the given measurements.

14) $1\frac{1}{4}$ in.　　　15) $1\frac{1}{8}$ in.　　　16) $2\frac{5}{16}$ in.

17) $4\frac{15}{16}$ in.　　　18) $2\frac{5}{8}$ in.

Write each fraction as a decimal. Give the exact answer and the approximation rounded to the nearest thousandth.

19) $\frac{1}{3}$　　　20) $\frac{1}{9}$　　　21) $-\frac{7}{9}$

22) $-\frac{8}{9}$　　　23) $\frac{5}{6}$　　　24) $\frac{1}{6}$

25) $\frac{29}{30}$　　　26) $\frac{8}{15}$　　　27) $-\frac{1}{11}$

28) $-\frac{10}{11}$

Objective 2: Write a Fraction as a Decimal Using Equivalent Fractions with a Denominator of 10, 100, etc.

29) What are two ways to write the fraction $\frac{17}{20}$ as a decimal?

30) Can we change $\frac{8}{13}$ to a decimal by writing it with a denominator that is a power of 10? Explain your answer.

Write each fraction or mixed number as a decimal by first getting a denominator of 10, 100, or 1000.

31) $\frac{3}{5}$　　　32) $\frac{1}{5}$　　　33) $-\frac{3}{4}$

34) $-\frac{1}{4}$　　　35) $-\frac{9}{20}$　　　36) $-\frac{31}{50}$

37) $\frac{16}{25}$　　　38) $\frac{17}{20}$　　　39) $\frac{189}{200}$

40) $\frac{7}{250}$　　　41) $-4\frac{1}{2}$　　　42) $-2\frac{3}{4}$

Objective 3: Compare Fractions and Decimals
Use =, >, or < to compare the decimals.

43) 0.4 ___ 0.5　　　44) 0.9 ___ 0.7

45) −2.5 ___ −2.4　　　46) −1.3 ___ −1.6

47) 0.35 ___ 0.350　　　48) 0.604 ___ 0.6040

49) 5.18 ___ 5.09　　　50) 9.42 ___ 9.06

51) −0.708 ___ −0.78　　　52) −0.39 ___ −0.309

53) −0.0500 ___ −0.05　　　54) −0.07 ___ −0.0700

55) 3.35 ___ 3.4　　　56) 1.2 ___ 1.19

Use =, >, or < to compare the numbers.

57) $\frac{1}{5}$ ___ 0.8　　　58) 0.6 ___ $\frac{1}{3}$

59) 0.8 ___ $\frac{1}{4}$　　　60) $\frac{2}{3}$ ___ 0.1

61) $-1\frac{7}{20}$ ___ −1.4　　　62) −4.3 ___ $-4\frac{6}{25}$

63) $-\frac{16}{25}$ ___ −0.64　　　64) −0.15 ___ $-\frac{3}{20}$

65) 2.67 ___ $2\frac{2}{3}$　　　66) 5.17 ___ $5\frac{1}{6}$

67) 3.004 ___ $3\frac{4}{500}$　　　68) $6\frac{3}{250}$ ___ 6.015

69) $0.\overline{2}$ ___ $\frac{2}{9}$　　　70) $1.\overline{6}$ ___ $1\frac{2}{3}$

Arrange each group of numbers in order from smallest to largest.

71) 4.26, $4\frac{3}{10}$, $4\frac{1}{5}$, 4.259　　　72) $5\frac{1}{2}$, 5.45, $5\frac{7}{10}$, 5.503

73) $2\frac{1}{8}$, 2.75, 2.7, $2\frac{1}{16}$　　　74) 3.68, $3\frac{1}{15}$, 3.6, $3\frac{1}{9}$

75) 0.97, $\frac{7}{8}$, $-\frac{5}{16}$, −0.3　　　76) $-\frac{3}{8}$, $\frac{15}{16}$, −0.8, 0.38

Mixed Exercises: Objectives 1–3
Write each fraction or mixed number as a decimal. Give the exact answer.

77) $1\frac{4}{5}$　　　78) $-\frac{19}{200}$

79) $-\frac{2}{11}$　　　80) $\frac{39}{8}$

81) $\frac{3}{500}$　　　82) $\frac{5}{9}$

83) $-\frac{41}{16}$　　　84) $-7\frac{3}{25}$

Write each fraction as a decimal. Give the exact answer and an approximation rounded to the nearest thousandth.

85) $\dfrac{7}{15}$

86) $\dfrac{3}{11}$

Use $=$, $>$, or $<$ to compare the numbers.

87) 1.50 _____ $1\dfrac{1}{2}$

88) -0.605 _____ -0.61

89) $-\dfrac{1}{10}$ _____ -0.01

90) $\dfrac{1}{1000}$ _____ 0.0001

91) $0.\overline{8}$ _____ $\dfrac{4}{5}$

92) -3.750 _____ $-3\dfrac{3}{4}$

Arrange each group of numbers in order from smallest to largest.

93) $-\dfrac{2}{3}, \dfrac{59}{100}, 0.7, -0.6$

94) $\dfrac{3}{4}, \dfrac{11}{16}, -0.5, -\dfrac{5}{8}$

R Rethink

R1) After completing the exercises, which procedure did you use most often to write a fraction as a decimal? Why?

R2) When would it be easier to understand what a fraction represents if it were written as a decimal?

R3) Summarize how to compare negative decimals. Include some examples.

5.7 Mean, Median, and Mode

P Prepare

O Organize

What are your objectives for Section 5.7?	How can you accomplish each objective?
1 Find the Mean	• Write the definition of a *mean* (*average*) in your own words. • Complete the given examples on your own. • Complete You Trys 1 and 2.
2 Find the Weighted Mean	• Write your own procedure for **Finding the Weighted Mean.** • Complete the given examples on your own. • Complete You Trys 3 and 4.
3 Find the Median	• Write a definition of *median*. • Write the procedure for **Finding a Median** in your own words. • Complete the given examples on your own. • Complete You Trys 5 and 6.
4 Find the Mode	• Write definitions of *mode* and *bimodal*. • Write the procedure for **Finding a Mode** in your own words. • Complete the given example on your own. • Complete You Try 7.

W Work   **Read the explanations, follow the examples, take notes, and complete the You Trys.**

In this section, we will learn about other ways that we use decimals in our everyday lives.

One way to analyze data or a list of numbers is to look for a **measure of central tendency.** This is a number that can be used to represent the entire list of numbers. In this section, we will learn about three measures of central tendency: the mean (or average), the median, and the mode. Let's start with the mean.

1 Find the Mean

What is the *mean,* or *average,* of a list of numbers?

Definition

The **mean (average)** is the sum of all the values in a list of numbers divided by the number of values in the list.

$$\text{Mean} = \frac{\text{Sum of all values}}{\text{Number of values}}$$

EXAMPLE 1

Four friends went bowling. Their scores were 183, 204, 162, and 195. Find the mean, or average, score.

Solution

To find the mean, add the scores and divide by 4, the *number* of scores.

$$\text{Mean} = \frac{183 + 204 + 162 + 195}{4} \quad \begin{array}{l}\text{Sum of the scores}\\ \hline \text{Number of scores}\end{array}$$

$$= \frac{744}{4} \qquad \text{Add.}$$

$$= 186 \qquad \text{Divide.}$$

The mean, or average, score was 186.

©Asia Images Group Pte Ltd/Alamy

Note

The mean will not necessarily be a value found in the original group of numbers!

[YOU TRY 1]

Tariq's driving times (in minutes) to work Monday through Friday last week were 21, 25, 27, 24, and 23. Find the mean, or average, driving time.

EXAMPLE 2

Estrella took six quizzes in her Psychology class this semester. Her scores were 95%, 82%, 88%, 93%, 91%, and 88%. What was her quiz average for the semester?

Solution

To find the average of her quiz scores, add the scores and divide by 6, the *number* of quizzes.

$$\text{Mean} = \frac{95 + 82 + 88 + 93 + 91 + 88}{6} \qquad \begin{array}{l}\text{Sum of the scores}\\ \text{Number of scores}\end{array}$$

$$= \frac{537}{6} \qquad\qquad \text{Add.}$$

$$= 89.5 \qquad\qquad \text{Divide.}$$

The average of Estrella's quiz scores was 89.5%.

> **[YOU TRY 2]** Enrique took four exams in his Political Science class this semester. His grades were 82%, 75%, 63%, and 69%. What was his exam average?

2 Find the Weighted Mean

Hint

How is this different from a "regular" mean?

If some values in a list of numbers appear more than once, then we can compute the *weighted mean* to represent that list of numbers. To find a **weighted mean,** we *weight* each value by multiplying it by the number of times it appears in the list. The number of times an item appears on the list is also called the **frequency.**

EXAMPLE 3

The table lists the high temperatures in Honolulu during May 2016 and the number of days that each was the high temperature. Find the weighted mean. (www.accuweather.com)

High Temperature, °F	Number of Days
81°	2
82°	2
83°	3
84°	8
85°	5
86°	2
87°	8
88°	1

Solution

The temperature is the value in the list, and the number of days is the number of times the temperature is counted, or the *weight*.

Multiply each temperature by the number of days that it was the high temperature. Then, find the weighted mean by taking the sum of all of these products and dividing by the total number of days. (Notice that this is 31 because May has 31 days.)

W Hint

When *you* are finding a weighted mean, make a table like this one to keep the information organized, neat, and easy to read.

High Temperature, °F	Number of Days	Product
81°	2	81 · 2 = **162**
82°	2	82 · 2 = **164**
83°	3	83 · 3 = **249**
84°	8	84 · 8 = **672**
85°	5	85 · 5 = **425**
86°	2	86 · 2 = **172**
87°	8	87 · 8 = **696**
88°	1	88 · 1 = **88**
Totals	**31**	**2628**

$$\text{Weighted mean} = \frac{2628}{31} \quad \begin{array}{l}\text{Sum of the products}\\ \text{Total number of days}\end{array}$$

$$\approx 84.8 \qquad \text{Divide and round to the nearest tenth.}$$

The weighted mean of the high temperatures in May 2016 was approximately 84.8°F.

YOU TRY 3

The table lists the number of students in each classroom at Prospect School and the number of classrooms containing that number of students. Find the weighted mean number of students per classroom. Round the answer to the nearest tenth.

Number of Students	Number of Classrooms
21	4
22	3
24	5
26	4
27	1

Have you ever wondered how a grade point average (GPA) is calculated? It is usually calculated using a weighted mean.

Let's look at how to find the weighted mean (or GPA) of a student's grades. The grade earned for the course is the value in the list, and the number of credits assigned to the course is the weight.

EXAMPLE 4

Bharavi's grades from last semester are listed in the table. Her school uses a 4-point scale so that an A is 4 points, a B is 3 points, a C is 2 points, a D is 1 point, and an F is 0 points. Find Bharavi's grade point average or GPA.

Course	Grade	Credits
Speech	C	3
Precalculus	A	5
Economics	B	3
Biology	B	4

Solution

The grade earned for the course is the value in the list, and the number of credits assigned to the course is the number of times the grade is counted, or the *weight*.

For each course, multiply the grade, in points, by the number of credits that course is worth. Then, find the weighted mean, or GPA, by taking the sum of all these numbers and dividing by the total number of credits.

Course	Grade	Grade, in Points	Credits	Points · Credits
Speech	C	2	3	$2 \cdot 3 = 6$
Precalculus	A	4	5	$4 \cdot 5 = 20$
Economics	B	3	3	$3 \cdot 3 = 9$
Biology	B	3	4	$3 \cdot 4 = 12$
Totals			15	47

$$\text{Weighted mean (GPA)} = \frac{47}{15} \quad \frac{\text{Total number of grade points}}{\text{Total number of credits}}$$

$$= 3.1\overline{3} \quad \text{Divide.}$$

Round $3.1\overline{3}$ to the nearest hundredth. Bharavi's GPA was 3.13.

[YOU TRY 4] Determine Wesley's GPA if he received this grade report. Round the answer to the nearest hundredth.

Course	Grade	Credits
French	C	4
Photography	A	3
U.S. History	D	3
Weight Training	B	1
Drawing	A	2

---

3 Find the Median

Another way to represent a list of numbers with a single number is to use the median.

Definition

The **median** of an ordered list of numbers is the middle number.

To find the median of a list of values, follow this three-step process.

Procedure Finding the Median of a List of Values

Step 1: Arrange the values from lowest to highest.

Step 2: Determine whether there is an even or odd number of values in the list.

Step 3: If there is an *odd number* of values, the median is the *middle number*. If there is an *even number* of values, the median is the *mean (average) of the middle two numbers*.

EXAMPLE 5

Here are the ages of Desmond's cousins: 16, 20, 14, 11, 22, 18, 15, 26, 21. Find the median age.

©Thomas Northcut/Getty Images

Solution

Follow the three-step process to find the median.

Step 1: Arrange the values from lowest to highest:
11, 14, 15, 16, 18, 20, 21, 22, 26

Step 2: Determine whether there is an even or odd number of values in the list. There are 9 values in the list; therefore, there is an *odd* number of values.

Step 3: Since there is an *odd* number of values, the median is the middle value, 18.

$$\underbrace{11,\ 14,\ 15,\ 16,}_{\substack{\text{First four}\\\text{numbers}}}\ \underset{\substack{\uparrow\\\text{Middle}\\\text{number}}}{18,}\ \underbrace{20,\ 21,\ 22,\ 26}_{\substack{\text{Last four}\\\text{numbers}}}$$

The median age of Desmond's cousins is 18.

[YOU TRY 5]

Here are the estimates for installing a new garage door at the O'Reillys' house: $385, $290, $350, $320, $340. Find the median estimate.

EXAMPLE 6

Jake finds the following airfares for a round-trip flight from Los Angeles to San Francisco: $155.40, $126.85, $119.38, $131.46, $179.22, $142.03. Find the median airfare.

Solution

Follow the three-step process to find the median.

Step 1: Arrange the values from lowest to highest:

$119.38, $126.85, $131.46, $142.03, $155.40, $179.22

Step 2: Determine whether there is an even or odd number of values in the list. There are 6 values in the list; therefore, there is an *even* number of values.

Step 3: Since there is an *even* number of values, the median is the *mean* of the middle two numbers.

$$\underbrace{\$119.38,\ \$126.85,}_{\substack{\text{First two}\\\text{numbers}}}\ \overbrace{\$131.46,\ \$142.03,}^{\text{Two middle numbers}}\ \underbrace{\$155.40,\ \$179.22}_{\substack{\text{Last two}\\\text{numbers}}}$$

$$\text{Median} = \frac{\$131.46 + \$142.03}{2} = \$136.745$$

Round the answer to the nearest cent, $136.75. The median airfare is $136.75.

[YOU TRY 6] Simone is a real estate agent and is compiling data on home sales. Here are the selling prices of the last 10 homes sold in her town: $265,400, $320,000, $318,500, $299,000, $305,900, $296,800, $332,000, $274,000, $260,000, $340,000. What was the median selling price?

4 Find the Mode

Another measure of central tendency is the *mode*.

Definition

The **mode** is the number that appears most frequently in a list of numbers.

Procedure Finding the Mode of a List of Numbers

Arrange the numbers from lowest to highest, and underline the numbers that are repeated.

1) If one number appears more often than the others, then that number is the **mode.**

2) If there are two numbers that appear most often in a list, then the list has two modes and is said to be **bimodal.** The numbers that appear most often are the modes.

3) The list has **no mode** if no number appears more often than any other number.

EXAMPLE 7

Find the mode for each list of numbers.

a) 40, 90, 70, 20, 50, 20, 60, 10, 80, 50, 30, 20

b) $1.05, $1.19, $1.09, $0.99, $1.19, $1.29, $1.17, $1.09

c) 67, 43, 118, 59, 96, 104, 80

Solution

a) If the list is long, it is helpful to arrange the numbers from lowest to highest and underline the numbers that are repeated:

$$10, \underline{20}, \underline{20}, \underline{20}, 30, 40, \underline{50}, \underline{50}, 60, 70, 80, 90$$

The *mode* is 20 because it is the number that appears most often.

b) Write the numbers from smallest to largest, and underline the numbers that are repeated:

$$\$0.99, \$1.05, \underline{\$1.09}, \underline{\$1.09}, \$1.17, \underline{\$1.19}, \underline{\$1.19}, \$1.29$$

Since $1.09 and $1.19 each appear twice, the list is *bimodal*. The *modes* are $1.09 and $1.19.

c) Writing the numbers in order, we get 43, 59, 67, 80, 96, 104, 118.

No numbers appear more often than any others, so there is *no mode*.

YOU TRY 7

Find the mode for each list of numbers.

a) 25, 15, 40, 75, 35

b) $2.49, $2.79, $2.50, $2.49, $2.59, $2.59, $2.69, $2.59

c) 9.3, 7.8, 10.2, 8.5, 7.8, 8.1, 9.4, 8.1, 8.7

ANSWERS TO [YOU TRY] EXERCISES

1) 24 min   2) 72.25%   3) 23.6 students   4) 2.62   5) $340   6) $302,450
7) a) no mode   b) The mode is $2.59.   c) The list is bimodal. The modes are 7.8 and 8.1.

Using Technology

We can use a calculator to find the mean (or average) of a list of numbers by first finding the sum of all the values and then dividing by the number of values in the list. For example, find the mean value of 127, 96, 85, and 116. Notice that we have four values in our list. First find the sum of the values by entering [1][2][7][+][9][6][+][8][5][+][1][1][6][=]. The display will show 424. Next we divide the sum by 4 by pressing [÷][4][=]. The display will show 106, which is the mean of our list of numbers.

E Evaluate **5.7** Exercises   Do the exercises, and check your work.

Objective 1: Find the Mean

1) What is another word for the *mean* of a list of numbers?

2) Explain, in your own words, how to find the mean of a list of numbers.

Solve each problem.

 3) The ages of the employees at a bakery are 43, 54, 16, 17, 23, and 21. Find the mean age.

4) The numbers of people in Pilates classes are 14, 19, 12, 11, 18, and 16. Find the mean number of people.

5) The number of laps Gerhard swam over a 5-day period was 50, 40, 40, 50, and 30. Find the mean number of laps.

6) The number of pounds Melinda lost each of the past seven months was 9, 8, 8, 6, 5, 7, and 6. Find the mean number of pounds.

©RuslanDashinsky/Getty Images

7) Scott's latest scores on *Halo: Reach* were 11,350, 10,670, 11,020, 12,690, 12,410, 11,880, and 12,160.

8) Over the past six months, Alma's daughter sent or received the following numbers of text messages each month: 3214, 3107, 3350, 3422, 3069, and 3386. Find the average number of texts per month.

9) In the six games of the 2011 NBA Championship Series against the Miami Heat, Dirk Nowitzki of the Dallas Mavericks scored 27 pts, 24 pts, 34 pts, 21 pts, 29 pts, and 21 pts, respectively. Find the mean number of points he scored per game in this series. (In basketball, this is called the *points-per-game* or *ppg*.) (www.nba.com)

10) See Exercise 9. The numbers of points scored in each championship game by Dwyane Wade of the Miami Heat were 22, 36, 29, 32, 23, and 17. Find the mean number of points he scored per game in this series. (www.nba.com)

11) Jagoda looked at several apartments, and these were the rents per month: $525, $480, $600, $570, $542, and $556. Find the average rent.

---

428   CHAPTER 5   **Signed Decimals**   www.mhhe.com/messersmith

12) The last several times Bjorn has filled up his car's gas tank, it cost: $47.20, $48.00, $51.35, $49.70, and $52.80. Find the average cost of filling up the tank.

13) Opening weekend ticket sales for the *Toy Story* movies were as follows: *Toy Story:* $29.1 million, *Toy Story 2:* $57.4 million, and *Toy Story 3:* $110.3 million. What was the mean revenue from ticket sales? (http://boxofficemojo.com)

14) The *Toy Story* movies were shown in the following numbers of theaters the first weekend of release: *Toy Story:* 2457, *Toy Story 2:* 3236, and *Toy Story 3:* 4028. What was the mean number of theaters? Round to the nearest whole number. (http://boxofficemojo.com)

Find the quiz average given the following scores. Round to the nearest tenth of a percent.

15) 79%, 71%, 73%, 64%

16) 85%, 92%, 81%, 87%

17) Adileh's exam scores are 75%, 61%, and 63%. What is her exam average? Round the answer to the nearest tenth.

18) Toby's exam scores are 80%, 84%, and 72%. Round the answer to the nearest tenth.

Objective 2: Find the Weighted Mean

Find the weighted mean. Round the answer to the nearest tenth, where necessary.

19)

Family Size	Number of Families
3	2
4	5
5	3
6	2

20)

Houses Sold per Month	Number of Months
1	3
2	6
3	2
5	1

21)

Cars Washed per Hour	Frequency
0	2
14	7
15	10
16	12
18	11

22)

Books Read per Year	Frequency
0	8
4	9
6	15
10	10
12	9

23) The table shows the cost of an adult ticket at several movie theaters and the number of theaters charging that price. Find the weighted mean.

Ticket Price	Number of Theaters
$5.50	2
$7.00	3
$9.00	1
$9.50	5
$10.00	8
$11.00	2

24) The table shows each amount that employees contributed to a going-away gift and the number of employees who contributed that amount. Find the weighted mean.

Amount	Number of Employees
$5.00	2
$8.00	1
$10.00	9
$12.00	2
$15.00	5
$20.00	3

Find the GPA for each student with the following grade report. Let A = 4 points, B = 3 points, C = 2 points, D = 1 point, and F = 0 points. Round the answer to the nearest hundredth.

25)

Course	Grade	Credits
Study Skills Strategies	B	2
Fundamentals of Writing	C	3
Fundamentals of Reading	A	3
Intro to Child Care	B	3
Aerobic Fitness	A	1

26)

Course	Grade	Credits
Intermediate Algebra	B	4
English Composition	A	3
Drafting	A	2
Sociology	C	3
Cardio Kickboxing	B	1

27)

Course	Grade	Credits
World History	C	3
Anatomy and Physiology	F	4
Precalculus	D	5
French	C	4

28)

Course	Grade	Credits
Automotive Engine Design	C	4
Automotive Electrical Systems	D	4
Technical Math	C	3
Speech	F	3

29)

Course	Grade	Credits
Art History	B	3
Website Design	A	3
Graphic Design II	B	4
Botany	C	4

30)

Course	Grade	Credits
Basic Phlebotomy	A	3
Medical Terminology	A	4
Noninvasive EKG	B	2
Speech	B	3

31)

Course	Grade	Credits
Child Psychology	C	3
Earth Science	C	4
Painting	B	2
English	A	3

32)

Course	Grade	Credits
Electronics	A	4
Algebra	B	5
Commercial Wiring	B	3
Technical Writing	C	3

Objective 3: Find the Median

33) What is the difference between the *mean* of a list of numbers and the *median* of the list?

34) Are the mean and the median of a list of numbers *always, sometimes,* or *never* the same value?

35) If a list has an *odd* number of numbers, how do you find the median?

36) If a list has an *even* number of numbers, how do you find the median?

Find the median of each list of numbers.

37) The cost of a manicure: $24, $20, $25, $28, $15

38) The cost of a facial: $60, $65, $58, $70, $72

©jakubzak/Getty Images

39) The ages of the members of a softball team: 21, 25, 20, 19, 24, 22, 25, 23, 20, 18, 24, 23, 19

40) Number of years of experience: 17, 12, 18, 11, 3, 5, 12, 2, 11

41) Amount Evelyn spent on coffee each week: $16, $14, $13, $10, $15, $17

42) Amount Stanislav spent on textbooks and supplies in previous semesters: $430, $380, $500, $440, $470, $320

43) The attendance at football games: 72,140, 73,632, 75,109, 76,008, 76,150, 75,413, 76,124, 76,235

44) Employees' salaries: $47,540, $49,316, $44,700, $46,215, $45,990, $48,625, $43,180, $48,900

45) The number of calls made to 911 per hour over a 24-hour period: 1291, 1056, 805, 639, 540, 522, 593, 781, 794, 836, 916, 1062, 1085, 964, 951, 1147, 1002, 1228, 1290, 1053, 997, 1178, 1260, 1359

46) The number of customers, per hour, at a 24-hour convenience store: 31, 17, 16, 16, 19, 52, 59, 68, 73, 75, 61, 63, 68, 60, 52, 51, 49, 67, 55, 51, 40, 40, 30, 27

47) The Bertagnoli family's last eight doctor bills: $74.00, $92.00, $120.00, $135.00, $97.00, $103.00, $84.00, $76.00

©Ingram Publishing

48) Hotel prices for Julia's trip to Florida: $169.00, $129.00, $109.00, $120.00, $139.00, $150.00, $130.00, $158.00

Objective 4: Find the Mode

49) What is the *mode* of a list of numbers?

50) Does every list of numbers have a mode?

Find the mode of each list of numbers. If the list has no mode or is bimodal, then say so.

51) $5.00, $4.80, $4.90, $5.00, $4.95, $5.00, $5.30, $5.00

52) $7.30, $9.10, $8.50, $9.10, $9.20, $8.70, $9.10, $9.10

53) 28.3, 28.7, 27.4, 29.0, 27.5, 28.6

54) 61.8, 60.5, 59.9, 60.2, 61.3, 59.4

55) 75, 71, 78, 71, 82, 80, 79, 78, 76

56) 41, 46, 46, 50, 49, 42, 53, 47, 42

57) The heights of the members of a college women's gymnastics team, in inches: 58, 61, 57, 59, 60, 62, 58, 59, 60, 59, 58, 59

©2009 Jupiterimages Corporation

58) The number of notebooks Levy bought each of the previous seven semesters: 4, 3, 5, 2, 3, 4, 3

59) Prices for the same laptop at different stores: $409.99, $399.99, $379.99, $389.99, $389.99

60) Prices for the same television at different stores: $549.99, $559.99, $539.99, $549.99, $569.99

61) The numbers of women job-sharing at several companies: 4, 8, 10, 16, 0, 2, 6

62) The numbers of cabins at several state parks: 16, 22, 18, 28, 34, 25, 20

63) The number of patients visiting a free clinic each day over the last week: 102, 114, 96, 114, 120, 102, 85

64) The number of students in each section of a PreAlgebra class: 30, 24, 27, 23, 23, 29, 31, 24, 28, 24, 23

65) The percent of students at each of twelve schools receiving financial aid: 65%, 80%, 85%, 70%, 80%, 60%, 80%, 85%, 70%, 85%, 65%, 60%

56) The number of children in each apartment in a building: 4, 2, 0, 3, 2, 1, 0, 5

57) The number of cars serviced each day for the last six days; 14, 11, 12, 9, 10, 15

58) The percent of customers who paid their restaurant bill with a credit card over each of the last seven days: 49%, 53%, 61%, 40%, 63%, 79%, 56%

Mixed Exercises: Objectives 1–4

69) The numbers of gold medals won by the United States in the Summer Olympics from 1968 in Mexico through 2016 in Rio de Janeiro are 45, 33, 34, 83, 41, 38, 44, 37, 36, 36, 46, and 46. Find the mean, median, and mode. (www.olympic.org)

©Datacraft Co Ltd/Getty Images

70) The numbers of runs scored by the Detroit Tigers each season from 2006 to 2016 are 822, 887, 821, 743, 751, 787, 726, 796, 757, 689, and 750. Find the mean, median, and mode. Round the answer to the nearest tenth, if necessary. (http://detroit.tigers.mlb.com)

71) Find the mean, median, and mode of Ms. Darvish's students' test grades: 74%, 85%, 61%, 73%, 92%, 85%, 40%, 78%, 93%, 85%, 73%, 72%, 54%, 66%, 70%, 89%, 81%, 52%, 76%, 65%

72) Find the mean, median, and mode of Mr. Cuesta's students' test grades: 82%, 59%, 73%, 70%, 88%, 65%, 91%, 94%, 67%, 73%, 82%, 39%, 75%, 68%, 80%, 73%, 54%, 82%, 59%

73) The table shows the number of milligrams (mg) of caffeine in a 12-oz serving of each soft drink. Find the average amount of caffeine.
(www.energyfiend.com)

Soft Drink	mg of Caffeine
Coca-Cola Classic	35
Pepsi-Cola	38
RC Cola	45
Mountain Dew	54
Barq's Root Beer	23

74) The table shows the percent of the recommended daily calcium in a 1-cup serving of different types of milk. Find the average percent of calcium.

Type of Milk	Percent of Recommended Daily Calcium
Whole Milk (cow)	30
Soy	20
Rice	30
Coconut	45
Almond	30

For Exercises 75 and 76, find the GPA. Let A = 4 points, B = 3 points, C = 2 points, D = 1 point, and F = 0 points. Round the answer to the nearest hundredth.

75)

Course	Grade	Credits
Architectural Design	B	4
Architectural Drafting	A	2
Calculus I	A	5
Art History	A	3

76)

Course	Grade	Credits
Technical Writing	B	3
Managerial Accounting	A	4
Computer Accounting	C	2
Finite Math	A	3

R Rethink

R1) You have divided a number by 2 many times now. What statement can you make regarding the quotient?

R2) Where do you often use measures of central tendency? Does one measure of central tendency make sense for certain data sets but not others? Why?

R3) Write a paragraph comparing and contrasting mean, median, and mode.

5.8 Solving Equations Containing Decimals

P Prepare

What are your objectives for Section 5.8?

O Organize

How can you accomplish each objective?

1 Solve an Equation Using One Property of Equality	• Review the properties of equality in Chapter 2, if necessary. • Be sure you can perform operations with decimals. • Complete the given examples on your own. • Complete You Trys 1 and 2.
2 Solve an Equation by Combining the Properties of Equality	• Review the steps for solving a linear equation in Chapter 2. • Complete the given example on your own. • Complete You Try 3.
3 Solve an Equation by First Eliminating the Decimals	• Know how many places to move the decimal point when you multiply a number by a power of 10. • Learn the procedure for **Eliminating Decimals from an Equation.** • Complete the given example on your own. • Complete You Try 4.
4 Solve Applied Problems Involving Decimals	• Review the **Five Steps for Solving Applied Problems** in Chapter 2. • Complete the given example on your own. • Complete You Try 5.

W Work

Read the explanations, follow the examples, take notes, and complete the You Trys.

1 Solve an Equation Using One Property of Equality

To solve $x - 4 = 9$, we use one property of equality: the addition property of equality. (The solution is $x = 13$.) We can solve some equations containing decimals using just one property of equality.

EXAMPLE 1

Solve each equation, and check the solution.

a) $x + 2.3 = 1.8$ b) $4.2 = k - 0.7$

Solution

a)   We use the *subtraction property of equality* to solve this equation.

$$x + 2.3 = 1.8$$
$$x + 2.3 - 2.3 = 1.8 - 2.3 \qquad \text{Subtract 2.3 from each side.}$$
$$x = -0.5 \qquad \text{Subtract.}$$

Check: $x + 2.3 = 1.8$

$-0.5 + 2.3 = 1.8$ Substitute -0.5 for x.

$1.8 = 1.8$ Add.

The solution is $x = -0.5$.

b) Notice that the variable is on the right side of the equation $4.2 = k - 0.7$. Use the *addition property of equality* to solve for k.

$$4.2 = k - 0.7$$

$$4.2 + 0.7 = k - 0.7 + 0.7 \qquad \text{Add 0.7 to each side.}$$

$$4.9 = k \qquad \text{Add.}$$

Check: $4.2 = k - 0.7$

$4.2 = 4.9 - 0.7$ Substitute 4.9 for k.

$4.2 = 4.2$ Subtract.

The solution is $k = 4.9$.

$\left[\text{YOU TRY 1}\right]$ Solve each equation, and check the solution.

a) $m + 7.1 = 4.3$ b) $0.8 = v - 5.9$

We use the *division property of equality* to solve $3y = 15$. Divide both sides of the equation by 3 to get the solution $y = 5$. We can use the division property of equality to solve some equations containing decimals.

EXAMPLE 2   Solve each equation, and check the solution.

a) $-2p = 1.3$ b) $-36 = -4.5n$

Solution

a) We use the *division property of equality* to solve $-2p = 1.3$.

$$-2p = 1.3$$

$$\frac{-2p}{-2} = \frac{1.3}{-2} \qquad \text{Divide each side by } -2.$$

$$p = -0.65 \qquad \text{Simplify.}$$

W Hint

Look at the earlier sections of this chapter if you need to review operations with decimals.

Check: $-2p = 1.3$

$-2(-0.65) = 1.3$ Substitute -0.65 for p.

$1.3 = 1.3$ Multiply.

The solution is $p = -0.65$.

b) The variable is on the right side of the equation $-36 = -4.5n$, so we will divide both sides by -4.5 to solve for n.

$$-36 = -4.5n$$

$$\frac{-36}{-4.5} = \frac{-4.5n}{-4.5} \qquad \text{Divide each side by } -4.5.$$

$$8 = n \qquad \text{Simplify.}$$

Check: $-36 = -4.5n$

$\qquad -36 = -4.5(8)$ Substitute 8 for n.

$\qquad -36 = -36$ Multiply.

Therefore, $n = 8$.

2 Solve an Equation by Combining the Properties of Equality

W Hint

Write out the steps for solving an equation.

Recall the steps for solving an equation from earlier chapters. The same steps apply to solving equations containing decimals.

EXAMPLE 3

Solve each equation.

a) $0.8t - 3.1 = 3.7$ b) $5(w + 0.28) = 8 - w$

Solution

a) To solve this equation, we have to use more than one property of equality. What do we do first? We must get the variable on one side of the equal sign and get the constants on the other side. So, start by adding 3.1 to each side of the equation.

$$0.8t - 3.1 = 3.7$$
$$0.8t - 3.1 + 3.1 = 3.7 + 3.1 \qquad \text{Add 3.1 to each side.}$$
$$0.8t = 6.8 \qquad \text{Simplify.}$$
$$\frac{0.8t}{0.8} = \frac{6.8}{0.8} \qquad \text{Divide both sides by 0.8.}$$
$$t = 8.5 \qquad \text{Simplify.}$$

The check is left to the student. The solution is $t = 8.5$.

b) What is the first step for solving $5(w + 0.28) = 8 - w$? Distribute the 5 to eliminate the parentheses.

$$5(w + 0.28) = 8 - w$$
$$5w + 1.4 = 8 - w \qquad \text{Distribute.}$$

There are variables and constants on both sides of the equal sign. We must get the variables on one side and the constants on the other side.

$$5w + w + 1.4 = 8 - w + w \qquad \text{Add } w \text{ to each side.}$$
$$6w + 1.4 = 8 \qquad \text{Combine like terms.}$$
$$6w + 1.4 - 1.4 = 8 - 1.4 \qquad \text{Subtract 1.4 from each side.}$$
$$6w = 6.6 \qquad \text{Subtract.}$$
$$\frac{6w}{6} = \frac{6.6}{6} \qquad \text{Divide both sides by 6.}$$
$$w = 1.1 \qquad \text{Simplify.}$$

Therefore, $w = 1.1$. The check is left to the student.

Solve each equation.

a) $4k - 1.9 = 12.1$ b) $-3.9 - n = 6(n + 0.75)$

3 Solve an Equation by First Eliminating the Decimals

Remember that we can solve an equation containing fractions by first multiplying the equation by the least common denominator (LCD) to eliminate the fractions. For example, we can solve $\frac{1}{6}x + \frac{1}{3} = \frac{1}{2}$ like this:

$$\frac{1}{6}x + \frac{1}{3} = \frac{1}{2}$$

$$6\left(\frac{1}{6}x + \frac{1}{3}\right) = 6\left(\frac{1}{2}\right) \qquad \text{Multiply both sides by 6, the LCD.}$$

$$x + 2 = 3 \qquad \text{Distribute.}$$

$$x = 1 \qquad \text{Subtract 2 from each side.}$$

We can do something similar to eliminate decimals from an equation.

Recall that multiplying a number by a power of 10 moves the decimal point in the number. For example,

$10(7.4) = 74$	$100(9.06) = 906$	$1000(8.152) = 8152$
Multiplying by 10 moves the decimal point one place.	Multiplying by 100 moves the decimal point two places.	Multiplying by 1000 moves the decimal point three places.

We can use this fact about multiplying by powers of 10 to eliminate decimals from an equation.

Procedure Eliminating Decimals from an Equation

To eliminate the decimals from an equation, multiply both sides of the equation by the smallest power of 10 that will eliminate all decimals. Then, solve using the usual steps.

EXAMPLE 4

Solve $0.2c + 0.18 = -0.5$.

Solution

The number containing a decimal place farthest to the right is 0.18. The 8 is in the **hundredths** place. So, multiply both sides of the equation by **100** and distribute to eliminate all of the decimals.

$$0.2c + 0.18 = -0.5$$

$$100(0.2c + 0.18) = 100(-0.5) \qquad \text{Multiply both sides of the equation by 100.}$$

$$20c + 18 = -50 \qquad \text{Distribute.}$$

W Hint

Remember to write out all of the steps on your paper as you are reading the example.

Now that all decimals have been eliminated, we solve as usual.

$$20c + 18 - 18 = -50 - 18 \quad \text{Subtract 18 from each side.}$$
$$20c = -68 \quad \text{Subtract.}$$
$$\frac{20c}{20} = \frac{-68}{20} \quad \text{Divide both sides by 20.}$$
$$c = -3.4 \quad \text{Simplify.}$$

Therefore, $c = -3.4$. The check is left to the student.

$\left[\textbf{YOU TRY 4}\right]$ Solve $0.15y + 0.4 = 0.28$.

---

4   Solve Applied Problems Involving Decimals

EXAMPLE 5

Write an equation and solve.

Gavin sells home theater systems and earns $12.40 per hour plus commission. One week, his commission was $291.87, and his total earnings were $663.87. How many hours did Gavin work?

Solution

Step 1: **Read** the problem carefully, and identify what we are being asked to find.

We must determine the number of hours Gavin worked.

Step 2: **Choose a variable** to represent the unknown.

$$x = \text{the number of hours Gavin worked}$$

W Hint

Review the Five Steps for Solving Applied Problems in Chapter 2, if necessary.

The amount Gavin earned from his hourly wage is $12.40x$.

Step 3: **Translate** the information that appears in English into an algebraic equation. Let's break up the problem slowly and restate it in our own words.

The amount earned from his hourly wage *plus* the amount of commission *equals* Gavin's total earnings.

Let's write this as an equation.

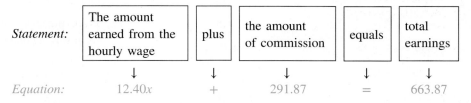

Statement:	The amount earned from the hourly wage	plus	the amount of commission	equals	total earnings
	↓	↓	↓	↓	↓
Equation:	12.40x	+	291.87	=	663.87

The equation is $12.40x + 291.87 = 663.87$.

Step 4: **Solve** the equation.

$$12.40x + 291.87 = 663.87$$

$$12.40x + 291.87 - 291.87 = 663.87 - 291.87 \qquad \text{Subtract 291.87 from each side.}$$

$$12.40x = 372 \qquad \text{Subtract.}$$

$$\frac{12.40x}{12.40} = \frac{372}{12.40} \qquad \text{Divide both sides by 12.40.}$$

$$x = 30 \qquad \text{Simplify.}$$

Step 5: **Check** the answer, and **interpret** the meaning of the solution as it relates to the problem.

Gavin worked 30 hours.

To check, multiply his hourly wage by 30 hours and add the amount of the commission:

$$\underset{\substack{\text{Amount earned} \\ \text{from hourly wage}}}{\$12.40(30)} \quad + \quad \underset{\substack{\text{Amount of} \\ \text{commission}}}{\$291.87} \quad = \quad \$372.00 + \$291.87 = \underset{\text{Total earnings}}{\$663.87} \checkmark$$

[YOU TRY 5]

Write an equation and solve.

Latoya sells advertising space on a website. She earns $10.30 per hour plus commission. One week, her commission was $240.68, and her total earnings were $601.18. How many hours did Latoya work?

ANSWERS TO [YOU TRY] EXERCISES

1) a) $m = -2.8$ b) $v = 6.7$ 2) a) $d = -0.85$ b) $r = 5$
3) a) $k = 3.5$ b) $n = -1.2$ 4) $y = -0.8$ 5) 35 hours

E Evaluate **5.8** Exercises   Do the exercises, and check your work.

Objective 1: Solve an Equation Using One Property of Equality

Solve each equation and check the solution.

1) $b - 0.6 = 0.8$ 2) $r + 0.9 = -0.3$

 3) $y + 1.2 = 4.7$ 4) $p - 5.1 = 1.8$

5) $-0.94 = k - 0.89$ 6) $0.43 = v + 0.61$

7) $c - 3.4 = -3.6$ 8) $p + 5.9 = 8.2$

9) $7 = m + 4.6$ 10) $-9 = t - 3.7$

11) $2a = 1.8$ 12) $3p = 2.1$

13) $-5u = 6.5$ 14) $-8t = -9.6$

15) $-66 = -1.1r$ 16) $-72 = 0.9w$

17) $0.3b = 5.4$ 18) $-0.2v = -7.8$

19) $-1.1 = -0.5n$ 20) $-0.64 = 0.4z$

Objective 2: Solve an Equation by Combining the Properties of Equality

21) Explain, in your own words, the steps for solving an equation.

22) Is solving an equation containing decimals different from solving one without decimals?

Solve each equation.

23) $0.2g - 0.3 = 0.5$ 24) $0.5f - 4.3 = -0.8$

25) $0.9c - 1.3 = -1.3$ 26) $0.8k - 0.5 = -0.5$

27) $-0.9 = 0.5n - 1.1$ 28) $1.3 = 0.4m + 0.5$

29) $5.6v - 21.9 = 3.2v + 38.1$

30) $7.9y - 58.2 = 6.3y + 13.8$

31) $2(b + 8.1) = b + 3.7$ 32) $4(z + 3.8) = 3z + 2.1$

33) $0.3(2p + 1) = p - 2$ 34) $0.2(3h + 4) = h - 5$

Objective 3: Solve an Equation by First Eliminating the Decimals

35) Explain, in your own words, how to eliminate the decimals from an equation.

36) Kavar multiplies the equation $0.4n + 5 = 1.7$ by 10 and gets $4n + 5 = 17$. What did he do wrong?

Solve each equation by first eliminating the decimals.

37) $0.2m + 5.7 = 9.3$ 38) $0.5x - 4.9 = 2.6$

39) $0.5 - 0.4n = 0.2$ 40) $0.5 - 0.2t = 0.15$

41) $4.9b + 1.3 = 5.8b + 3.1$

42) $5.7n + 1.2 = 6.3n + 5.4$

43) $1.4k + 0.73 = 1.8k + 1.61$

44) $1.3y + 0.67 = 1.7y + 2.31$

Objective 4: Solve Applied Problems Involving Decimals

Write an equation and solve.

45) One month, Phoebe earned a commission of $592.74 as well as $18.70 per hour. If her total income was $3322.94, how many hours did she work?

46) One week, Rick's commission was $183.22 and he also earned $11.90 per hour. If his total income was $647.30 that week, how many hours did he work?

47) A rectangular piece of poster board has an area of 7.5 ft². Find the length if it is 2.5 ft wide.

48) A rectangular table has an area of 14 ft². Find the length if it is 3.5 ft wide.

49) One side of a triangular sign is 8.4 in. long. Of the remaining two sides, one side is 5.1 in. longer than the other. If the perimeter of the triangle is 27.9 in., how long is each of the unknown sides?

50) A triangular flag has a perimeter of 51.3 in. One side is 18.5 in. long. Of the two remaining sides, one side is 4.2 in. shorter than the other. How long is each of the remaining sides?

Mixed Exercises

This section contains expressions and equations with integers, fractions, and decimals. Look at each exercise and determine whether it is an expression or an equation. If it is an expression, simplify it by combining like terms. If it is an equation, solve it.

51) $-6p + 7 = 19$ 52) $r - \dfrac{3}{8} - \dfrac{10}{7}r + \dfrac{1}{9}$

53) $11m + 2m - 17m$ 54) $3v - 4.61 = 2.11$

55) $18 - \dfrac{3}{5}a = \dfrac{1}{5}a - 30$ 56) $11 = -\dfrac{y}{8}$

57) $2.3h - 5.8 - 8.4h + 2.6$

58) $\dfrac{3}{8}w = 18$ 59) $11.5 = -2.5c$

60) $14x - y + 5(x + 8) - 3y + 1$

R Rethink

R1) How is solving an equation containing decimals similar to solving an equation containing fractions? How are they different?

R2) Write a paragraph describing how to solve all linear equations. Include information on how to solve equations containing decimals and fractions.

5.9 Square Roots and the Pythagorean Theorem

P Prepare

P Prepare

O Organize

What are your objectives for Section 5.9?	How can you accomplish each objective?
1 Approximate Square Roots	• Refresh your memory on **The First 12 Perfect Square Roots,** and write them in your notes. • Write a procedure for approximating a square root by following Example 1. • Complete You Try 1.
2 Find an Unknown Length in a Right Triangle	• Write the definition of a *hypotenuse* and *legs* of a right triangle in your notes. • Understand and memorize the **Pythagorean Theorem.** • Write the formula for **Finding an Unknown Side of a Right Triangle** in your notes. • Complete the given example on your own. • Complete You Try 2.
3 Apply the Pythagorean Theorem	• Be able to identify when a word problem requires use of the Pythagorean theorem. • Complete the given example on your own. • Complete You Try 3.

W Work

Read the explanations, follow the examples, take notes, and complete the You Trys.

In Section 1.8, we first learned how to find the square root of a number. For example,

$$\sqrt{9} = 3 \qquad \text{because} \qquad 3^2 = 9$$

We can represent this geometrically with a square.

3 in.

3 in.

Area = 3 in. · 3 in. = 9 in^2

We say that 9 is a *perfect square* and that $\sqrt{9}$ is a *perfect square root* because $\sqrt{9} = 3$. Here is a list of the first 12 perfect square roots.

> **Summary** The First 12 Perfect Square Roots
>
> $\sqrt{1} = 1 \qquad \sqrt{16} = 4 \qquad \sqrt{49} = 7 \qquad \sqrt{100} = 10$
>
> $\sqrt{4} = 2 \qquad \sqrt{25} = 5 \qquad \sqrt{64} = 8 \qquad \sqrt{121} = 11$
>
> $\sqrt{9} = 3 \qquad \sqrt{36} = 6 \qquad \sqrt{81} = 9 \qquad \sqrt{144} = 12$
>
> Note: $\sqrt{0} = 0$

1 Approximate Square Roots

If a number is *not* a perfect square, its square root will *not* be a whole number. We can approximate its value using perfect squares that we *do* know.

EXAMPLE 1

Approximate each square root to the nearest tenth, and plot it on a number line.

a) $\sqrt{5}$ b) $\sqrt{22}$

Solution

a) The number 5 is *not* a perfect square. Let's ask ourselves two questions:

1) What is the largest perfect square that is *less than* 5?   **4**

2) What is the smallest perfect square that is *greater than* 5?   **9**

Listing these numbers from smallest to largest, we get

$$4 \qquad 5 \qquad 9$$
$$\uparrow \qquad\qquad\qquad \uparrow$$
$$\text{Perfect square} \qquad \text{Perfect square}$$
$$\sqrt{4} = 2 \qquad\qquad \sqrt{9} = 3$$

Because 5 *is between* 4 and 9, $\sqrt{5}$ *is between* $\sqrt{4}$ and $\sqrt{9}$. In order from smallest to largest, we get

$$\sqrt{4} = 2 \qquad \sqrt{5} = ? \qquad \sqrt{9} = 3$$

$\sqrt{5}$ must be between 2 and 3. Because 5 is closer to 4 than to 9, $\sqrt{5}$ will be closer to $\sqrt{4}$ than to $\sqrt{9}$. So, $\sqrt{5}$ will be closer to 2 than to 3. Let's see whether 2.2 is a good approximation of $\sqrt{5}$.

If $\sqrt{5} \approx 2.2$, then $(2.2)^2 \approx 5$.

$$(2.2)^2 = 2.2 \times 2.2 = 4.84 \qquad \text{Close to 5 but } less\ than\ 5$$

Let's see whether 2.3 is a *better* approximation of $\sqrt{5}$.

If $\sqrt{5} \approx 2.3$, then $(2.3)^2 \approx 5$.

$$(2.3)^2 = 2.3 \times 2.3 = 5.29 \qquad \text{Close to 5 but } more\ than\ 5$$

Since 4.84 is closer to 5 than 5.29, the better approximation is 2.2.

$$\sqrt{5} \approx 2.2$$

b) The number 22 is *not* a perfect square. Ask ourselves,

1) What is the largest perfect square that is *less than* 22?   **16**

2) What is the smallest perfect square that is *greater than* 22?   **25**

List the numbers from smallest to largest.

$$16 \qquad 22 \qquad 25$$
$$\uparrow \qquad\qquad\qquad \uparrow$$
$$\text{Perfect square} \qquad \text{Perfect square}$$
$$\sqrt{16} = 4 \qquad\qquad \sqrt{25} = 5$$

Because 22 *is between* 16 and 25, $\sqrt{22}$ *is between* $\sqrt{16}$ and $\sqrt{25}$. In order from smallest to largest, we get

$$\sqrt{16} = 4 \qquad \sqrt{22} = ? \qquad \sqrt{25} = 5$$

$\sqrt{22}$ *is between* 4 and 5. Because 22 is closer to 25 than to 16, $\sqrt{22}$ will be closer to 5 than to 4. Let's try 4.6 as an approximation of $\sqrt{22}$.

If $\sqrt{22} \approx 4.6$, then $(4.6)^2 \approx 22$.

$$(4.6)^2 = 4.6 \times 4.6 = 21.16 \qquad \text{Close to 22 but \textit{less than} 22}$$

Let's see whether 4.7 is a better approximation.

If $\sqrt{22} \approx 4.7$, then $(4.7)^2 \approx 22$.

$$(4.7)^2 = 4.7 \times 4.7 = 22.09 \qquad \text{This is closer to 22 than the other approximation.}$$

Therefore, $\sqrt{22} \approx 4.7$.

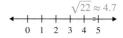

$\sqrt{22} \approx 4.7$

0 1 2 3 4 5

[**YOU TRY 1**] Approximate each square root to the nearest tenth, and plot it on a number line.

a) $\sqrt{13}$ b) $\sqrt{59}$

We can use a calculator to approximate square roots.

Using Technology

A regular or scientific calculator will have a $\sqrt{\ }$ key or $\sqrt{x}$ key. To approximate $\sqrt{5}$ and $\sqrt{22}$ from Example 1 on a calculator, follow the steps here. Depending on your calculator, you may not have to press the = key.

To find $\sqrt{5}$, press 5 $\sqrt{x}$. Answer: 2.236067977
    Round this to 2.2.
To find $\sqrt{22}$, press 22 $\sqrt{x}$. Answer: 4.69041576
    Round this to 4.7.

The long answer that the calculator gives us is also an approximation.

2 Find an Unknown Length in a Right Triangle

We have learned that a right triangle contains a right, or 90°, angle. Right triangles have special properties that other triangles do not. We can label a right triangle as shown in the figure.

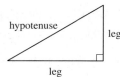

hypotenuse   leg

leg

The side opposite the right angle is the longest side of the triangle and is called the **hypotenuse.** The other two sides are called the **legs.** The *Pythagorean theorem* states a relationship between the lengths of the sides of a right triangle.

Property Pythagorean Theorem

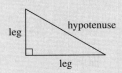

For any **right** triangle,

$$(\text{leg})^2 + (\text{leg})^2 = (\text{hypotenuse})^2$$

We can use the Pythagorean theorem to find an unknown side of a right triangle. From the Pythagorean theorem, we get these two formulas.

Formulas Finding an Unknown Side of a Right Triangle

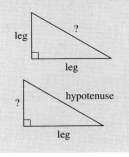

1) To find the **hypotenuse** of a right triangle, use this formula:

$$\text{hypotenuse} = \sqrt{(\text{leg})^2 + (\text{leg})^2}$$

2) To find a **leg** of a right triangle, use this formula:

$$\text{leg} = \sqrt{(\text{hypotenuse})^2 - (\text{known leg})^2}$$

 The Pythagorean theorem and the given formulas apply *only* to **right** triangles.

EXAMPLE 2

Find the length of the missing side. Give the exact answer and the answer to the nearest tenth, if appropriate.

a)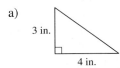

b)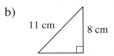

Solution

a) First ask yourself, *"Which part of the right triangle is unknown?"* The side *across from* the right angle is unknown, and that is the *hypotenuse*. Use the formula for finding the hypotenuse.

$$\begin{aligned}
\text{hypotenuse} &= \sqrt{(\text{leg})^2 + (\text{leg})^2} \\
&= \sqrt{(3)^2 + (4)^2} \qquad \text{Substitute the values for the legs.} \\
&= \sqrt{9 + 16} \qquad 3^2 = 9;\ 4^2 = 16 \\
&= \sqrt{25} \qquad \text{Add.} \\
&= 5 \qquad \text{Evaluate } \sqrt{25}.
\end{aligned}$$

The length of the hypotenuse is 5 in.

Hint

Can you think of a quick way to check your work when finding the hypotenuse?

 Hint

Can you think of a quick way to check your work when finding a leg of a right triangle?

b) Ask yourself, *"Which part of the right triangle is unknown?"* The unknown side is *next to* the right angle, so it is a *leg*. Use the formula for finding an unknown leg.

$$\text{leg} = \sqrt{(\text{hypotenuse})^2 - (\text{known leg})^2}$$

Identify the hypotenuse and the leg. Remember, the hypotenuse is the side *across from* the right angle.

$$\text{hypotenuse} = 11 \text{ cm} \qquad \text{leg} = 8 \text{ cm}$$

$$\text{leg} = \sqrt{(\text{hypotenuse})^2 - (\text{known leg})^2}$$
$$\text{leg} = \sqrt{(11)^2 - (8)^2} \qquad \text{Substitute the values for the hypotenuse and leg.}$$
$$= \sqrt{121 - 64} \qquad 11^2 = 121; \ 8^2 = 64$$
$$= \sqrt{57} \qquad \text{Subtract.}$$
$$\approx 7.5 \qquad \text{Round to the nearest tenth.}$$

The length of the leg is *exactly* $\sqrt{57}$ cm, and it is *approximately* 7.5 cm.

[YOU TRY 2] Find the length of the missing side. Give the exact answer and the answer to the nearest tenth, if appropriate.

a)
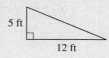
5 ft

12 ft

b)
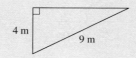
4 m

9 m

---

3 Apply the Pythagorean Theorem

The Pythagorean theorem can be used in many ways.

EXAMPLE 3

A wire is attached to the top of a pole as shown here. Find the length of the wire.

Solution

Notice that the pole, the wire, and the ground form a right triangle. Ask yourself, *"Which part of the triangle is unknown?"* The unknown side is *across from* the right angle, so this is the *hypotenuse*. Use the formula for finding the hypotenuse.

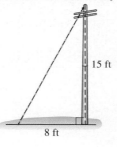
15 ft

8 ft

$$\text{hypotenuse} = \sqrt{(\text{leg})^2 + (\text{leg})^2}$$
$$= \sqrt{(8)^2 + (15)^2} \qquad \text{Substitute the values of the legs.}$$
$$= \sqrt{64 + 225} \qquad 8^2 = 64; \ 15^2 = 225$$
$$= \sqrt{289} \qquad \text{Add.}$$
$$= 17 \qquad \sqrt{289} = 17$$

The wire is 17 ft long.

YOU TRY 3

A computer screen is 13 in. long and 8 in. wide. Find the exact length of the diagonal and the approximation to the nearest tenth.

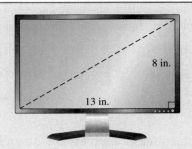

8 in.

13 in.

ANSWERS TO YOU TRY EXERCISES

1) a) 3.6; $\sqrt{13} \approx 3.6$ b) 7.7; $\sqrt{59} \approx 7.7$

0 1 2 3 4 5   0 1 2 3 4 5 6 7 8 9

2) a) 13 ft   b) exact: $\sqrt{65}$ m; approximation: 8.1 m   3) exact: $\sqrt{233}$ in.; approximation: 15.3 in.

Using Technology

We can use a calculator to help us find an unknown leg of a triangle using the formula leg $= \sqrt{(\text{hypotenuse})^2 - (\text{known leg})^2}$. Remember that we must perform the exponent operations before we subtract and take the square root. To instruct the calculator to do this, we must use the parenthesis keys.

Suppose we have a right triangle with a hypotenuse of length 13 mm and a known leg length of 5 mm. To find the length of the missing leg, we enter $\boxed{1}\boxed{3}\boxed{y^x}\boxed{2}\boxed{=}\boxed{-}\boxed{(}\boxed{5}\boxed{y^x}\boxed{2}\boxed{)}\boxed{=}\boxed{\sqrt{}}$ into the calculator. The display screen will show 12 for the result. This means that the length of the unknown leg is 12 mm.

E Evaluate 5.9 Exercises   Do the exercises, and check your work.

Objective 1: Approximate Square Roots

1) Write down the first four perfect squares and their square roots.

2) What makes a number a perfect square?

Approximate each square root to the nearest tenth and plot it on a number line.

3) $\sqrt{2}$ 0 1 2 3 4 5

4) $\sqrt{3}$ 0 1 2 3 4 5

5) $\sqrt{8}$ 0 1 2 3 4 5

6) $\sqrt{6}$ 0 1 2 3 4 5

7) $\sqrt{20}$ 0 1 2 3 4 5

8) $\sqrt{12}$ 0 1 2 3 4 5

9) $\sqrt{10}$ 0 1 2 3 4 5

10) $\sqrt{27}$ 0 1 2 3 4 5 6

11) $\sqrt{45}$ 4 5 6 7 8 9

12) $\sqrt{66}$ 4 5 6 7 8 9

Use a calculator and its square root key to approximate the following square roots to the nearest thousandth.

13) $\sqrt{20}$

14) $\sqrt{12}$

15) $\sqrt{45}$

16) $\sqrt{66}$

17) $\sqrt{157}$

18) $\sqrt{300}$

Objective 2: Find an Unknown Length in a Right Triangle

19) Write the formula to find the leg of a right triangle.

20) Write the formula to find the hypotenuse of a right triangle.

Find the length of the missing side. Give the exact answer and the answer to the nearest tenth, if appropriate.

21)
3 km
5 km

22)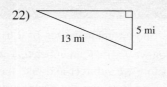
13 mi
5 mi

23)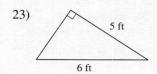
5 ft
6 ft

24)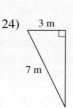
3 m
7 m

25)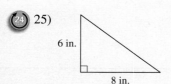
6 in.
8 in.

26)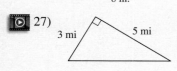
12 dm
5 dm

27)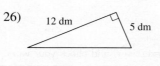
3 mi
5 mi

28)
3 km
2 km

29)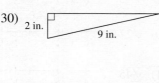
4 cm
10 cm

30)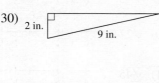
2 in.
9 in.

31)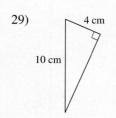
16 cm
20 cm

32)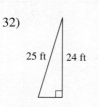
25 ft
24 ft

Use a calculator to find the unknown side length. Approximate your answer to the nearest thousandth.

33)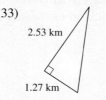
2.53 km
1.27 km

34)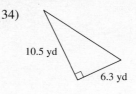
10.5 yd
6.3 yd

35)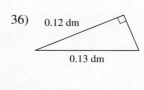
1.6 m
3.5 m

36)
0.12 dm
0.13 dm

Objective 3: Apply the Pythagorean Theorem

37) Does the Pythagorean theorem apply to any type of triangle?

38) Answer *true* or *false*: $\sqrt{3^2 + 4^2} = 3 + 4$.

Solve each problem. Give the exact answer and an answer approximated to the nearest tenth, if appropriate.

39) Trang flies his kite using 20 m of kite string. How high above Trang is his kite when he uses all 20 m of his kite string and he is standing 15 m away from directly below the kite?

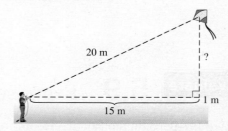
20 m
?
15 m
1 m

40) A flat-screen TV is 38 in. wide and 13 in. high. Find the length of the diagonal.

41) In the middle of a baseball game, a player tried to steal second base. The catcher threw the ball from home base to second base in time to get the player out. How far did the catcher have to throw the ball?

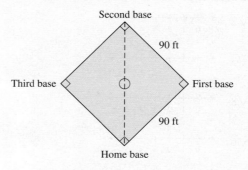
Second base
90 ft
Third base
First base
90 ft
Home base

42) Find the height of an equilateral triangle with side lengths of 6 dm. Give both the exact height and an approximation rounded to the nearest tenth.

43) Ethan and Trevin have a tin-can phone with a string of length 9 yd. In order for the phone to work, the string has to be tight. If Ethan is in a tree house 4 yd off the ground, how far does Trevin have to be from the base of the tree to make the phone work? Assume Ethan and Trevin are the same height.

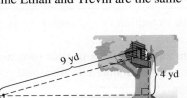

44) During a rescue exercise, a ladder truck parks 15 m from the base of a building. The truck must place its ladder at a fourth-story window, 15 m above the base of the ladder. How far should the ladder be extended to reach the window?

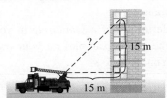

Mixed Exercises: Objectives 1–3

Approximate each square root to the nearest tenth and plot it on a number line.

45) $\sqrt{15}$

46) $\sqrt{32}$

Use a calculator and its square root key to approximate the following square roots to the nearest thousandth.

47) $\sqrt{17}$

48) $\sqrt{59}$

Find the length of the missing side. Give the exact answer and the answer to the nearest tenth, if appropriate.

49)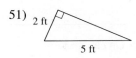

50)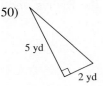

51)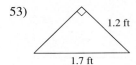

52)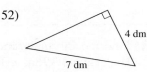

Use a calculator to find the unknown side length. Approximate your answer to the nearest thousandth.

53)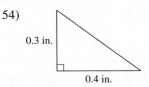

54)

Solve each problem. Give the exact answer and an answer approximated to the nearest tenth, if appropriate.

55) A ladder 13 ft long is leaning against a wall, forming a right triangle. The bottom of the ladder is 5 ft away from the wall. How far off the ground does the top of the ladder touch the wall?

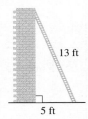

56) A news helicopter travels directly from the news station to a warehouse fire in a neighboring city. The news van leaves the same location but must travel a rectangular path through city blocks to get to the fire. The van first travels 7 km south and then 3 km east. How much further than the helicopter did the news van travel to get to the scene of the fire?

R Rethink

R1) After completing these exercises and taking a closer look at Exercise 41, what general statement could you make about a diagonal of a square? How could you find the length of a diagonal?

R2) When solving applications involving geometry, explain how you begin to solve the problem.

5.10 Circles, Spheres, Cylinders, and Cones

What are your objectives for Section 5.10?	How can you accomplish each objective?
1 Find the Radius and Diameter of a Circle	• Write the definition of a *circle* in your own words, draw an example, and note its main characteristics. • Understand what a *radius* and a *diameter* are. • Write the formulas relating the **Diameter and Radius of a Circle** in your notes. • Complete the given examples on your own. • Complete You Trys 1 and 2.
2 Find the Circumference of a Circle and Understand What π Represents	• Write the definition of *circumference* in your own words. • Write the definition of π in your own words. • Understand the decimal and fractional **Approximations for π.** • Write the formulas for determining the **Circumference of a Circle** in your notes. • Complete the given examples on your own. • Complete You Trys 3 and 4.
3 Find the Area of a Circle	• Understand how the area of a circle is derived from using information known about a parallelogram. • Write the formula for determining the **Area of a Circle** in your notes. • Write the definition of a *semicircle* in your notes. • Complete the given examples on your own. • Complete You Trys 5–8.
4 Find the Volume and Surface Area of Solids	• Write the definitions for *volume* and *surface area* in your own words. • Learn the formulas in the table. • Complete the given examples on your own. • Complete You Trys 9–11.

W Work Read the explanations, follow the examples, take notes, and complete the You Trys.

1 Find the Radius and Diameter of a Circle

Definition

A **circle** is a two-dimensional, or flat, figure in which all points are the same distance from the fixed center point.

Example: Each point on this circle is the same distance from the center.

(Continued)

The **radius,** *r*, is the distance from the center of the circle to any point on the circle.

The **diameter,** *d*, is the distance across the circle passing through the center.

Notice that the diameter is *twice* the length of the radius. (We can also say that the radius is half the length of the diameter.)

W Hint

If you memorize only one of the formulas, you can use it to find the other value.

Formula Diameter and Radius of a Circle

1) diameter = 2 · radius   or $d = 2 \cdot r$ or $d = 2r$

2) radius = $\dfrac{\text{diameter}}{2}$ or $r = \dfrac{d}{2}$

EXAMPLE 1   Find the diameter of a circle with a radius of 3 in.

Solution

The diameter is *two times* the radius.

$$d = 2r$$
$$d = 2(3 \text{ in.}) \qquad \text{Substitute the value.}$$
$$d = 6 \text{ in.}$$

[YOU TRY 1]   Find the diameter of a circle with a radius of 11 m.

EXAMPLE 2   Find the radius of a circle with a diameter of 15 cm.

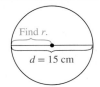

Solution

The radius is *half* the diameter.

$$r = \frac{d}{2}$$

$$r = \frac{15 \text{ cm}}{2} \qquad \text{Substitute the value.}$$

$$r = \frac{15}{2} \text{ cm or } 7\frac{1}{2} \text{ cm or } 7.5 \text{ cm}$$

We can write the answer in any of these forms.

$\left[\text{YOU TRY 2}\right]$ Find the radius of a circle with a diameter of 21 in.

$d = 21$ in.

2 Find the Circumference of a Circle and Understand What π Represents

We know that the perimeter of a figure is the distance around the figure. This is true for circles as well. The perimeter of a circle, however, has a special name. It is called the *circumference*.

Definition

The **circumference** of a circle is the distance around the circle. (The circumference is the *perimeter* of the circle.) It is usually abbreviated with C.

Let's look at several circles of different sizes and do some calculations.

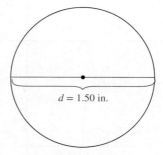

$d = 1.50$ in.

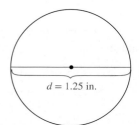

$d = 0.75$ in.

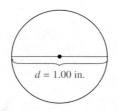

$d = 1.25$ in.

$d = 1.00$ in.

diameter = 1.50 in.
Circumference:
$C \approx 4.71$ in.
$\dfrac{C}{d} = \dfrac{4.71 \text{ in.}}{1.50 \text{ in.}} = 3.14$

diameter = 0.75 in.
Circumference:
$C \approx 2.36$ in.
$\dfrac{C}{d} = \dfrac{2.36 \text{ in.}}{0.75 \text{ in.}} = 3.14\overline{6}$

diameter = 1.25 in.
Circumference:
$C \approx 3.93$ in.
$\dfrac{C}{d} = \dfrac{3.93 \text{ in.}}{1.25 \text{ in.}} = 3.144$

diameter = 1.00 in.
Circumference:
$C \approx 3.14$ in.
$\dfrac{C}{d} = \dfrac{3.14 \text{ in.}}{1.00 \text{ in.}} = 3.14$

Even though the circles are different sizes, whenever we divide the circumference of a circle by its diameter, we *always* get a number that is close to 3.14. This is *not* a coincidence. *The circumference of a circle divided by its diameter is always the same value, and that value is called π (pi).*

Definition

π is the ratio of any circle's circumference to its diameter. That is, $\pi = \dfrac{C}{d}$.
(π is the Greek letter pi and is read as "pie.")

There is no exact decimal equivalent of π. An approximate decimal value is $\pi = 3.14159265$, but it is most common to approximate π as 3.14 or as the fraction $\dfrac{22}{7}$.

W Hint

This section contains many new terms and formulas. Make and use flash cards to help you learn them.

Property Approximations for π

$$\pi \approx 3.14 \quad \text{and} \quad \pi \approx \frac{22}{7}$$

Because these are *approximate* values, calculations using π will give *approximate* answers. Therefore, we should use the $\approx$ symbol instead of $=$.

We use π to find the circumference of a circle.

Formula Circumference of a Circle

1) Circumference $= \pi \cdot$ diameter   or $C = \pi \cdot d$ or $C = \pi d$

   Because $d = 2 \cdot$ radius, we can also write $C = \pi \cdot 2 \cdot$ radius

2) Circumference $= 2 \cdot \pi \cdot$ radius   or $C = 2 \cdot \pi \cdot r$ or $C = 2\pi r$

Note

Remember, because the *circumference* of a circle is the same as the *perimeter* of a circle, the units for circumference are linear units like ft, cm, in., etc. They are *not* square units.

EXAMPLE 3   Find the circumference of each circle. Give the *exact* value and an *approximate* value using $\pi \approx 3.14$.

a)

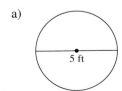

5 ft

b)

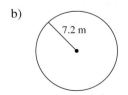

7.2 m

Solution

a) Are we given the diameter or the radius of this circle? We are given that the *diameter* = 5 ft. So, use the formula containing the diameter.

$$C = \pi \cdot d$$
$$C = \pi \cdot 5 \text{ ft} \qquad \text{Substitute the diameter.}$$
$$C = 5\pi \text{ ft} \qquad \text{This is the exact value.}$$

(We usually write π between the number and the units.) To find an *approximate* value for the circumference, substitute 3.14 for π and multiply.

$$C \approx 5(3.14) \text{ ft} \qquad \text{Substitute 3.14 for } \pi.$$
$$C \approx 15.7 \text{ ft} \qquad \text{Multiply.}$$

The circumference is *exactly* 5π ft. The *approximate* value is 15.7 ft.

(Notice that we write 5π ft *without* a multiplication symbol.)

b) We are given that the *radius* = 7.2 m. Use the formula containing the radius.

$$C = 2 \cdot \pi \cdot r$$
$$C = 2 \cdot \pi \cdot (7.2 \text{ m}) \qquad \text{Substitute the radius.}$$
$$C = 14.4\pi \text{ m} \qquad \text{Multiply.}$$

The circumference is *exactly* equal to 14.4π m. Find the *approximation* of the circumference by substituting 3.14 for π.

$$C \approx 14.4(3.14) \text{ m} \qquad \text{Substitute 3.14 for } \pi.$$
$$C \approx 45.216 \text{ m} \qquad \text{Multiply.}$$

The circumference is *approximately* equal to 45.216 m.

[YOU TRY 3] Find the circumference of each circle. Give the exact value and an approximate value using $\pi \approx 3.14$.

a)
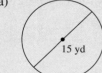
15 yd

b)
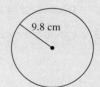
9.8 cm

EXAMPLE 4   Find the circumference of the circle. Give the exact value and an approximate value using $\pi \approx \dfrac{22}{7}$.

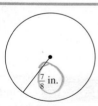
$\frac{7}{8}$ in.

Solution

We are given the *radius* of this circle, so we will use the formula containing the radius.

$$C = 2 \cdot \pi \cdot r$$
$$C = 2 \cdot \pi \cdot \frac{7}{8} \text{ in.} \qquad \text{Substitute the radius.}$$
$$C = \frac{\overset{1}{2}}{1} \cdot \pi \cdot \frac{7}{\underset{4}{8}} \text{ in.} \qquad \text{Divide out 2.}$$
$$C = \frac{7}{4}\pi \text{ in.} \qquad \text{Multiply.}$$

Exact value of the circumference: $C = \dfrac{7}{4}\pi$ in. Find an *approximate* value of the circumference by substituting $\dfrac{22}{7}$ for π.

$$C \approx \frac{7}{4}\left(\frac{22}{7}\right) \text{ in.}$$ Substitute $\dfrac{22}{7}$ for π.

$$C \approx \frac{\overset{1}{7}}{\underset{2}{4}}\left(\frac{\overset{11}{22}}{\underset{1}{7}}\right) \text{ in.}$$ Divide out common factors.

$$C \approx \frac{11}{2} \text{ in. or } 5\frac{1}{2} \text{ in. or 5.5 in.}$$ Multiply.

The *approximation* of the circumference is $C \approx \dfrac{11}{2}$ in. or $5\dfrac{1}{2}$ in. or 5.5 in.

[YOU TRY 4]   Find the circumference of the circle. Give the exact value and an approximate value using $\pi \approx \dfrac{22}{7}$.

---

3 Find the Area of a Circle

We can understand where the formula for the area of a circle comes from if we relate it to the area of a parallelogram. Let's begin by cutting a circle in half and then dividing it into pie-shaped pieces.

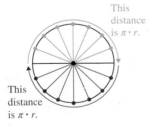

This distance is $\pi \cdot r$.

This distance is $\pi \cdot r$.

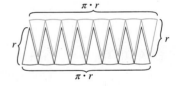

$\pi \cdot r$

r r

$\pi \cdot r$

Since the distance around a whole circle is $2 \cdot \pi \cdot r$, the distance around *half* the circle (or the *semicircle*) is $\dfrac{1}{2} \cdot 2 \cdot \pi \cdot r$ or $\pi \cdot r$.

If we cut out the pie-shaped pieces and put them together as seen here, we get a figure that is approximately a parallelogram with **base $= \pi \cdot r$** and **height $= r$.**

The formula for the area of a parallelogram is Area = base · height. Use this formula to find the area of the parallelogram that was formed from the circle.

$$\text{Area} = \text{base} \cdot \text{height}$$ Area of a parallelogram
$$\text{Area} = \pi \cdot r \cdot r$$ Substitute for base and height.
$$\text{Area} = \pi \cdot r^2$$ Multiply; $r \cdot r = r^2$

This is the formula for the area of a circle.

Formula Area of a Circle

The **area**, A, of a circle with radius r is

$$A = \pi \cdot r^2 \quad \text{or} \quad A = \pi r^2$$

Remember that we use *square units* for area.

EXAMPLE 5

Find the area of a circle with a radius of 1.5 mi. Give an exact value and an approximation using $\pi \approx 3.14$.

Solution

If we are not given a picture, it can be helpful if we sketch the circle.

1.5 mi

Use the area formula with $r = 1.5$ mi.

$A = \pi \cdot r^2$	Area formula
$A = \pi \cdot (1.5 \text{ mi})^2$	Substitute 1.5 mi for r.
$A = \pi \cdot 2.25 \text{ mi}^2$	$(1.5 \text{ mi})^2 = 1.5 \text{ mi} \cdot 1.5 \text{ mi} = 2.25 \text{ mi}^2$
$A = 2.25\pi \text{ mi}^2$	Use the commutative property to rewrite the answer.

The exact area is $2.25\pi \text{ mi}^2$.

Find the *approximate* area by substituting 3.14 for π.

$A \approx 2.25(3.14) \text{ mi}^2$	Substitute 3.14 for π.
$A \approx 7.065 \text{ mi}^2$	Multiply.

The area is approximately 7.065 mi^2.

[YOU TRY 5]

Find the area of a circle with a radius of 70 mm. Give an exact value and an approximation using $\pi \approx 3.14$.

Remember, we must know the *radius* to find the area of a circle.

EXAMPLE 6

Find the area of a circle with a diameter of 20 cm. Give an exact value and an approximation using $\pi \approx 3.14$.

Solution

Let's sketch the circle and label it with the information we are given.

20 cm

We can see from our drawing that the *diameter* is 20 cm, but we need the *radius*. The radius is *half* the diameter, so $r = \dfrac{d}{2} = \dfrac{20 \text{ cm}}{2} = 10$ cm. Use $r = 10$ cm in the area formula.

$$A = \pi \cdot r^2 \qquad \text{Area formula}$$
$$A = \pi \cdot (10 \text{ cm})^2 \qquad \text{Substitute 10 cm for } r.$$
$$A = \pi \cdot 100 \text{ cm}^2 \qquad (10 \text{ cm})^2 = 10 \text{ cm} \cdot 10 \text{ cm} = 100 \text{ cm}^2$$
$$A = 100\pi \text{ cm}^2 \qquad \text{Use the commutative property.}$$

The area of the circle is *exactly* 100π cm². Substitute 3.14 for π to find the approximate area.

$$A \approx 100(3.14) \text{ cm}^2 \qquad \text{Substitute 3.14 for } \pi.$$
$$A \approx 314 \text{ cm}^2 \qquad \text{Multiply.}$$

The area of the circle is *approximately* 314 cm².

[**YOU TRY 6**] Find the area of a circle with a diameter of 200 yd. Give an exact value and an approximation using $\pi \approx 3.14$.

Note

If you are working with geometric figures and a picture is *not* given in the problem, it can be helpful to draw a picture and label it with the information in the problem. Sometimes, looking at a *picture* can help us to better understand what is happening in the problem so that we can solve it.

A **semicircle** is half a circle. To find the area of a semicircle, we use the formula for the area of a circle and divide by 2.

EXAMPLE 7

Find the area of this semicircle. Give the exact value and an approximation using 3.14 for π.

16 in.

Solution

First, we will find the area of a *whole* circle that has a radius of 16 in. Then, we will divide by 2 since a semicircle is *half* a circle.

Whole Circle

$$A = \pi \cdot r^2$$
$$A = \pi \cdot (16 \text{ in.})^2 \qquad \text{Substitute 16 in. for } r.$$
$$A = 256\pi \text{ in}^2 \qquad (16 \text{ in.})^2 = 16 \text{ in.} \cdot 16 \text{ in.} = 256 \text{ in}^2$$

Divide by 2 to find the area of the semicircle.

Semicircle

$$A = \frac{\text{Area of whole circle}}{2} = \frac{256\pi \text{ in}^2}{2} = 128\pi \text{ in}^2$$

W Hint

Would it be helpful to write a procedure for finding the area of a semicircle?

The *exact* area of the semicircle is 128π in². To find the *approximate* area, substitute 3.14 for π.

$$A \approx 128(3.14) \text{ in}^2 \qquad \text{Substitute 3.14 for } \pi.$$
$$A \approx 401.92 \text{ in}^2 \qquad \text{Multiply.}$$

[YOU TRY 7] Find the area of the semicircle. Give the exact value and an approximation using 3.14 for π.

We can use what we have learned about circles to solve problems.

EXAMPLE 8

Tiffany wants a custom teak table made for her dining room. The tabletop will be a circle with a diameter of 4 ft. If the wood for the tabletop costs \$4.20/ft^2, find the total cost of the wood. Use 3.14 for π.

Solution

Let's draw and label a picture first.

We need to determine the *amount* of wood that is needed to make the tabletop. *This is the area.* Since $A = \pi \cdot r^2$, we need to know the *radius* of the table.

The diameter $= 4$ ft, so the radius $= \dfrac{4 \text{ ft}}{2} = 2$ ft.

$A = \pi \cdot r^2$	Area of a circle
$A = \pi \cdot (2 \text{ ft})^2$	Substitute 2 ft for r.
$A = 4\pi \text{ ft}^2$	$(2 \text{ ft})^2 = 2 \text{ ft} \cdot 2 \text{ ft} = 4 \text{ ft}^2$
$A \approx 4(3.14) \text{ ft}^2$	$\pi \approx 3.14$
$A \approx 12.56 \text{ ft}^2$	Multiply.

To find the total cost of the wood, multiply the area and the unit cost of the wood.

$$\text{Cost} = 12.56 \text{ ft}^2 \cdot \frac{\$4.20}{\text{ft}^2} = \$52.752$$

Round the answer to the nearest cent: \$52.75. The cost of the wood for the tabletop is \$52.75.

[YOU TRY 8] Esteban will make a cover for the circular swimming pool in his backyard. How much will it cost if the material for the cover costs \$0.09/ft^2 and the diameter of the cover will be 22 ft? Use 3.14 for π.

4 Find the Volume and Surface Area of Solids

Let's review the definitions of volume and surface area.

Definition

The **volume** of a three-dimensional object is a measure of the amount of space occupied by the object or the amount of space inside the object. Volume is measured in *cubic* units.

Definition

The **surface area** of a three-dimensional object is a measure of the total area on the surface of the object. Surface area is measured in *square* units.

In the table, we summarize the formulas for volume, *V*, and surface area, *SA*. We have included the formulas we learned in Chapter 4 so that all of the formulas can be found in one place.

Solid		Formulas
Rectangular solid		$V = lwh$ $SA = 2lh + 2wh + 2lw$ where l = length, w = width, h = height
Cube		$V = s^3$ $SA = 6s^2$ where s = length of a side
Sphere		$V = \dfrac{4}{3}\pi r^3$ $SA = 4\pi r^2$ where r = radius
Right circular cylinder		$V = \pi r^2 h$ $SA = 2\pi rh + 2\pi r^2$ where r = radius, h = height
Rectangular pyramid		$V = \dfrac{1}{3}lwh$ $SA = lw + ls_1 + ws_2$ where l = length, w = width, s_1 = slant height of side with length l s_2 = slant height of side with width w
Right circular cone		$V = \dfrac{1}{3}\pi r^2 h$ $SA = \pi r\sqrt{r^2 + h^2} + \pi r^2$ where r = radius, h = height

A **sphere** is a round, three-dimensional object like a basketball or the Earth. The radius of a sphere is the distance from the center to the edge of the sphere.

EXAMPLE 9

Find the volume and surface area of a sphere with radius 2 ft. Give an exact value and an approximation using $\frac{22}{7}$ for π.

Solution

Use the volume formula with $r = 2$ ft.

$$V = \frac{4}{3}\pi r^3$$

$$V = \frac{4}{3}\pi(2 \text{ ft})^3 \qquad \text{Substitute 2 ft for } r.$$

$$V = \frac{4}{3}\pi(8 \text{ ft}^3) \qquad (2 \text{ ft})^3 = 2 \text{ ft} \cdot 2 \text{ ft} \cdot 2 \text{ ft} = 8 \text{ ft}^3$$

$$V = \frac{32}{3}\pi \text{ ft}^3 \qquad \text{Multiply.}$$

The *exact* volume is $\frac{32}{3}\pi$ ft³.

To find an approximation, substitute $\frac{22}{7}$ for π.

$$V \approx \frac{32}{3}\left(\frac{22}{7}\right) \text{ ft}^3 \qquad \text{Substitute } \frac{22}{7} \text{ for } \pi.$$

$$V \approx \frac{704}{21} \text{ ft}^3 \qquad \text{Multiply.}$$

The volume is approximately $\frac{704}{21}$ ft³ or $33\frac{11}{21}$ ft³ or 33.52 ft³, rounded to the nearest hundredth.

To find the surface area, use the formula $SA = 4\pi r^2$.

$$SA = 4\pi r^2 = 4\pi(2 \text{ ft})^2 = 4\pi(4 \text{ ft}^2) = 16\pi \text{ ft}^2$$

To find an approximation, substitute $\frac{22}{7}$ for π.

$$SA = 16\pi \text{ ft}^2 \approx 16\left(\frac{22}{7}\right)\text{ft}^2 = \frac{352}{7} \text{ ft}^2$$

The surface area is exactly 16π ft². It is approximately $\frac{352}{7}$ ft² or $50\frac{2}{7}$ ft². Rounded to the nearest hundredth, the approximation is 50.29 ft².

YOU TRY 9

Find the volume and surface area of a sphere with radius 6 m.

Give an exact value and an approximation using $\dfrac{22}{7}$ for π.

6 m

Note

Remember to use the symbol $\approx$ when *approximating* a value.

EXAMPLE 10

A hockey puck is 3 in. in diameter, and it is 1 in. thick. Find the exact volume of the hockey puck and the approximate volume to the nearest hundredth. Use 3.14 for π.

3 in.

1 in.

Solution

The formula for the volume of a cylinder is $V = \pi \cdot r^2 \cdot h$. The height of the hockey puck, h, is 1 in. But, we are given the diameter, *not* the radius. How do we find the radius?

W Hint

Do you see a relationship between the volume of a cylinder and a formula you learned before?

$$\text{radius} = \frac{\text{diameter}}{2} = \frac{3 \text{ in.}}{2} = 1.5 \text{ in.}$$

Find the volume of the hockey puck using $r = 1.5$ in. and $h = 1$ in.

$V = \pi r^2 h$

$V = \pi(1.5 \text{ in.})^2(1 \text{ in.})$ Substitute the values.

$V = \pi(2.25 \text{ in}^2)(1 \text{ in.})$ $(1.5 \text{ in.})^2 = 2.25 \text{ in}^2$

$V = 2.25\pi \text{ in}^3$ Multiply.

The volume of the hockey puck is exactly $2.25\pi \text{ in}^3$. Find the approximate value by substituting 3.14 for π.

$$V \approx 2.25(3.14) \text{ in}^3 \approx 7.065 \text{ in}^3 \approx 7.07 \text{ in}^3$$

The volume of the hockey puck is *approximately* 7.07 in^3.

YOU TRY 10

Find the volume of this can. Give the exact value and an approximation to the nearest hundredth. Use 3.14 for π.

3 in.

4 in.

EXAMPLE 11

Find the volume of the cone. Give the exact value and an approximation to the nearest hundredth. Use 3.14 for π.

Solution

Identify the radius and height: $r = 2$ m, $h = 3$ m. Use the volume formula.

$$V = \frac{1}{3}\pi r^2 h$$

$$V = \frac{1}{3}\pi (2 \text{ m})^2 3 \text{ m} \qquad \text{Substitute the values.}$$

$$V = \frac{1}{3}\pi (4 \text{ m}^2)(3 \text{ m}) \qquad (2 \text{ m})^2 = 4 \text{ m}^2$$

$$V = \frac{1}{\cancel{3}}\pi (4 \text{ m}^2)(\cancel{3} \text{ m}) \qquad \text{Divide out 3.}$$

$$V = 4\pi \text{ m}^3 \qquad \text{Multiply.}$$

The volume is *exactly* 4π m³.

Substitute 3.14 for π to find the approximate volume.

$$V \approx 4(3.14) \text{ m}^3 \approx 12.56 \text{ m}^3$$

The volume is *approximately* 12.56 m³.

[YOU TRY 11]

Find the volume of the cone. Give an exact value and an approximation to the nearest hundredth. Use 3.14 for π.

Note

The formula for the volume of a cone, $V = \frac{1}{3}\pi r^2 h$, can also be thought of as $V = \frac{1}{3} \cdot Ah$ where $A = \pi r^2$, the area of the base of the cone.

ANSWERS TO [YOU TRY] EXERCISES

1) 22 m   2) $\frac{21}{2}$ in. or $10\frac{1}{2}$ in. or 10.5 in.

3) a) exact: 15π yd; approximation: 47.1 yd   b) exact: 19.6π cm; approximation: 61.544 cm

4) exact: $\frac{3}{2}\pi$ km; approximation: $\frac{33}{7}$ km or $4\frac{5}{7}$ km   5) exact: 4900π mm²; approximation: 15,386 mm²

6) exact: $10{,}000\pi$ yd²; approximation: 31,400 yd²   7) exact: 288π cm²; approximation: 904.32 cm²

8) \$34.19

9) Volume: exact: 288π m³; approximation: $\frac{6336}{7}$ m³ or $905\frac{1}{7}$ m³ or 905.14 m³.

   Surface area: exact: 144π m²; approximation: $\frac{3168}{7}$ m² or 452.57 m².

10) exact: 9π in³; approximation: 28.26 in³   11) exact: 32π in³; approximation: 100.48 in³

We can use a calculator to help us calculate the area of a circle using the formula $A = \pi r^2$. Remember that we must perform the exponent operation before we multiply. To instruct the calculator to do this, we must use the parenthesis keys.

Suppose we have a circle with radius 3 m. We can approximate the area using either $\pi \approx 3.14$ or $\pi \approx \frac{22}{7}$. If we approximate the area using $\pi \approx 3.14$, we enter $\boxed{3}\boxed{.}\boxed{1}\boxed{4}\boxed{\times}\boxed{(}\boxed{3}\boxed{y^x}\boxed{2}\boxed{)}\boxed{=}$ into the calculator. The display screen will show 28.26. This means that the area is approximately 28.26 m².

If we approximate the area using $\pi \approx \frac{22}{7}$, we enter $\boxed{2}\boxed{2}\boxed{\div}\boxed{7}\boxed{\times}$ $\boxed{(}\boxed{3}\boxed{y^x}\boxed{2}\boxed{)}\boxed{=}$ into the calculator. The display screen will show 28.285714. Rounded to the nearest hundredth, the area is approximately 28.29 m².

E Evaluate 5.10 Exercises

Do the exercises, and check your work.

Objective 1: Find the Radius and Diameter of a Circle

1) What is the difference between the radius and the diameter of a circle?

2) How are the radius and diameter of a circle related?

Find the diameter of the circle.

3)

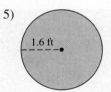
7 km

4)

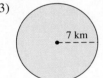
4 mi

5)

1.6 ft

6)

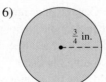
$\frac{3}{4}$ in.

Find the radius of the circle.

7)

18 cm

8)

12 in.

9)

$\frac{1}{6}$ cm

10)

$\frac{5}{8}$ in.

11)

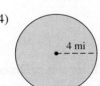
14 in.
14 in.

12)

6.5 cm
8 cm
10 cm
11.5 cm

Objective 2: Find the Circumference of a Circle and Understand What π Represents

13) What do we call the perimeter of a circle?

14) Where does the number π come from? What are the two values we can use for the approximation of π?

15) If we are given the diameter of a circle, what formula should we use to find the circumference?

16) If we are given the radius of a circle, what formula should we use to find the circumference?

Find the circumference of each circle. Give the *exact* value and an *approximate* value using $\pi \approx 3.14$.

17)

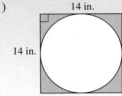
5 in.

18)

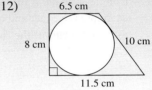
4 yd

19)

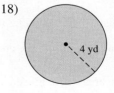
8 cm

20)

7 m

21) 6.8 in.

22) 3.5 yd

35) 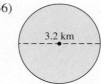 18 cm

36) 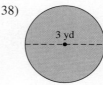 3.2 km

23) $1\frac{1}{2}$ ft

24) $2\frac{2}{5}$ ft

37) 4.8 ft

38) 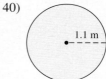 3 yd

25) Celine has a garden shaped as a semicircle. Use the figure to calculate the exact and approximate perimeter of the garden. Use $\pi \approx 3.14$.

 4 m

26) A Little League baseball field has a shape equivalent to one-fourth of a circle. Use the figure to calculate the exact and approximate perimeter of the field shown in the figure. Use $\pi \approx 3.14$.

 100 yd   100 yd

39) 50 ft

40) 1.1 m

Find the area of the semicircle. Give the *exact* value and an *approximate* value using $\pi \approx 3.14$.

41) 0.8 m

42) 6 ft

Find the circumference of each circle. Give the *exact* value and an *approximate* value using $\pi \approx \frac{22}{7}$.

43) 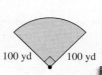 16 yd

44) 60 cm

27) 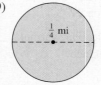 7 m

28) 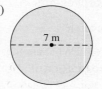 5 cm

29) 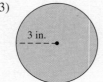 $\frac{1}{4}$ mi

30) 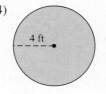 $\frac{14}{15}$ cm

Mixed Exercises: Objectives 1–3
Solve each problem. Use $\pi \approx 3.14$.

45) The General Sherman Tree in Sequoia National Park is the largest tree (by volume) in the world. The maximum diameter at its base is 36.5 ft. What is the circumference of the tree at this diameter? Using an average arm span of a human being as 67 in., estimate the number of people required to make a human chain around the tree. (www.nps.gov)

©SuperStock/Alamy

Objective 3: Find the Area of a Circle

31) What is the formula for the area of a circle?

32) How do you find the area of a semicircle?

Find the area of the given circle. Give the *exact* value and an *approximate* value using $\pi \approx 3.14$.

33) 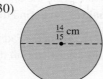 3 in.

34) 4 ft

46) Ji-Min needs to replace a circular window having a diameter of 5 ft. She decides to use double-strength glass costing $3 per square foot. Find the total cost for the glass.

47) The Johnsons have a round, 8-ft-diameter pool that needs a new custom-fit insulated cover. If the cover costs $4.50 per square foot, what is the total cost?

48) The playground pictured here is having rubber flooring installed to help protect children against injuries from falls. If the cost, including installation, is $8.00 per square foot, what is the total cost of installing the surface?

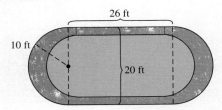

49) A playing field will have sod installed at a cost of $0.50 per square foot. In addition, the field will have a fence installed around its perimeter for $5.00 per foot. Find the combined cost for both jobs.

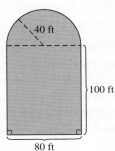

50) Thaddeus wants to cover an area around a pond with natural slate stone. If the cost of the slate is $6.50 per square foot, how much will it cost to surround the pond?

Objective 4: Find the Volume and Surface Area of Solids

51) Write the formulas for the volume and surface area of a sphere with radius r.

52) Because a hemisphere is half a sphere, what is the formula for the volume of a hemisphere with radius r?

Find the volume and surface area of each sphere. Give an exact answer and an approximation using the indicated value for π. Round the approximation to the nearest hundredth.

53) Use 3.14 for π.

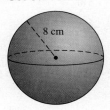

54) Use 3.14 for π.

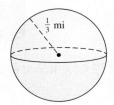

55) Use $\frac{22}{7}$ for π.

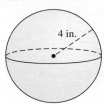

56) Use $\frac{22}{7}$ for π.

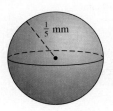

57) Use 3.14 for π.

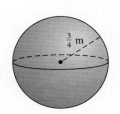

58) Use $\frac{22}{7}$ for π.

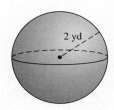

A **hemisphere** is half a sphere. Find the volume of each hemisphere. Give an exact answer and an approximation using the indicated value for π. Round the approximation to the nearest hundredth.

59) Use 3.14 for π.

60) Use $\frac{22}{7}$ for π.

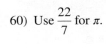

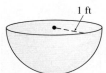

61) Write down the formulas for the volume and surface area of a right circular cylinder with radius r and height h.

62) If you double the height of a right circular cylinder and keep the radius the same, by what factor does the volume change?

Find the volume and surface area of each right circular cylinder. Give an exact value and an approximation using the indicated value for π.

63) Use 3.14 for π.

5 m
0.2 m

64) Use 3.14 for π.

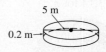
4 in.
1.5 in

65) Use $\frac{22}{7}$ for π. Round to the nearest hundredth.

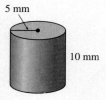
5 mm
10 mm

66) Use $\frac{22}{7}$ for π. Round to the nearest hundredth.

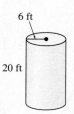
6 ft
20 ft

67) Use 3.14 for π.

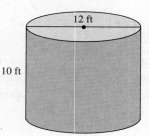
12 ft
10 ft

68) Use $\frac{22}{7}$ for π.

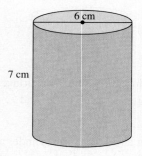
6 cm
7 cm

69) Write down the formula that represents the volume of a cone with a base of radius r and height h.

70) If you double the radius of a cone and keep the height the same, by what factor does the volume change?

Find the volume of each cone. Give an exact value and an approximation using $\pi = 3.14$.

71)

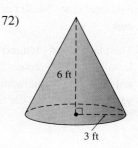
6 in.
5 in.

72)

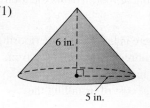
6 ft
3 ft

73)

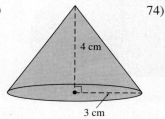
4 cm
3 cm

74)

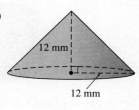
12 mm
12 mm

75) The radius of the Earth is approximately 4000 mi. Using this dimension, calculate the circumference of the Earth. (nssdc.gsfc.nasa.gov)

76) Ethanol is an alcohol that is currently blended into 50% of the nation's fuel supply. The primary source for ethanol is corn, which is sometimes stored in a corn silo shaped like a right circular cylinder. If the silo has a diameter of 40 ft and a height of 160 ft, what is its volume? (www.afdc.energy.gov)

77) A dime is 1.35 mm thick, and its diameter is approximately 18 mm. What is the volume of a dime? (www.usmint.gov)

78) A waffle cone maker lets you make your own waffle cone having a diameter of approximately 3 in. and a height of approximately 8 in. Using these dimensions, find the volume of the waffle cone.

79) The Millers' farm has a corn storage silo shaped as a right circular cylinder with a hemispherical top. If the cylinder has a diameter of 6 m and a height of 40 m, find the volume of the silo.

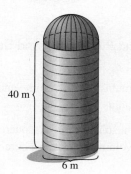
40 m
6 m

80) The Epcot Center geodesic sphere has a diameter of approximately 50 m. What is the volume of the sphere rounded to the nearest cubic meter?

Mixed Exercises: Objectives 1–4

Exercises 81–92 contain figures from Chapters 4 and 5.

Find the area and perimeter (or circumference) of each figure. In the case of a circle, find the exact values and the approximations using 3.14 for π. Round the answers to the nearest hundredth, where appropriate.

81)

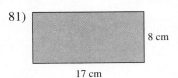

8 cm

17 cm

82)

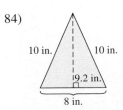

15 ft

83)

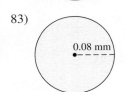

0.08 mm

84)

10 in.   10 in.

9.2 in.

8 in.

85) Find the area of the shaded region. Use 3.14 for π.

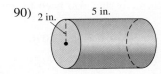

6 cm

12 cm

86) Marla needs to fence in an area of her yard that has a shape equivalent to three-fourths of an entire circle. What length of fence does she need to completely enclose this region? Use 3.14 for π.

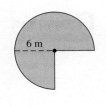

6 m

Find the exact volume of each solid. Use $\pi \approx 3.14$, if required, to also approximate the volume to the nearest hundredth.

87)

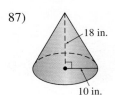

18 in.

10 in.

88)

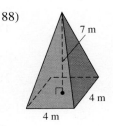

7 m

4 m

4 m

89)

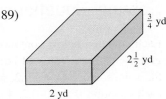

$\frac{3}{4}$ yd

$2\frac{1}{2}$ yd

2 yd

90)

2 in.   5 in.

91) A seventeenth-century cannonball found at Charles Fort in Kinsale, Ireland, has a diameter of 15.5 cm. Find the volume of the cannonball. (Information on display at Charles Fort)

92) The Great Pyramid of Giza is the world's largest pyramid. The base of the pyramid has four side lengths, all approximately equal to 230 m. The height of the pyramid is approximately 147 m. Calculate the volume of the Great Pyramid of Giza using these dimensions. (www.nationalgeographic.com)

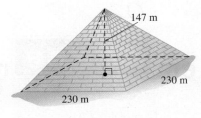

147 m

230 m

230 m

R Rethink

R1) Which formulas do you find easier to use and why?

R2) When is it easier to use the decimal approximation for π, and when is it easier to use the fractional approximation?

R3) Which two formulas would you need to use to find the volume of ice cream in a waffle cone, assuming that the cone is completely full and that there is one round scoop on the top?

Group Activity – Signed Decimals

Students should work in groups of 3 or 4 to complete the following exercises.

Evaluating Expressions

Evaluate each expression below for $a = -0.2$, $b = 4.05$, $c = -1.3$, and $d = 0.85$.

1) $ab - cd$

2) $\dfrac{b - c}{a}$

3) $-a + 3c$

4) $5a^2 - bc + d$

Solving Equations

Each equation below contains decimals. Solve each equation with the help of your group members. You will know that you have the correct solution if it matches one of the values above ($a = -0.2$, $b = 4.05$, $c = -1.3$, or $d = 0.85$). If your final solution to each equation is not -0.2, 4.05, -1.3, or 0.85, then you have solved the equation incorrectly and you need to review your work with your group members.

1) $0.6(5 - 0.1x) = 3.078$

2) $-0.05(3 + x) = -0.14$

3) $0.8x = 0.4(3x - 0.85)$

4) $6.25 - 2x = -1.85$

em POWER me Is Anxiety the Hardest Problem on the Test?

Do you feel anxious at the very thought of a math test, or are you cool and calm in the face of this situation? Get a sense of your test-taking style by checking off every statement below that applies to you.

☐ 1. The closer a test date approaches, the more nervous I get.

☐ 2. I am sometimes unable to sleep the night before a test.

☐ 3. I have "frozen up" during a math test, finding myself unable to think or respond.

☐ 4. I can feel my hands shaking as I pick up my pencil to begin the first problem on a test.

☐   5. The minute I get to a tough question, I forget all the skills I've learned and have no idea how to begin.

☐   6. I have become physically ill before or during a test.

☐   7. Nervousness prevents me from studying immediately before a test.

☐   8. I often dream about an upcoming test.

☐   9. Even if I successfully answer a number of questions, my anxiety stays with me throughout the test.

☐   10. I check and recheck the calculations on problems I know how to do, rather than deal with the questions I don't know how to answer.

If you checked off more than four statements, you have experienced fairly serious test anxiety. If you checked off more than six statements, your anxiety is probably interfering with your test performance. In particular, statements 3, 5, 6, and 7 may indicate serious test anxiety.

If, based on your responses to this questionnaire and your previous experience, your level of test anxiety is high, there are several things you can do.

- *Prepare thoroughly*. Good preparation can give you a sense of control and mastery, and it will prevent test anxiety from overwhelming you.
- *Take a realistic view of the test*. Remember that your future success does not hinge on your performance on any single exam. Think of the big picture: Put the task ahead in context, and remind yourself of all the hurdles you've passed so far.
- *Eat right and get enough sleep*. Good mental preparation can't occur without your body being well prepared.
- *Learn relaxation techniques*. You can learn to reduce or even eliminate the jittery physical symptoms of test anxiety by using relaxation techniques. The basic process is straightforward: Breathe evenly, gently inhaling and exhaling. Focus your mind on a pleasant, relaxing scene, or on a restful sound such as that of ocean waves breaking on the beach.
- *Visualize success*. Think of an image of your instructor handing back your test marked with an "A." Positive visualizations that highlight your potential success can help replace images of failure that may fuel test anxiety.

Chapter 5: Summary

Definition/Procedure	Example

5.1 Reading and Writing Decimals

A **decimal** is a number containing a *decimal point* that is another way to represent a fraction with a denominator that is a power of 10.

Use a decimal to represent the shaded part of the figure.

The square is divided into 100 equal parts. 27 parts are shaded. As a *fraction,* we say that $\frac{27}{100}$ (twenty-seven hundredths) is shaded. In *decimal* form, we say that 0.27 (twenty-seven hundredths) is shaded.

The Place-Value Chart

1) The *ones* column is in the middle of the chart.

2) The place values to the *left* of the decimal point (the whole number part) end in **s.**

3) The place values to the *right* of the decimal point (the fractional part) end in **ths.**

Identify the place value of each digit of 508.3192.

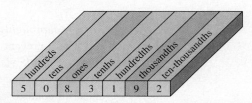

Reading a Decimal Number

1) Read the whole-number part first.

2) Read the decimal point as "*and.*"

3) Read the fractional part, the digits to the *right* of the decimal point, last.

Write 43.78 in words.

Read the whole-number part first, read the decimal point as "*and,*" then read the part to the right of the decimal point.

$$43.78$$
$$\uparrow$$
and

Read 43.78 as "forty-three and seventy-eight hundredths."

Write Decimals as Fractions or Mixed Numbers

1) If the decimal has a whole number, it will convert to a mixed number. Write the whole number in front of the fraction.

2) The numerator of the fraction will be the digits to the right of the decimal point.

3) The denominator will be the place value of the digit farthest to the right of the decimal point.

4) Write the fraction in lowest terms.

Write 5.0789 as a mixed number in lowest terms.

numerator = 789

$$5.\widehat{0789} = 5\frac{789}{10,000}$$

9 is in the *ten-thousandths* place, so the denominator = 10,000.

Definition/Procedure	Example

5.2 Rounding Decimals

Rounding Decimals

Step 1: Find the place to which we are asked to round. Underline the digit in that place and draw a vertical line after it.

Step 2: Look at the digit to the right of the vertical line.

    a) If the digit to the right of the vertical line is **less than 5**, "drop off" the digits to the right of the vertical line and leave the underlined digit as it is.

    b) If the digit to the right of the vertical line is **5 or more**, "drop off" the digits to the right of the vertical line and increase the underlined digit by 1.

Round 45.90825 to the nearest thousandth.

Step 1: Underline the digit in the thousandths place and draw a vertical line after it.

$$45.908|25$$

Step 2: Look at the digit to the right of the vertical line. Because 8 is more than 5, "drop off" the digits to the right of the vertical line and increase the underlined digit by 1.

Keep this digit the same.
2 is less than 5.

$$45.908|25$$

Drop these.

Round to 45.908

Rounded to the nearest thousandth, 45.90825 is 45.908.

Round Money Amounts to the Nearest Cent

The value of a penny is 1¢ (1 cent). We can also write it in terms of a dollar as $0.01 or 0.01 dollar. $\left(\text{We can also say that it is } \frac{1}{100} \text{ dollar.}\right)$ Because it is the smallest denomination of money in the United States, most everyday money amounts are rounded *to the nearest cent*. This is the same as rounding *to the nearest hundredth of a dollar*.

Brittany figured out that the exact amount of sales tax for the card she bought for her friend Zoe was $0.2625. Round the sales tax to the nearest cent.

Underline the 6 in the hundredths (or cents) place, and round.

$$\$0.26|25 \qquad \text{Round to } \$0.26$$

Brittany will spend $0.26 on sales tax. This can also be written as 26¢.

Round Money Amounts to the Nearest Dollar

Rounding to the nearest *dollar* means rounding to the *ones* place.

Round $368.83 to the nearest dollar.

Underline the 8 in the ones place in $368.83 and round.

$$\$368.|83 \qquad \text{Round to } \$369$$

Or ask yourself, *"Is $368.83 closer to $368 or $369?"* It is closer to $369.

Definition/Procedure	Example

5.3 Adding and Subtracting Signed Decimals

Add or Subtract Decimals

1) Write the numbers vertically so that the decimal points are lined up.

2) If any numbers are missing digits to the right of the decimal point, insert zeros. Then, add or subtract the same way we add or subtract whole numbers.

3) Place the decimal point in the answer *directly below* the decimal point in the problem.

Add or subtract as indicated.

a) $9.4 + 7.3$ b) $47 - 33.395$

Solution

a) Write the numbers vertically so that the decimal points are lined up. Then, add just like we add whole numbers.

```
    9.4
 +  7.3
 1  6.7
```
↑
Line up the decimal points.

b) First, rewrite 47 with a decimal point: $47 = 47.$
Line up the decimal points. Insert zeros and subtract.

Insert zeros.

```
        9  9
     6 10 10 10
   4 7. 0  0  0  ←  Insert zeros.
 - 3 3. 3  9  5
   1 3. 6  0  5
```

We add and subtract signed decimal numbers using the same rules we use for integers.

Subtract $-25.87 - 12.05$.

Solution

Write $-25.87 - 12.05$ as $-25.87 + (-12.05)$. Add the absolute values of the numbers, then make the answer negative.

```
        1
     2 5. 8 7
  +  1 2. 0 5
     3 7. 9 2
```
$-25.87 - 12.05 = -37.92$

5.4 Multiplying Signed Decimals

Multiplying Decimals

1) **Multiply the numbers (factors) just like you would multiply whole numbers.** (Line up the numbers on the right; the decimal points do *not* have to be lined up.)

2) **Determine the total number of decimal places in the answer (the product).** This will be the *total* number of decimal places in the factors.

3) **Insert the decimal point in the answer.** Start at the right side of the product and count the *total* number of places you determined in 2). Sometimes, you may need to insert zeros as placeholders on the left side of the answer.

4) **Determine the sign of the product.** If the two factors have the *same sign*, the product is positive. If the two factors have *different signs*, the product is negative.

Find $-65.89 \times (-5.4)$.

Multiply the numbers just like they were whole numbers. Determine the total number of decimal places in the answer. Insert the decimal point in the answer.

```
     6 5. 8 9  →   2 decimal places
  ×     5. 4  →  + 1 decimal place
   2 6 3 5 6      3 decimal places in
 3 2 9 4 5        the answer.
 3 5 5. 8 0 6
```

The factors are both negative, so the product is positive.

$-65.89 \times (-5.4) = 355.806$

Definition/Procedure	Example
Multiplying a Number by a Power of 10 1) Count the number of zeros in the power of 10. 2) Move the decimal point in the number *to the right* the same number of spaces as the number of zeros in the power of 10. 3) If necessary, add zeros as placeholders on the right.	Multiply 34.59×1000. $$34.59 \times 1\underline{000} = 34.590 = 34{,}590$$ 3 decimal places — Move the decimal point 3 places to the right.
Multiplying a Number by 0.1, 0.01, 0.001, etc. 1) 0.1, move the decimal point in the number 1 place to the left; 2) 0.01, move the decimal point 2 places to the left; 3) 0.001, move the decimal point 3 places to the left; and so on.	Multiply 27.1×0.01. Write a 0 in the ones place. ↓ $$27.1 \times 0.01 = .271 = 0.271$$ 2 decimal places — Move the decimal point 2 places to the left.

5.5 Dividing Signed Decimals and Order of Operations

Divide a Decimal by a Whole Number 1) Write the problem in long division form. 2) Write the decimal point in the quotient directly above the decimal point in the dividend. 3) Perform the division as if the numbers were whole numbers. 4) Determine the sign of the quotient. If both numbers have the *same sign,* the quotient is *positive.* If the numbers have *different signs,* the quotient is *negative.*	Divide $2.748 \div 6$. **Solution** $$\begin{array}{r} 0.458 \\ 6\overline{)2.748} \\ \underline{-2\,4} \\ 3\,4 \\ \underline{-3\,0} \\ 4\,8 \\ \underline{-4\,8} \\ 0 \end{array}$$ ← Remainder = 0 Therefore, $2.748 \div 6 = 0.458$.
Divide a Number by a Decimal 1) Move the decimal point to the right end of the divisor. This makes the divisor a whole number. Count the number of decimal places you have moved. 2) Move the decimal point in the dividend the *same* number of places to the right. (If necessary, write in zeros.) 3) Write the decimal point in the quotient directly above the decimal point in the dividend. Then, divide.	Divide 61.6 by 0.011. Move the decimal point 3 places to the right. $$0.011\overline{)61.6} \longrightarrow 011.\overline{)61600.}$$ Move decimal points 3 places to the right. Write the decimal point in the quotient directly above the decimal point in the dividend. $$\begin{array}{c} 5600. \\ 11\overline{)61600.} \end{array}$$ Check by multiplying.
Use the Order of Operations with Decimals We use the order of operations with decimals just like we did with whole numbers and fractions.	Simplify $8.1 \div 9 - 0.1(1.3)^2$ using the order of operations. $8.1 \div 9 - 0.1(1.3)^2$ $= 8.1 \div 9 - 0.1(1.69)$ Evaluate exponents first. $= 0.9 - 0.169$ Divide and multiply before subtracting. $= 0.731$ Subtract.

Definition/Procedure	Example

5.6 Writing Fractions as Decimals

Write a Fraction as a Decimal Using Division

To write a fraction as a decimal, divide the numerator by the denominator.

Write $\frac{1}{4}$ as a decimal.

$\frac{1}{4}$ means $1 \div 4$. Divide.

$$
\begin{array}{r}
0.25 \\
4\overline{)1.00} \\
\underline{-8} \\
20 \\
\underline{-20} \\
0
\end{array}
$$

Therefore, $\frac{1}{4} = 0.25$

Write a Fraction as a Decimal Using Equivalent

Fractions with a Denominator of 10, 100, etc.

If we can write a fraction as an equivalent fraction with a denominator of 10, 100, 1000, or another power of 10, then we can convert it to a decimal.

Write $-\frac{7}{50}$ as a decimal.

We can write $-\frac{7}{50}$ with a denominator of 100.

$$-\frac{7}{50} \cdot \frac{2}{2} = -\frac{14}{100}$$

Write $-\frac{14}{100}$ as a decimal: -0.14. Therefore, $-\frac{7}{50} = -0.14$

Comparing Fractions and Decimals

To compare two decimals or a fraction and a decimal, we can think about where they appear on a number line. Or, we can write both numbers as fractions or both as decimals.

Use $=$, $>$, or $<$ to compare 0.45 _____ $\frac{7}{20}$.

Let's change 0.45 to a fraction to compare it to $\frac{7}{20}$.

$$0.45 = \frac{45}{100} = \frac{45 \div 5}{100 \div 5} = \frac{9}{20}$$

Because $0.45 = \frac{9}{20}$, compare $\frac{9}{20}$ _____ $\frac{7}{20}$.

$$\frac{9}{20} > \frac{7}{20}, \text{ so } 0.45 > \frac{7}{20}$$

5.7 Mean, Median, and Mode

Find the Mean

The **mean (average)** is the sum of all the values in a list of numbers divided by the number of values in the list.

$$\text{Mean} = \frac{\text{Sum of all values}}{\text{Number of values}}$$

The mean will not necessarily be a value found in the list.

Find the mean, or average, of 37, 42, 31, and 34.

Solution

To find the mean, add the scores and divide by 4, the *number* of scores.

$$
\begin{aligned}
\text{Mean} &= \frac{37 + 42 + 31 + 34}{4} \quad &&\text{Sum of the scores} \\
&&&\text{Number of scores} \\
&= \frac{144}{4} \quad &&\text{Add.} \\
&= 36 \quad &&\text{Divide.}
\end{aligned}
$$

The mean, or average, is 36.

Definition/Procedure	Example

Find the Weighted Mean

If some values in a list of numbers appear more than once, then we can compute the *weighted mean* to represent that list of numbers. To find a **weighted mean,** we *weight* each value by multiplying it by the number of times it appears in the list.

Find a Grade Point Average (GPA)

Computing a student's GPA is the same as finding a weighted mean. For each course, multiply the grade, in points, by the number of credits that course is worth. Then, find the weighted mean, or GPA, by taking the sum of all of these numbers and dividing by the total number of credits.

Athena's grades from last semester are listed in the table. Her school uses a 4-point scale so that an A is 4 points, a B is 3 points, a C is 2 points, a D is 1 point, and an F is 0 points. Find Athena's grade point average or GPA.

Course	Grade	Credits
Human Services	B	4
Crisis Intervention	A	2
Sociology	C	3
Addictions Counseling	A	4

Solution

Multiply the grade, in points, by the number of credits. Then, add those products and divide by the total number of credits.

Course, Grade, Grade Points	Credits	Points · Credits
Human Services B = 3	4	$3 \cdot 4 = 12$
Crisis Intervention A = 4	2	$4 \cdot 2 = 8$
Sociology C = 2	3	$2 \cdot 3 = 6$
Addictions Counseling, A = 4	4	$4 \cdot 4 = 16$
Totals	13	**42**

$$\text{GPA} = \frac{42}{13} \qquad \frac{\text{Total number of grade points}}{\text{Total number of credits}}$$

$$= 3.23 \qquad \text{Divide. Round to the nearest hundredth.}$$

Athena's GPA was 3.23.

Find the Median

The **median** of an ordered list of numbers is the middle number. To find the median of a list of values, follow this three-step process.

Procedure for Finding the Median of a List of Values

Step 1: Arrange the values from lowest to highest.

Step 2: Determine whether there is an even or odd number of values in the list.

Step 3: If there is an *odd number* of values, the median is the *middle number*. If there is an *even number* of values, the median is the *mean (average) of the middle two numbers.*

Find the median of each list of numbers.

a) 19, 25, 27, 20, 26, 18, 23

b) 8.5, 5.4, 6.7, 8.8, 9.2, 7.9

Solution

a) Arrange the numbers from lowest to highest.

18, 19, 20, 23, 25, 26, 27

Because there is an *odd* number of values in the list (7 values), the median is the *middle number* in the list. The median is 23.

b) Arrange the numbers from lowest to highest.

5.4, 6.7, 7.9, 8.5, 8.8, 9.2

Because there is an *even* number of values in the list (6 values), the median is the *mean of the middle two numbers* in the list.

$$\text{Median} = \frac{7.9 + 8.5}{2} = 8.2$$

The median is 8.2.

Definition/Procedure	Example

Find the Mode

The **mode** is the number that appears most frequently in a list of numbers.

Procedure for Finding the Mode of a List of Numbers
Arrange the numbers from lowest to highest, and underline the numbers that are repeated.

1) If one number appears more often than the others, then that number is the **mode.**

2) If there are two numbers that appear most often in a list, then the list has two modes and is said to be **bimodal.** The numbers that appear most often are the modes.

3) The list has **no mode** if no number appears more often than any other number.

Find the mode of each list of numbers.

a) 74%, 71%, 76%, 87%, 90%, 71%, 64%, 76%, 71%

b) $2.99, $2.69, $2.99, $3.09, $2.89, $3.09, $3.29

c) 51, 58, 49, 56, 50, 62, 68, 54, 57

Solution

a) It is helpful to arrange the numbers from lowest to highest and underline the numbers that are repeated:

$$64\%, \underline{71\%}, \underline{71\%}, \underline{71\%}, 74\%, \underline{76\%}, \underline{76\%}, 87\%, 90\%$$

The number that appears most often is 71%, so the mode is 71%.

b) Arrange the numbers from lowest to highest and underline the numbers that are repeated:

$$\$2.69, \$2.89, \underline{\$2.99}, \underline{\$2.99}, \underline{\$3.09}, \underline{\$3.09}, \$3.29$$

Both $2.99 and $3.09 appear twice. The list is *bimodal,* and the modes are $2.99 and $3.09.

c) Arrange the numbers from lowest to highest.

$$49, 50, 51, 54, 56, 57, 58, 62, 68$$

No number appears more often than any other. The list has **no mode.**

5.8 Solving Equations Containing Decimals

To solve equations containing decimals, we can use the same steps for solving a linear equation that we learned in Chapter 2.	Solve $1.8x + 0.4 = 1.3x + 1.4$.

Solution

$1.8x + 0.4 = 1.3x + 1.4$.

$1.8x + 0.4 - 0.4 = 1.3x + 1.4 - 0.4$	Subtract 0.4 from each side.
$1.8x = 1.3x + 1$	Subtract.
$1.8x - 1.3x = 1.3x - 1.3x + 1$	Subtract $1.3x$ from each side.
$0.5x = 1$	Combine like terms.
$\dfrac{0.5x}{0.5} = \dfrac{1}{0.5}$	Divide both sides by 0.5.
$x = 2$	Simplify.

Definition/Procedure	Example
Eliminating Decimals from an Equation To eliminate decimals from an equation, multiply both sides by the smallest power of 10 that will eliminate all decimals.	Solve $1.8x + 0.4 = 1.3x + 1.4$ by first eliminating the decimals. **Solution** $10(1.8x + 0.4) = 10(1.3x + 1.4)$ — Multiply both sides by 10. $18x + 4 = 13x + 14$ — Distribute. $18x + 4 - 4 = 13x + 14 - 4$ — Subtract 4 from each side. $18x = 13x + 10$ — Subtract. $18x - 13x = 13x - 13x + 10$ — Subtract 13x from each side. $5x = 10$ — Combine like terms. $\dfrac{5x}{5} = \dfrac{10}{5}$ — Divide both sides by 5. $x = 2$ — Simplify. Notice that this is the same equation and solution as in the previous box.

5.9 Square Roots and the Pythagorean Theorem

Perfect Squares and Square Roots A number such as 16 is a **perfect square** because it is the square of a whole number: $16 = 4^2$ We also say that $\sqrt{16}$ is a **perfect square root** because $\sqrt{16} = 4$.	**The First 12 Perfect Square Roots** $\sqrt{1} = 1 \quad \sqrt{16} = 4 \quad \sqrt{49} = 7 \quad \sqrt{100} = 10$ $\sqrt{4} = 2 \quad \sqrt{25} = 5 \quad \sqrt{64} = 8 \quad \sqrt{121} = 11$ $\sqrt{9} = 3 \quad \sqrt{36} = 6 \quad \sqrt{81} = 9 \quad \sqrt{144} = 12$ Note: $\sqrt{0} = 0$
Characteristics of Right Triangles Right triangles have special properties that other triangles do not. The side opposite the right angle is the longest side of the triangle and is called the **hypotenuse.** The other two sides are called the **legs.**	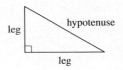
The Pythagorean Theorem The **Pythagorean theorem** states the following relationship between the lengths of the sides of a right triangle. For any **right** triangle, $$(\text{leg})^2 + (\text{leg})^2 = (\text{hypotenuse})^2$$ The Pythagorean theorem applies **only** to right triangles.	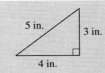 $(3 \text{ in.})^2 + (4 \text{ in.})^2 \overset{?}{=} (5 \text{ in.})^2$ $9 \text{ in}^2 + 16 \text{ in}^2 = 25 \text{ in}^2 \quad \checkmark$
Finding an Unknown Side of a Right Triangle 1) To find the **hypotenuse** of a right triangle, use this formula: $$\text{hypotenuse} = \sqrt{(\text{leg})^2 + (\text{leg})^2}$$ 2) To find a **leg** of a right triangle, use this formula: $$\text{leg} = \sqrt{(\text{hypotenuse})^2 - (\text{known leg})^2}$$	Find the length of the missing side. Give the exact answer and the answer to the nearest tenth. 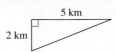

Definition/Procedure	Example
	First ask yourself, *"Which part of the right triangle is unknown?"* The side *across from* the right angle is unknown, and that is the *hypotenuse*. Use the formula for finding the hypotenuse.

$$\text{hypotenuse} = \sqrt{(\text{leg})^2 + (\text{leg})^2}$$
$$= \sqrt{(5)^2 + (2)^2} \qquad \text{Substitute the values.}$$
$$= \sqrt{25 + 4} \qquad 5^2 = 25;\ 2^2 = 4$$
$$= \sqrt{29} \qquad \text{Add.}$$

The length of the hypotenuse is exactly $\sqrt{29}$ km.

$$\sqrt{29} \approx 5.4 \qquad \text{Approximate by hand or using a calculator}$$

The length of the hypotenuse is approximately $\sqrt{29} \approx 5.4$ km. |

5.10 Circles, Spheres, Cylinders, and Cones

Radius and Diameter of a Circle

A **circle** is a two-dimensional, or flat, figure in which all points are the same distance from the fixed center point.

The **radius**, r, is the distance from the center of the circle to any point on the circle.

The **diameter**, d, is the distance across the circle passing through the center.

The diameter and the radius are related in this way:

1) Diameter $= 2 \cdot$ radius   or $d = 2 \cdot r$ or $d = 2r$

2) Radius $= \dfrac{\text{Diameter}}{2}$ or $r = \dfrac{d}{2}$

What Is π?

π is the ratio of the circle's circumference to its diameter. That is, $\pi = \dfrac{C}{d}$. (π is the Greek letter pi and is read as "pie.")

There is no exact, decimal equivalent of π. An approximate decimal value is $\pi \approx 3.14159265$, but we usually use one of the following approximations for π.

$$\pi \approx 3.14 \qquad \text{and} \qquad \pi \approx \frac{22}{7}$$

Because these are *approximate* values, calculations using π will give *approximate* answers. Therefore, we should use the $\approx$ symbol instead of $=$.

This is how the value of π is computed.

$d = 1$ in.

If we measure the distance around the circle, its circumference, we get approximately 3.14 in.

$$\frac{\text{circumference}}{\text{diameter}} \approx \frac{3.14 \text{ in.}}{1 \text{ in.}} = 3.14$$
$$\pi \approx 3.14$$

Definition/Procedure	Example

Circumference of a Circle
 	
The **circumference** of a circle is the distance around the circle. (The circumference is the *perimeter* of the circle.) It is usually abbreviated with *C*.
 	
These are formulas for the circumference of a circle.
 	
1) Circumference = $\pi \cdot$ diameter   or $C = \pi d$
 	
    Because $d = 2 \cdot$ radius, we can also write
    $C = \pi \cdot 2 \cdot$ radius
 	
2) Circumference = $2 \cdot \pi \cdot$ radius   or $C = 2\pi r$
	Find the circumference of this circle. Give the *exact* value and an *approximate* value using $\pi \approx \dfrac{22}{7}$.
	
	We are given the diameter, so we will use the formula containing the diameter of the circle.
	
	$C = \pi \cdot d$
	$C = \pi \cdot 14$ cm Substitute the diameter.
	$C = 14\pi$ cm This is the *exact* value.
	
	To find the *approximate* circumference, substitute $\dfrac{22}{7}$ for π.
	
	$C \approx 14\left(\dfrac{22}{7}\right)$ cm Substitute $\dfrac{22}{7}$ for π.
	$C \approx 44$ cm Multiply.
	
	The *circumference* is exactly 14π cm, and it is *approximately* 44 cm.

Area of a Circle The **area**, *A*, of a circle with radius *r* is $A = \pi \cdot r^2$ or $A = \pi r^2$	Find the area of the circle above. Give the *exact* value and an *approximate* value using 3.14 for π. In the circle above, we are given the diameter. To use the area formula, we need to find the radius. $r = \dfrac{d}{2} = \dfrac{14 \text{ cm}}{2} = 7$ cm radius = 7 cm $A = \pi r^2$ $A = \pi (7 \text{ cm})^2$ Substitute the value of the radius. $A = 49\pi$ cm^2 Evaluate the exponent. The *exact* area is 49π cm^2. This time, we will substitute 3.14 for π to *approximate* the area. $A \approx 49\pi$ cm^2 $A \approx 49(3.14)$ cm^2 Substitute 3.14 for π. $A \approx 153.86$ cm^2 Multiply. The area is *approximately* 153.86 cm^2.
The **volume** of a three-dimensional object is a measure of the amount of space occupied by the object or the amount of space inside the object. Volume is measured in *cubic* units, the number of *cubic* units it takes to fill the object.	For the volume formulas of all solids we have studied so far, see the table in this section.
The **surface area** of a three-dimensional object is a measure of the total area on the surface of the object. Surface area is measured in *square* units.	For the surface area formulas for all solids we have studied so far, see the table in this section.

Definition/Procedure	Example
Using Volume and Surface Area Formulas A **sphere** is a round, three-dimensional object like a basketball or the Earth. The radius of a sphere is the distance from the center to the edge of the sphere. The **volume**, V, of a sphere with radius r is $$\text{Volume} = \frac{4}{3}\pi(\text{radius})^3 \quad \text{or}$$ $$V = \frac{4}{3}\pi r^3$$ 	Find the volume of a sphere with radius 6 m. Give an exact value and an approximation using $\frac{22}{7}$ for π. $$V = \frac{4}{3}\pi r^3$$ $$V = \frac{4}{3}\pi(6\text{ m})^3 \qquad \text{Substitute 6 m for } r.$$ $$V = \frac{4}{3}\pi(216\text{ m}^3) \qquad \begin{array}{l}(6\text{ m})^3 = 6\text{ m} \cdot 6\text{ m} \cdot 6\text{ m}\\ = 216\text{ m}\end{array}$$ $$V = \frac{4}{3}\pi\left(\frac{\overset{72}{216}\text{ m}^3}{1}\right) \qquad \text{Divide 3 and 216 by 3.}$$ $$V = 288\pi\text{ m}^3 \qquad \text{Multiply.}$$ The *exact* volume is 288π m^3. To find an approximation, substitute $\frac{22}{7}$ for π. $$V \approx 288\left(\frac{22}{7}\right)\text{m}^3 \qquad \text{Substitute } \frac{22}{7} \text{ for } \pi.$$ $$V \approx \frac{6336}{7}\text{ m}^3 \qquad \text{Multiply.}$$ The volume is approximately $\frac{6336}{7}$ m^3 or $905\frac{1}{7}$ m^3 or 905.14 m^3, rounded to the nearest hundredth. To find the surface area, use $SA = 4\pi r^2$. $$SA = 4\pi(6\text{ m})^2 = 4\pi(36\text{ m}^2) = 144\pi\text{ m}^2$$ The *exact* surface area is 144π m^2. To find an approximation, substitute $\frac{22}{7}$ for π. $$SA \approx 144\left(\frac{22}{7}\right)\text{m}^2 = \frac{3168}{7}\text{ m}^2 \approx 452.57\text{ m}^2$$

Chapter 5: Review Exercises

(5.1)

1) Use a fraction with a denominator of 10 as well as a decimal to represent the shaded part of the rectangle. Then, represent the fraction as a decimal on a number line.

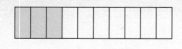

2) Use a fraction and a decimal to represent the shaded part of the figure. Write the fraction in lowest terms.

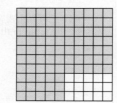

Approximate the location of the decimal on the given number line.

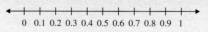

3) 0.93

4) 0.17

Write each fraction or mixed number as a decimal.

5) $\frac{97}{1000}$

6) $-4\frac{1}{10}$

7) $-\frac{3867}{1000}$

8) $\frac{1724}{100}$

9) Identify the place value of each digit.

52.406798

10) Write 0.029 in words.

Write each word statement as a decimal number.

11) Fifty thousand seventy-two and thirty-six hundredths

12) Four hundred nineteen ten-thousandths

Write each decimal as a fraction or mixed number in lowest terms.

13) −0.98

14) 0.575

15) 1.5

16) −6.0072

(5.2)

17) Round 39,604.9951 to the indicated place.

   a) thousands   b) ten-thousands

   c) tenths   d) hundredths

18) Round $3.78195 to the nearest a) cent and b) dollar.

Solve each problem.

19) Tangaroa bought a surfboard for $549.99, and he computed the amount of sales tax to be exactly $35.74935. How much will Tangaroa pay at the cash register including the sales tax?

©John Lund/Drew Kelly/Blend Images LLC RF

20) Greta takes her son shopping for winter clothes and puts the following items in her shopping cart. Round the price of each item to the nearest dollar, then add the rounded numbers to estimate the total cost of the items.

   Coat: $49.99

   Hat: $6.09

   Gloves: $4.85

   Boots: $23.49

(5.3)

21) Explain how to add decimals.

22) What is the first step in performing this subtraction? Do *not* subtract.

$$\begin{array}{r} 19 \\ -\ 2.584 \\ \hline \end{array}$$

Add or subtract as indicated.

23) Find the sum of 4.19 and 3.05.

24) $\begin{array}{r} 50.042 \\ -\ 39.619 \\ \hline \end{array}$

25) −405.293 + 87.55

26) 53.1 + (−43.908)

27) 19 minus 2.584

28) Subtract 0.0397 from 1.

29) 16,001.5 + 42.936 + 579 + 0.02

Solve each problem.

30) Amanda has $817.32 in her checking account. When she uses her debit card, the money comes out of this account. She makes the following purchases with her debit card:

   Boots: $139.87

   Make-Up: $25.09

   Scarf: $32.46

   Jeans: $53.71

   Leather Coat: $206.50

   Purse: $74.83

   a) How much did Amanda spend on this shopping trip?

   b) How much money is left in her account?

   c) If Amanda wants to bring her account balance up to $500, how much must she deposit?

31) At a pet supply store, Anand buys a dog collar and some treats for $16.28. If he gives the cashier a $20 bill and three pennies, how much change will he receive?

(5.4)

32) Is it necessary to line up the decimal points when multiplying decimal numbers?

Multiply.

33) 43.57(−1.08)

34) 16 · 62.195

35) $\begin{array}{r} 0.064 \\ \times\ 0.03 \\ \hline \end{array}$

36) (−971.8)(−250)

37) $\begin{array}{r} 31.059 \\ \times\ 2.76 \\ \hline \end{array}$

38) $(1.2)^2$

39) $(−4.7)^2$

40) 3.8 × 1000

41) −200.5 · 0.0001

42) (0.00001)(3505)

Solve each problem.

43) A rugby field is in the shape of a rectangle and is 157.5 yards long and 76.6 yards wide. What is the area of the field?

©Don Hammond/Design Pics

44) For his daughter's birthday party, Dashiell bought fourteen birthday cupcakes for $2.55 each. Sales tax on this purchase was $2.29. What was the total cost of the cupcakes? If Dashiell pays for the cupcakes by giving the cashier two $20 bills, how much change does he receive?

(5.5) Divide.

45) 45.48 ÷ 6

46) $\dfrac{-15}{0.05}$

47) 0.03596 ÷ 6.2

48) Divide 2146 by 0.029.

49) Divide $\dfrac{59.2}{0.9}$. Give the exact answer and an approximation to the nearest thousandth.

Simplify each expression using the order of operations.

50) $410 - 67.48 \times 3.5 + 7.625$

51) $(1.6)^2 + 9.6 \div (-0.3)$

52) $100(51 - 50.328) \div (0.2)^2$

Solve each problem.

53) Savannah finances $11,654.40 to purchase her new car. Her loan is a 48-month, interest-free loan. If Savannah pays equal amounts per month for 48 months, what is her monthly payment?

54) The total weight of a shipment of boxes was 317.3 pounds, and each box weighed 16.7 pounds. How many boxes were in the shipment?

(5.6) Use division to write each fraction or mixed number as a decimal.

55) $\dfrac{4}{5}$

56) $1\dfrac{1}{8}$

57) $-5\dfrac{1}{16}$

58) $-\dfrac{13}{20}$

Write each fraction as a decimal. Give the exact answer and an approximation rounded to the nearest hundredth.

59) $\dfrac{2}{11}$

60) $\dfrac{4}{15}$

Write each fraction as a decimal by first writing the fraction with a denominator of 10, 100, or 1000.

61) $\dfrac{9}{250}$

62) $-\dfrac{35}{4}$

Use =, >, or < to compare the numbers.

63) $6.05 \underline{\hphantom{xx}} 6.049$

64) $3\dfrac{4}{500} \underline{\hphantom{xx}} 3.004$

65) $\dfrac{2}{3} \underline{\hphantom{xx}} 0.\overline{6}$

66) $-0.27 \underline{\hphantom{xx}} -\dfrac{13}{50}$

Arrange each group of numbers in order from smallest to largest.

67) $-0.76, \dfrac{1}{2}, -0.7, \dfrac{3}{4}$

68) $0.\overline{8}, 1.01, -\dfrac{4}{5}, -\dfrac{9}{10}$

(5.7)

69) Hillary is looking at four used cars with the following mileages: 51,427, 49,058, 60,229, 55,694. Find the average number of miles on these cars.

70) A restaurant has seven appetizers on its menu, with the following prices: $6.99, $8.49, $7.89, $7.99, $9.29, $10.99, $8.99. What is the mean price of an appetizer?

71) Find Paolo's GPA. Let A = 4 points, B = 3 points, C = 2 points, D = 1 point, and F = 0 points. Round the answer to the nearest hundredth.

Course	Grade	Credits
Constitutional Law	C	3
Criminal Law	A	3
Business Math	A	4
Word Processing	B	2

72) The table shows the number of miles employees drive to work each day and the number of employees who drive that distance. Find the weighted mean rounded to the nearest tenth.

Distance	Number of Employees
3	1
8	3
15	9
20	6
23	2

73) The numbers of miles Maurice rode his bike each day last week were 18, 17, 25, 20, 28, 25, and 16. Find the median.

©Roberto Caucino/Shutterstock

74) The attendances at a Broadway play for each of its previous ten shows were 1430, 1392, 1571, 1580, 1465, 1609, 1718, 1696, 1381, and 1628. Find the median.

Find the mode of each list of numbers. If the list is bimodal or has no mode, then say so.

75) 12.5, 9.4, 11.6, 12.1, 11.6, 8.9

76) 70%, 91%, 64%, 52%, 73%, 85%, 73%, 66%, 52%, 73%

77) 38, 45, 49, 42, 37, 40

78) $6.29, $6.79, $6.39, $6.99, $6.89, $6.99, $7.19, $6.79

Mixed Exercises: Sections 5.1–5.7

Perform the indicated operations.

79) $(-0.098)(-72.1)$

80) Subtract 14.188 from 23.

81) $1135.7 + 4.092 + 1135.688 + 730$

82) $2.4\overline{)73.68}$

83) $-46.5 \times 1.2 + 0.01(63.94 - 87)$

84) $(1.2)^2 - 4.5 \div 15 + 2.8$

85) Round $73.8975 to the nearest

    a) dollar   b) cent

86) Write $\dfrac{39}{10,000}$ as a decimal.

87) Write 7.528 as a fraction or mixed number in lowest terms.

Solve each problem.

88) Here is a list of Mr. Carson's students' grades: 62%, 75%, 69%, 81%, 90%, 69%, 72%, 74%, 86%, 72%, 83%, 72%, 51%, 43%, 69%, 79%, 74%, 48%. Find the mean, median, and mode.

89) A sandwich shop is open for business from 11:00 A.M. to 9:00 P.M. One day, the tips totaled $61.20. The manager does calculations each day to determine how much each employee earns in tips. Here is the list of employees and the hours they worked:

    Vernon   10:00 A.M.–5:30 P.M.

    Jane   10:00 A.M.–2:00 P.M.

    Aliyah   11:00 A.M.–2:30 P.M.

    Domingo   12:00 P.M.–8:00 P.M.

    Nayana   3:00 P.M.–10:00 P.M.

    Steve   6:00 P.M.–10:00 P.M.

    a) Determine the total number of hours worked by all the employees.

    b) Determine the average amount of tips earned per hour.

    c) The manager gives each of the employees his or her tips in cash. How much will she give each employee? (Multiply the number of hours worked by each employee by the average amount of tips earned per hour.)

90) If the perimeter of the square is 12.224 in., what is the measure of one side length?

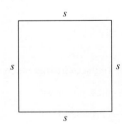

(5.8) Solve each equation.

91) $-4z = 3.08$

92) $p + 5.9 = 14.2$

93) $2.9 = 8.1 - 1.3x$

94) $12a - 11.4 = 6(2.5a - 0.7)$

Solve each equation by first eliminating the decimals.

95) $0.2n - 0.33 = -0.1$

96) $0.08k + 0.24 = 0.1k + 0.4$

Write an equation, and solve.

97) One month, Ramone earned a commission of $607.42 in addition to $13.70 per hour. If his total income was $1128.02, how many hours did he work?

98) A rectangular window is 2.4 ft long and has an area of 3.84 ft². What is the width?

(5.9) Approximate each square root to the nearest tenth and plot it on a number line.

99) $\sqrt{32}$

100) $\sqrt{50}$

Find the length of the missing side. Give the exact answer and the answer to the nearest tenth, if appropriate.

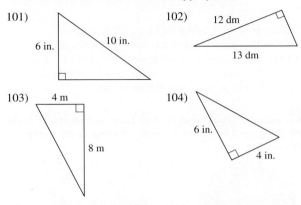

101)   6 in.   10 in.

102)   12 dm   13 dm

103)   4 m   8 m

104)   6 in.   4 in.

105) A ladder 13 ft long is leaning against a wall, forming a right triangle. The bottom of the ladder is 5 ft away from the wall. Find the area of the formed triangle.

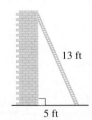

13 ft

5 ft

(5.10) Find the diameter of each circle.

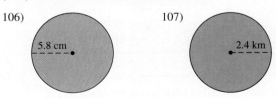

106)   5.8 cm

107)   2.4 km

Find the radius of each circle.

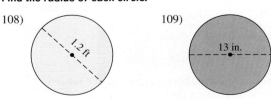

108)   1.2 ft

109)   13 in.

Find the area and circumference of each circle. Give the *exact* value and an *approximate* value using π ≈ 3.14. Round the approximation to the nearest hundredth.

110)
$1\frac{1}{2}$ m

111)
13 mm

112) Find the area of the semicircle. Give the *exact* value and an *approximate* value using $\pi \approx \frac{22}{7}$.

7 ft

113) At one point as it moved through the Caribbean in September of 2017, Hurricane Irma (in the shape of a circle) had a diameter of about 400 miles. Find the area covered by this hurricane. Give an approximate answer using 3.14 for π. (www.washingtonpost.com)

Find the volume of each solid. Give an exact and an approximate answer using π ≈ 3.14. Round the approximation to the nearest hundredth.

114)
30 cm

115)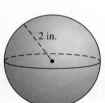
2 in.

116)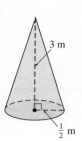
3 m
$\frac{1}{2}$ m

117)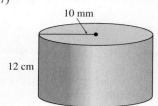
10 mm
12 cm

118)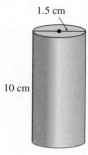
1.5 cm
10 cm

119) Find the surface area of the sphere in Exercise 115. Give an exact and an approximate answer using π ≈ 3.14. Round the approximation to the nearest hundredth.

120) A soup can in the shape of a right circular cylinder has a radius of 3.5 cm, and it is 11 cm tall. How much soup will it hold? Use 3.14 for π and give an approximate answer to the nearest hundredth.

Chapter 5:   Test

1) Use a decimal to represent the shaded part of the figure, and represent the decimal on a number line.

2) Write each decimal as a fraction or mixed number in lowest terms.

   a) 0.73   b) −0.8   c) 2.075

3) Write each decimal in words.

   a) 4.09   b) 0.0614

4) Write as a decimal number: *sixteen and five hundred seventy-three thousandths*

Round each number to the indicated place.

5) 9.4683 to the nearest hundredth

6) 310.974 to the nearest tenth

7) $2375.46 to the nearest dollar

8) $62.1352 to the nearest cent

9) Explain, in your own words, how to multiply decimals. Do you need to line up the decimal points when you write out the problem?

Perform the indicated operations.

10) −3.86(−5.9)   11) 87 ÷ 0.012

12) 63.14 + 5.9362 + 158.791

13) −74.3 + 28.96   14) $\dfrac{0.8785}{-35}$

15) $3.5(12.7 - 10.1) + (0.9)^2 - 1.2 \times 5$

16) $1000(0.000829)$ 17) $273 - 0.406$

18) Find $0.78\overline{)5.2}$. Give the exact answer and an approximation rounded to the nearest thousandth.

19) Write $9\dfrac{3}{8}$ as a decimal.

Use =, >, or < to compare the numbers.

20) -0.82 ____ -0.8256 21) $2\dfrac{3}{5}$ ____ 2.6

22) $\dfrac{32}{25}$ ____ 1.44

23) Write in order from smallest to largest.

   -1.207 1.073 0.73 1.25 -1.2

Solve each problem.

24) Rodrigo had $653.28 in his checking account and deposited his paycheck of $807.14. Then, he paid $66.40 for the electric bill, $120.79 for the cable bill, $195.82 for his credit card, and $19.60 for his water bill. How much money remains in Rodrigo's account?

25) A 20-pound bag of dog food costs $18.99. Find the cost per pound to the nearest cent.

©Ingram Publishing

26) A coffee shop is open from 5 A.M. to 7 P.M., and each hour on Thursday it served the following number of customers: 11, 47, 54, 63, 61, 34, 26, 25, 26, 29, 33, 31, 20, 15. Find the mean, median, and mode. Round the answer to the nearest tenth, if necessary.

27) Find Ava's GPA. Let A = 4 points, B = 3 points, C = 2 points, D = 1 point, and F = 0 points. Round the answer to the nearest hundredth.

Course	Grade	Credits
English Composition	A	4
Advertising	B	3
E-Marketing	A	3
Management	C	4

28) Solve each equation.

   a) $-11.6 = n - 3.9$

   b) $3(0.4k + 2) - 1 = 9.2$

   c) $0.9w + 0.4 = 0.7w + 1.6$

29) Write an equation and solve.

   A rectangular lapdesk is 1.2 ft wide and has an area of 2.16 ft^2. What is the length?

30) Can the Pythagorean theorem be used to find the lengths of the sides of any kind of triangle? Explain your answer.

31) Find the length of the missing side. Give the exact value and an approximation to the nearest tenth.

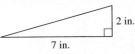
2 in.
7 in.

32) Romeo brings a 13-ft ladder to Juliet's window. He places the bottom of the ladder 5 ft from her house, and the top of the ladder rests at the bottom of her window. Find the height of Juliet's window.

13 ft
5 ft

33) What term do we use for the perimeter of a circle?

34) For this circle,

10 in.

   a) find the area.

   b) find the circumference.

   Give the exact value and an approximation using 3.14 for π.

35) For this figure,

5 ft

   a) identify the figure.

   b) find the volume of the solid.

   c) find the surface area of the figure.

   Give the exact value and, where appropriate, an approximate value rounded to the nearest hundredth using 3.14 for π.

36) A Tootsie Roll is in the shape of a cylinder. It is 21 mm long, and its diameter is 10 mm. What is the volume of the Tootsie Roll? Use 3.14 as an approximation for π.

Perform the indicated operations.

1) $74{,}596 + 103 + 8449 + 62$

2) $43{,}000 - 599$

3) $\begin{array}{r} 778 \\ \times\, 243 \\ \hline \end{array}$

4) $0.12\overline{)87.6}$

5) $-\dfrac{8}{15} \div 6$

6) $-\dfrac{1}{6} + \left(-\dfrac{5}{12}\right) + \dfrac{1}{8}$

7) If the perimeter of the figure is 60 cm, find the missing side length.

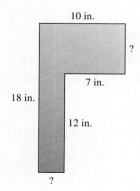

8) Find the two missing side lengths, then find the perimeter of the figure.

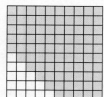

9) Use a fraction and a decimal to represent the unshaded part of the figure. Reduce the fraction if possible.

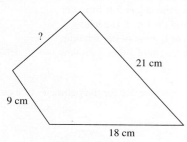

10) Approximate the location of the decimal value on the given number line.

0.52

```
    0  0.1 0.2 0.3 0.4 0.5 0.6 0.7 0.8 0.9  1
```

11) Write *eighty-three and twenty-three ten-thousandths* using numbers.

Round each number as indicated.

12) 6,399,699 to the nearest thousand

13) 0.549873 to the nearest ten-thousandth

Write each decimal as a fraction or mixed number in lowest terms.

14) 11.8

15) 0.0016

Simplify each expression using the order of operations.

16) $9 \cdot \sqrt{16} - 3 + \sqrt{25}$

17) $\left(-\dfrac{3}{4}\right)^2 \div \left(\dfrac{3}{2} - \dfrac{5}{4}\right) + \dfrac{3}{8}$

18) $0.96 - 0.001(1002.5 - 32.9)$

19) Find the mode of this list of numbers: 61, 59, 48, 92, 70, 59, 66, 83, 61, 74, 88, 59, 49

Solve each equation.

20) $\dfrac{1}{3}t - \dfrac{5}{12} = \dfrac{1}{2}t + \dfrac{1}{6}$

21) $0.2n + 4.9 = 10.7$

Solve each problem.

22) A soup recipe uses $3\dfrac{1}{2}$ cups of milk. If Elijah wants to cut the recipe in half, how much milk should he use?

23) Korina and three of her friends purchase a four-pack of 16-GB USB 2.0 flash drives for $50.68. If they are going to split the cost equally among themselves, how much does each person pay?

24) Opening weekend ticket sales in the United States for the first three *Twilight Saga* movies were $69.6 million, $142.8 million, and $68.5 million. What was the mean revenue from ticket sales? Round the answer to the nearest tenth of a million. (www.imdb.com)

Write an equation and solve.

25) Alec and his cousin Kullen were playing Nintendo. Kullen's score was 17.833 points less than Alec's score. If they had 384,241 points combined, what was each boy's score?

©Ingram Publishing

Design elements: Yellow diamond road sign ©Nora Development; Texas Instruments TI-83 graphing calculator ©mbbirdy/Getty Images; Yellow pencil ©McGraw-Hill Education.

Ratios, Rates, and Proportions

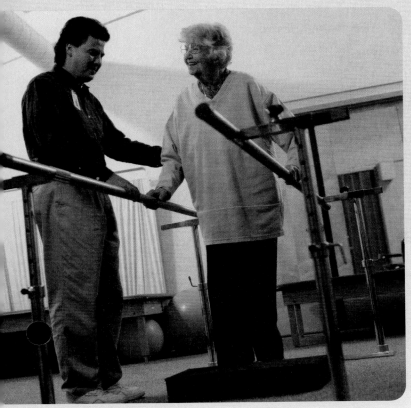

©Keith Brofsky/Getty Images

Math at Work:

Physical Therapist

Sam Blekicki's days are filled with broken legs, torn ligaments, and dislocated shoulders. And he wouldn't have it any other way. As a physical therapist, Sam's job is to help people who have suffered such injuries recover their strength and mobility.

"It is very rewarding work," Sam says. "Seeing someone who has been in a wheelchair since a car accident walk across the room . . . well, there's just nothing like it."

Even though Sam's work is focused on muscles and joints, he couldn't do his job without math. In order to understand his patients' progress, he determines the proportion of muscle loss and considers the ratio of time spent on therapy to degree of injury. Sometimes he tracks recovery rates so closely that he examines the number of steps per day a patient makes. "In physical therapy, you know that nothing happens overnight," Sam explains. "What you hope for is a rate of slow and steady progress."

The field of physical therapy is changing rapidly. Sam says that if he didn't regularly read medical journals during his off-hours, he wouldn't be able to keep up with the latest advances in treatment. "I always imagined that when I finished school, I'd never have to do homework again," Sam says. "The truth is that if you want to be good at your job, you need to spend time working at it even when you're not in the office."

In this chapter, we explore ratios, rates, and proportions, the same type of math Sam uses to help his patients on the road to recovery. We'll also look at some strategies to help you do your best on your math homework.

 **Study Strategies** Doing Math Homework

If you're like most students, homework is probably not your favorite part of being in college. Yet homework is essential to college success. Let's look at how you can use the steps of P.O.W.E.R. to *effectively and efficiently* do math homework.

- **I will do my homework effectively and efficiently.**

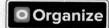

- Complete the emPOWERme survey that appears before the Chapter Summary to learn about your attitudes toward homework.
- Realize that you should be doing the homework to *truly learn the material* and not just to get it done.
- Decide what you will need to do your homework: your class notes, book, pencils, pens, ruler, and paper, and have those materials with you.
- Know your instructor's expectations: Will you be able to use a calculator on your quizzes and exams? If not, do not use one while doing your homework.
- Think about *how* you learn best: Is it in a completely quiet room? Is it in a room with ambient noise, like a coffee shop or a tutoring center? Is it while listening to music? Really think about how you learn best *and* about what does *not* work well for you. Then, choose the appropriate environment for doing your homework.
- Plan to study in a place with a comfortable chair and a table that has enough space for you to open your book and notes *in addition to* having enough space to write.
- Set aside enough time to do your homework, and do it soon after it is assigned so that you do not forget what you learned in class. Research shows that you are much more likely to remember what you have learned if you do your homework within 24 hours of leaving class than if you wait longer to start it. At the very least, do the 24-hr problems noted with this icon ⓬ in the exercise set within 24 hours of leaving class if you must do the rest of the homework at a later time.

- When it is time to do your homework, sit at a table and spread out everything you need.
- Turn off your cell phone and put it away to minimize distractions.
- Read over your notes and examples in the book to remind you of what you learned in class. *Then*, start the homework your instructor has assigned.
- Do your homework in a designated math notebook or on loose-leaf paper that you keep in a three-ring binder or folder that is just for your math course.
- Write neatly and show all of your work. This allows your instructor to see your thought processes if you made a mistake and need help.
- Don't give up if you get stuck. Try a different approach. Look back at your notes or the book if you need help. Circle the problem and come back to it later. Consider going to the tutoring center for help.
- If you cannot use a calculator on your quizzes and exams, do not use one while doing your homework. The same is true for homework websites. Do not use them just to complete the homework. Remember, the purpose of doing homework is for learning the math! You will not learn and you will not be ready for quizzes and tests if you do not do the homework yourself.
- Circle any problems that you could not do, and ask about them in class.

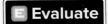 **Evaluate**

- After going over the homework in class, think about how well you did. Do you have a true understanding of the concepts in the assignment?
- Make corrections and notes on your paper for any problems that you did wrong. Be sure you understand your mistakes and the correct way to do the problems.

 **Rethink**

- Keep all of your corrected homework in the same place so that you can refer to it later when studying for a quiz or exam.
- Even after making corrections, do you still have questions about anything in the assignment? If so, see your instructor during office hours or go to the tutoring center. Be sure you understand everything because you will need to know it for the quizzes and test and for learning future concepts.
- If you feel like you had a very good understanding of how to do the exercises in the homework, think about what you did well to learn the material. Would you use the same approach next time? What might you do differently?
- If you had difficulty completing the homework, ask yourself why. Should you have chosen a different place to do your homework? Were your notes difficult to understand? Did you wait too long before you started to do your homework? Think about adjustments that need to be made.

©Fuse/Getty Images

Chapter 6  Plan

What are your goals for Chapter 6?	How can you accomplish each goal?
1 Be prepared before and during class.	• Don't stay out late the night before class, and be sure to set your alarm! • Bring a pencil, notebook paper, and textbook to class. • Avoid distractions by turning off your cell phone during class. • Pay attention, take good notes, and ask questions. • Complete your homework on time, and ask questions on problems you do not understand.
2 Understand the homework to the point where you could do it without needing any help or hints.	• Read the directions, and show all of your steps. • Go to the instructor's office for help. • Rework homework and quiz problems, and find similar problems for practice.
3 Use the P.O.W.E.R. framework to learn some strategies for doing homework.	• Complete the emPOWERme that appears before the Chapter Summary. • Read the Study Strategies that explain how you can use P.O.W.E.R. to do your homework effectively. • Think of ways you can apply this strategy beyond your math courses.
4 Write your own goal. _____ _____	• _____ _____

What are your objectives for Chapter 6?	How can you accomplish each objective?
1 Learn how to recognize and write ratios and rates.	• Understand the definitions as well as the similarities and differences between ratios and rates. • Know how to write and simplify ratios and rates. • Be able to use ratios and rates in an applied problem.
2 Determine whether a proportion is true or false.	• Know the definition of a *proportion*. • Know how to use cross products, and determine whether a proportion is true.
3 Learn to solve a proportion.	• Use the division property of equality, and combine it with using cross products to solve a proportion. • Use the Five Steps for Solving Applied Problems.
4 Understand the relationships between angles.	• Learn the definitions. • Apply the definitions to solve problems.
5 Identify congruent and similar triangles and solve applied problems.	• Learn the definitions. • Summarize the SSS, SAS, and ASA theorems and apply them. • Apply your knowledge of similarity to solve similar triangle problems.
6 Write your own goal. _____ _____	• _____ _____

E Evaluate Complete the Chapter Review and Chapter Test. How did you do?	**R Rethink** • Could you apply ratios and proportions to situations outside of a math class? Which concepts do you need to review?
	• How do you feel about your understanding of the geometry topics in this chapter? What do you understand well, and what do you need to work on more?
	• Did you use the P.O.W.E.R. framework to help you do your homework? If so, what worked well and what did not? If you did not use it, why not?

6.1 Ratios

P Prepare　　　　　　　　　　　**O Organize**

What are your objectives for Section 6.1?	How can you accomplish each objective?
1 Write Basic Ratios	• Write the definition of a *ratio* and your own procedure for **Writing a Ratio as a Fraction.**
	• Complete the given examples on your own.
	• Complete You Trys 1 and 2.
2 Write Ratios Comparing Fractions, Mixed Numbers, or Decimals	• Create a list of the different ways to simplify a ratio.
	• Use the Five Steps for Solving Applied Problems.
	• Complete the given examples on your own.
	• Complete You Trys 3 and 4.
3 Write Ratios After Converting Units	• Follow Example 5, and write your own procedure for **Converting Measurements.**
	• Examine the given relationships between measurements, and memorize them.
	• Complete the given examples on your own.
	• Complete You Trys 5–7.

 Work　　　**Read the explanations, follow the examples, take notes, and complete the You Trys.**

1　Write Basic Ratios

We hear about *ratios* and use them in many ways in everyday life. But, what is a ratio? A **ratio** is a comparison of two quantities. It can compare numbers or measurements with the same units. (*Rates* compare measurements with different units.) Ratios can be written in several different ways. For example, the ratio of 3 to 4 can be written as

$$
3 \text{ to } 4 \qquad \text{or} \qquad 3 \underset{\uparrow}{\overset{\downarrow}{:} } 4 \qquad \text{or} \qquad \frac{3}{4} \begin{array}{l} \leftarrow \text{First number} \\ \leftarrow \text{Second number} \end{array}
$$

Second number ↓ ... First number ↑

> **Note**
>
> When a ratio is to be written as a fraction, the number that comes first is the numerator and the number that comes second is the denominator.

Let's practice writing ratios.

EXAMPLE 1

Write each ratio as a fraction.

a) $7 to $16   b) 40 min to 15 min   c) 8 ft to 2 ft

Solution

a) The quantity that comes first is the numerator, and the second quantity is the denominator.

The ratio of $7 to $16 is $\dfrac{\$7}{\$16}$ ← First number ← Second number

$$= \dfrac{7}{16} \qquad \text{Divide out common units just like common factors.}$$

The ratio of $7 to $16 is $\dfrac{7}{16}$. Notice that we do not write units in a ratio.

b) Remember, the quantity that comes first is the numerator, and the second quantity is the denominator.

The ratio of 40 min to 15 min is

$$\dfrac{40\ \text{min}}{15\ \text{min}} = \dfrac{40}{15} \qquad \text{Divide out common units just like common factors.}$$

$$= \dfrac{40 \div 5}{15 \div 5} = \dfrac{8}{3} \qquad \text{Write the ratio in lowest terms.}$$

The ratio of 40 min to 15 min is $\dfrac{8}{3}$. We do *not* write a ratio as a mixed number.

c) The ratio of 8 ft to 2 ft is

$$\dfrac{8\ \text{ft}}{2\ \text{ft}} = \dfrac{8}{2} \qquad \text{Divide out common units just like common factors.}$$

$$= \dfrac{8 \div 2}{2 \div 2} = \dfrac{4}{1} \qquad \text{Write the ratio in lowest terms.}$$

The ratio of 8 ft to 2 ft is $\dfrac{4}{1}$. Notice that even though $\dfrac{4}{1}$ simplifies to 4, we do *not* write the ratio this way. Because a ratio compares two quantities, we write it as the fraction $\dfrac{4}{1}$.

Note

Keep in mind the following important facts about ratios.

1) Ratios are not written with units.
2) Ratios are usually written in lowest terms.
3) Ratios are not written as mixed numbers.
4) Ratios are not written as whole numbers.

$\begin{bmatrix}\textbf{YOU TRY 1}\end{bmatrix}$ Write each ratio as a fraction.

a) $5 to $8   b) 36 ft to 24 ft   c) 9 hr to 3 hr

EXAMPLE 2

The ratio of potassium to sodium in a single serving of oatmeal is 105 mg to 80 mg. Write this ratio as a fraction, and explain what it means.

©John A. Rizzo/Getty Images

Solution

The quantity that comes first is the numerator, and the second quantity is the denominator:

$$\frac{\text{Potassium}}{\text{Sodium}} = \frac{105 \text{ mg}}{80 \text{ mg}}$$

$$= \frac{105}{80} \qquad \text{Divide out the common units.}$$

$$= \frac{105 \div 5}{80 \div 5} = \frac{21}{16} \qquad \text{Write the ratio in lowest terms.}$$

W Hint

Why would it be incorrect to write a ratio as a mixed number?

The ratio of potassium to sodium is $\frac{21}{16}$. This means that for every 21 mg of potassium in the oatmeal, there are 16 mg of sodium. We do *not* write the ratio as a mixed number.

[YOU TRY 2]

The ratio of the number of girls in the choir to the number of boys in the choir is 18 to 10. Write this ratio as a fraction, and explain what it means.

2 Write Ratios Comparing Fractions, Mixed Numbers, or Decimals

If a ratio compares fractions, mixed numbers, or decimals, we can rewrite the ratio so that it compares whole numbers.

EXAMPLE 3

Write each ratio so that it compares two whole numbers in lowest terms.

a) $\frac{7}{9}$ to $\frac{1}{3}$ b) 2 hr to $2\frac{1}{2}$ hr   c) $1.25 to $4.50

Solution

a) Begin by writing the ratio $\frac{7}{9}$ to $\frac{1}{3}$ as $\dfrac{\frac{7}{9}}{\frac{1}{3}}$. To simplify $\dfrac{\frac{7}{9}}{\frac{1}{3}}$, rewrite it as a division problem.

$$\frac{\frac{7}{9}}{\frac{1}{3}} = \frac{7}{9} \div \frac{1}{3} \qquad \text{Rewrite as a division problem.}$$

W Hint

You have simplified these types of fractions, or ratios, before!

$$= \frac{7}{9} \cdot \frac{3}{1} \qquad \text{Multiply by the reciprocal.}$$

$$= \frac{7}{\underset{3}{9}} \cdot \frac{\overset{1}{3}}{1} \qquad \text{Divide out common factors.}$$

$$= \frac{7}{3} \qquad \text{Multiply.}$$

The ratio of $\frac{7}{9}$ to $\frac{1}{3}$ is $\frac{7}{3}$.

b) Write the ratio 2 hr to $2\frac{1}{2}$ hr as a fraction, and divide out the common units.

$$\frac{2 \text{ hr}}{2\frac{1}{2}\text{ hr}} = \frac{2}{2\frac{1}{2}}$$

Rewrite $\dfrac{2}{2\frac{1}{2}}$ as a division problem and perform the division.

$$\frac{2}{2\frac{1}{2}} = 2 \div 2\frac{1}{2} \qquad \text{Rewrite as a division problem.}$$

$$= 2 \div \frac{5}{2} \qquad \text{Change the mixed number to an improper fraction.}$$

$$= 2 \cdot \frac{2}{5} \qquad \text{Multiply by the reciprocal.}$$

$$= \frac{2}{1} \cdot \frac{2}{5} = \frac{4}{5} \qquad \text{Rewrite 2 as } \frac{2}{1}, \text{ and multiply the fractions.}$$

The ratio of 2 hr to $2\frac{1}{2}$ hr is $\dfrac{4}{5}$.

c) Begin by writing the ratio of $1.25 to $4.50 as a fraction, and divide out the common units.

$$\frac{\$1.25}{\$4.50} = \frac{1.25}{4.50}$$

If we multiply the numerator and denominator by 100, we will eliminate the decimal *and* we will not change the value of the fraction.

$$\frac{1.25}{4.50} = \frac{1.25}{4.50} \cdot \frac{100}{100} \qquad \text{Multiply numerator and denominator by 100.}$$

$$= \frac{125}{450} \qquad \text{Multiply to eliminate the decimals.}$$

$$= \frac{125 \div 25}{450 \div 25} = \frac{5}{18} \qquad \text{Write the ratio in lowest terms.}$$

The ratio of $1.25 to $4.50 is $\dfrac{5}{18}$.

[YOU TRY 3] Write each ratio so that it compares two whole numbers in lowest terms.

a) $\dfrac{9}{10}$ to $\dfrac{4}{5}$ b) 3 days to $3\frac{1}{2}$ days   c) $2.40 to $3.20

Let's solve another applied problem.

EXAMPLE 4 The regular price of a phone case was $28.00, and now it is on sale for $22.40. Find the ratio of the decrease in the price to the regular price.

Solution

We must find the decrease in the price, then write a *ratio* of the decrease in the price to the regular price. Rewrite the ratio with whole numbers.

Find the decrease in the price of the DVD.

$$\text{Original price} - \text{Sale price} = \text{Decrease in price}$$
$$\$28.00 \quad - \quad \$22.40 \quad = \$5.60$$

Write the ratio of the decrease in price to the original price.

$$\frac{5.60}{28.00} = \frac{5.60}{28.00} \cdot \frac{10}{10}$$ 　　Multiplying by $\frac{10}{10}$ will eliminate the decimals.

$$= \frac{56}{280}$$ 　　Multiply.

$$= \frac{56 \div 56}{280 \div 56} = \frac{1}{5}$$ 　　Write the ratio in lowest terms.

W Hint

You'll see problems similar to these in Chapter 8. How do you normally see discounts represented?

The ratio of the decrease in price to the regular price is $\frac{1}{5}$.

[**YOU TRY 4**] The regular price of a shirt was $24.00, and now it is on sale for $16.80. Find the ratio of the decrease in the price to the regular price.

3 Write Ratios After Converting Units

At the beginning of this section, we said that a ratio comparing two measurements must have the same units. If the units are different, we must begin by changing one of the units so that it is the same as the other one.

EXAMPLE 5　Write the ratio of 4 ft to 2 yd.

Solution

Write each quantity with the same units. Let's change yards to feet since a foot is smaller than a yard. (Converting to the smaller unit can help avoid fractions in the conversion.) Since there are 3 ft in 1 yd,

$$2 \text{ yd} = 2 \cdot 3 \text{ ft} = 6 \text{ ft}$$

Then the ratio of 4 ft to 2 yd is

$$\frac{4 \text{ ft}}{2 \text{ yd}} = \frac{4 \text{ ft}}{6 \text{ ft}} = \frac{4 \text{ ft}}{6 \text{ ft}} = \frac{4}{6} = \frac{2}{3}$$

Note

If we had changed feet to yards, the answer would have been the same but the calculations would have been more complicated.

$$4 \text{ ft} = 1\frac{1}{3} \text{ yd}$$

Then the ratio of 4 ft to 2 yd is

$$\frac{4 \text{ ft}}{2 \text{ yd}} = \frac{1\frac{1}{3} \text{ yd}}{2 \text{ yd}} = \frac{1\frac{1}{3} \text{ yd}}{2 \text{ yd}} = 1\frac{1}{3} \div 2 = \frac{4}{3} \cdot \frac{1}{2} = \frac{\overset{2}{4}}{3} \cdot \frac{1}{\underset{1}{2}} = \frac{2}{3}$$

The result is the same, but changing from the larger units to the smaller ones will help avoid these more complicated calculations.

[YOU TRY 5] Write the ratio of 3 ft to 24 in.

EXAMPLE 6   Write the ratio of 8 weeks to 14 days.

Solution

Write each quantity with the same units. Since a day is smaller than a week, we will change weeks to days. There are 7 days in 1 week, so

$$8 \text{ weeks} = 8 \cdot 7 \text{ days} = 56 \text{ days}$$

Then the ratio of 8 weeks to 14 days is

$$\frac{8 \text{ weeks}}{14 \text{ days}} = \frac{56 \text{ days}}{14 \text{ days}} = \frac{56 \text{ days}}{14 \text{ days}} = \frac{4}{1}$$

[YOU TRY 6]   Write the ratio of 20 hr to 2 days.

We can use the following list of relationships between measurements to help us write measurements with the same units.

W Hint

Does this chart look familiar? If not, spend some time memorizing these relationships.

Relationships Between Measurements

Length	Time
12 inches = 1 foot	60 seconds = 1 minute
3 feet = 1 yard	60 minutes = 1 hour
5280 feet = 1 mile	24 hours = 1 day
	7 days = 1 week

Volume (Capacity)	Weight
2 cups = 1 pint	16 ounces = 1 pound
2 pints = 1 quart	2000 pounds = 1 ton
4 quarts = 1 gallon	

EXAMPLE 7   The ratio of butternut squash puree to cream in a butternut squash soup recipe is 3 pints to 1 cup. Write the ratio of the puree to the cream.

Solution

Write each quantity with the same units. A cup is smaller than a pint, so let's convert 3 pints to cups.

Then the ratio of 3 pints to 1 cup is

$$\frac{3 \text{ pints}}{1 \text{ cup}} = \frac{6 \text{ cups}}{1 \text{ cup}} = \frac{6 \text{ cups}}{1 \text{ cup}} = \frac{6}{1}$$

The ratio of butternut squash puree to cream is $\frac{6}{1}$.

©James Gathany/CDC

YOU TRY 7	The ratio of ginger ale to orange juice in a fruit punch recipe is 3 quarts to 4 pints. Write the ratio of the ginger ale to orange juice.

ANSWERS TO [YOU TRY] **EXERCISES**

1) a) $\frac{5}{8}$ b) $\frac{3}{2}$ c) $\frac{3}{1}$ 2) $\frac{9}{5}$; for every 9 girls in the choir, there are 5 boys. 3) a) $\frac{9}{8}$ b) $\frac{6}{7}$ c) $\frac{3}{4}$

4) $\frac{3}{10}$ 5) $\frac{3}{2}$ 6) $\frac{5}{12}$ 7) $\frac{3}{2}$

E Evaluate **6.1** Exercises   Do the exercises, and check your work.

Get Ready
Fill in the blank with the correct number.

1) _____ weeks = 1 year

2) _____ feet = 1 yard

3) 1 pint = _____ cups

4) 1 pound = _____ ounces

Objective 1: Write Basic Ratios

5) What is a ratio?

6) How are ratios and rates different?

Write each ratio as a fraction in lowest terms.

7) 11 min to 16 min   8) 8 oz to 11 oz

9) $13 to $26   10) 6¢ to 36¢

11) 75 in. to 50 in.   12) 300 mi to 450 mi

13) 72¢ to 18¢   14) $80 to $40

15) At a fast-food restaurant, the ratio of part-time workers to full-time employees is 16 to 6. Write this ratio as a fraction, and explain what it means.

16) In a classroom, the ratio of students who play *Pokemon GO* on their phones to those who do not is 14 to 10. Write this ratio as a fraction, and explain what it means.

The table shows the number of music album sales, by type, in the United States during the first six months of 2016. Use this information for Exercises 17–24. Write each ratio as a fraction in lowest terms. (www.billboard.com)

Type of Album	Number Sold
CD	50.0 million
Digital	43.8 million
Vinyl	6.2 million

17) Find the ratio of vinyl albums to CDs.

18) Find the ratio of CDs to digital albums.

19) Find the ratio of digital albums to vinyl albums.

20) Find the ratio of the number of vinyl albums sold to the total number of albums sold.

21) Find the ratio of the number of CDs sold to the total number of albums sold.

22) Find the ratio of the number of digital albums sold to the total number of albums sold.

23) Explain the meaning of the answer to Exercise 19.

24) Explain the meaning of the answer to Exercise 22.

Objective 2: Write Ratios Comparing Fractions, Mixed Numbers, or Decimals
Write each ratio so that it compares two whole numbers in lowest terms.

25) $\frac{8}{9}$ to $\frac{1}{3}$ 26) $\frac{11}{12}$ to $\frac{5}{6}$

27) $\frac{1}{10}$ to $\frac{11}{15}$ 28) $\frac{1}{6}$ to $\frac{3}{4}$

29) 3 to $3\frac{3}{4}$ 30) 2 to $2\frac{1}{2}$

31) 4 min to $6\frac{1}{2}$ min   32) 7 min to $10\frac{1}{2}$ min

33) $2\frac{5}{6}$ yd to $1\frac{3}{4}$ yd   34) $3\frac{1}{3}$ hr to $2\frac{3}{4}$ hr

35) $4\frac{2}{3}$ mi to $1\frac{1}{6}$ mi   36) $5\frac{1}{2}$ days to $1\frac{5}{6}$ days

37) 7.6 to 4.4   38) 15.6 to 8.8

39) $4.90 to $5.60

40) $6.00 to $8.40

41) $1.05 to $0.35

42) $3.25 to $0.65

43) 6.4 mi to 9.2 mi

44) 2.7 lb to 14.4 lb

Solve each problem.

45) The regular price of a soccer ball was $42.00, and now it is on sale for $31.50. Find the ratio of the decrease in the price to the regular price.

©Mike Kemp/Blend Images

46) A pair of shoes that normally sells for $36.00 is marked down to $24.00. Find the ratio of the decrease in price to the original price.

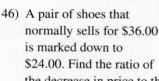 47) In Fall 2016, a community college had an enrollment of 5400. In Fall 2017, that number rose to 5514. Find the ratio of the increase in enrollment to the enrollment in Fall 2016.

48) Due to a slowing economy, a company must decrease its normal 40-hr workweek to 32 hr. Find the ratio of the decrease in new hours worked to the normal hourly workweek.

49) At birth, a baby weighed 6.8 lb. When she was 4 months old, she weighed 13.2 lb. Find the ratio of the increase in weight to her original weight.

©John Lund/Sam Diephuis/ Blend Images LLC

50) In 2000, approximately 80.4% of the U.S. population had a high school diploma or higher. In 2015, this percentage rose to 88.0%. Find the ratio of the increase in the percentage to the percentage earning a high school diploma or higher in 2015. (www.census.gov)

Objective 3: Write Ratios After Converting Units

Write each ratio as a fraction in lowest terms.

51) 14 in. to 2 ft

52) 20 in. to 2 yd

53) 8 ft to 4 yd

54) 6 cups to 8 pints

55) 12 min to 2 hr

56) 36 min to 4 hr

57) 6 qt to 8 gal

58) 8 oz to 8 lb

59) 15 weeks to 21 days

60) 9 min to 180 sec

61) 14 gal to 8 qt

62) 16 qt to 4 gal

Solve each problem.

63) The ratio of the height of a bookshelf to the width of the bookshelf is 6 ft to 32 in. Write the ratio of the height to the width.

64) The ratio of the height of a man to the length of his arm is 6 ft to 30 in. Write the ratio of his height to the length of his arm.

 65) The ratio of heavy cream to whole milk in a vanilla ice cream recipe is 3 cups to 1 pint. Write the ratio of heavy cream to whole milk.

66) The ratio of white wine to fish broth in a bouillabaisse recipe is 2 cups to 1.5 pints. Write the ratio of white wine to fish broth.

67) The ratio of the width of an oil painting to the length is 10 in. to 1.5 ft. Write the ratio of the width to the length.

68) The ratio of the length of a stained glass window to the width is 4 ft to 30 in. Write the ratio of the length to the width.

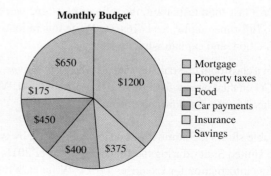

©Apis/Abramis/Alamy

Mixed Exercises: Objectives 1–3

The circle graph shows a family's monthly budget, where the total amount of money shown equals their monthly income. Use the graph for Exercises 69–74. Write each ratio as a fraction in lowest terms.

Monthly Budget

$650

$175

$450

$400

$375

$1200

☐ Mortgage
☐ Property taxes
☐ Food
☐ Car payments
☐ Insurance
☐ Savings

69) What is the family's total monthly income?

70) What is the family's combined monthly mortgage, insurance, and property tax payment?

71) Find the ratio of the family's total monthly income to savings.

72) Find the ratio of the family's monthly food budget to car payment.

73) Find the ratio of the family's combined monthly mortgage and property tax payment to insurance payment.

74) What is the family's yearly savings?

75) Today, a gallon of unleaded gasoline sells for $3.36. Last year, one gallon cost $3.20. Find the ratio of the increase in price to the price last year.

76) The regular price of a hair dryer is $22.00, and now it is on sale for $16.40. Find the ratio of the decrease in price to the original price.

Write each ratio as a fraction in lowest terms.

77) 3 gal to 8 qt

78) 21 hr to 2 days

For Exercises 79 and 80, write the ratio of the longest side to the shortest side. Reduce to lowest terms.

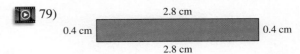
79)

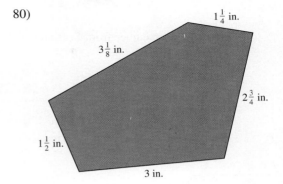
80)

R **Rethink**

R1) How often do you encounter ratios every day? Give some examples.

R2) How could you use a ratio to help make a generalization about a bigger number?

6.2 Rates

P **Prepare**

O **Organize**

What are your objectives for Section 6.2?	How can you accomplish each objective?
1 Write a Rate as a Fraction	• Write the definition of *rate* in your notes, and be sure to include the units. • Complete the given example on your own. • Complete You Try 1.
2 Find a Unit Rate	• Write the definition of *unit rate*. • Write your own procedure for **Finding a Unit Rate.** • Complete the given example on your own. • Complete You Try 2.
3 Solve Applied Problems Involving Unit Rates	• Complete the given examples on your own. • Complete You Trys 3 and 4.

W **Work** **Read the explanations, follow the examples, take notes, and complete the You Trys.**

Rates appear in many different situations. A car's mileage might be 28 miles per gallon. A part-time worker might earn $9.50/hour. Each of these is an example of a *rate*.

Definition

A **rate** compares quantities with *different* units.

Remember from Section 6.1 that a ratio compares quantities with the *same* units.

1 Write a Rate as a Fraction

Suppose that you drove 170 miles in 4 hours. We can write the *rate* at which you drove as a fraction in lowest terms:

$$\frac{170 \text{ mi}}{4 \text{ hr}} = \frac{170 \text{ mi} \div 2}{4 \text{ hr} \div 2} = \frac{85 \text{ mi}}{2 \text{ hr}}$$

Hint
Rates are different from ratios!

The rate at which you drove was $\frac{85 \text{ mi}}{2 \text{ hr}}$ or 85 miles in 2 hours. When writing a rate, we *always* include the units.

Note
We can use different words to indicate rate:

in   per   for   from   on

Rates are usually written in lowest terms.

EXAMPLE 1

Write each rate as a fraction in lowest terms.

a)   600 mi on 21 gal of gas   b)   18 lb of dog food for $15

c) $960 in 4 weeks

Solution

a)   Set this up like we set up ratios. The quantity that comes first goes in the numerator, and the second quantity is the denominator.

$$\frac{600 \text{ mi}}{21 \text{ gal}} = \frac{600 \text{ mi} \div 3}{21 \text{ gal} \div 3} = \frac{200 \text{ mi}}{7 \text{ gal}}$$

b) $\dfrac{18 \text{ lb}}{15 \text{ dollars}} = \dfrac{18 \text{ lb} \div 3}{15 \text{ dollars} \div 3} = \dfrac{6 \text{ lb}}{5 \text{ dollars}}$

c) $\dfrac{960 \text{ dollars}}{4 \text{ weeks}} = \dfrac{960 \text{ dollars} \div 4}{4 \text{ weeks} \div 4} = \dfrac{240 \text{ dollars}}{1 \text{ week}}$

[YOU TRY 1]

Write each rate as a fraction in lowest terms.

a)   450 mi on 20 gal of gas   b)   10 lb of potatoes for $4

c) $1092 in 3 weeks

2 Find a Unit Rate

In Example 1c, we wrote "$960 in 4 weeks" in fractional form as $\dfrac{240 \text{ dollars}}{1 \text{ week}}$. Notice that the denominator is 1. This is an example of a *unit rate*.

Definition

A **unit rate** is a rate with a denominator of 1.

The rate $\dfrac{240 \text{ dollars}}{1 \text{ week}}$ can also be written as $240/week or $240 per week. The slash mark, /, means *per*.

EXAMPLE 2

Find each unit rate.

a) 192 mi in 3 hr

b) 336 mi on 15 gal

c) $13.20 for 20 lb of charcoal

Solution

a) Write the rate as a fraction, then simplify so that the denominator is 1.

$$\frac{192 \text{ mi}}{3 \text{ hr}} = \frac{192 \text{ mi} \div 3}{3 \text{ hr} \div 3} = \frac{64 \text{ mi}}{1 \text{ hr}}$$

The unit rate is 64 mi/hr or 64 miles per hour. This is also called a vehicle's **average speed.** It can also be written as 64 mph.

W Hint

This looks like a procedure to be used for all unit rates!

b) $\dfrac{336 \text{ mi}}{15 \text{ gal}} = \dfrac{336 \text{ mi} \div 3}{15 \text{ gal} \div 3} = \dfrac{112 \text{ mi}}{5 \text{ gal}}$

Next, divide 112 by 5 to find the unit rate: $5)\overline{112.0}$ with quotient 22.4

The unit rate is 22.4 mi/gal or 22.4 miles per gallon. This is also called a vehicle's **gas mileage** and can be written as 22.4 mpg.

c) Write the rate as $\dfrac{\$13.20}{20 \text{ lb}}$, then divide: $20)\overline{13.20}$ with quotient 0.66

The unit rate is $0.66/lb or $0.66 per pound. This is also called **unit price.** (This is the same as 66¢ per pound.)

[YOU TRY 2]

Find each unit rate.

a) 504 mi on 14 gal of gas

b) $198 for 12 hr of work

c) $3.84 for 64 oz of grapefruit juice

---

3 Solve Applied Problems Involving Unit Rates

We can use unit rates to help us solve many different problems.

EXAMPLE 3

Jim earned $253.00 for working 22 hr one week. What is his hourly wage? (This is his **unit pay.**)

Solution

Jim's hourly wage is the amount of money he earns *per* hour. Set up the rate with his total earnings on the top and hours worked on the bottom.

> **W Hint**
> It is helpful to know the everyday terminology that is used for unit rates.

$$\frac{\$253.00}{22 \text{ hr}} \qquad \text{Divide:} \quad 22)\overline{253.00}^{\,11.50}$$

Jim's hourly wage is $11.50/hr.

[YOU TRY 3] Oksana earned $433.50 for working 34 hr one week. What is her hourly wage?

A unit rate that is used often is the *unit price*. The **unit price** of an item is the cost of the item per unit.

> **Note**
> Some examples of unit rates are $0.25 per ounce, $1.98 per pound, $3.00 per gallon, and $49.99 per video game.

We can use the unit price to figure out which item in a store gives us the most value for our money. We call this the **best buy.** The item with the lowest unit price is the best buy.

EXAMPLE 4

A store sells cornflakes cereal in three different sizes. The sizes and prices are listed here. Which size is the best buy?

Size	Price
12 oz	$3.69
18 oz	$4.89
24 oz	$6.57

©McGraw-Hill Education/John Thoeming, photographer

Solution

For each box of cereal, we must find the unit price, or how much the cereal costs per ounce. We will find the unit price by dividing.

$$\text{Unit price} = \frac{\text{Price of a box of cereal}}{\text{Number of ounces in the box}} = \text{Cost per ounce}$$

Size	Unit Price
12 oz	$\dfrac{\$3.69}{12 \text{ oz}} \approx \$0.308$ per oz
18 oz	$\dfrac{\$4.89}{18 \text{ oz}} \approx \$0.272$ per oz
24 oz	$\dfrac{\$6.57}{24 \text{ oz}} \approx \$0.274$ per oz

The 18-oz box of cereal has the lowest unit price of $0.272/oz, so it is the best buy.

Note

Round the answers to the thousandths place because, as you can see, if we rounded the unit price of the 18-oz size and the 24-oz size to the nearest hundredth (or nearest cent), they would be the same. Rounding to the thousandths place, however, shows us that the 18-oz size is the best buy.

YOU TRY 4

A store sells cheddar fish crackers in three sizes. A 7-oz box costs $2.25, a 15-oz box costs $3.96, and the price of a 34-oz box is $8.89. Which size is the best buy, and what is its unit price?

ANSWERS TO [YOU TRY] **EXERCISES**

1) a) $\dfrac{45 \text{ mi}}{2 \text{ gal}}$ b) $\dfrac{5 \text{ lb}}{\$2}$ c) $\dfrac{364 \text{ dollars}}{1 \text{ week}}$

2) a) 36 mi/gal b) $16.50/hr c) $0.06/oz   3) $12.75/hr

4) The 34-oz box has the lowest unit price of $0.261/oz, so it is the best buy.

Using Technology

We can use a calculator to find a vehicle's gas mileage in miles per gallon or mpg.

The next time you fill up your tank, record the number of gallons you purchased and the number of miles you have driven since your last fill-up. For example, suppose you drove 284.2 mi using 11.6 gal of gas. We find gas mileage by dividing the number of miles driven by the number of gallons of gas used.

To calculate the gas mileage, enter ⟨2⟩⟨8⟩⟨4⟩⟨.⟩⟨2⟩⟨÷⟩⟨1⟩⟨1⟩⟨.⟩⟨6⟩⟨=⟩ into the calculator.

Your display will show 24.5, which means that the car's gas mileage was 24.5 miles per gallon or 24.5 mpg.

E Evaluate **6.2** Exercises   Do the exercises, and check your work.

Objective 1: Write a Rate as a Fraction

1) What is the difference between a rate and a ratio?

2) Is $\dfrac{97 \text{ mi}}{2 \text{ hr}}$ a rate or a ratio? Explain your answer.

Write each rate as a fraction in lowest terms.

 3) 96 ft in 36 sec

4) 150 ft in 20 sec

 5) 270 mi on 12 gal

6) 504 mi on 16 gal

7) 12 cups for 10 servings

8) 10 cups for 6 servings

9) 280 mi in 6 hr

10) 22 ft in 8 sec

11) $12 for 48 daisies

12) $20 for 12 tulips

13) 8 oz of tea for $60

14) 30 boxes of cookies for $110

15) $597 in 3 weeks

16) $2275 in 5 weeks

Objective 2: Find a Unit Rate

17) What is a unit rate?

18) Write the rate $\dfrac{\$523}{1 \text{ wk}}$ in two other ways.

Find each unit rate. Round the answer to the nearest tenth where appropriate.

19) 316 mi in 4 hr

20) 1143 mi in 3 hr

21) 780 mi on 30 gal

22) 425 mi on 25 gal

23) $120 in 3 hr

24) $540 in 30 days

25) $360.75 for 6.5 days

26) $305.25 in 16.5 hr

27) 421 mi on 18 gal

28) 595 mi on 16 gal

Find the unit price of each item. Round the answer to the nearest thousandth, if necessary.

29) $5.49 for 4.5 oz of chili powder

©Ingram Publishing

30) $3.15 for 2.5 oz of cinnamon

31) $3.19 for 15 oz of pasta sauce

32) $3.39 for 12 oz of molasses

33) $8.99 for 24 cans of soda

34) $2.59 for 6 individual packs of applesauce

Objective 3: Solve Applied Problems Involving Unit Rates

Solve each problem. For problems involving money, round to the nearest cent. Otherwise, round to the nearest tenth, where appropriate.

35) Arturo rode 91 mi in 5 hr. What was his average speed?

36) Monique earns $1424 for a 40-hr workweek. What is her hourly wage?

37) A minivan can travel 374 mi on 20 gal of gasoline. Find its gas mileage.

38) During the 2016-2017 regular season, the Chicago Blackhawks' Duncan Keith played 2049 min in 80 games. Find the average number of minutes he played per game. (www.foxsports.com)

39) The Vespa® GTS 300 scooter can travel approximately 190 mi on its 2.5-gal gas tank. What is the gas mileage for this scooter? (www.vespa.com)

©Ingram Publishing/SuperStock

40) Marilu drove 234.5 mi on a highway for 3.5 hr. What was her average speed?

41) Tuan recorded the number of hours he studied over the past 5 weeks. If he studied a total of 98 hr, what was his rate of study in hours per week? What was his rate of study in hours per day?

42) Brandy drove her car 11,830 mi over the past year. What was her driving rate in miles per week? What was her driving rate in miles per day?

43) Fumiko has a text messaging plan that allows her 3000 text messages for $9.99/month. For every message above the cap, she is charged $0.18. Last month she sent 3172 text messages. What is her total monthly text message charge?

44) Jerry has a text messaging plan that allows him 2000 text messages for $7.99/month. For every message above the cap, he is charged $0.15. Last month he sent 2198 text messages. What is his total monthly text message charge?

45) Alicia buys a two-year health club membership and agrees to pay $1896. What is her monthly payment?

©Dave and Les Jacobs/Blend Images

46) Siyamak buys a used car and agrees to pay $10,800 over four years. What is his monthly payment?

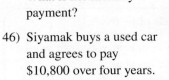

47) Devin Hester broke the National Football League record for number of touchdowns on kickoffs or punt returns when the Chicago Bears played the Minnesota Vikings on December 20, 2010. In that game, Hester ran back three kickoffs or punts for 146 yd. Find the average number of yards he ran per return. (http://espn.go.com)

48) During his first five seasons in the National Hockey League, Sidney Crosby played a total of 371 games with the Pittsburgh Penguins. Find the average number of games he played per season. (www.nhl.com)

49) Tessa's time sheet is shown below. If she was paid $391.50 for working these hours, what is her hourly wage?

Mon.	Tues.	Wed.	Thurs.	Fri.
6.0 hr	4.5 hr	3.5 hr	8.0 hr	7.0 hr

50) A nursing student sets an intravenous drip rate at 2880 drops per 1 hr. What is the unit drip rate in drops per minute?

51) The California high-speed rail authority claims that its proposed high-speed train will travel 600 mi from San Diego to San Francisco in $3\frac{3}{4}$ hr. Find the average speed of the train. (www.cahighspeedrail.ca.gov)

52) Corey's morning commute to work is 34.5 mi. In heavy traffic, it took him $1\frac{1}{4}$ hr to get to work. What was his average speed?

©Philip & Karen Smith/Getty Images

R Rethink

R1) How did these exercises compare with those you completed for Section 6.1?

For each item, determine which size is the best buy, and list its unit price.

53) Soy sauce

Size	Price
10 oz	$2.69
24 oz	$4.73
36 oz	$6.39

54) Cheese

Size	Price
8 oz	$2.99
12 oz	$3.99
16 oz	$6.09

55) Cups of coffee

Size	Price
12 oz	$1.50
16 oz	$1.95
20 oz	$2.25

56) Packages of diapers

Number of Diapers in the Package	Price
31	$10.99
96	$23.99
132	$41.99

57) Brand A lightbulbs cost $7.38 for 6 bulbs, and Brand B lightbulbs cost $11.34 for 9 bulbs. Which is the better buy, and what is its unit cost?

58) A 4-pack of AA batteries costs $4.99 while a package of 6 costs $7.39. Which is the better buy, and what is its unit cost?

The table lists the monthly rental cost for storage units in a city. Use the table for Exercises 59 and 60.

Dimensions in feet	Monthly Cost	Floor Area in square feet	Cost/ft^2
15 × 15	$281.25		
10 × 10	$150.00		
8 × 8	$112.00		
8 × 6	$88.80		
5 × 5	$48.75		

59) Find the floor area for each unit and the cost per square foot.

60) How much does it cost to rent the largest storage unit for one year?

R2) After completing the exercises involving price, have you noticed where you might be able to make better decisions at the store? Explain.

6.3 Proportions

What are your objectives for Section 6.3?	How can you accomplish each objective?
1 Write Proportions	• Write the definition of *proportion* in your own words. • Complete the given example on your own. • Complete You Try 1.
2 Determine Whether a Proportion Is True or False	• Know the two different ways to determine whether a proportion is true. • Write the procedure for **Using Cross Products** in your own words. • Complete the given examples on your own. • Complete You Trys 2 and 3.

W Work   Read the explanations, follow the examples, take notes, and complete the You Trys.

1 Write Proportions

A **proportion** is a statement that two ratios or two rates are equal. For example, since $\frac{1}{2}$ and $\frac{3}{6}$ are equivalent fractions, we can write the proportion $\frac{1}{2} = \frac{3}{6}$. This can be read as

$$
\underset{\substack{\uparrow \\ \text{fraction bar}}}{1 \text{ is to } 2} \quad \overset{\substack{= \\ \downarrow}}{\text{as}} \quad \underset{\substack{\uparrow \\ \text{fraction bar}}}{3 \text{ is to } 6}
$$

Note
The words *is to* represent the fraction bar, and the word *as* represents =.

Similarly, the proportion $\frac{\$15}{1 \text{ hr}} = \frac{\$30}{2 \text{ hr}}$ says that the two rates are equal. This is true because if the amount of money is multiplied by 2, the hours are also multiplied by 2. We read this proportion as

$$
\underset{\substack{\uparrow \\ \text{fraction bar}}}{\$15 \text{ is to } 1 \text{ hr}} \quad \overset{\substack{= \\ \downarrow}}{\text{as}} \quad \underset{\substack{\uparrow \\ \text{fraction bar}}}{\$30 \text{ is to } 2 \text{ hr}}
$$

EXAMPLE 1   Write each statement as a proportion.

a)   2 in. is to 9 in.   as   14 in. is to 63 in.

b)   100 mi is to 2 hr   as   500 mi is to 10 hr.

Solution

a) 2 in. is to 9 in. as 14 in. is to 63 in.

$$\text{is to} \to \frac{2 \text{ in.}}{9 \text{ in.}} \underset{\text{as}}{\overset{\uparrow}{=}} \frac{14 \text{ in.}}{63 \text{ in.}} \leftarrow \text{is to}$$

b) 100 mi is to 2 hr as 500 mi is to 10 hr

$$\frac{100 \text{ mi}}{2 \text{ hr}} = \frac{500 \text{ mi}}{10 \text{ hr}}$$

[YOU TRY 1] Write each statement as a proportion.

a) 4 in. is to 5 in. as 12 in. is to 15 in.

b) $39 is to 3 hr as $78 is to 6 hr.

---

2 Determine Whether a Proportion Is True or False

One way to determine whether a proportion is true or false is by writing each fraction in lowest terms.

EXAMPLE 2 Determine whether each proportion is true or false by writing each fraction in lowest terms.

a) $\dfrac{4}{7} = \dfrac{24}{40}$ b) $\dfrac{10}{6} = \dfrac{15}{9}$

Solution

a) The fraction $\dfrac{4}{7}$ is already in lowest terms. Write $\dfrac{24}{40}$ in lowest terms.

$$\frac{24}{40} = \frac{24 \div 8}{40 \div 8} = \frac{3}{5}$$

Since $\dfrac{4}{7} \neq \dfrac{3}{5}$, the proportion is false.

b) Write each fraction in lowest terms.

$$\frac{10}{6} = \frac{10 \div 2}{6 \div 2} = \frac{5}{3} \qquad \frac{15}{9} = \frac{15 \div 3}{9 \div 3} = \frac{5}{3}$$

Since each fraction simplifies to $\dfrac{5}{3}$, the proportion is true.

[YOU TRY 2] Determine whether each proportion is true or false by writing each fraction in lowest terms.

a) $\dfrac{42}{56} = \dfrac{6}{8}$ b) $\dfrac{16}{12} = \dfrac{28}{20}$

Another way to determine whether a proportion is true or false is by finding the *cross products*. The **cross products** are the numbers we get when we *multiply along the diagonals* of the proportion.

For example, the cross products of the proportion $\dfrac{3}{4} = \dfrac{18}{24}$ are

One cross product

$4 \cdot 18 = 72$

$$\dfrac{3}{4} \diagup\hspace{-0.9em}= \dfrac{18}{24}$$

$3 \cdot 24 = 72$

The other cross product

Procedure Using Cross Products to Determine Whether a Proportion Is True or False

1) If the cross products in a proportion are equal, then the proportion is true.

   Example: $\dfrac{3}{4} \diagup\hspace{-0.9em}= \dfrac{18}{24}$ $\begin{array}{l} 4 \cdot 18 = 72 \\ 3 \cdot 24 = 72 \end{array}$ The cross products are equal. The proportion is true.

2) If the cross products in a proportion are *not* equal, then the proportion is false.

   Example: $\dfrac{8}{12} \diagup\hspace{-0.9em}= \dfrac{10}{16}$ $\begin{array}{l} 12 \cdot 10 = 120 \\ 8 \cdot 16 = 128 \end{array}$ The cross products are *not* equal. The proportion is false.

Using cross products is especially good when one or both of the fractions contain mixed numbers or decimals.

EXAMPLE 3

Determine whether each proportion is true or false by finding the cross products.

a) $\dfrac{16}{6} = \dfrac{10}{4}$

b) $\dfrac{2\frac{3}{4}}{4\frac{1}{2}} = \dfrac{11}{18}$

Solution

a) Find the cross products.

$$\dfrac{16}{6} \diagup\hspace{-0.9em}= \dfrac{10}{4}$$ $\begin{array}{l} 6 \cdot 10 = 60 \\ 16 \cdot 4 = 64 \end{array}$ False. The cross products are *not* equal.

Since the cross products are *not* equal, the proportion is false.

W Hint

Which procedure do you prefer?

b) Find the cross products.

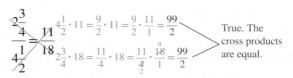

$$\dfrac{2\frac{3}{4}}{4\frac{1}{2}} \diagup\hspace{-0.9em}= \dfrac{11}{18}$$ $\begin{array}{l} 4\frac{1}{2} \cdot 11 = \frac{9}{2} \cdot 11 = \frac{9}{2} \cdot \frac{11}{1} = \frac{99}{2} \\ 2\frac{3}{4} \cdot 18 = \frac{11}{4} \cdot 18 = \frac{11}{4} \cdot \frac{\overset{9}{\cancel{18}}}{1} = \frac{99}{2} \end{array}$ True. The cross products are equal.

Since the cross products are equal, the proportion is true.

[**YOU TRY 3**] Determine whether each proportion is true or false by finding the cross products.

a) $\dfrac{45}{18} = \dfrac{20}{8}$ b) $\dfrac{1\frac{4}{7}}{3\frac{2}{3}} = \dfrac{2}{7}$

BE CAREFUL We use cross products only when we are working with proportions. Cross products are *not* used to multiply, divide, add, or subtract fractions.

ANSWERS TO [YOU TRY] EXERCISES

1) a) $\dfrac{4 \text{ in.}}{5 \text{ in.}} = \dfrac{12 \text{ in.}}{15 \text{ in.}}$ b) $\dfrac{\$39}{3 \text{ hr}} = \dfrac{\$78}{6 \text{ hr}}$ 2) a) true b) false   3) a) true b) false

Using Technology

A calculator can help us determine whether a proportion is true or false by finding cross products. Suppose we want to know whether the proportion $\dfrac{52}{98} = \dfrac{13}{24.5}$ is true. To find one cross product, we enter $\boxed{5}\boxed{2}\boxed{\times}\boxed{2}\boxed{4}\boxed{.}\boxed{5}\boxed{=}$ into the calculator. The display will show 1274. To calculate the other cross product, we enter $\boxed{9}\boxed{8}\boxed{\times}\boxed{1}\boxed{3}\boxed{=}$ into the calculator. The display will show 1274 again. Since the cross products are equal, the proportion is true.

E Evaluate   **6.3** Exercises   Do the exercises, and check your work.

Objective 1: Write Proportions

1) What is a proportion?

2) Is the following statement true or false? Because $\dfrac{3}{4}$ and $\dfrac{12}{16}$ are equivalent fractions, we can write the proportion $\dfrac{3}{4} = \dfrac{12}{16}$.

Write each statement as a proportion.

3) 8 is to 11 as 24 is to 33.

4) 2 is to 7 as 10 is to 35.

5) 14 is to 8 as 35 is to 20.

6) 18 is to 12 as 15 is to 10.

7) $50 is to $18 as $25 is to $9.

8) $84 is to $28 as $36 is to $12.

9) 2 ft is to 5 ft as 8 ft is to 20 ft.

10) 3 yd is to 8 yd as 6 yd is to 16 yd.

11) 480 mi is to 8 hr as 180 mi is to 3 hr.

12) 128 ft is to 8 sec as 48 ft is to 3 sec.

13) 2 in. is to 5 ft as 10 in. is to 25 ft.

14) 3 cm is to 4 m as 12 cm is to 16 m.

15) $8 is to 12 songs as $16 is to 24 songs.

16) 120 trees is to 1 park as 1320 trees is to 11 parks.

Objective 2: Determine Whether a Proportion Is True or False

17) What are two ways to determine whether a proportion is true or false?

18) Explain how to use the cross products to determine whether a proportion is true or false.

Determine whether each proportion is true or false by writing each fraction in lowest terms.

19) $\dfrac{9}{27} = \dfrac{1}{3}$

20) $\dfrac{18}{24} = \dfrac{3}{4}$

21) $\dfrac{24}{21} = \dfrac{18}{14}$

22) $\dfrac{32}{12} = \dfrac{60}{24}$

23) $\dfrac{150}{225} = \dfrac{20}{36}$

24) $\dfrac{45}{150} = \dfrac{30}{105}$

25) $\dfrac{124}{36} = \dfrac{155}{45}$

26) $\dfrac{220}{120} = \dfrac{55}{30}$

27) $\dfrac{18}{4} = \dfrac{63}{14}$

28) $\dfrac{15}{9} = \dfrac{20}{12}$

Determine whether each proportion is true or false by finding the cross products.

29) $\dfrac{1}{4} = \dfrac{4}{16}$

30) $\dfrac{7}{21} = \dfrac{1}{3}$

31) $\dfrac{6}{20} = \dfrac{2}{7}$

32) $\dfrac{12}{16} = \dfrac{3}{5}$

33) $\dfrac{19}{5} = \dfrac{76}{20}$

34) $\dfrac{13}{8} = \dfrac{91}{56}$

35) $\dfrac{8}{12} = \dfrac{18}{24}$

36) $\dfrac{4}{30} = \dfrac{6}{50}$

37) $\dfrac{15}{4.3} = \dfrac{17}{6.9}$

38) $\dfrac{23}{9.8} = \dfrac{18}{5.7}$

39) $\dfrac{3.2}{12} = \dfrac{1.6}{6}$

40) $\dfrac{9.8}{21} = \dfrac{1.4}{3}$

41) $\dfrac{2\frac{5}{6}}{5\frac{3}{8}} = \dfrac{68}{129}$

42) $\dfrac{1\frac{4}{9}}{3\frac{1}{6}} = \dfrac{26}{57}$

43) $\dfrac{3\frac{1}{8}}{5} = \dfrac{2}{4\frac{3}{4}}$

44) $\dfrac{4\frac{3}{5}}{9} = \dfrac{3}{6\frac{2}{3}}$

45) $\dfrac{2.5}{6} = \dfrac{5\frac{1}{3}}{12.8}$

46) $\dfrac{2\frac{2}{5}}{1.2} = \dfrac{4\frac{1}{4}}{2.25}$

47) Do you multiply fractions and determine whether a proportion is true in the same way? Explain.

48) a) Determine whether the proportion $\dfrac{12}{18} = \dfrac{8}{10}$ is true or false.

   b) Multiply $\dfrac{12}{18} \cdot \dfrac{8}{10}$.

49) a) Determine whether the proportion $\dfrac{9}{21} = \dfrac{21}{49}$ is true or false.

   b) Multiply $\dfrac{9}{21} \cdot \dfrac{21}{49}$.

50) a) Determine whether the proportion $\dfrac{8}{20} = \dfrac{18}{45}$ is true or false.

   b) Multiply $\dfrac{8}{20} \cdot \dfrac{18}{45}$.

51) a) Divide $\dfrac{18}{28} \div \dfrac{12}{16}$.

   b) Determine whether the proportion $\dfrac{18}{28} = \dfrac{12}{16}$ is true or false.

52) a) Divide $\dfrac{12}{32} \div \dfrac{8}{18}$.

   b) Determine whether the proportion $\dfrac{12}{32} = \dfrac{8}{18}$ is true or false.

53) a) Multiply $\dfrac{11}{14} \cdot \dfrac{6}{13}$.

   b) Determine whether the proportion $\dfrac{11}{14} = \dfrac{6}{13}$ is true or false.

54) a) Multiply $\dfrac{15}{7} \cdot \dfrac{23}{10}$.

   b) Determine whether the proportion $\dfrac{15}{7} = \dfrac{23}{10}$ is true or false.

55) Doug got 15 out of 18 problems correct on his quiz and 35 out of 42 problems correct on his exam. He claims that he did equally well on both. Use a proportion and cross products to determine whether Doug's claim is correct.

©Wavebreakmedia Ltd/Getty Images

56) Janice and Raymond each go to their favorite MP3 download website to get songs. During a holiday sale, Janice downloads 8 songs for $5.20 and Raymond downloads 12 songs for $7.80. Raymond claims that Janice paid more per song than he did. Use a proportion and cross products to determine whether Raymond's claim is correct.

R Rethink

R1) How have you done on the exercises?

R2) Where could you go or who could help you with any questions you have?

6.4 Solve Proportions

P Prepare

O Organize

What are your objectives for Section 6.4?	How can you accomplish each objective?
1 Solve a Proportion	• Review **Using Cross Products** from Section 6.3 if you need to refresh your skills. • Write the procedure for **Solving a Proportion** in your own words. • Complete the given example on your own, and check the answer. • Complete You Try 1.
2 Solve a Proportion Containing Mixed Numbers or Decimals	• Use the same procedure you outlined for Objective 1. • Complete the given examples on your own. • Complete You Try 2.

W Work

Read the explanations, follow the examples, take notes, and complete the You Trys.

1 Solve a Proportion

We can use what we know about solving equations to solve proportions. If a proportion contains one variable and we have to find the number that makes the proportion true, we use cross products and the division property of equality to find that missing value.

> **Procedure** How to Solve a Proportion
>
> **Step 1:** Find the cross products.
>
> **Step 2:** Set the cross products equal to each other, and solve the equation.
>
> **Step 3:** Check the solution by substituting it into the original proportion and finding the cross products.

Let's begin with an example.

EXAMPLE 1

Solve each proportion.

a) $\dfrac{2}{5} = \dfrac{x}{15}$ b) $\dfrac{x}{10} = \dfrac{12}{15}$

Solution

a) **Step 1:** Find the cross products.

$$\dfrac{2}{5} \bowtie \dfrac{x}{15} \qquad \begin{aligned} 5 \cdot x &= 5x \\ 2 \cdot 15 &= 30 \end{aligned}$$

Hint

Write out the example as you are reading it.

Step 2: Set the cross products equal to each other, and solve the equation.

$$5x = 30 \qquad \text{Set the cross products equal to each other.}$$

$$\dfrac{5x}{5} = \dfrac{30}{5} \qquad \text{Divide each side by 5.}$$

$$\dfrac{\overset{1}{\cancel{5}}x}{\underset{1}{\cancel{5}}} = \dfrac{30}{5} \qquad \text{Divide out the common factor.}$$

$$x = 6 \qquad \text{Simplify.}$$

Step 3: Check the solution in the original proportion, $\dfrac{2}{5} = \dfrac{x}{15}$, by checking to see whether the cross products are equal.

$$\dfrac{2}{5} \bowtie \dfrac{6}{15} \qquad \begin{aligned} 5 \cdot 6 &= 30 \qquad \text{Substitute 6 for } x. \\ 2 \cdot 15 &= 30 \end{aligned}$$

Since the cross products are equal, the solution of the proportion is $x = 6$.

The solution of the equation is 6, *not* 30, the value of each cross product!

b) **Step 1:** Find the cross products.

$$\frac{x}{10} \diagup\!\!\!\!= \frac{12}{15}$$

$10 \cdot 12 = 120$

$x \cdot 15 = x15 = 15x$ Commutative property

Step 2: Set the cross products equal to each other, and solve the equation.

$$120 = 15x \qquad \text{Set the cross products equal to each other.}$$

$$\frac{120}{15} = \frac{15x}{15} \qquad \text{Divide each side by 15.}$$

$$\frac{120}{15} = \frac{\overset{1}{\cancel{15}}x}{\underset{1}{\cancel{15}}} \qquad \text{Divide out the common factor.}$$

$$8 = x \qquad \text{Simplify.}$$

Step 3: The check is left to the student. The solution of $\dfrac{x}{10} = \dfrac{12}{15}$ is $x = 8$.

[**YOU TRY 1**] Solve each proportion.

a) $\dfrac{4}{9} = \dfrac{x}{27}$ b) $\dfrac{x}{12} = \dfrac{6}{9}$

---

2 Solve a Proportion Containing Mixed Numbers or Decimals

Sometimes, proportions contain mixed numbers or decimals.

EXAMPLE 2 Solve each proportion.

a) $\dfrac{1\frac{2}{3}}{x} = \dfrac{8}{12}$ b) $\dfrac{0.2}{0.8} = \dfrac{3}{x}$

Solution

a) **Step 1:** Find the cross products.

> **W Hint**
> Don't be intimidated by how these problems look. You will use the same procedure as in Objective 1, and you have worked with mixed numbers and decimals before!

$$\frac{1\frac{2}{3}}{x} \diagup\!\!\!\!= \frac{8}{12}$$

$x \cdot 8 = x8 = 8x$ Commutative property

$1\dfrac{2}{3} \cdot 12 = \dfrac{5}{3} \cdot 12 = \dfrac{5}{\cancel{3}_1} \cdot \dfrac{\cancel{12}^4}{1} = \dfrac{20}{1} = 20$

Step 2: Set the cross products equal to each other, and solve the equation.

$$8x = 20 \qquad \text{Set the cross products equal to each other.}$$

$$\frac{8x}{8} = \frac{20}{8} \qquad \text{Divide each side by 8.}$$

$$\frac{\overset{1}{\cancel{8}}x}{\cancel{8}} = \frac{20 \div 4}{8 \div 4} \qquad \text{Divide out the common factor.}$$

$$x = \frac{5}{2} = 2\frac{1}{2} \qquad \text{Simplify.}$$

We can write the solution either as an improper fraction or as a mixed number. Let's write it as a mixed number because there is a mixed number in the original proportion.

Step 3: Check the solution in the original proportion, $\dfrac{1\frac{2}{3}}{x} = \dfrac{8}{12}$, by checking to see whether the cross products are equal.

Substitute $2\frac{1}{2}$ for x.

$$2\frac{1}{2} \cdot 8 = \frac{5}{2} \cdot 8 = \frac{5}{\cancel{2}} \cdot \frac{\cancel{8}^{4}}{1} = \frac{20}{1} = 20$$

$$1\frac{2}{3} \cdot 12 = \frac{5}{3} \cdot 12 = \frac{5}{\cancel{3}} \cdot \frac{\cancel{12}^{4}}{1} = \frac{20}{1} = 20$$

Since the cross products are equal, the solution is $x = 2\frac{1}{2}$.

b) **Step 1:** Find the cross products.

$$0.8 \cdot 3 = 2.4$$
$$0.2 \cdot x = 0.2x$$

Step 2: Set the cross products equal to each other, and solve the equation.

$$2.4 = 0.2x \qquad \text{Set the cross products equal to each other.}$$

$$\frac{2.4}{0.2} = \frac{0.2x}{0.2} \qquad \text{Divide each side by 0.2.}$$

$$\frac{2.4}{0.2} = \frac{\overset{1}{\cancel{0.2}}x}{\cancel{0.2}} \qquad \text{Divide out the common factor.}$$

$$12 = x \qquad \text{Simplify.}$$

Step 3: The check is left to the student. The solution of $\dfrac{0.2}{0.8} = \dfrac{3}{x}$ is $x = 12$.

[**YOU TRY 2**] Solve each proportion.

a) $\dfrac{3\frac{1}{8}}{x} = \dfrac{20}{16}$ b) $\dfrac{0.4}{0.6} = \dfrac{7}{x}$

[W] **Hint**

Be sure you have manually checked your answers first.

ANSWERS TO [YOU TRY] **EXERCISES**

1) a) $x = 12$ b) $x = 8$ 2) a) $x = 2\frac{1}{2}$ b) $x = 10.5$

Using Technology

We can use a calculator to solve a proportion in the way we have already learned in this section. Or, we can use a shortcut, along with a calculator, to solve a proportion of the form $\dfrac{x}{b} = \dfrac{c}{d}$. To solve for x, we first find the product of b and c, and then divide by d.

Using this shortcut, we can solve $\dfrac{x}{7} = \dfrac{5.4}{6}$ by multiplying 7 and 5.4 and then dividing the product by 6: Enter $\boxed{7}\ \boxed{\times}\ \boxed{5}\ \boxed{.}\ \boxed{4}\ \boxed{\div}\ \boxed{6}\ \boxed{=}$ into the calculator.

The display will show 6.3. Therefore, the solution is $x = 6.3$. Now, solve $\dfrac{x}{7} = \dfrac{5.4}{6}$ by hand. Did you get the same answer?

E Evaluate **6.4** Exercises   Do the exercises, and check your work.

Get Ready

1) $4\dfrac{2}{3} \cdot 9$

2) $8 \cdot 1\dfrac{1}{2}$

3) 7×5.6

4) 3.7×4

5) $2\dfrac{3}{4} \cdot \dfrac{8}{9}$

6) $4\dfrac{1}{6} \cdot \dfrac{4}{5}$

7) $\dfrac{7\frac{4}{5}}{1.2}$

8) $\dfrac{5\frac{7}{10}}{1.9}$

Objective 1: Solve a Proportion

9) What does it mean to solve a proportion like $\dfrac{5}{9} = \dfrac{x}{63}$?

10) Check the solution $x = 8$ in the proportion $\dfrac{4}{7} = \dfrac{x}{21}$ to determine whether it is the correct answer. Explain.

11) Explain how to solve a proportion.

12) Is the solution to a proportion always a whole number?

Solve each proportion, and check the answer using the cross products.

13) $\dfrac{x}{16} = \dfrac{1}{8}$

14) $\dfrac{x}{12} = \dfrac{1}{4}$

 15) $\dfrac{3}{7} = \dfrac{6}{x}$

16) $\dfrac{5}{12} = \dfrac{10}{x}$

17) $\dfrac{3}{4} = \dfrac{x}{24}$

18) $\dfrac{2}{3} = \dfrac{x}{24}$

19) $\dfrac{1}{100} = \dfrac{x}{1000}$

20) $\dfrac{1}{100} = \dfrac{x}{10,000}$

21) $\dfrac{6}{x} = \dfrac{4}{15}$

22) $\dfrac{12}{x} = \dfrac{8}{5}$

23) $\dfrac{81}{6} = \dfrac{8}{x}$

24) $\dfrac{57}{9} = \dfrac{14}{x}$

Objective 2: Solve a Proportion Containing Mixed Numbers or Decimals

Solve each proportion, and check the answer using the cross products.

25) $\dfrac{1\frac{4}{5}}{x} = \dfrac{9}{10}$

26) $\dfrac{2\frac{1}{4}}{x} = \dfrac{3}{8}$

27) $\dfrac{2\frac{3}{4}}{x} = \dfrac{7}{8}$

28) $\dfrac{2\frac{1}{2}}{x} = \dfrac{3}{4}$

29) $\dfrac{1\frac{4}{9}}{5} = \dfrac{x}{3}$

30) $\dfrac{3\frac{1}{4}}{10} = \dfrac{x}{2}$

 31) $\dfrac{x}{5\frac{1}{2}} = \dfrac{7}{9}$

32) $\dfrac{x}{4\frac{3}{8}} = \dfrac{3}{11}$

33) $\dfrac{x}{1.4} = \dfrac{6}{2.5}$

34) $\dfrac{x}{3.1} = \dfrac{7}{2.8}$

35) $\dfrac{0.5}{1.2} = \dfrac{0.7}{x}$

36) $\dfrac{0.4}{2.3} = \dfrac{0.6}{x}$

37) $\dfrac{6}{x} = \dfrac{0.3}{5.2}$

38) $\dfrac{8}{x} = \dfrac{0.4}{6.1}$

39) $\dfrac{1\frac{1}{6}}{x} = \dfrac{\frac{2}{5}}{1\frac{1}{2}}$

40) $\dfrac{2\frac{1}{4}}{x} = \dfrac{\frac{2}{3}}{3\frac{1}{3}}$

41) $\dfrac{1\frac{1}{2}}{4} = \dfrac{2\frac{1}{3}}{x}$

42) $\dfrac{4\frac{1}{4}}{6} = \dfrac{3\frac{2}{3}}{x}$

43) $\dfrac{x}{6\frac{1}{4}} = \dfrac{0.8}{1.5}$

44) $\dfrac{x}{7\frac{1}{2}} = \dfrac{0.6}{2.5}$

49) $\dfrac{x}{16} = \dfrac{9}{6}$

50) $9x = 63$

45) $\dfrac{5}{2\frac{1}{3}} = \dfrac{x}{1.4}$

46) $\dfrac{4}{2\frac{1}{5}} = \dfrac{x}{5.5}$

51) $18 = 3x$

52) $\dfrac{35}{20} = \dfrac{x}{8}$

Mixed Exercises: Objectives 1 and 2
Solve.

47) $8x = 40$

48) $\dfrac{0.7}{1.2} = \dfrac{x}{6}$

53) $\dfrac{14}{9} = \dfrac{x}{1\frac{5}{7}}$

54) $16 = 2x$

55) $\dfrac{x}{2} = \dfrac{2.1}{3.5}$

56) $\dfrac{15}{4} = \dfrac{x}{3\frac{1}{3}}$

R Rethink

R1) There is another way to check whether your answer is correct after solving a proportion. Think of what it takes to compare fractions and what you know about common denominators to develop a new way to check your answer.

R2) Try this new method of checking your answer on a few of the exercises you just completed.

6.5 Solve Applied Problems Involving Proportions

P Prepare

O Organize

What are your objectives for Section 6.5?	How can you accomplish each objective?
1 Solve Applied Problems Involving Proportions	• Use the Five Steps for Solving Applied Problems that you used previously, and add some details to Step 3 regarding proportions and cross multiplication. • Complete the given examples on your own. • Complete You Trys 1 and 2.

W Work

Read the explanations, follow the examples, take notes, and complete the You Trys.

1 Solve Applied Problems Involving Proportions

Proportions are used to solve many applied problems. Since a proportion is a statement that two ratios or rates are equal, look for the key words we learned in Sections 6.2 and 6.3 as well as some new ones: *in, per, for, from, on, out of, is to,* and *as.*

To solve an application involving a proportion, we will use the Five Steps for Solving Applied Problems that we first learned in Section 2.6.

Procedure Five Steps for Solving Applied Problems

Step 1: **Read** the problem carefully, more than once if necessary, until you understand it. Draw a picture, if applicable. Identify what you are being asked to find.

Step 2: **Choose a variable** to represent an unknown quantity. If there are any other unknowns, define them in terms of the variable.

Step 3: **Translate** the problem from English into an equation using the chosen variable.

Step 4: **Solve** the equation.

Step 5: **Check** the answer in the original problem, and **interpret** the solution as it relates to the problem. Be sure your answer makes sense in the context of the problem.

Let's solve a problem using the scale on a map.

EXAMPLE 1

On a map of the United States, 1 inch represents 120 miles. If two cities are 3.5 inches apart on the map, what is the actual distance between the two cities?

0   120 mi
1 in.

Solution

Step 1: **Read** the problem carefully, and identify what we are being asked to find.

We must find the actual distance between the two cities.

We can think of this problem as a proportion in this way:

1 in. *is to* 120 mi *as* 3.5 in. *is to* the actual distance.

Step 2: **Choose a variable** to represent the unknown.

Let x = the actual distance between the two cities

Step 3: **Translate** the information that appears in English into an algebraic equation.

Write a proportion using x for the unknown quantity. We will write our rates in the form $\dfrac{\text{number of inches}}{\text{number of miles}}$ so that the numerators contain the same quantities and the denominators contain the same quantities:

$$\frac{1 \text{ in.}}{120 \text{ mi}} = \frac{3.5 \text{ in.}}{x}$$

1 in. *is to* 120 mi *as* 3.5 in. *is to* the actual distance, x miles.

Step 4: Solve the problem. We do not need the units when solving for x.

$$\text{Number of inches} \rightarrow \frac{1 \text{ in.}}{120 \text{ mi}} = \frac{3.5 \text{ in.}}{x} \leftarrow \text{Number of inches} \\ \text{Number of miles} \rightarrow \phantom{\frac{1}{120}} \phantom{\frac{3.5}{x}} \leftarrow \text{Number of miles}$$

$$\frac{1}{120} = \frac{3.5}{x} \qquad \text{Write the equation without the units.}$$

$$120 \cdot 3.5 = 1x \qquad \text{Find the cross products.}$$

$$420 = x \qquad \text{Solve the equation.}$$

Step 5: Check the answer, and **interpret** the solution as it relates to the problem.

The actual distance between the two cities is 420 mi.

Use the cross products to check $x = 420$ in the proportion $\dfrac{1}{120} = \dfrac{3.5}{x}$.

$$120 \cdot 3.5 = 420$$

$$\frac{1}{120} \diagdown \frac{3.5}{420} \qquad \text{Substitute 420 for } x. \qquad \begin{array}{l}\text{The cross products} \\ \text{are equal.}\end{array}$$

$$1 \cdot 420 = 420$$

The answer is correct.

[YOU TRY 1] If 4 granola bars cost \$4.76, find the cost of 6 granola bars.

 BE CAREFUL Be sure you write the quantities in the right places in the proportion. The units in the numerators must be the same, and the units in the denominators must be the same.

EXAMPLE 2

According to the American Lung Association, approximately 7 out of 50 premature births are the result of smoking during pregnancy. At that rate, how many of a hospital's 292 premature births were due to smoking? Round the answer to the nearest whole number. (www.lungusa.org)

Solution

Step 1: Read the problem carefully, and identify what we are being asked to find.

We must find the number of babies born early because of smoking out of a group of 292 who are born prematurely.

Step 2: Choose a variable to represent the unknown.

Let $x = $ the number of babies born prematurely because of smoking out of 292 babies

---

W Hint

Do you think many applied problems involve ratios or rates?

Step 3: **Translate** the information that appears in English into an algebraic equation.

Write a proportion using x for the unknown quantity. We will write our rates in the form $\dfrac{\text{number born early due to smoking}}{\text{total number born early}}$:

Number born early due to smoking $\rightarrow$ $\dfrac{7 \text{ babies}}{50 \text{ babies}} = \dfrac{x}{292 \text{ babies}}$ $\leftarrow$ Number born early due to smoking
Total number born early $\rightarrow$ $\leftarrow$ Total number born early

7 babies *are to* 50 babies *as x is to* 292 babies.

Step 4: **Solve** the problem. We do not need the units when solving for x.

$$\frac{7 \text{ babies}}{50 \text{ babies}} = \frac{x}{292 \text{ babies}}$$

$$\frac{7}{50} = \frac{x}{292} \qquad \text{Write the equation without the units.}$$

$$50x = 7 \cdot 292 \qquad \text{Find the cross products.}$$

$$50x = 2044 \qquad \text{Multiply.}$$

$$\frac{50x}{50} = \frac{2044}{50} \qquad \text{Divide by 50.}$$

$$x = 40.88 \qquad \text{Perform the division.}$$

$$x \approx 41 \qquad \text{Round to the nearest whole number.}$$

Step 5: **Check** the answer, and **interpret** the solution as it relates to the problem.

Approximately 41 babies out of 292 were born prematurely due to smoking.

Use the cross products to check $x = 40.88$ in the proportion $\dfrac{7}{50} = \dfrac{x}{292}$.

$$\frac{7}{50} = \frac{40.88}{292} \qquad \text{Substitute 40.88 for } x.$$

When we round 40.88 to the nearest whole number, we get 41. The answer is correct.

[**YOU TRY 2**] A survey at Harrington High School revealed that 5 out of 8 of its seniors would be going away for college. How many of the 424 seniors were planning to go away for college?

ANSWERS TO [YOU TRY] **EXERCISES**

1) $7.14   2) 265

Objective 1: Solve Applied Problems Involving Proportions

Solve each problem using a proportion. Be sure to keep the units organized.

1) On a highway map, 2 in. represents 140 mi. If two cities are 8.5 in. apart on the map, what is the actual distance between the two cities?

2) In a scaled drawing, a 200-ft building is drawn 5 in. high. What is the actual height of a building that is drawn 3 in. high?

3) If 120 g of frozen yogurt contains 12 g of fat, how much fat is in 200 g of frozen yogurt?

4) A 12-oz serving of Pepsi contains 38 mg of caffeine. How much caffeine is in 20 oz of Pepsi? Round the answer to the nearest whole number. (pepsiproductfacts.com)

5) A high-speed train can travel 240 mi in 1.5 hr. At the same rate, how far can the train travel in 4 hr?

©Ingram Publishing

6) On average, 5 gal of paint will cover 1750 sq ft of wall space. How many gallons of paint will be required to paint a home if the wall space is 1050 sq ft?

7) If Gary earns $435.75 in 5 days, how much does he earn in 3 days?

8) If the cost to deliver 30 newspapers to a person's home is $48.00, what is the cost to deliver 8 newspapers?

9) In his pickup, Daryl can drive 224 mi on 14 gal of gas. How far could he drive on 20 gal?

10) If a tortoise can walk 7 ft in 20 sec, how long would it take to walk 50 ft? Round to the nearest second.

11) The number of calories a person burns is proportional to his weight.
©George Brits/Gallo Images/ Getty Images
A 160-lb person burns 219 calories during a 30-min hike. How many calories will a 200-lb person burn on the same 30-min hike? Round to the nearest calorie. (www.mayoclinic.com)

12) A recent study at a university found that 17 out of 20 American youths aged 8 to 18 who play video games show multiple signs of behavioral addiction. In a group of 260 youths in this age group, how many are expected to have this addiction?

13) In April 2017, the U.S. Bureau of Labor Statistics reported that 4.4 out of 100 citizens were unemployed. In a group of 2500 citizens, how many were expected to be unemployed? (www.bls.gov)

14) At a campus election, it is expected that 7 out of 10 students will vote. On a campus of 12,318 students, how many students are expected to vote? Round the answer to the nearest whole number.

Use the floor plan shown for Exercises 15–22. One inch represents 4 feet on the plan.

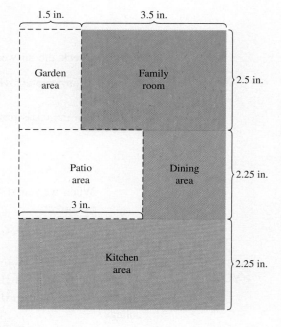

15) What are the actual length and width of the kitchen?

16) What are the actual width and length of the dining area?

17) What are the actual width and length of the family room?

18) What are the actual width and length of the garden area?

19) What are the actual width and length of the entire floor plan?

20) What is the actual area of the patio area?

21) What is the actual area of the entire floor plan?

22) What is the combined area of the dining, family, and kitchen areas?

Solve each problem using a proportion.

23) A nursing student observes an intravenous drip rate of 8 drops per 10 sec. How many drops should be observed in 1 min?

24) Nadia teaches two classes, and each contains 35 students. If she grades the final exam at a rate of 15 min per exam, how many minutes will be required to grade all the finals? How many hours is this?

25) To safely thaw a frozen turkey in the refrigerator takes 24 hr per every 4.5 lb. How many hours will it take to thaw a 15-lb turkey in this manner?
(www.fsis.usda.gov)

26) Suppose that a bicyclist pedals a bicycle at 40 revolutions per minute, resulting in a speed of 7 miles per hour. How fast will the bicyclist go if she pedals 60 revolutions per minute?

27) At 2 P.M., Maria's shadow is 2 ft long, and she is 5 ft tall. If the flagpole casts a shadow that is 8.4 ft long, how tall is the flagpole?

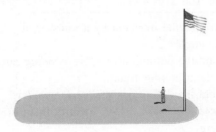

28) Refer to Exercise 27. At 5 P.M., Maria's shadow is 8 ft long. How long is the flagpole's shadow at this time?

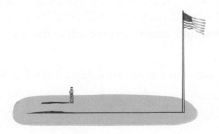

The standard wide-screen format ratio is 16:9, width to height. A new movie theatre is being constructed with four different screen sizes. Fill in the missing dimensions in the table.

	Width in Feet	Height in Feet
29)	32	
30)	40	
31)		36
32)		45

33) Rosa Vasquez won an election over her opponent by a ratio of 6:5. If her opponent received 2820 votes, how many did Rosa receive?

34) The ratio of women to men in a college physics lecture course is 4:17. Find the number of men in this course if it contains 32 women.

35) Suppose the gas-to-oil ratio for a two-stroke engine is 32:1. How much oil, in gallons, must be mixed with 8 gal of gasoline? If 1 gal equals 128 fl oz, how many ounces of oil must be added to the 8 gal?

36) A particular lawn mower requires a 40:1 gas-to-oil ratio. How much oil, in gallons, must be mixed with 2 gal of gasoline? If 1 gal equals 128 fl oz, how many ounces of oil must be added to the 2 gal?

37) A high school wishes to have a student-to-teacher ratio of 22 to 1. If there are 2315 students, how many teachers are needed? Round to the nearest whole number.

38) If the ratio of cats to dogs at an animal shelter is 3 to 5 and there are 45 dogs at the shelter, how many cats are in the shelter?

39) Approximately 9 out of 10 students believe that studying on a daily basis will improve one's grade. But of those who believe, only 1 out of 7 students study on a daily basis. Out of 8760 students, how many believe that studying on a daily basis will improve one's grade and do in fact study on a daily basis? Round to the nearest whole number.

40) A school survey reveals that 4 out of 5 students eat at a fast-food restaurant every Friday after school. Only 1 out of 12 students who do this order a healthy food item from the menu. Out of 3850 students, how many students will go to a fast-food restaurant on Friday and order a healthy food item? Round to the nearest whole number.

R1) Did you remember to include units in your answers? Why is it important to include the units?

R2) Where have you previously encountered a ratio or rate that you could use to set up a proportion to solve for an unknown?

6.6 Angles

P Prepare

O Organize

What are your objectives for Section 6.6?	How can you accomplish each objective?
1 Understand Complementary and Supplementary Angles	• Write the definition of *complementary angles* in your own words, and draw examples. • Write the definition of *supplementary angles* in your own words, and draw examples. • Complete the given examples on your own. • Complete You Trys 1–3.
2 Understand Congruent and Vertical Angles	• Write the definition of *congruent angles* in your own words, and draw examples. • Write the definition of *vertical angles* in your own words, and draw examples. • Write the definition of *adjacent angles* in your own words, and draw examples. • Complete the given examples on your own. • Complete You Trys 4–8.
3 Use the Relationships of Corresponding Angles and Alternate Interior Angles	• Draw two parallel lines cut by a transversal. Label the angles. • Understand the definition of *corresponding angles*. Identify them in your figure. • Understand the definitions of *interior angles* and *alternate interior angles*. Identify them in your figure. • Complete the given example on your own. • Complete You Try 9.

W Work

Read the explanations, follow the examples, take notes, and complete the You Trys.

In this section, we will learn more about angles and their relationships. Let's begin with complementary and supplementary angles.

1   Understand Complementary and Supplementary Angles

 Hint

This section contains many new terms. Make and use flashcards to learn them.

Definition

Two angles are **complementary angles** if their measures add to 90°.

Example:

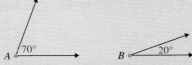

∠A and ∠B are complementary angles
because m∠A + m∠B = 70° + 20° = 90°.

We say that ∠A and ∠B are *complements* of each other.

 Hint

Write a procedure for finding the complement of an angle in your notes.

In the definition box, ∠A and ∠B are complementary because 70° + 20° = 90°. But, it is also true that 90° − 70° = 20° and 90° − 20° = 70°. So, if we know the measure

$$\uparrow \qquad \uparrow \qquad\qquad \uparrow \qquad \uparrow$$
$$\text{m}\angle A \quad \text{m}\angle B \qquad\quad \text{m}\angle B \quad \text{m}\angle A$$

of one angle and want to find its complement, we subtract the measure of the given angle from 90°.

EXAMPLE 1

If m∠R = 31°, find the measure of its complement.

Solution

To find the measure of the complement of ∠R, subtract 31° from 90°:

$$90° − 31° = 59°$$

The measure of the complement of ∠R is 59°.

Check by adding:

$$\text{m}\angle R + \text{Measure of its complement} = 31° + 59° = 90°. \ \checkmark$$

[YOU TRY 1]

If m∠Z = 16°, find the measure of its complement.

We use a special term for two angles whose sum is 180°.

Definition

Two angles are **supplementary angles** if their measures add to 180°.

Example:

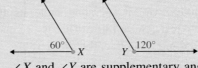

∠X and ∠Y are supplementary angles
because m∠X + m∠Y = 60° + 120° = 180°.

We say that ∠X and ∠Y are *supplements* of each other.

In the definition box for supplementary angles, $\angle X$ and $\angle Y$ are supplementary because $60° + 120° = 180°$. It is also true that $180° - 60° = 120°$ and $180° - 120° = 60°$. So, if

$$\underset{\text{m}\angle X}{\uparrow} \quad \underset{\text{m}\angle Y}{\uparrow} \qquad\qquad \underset{\text{m}\angle Y}{\uparrow} \quad \underset{\text{m}\angle X}{\uparrow}$$

we know the measure of one angle and we want to find its supplement, we subtract the measure of the given angle from 180°.

EXAMPLE 2

If $m\angle C = 114°$, find the measure of its supplement.

Solution

To find the measure of the supplement of $\angle C$, subtract 114° from 180°:

$$180° - 114° = 66°$$

The measure of the supplement of $\angle C$ is 66°.

Check by adding:

$$m\angle C + \text{Measure of its supplement} = 114° + 66° = 180°. \quad \checkmark$$

$\Big[$ **YOU TRY 2** $\Big]$ If $m\angle H = 35°$, find the measure of its supplement.

We can think about complementary and supplementary angles in another way.

EXAMPLE 3

Find each missing angle measure.

a) $\angle PQR$ is a right angle.
Find $m\angle x$.

b) $\angle ABC$ is a straight angle.
Find $m\angle y$.

 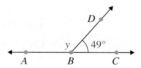

Solution

a)   Because $\angle PQR$ is a right angle, its measure is 90°. The figure shows that $m\angle TQR + m\angle x = 90°$. (That is, $\angle TQR$ and $\angle x$ are *complementary* angles.) To find $m\angle x$, subtract 68° from 90°:

$$90° - m\angle TQR = m\angle x$$
$$90° - 68° = 22°$$
$$m\angle x = 22°$$

b)   Because $\angle ABC$ is a straight angle, its measure is 180°. The figure shows that $m\angle y + m\angle DBC = 180°$. (That is, $\angle y$ and $\angle DBC$ are *supplementary* angles.) To find $m\angle y$, subtract 49° from 180°:

$$180° - m\angle DBC = m\angle y$$
$$180° - 49° = 131°$$
$$m\angle y = 131°$$

Find each missing angle measure.

a)   ∠*CDF* is a right angle.
    Find m∠*n*.

b)   ∠*RTV* is a straight angle.
    Find m∠*w*.

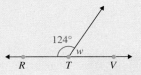

2 Understand Congruent and Vertical Angles

Definition

Two angles are **congruent angles** if their measures are the same.

Example:

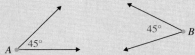

∠*A* and ∠*B* are *congruent angles* because their
angle measures are the same.

We can also say that ∠*A* and ∠*B* are *congruent*. We can write this as ∠*A* ≅ ∠*B*.
Read this as "Angle *A* *is congruent to* angle *B*." The symbol ≅ means "is
congruent to."

When two lines intersect, four angles are formed.

Definition

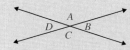

Vertical angles are the angles opposite each other—the angles that do not share a
common side.

Example: ∠*A* and ∠*C* are vertical angles.

∠*D* and ∠*B* are vertical angles.

Vertical angles are congruent. That is, their angle measures are the same.

Example: ∠*A* ≅ ∠*C* and ∠*D* ≅ ∠*B*

W Hint

Would it be helpful to
make notes on the
example?

EXAMPLE 4   Identify the vertical angles, and make a statement about their congruence.

Solution

$\angle W$ and $\angle Y$ are vertical angles. Therefore, $\angle W \cong \angle Y$. ($\angle W$ is congruent to $\angle Y$.)

$\angle X$ and $\angle Z$ are vertical angles. Therefore, $\angle X \cong \angle Z$. ($\angle X$ is congruent to $\angle Z$.)

[YOU TRY 4]   Identify the vertical angles, and make a statement about their congruence.

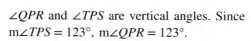

We can use what we know about vertical angles to find missing angle measures.

EXAMPLE 5   Find m$\angle QPT$ and m$\angle QPR$.

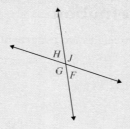

Solution

Begin by identifying the vertical angles because *vertical angles have the same measure.*

$\angle QPT$ and $\angle RPS$ are vertical angles. Since m$\angle RPS = 57°$, m$\angle QPT = 57°$.

$\angle QPR$ and $\angle TPS$ are vertical angles. Since m$\angle TPS = 123°$, m$\angle QPR = 123°$.

[YOU TRY 5]   Find m$\angle QWY$ and m$\angle ZWY$.

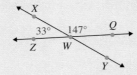

Two angles that share a common side also have a special relationship.

Definition

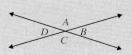

Adjacent angles are angles that share a common side and a common vertex.

 Example: ∠A and ∠B are adjacent angles. ∠B and ∠C are adjacent angles.

     ∠C and ∠D are adjacent angles. ∠D and ∠A are adjacent angles.

Note

When angles are formed with two intersecting lines as pictured in the definition box, the measures of adjacent angles add up to 180°. (Can you see that two adjacent angles form a straight angle?)

EXAMPLE 6 Identify the adjacent angles.

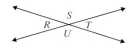

Solution

  ∠S and ∠T are adjacent angles. ∠T and ∠U are adjacent angles.

  ∠U and ∠R are adjacent angles. ∠R and ∠S are adjacent angles.

[YOU TRY 6] Identify the adjacent angles.

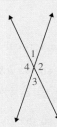

Now let's find some angle measures based on what we have learned about vertical and adjacent angles.

EXAMPLE 7 Find the measures of angles *A*, *B*, and *C*.

W Hint

Write down whatever relationship you see first.

Solution

Look for any relationship between the 41° angle and *any other angle*: **The 41° angle and ∠B are vertical angles, so they have the same angle measure.**

$$m\angle B = 41°$$

What is the relationship between the 41° angle and ∠A? They are not only adjacent angles, but *they also form a straight angle*. Therefore, **the sum of their measures is 180°.**

$$41° + m\angle A = 180° \quad \text{or} \quad 180° - 41° = m\angle A = 139°$$

$$m\angle A = 139°$$

∠A and ∠C are vertical angles, so m∠A = m∠C.

$$m\angle C = 139°$$

We can fill in the missing angle measures:

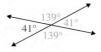

[**YOU TRY 7**] Find the measures of angles *T*, *V*, and *W*.

EXAMPLE 8   Find each missing angle measure.

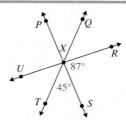

Solution

Begin by finding the vertical angles because they are congruent.

∠RXS ≅ ∠UXP so m∠RXS = m∠UXP

m∠UXP = 87°

∠TXS ≅ ∠PXQ so m∠TXS = m∠PXQ

m∠PXQ = 45°

Fill in these angle measures:

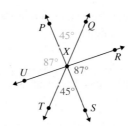

∠TXQ is a straight angle, so its measure is 180°.

Therefore, m∠TXU + m∠UXP + m∠PXQ = 180°. To find m∠TXU, add the measures of ∠UXP and ∠PXQ, then subtract from 180°.

$$m\angle UXP + m\angle PXQ = 87° + 45° = 132°$$

Then, **m∠*TXU* = 180° − 132° = 48°.**

∠*TXU* and ∠*QXR* are vertical angles, so their measures are the same:
m∠*QXR* = 48°.

Fill in all the missing angle measures.

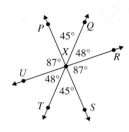

Find each missing angle measure.

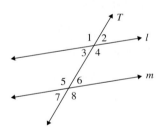

3 Use the Relationships of Corresponding Angles and Alternate Interior Angles

Now let's look at the angles we get when we intersect two parallel lines with another line called a *transversal.*

Here is a line, *T*, that intersects two parallel lines, *l* and *m*. The line *T* is called a **transversal.** We say that lines *l* and *m* are *cut* by the transversal. The angles formed have special names and properties.

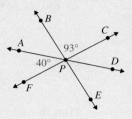

∠1 and ∠5 are examples of **corresponding angles.** They are on the same side of the transversal and in the same position with respect to line *l* and line *m*. **Corresponding angles are congruent.** So, we can write ∠1 ≅ ∠5. We can list the other pairs of corresponding angles:

∠2 and ∠6 are corresponding angles, so ∠2 ≅ ∠6.

∠3 and ∠7 are corresponding angles, so ∠3 ≅ ∠7.

∠4 and ∠8 are corresponding angles, so ∠4 ≅ ∠8.

Notice that angles 3, 4, 5, and 6 are inside the parallel lines. These angles are called **interior angles.** But, we can be more specific and say that ∠3 and ∠6 are **alternate interior angles,** and ∠4 and ∠5 are **alternate interior angles.** (Alternate interior angles are on opposite sides of the transversal.) **Alternate interior angles are congruent,** so we can say that

∠3 and ∠6 are alternate interior angles, so ∠3 ≅ ∠6.

∠4 and ∠5 are alternate interior angles, so ∠4 ≅ ∠5.

EXAMPLE 9

Lines *l* and *m* are parallel lines cut by transversal *T*. If m∠2 = 125°, find the measures of the other angles.

Solution

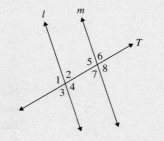

As you find the measure of each angle, write it on the figure.

> **W Hint**
>
> Make a note of *whatever* relationship you see. You don't have to go in a certain order.

Find m∠1.

∠1 and ∠2 are *supplementary angles,* so m∠1 + m∠2 = 180°. Because m∠2 = 125°,

$$m∠1 = 180° − 125° = 55° \qquad m∠1 = 55°$$

Find m∠3.

∠2 and ∠3 are *vertical angles,* so m∠2 = m∠3: m∠3 = 125°.

Find m∠4.

∠1 and ∠4 are *vertical angles,* so m∠1 = m∠4: m∠4 = 55°.

Now, we can substitute the angle measures this way:

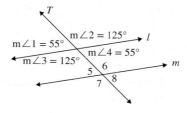

Find m∠5.

∠1 and ∠5 are *corresponding angles,* so m∠1 = m∠5: m∠5 = 55°.

Find m∠6.

∠2 and ∠6 are *corresponding angles,* so m∠2 = m∠6: m∠6 = 125°.

Find m∠7.

∠6 and ∠7 are *vertical angles,* so m∠6 = m∠7: m∠7 = 125°.

Find m∠8.

m∠5 = m∠8 are *vertical angles,* so m∠5 = m∠8: m∠8 = 55°.

[YOU TRY 9]

l and *m* are parallel lines cut by transversal *T*. If m∠2 = 78°, find the measures of the other angles.

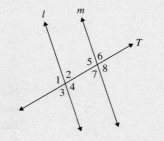

E Evaluate 6.6 Exercises

Do the exercises, and check your work.

Objective 1: Understand Complementary and Supplementary Angles

1) What are supplementary angles?

2) What are complementary angles?

Find the measure of the complement of each of the given angles.

3) m∠V = 61°   4) m∠Y = 19°

5) m∠Q = 23°   6) m∠H = 38°

7) m∠K = 7°   8) m∠T = 46°

Find the measure of the supplement of each of the given angles.

9) m∠C = 103°   10) m∠X = 166°

 11) m∠J = 54°   12) m∠U = 21°

13) m∠B = 117°   14) m∠D = 8°

For Exercises 15–24, find the missing angle measure.

15) ∠ATB is a right   16) ∠ZXV is a right
     angle. Find m∠b.   angle. Find m∠a.

 17) ∠BGN is a straight angle. Find m∠z.

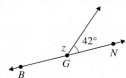

18) ∠RYQ is a straight angle. Find m∠a.

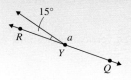

19) ∠ATB is a right   20) ∠KMR is a right
     angle. Find m∠y.   angle. Find m∠n.

21) ∠WTV is a straight angle. Find m∠x.

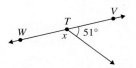

22) ∠KZD is a straight angle. Find m∠y.

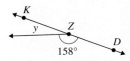

23) ∠JZE is a right angle. Find m∠q.

24) ∠CWQ is a straight angle. Find m∠k.

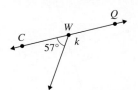

Fill in the blank with *always*, *sometimes*, or *never* to make the statement true.

25) The measure of the supplement of an acute angle is _____ an obtuse angle.

26) The measure of the complement of an acute angle is _____ an obtuse angle.

Objective 2: Understand Congruent and Vertical Angles

27) What does it mean if two angles are congruent?

28) Use the correct notation to write "∠T is congruent to ∠K."

29) What do we know about the measures of vertical angles?

30) What is the difference between vertical angles and adjacent angles?

Identify the vertical angles, and make a statement about their congruence.

31)

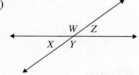

32)

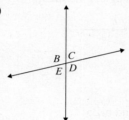

33)

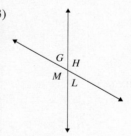

Find the measures of the indicated angles.

34) ∠SRN and ∠BRN

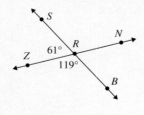

35) ∠TJQ and ∠DJQ

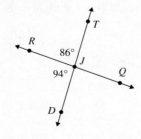

36) ∠PQN and ∠PQV

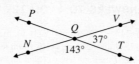

37) ∠ADF and ∠FDB

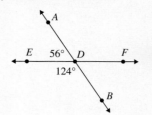

38) ∠WKR and ∠ZKR

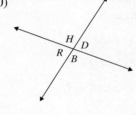

Identify the adjacent angles.

39)

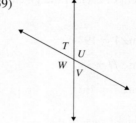

40)

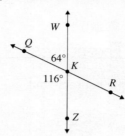

41) Draw two intersecting lines with vertical angles P and R.

42) Draw two intersecting lines with adjacent angles G and H.

43) Find the measures of angles A, C, and F.

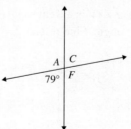

44) Find the measures of angles K, M, and P.

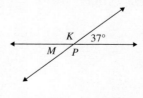

45) Find the measures of angles *T*, *U*, and *V*.

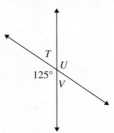

46) Find the measures of angles *H*, *D*, and *B*.

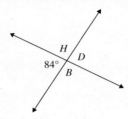

Find each missing angle measure.

47)

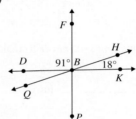

48)

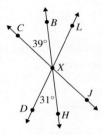

49)

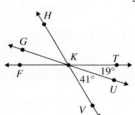

50)

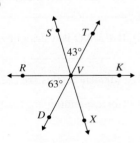

51) When two lines intersect, four angles are formed. When three lines intersect at the same point, six angles are formed. How many angles are formed when five lines intersect at the same point, and what is the sum of the measures of all the angles?

52) What angle is congruent to its complement?

Objective 3: Use the Relationships of Corresponding Angles and Alternate Interior Angles

Use this figure for Exercises 53–58.

Lines *l* and *m* are parallel. Fill in the blank with the appropriate term.

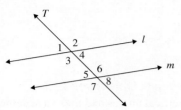

Terms

| Alternate interior | Vertical |
| Corresponding | Transversal |

53) Line *T* is a _____.

54) ∠3 and ∠7 are _____ angles.

55) ∠4 and ∠5 are _____ angles.

56) ∠5 and ∠8 are _____ angles.

57) ∠1 ≅ ∠5. Why?

58) ∠3 ≅ ∠6. Why?

In Exercises 59 and 60, lines *l* and *m* are parallel. Identify all pairs of corresponding angles and all pairs of alternate interior angles.

59)

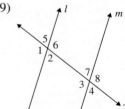

60)

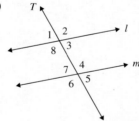

In Exercises 61–64, lines *l* and *m* are parallel, and each figure is labeled with one angle measure. Find the measures of the other angles.

61)

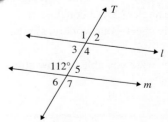

62)

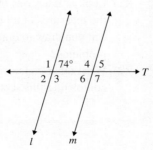

63)

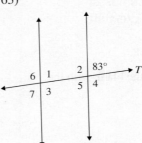

64)

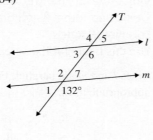

70) Find the measure of angles *F*, *G*, and *H*.

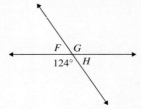

In Exercises 71 and 72, lines *m* and *n* are parallel, and each figure is labeled with one angle measure. Find the measure of the other angles.

Mixed Exercises: Objectives 1–3

Identify the vertical angles and make a statement about their congruence.

71)   **72)**

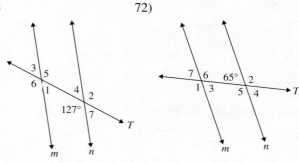

 65)   **66)**

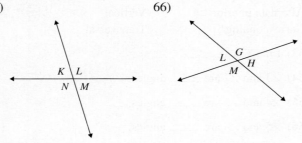

67) If m∠*G* = 32°, find the measure of its supplement.

68) If m∠*R* = 74°, find the measure of its complement.

69) Find each missing angle measure.

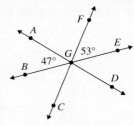

Fill in the blank with *always*, *sometimes*, or *never* to make the statement true.

73) Congruent angles _____ have the same measure.

74) Vertical angles are _____ congruent.

75) The sum of the measures of supplementary angles is _____ 90°.

76) If ∠*MNQ* is a straight angle, then m∠*MNQ* is _____ 360°.

77) Adjacent angles are _____ congruent.

78) Corresponding angles are _____ congruent.

R Rethink

R1) Look at Exercises 69 and 70. What is the sum of the measures of all the angles? What conclusion can you draw to quickly check your work on similar problems?

R2) What other conclusions were you able to draw after completing this section?

R3) This section contained many new terms and concepts. Are you confident that you could explain them to someone without looking at the book or your notes? What do you know well, and what do you need to review more?

6.7 Solve Applied Problems Involving Congruent and Similar Triangles

What are your objectives for Section 6.7?	How can you accomplish each objective?
1 Identify Corresponding Angles and Corresponding Sides of Congruent Triangles	• In your own words, explain the difference between *congruent* figures and *similar* figures. • Complete the given example on your own. • Complete You Try 1.
2 Use SSS, SAS, and ASA to Prove That Triangles Are Congruent	• Summarize the three theorems. Include figures with your summary. • Complete the given example on your own. • Complete You Try 2.
3 Find Unknown Lengths in Similar Triangles	• Understand how the angles and sides of similar triangles are marked. • Write the definitions of *corresponding angles* and *corresponding sides* in your own words. • Write the definition of *similar triangles* in your own words. • Write a procedure for finding an unknown length in similar triangles. • Complete the given examples on your own. • Complete You Trys 3–5.
4 Solve Applied Problems Involving Similar Triangles	• Use the five-step process to solve similar triangles applications. • Complete the given example on your own. • Complete You Try 6.

 W Work   **Read the explanations, follow the examples, take notes, and complete the You Trys.**

In this section, we will learn about congruent and similar triangles.

Two figures are **congruent** if they are *exactly* the same; that is, they are the *same shape* and the *same size*. Two figures are **similar** if they are the *same shape* but they are *different in size*. First, we will learn about congruent triangles.

1 Identify Corresponding Angles and Corresponding Sides of Congruent Triangles

Look at these triangles:

They are exactly the same shape *and* the same size. Therefore, they are **congruent triangles.**

If we picked up ΔABC and put it on top of ΔXYZ, ∠A would be on top of ∠X, ∠B would be on top of ∠Y, and ∠C would be on top of ∠Z.

∠A and ∠X are *corresponding angles,* so m∠A = m∠X.

∠B and ∠Y are *corresponding angles,* so m∠B = m∠Y.

∠C and ∠Z are *corresponding angles,* so m∠C = m∠Z.

What can we say about the *sides* of the triangles when we put one triangle on top of the other? Side AB would be on top of side XY, so their lengths are the same. (Writing AB = XY means that the lengths of the sides are the same.) Also, side BC would be on top of side YZ, and side AC would be on top of XZ.

$\overline{AB}$ and $\overline{XY}$ are *corresponding sides,* and AB = XY.

$\overline{BC}$ and $\overline{YZ}$ are *corresponding sides,* and BC = YZ.

$\overline{AC}$ and $\overline{XZ}$ are *corresponding sides,* and AC = XZ.

Because all of their corresponding angles have the same measure and all of their corresponding sides have the same length, ΔABC and ΔXYZ are congruent triangles. We write this as ΔABC ≅ ΔXYZ.

EXAMPLE 1 For each pair of congruent triangles, identify the corresponding angles and corresponding sides.

a)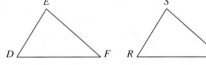

b)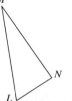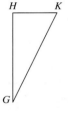

Solution

a) If you picked up ΔDEF and put it on top of ΔRST, you would not have to change the position of the triangles at all to get their corresponding parts to match up.

Corresponding Angles	**Corresponding Sides**
∠D and ∠R	$\overline{DE}$ and $\overline{RS}$
∠E and ∠S	$\overline{EF}$ and $\overline{ST}$
∠F and ∠T	$\overline{DF}$ and $\overline{RT}$

b) These triangles are not positioned in the same way. Let's redraw the second triangle, ΔGHK, by rotating it clockwise so that it matches ΔLMN. This makes it easier for us to see the corresponding parts.

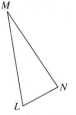 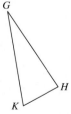

Corresponding Angles	**Corresponding Sides**
∠L and ∠K	$\overline{LM}$ and $\overline{KG}$
∠M and ∠G	$\overline{MN}$ and $\overline{GH}$
∠N and ∠H	$\overline{LN}$ and $\overline{KH}$

For each pair of congruent triangles, identify the corresponding angles and corresponding sides.

a)

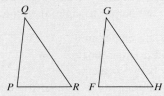

b)

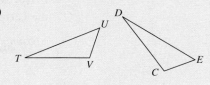

2 Use SSS, SAS, and ASA to Prove That Triangles Are Congruent

We can show that two triangles are congruent by showing that all of their corresponding angles are congruent *and* that all of their corresponding sides are congruent. Or, we can use these rules.

Procedure Using SSS, SAS, and ASA to Prove That Triangles Are Congruent

We can use these rules, or *theorems,* to prove that two triangles are congruent.

1) Side-Side-Side (SSS) Theorem

If all of the sides of one triangle are the same length as the corresponding sides of another triangle, then the triangles are congruent.

Example: Let *a, b,* and *c* be the lengths of the sides in the first triangle. Let *x, y,* and *z* be the lengths of the sides in the second triangle.

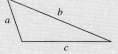
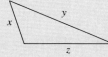

If $a = x$, $b = y$, and $c = z$, then the triangles are congruent.

2) Side-Angle-Side (SAS) Theorem

If two sides of a triangle and the angle between them are the same measure as the corresponding parts on another triangle, then the triangles are congruent.

Example: Let *a* and *b* be the lengths of two sides of a triangle and let 1 be the angle between them. Let *x* and *y* be the lengths of two sides of a second triangle and let 2 be the angle between them.

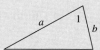
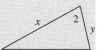

If $a = x$, $b = y$, and $m\angle 1 = m\angle 2$, then the two triangles are congruent.

(Continued)

3) Angle-Side-Angle (ASA) Theorem

If two angles of a triangle and the side between them are the same measure as the corresponding parts on another triangle, then the triangles are congruent.

Example: ∠1 and ∠2 are angles in a triangle and *a* is the length of the side between them.

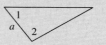 

∠3 and ∠4 are angles in another triangle and *x* is the length of the side between them.

If m∠1 = m∠3, m∠2 = m∠4, and *a* = *x*, then the two triangles are congruent.

EXAMPLE 2

For each pair of triangles, determine which theorem proves that the triangles are congruent. Use SSS, SAS, or ASA. Explain your answer.

a)

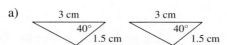

b)

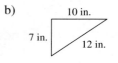

W Hint

Think about how you could "pick up" one triangle and put it on top of the other so that their corresponding parts will match.

c)

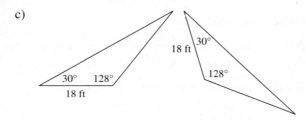

Solution

a) In both triangles, the corresponding sides and the angle between them have the same measure. Therefore, the **SAS theorem** (or Side-Angle-Side) proves that the triangles are congruent.

b) The corresponding sides of the triangles are the same length. Therefore, the **SSS theorem** (or Side-Side-Side) proves that the triangles are congruent.

c) In the first triangle, two angles and the side between them have the same measure as the corresponding parts of the second triangle. Therefore, the **ASA theorem** (or Angle-Side-Angle) proves that the triangles are congruent.

For each pair of triangles, determine which theorem proves that the triangles are congruent. Use SSS, SAS, or ASA. Explain your answer.

a)   b)

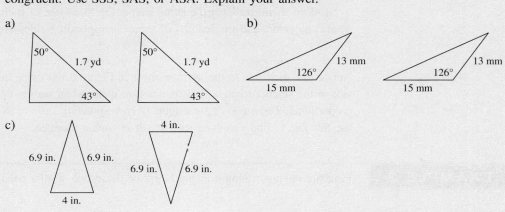

c)

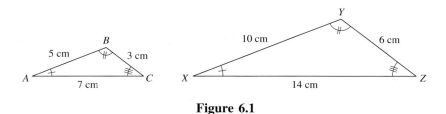

Next we will learn about similar triangles.

3 Find Unknown Lengths in Similar Triangles

At the beginning of this section, we said that two figures are congruent if they are *exactly* the same *shape* and *size*. Two figures are **similar,** however, if they are the *same shape* but they are *different in size.*

Look at these triangles in Figure 6.1:

Y
10 cm   6 cm

B
5 cm   3 cm

A   7 cm   C   X   14 cm   Z

Figure 6.1

They are the same *shape,* but they are not the same size. We can also say something about their angles:

$m\angle A = m\angle X$ which is indicated by   ╬

$m\angle B = m\angle Y$ which is indicated by   ╬

$m\angle C = m\angle Z$ which is indicated by   ╬

The angles with the same measure are corresponding angles. The sides opposite the corresponding angles are the corresponding sides. Although corresponding angles have the same measure, corresponding sides are *not* necessarily the same length.

Corresponding Angles	**Corresponding Sides**
$\angle A$ and $\angle X$	$\overline{BC}$ and $\overline{YZ}$
$\angle B$ and $\angle Y$	$\overline{AC}$ and $\overline{XZ}$
$\angle C$ and $\angle Z$	$\overline{AB}$ and $\overline{XY}$

Triangles like these are called *similar triangles.*

Definition

Similar triangles have the same shape, the measures of their corresponding angles are the same, and the lengths of their corresponding sides are proportional.

Notice that each side in the larger triangle in Figure 6.1 is twice the length of the corresponding side in the smaller triangle. This is what we mean when we say that corresponding sides are proportional: *The ratios of the lengths of corresponding sides of the triangles are the same.* We use this fact to find a missing side length in similar triangles.

EXAMPLE 3   Find the unknown length in the larger of these two similar triangles.

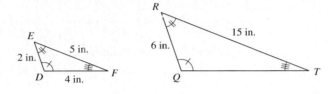

Solution

We can find the missing length, $\overline{QT}$, just by thinking about how the sides of the first triangle are related to the sides of the second triangle.

Corresponding Sides	Lengths of Those Sides
$\overline{DE}$ and $\overline{QR}$	2 in. and 6 in.
$\overline{EF}$ and $\overline{RT}$	5 in. and 15 in.

How are the corresponding sides related? **The sides in the larger triangle are 3 times the length of the corresponding sides in the smaller triangle.**

So, the length of side $\overline{QT}$ is 3 *times* the length of $\overline{DF}$.

$$\text{Length of } \overline{QT} = 3 \cdot \text{length of } \overline{DF}$$
$$= 3 \cdot (4 \text{ in.}) \qquad \text{Substitute the value.}$$
$$= 12 \text{ in.} \qquad \text{Multiply.}$$

[YOU TRY 3]   Find the unknown length in the larger of these two similar triangles.

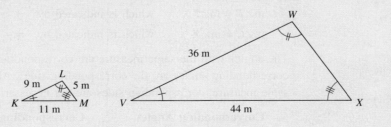

If we cannot figure out the relationship between the sides just by looking at the lengths of the sides, then we can use a proportion. Let's solve Example 3 using a proportion.

EXAMPLE 4

Use a proportion to find the unknown length in Example 3.

Solution

Label the unknown length x.

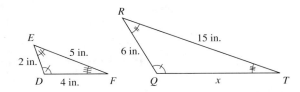

Choose either of the two corresponding side pairs that we know; let's choose 2 in. and 6 in. The corresponding sides that include the side we do not know are 4 in. and x.

The ratios of the corresponding sides are equal, so we can set up a proportion. Read it as

$$\text{Corresponding sides} \left\langle \begin{array}{c} \nearrow \\ \searrow \end{array} \frac{2}{6} \uparrow \frac{4}{x} \begin{array}{c} \nwarrow \\ \swarrow \end{array} \right\rangle \text{Corresponding sides}$$

2 is to 6 as 4 is to x.

Our proportion is $\dfrac{2}{6} = \dfrac{4}{x}$. Set the cross products equal to each other, and solve.

> **W Hint**
>
> Could this example be solved by setting up a different proportion?

$$\frac{2}{6} \times \frac{4}{x} \quad \begin{array}{l} 6 \cdot 4 = 24 \\ 2 \cdot x = 2x \end{array}$$

$2x = 24$ Set the cross products equal to each other.

$\dfrac{2x}{2} = \dfrac{24}{2}$ Divide by 2.

$x = 12$ Simplify.

The length of the missing side is 12 in. This is the same result we obtained in Example 3.

[YOU TRY 4]   Use a proportion to find the unknown length in You Try 3.

Sometimes, you must look at the triangles carefully to recognize the corresponding sides.

EXAMPLE 5

Find the length of side x, and find the perimeter of the smaller of these similar triangles.

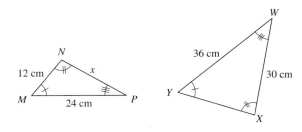

Solution

These triangles are not positioned in the same way. Let's redraw the second triangle so that we can identify the corresponding sides more easily.

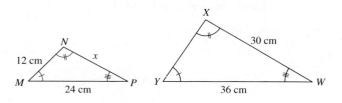

Identify two corresponding sides that we *do* know: 24 cm and 36 cm

Identify the corresponding sides that include the side we must find: x and 30 cm

Write a proportion:

$$\text{Corresponding sides} \left\langle \frac{24}{36} = \frac{x}{30} \right\rangle \text{Corresponding sides}$$

24 is to 36 as x is to 30.

Set the cross products equal to each other, and solve.

$$\frac{24}{36} \diagdown\diagup \frac{x}{30} \quad \begin{array}{l} 36 \cdot x = 36x \\ 24 \cdot 30 = 720 \end{array}$$

$36x = 720$ Set the cross products equal to each other.

$\dfrac{36x}{36} = \dfrac{720}{36}$ Divide by 36.

$x = 20$ Simplify.

The length of the missing side is 20 cm. Now, we can find the perimeter of the smaller triangle.

$$\text{Perimeter} = 12 \text{ cm} + 20 \text{ cm} + 24 \text{ cm} = 56 \text{ cm}$$

[YOU TRY 5]

Find the length of side x, and find the perimeter of the smaller of these similar triangles.

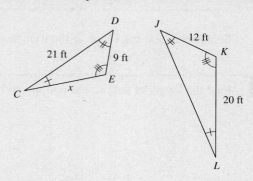

4 Solve Applied Problems Involving Similar Triangles

Many applications involve similar triangles. Let's use the five-step process to solve this problem.

EXAMPLE 6   Josh is 6 ft tall, and he casts a shadow that is 8 ft long. At the same time, a tree nearby casts a shadow that is 60 ft long. How tall is the tree?

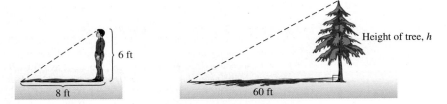

Height of tree, h

Solution

Step 1: **Read** the problem carefully, and identify what we are being asked to find.

We must find the height of the tree.

Step 2: **Choose a variable** to represent the unknown.

Let h = the height of the tree.

Step 3: **Translate** the information that appears in English into an algebraic equation.

8 ft and 60 ft are corresponding sides. 6 ft and h are corresponding sides.

Corresponding sides $\Big\langle \dfrac{8}{60} = \dfrac{6}{h} \Big\rangle$ Corresponding sides

8 is to 60 as 6 is to h.

Step 4: **Solve** the problem.

$$\dfrac{8}{60} \diagdown \dfrac{6}{h} \qquad 60 \cdot 6 = 360$$
$$8 \cdot h = 8h$$

$$360 = 8h \qquad \text{Set the cross products equal.}$$

$$\dfrac{360}{8} = \dfrac{8h}{8} \qquad \text{Divide by 8.}$$

$$45 = h \qquad \text{Simplify.}$$

Step 5: **Check** the answer, and **interpret** the solution as it relates to the problem.

The tree is 45 ft tall.

Check that the ratios of the corresponding sides are the same.

Corresponding sides $\Big\langle \dfrac{8}{60} = \dfrac{8 \div 4}{60 \div 4} = \dfrac{2}{15}$

Corresponding sides $\Big\langle \dfrac{6}{45} = \dfrac{6 \div 3}{45 \div 3} = \dfrac{2}{15}$

The ratios are the same, so the answer is correct.

A girl is 4 ft tall and casts a shadow that is 6 ft long. At the same time, a telephone pole casts a shadow that is 27 ft long. How tall is the pole?

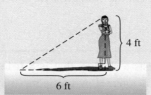

4 ft

6 ft

27 ft

ANSWERS TO YOU TRY EXERCISES

1) a) Corresponding angles: $\angle P$ and $\angle F$, $\angle Q$ and $\angle G$, $\angle R$ and $\angle H$; corresponding sides: $\overline{PQ}$ and $\overline{FG}$, $\overline{QR}$ and $\overline{GH}$. $\overline{PR}$ and $\overline{FH}$
   b) Corresponding angles: $\angle T$ and $\angle D$, $\angle U$ and $\angle E$, $\angle V$ and $\angle C$; corresponding sides: $\overline{TU}$ and $\overline{DE}$, $\overline{UV}$ and $\overline{EC}$, $\overline{TV}$ and $\overline{DC}$
2) a) ASA   b) SAS   c) SSS   3) 20 m   4) 20 m
5) $x = 15$ ft; perimeter = 45 ft   6) The pole is 18 ft tall.

E Evaluate **6.7** Exercises   Do the exercises, and check your work.

Objective 1: Identify Corresponding Angles and Corresponding Sides of Congruent Triangles

1) What does it mean if two figures are congruent?

2) What does it mean if two figures are similar?

For each pair of congruent triangles, identify the corresponding angles and corresponding sides.

3)
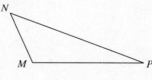

4)
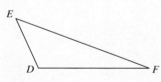

5)
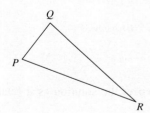

6)
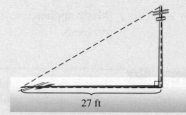

Objective 2: Use SSS, SAS, and ASA to Prove That Triangles Are Congruent

For each pair of triangles, determine which theorem proves that the triangles are congruent. Use SSS, SAS, or ASA.

7)
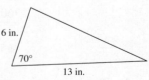

8)
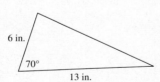

9)

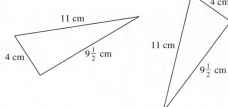

10)

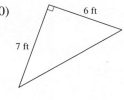

11)

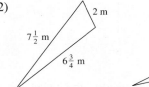

12)

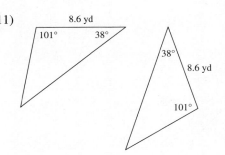

Objective 3: Find Unknown Lengths in Similar Triangles

13) Write the definition of similar triangles and then look up the definition of the word *similar* in a dictionary. Compare the two definitions.

14) What does it mean when we say that corresponding sides are proportional?

15) Are all equilateral triangles similar? In your own words, explain why or why not.

16) Are all right triangles similar? In your own words, explain why or why not.

Write the corresponding angles and the corresponding sides for each pair of similar triangles.

17)

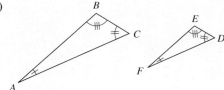

18)

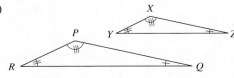

19)

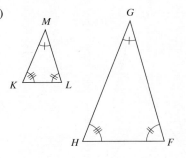

20)

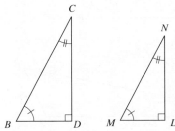

21)

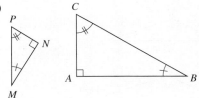

22)

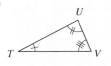

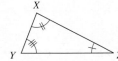

Find the unknown length in the larger of each pair of similar triangles; then, find its perimeter.

23)

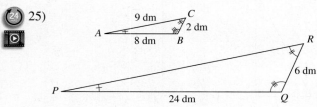

24)

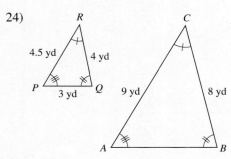

25)

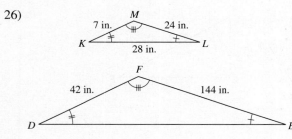

26)

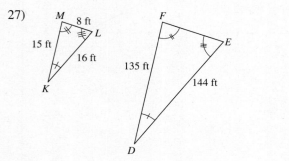

27)

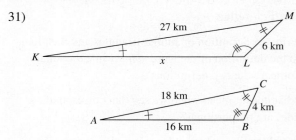

28)

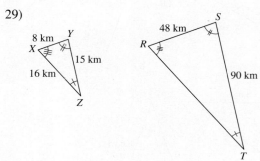

29)

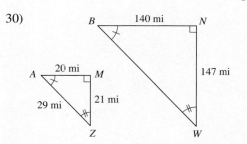

30)

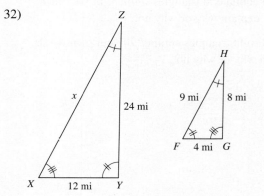

Use a proportion to find the unknown length labeled *x*. Then, find the perimeter of the triangle with the unknown side length labeled *x*.

31)

32)

33)

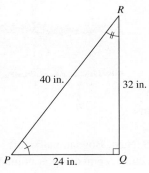

34)

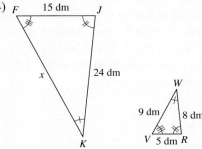

35)

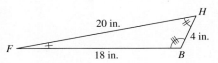

T 3 in. U
x
15 in.
P

36)

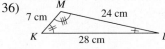

F
10.5 cm x
D   42 cm   E

37)

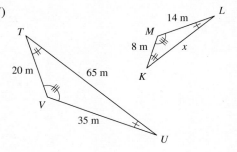

38)

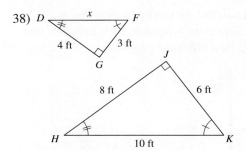

Fill in the blank with the correct term from this section.

39) The lengths of the corresponding sides of similar triangles are _____.

40) Two figures are _____ if they have the same shape but they are different in size.

41) Two figures are _____ if they are exactly the same; that is, they are the same shape *and* the same size.

42) The measures of the _____ angles of similar triangles are the same.

Objective 4: Solve Applied Problems Involving Similar Triangles

Solve each problem.

43) A college had a ribbon-cutting ceremony for its new Math and Science building, and the school's band was asked to participate. To find the height of the building, a math student measured the shadows cast by the new building and the drum major's baton. The shadow of the 3.5-ft baton was 4 ft. The shadow of the new building was 52 ft. Find the height of the building.

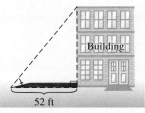

44) The height of a tree can be found by lining up the top of the tree with the top of a 2-m stick. Use similar triangles to find the height of the tree.

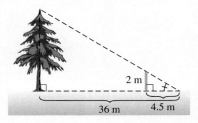

45) Maria is 5 ft tall, and she casts a shadow that is 2 ft long. At the same time, a flagpole casts a shadow that is 8.4 ft long. How tall is the flagpole?

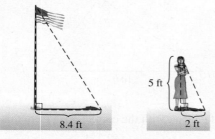

5 ft

8.4 ft   2 ft

46) Refer to Exercise 45. Later in the day, Maria's shadow is 8 ft long. How long is the flagpole's shadow at this time?

47) The height of a statue can be found by comparing its shadow to a shadow cast by a 1.5-m stick at the same time of day. Suppose a statue casts a 5-m shadow on the ground while the stick casts a 2-m shadow. Find the height of the statue.

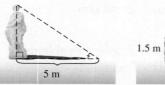

1.5 m

5 m   2 m

48) Jaromir and Bogdan are on opposite sides of a river and want to calculate its width. On one side of the river, Jaromir places a marker at point A. On the other side, Bogdan places a marker at point B, creating line segment $\overline{AB}$. Next, Bogdan marks off line segment $\overline{CD}$, intersecting segment $\overline{AB}$ at point M, forming two similar right triangles as shown in the figure to the right. Note that point D is directly across the river from point A. With his tape measure, Bogdan finds the length of segments $\overline{CM}$, $\overline{MD}$, and $\overline{BC}$ to be 3 ft, 6 ft, and 5 ft, respectively. Find the width of the river using Bogdan's measurements.

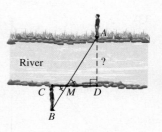

River   ?

C   M   D

B

R Rethink

R1) When wouldn't be a good time of day to try your own experiment similar to Exercise 45? Why?

R2) Which objectives do you still need to master? Do you need to review material from previous sections or chapters to get up to speed?

Group Activity – Ratios and Proportions

Is there more than one way to write a proportion?

In this chapter, you learned how to write proportions to solve problems like the one below. "If 4 granola bars cost $4.76, find the cost of 6 granola bars."

$$\frac{4 \text{ granola bars}}{\$4.76} = \frac{6 \text{ granola bars}}{x}$$

$$\frac{4}{4.76} = \frac{6}{x}$$

$$4x = (4.76)(6)$$

$$4x = 28.56$$

$$x = 7.14$$

The cost of 6 granola bars is $7.14.

Work with a partner on the rest of this activity.

1) Is the proportion shown above the only proportion that could be used to solve the granola bar problem? Could we use the proportion below to solve the granola bar problem? Explain why or why not.

$$\frac{4}{6} = \frac{4.76}{x}$$

2) "A survey at Harrington High School revealed that 5 out of 8 of its seniors would be going away for college. How many of the 424 seniors are planning to go away for college?" Two of the proportions below can be used to solve this problem and one cannot. Determine which proportion cannot be used. Give a reason to support your choice. You do not have to actually solve the proportion.

   a) $\dfrac{5}{8} = \dfrac{x}{424}$ b) $\dfrac{5}{8} = \dfrac{424}{x}$ c) $\dfrac{5}{x} = \dfrac{8}{424}$

3) "To safely thaw a frozen turkey in the refrigerator takes 24 hours per every 4.5 lb. How many hours will it take to thaw a 15-lb turkey in this manner?" Two of the proportions below can be used to solve this problem and one cannot. Determine which proportion cannot be used. Give a reason to support your choice. You do not have to actually solve the proportion.

   a) $\dfrac{24}{4.5} = \dfrac{15}{x}$ b) $\dfrac{4.5}{24} = \dfrac{15}{x}$ c) $\dfrac{24}{x} = \dfrac{4.5}{15}$

Write two different proportions that could be used to solve Problems 4 through 6. Solve each proportion and verify that both proportions yield the correct answer to the problem.

4) In the last three days, George signed into his Facebook account 8 times. At this rate, how many times will he sign into his account in the next 90 days?

5) Approximately 16 out of every 25 people who own a mobile device with Internet access will use their mobile devices to make an online purchase. A group of web-accessible mobile device owners was surveyed and it was found that 800 people in the group used their devices to make an online purchase. How many people were in the group? (http://ecommerce-news.internetretailer.com/)

6) There are 110 calories in $\frac{3}{4}$ cup of Lucky Charms cereal. How many calories are in 2 cups of Lucky Charms cereal? Round to the nearest whole number.

emPOWERme The Right Approach to Homework

Let's face it: Homework has a bad reputation. Nonetheless, it is a fact of college life, and more importantly, it is an excellent opportunity to learn. Too many students sacrifice this opportunity either by not doing their homework, or, almost as harmfully, by allowing their negative attitudes to interfere with their performance.

To assess your own attitudes toward homework, consider each of the following statements. On a scale of 1 (Does not apply to me) to 5 (Strongly applies to me), mark how well each statement describes you.

_____ Usually I throw away my homework as soon as my instructor returns it.

_____ I put off my homework as long as possible.

_____ I don't write down my homework assignments; I just try to remember them.

_____ I like to have music and the TV on to keep me entertained while I'm doing homework.

_____ I often do my homework during class.

_____ I usually get so frustrated with homework that I just give up after a few questions.

_____ My goal with homework is to finish as quickly as possible, so that I can move on to something fun.

_____ I don't think I've ever completed a homework assignment early.

_____ I always keep my phone on while I'm doing my homework, in case anyone wants to talk.

_____ If there are homework problems I can't figure out, I just ask my friends to tell me the answers.

Find your average response to the statements. Do this by adding up all the numbers you wrote and dividing by 10. If your average is more than 4, it is likely that your feelings about homework are negatively impacting your performance. Remember, homework benefits you in the long run. It will help bring you closer to your educational and career goals. Make the most of your time doing homework by using the strategies described at the beginning of the chapter.

Chapter 6:   **Summary**

Definition/Procedure	Example

6.1 Ratios

A **ratio** is a comparison of two quantities. It can compare numbers or measurements with the same units. Ratios can be written in several different ways.	The ratio of 5 to 9 can also be written as $$5:9 \quad \text{or} \quad \frac{5}{9}$$
Writing a Ratio in Lowest Terms When writing a ratio as a fraction in lowest terms, we divide out common units.	Write the ratio 4 ft to 10 ft as a fraction in lowest terms. $\dfrac{4 \cancel{ft}}{10 \cancel{ft}} = \dfrac{4}{10}$ Divide out common units. $= \dfrac{4 \div 2}{10 \div 2} = \dfrac{2}{5}$ Write the ratio in lowest terms.
If the units in a ratio are different, we must begin by changing one of the units so that it is the same as the other one.	Write the ratio as a fraction in lowest terms: 2 hr to 50 min Change hours to minutes: 2 hr = 2 · 60 min = 120 min Then, the ratio of 2 hr to 50 min is $\dfrac{2\text{ hr}}{50\text{ min}} = \dfrac{120\text{ min}}{50\text{ min}} = \dfrac{120\ \cancel{min}}{50\ \cancel{min}} = \dfrac{120}{50} = \dfrac{12}{5}$

6.2 Rates

A **rate** compares quantities with different units. Rates are usually written in lowest terms.	Write the rate 190 mi in 4 hr as a fraction in lowest terms. $\dfrac{190\text{ mi}}{4\text{ hr}} = \dfrac{190\text{ mi} \div 2}{4\text{ hr} \div 2} = \dfrac{95\text{ mi}}{2\text{ hr}}$
Unit Rate A **unit rate** is a rate with a denominator of 1.	Find the unit rate: 376 mi on 16 gal $\dfrac{376\text{ mi}}{16\text{ gal}} = \dfrac{376\text{ mi} \div 8}{16\text{ gal} \div 8} = \dfrac{47\text{ mi}}{2\text{ gal}}$ Next, divide 47 by 2 to find the unit rate: $2)\overline{47.0}$ gives 23.5 The unit rate is 23.5 mi/gal or 23.5 miles per gallon.
Unit Price and Best Buy The **unit price** of an item is the cost of the item per unit. We can use the unit price to figure out which item in a store gives us the most value for our money. This is called the **best buy.** Round answers to the thousandths place.	A grocery store sells an 8-oz container of cream cheese for $2.89 and a 12-oz container for $4.39. Which is the better buy? Find the unit price of each item. 8-oz container: $\dfrac{\$2.89}{8\text{ oz}} \approx \$0.361$ per ounce 12-oz container: $\dfrac{\$4.39}{12\text{ oz}} \approx \$0.366$ per ounce The 8-oz container has the lower unit price, so it is the better buy.

Definition/Procedure	Example

6.3 Proportions

A **proportion** is a statement that two ratios or two rates are equal.

Write the statement as a proportion:

1 in. is to 3 ft as 4 in. is to 12 ft.

$$\frac{1 \text{ in.}}{3 \text{ ft}} = \frac{4 \text{ in.}}{12 \text{ ft}}$$

Determining Whether a Proportion Is True or False by Writing the Ratios in Lowest Terms

One way to determine whether a proportion is true or false is by writing each ratio in lowest terms.

Determine whether each proportion is true or false by writing each ratio in lowest terms.

a) $\frac{24}{18} = \frac{12}{9}$ b) $\frac{2}{7} = \frac{14}{35}$

a) Write each fraction in lowest terms.

$$\frac{24}{18} = \frac{24 \div 6}{18 \div 6} = \frac{4}{3} \qquad \frac{12}{9} = \frac{12 \div 3}{9 \div 3} = \frac{4}{3}$$

Since each fraction simplifies to $\frac{4}{3}$, the proportion is true.

b) The fraction $\frac{2}{7}$ is already in lowest terms. Write $\frac{14}{35}$ in

lowest terms: $\frac{14}{35} = \frac{14 \div 7}{35 \div 7} = \frac{2}{5}$

Since $\frac{2}{7} \neq \frac{2}{5}$, the proportion is false.

Determining Whether a Proportion Is True or False Using Cross Products

Another way to determine whether a proportion is true or false is by finding the *cross products*. The **cross products** are the numbers we get when we *multiply along the diagonals* of the proportion.

Using Cross Products to Determine Whether a Proportion Is True or False

1) If the cross products in a proportion are equal, then the proportion is true.

2) If the cross products in a proportion are not equal, then the proportion is false.

Determine whether each proportion is true or false by finding the cross products.

a) $\frac{10}{15} = \frac{4}{6}$ b) $\frac{5.8}{3} = \frac{4.2}{2}$

a) Find the cross products.

$\frac{10}{15} = \frac{4}{6}$ $15 \cdot 4 = 60$ True. The cross products are equal.
$10 \cdot 6 = 60$

The cross products are equal, so the proportion is true.

b) Find the cross products.

$\frac{5.8}{3} = \frac{4.2}{2}$ $3 \cdot 4.2 = 12.6$ False. The cross products are *not* equal.
$5.8 \cdot 2 = 11.6$

The cross products are *not* equal, so the proportion is false.

Definition/Procedure	Example

6.4 Solve Proportions

Solving a Proportion

If a proportion contains one variable, we use cross products and the division property of equality to find that missing value. This is called **solving the proportion.**

How to Solve a Proportion

Step 1: Find the cross products.

Step 2: Set the cross products equal to each other, and solve the equation.

Step 3: Check the solution by substituting it into the original proportion and finding the cross products.

Solve $\dfrac{3}{8} = \dfrac{x}{24}$.

Step 1: Find the cross products.

$$\frac{3}{8} \diagup\!\!\!\!\diagdown \frac{x}{24} \qquad \begin{array}{l} 8 \cdot x = 8x \\ 3 \cdot 24 = 72 \end{array}$$

Step 2: Set the cross products equal to each other and solve the equation.

$$8x = 72 \qquad \text{Set the cross products equal to each other.}$$
$$\frac{8x}{8} = \frac{72}{8} \qquad \text{Divide each side by 8.}$$
$$\frac{\overset{1}{\cancel{8}}x}{\underset{1}{\cancel{8}}} = \frac{72}{8} \qquad \text{Divide out the common factor.}$$
$$x = 9 \qquad \text{Simplify.}$$

Step 3: Check the solution in the original proportion,
$\dfrac{3}{8} = \dfrac{x}{24}$, by checking to see whether the cross products are equal.

$$\frac{3}{8} \diagup\!\!\!\!\diagdown \frac{9}{24} \qquad \begin{array}{l} 8 \cdot 9 = 72 \\ \text{Substitute 9 for } x. \\ 3 \cdot 24 = 72 \end{array}$$

Since the cross products are equal, the solution is $x = 9$.

6.5 Solve Applied Problems Involving Proportions

We can solve many applied problems using proportions. Look for key words like *in, per, for, from, on, out of, is to,* and *as.*

Use the Five Steps for Solving Applied Problems that were first presented in Section 2.6.

Steps for Solving Applied Problems

Step 1: **Read** the problem carefully, more than once if necessary, until you understand it. Draw a picture, if applicable. Identify what you are being asked to find.

Step 2: **Choose a variable** to represent an unknown quantity. If there are any other unknowns, define them in terms of the variable.

Step 3: **Translate** the problem from English into an equation using the chosen variable.

Step 4: **Solve** the equation.

Step 5: **Check** the answer in the original problem, and **interpret** the solution as it relates to the problem. Be sure your answer makes sense in the context of the problem.

In her hybrid car, Emilie can drive 336 mi on 8 gal of gas. How far can she drive on 12 gal?

Step 1: **Read** the problem carefully, and identify what we are being asked to find.

We must find the number of miles Emilie can drive on 12 gal of gas.

Step 2: **Choose a variable** to represent the unknown.

Let x = number of miles Emilie can drive on 12 gal of gas

Step 3: **Translate** the information that appears in English into an algebraic equation.

Write a proportion using x for the unknown quantity.

$$\frac{336 \text{ mi}}{8 \text{ gal}} = \frac{x}{12 \text{ gal}}$$

Definition/Procedure	Example
	Step 4: **Solve** the problem. We do not need the units when solving for x.
	$\dfrac{336}{8} = \dfrac{x}{12}$ Write the equation without the units.
	$8x = 336 \cdot 12$ Find the cross products.
	$8x = 4032$ Multiply.
	$x = 504$ Divide by 8 to solve the equation.
	Step 5: **Check** the answer, and **interpret** the solution as it relates to the problem.
	Emilie can drive 504 mi on 12 gal of gas.
	Use the cross products to check $x = 504$ in the proportion. The check is left to the student.

6.6 Angles

Definition/Procedure	Example
Complementary Angles Two angles are **complementary angles** if their measures add to 90°.	If $m\angle B = 67°$, find the measure of its complement. To find the *complement* of $\angle B$, subtract 67° from 90°: $90° - 67° = 23°$. The measure of the complement of $\angle B$ is 23°.
Supplementary Angles Two angles are **supplementary angles** if their measures add to 180°.	If $m\angle R = 95°$, find the measure of its supplement. To find the *supplement* of $\angle R$, subtract 95° from 180°: $180° - 95° = 85°$. The measure of the supplement of $\angle R$ is 85°.
Congruent Angles Two angles are **congruent angles** if their measures are the same.	$\angle A$ and $\angle B$ are *congruent angles* because their angle measures are the same. 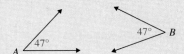
Vertical and Adjacent Angles When two lines intersect, **vertical angles** are the angles opposite each other, the angles that do not share a common side. Vertical angles are congruent. That is, their angle measures are the same. **Adjacent angles** are angles that share a common side and a common vertex.	Identify the vertical and adjacent angles, and find the missing angle measures. **Vertical Angles** $\angle WMZ$ and $\angle XMY$, $\angle XMW$ and $\angle YMZ$ **Adjacent Angles** $\angle WMZ$ and $\angle YMZ$, $\angle WMZ$ and $\angle XMW$ $\angle XMW$ and $\angle XMY$, $\angle XMY$ and $\angle YMZ$ $m\angle XMY = 119°$ $m\angle YMZ = m\angle XMW = 180° - 119° = 61°$

Definition/Procedure	Example
Corresponding Angles and Alternate Interior Angles Refer to the figure at the right. Line T is a **transversal** that intersects the parallel lines l and m. **Corresponding angles** are on the same side of the transversal and in the same position with respect to line l and line m. **Corresponding angles are congruent.** **Alternate interior angles** are inside the parallel lines and are on opposite sides of the transversal. **Alternate interior angles are congruent.**	Lines l and m are parallel. Identify all pairs of corresponding angles and all pairs of alternate interior angles. 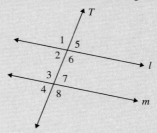 Corresponding angles: $\angle 1$ and $\angle 3$, $\angle 5$ and $\angle 7$, $\angle 2$ and $\angle 4$, $\angle 6$ and $\angle 8$. Alternate interior angles: $\angle 2$ and $\angle 7$, $\angle 3$ and $\angle 6$.

6.7 Solve Applied Problems Involving Congruent and Similar Triangles

Congruent and Similar Triangles Two figures are **congruent** if they are the *same shape* and the *same size*. Two figures are **similar** if they are the *same shape* but *different in size*. If the corresponding angles of two triangles have the same measure and if all of the corresponding sides have the same length, then the triangles are congruent.	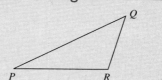 The corresponding angles are $\angle P$ and $\angle D$, $\angle Q$ and $\angle E$, and $\angle R$ and $\angle F$. The corresponding sides are $\overline{PQ}$ and $\overline{DE}$, $\overline{QR}$ and $\overline{EF}$, and $\overline{PR}$ and $\overline{DF}$. The corresponding angles have the same measure, and the corresponding sides have the same length. Therefore, $\triangle PQR \cong DEF$.

Using SSS, SAS, and ASA to Prove That Triangles Are Congruent
We can use these rules, or *theorems,* to prove that two triangles are congruent.
 	
1) Side-Side-Side (SSS) Theorem
 	
   If all of the sides of one triangle are the same length as the corresponding sides of another triangle, then the triangles are congruent.
 	
2) Side-Angle-Side (SAS) Theorem
 	
   If two sides of a triangle and the angle between them are the same measure as the corresponding parts on another triangle, then the triangles are congruent.
 	
3) Angle-Side-Angle (ASA) Theorem
 	
   If two angles of a triangle and the side between them are the same measure as the corresponding parts on another triangle, then the triangles are congruent.
	For each pair of triangles, determine which theorem proves that the triangles are congruent.
	
	a) 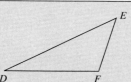
	
	b) 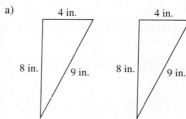 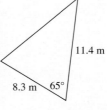

Definition/Procedure	Example

c)

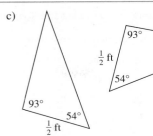

Solution

a) The corresponding sides of the triangles are the same length. Therefore, the **SSS theorem** (or Side-Side-Side) proves that the triangles are congruent.

b) In both triangles, the corresponding sides and the angle between them have the same measure. Therefore, the **SAS theorem** (or Side-Angle-Side) proves that the triangles are congruent.

c) In the first triangle, two angles and the side between them have the same measure as the corresponding parts of the second triangle. Therefore, the **ASA theorem** (or Angle-Side-Angle) proves that the triangles are congruent.

Similar Triangles

The angles with the same measure are called **corresponding angles.**

The sides opposite the corresponding angles are called the **corresponding sides.**

Although corresponding angles have the same measure, corresponding sides are *not* necessarily the same length.

Similar triangles have the same shape, the measures of their corresponding angles are the same, and the lengths of their corresponding sides are proportional.

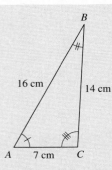 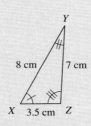

Corresponding Angles	Corresponding Sides
$\angle A$ and $\angle X$	$\overline{BC}$ and $\overline{YZ}$
$\angle B$ and $\angle Y$	$\overline{AC}$ and $\overline{XZ}$
$\angle C$ and $\angle Z$	$\overline{AB}$ and $\overline{XY}$

These are similar triangles. The ratio of the corresponding sides of the first triangle to those of the second triangle is $\frac{2}{1}$.

Definition/Procedure	Example

Finding the Length of an Unknown Side in Similar Triangles

We can use a proportion to find the length of an unknown side in similar triangles.

Use a proportion to find the unknown length, x.

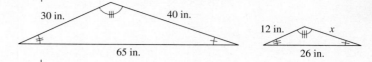

Choose *either* of the two corresponding side pairs that we know; let's choose 30 in. and 12 in. The corresponding sides that include the side we do not know are 40 in. and x. *The ratios of the corresponding sides are equal, so we can set up a proportion and solve for x.*

$$\frac{30}{12} \diagdown \frac{40}{x} \quad \begin{array}{l} 12 \cdot 40 = 480 \\ \text{Set up the proportion.} \\ 30 \cdot x = 30x \end{array}$$

$30x = 480$ Set the cross products equal to each other.

$$\frac{30x}{30} = \frac{480}{30} \quad \text{Divide by 30.}$$

$x = 16$ Simplify.

The unknown length is 16 in.

Chapter 6:   Review Exercises

(6.1)

1) What is a ratio?

2) The child to teacher ratio in the toddler room at a daycare is 4 to 1. What does this mean?

Write each ratio as a fraction in lowest terms.

3) 175 yd to 300 yd

4) 60 oz to 160 oz

5) \$1.25 to \$0.25

6) 3.8 lb to 1.2 lb

7) $\frac{5}{12}$ to $\frac{7}{16}$

8) $2\frac{1}{4}$ to $1\frac{1}{2}$

9) 10 oz to 2 lb

10) 12 hr to 12 min

The table shows the approximate lifetime theatre ticket sales for the top-grossing 3D movies. Use this information for Exercises 11–16 to write each ratio as a fraction in lowest terms.

3D Movie Title	Lifetime Gross in (Theatres)
Star Wars: The Force Awakens	\$940 million
Avatar	\$760 million
Jurassic World	\$650 million
Marvel's The Avengers	\$620 million
Rogue One: A Star Wars Story	\$530 million
Beauty and the Beast (2017)	\$500 million
Finding Dory	\$490 million
Avengers: Age of Ultron	\$460 million

(www.boxofficemojo.com)

11) Find the ratio that compares the number one 3D movie to the eighth-best seller.

12) Find the ratio that compares the ticket sales for *Marvel's The Avengers* to *Avengers: Age of Ultron.*

13) What is the ratio that compares the total ticket sales for *Finding Dory* to *Beauty and the Beast?*

14) What is the ratio that compares the ticket sales for *Rogue One: A Star Wars Story* to *Star Wars: The Force Awakens?*

15) The gross earnings of which two movies give a ratio of $\frac{13}{10}$?

16) The gross earnings of which two movies give a ratio of $\frac{38}{47}$?

17) The year before a college football team went to a bowl game, the team sold 2800 season tickets. The year after the bowl game, they sold 4000 season tickets. Find the ratio of the increase in the number of tickets sold to the number sold before the bowl game.

18) Write the ratio of the longest side to the shortest side.

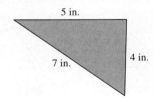

5 in.

7 in.   4 in.

(6.2)

19) What is the difference between a ratio and a rate?

Write each rate as a fraction in lowest terms.

20) 140 mi in 30 hr

21) 2 cups for 6 servings

22) $48 for 16 gal

23) What is a unit rate?

24) Write the rate $\frac{\$15}{1 \text{ hr}}$ in two other ways.

Find each unit rate. Round the answer to the nearest hundredth where appropriate.

25) 180 text messages in 3 hr

26) 384 desks for 12 classrooms

27) $90 in 60 days

28) $131.25 in 15 hr

29) Determine which package of paper napkins is the best buy, and give its unit price.

Number of Napkins	Price
100	$1.49
250	$3.29
400	$4.99

30) Daryl drove his pickup truck 290 mi on 22.3 gal of gas. Find the vehicle's gas mileage to the nearest tenth.

31) A group of hikers in Yosemite National Park hiked the 14-mi round-trip trail to the top of Half-Dome and back. If it took $11\frac{1}{2}$ hr to do the hike, what is the group's average speed in miles per hour? Round the answer to the nearest tenth.

©rubberball/Getty Images

32) Selena worked 24 hr and earned $468.00. Find her hourly wage.

(6.3) Write each statement as a proportion.

33) 16 is to 14 as 24 is to 21.

34) 9 mi is to 11 mi as 36 mi is to 44 mi.

35) 350 cell phones is to 450 adults as 7 cell phones is to 9 adults.

36) $80 is to 4 hr as $240 is to 12 hr.

37) What are two ways to determine whether a proportion is true or false?

38) Explain how to use the cross products to determine whether a proportion is true or false.

Determine whether each proportion is true or false by writing each ratio in lowest terms.

39) $\frac{28}{6} = \frac{42}{9}$

40) $\frac{30}{48} = \frac{16}{28}$

41) $\frac{120}{340} = \frac{200}{460}$

42) $\frac{225}{300} = \frac{63}{84}$

Determine whether each proportion is true or false by finding the cross products.

43) $\frac{14}{24} = \frac{4}{9}$

44) $\frac{1.4}{16} = \frac{1.8}{17}$

45) $\frac{5\frac{1}{4}}{6} = \frac{4\frac{1}{2}}{4\frac{2}{3}}$

46) $\frac{3\frac{1}{4}}{2.02} = \frac{4\frac{1}{2}}{3.04}$

47) a) Add $\frac{8}{9} + \frac{13}{15}$.

b) Determine whether the proportion $\frac{8}{9} + \frac{13}{15}$ is true or false.

(6.4)

48) What does *solve a proportion* mean?

Solve each proportion, and check the answer using the cross products.

49) $\dfrac{4}{3} = \dfrac{x}{9}$

50) $\dfrac{5}{x} = \dfrac{11}{55}$

51) $\dfrac{8}{15} = \dfrac{2}{x}$

52) $\dfrac{x}{3} = \dfrac{63}{18}$

53) $\dfrac{x}{2\frac{1}{4}} = \dfrac{1\frac{2}{3}}{10}$

54) $\dfrac{\frac{5}{9}}{3\frac{1}{3}} = \dfrac{\frac{5}{6}}{x}$

55) $\dfrac{5.7}{x} = \dfrac{2.1}{1.4}$

56) $\dfrac{7}{4.5} = \dfrac{x}{3.6}$

(6.5) Solve each problem using a proportion.

57) In an architectural drawing, a 300-ft-tall building is drawn 4 in. high. What is the actual height of a building that is drawn 6 in. high?

58) On average, one tree produces nearly 260 lb of oxygen each year. How many pounds of oxygen will be produced by 12 trees in one year? (www.treesaregood.com)

59) A nursing student observes an intravenous drip rate of 10 drops per 12 sec. How many drops should be observed in 1 min?

60) A university finds that, on average, 2 out of 9 of its students receive credit for English composition due to a high Advanced Placement test score. If the most recent freshman class contains 2548 students, approximately how many will receive credit for English composition through their AP tests? Round the answer to the nearest whole number.

(6.6) Find the measure of the complement of each of the given angles.

61) $m\angle V = 47°$

62) $m\angle Y = 56°$

Find the measure of the supplement of each of the given angles.

63) $m\angle C = 106°$

64) $m\angle X = 23°$

Find the missing angle measure.

65) $\angle DNA$ is a right angle. Find $m\angle r$.

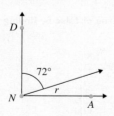

66) $\angle ABD$ is a straight angle. Find $m\angle d$.

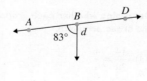

67) $\angle ZXV$ is a straight angle. Find $m\angle p$.

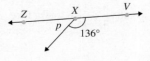

68) $\angle EZJ$ is a right angle. Find $m\angle q$.

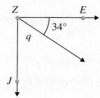

69) Find the measures of angles N, T, and Z.

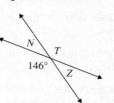

70) Find the measures of angles 1, 2, and 3.

71) Lines l and m are parallel. Identify all pairs of corresponding angles and all pairs of alternate interior angles.

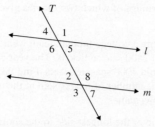

72) Lines l and m are parallel. Find the missing angle measures.

(6.7)

73) For the given pair of congruent triangles, identify the corresponding angles and corresponding sides.

For each pair of triangles, determine which theorem proves that the triangles are congruent. Use SSS, SAS, or ASA.

74)

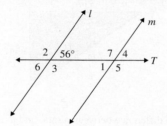

75)

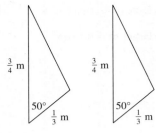

76)

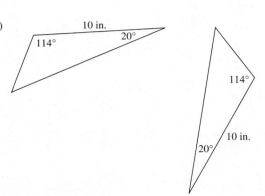

Find the unknown length in the larger of each pair of similar triangles.

77)

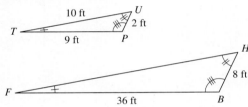

78)

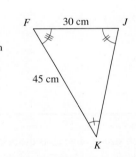

Use a proportion to find the unknown length labeled x.

79)

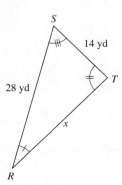

80)

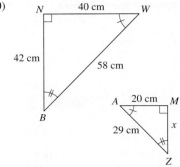

Solve each problem.

81) The height of a rocket can be found by comparing its shadow to a shadow cast by a 1.5-m stick at the same time of day. Suppose a rocket casts a 112-m shadow on the ground while the stick casts a 4-m shadow. Find the height of the rocket by writing a proportion and solving it.

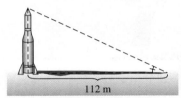

82) Refer to the previous exercise. Suppose at the same time, a tree nearby casts a 16-m shadow on the ground. How tall is the tree?

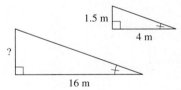

Mixed Exercises: Sections 6.1–6.7

Fill in the blank with *always*, *sometimes*, or *never* to make the statement true.

83) Vertical angles _____ have the same measure.

84) Two figures are _____ congruent if they have the same shape but are different in size.

85) The sum of the angles in a triangle is _____ 360°.

86) The sum of two complementary angles is _____ 90°.

Write each ratio or rate as a fraction in lowest terms.

87) 27 books to 45 books   88) 2850 ft in 90 sec

Find each unit rate.

89) 224 miles in 3.5 hours   90) $31.50 for 6 people

Determine whether each proportion is true or false.

91) $\dfrac{5.5}{10} = \dfrac{4.2}{8}$

92) $\dfrac{16}{32} = \dfrac{4}{8}$

Solve each proportion.

93) $\dfrac{4}{x} = \dfrac{11}{44}$

94) $\dfrac{x}{2\frac{1}{2}} = \dfrac{3\frac{1}{3}}{4}$

Solve each problem.

95) If the ratio of potato starch flour to tapioca flour in a gluten-free cookie recipe is 3 to 2, how much tapioca flour is used if the recipe calls for 1 cup of potato starch flour?

96) If Faviana slept a total of 511 hr over the past 10 weeks, what was her rate of sleep in hours per week?

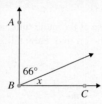

©RyanKing999/Getty Images

97) $\angle ABC$ is a right angle. Find m$\angle x$.

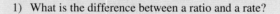

98) If m$\angle R = 53°$, find the measure of its supplement.

Find the unknown length in the larger of these two similar triangles.

99)

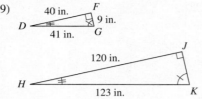

100)

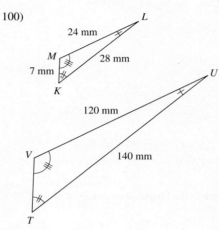

Chapter 6:   Test

1) What is the difference between a ratio and a rate?

Write each ratio as a fraction in lowest terms.

2) 63 points to 35 points

3) 24 in. to 3 ft

4) In 12 oz of yogurt, the ratio of cholesterol to protein is 10 g to 28 g. Write the ratio as a fraction in lowest terms, and explain what it means.

Find each unit rate.

5) $12.45 for 5 lb

6) 170 mi in 4 hr

7) Thomas drove his motorcycle 5752 miles over the past year. What was his driving rate in miles per week? Round the answer to the nearest tenth.

8) A nursing student sets an intravenous drip rate at 1500 drops per $\dfrac{1}{2}$ hour. What is the unit drip rate in drops per minute?

9) Determine which size of olive oil is the best buy, and give its unit price.

Size	Price
8 oz	$3.99
17 oz	$8.19
26 oz	$12.69

10) Write the statement as a proportion. $12 is to 9 lb as $20 is to 15 lb.

11) Determine whether each proportion is true or false.

a) $\dfrac{25}{35} = \dfrac{10}{14}$

b) $\dfrac{3\frac{1}{4}}{2} = \dfrac{9\frac{1}{2}}{6}$

12) Elsa multiplies $\dfrac{3}{10} \cdot \dfrac{4}{9}$ like this:

$$\dfrac{3}{10} \diagup \dfrac{4}{9} \quad \begin{array}{l} 10 \cdot 4 = 40 \\ 3 \cdot 9 = 27 \end{array} \quad \left(\dfrac{40}{27}\right)$$

Did she multiply these fractions correctly? Explain your answer. If Elsa's answer is wrong, find the correct answer.

Solve each proportion. Check the answer using the cross products.

13) $\dfrac{5}{2} = \dfrac{x}{6}$

14) $\dfrac{2\frac{5}{8}}{x} = \dfrac{5}{8}$

Solve each problem using a proportion.

15) On an architectural drawing, 1 in. represents an actual length of 6 ft. How many inches long should an architect draw a wall that is 27 ft long?

16) In a pharmacy, Dara reads that a medication is to be given to a patient at a rate of 2.7 mg for every 40 lb. How much medicine should be given to a patient weighing 140 lb?

17) In 2015, approximately 15 out of 100 injury car crashes were the result of distracted driving. In a county that had 40 crashes resulting in injury in 2015, how many would have been expected to involve distracted driving? (crashstats.nhtsa.dot.gov)

©Mikael Karlsson/Alamy Images

18) Find
   a) the supplement of 21°.
   b) the complement of 21°.

19) Find each missing angle measure.
   m∠RAT = _____   m∠WAX = _____
   m∠TAW = _____   m∠YAZ = _____

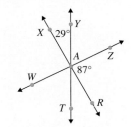

20) Lines l and m are parallel. Find the measures of the missing angles.

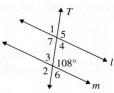

21) Which theorem proves that the triangles are congruent? Use SSS, SAS, or ASA. Explain your answer.

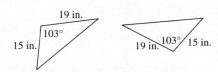

22) Given these two similar triangles, find the length of the side labeled x.

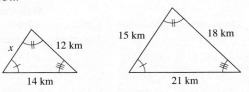

Determine whether each statement is true or false.

23) Adjacent angles are always congruent.

24) Corresponding angles of similar triangles are congruent.

25) 1 year = 48 weeks

Chapter 6:   Cumulative Review for Chapters 1–6

Perform the indicated operations. Write all answers in lowest terms.

1) $-\dfrac{8}{35} \div \dfrac{4}{21}$

2) $12 - 3.61$

3) $\begin{array}{r} 4073 \\ \times\, 809 \end{array}$

4) $5\dfrac{1}{6} + 2\dfrac{3}{4}$

5) $-\dfrac{7}{12} - \dfrac{3}{8}$

6) 65.7×0.03

7) $7.84 \div 1.4$

8) $\dfrac{7}{10} - \dfrac{11}{15}$

9) 4720×1000

10) $-56{,}000{,}000 \div 10{,}000$

11) $100 \div 4 + 2^3 \cdot 5 - (8 - 2)^2$

12) $-\sqrt{144} + 3(2 - 10) \div 6$

For Exercises 13 and 14, round 174.8156 to the indicated place.

13) hundredths   14) tens

15) Write 0.64 as a fraction in lowest terms.

16) Write as a ratio in lowest terms: 50 minutes to 2 hours

Solve each equation.

17) $-5n + 8 = 9 - 4n$

18) $0.6c - 1.7 + 0.3c = 0.5c - 0.6$

19) Solve the proportion $\dfrac{x}{3} = \dfrac{15}{9}$.

20) Cantu records her gas purchases and odometer readings for each gas station visit. Complete the table below by finding the number of miles traveled and the gas mileage.

Date	Odometer Start	Odometer End	Miles Traveled	Gallons Pur- chased	Miles per Gallon
8/24	15,352.8	15,625.2		12.0	
9/2	15,625.2	15,958.9		14.2	
9/13	15,958.9	16,248.7		13.8	
9/22	16,248.7	16,457.8		8.5	

21) If $m\angle X = 59°$, then what is the measure of its supplement?

22) Lines P and Q are parallel. Fill in each blank to make the statement true.

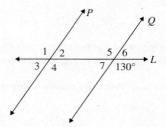

a) $m\angle 5 =$ _____

b) $m\angle 2 =$ _____

c) $\angle 1$ and $\angle$ _____ are vertical angles.

d) $m\angle 5 + m\angle 6 =$ _____

Solve each problem.

23) A sign in a clothing store says, "All merchandise is $\dfrac{1}{3}$ off."

A shirt with a regular price of $44.99 is marked with a sale price of $32.99. Is this correct?

24) Three friends go out to dinner and decide to split the bill evenly. Their dinner amounts to $47.74, and they leave a $9.50 tip. How much does each person owe?

25) In an area designated as a flood zone, approximately 5 out of 7 homeowners have flood insurance. If there are 1852 homes in this area, how many have flood insurance? Round the answer to the nearest whole number.

Measurement and Conversion

©Comstock/Getty Images

Math at Work:

Technical Support Specialist

Mitch Kramer is a technical support specialist. His job is to provide assistance to companies that are having problems with their computers.

When his company takes on a new client, Mitch has a lot to learn. "I need to know what type of work the company does, the types of computers they use, and what software they use," he explains. "If I don't know the answers to these basic questions, I can't fix their computers when they aren't working. The more I know about the company's needs and their computers, the better equipped I am to help them."

Mitch is confident that if he can learn about his client's operations and their computers' hardware and software, he will be ready to fix any problem they may have. The same is true for students attending a new school; the more students know about their school's operation and culture, the more tools they have to succeed. In this chapter, we will learn about measurement and conversion and offer strategies for learning about your school.

 **Study Strategies** Know Your School

Understanding how your college or university works and where to find resources are critical components of being successful in college. Where do you go if you have a question about financial aid? Where can you find out about clubs that are of interest to you? Where do you go if you have questions about which classes you still need to take in order to graduate? Let's use the P.O.W.E.R. framework to help you get to know your school.

 **Prepare**

- **I will learn about the different offices, resources, and services at my school as well as how my school works.**

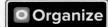 **Organize**

- Complete the emPOWERme survey that appears before the Chapter Summary to learn how well you know your school. Think about what you know and what you still need to learn.
- Know your school's website. Identify important places on the website such as where you can register for classes, pay your tuition online, how to make an appointment to see an adviser, and where you can find information about the tutoring center.
- Identify places you need to know at your school: important offices where you can get questions answered, services offered at your school, and organizations or clubs that you can join.
- Identify important deadlines you should know such as registering for classes, paying tuition, and submitting financial aid applications.
- Identify academic services that you may want to use such as the library, a tutoring center, and a testing center.
- Identify support services that may be helpful to you such as the campus health center, a veterans' support office, an office to help students with disabilities, and a child care center.
- Be sure you have your syllabus for each class with information about the class as well as the location of your instructors' offices and their office hours.

 **Work**

- It's time to learn about some of the offices, resources, and support services at your school and to learn about some of the procedures. Some important, common offices and resources are listed here, but different schools have different names for these services. Some apply to all students, and some may not apply to you. For the items that apply to you, fill in the blanks with the information for *your* school.
- *My school's website is* _____. Among other things, know how you can register for classes and pay your tuition online.
- *The office where I can go to register for classes is called the* _____ *office. Its location is* _____. Know the deadlines for registering for classes.
- When you are choosing your classes, know where they are located. If the campus is large, be sure you have enough time to get from one class to the next.
- *The office where I can go to ask questions about my tuition bill or to pay my bill in person is called the* _____ *office. Its location is* _____. Know when tuition is due.
- *The office where I can go to ask questions about financial aid, get financial aid forms, or turn in forms is called the* _____ *office. Its location is* _____. Know when to get the financial aid process started and when the forms are due.
- *The office where I can talk with someone about choosing a major or get help choosing classes is called the* _____ *office. Its location is* _____.

- *The office where I can go to talk with someone about personal problems that are interfering with my success in college is called the _____ office. Its location is _____.*
- Does your school have an emergency alert system? If so, know how to sign up for it. You may have options of receiving an email, text, or automated phone call.
- Most institutions have a Tutoring Center or a Math Lab that is available to students free of charge. *The place I can go to get free help with my math class is called the _____. Its location is _____, and its hours are _____.* What are the procedures for seeing a tutor?
- *My math instructor's office is located in _____, and his/her office hours are _____ _____.*
- Most schools have a testing center where students can go if they are allowed extra time on their tests or if they need to take a make-up test. *At my school, this center is called the _____. Its hours are _____ _____.* What are the procedures for taking a test in their center?
- *The office where I can learn about the resources available to me as a student with a disability is called the _____ office. Its location is _____.*
- *Most campuses have an office that helps veterans. On my campus, that office is called the _____. Its location is _____.*
- Many campuses have a child care center for students with children. *I can get information about the child care center at _____.*
- *The place I can go to find out about campus clubs, events, and organizations is called the _____.*
- Most institutions have a place where students can get help finding a job or preparing for an interview. *At my school, this office is called the _____ _____ office. Its location is _____.*

Evaluate
- Do you feel like you have a good understanding of how your school works and where to go for help? Did you learn about the names and locations of the offices listed in the Work section?

Rethink
- If you feel like you learned about the places and services that will be beneficial to you, be sure to use them when you need them! Also, guide classmates to the proper place when they need help.
- If you did not learn everything you think you need to know, ask one of your instructors or go to an information office at your school. Search the institution's website.
- Are there other services or resources that you feel like you need but did not learn about? Talk to one of your instructors and ask if there is such a place at your school.

©MBI/Alamy

Chapter 7 **P·O·W·E·R** Plan

P Prepare

What are your goals for Chapter 7?

1 Be prepared before and during class.

2 Understand the homework to the point where you could do it without needing any help or hints.

3 Use the P.O.W.E.R. framework to learn about your school.

4 Write your own goal.

_____

_____

O Organize

How can you accomplish each goal? (Write in the steps you will take to succeed.)

- _____
- _____
- _____
- _____

- _____
- _____
- _____

- Complete the emPOWERme that appears before the Chapter Summary to determine how much you know about your school.
- Read the Study Strategies to learn how to use the P.O.W.E.R. framework to learn about the procedures and resources at your school.
- _____

- _____

What are your objectives for Chapter 7?	**How can you accomplish each objective?**
1 Learn how to use U.S. customary measurements and the metric system.	<ul><li>Understand the relationships between the different ways to represent length, weight (mass), and capacity.</li><li>Be able to reproduce a chart that will help you convert between units.</li><li>Know how to correctly decide which unit would best describe a measurement.</li></ul>
2 Know how to convert units using multiplication and division or unit analysis.	<ul><li>Understand that the procedure for multiplying or dividing to convert between units will work for U.S. customary measurements as well as for the metric system.</li><li>Understand unit fractions, and be able to apply the procedure for using unit analysis for both systems.</li></ul>
3 Know how to solve applied problems that involve measurements.	<ul><li>Understand what you are being asked to find, and use the correct unit fractions.</li><li>Use the Five Steps for Solving Applied Problems, and add steps in the solving process that involve unit analysis.</li></ul>
4 Be able to convert between metric and U.S. customary measurements.	<ul><li>Memorize the relationships between metric and U.S. customary measurements for length, weight (mass), and capacity.</li><li>Know how to use the formulas to convert between Fahrenheit and Celsius.</li></ul>
5 Write your own goal. _____ _____	<ul><li>_____</li></ul>

| **E** Evaluate | Complete the Chapter Review and Chapter Test. How did you do? | **R** Rethink | • How well do you know all of the relationships presented in this chapter? Can you convert from one unit to another without looking at your book or notes?
• After doing the emPOWERme and Study Strategy, do you know more about your school than you did before? Is there anything else you would like to know? |

7.1 Using U.S. Customary Measurements

P Prepare   **O** Organize

What are your objectives for Section 7.1?	How can you accomplish each objective?
1 Learn U.S. Customary Measurements	• Memorize the **Relationships Between U.S. Customary Measurements.** • Complete the given example on your own. • Complete You Try 1.
2 Use Multiplication or Division to Convert Between Units	• Write the procedure for **Using Multiplication or Division to Convert Between Units** in your own words, and add necessary intermediary steps after following the examples. • Complete the given example on your own. • Complete You Try 2.
3 Use Unit Analysis to Convert Between Units	• Write the definition of a *unit fraction* in your notes. • Write the equivalencies found in the **Relationships Between U.S. Customary Measurements** table as unit fractions. • Write the procedure for **How to Convert Between Units Using Unit Analysis** in your own words. • Complete the given examples on your own. • Complete You Trys 3–6.
4 Solve Applied Problems Using Unit Analysis	• Use the Five Steps for Solving Applied Problems and add additional steps, as needed, from this section. • Complete the given examples on your own. • Complete You Trys 7 and 8.

 Read the explanations, follow the examples, take notes, and complete the You Trys.

1 Learn U.S. Customary Measurements

We use measurements every day: inches to measure the diagonal of a television screen, miles to measure the distance between two cities, quarts or gallons to measure a quantity of milk, pounds to measure our weight, and so on. Each of these is an example of a *U.S. customary measurement unit*. In the United States, we use **U.S. customary units** while most of the rest of the world uses the *metric system*. (The **metric system** uses units such as meters, liters, and grams. We will study the metric system in future sections.) In both systems, time is measured with the same units, such as: seconds, minutes, hours, and days.

Sometimes, we must convert between different units like yards and feet. Therefore, we need to memorize the relationships between the units and their abbreviations given in the table here.

Relationships Between U.S. Customary Measurements

Length	Time
12 inches (in.) = 1 foot (ft)	60 seconds (sec) = 1 minute (min)
3 feet (ft) = 1 yard (yd)	60 minutes (min) = 1 hour (hr)
5280 feet (ft) = 1 mile (mi)	24 hours (hr) = 1 day
	7 days = 1 week (wk)

Capacity	Weight
8 fluid ounces (fl oz) = 1 cup (c)	16 ounces (oz) = 1 pound (lb)
2 cups (c) = 1 pint (pt)	2000 pounds (lb) = 1 ton
2 pints (pt) = 1 quart (qt)	
4 quarts (qt) = 1 gallon (gal)	

Let's practice using these basic relationships.

EXAMPLE 1

Fill in the blank.

a) 1 ft = _____ in.   1 ft = 12 in.

b) _____ wk = 1 yr   52 wk = 1 yr

c) 1 gal = _____ qt   1 gal = 4 qt

YOU TRY 1

Fill in the blank.

a) 1 yd = _____ ft   b) _____ lb = 1 ton   c) _____ c = 1 pt

Very often we have to convert between the units given in the table. A recipe, for example, might call for $2\frac{1}{2}$ cups of milk, but at the grocery store we buy milk in pints, quarts, or gallons. Do we buy a pint, a quart, or a gallon? Let's learn how to change from one U.S. unit to another. We can do this in two different ways: using multiplication or division, or using unit analysis.

2 Use Multiplication or Division to Convert Between Units

How do we know whether we multiply or divide to change units? Use these guidelines.

W Hint

Look at the relationships between measurements in Objective 1. What do you notice that helps this make sense?

Procedure Use Multiplication or Division to Convert Between Units

1) When converting from a *larger unit* to a *smaller unit,* use *multiplication.*

2) When converting from a *smaller unit* to a *larger unit,* use *division.*

EXAMPLE 2 Convert each measurement to the indicated unit.

a) 4 ft to inches   b) 1.5 days to hours   c) $2\frac{1}{2}$ c to pints

Solution

a) First, we need to know that 1 ft = 12 in. Next, notice that we are changing from a *larger unit,* feet, to a *smaller unit,* inches. So to change 4 ft to inches, we *multiply.*

$$4 \cdot 12 = 48 \text{ in.} \qquad 12 \text{ in.} = 1 \text{ ft}$$

Therefore, 4 ft = 48 in.

b) Since we are asked to convert 1.5 days to hours, we need to know that 1 day = 24 hr. We are changing from a larger unit, *days,* to a smaller unit, *hours.* To change 1.5 days to hours, we will multiply.

$$1.5 \cdot 24 = 36 \qquad 24 \text{ hr} = 1 \text{ day}$$

So, 1.5 days = 36 hr.

c) First ask yourself, *"How are cups and pints related?"* 2 c = 1 pt. This time we are asked to convert from a *smaller unit,* cups, to a *larger unit,* pints. Therefore, we will *divide.*

$$\text{Change } 2\tfrac{1}{2} \text{ to } \tfrac{5}{2}. \searrow$$
$$2\frac{1}{2} \div 2 = \frac{5}{2} \div \frac{2}{1} = \frac{5}{2} \cdot \frac{1}{2} = \frac{5}{4} = 1\frac{1}{4} \text{ or } 1.25$$
$$\text{Write 2 as } \tfrac{2}{1}. \nearrow \qquad \uparrow \text{ Change division to multiplication by the reciprocal.}$$

Therefore, $2\frac{1}{2}$ c = $1\frac{1}{4}$ pt or 1.25 pt.

Note

If we get an improper fraction for a unit of measurement, we usually write it as a mixed number or as a decimal.

[YOU TRY 2] Convert each measurement to the indicated unit.

a) 8000 lb to tons b) $1\frac{1}{2}$ qt to pints

c) 12 sec to minutes d) 4.5 yd to feet

3 Use Unit Analysis to Convert Between Units

Another way to convert between units is to use *unit analysis*. This method is commonly used in chemistry, nursing, and other science courses. Recall from our work with ratios that we can divide out common units that appear in the numerator and denominator of a fraction. For example, we can write the ratio 9 feet to 12 feet as

$$\frac{9\text{ ft}}{12\text{ ft}} = \frac{9\text{ ft}}{12\text{ ft}} = \frac{3}{4}$$ Divide 9 and 12 by 3.

because the units divide out. *In unit analysis, we divide out like units to help us convert between units and to guide us in performing the conversion correctly.*

Another concept we need to understand is that of a *unit fraction*. A **unit fraction** is a fraction that is equivalent to 1. The measurement equivalencies can be written as unit fractions. For example, we can think of the relationship 12 in. = 1 ft as the ratio 12 in. to 1 ft. Then, we can write the ratio

> **Hint**
> Write the equivalencies as unit fractions.

$$\frac{12\text{ in.}}{1\text{ ft}} = \frac{12\text{ in.}}{12\text{ in.}} = \frac{12\text{ in.}}{12\text{ in.}} = \frac{1}{1} = 1$$

So, $\frac{12\text{ in.}}{1\text{ ft}}$ is a unit fraction because it is equal to 1. All of the equivalencies in the table of measurement relationships can be written as unit fractions. To convert from one unit of measurement to another, we will multiply the measurement we are given by the appropriate unit fraction.

EXAMPLE 3 Use unit analysis to convert 36 in. to feet.

Solution

First, identify the units we are given and the units we want to get. Since we are given *inches* and we want to get *feet*, write down the relationship between inches and feet:
12 in. = 1 ft.

Next, multiply the measurement we are given, 36 in., by the unit fraction relating 12 in. and 1 ft *so that the inches divide out* and we will be left with feet.

> **Hint**
> Write out the example on your own paper as you read it.

$$36\text{ in.} \cdot \frac{1\text{ ft}}{12\text{ in.}} = \frac{36\text{ in.}}{1} \cdot \frac{1\text{ ft}}{12\text{ in.}}$$ Divide 36 and 12 by 12; divide out the unit of inches.

$$= \frac{3\text{ ft}}{1}$$ Multiply.

$$= 3\text{ ft}$$ Simplify.

Therefore, 36 in. = 3 ft.

Let's summarize the steps for converting from one unit to another using unit analysis.

Procedure How to Convert Between Units Using Unit Analysis

Step 1: **Identify** the units given and the units we want to get. **Write down** the relationship between those units.

Step 2: **Multiply** the given measurement by the unit fraction relating the unit given and the unit we want to get so that the given unit will divide out and leave us with the unit we want.

Note

The units tell us whether we are doing the conversion correctly. If the units do not divide out, then we are doing the problem incorrectly.

[**YOU TRY 3**]   Use unit analysis to convert 120 min to hours.

EXAMPLE 4   Use unit analysis to convert each measurement to the indicated unit.

a)   5 wk to days

b)   7 pt to quarts

c)   1800 lb to tons

d) $5\frac{2}{3}$ yd to feet

Solution

a)   *Step 1:* **Identify** the units given and the units we want to get. **Write down** the relationship between those units.

We are given *weeks* and want to convert to *days*. The relationship is

$$1 \text{ week} = 7 \text{ days}$$

Step 2: **Multiply** the given measurement by the unit fraction relating the unit given and the unit we want to get so that the given unit will divide out and leave us with the unit we want.

$$5 \text{ wk} \cdot \frac{\textbf{7 days}}{\textbf{1 wk}} = \frac{5 \text{ wk}}{1} \cdot \frac{7 \text{ days}}{1 \text{ wk}} \qquad \text{Divide out the unit of weeks.}$$

$$= \frac{35 \text{ days}}{1} \qquad \text{Multiply.}$$

$$= 35 \text{ days} \qquad \text{Simplify.}$$

So, 5 weeks = 35 days.

b) *Step 1:* **Identify** the units given and the units we want to get. **Write down** the relationship between those units.

We are given 7 *pints* and want to convert to *quarts*. The relationship is

$$2 \text{ pt} = 1 \text{ qt}$$

Step 2: **Multiply** the given measurement by the unit fraction relating the unit given and the unit we want to get so that the given unit will divide out and leave us with the unit we want.

$$7 \text{ pt} \cdot \frac{1 \text{ qt}}{2 \text{ pt}} = \frac{7 \text{ p\!\!/t}}{1} \cdot \frac{1 \text{ qt}}{2 \text{ p\!\!/t}} \qquad \text{Divide out the unit of pints.}$$

$$= \frac{7 \text{ qt}}{2} \qquad \text{Multiply.}$$

$$= \frac{7}{2} \text{ qt} \qquad \text{Simplify.}$$

$$= 3\frac{1}{2} \text{ qt or } 3.5 \text{ qt} \qquad \text{Write the final answer as a mixed number or decimal.}$$

Therefore, $7 \text{ pt} = 3\dfrac{1}{2}$ qt or 3.5 qt.

c) *Step 1:* **Identify** the units given and the units we want to get. **Write down** the relationship between those units.

We are given 1800 *lb* and want to convert to *tons*. The relationship is

$$2000 \text{ lb} = 1 \text{ ton}$$

Step 2: **Multiply** the given measurement by the unit fraction relating the unit given and the unit we want to get so that the given unit will divide out and leave us with the unit we want.

$$1800 \text{ lb} \cdot \frac{1 \text{ ton}}{2000 \text{ lb}} = \frac{\overset{9}{\cancel{1800}} \text{ l\!\!/b}}{1} \cdot \frac{1 \text{ ton}}{\underset{10}{\cancel{2000}} \text{ l\!\!/b}} \qquad \text{Divide the numbers by 2000; divide out the unit of pounds.}$$

$$= \frac{9 \text{ ton}}{10} \qquad \text{Multiply.}$$

$$= \frac{9}{10} \text{ ton or } 0.9 \text{ ton} \qquad \text{Simplify.}$$

Therefore, $1800 \text{ lb} = \dfrac{9}{10}$ ton or 0.9 ton.

d) **Step 1:** **Identify** the units given and the units we want to get. **Write down** the relationship between those units.

We are given $5\frac{2}{3}$ *yd* and want to convert to *feet*. The relationship is

1 yd = 3 ft

Step 2: **Multiply** the given measurement by the unit fraction relating the unit given and the unit we want to get so that the given unit will divide out and leave us with the unit we want.

$$5\frac{2}{3}\ \text{yd} \cdot \frac{\textbf{3 ft}}{\textbf{1 yd}} = \frac{17}{3}\ \text{yd} \cdot \frac{3\ \text{ft}}{1\ \text{yd}} \qquad \text{Change the mixed number to an improper fraction.}$$

$$= \frac{17}{\cancel{3}_{1}}\ \cancel{\text{yd}} \cdot \frac{\cancel{3}^{1}\ \text{ft}}{1\ \cancel{\text{yd}}} \qquad \text{Divide out 3; divide out the unit of yards.}$$

$$= \frac{17\ \text{ft}}{1} \qquad \text{Multiply.}$$

$$= 17\ \text{ft} \qquad \text{Simplify.}$$

So, $5\frac{2}{3}\ \text{yd} = 17\ \text{ft}$.

[YOU TRY 4] Use unit analysis to convert each measurement to the indicated unit.

a) 3 lb to ounces   b) 9 c to pints

c) 48 min to hours   d) $2\frac{1}{4}$ mi to feet

Sometimes, we have to use more than one unit fraction to convert to the desired unit. This is often done in science classes.

EXAMPLE 5   Convert 96 in. to yards.

Solution

We are given *inches* and want to convert to *yards*. We will use two relationships to do this conversion. They are

$$12\ \text{in.} = 1\ \text{ft} \qquad \text{and} \qquad 3\ \text{ft} = 1\ \text{yd}$$

Use this unit fraction to change feet to yards.

$$\frac{\overset{8}{\cancel{96}}\ \cancel{\text{in.}}}{1} \cdot \frac{1\ \cancel{\text{ft}}}{\underset{1}{\cancel{12}}\ \cancel{\text{in.}}} \cdot \frac{1\ \text{yd}}{3\ \cancel{\text{ft}}} = \frac{8\ \text{yd}}{3} \qquad \text{Multiply.}$$

Use this unit fraction to change inches to feet.

$$= \frac{8}{3}\ \text{yd} \ \text{or} \ 2\frac{2}{3}\ \text{yd}$$

[YOU TRY 5]   Convert 1200 sec to hours.

EXAMPLE 6 Use unit analysis to convert 3 days to seconds.

Solution

We are given *days* and want to convert to *seconds*. The relationships are

$$1 \text{ day} = 24 \text{ hr} \qquad 1 \text{ hr} = 60 \text{ min} \qquad 1 \text{ min} = 60 \text{ sec}$$

We will do this conversion in a single step using more than one unit fraction.

Use this unit fraction to change hours to minutes.

$$\frac{3 \text{ days}}{1} \cdot \frac{24 \text{ hr}}{1 \text{ day}} \cdot \frac{60 \text{ min}}{1 \text{ hr}} \cdot \frac{60 \text{ sec}}{1 \text{ min}} \qquad \text{Divide out the correct units.}$$

Use this unit fraction to change days to hours.

Finally, use this unit fraction to change minutes to seconds.

$$= \frac{259{,}200 \text{ sec}}{1} = 259{,}200 \text{ sec} \qquad \text{Multiply and simplify.}$$

Therefore, 3 days = 259,200 sec.

[YOU TRY 6] Use unit analysis to convert 3 gal to cups.

---

4 Solve Applied Problems Using Unit Analysis

Unit analysis can be used to solve many different types of applied problems.

EXAMPLE 7 Larry is a chef, and his company is catering a large party. One batch of cheese fondue uses $\frac{3}{4}$ cup heavy cream. If he needs to make 12 times that amount of cheese fondue, how many quarts of heavy cream does he need?

Solution

Let's restate the problem in our own words to be sure we understand it.

One batch of cheese fondue uses $\frac{3}{4}$ c heavy cream. We must determine the number of *quarts* of heavy cream needed to make 12 batches of fondue.

To solve the problem, first multiply $\frac{3}{4}$ c by 12 to determine the number of *cups* of heavy cream needed. Then, change the number of cups to quarts.

©Pixtal/Age Fotostock

Total cups of heavy cream needed: $\frac{3}{4} \text{ c} \cdot 12 = \frac{3}{4} \text{ c} \cdot \frac{12}{1} = 9 \text{ c}$

Change 9 *cups* to *quarts:*

This unit fraction changes
cups to pints.

$$\frac{9 \cancel{c}}{1} \cdot \frac{1 \ \cancel{pt}}{2 \ \cancel{c}} \cdot \frac{1 \ qt}{2 \ \cancel{pt}}$$ Divide out the correct units.

This unit fraction changes
pints to quarts.

$$= \frac{9 \ qt}{4} = \frac{9}{4} \ qt = 2\frac{1}{4} \ qt$$

Larry needs $2\frac{1}{4}$ qt heavy cream for the cheese fondue.

Check the answer by working backward. Change quarts to cups, then divide by 12.

$$2\frac{1}{4} \ qt = \frac{9}{4} \ qt$$ Change the mixed number to a fraction.

$$\frac{9}{4} \ \cancel{qt} \cdot \frac{2 \ \cancel{pt}}{1 \ \cancel{qt}} \cdot \frac{2 \ c}{1 \ \cancel{pt}}$$ Divide out the correct factors and units.

$$= 9 \ c$$

This 9 c heavy cream is for 12 batches of cheese fondue. Divide 9 by 12 to determine how many cups of heavy cream are used for one batch.

$$\frac{9 \ c}{12} = \frac{9}{12} \ c = \frac{3}{4} \ c \text{ heavy cream}$$

This is the same as the amount stated in the original problem, so the answer is correct.

[YOU TRY 7]

Julia's frozen lemonade recipe uses $\frac{2}{3}$ c lemon juice. If she needs to make 6 times the amount of the recipe, how many quarts of lemon juice will she need?

EXAMPLE 8

Kate has to buy paprika at the grocery store. The label on the shelf for Brand A says that its unit cost is $1.79/oz. The label for Brand B says that the unit cost is $22.59/lb. Which is the better buy?

Solution

The units on each unit cost must be the same in order to compare them. Therefore, we will change *ounces* to *pounds* in the unit cost of Brand A.

Unit cost of Brand A in dollars/lb: $\dfrac{\$1.79}{\cancel{oz}} \cdot \dfrac{16 \ \cancel{oz}}{1 \ lb} = \dfrac{\$28.64}{lb} = \$28.64/lb$

The unit cost of Brand B is $22.59/lb. Brand B is the better buy because it has the lower unit cost.

Check the answer by working backward. The check is left to the student.

Note

We could have changed the units in the unit cost of Brand B from $22.59/lb to dollars/oz so that the units matched those of Brand A.

[YOU TRY 8]

Sonja has to buy basil leaves at the grocery store. The label on the shelf for Brand A says that its unit cost is $3.59/oz. The label for Brand B says that the unit cost is $49.89/lb. Which is the better buy?

ANSWERS TO [YOU TRY] EXERCISES

1) a) 3   b) 2000   c) 2   2) a) 4 tons   b) 3 pt   c) $\frac{1}{5}$ min   d) 13.5 ft   3) 2 hr

4) a) 48 oz   b) $4\frac{1}{2}$ pt or 4.5 pt   c) $\frac{4}{5}$ hr or 0.8 hr   d) 11,880 ft   5) $\frac{1}{3}$ hr   6) 48 c

7) 1 qt   8) Brand B is the better buy because the unit cost of Brand A is $57.44/lb.

Using Technology

When using more than one unit fraction to convert to a desired unit, we can use a calculator to perform the required calculation. Suppose we wanted to convert 100 c to gallons. In this case, our expression of unit fractions would be

$$\frac{100 \text{ c}}{1} \cdot \frac{1 \text{ pt}}{2 \text{ c}} \cdot \frac{1 \text{ qt}}{2 \text{ pt}} \cdot \frac{1 \text{ gal}}{4 \text{ qt}}.$$

(Note: Because dividing or multiplying by 1 does not affect the final answer, we can disregard them in the calculation.)

To perform the calculation with a calculator (disregarding the 1's) we enter
[1] [0] [0] [÷] [2] [÷] [2] [÷] [4] [=]. The display will likely show 6.25. This means 100 c = 6.25 gal.

It is a common mistake to enter [1] [0] [0] [÷] [2] [×] [2] [×] [4] [=] into the calculator for this problem. Why is this incorrect?

E Evaluate   7.1 Exercises   Do the exercises, and check your work.

Objective 1: Learn U.S. Customary Measurements

In U.S. customary units,

1) name three units used to measure length.

2) name three units used to measure capacity.

3) what is the relationship between inches and feet?

4) what is the relationship between ounces and pounds?

Fill in the blank.

5) 1 min = _____ sec; _____ min = 1 hr

6) 1 day = _____ hr; _____ sec = 1 min

7) _____ ft = 1 yd; _____ ft = 1 mi

8) _____ fl oz = 1 c; 1 pt = _____ c

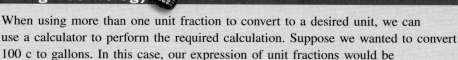

9) _____ pt = 1 qt; 1 c = _____ fl oz

10) 1 ton = _____ lb; _____ oz = 1 lb

Objective 2: Use Multiplication or Division to Convert Between Units

In U.S. customary units,

11) when do you use division to convert between units?

12) when do you use multiplication to convert between units?

Use multiplication or division to convert to the indicated unit.

 13) 3 hr to minutes

14) 5 hr to minutes

15) 36 in. to feet

16) 8 c to pints

17) 6 qt to gallons

18) 2 qt to gallons

19) 1500 lb to tons

20) 800 lb to tons

21) $2\frac{1}{2}$ hr to minutes

22) $3\frac{1}{2}$ lb to ounces

23) $1\frac{3}{4}$ ft to inches

24) $4\frac{2}{3}$ ft to inches

25) 1.25 gal to quarts

26) 2.75 days to hours

27) 3.5 pt to quarts

28) 6.5 pt to quarts

29) A snow leopard can pounce on its prey from 45 ft away. How many yards is this? (www.sandiegozoo.com)

©Moment Open/Getty Images

30) A baby elephant can weigh up to $\frac{1}{8}$ ton at birth. How many pounds is this? (www.sandiegozoo.com)

Objective 3: Use Unit Analysis to Convert Between Units

31) Write two different unit fractions for the relationship 60 min = 1 hr.

32) Write two different unit fractions for the relationship 3 ft = 1 yd.

33) What are the steps we use to convert between units using unit analysis?

34) To convert 4.2 min to seconds using unit analysis, Nisha sets up the problem this way:

$$4.2 \text{ min} \cdot \frac{1 \text{ min}}{60 \text{ sec}} = \frac{4.2 \text{ min}}{1} \cdot \frac{1 \text{ min}}{60 \text{ sec}}$$

a) How do we know this is wrong?

b) What is the correct way to set up this problem and why?

c) Finish the correct work in part b) to convert 4.2 min to seconds.

Use unit analysis to convert to the indicated unit.

35) 7 ft to inches

36) 4 hr to minutes

37) 10 qt to pints

38) 8 yd to feet

39) 52 oz to pounds

40) 5 c to pints

41) $3\frac{1}{2}$ days to hours

42) $4\frac{1}{3}$ yd to feet

Use unit analysis to convert to the indicated unit.

43) 5 yd = _____ ft

44) 12 qt = _____ pt

45) 72 oz = _____ lb

46) 44 oz = _____ lb

47) 1.75 hr = _____ min

48) 3.25 hr = _____ min

49) $3\frac{1}{2}$ pt = _____ c

50) $5\frac{1}{2}$ pt = _____ c

51) $1\frac{1}{4}$ tons = _____ lb

52) $2\frac{3}{4}$ tons = _____ lb

53) 2640 ft = _____ mi

54) 7920 ft = _____ mi

55) a) Convert 2 days to minutes in two separate steps, first by changing days to hours and then by changing hours to minutes.

b) Convert 2 days to minutes in a single step, using more than one unit fraction.

56) a) Convert 2 hours to seconds in two separate steps, first by changing hours to minutes and then by changing minutes to seconds.

b) Convert 2 hours to seconds in a single step, using more than one unit fraction.

Use more than one unit fraction to convert to the indicated unit.

57) 1 day = _____ sec

58) 2.5 days = _____ sec

59) 18 c = _____ qt

60) 15 c = _____ qt

61) 504 hr = _____ wk

62) 336 hr = _____ wk

63) 2.5 gal = _____ c

64) 1.25 gal = _____ c

65) 6600 yd = _____ mi

66) 3740 yd = _____ mi

Objective 4: Solve Applied Problems Using Unit Analysis

Solve each problem.

67) One batch of homemade vanilla ice cream uses $2\frac{1}{2}$ c half-and-half. How many pints of half-and-half are needed for four batches of ice cream?

68) Irina will make buttermilk pancakes for her daughter's pajama party. She has to make six batches, and each batch uses $2\frac{1}{2}$ c buttermilk. How many pints of buttermilk should she buy?

69) Melvina babysat four days in a row, 3.75 hr each time. Find the total number of minutes she babysat.

70) Susan is a restaurant manager and scheduled five employees to work 7.5 hr on Friday. Find the total number of minutes the employees worked.

©John Lund/Marc Romanelli/
Blend Images LLC

71) Mohsin purchased a 30-oz bag of dried Turkish figs for $11.40. What is the cost per pound?

72) Shufan bought an 8-oz jar of blackberry jam for $3.99. What is the cost per pound?

73) Roberto has to buy saltine crackers at the store. The label on the shelf for Brand A says that its unit cost is $3.19/lb. Brand B says that its unit cost is $0.18/oz. Which is the better buy?

74) Jung-Su is having a dinner party and wants to make enough cheesecake for everyone. If he makes eight cheesecakes and each one uses $\frac{3}{4}$ c milk, how many quarts of milk should he buy?

75) Valerie compares two brands of jars of pickles. Brand A sells for $0.19/oz while Brand B sells for $2.99/lb. Which is the better buy?

76) Logan wants to buy the box of cornflakes with the lower unit cost. The label on the shelf for Brand A says that its unit cost is $0.31/oz, and the unit cost of Brand B is $5.09/lb. Which brand should Logan buy?

77) When ladybugs fly, they flap their wings approximately 85 times per second. How many times does a ladybug flap its wings during a 5-min flight?
(www.sandiegozoo.com)

78) A television has a screen size that measures $4\frac{1}{3}$ ft diagonally. How many inches is this?

79) Joseph wants to line his garden with a fence that costs $2.25 per foot. Use the diagram below to calculate the total cost.

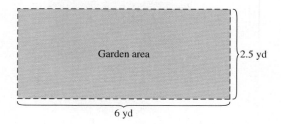
Garden area
2.5 yd
6 yd

80) A high school wants to put a fence around its new football field. The length of the area that will be enclosed is 138 yd, and the width is 72 yd. If the fence costs $8.40 per foot, find the cost of the fence.

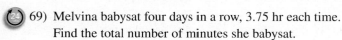

R Rethink

R1) Which process do you prefer: multiplication and division, or unit analysis? Why?

R2) Why do you never have to "divide" when using unit analysis?

R3) Think of a situation where you needed to convert between U.S. customary units, and write an application problem similar to those you just solved.

7.2 The Metric System: Length

P Prepare

What are your objectives for Section 7.2?	How can you accomplish each objective?
1 Learn the Basic Units of Length	• Know what a *meter* represents by looking at the pictures of common objects, and understand the different associated prefixes. • Memorize the **Relationships Between Metric Units of Length.** • Complete the given examples on your own. • Complete You Trys 1 and 2.
2 Use Unit Fractions to Convert Between Units	• Use the same procedure for using unit fractions as done previously. • Complete the given example on your own. • Complete You Try 3.
3 Convert Between Units Using the Metric Conversion Chart	• Understand the *metric conversion chart,* and be able to reproduce it on your own. • Write the procedure for **How to Use the Metric Conversion Chart** in your own words. • Follow the example on your own. • Complete You Try 4.

O Organize

Read the explanations, follow the examples, take notes, and complete the You Trys.

1 Learn the Basic Units of Length

Before the late 1700s, different countries used different units of measurement, which led to difficulties trading between countries. In the 1790s, the French Academy of Science developed a simpler system of measurement, based on multiples of 10, called the **metric system.** This system was adopted by almost all countries around the world. It made both measuring and trade easier. Although the United States still uses its own system of feet, cups, pounds, and so on, most scientific measurement in the United States is done in metric units.

Note

The metric system is also called the *International System of Units,* or *SI* for short.

The basic unit of length in the metric system is the **meter.** A meter is about 39.37 inches long, so it is a little longer than a yard.

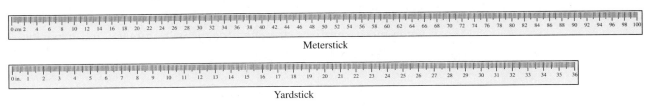

Meterstick

Yardstick

The other units of measurement in the decimal system are based on powers of 10. Because they are all related to the meter, these units are described by using a prefix on the word *meter*. The table shows the different units of measurement and their relationships to the meter.

Metric Units of Length

1 kilometer (km) = 1000 m
1 hectometer (hm) = 100 m
1 dekameter (dam) = 10 m
1 meter (m) = 1 m
1 decimeter (dm) = $\frac{1}{10}$ m or 0.1 m
1 centimeter (cm) = $\frac{1}{100}$ m or 0.01 m
1 millimeter (mm) = $\frac{1}{1000}$ m or 0.001 m

Note

The units used most often are kilometer, meter, centimeter, and millimeter.

We have already compared the length of a meter with the length of a yard. Here are some examples that will give you an idea of the lengths the metric units represent.

A dime is about 1 mm thick.

A USB plug is about 1 cm wide.

Distance is usually measured in kilometers.

A kilometer is about $\frac{2}{3}$ of a mile.

Height is usually measured in centimeters.

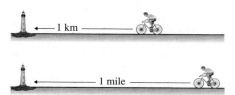

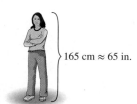

165 cm ≈ 65 in.

Note

These same metric prefixes are used for measuring mass and capacity, as we will see later in this chapter.

Now let's look closely at a meterstick to see how the meter, decimeter, centimeter, and millimeter are related.

The meterstick above shows us the following relationships:

**Relationships Between
Metric Units of Length**

10 decimeters = 1 meter
100 centimeters = 1 meter
1000 millimeters = 1 meter
10 millimeters = 1 centimeter
10 centimeters = 1 decimeter

Let's practice deciding which metric unit is appropriate for measuring a particular length.

EXAMPLE 1   Fill in the blank with the appropriate metric unit. Use km, m, cm, or mm.

a)   A pen is 14 _____ long.

b)   Gisele drives 8 _____ to work every day.

c)   The lens of a camera on a phone is 3 _____ wide.

d)   The width of a soccer field is 72 _____.

Solution

a)   A pen is 14 <u>cm</u> long. (This is about 5.5 in.) Usually, we use centimeters instead of inches.

b)   Gisele drives 8 <u>km</u> to work everyday. (8 km ≈ 5 mi) Kilometers are used instead of miles.

c)   The lens of a camera on a phone is 3 <u>mm</u> wide. (3 mm ≈ 0.12 in.)

d)   The width of a soccer field is 72 <u>m</u>. (This is about 78.74 yd.) Usually, we use meters instead of yards.

[YOU TRY 1]   Fill in the blank with the appropriate metric unit. Use km, m, cm, or mm.

a)   The width of a bedroom is 4.5 _____.

b)   A computer screen is 33 _____ wide.

c)   Boston and New York City are approximately 306 _____ apart.

d)   An ant is 5 _____ long.

EXAMPLE 2   Find the length of the paper clip in the indicated unit.

a) mm   b) cm

Solution

a) Each of the smallest marks on the ruler is a millimeter. Each of the larger marks on the ruler is a centimeter. Remember, 1 cm = 10 mm, so we can begin by counting the 10 millimeters in each centimeter. Then, count the individual millimeters.

The length of the paper clip is 46 mm.

b) To measure the paper clip in centimeters, think of the ruler in terms of centimeters. Each of the larger marks on the ruler is a centimeter. Since it takes 10 mm to make a centimeter, each smaller mark, or millimeter, is $\frac{1}{10}$ cm or 0.1 cm. Count the 4 cm, then count the additional 6 mm or 0.6 cm.

The length of the paper clip is 4.6 cm.

[YOU TRY 2]   Find the length of the nail clippers in the indicated unit.

a) mm   b) cm

Often, measurements must be converted from one unit to another. We can do this by using unit fractions or by moving the decimal point. Let's use unit fractions first.

2 Use Unit Fractions to Convert Between Units

Since we know the relationships between the units of measurement in the metric system, we can write them as unit fractions to convert from one unit to another. We will use the unit fractions just as we did in Section 7.1.

EXAMPLE 3

Use unit fractions to change to the indicated unit.

a) 83 km to m   b) 162 mm to m   c) 1.5 m to cm

Solution

a) *Step 1:* **Identify** the units given and the units we want to get. **Write down** the relationship between those units.

We are given *km* and want to convert to *m*. The relationship is

$$1 \text{ km} = 1000 \text{ m}$$

Step 2: **Multiply** the given measurement by the unit fraction relating the unit given and the unit we want to get so that the given unit will divide out and leave us with the unit we want.

$$83 \text{ km} \cdot \frac{1000 \text{ m}}{1 \text{ km}} = \frac{83 \text{ km}}{1} \cdot \frac{1000 \text{ m}}{1 \text{ km}} \qquad \text{Divide out the unit of kilometers.}$$

$$= \frac{83{,}000 \text{ m}}{1} \qquad \text{Multiply.}$$

$$= 83{,}000 \text{ m} \qquad \text{Simplify.}$$

So, 83 km = 83,000 m.

W Hint

Does this seem similar to Objective 3 of Section 7.1?

b) We are given 162 *mm* and want to convert to *m*. The relationship is

$$1000 \text{ mm} = 1 \text{ m}$$

Multiply the given measurement by the unit fraction so that millimeters divide out and meters remain.

$$162 \text{ mm} \cdot \frac{1 \text{ m}}{1000 \text{ mm}} = \frac{162 \text{ mm}}{1} \cdot \frac{1 \text{ m}}{1000 \text{ mm}} \qquad \text{Divide out the unit of millimeters.}$$

$$= \frac{162 \text{ m}}{1000} \qquad \text{Multiply.}$$

$$= 0.162 \text{ m} \qquad \text{Simplify.}$$

Therefore, 162 mm = 0.162 m. Notice that when we multiplied by the unit fraction, we did not divide out common factors of 162 and 1000. This is because we want to express the answer as a decimal.

c)   We are given 1.5 *m* and want to convert to *cm*. The relationship is

$$1 \text{ m} = 100 \text{ cm}$$

Multiply the given measurement by the unit fraction so that meters divide out and centimeters remain.

$$1.5 \text{ m} \cdot \frac{\textbf{100 cm}}{\textbf{1 m}} = \frac{1.5 \ \cancel{\text{m}}}{1} \cdot \frac{100 \text{ cm}}{1 \ \cancel{\text{m}}} \qquad \text{Divide out the unit of meters.}$$

$$= \frac{150 \text{ cm}}{1} \qquad\qquad\qquad \text{Multiply.}$$

$$= 150 \text{ cm} \qquad\qquad\qquad \text{Simplify.}$$

So, 1.5 m = 150 cm.

[YOU TRY 3]   Use unit fractions to change to the indicated unit.

a) 962 cm to m   b) 8.3 m to mm   c) 174 mm to m

Another way to change from one unit to another is by moving the decimal point.

3 Convert Between Units Using the Metric Conversion Chart

The previous examples have shown that changing from one unit to another in the metric system involves multiplying or dividing by powers of 10. Therefore, we can convert between metric units by moving the decimal point. How do we know which way to move the decimal point to convert between units? We can use a *metric conversion chart*.

Metric Conversion Chart

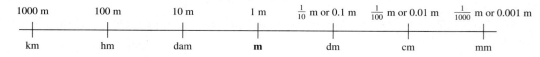

1000 m	100 m	10 m	1 m	$\frac{1}{10}$ m or 0.1 m	$\frac{1}{100}$ m or 0.01 m	$\frac{1}{1000}$ m or 0.001 m
km	hm	dam	**m**	dm	cm	mm

To make the chart, list the metric units from largest to smallest.

Procedure How to Use the Metric Conversion Chart

1)   Find the unit that you are given.

2)   Count the number of places you must move to go from the unit you are given to the unit you want to get.

3)   Move the decimal point the same number of places and in the same direction that you did on the metric conversion chart.

Note

Moving the decimal point to the *right* is like multiplying by a power of 10.
Moving the decimal point to the *left* is like dividing by a power of 10.

EXAMPLE 4

Convert each measurement to the indicated unit.

a)   8.416 km to meters   b)   278.5 mm to meters

c)   9 m to centimeters   d)   120 mm to kilometers

Solution

a)   Find km on the metric conversion chart. Since we have to change 8.416 km to meters, move *three places to the right* to get to m.

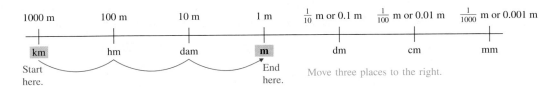

Since we moved *three places to the right* to get to m, move the decimal point *three places to the right* to change to meters.

$$8.416 \text{ km} = 8416 \text{ m}$$

Move the decimal point three places to the right.

Therefore, 8.416 km = 8416 m.

b)   To change 278.5 mm to meters, find mm on the metric conversion chart. Move *three places to the left* to get to m.

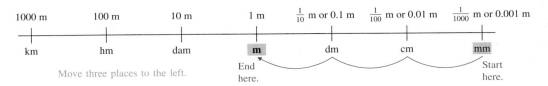

Since we moved *three places to the left* to get to m, move the decimal point *three places to the left* to change 278.5 mm to meters.

$$278.5 \text{ mm} = .2785 \text{ m}$$

Move the decimal point three places to the left.

So, 278.5 mm = 0.2785 m.

c)   To change 9 m to centimeters, find m on the metric conversion chart. Move *two places to the right* to get to cm.

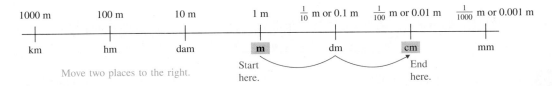

Put a decimal point at the end of the number 9. When you move the decimal point *two places to the right,* you must put in zeros as placeholders.

$$9. \text{ m} = 900 \text{ cm}$$ Put in zeros as placeholders.

Put the decimal point at
the end of the number.   Move the decimal point two places
to the right.

Therefore, 9 m = 900 cm.

d) To change 120 mm to kilometers, find mm on the metric conversion chart. Move *six places to the left* to get to km.

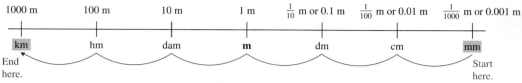

Move six places to the left.

Put a decimal point at the end of the number 120. When you move the decimal point *six places to the left,* you must put in zeros as placeholders.

$$120. \text{ m} = \underset{\text{Move the decimal point six places to the left.}}{\underbrace{000}120 \text{ km}} \qquad \text{Put in zeros as placeholders.}$$

Put the decimal point at the end of the number.

So, 120 m = 0.000120 km or 0.00012 km.

[**YOU TRY 4**] Convert each measurement to the indicated unit.

a) 72.9 cm to meters   b) 3.881 m to centimeters

c) 645 cm to kilometers   d) 2 m to millimeters

ANSWERS TO [**YOU TRY**] **EXERCISES**

1) a) m   b) cm   c) km   d) mm   2) a) 53 mm   b) 5.3 cm   3) a) 9.62 m   b) 8300 mm
c) 0.174 m   4) a) 0.729 m   b) 388.1 cm   c) 0.00645 km   d) 2000 mm

---

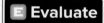 **Evaluate** **7.2** Exercises   Do the exercises, and check your work.

Objective 1: Learn the Basic Units of Length
Use the meaning of metric prefixes to fill in the blanks.

1) *milli* means _____ so

   1 mm = _____ m

2) *deci* means _____ so

   1 dm = _____ m

3) *kilo* means _____ so

   1 km = _____ m

4) *hecto* means _____ so

   1 hm = _____ m

5) *centi* means _____ so

   1 cm = _____ m

6) *deka* means _____ so

   1 dam = _____ m

Use the ruler to find the measurements in Exercises 7–14.

7) The length of a dollar bill,

   a) in centimeters.

   b) in millimeters.

9) The diameter of a quarter,

   a) in millimeters.

   b) in centimeters.

8) The width of a dollar bill,

   a) in centimeters.

   b) in millimeters.

10) The diameter of a nickel,

   a) in millimeters.

   b) in centimeters.

11) Find the length of a cell phone, in centimeters.

12) Find the width of a driver's license, in millimeters.

13) Find the length of your pen or pencil, in centimeters.

14) Find the length of a key, in centimeters.

15) Arrange the following lengths in order from smallest to largest.

   0.5 dm   3 cm   40 mm   0.06 m

16) Arrange the following lengths in order from largest to smallest.

   0.7 dm   80 mm   6 cm   0.05 m

Fill in the most reasonable metric length unit. Choose from cm, km, m, and mm.

17) The textbook is 4 _____ thick.

18) The teacher is 1.7 _____ tall.

19) The yardstick is approximately 0.9144 _____ in length.

20) The diameter of a penny is approximately 19 _____.

21) The diameter of a dime is approximately 1.8 _____.

22) A CD-ROM disk has a thickness of 1.2 _____.

23) 65 mph is equivalent to approximately 104.6 _____ per hour.

24) Joseph can run the 100-_____ dash in 14.2 sec.

25) 10 ft per second is equivalent to approximately 3 _____ per second.

26) Carol drives 19 _____ to get to school.

Objective 2: Use Unit Fractions to Convert Between Units

Use unit fractions to change to the indicated unit.

27) 6 m to centimeters
28) 4 m to millimeters
29) 3.8 m to millimeters
30) 1.25 m to centimeters
31) 7300 m to kilometers
32) 300 m to kilometers
33) 250 mm to meters
34) 425 cm to meters
35) 18.2 m to kilometers
36) 21.9 m to kilometers
37) 13 dam to meters
38) 47 dam to meters
39) 2500 m to kilometers
40) 1800 m to kilometers
41) 342.8 dm to meters
42) 4.6 hm to meters
43) 0.75 cm to millimeters
44) 342.8 dm to centimeters

Objective 3: Convert Between Units Using the Metric Conversion Chart

Metric Conversion Chart

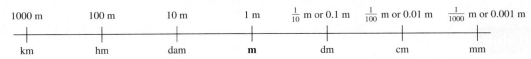

1000 m	100 m	10 m	1 m	$\frac{1}{10}$ m or 0.1 m	$\frac{1}{100}$ m or 0.01 m	$\frac{1}{1000}$ m or 0.001 m
km	hm	dam	**m**	dm	cm	mm

Use the metric conversion chart to convert each measurement to the indicated unit.

45) 15 m to centimeters
46) 64 m to decimeters
47) 7.5 cm to millimeters
48) 8.3 cm to millimeters
49) 970 m to kilometers
50) 510 m to kilometers
51) 820 mm to meters
52) 475 cm to meters
53) 1600 cm to kilometers
54) 2400 mm to kilometers
55) 8.6 cm to millimeters
56) 11.7 cm to millimeters
57) 3.49 km to meters
58) 8.25 km to meters
59) 2.34 m to decimeters
60) 0.77 km to dekameters
61) 43,000 cm to hectometers
62) 26,000 dm to hectometers

63) 9.44 km to millimeters
64) 5.83 km to centimeters
65) Arrange the following lengths in order from smallest to largest.

   1.01 m   100 cm   1400 mm   12 dm

66) Arrange the following lengths in order from largest to smallest.

   0.8 m   9.5 dm   75 cm   825 mm

67) The system of blood vessels (arteries, veins, and capillaries) in the human body is over 96,000 km long. How many meters is this? (www.webmd.com)

68) The width of a balance beam used by female gymnasts is 10 cm. How many millimeters is this?

69) American Veterinary Identification Devices (AVID) microchips measure 1.2 cm in length. Convert this number to millimeters. (www.avidplc.com)

70) The average height of an adult female in the United States is 1.62 m. Convert the height to centimeters and millimeters. (National Center for Health Statistics)

©Jack Hollingsworth/Getty Images

Mixed Exercises: Objectives 1–3

Fill in the most reasonable metric length unit. Choose from cm, km, m, and mm.

71) The height of the door is 200 _____.

72) The width of the door is 760 _____.

73) A bicycle wheel has a radius of 0.8 _____.

74) The length of a goldfish is 3 _____.

Convert to the indicated unit using either unit fractions or the metric conversion chart.

75) 340 cm to meters

76) 7 m to millimeters

77) 1.5 km to meters

78) 86 cm to meters

79) 20 km to centimeters

80) 4310 mm to kilometers

81) The Eiffel Tower in Paris is approximately 321 m tall. Change this number to centimeters. (www.discoverfrance.net)

©Sylvain Sonnet/Getty Images

82) A calculator is 7.2 cm wide. Change this number to millimeters.

R Rethink

R1) What other common items are measured using a form of the meter?

R2) Compare converting the U.S. customary measurements to converting using the metric system. Which is easier to do? Why?

7.3 The Metric System: Capacity and Weight (Mass)

What are your objectives for Section 7.3?	How can you accomplish each objective?
1 Learn the Basic Units of Capacity	• Understand what a *liter* represents, and memorize the relationships between *metric units of capacity*. • Complete the given example on your own. • Complete You Try 1.
2 Convert Between Metric Units of Capacity	• Use the same methods from Section 7.2. • Complete the given example on your own. • Complete You Try 2.
3 Learn the Basic Units of Weight (Mass)	• Understand what a *gram* represents, and memorize the relationships between *metric units of weight*. • Complete the given example on your own. • Complete You Try 3.
4 Convert Between Metric Units of Weight (Mass)	• Use the same methods from Section 7.2. • Complete the given example on your own. • Complete You Try 4.
5 Distinguish Between the Different Metric Units	• Develop a series of questions to help decide which metric unit to use to describe a measurement. • Complete the given example on your own. • Complete You Try 5.

Read the explanations, follow the examples, take notes, and complete the You Trys.

1 Learn the Basic Units of Capacity

In Section 7.1, we learned that the basic units of capacity in the U.S. customary system are cups, pints, quarts, and gallons. We use units of capacity to measure quantities of liquids. For example, at the grocery store, we might buy a *quart* of milk. At the gas station, we put *gallons* of gasoline in our cars. But what are the units of capacity in the metric system?

The basic unit of capacity in the metric system is the **liter,** sometimes spelled *litre.* Liter can be abbreviated as either l or L. We will use capital L so that we do not confuse the abbreviation with the number 1. The liter is related to the metric length.

A box measuring 10 cm on each side has a volume of 10 cm · 10 cm · 10 cm = 1000 cubic centimeters, and **1000 cubic cm = 1 liter.** **This box holds exactly 1 L.**

A liter is a little more than a quart.

10 cm

10 cm

10 cm

Soda is sold in 2-L bottles, and the amount of water in a swimming pool can be measured in liters.

2-L bottle of soda

This pool contains 2000 L of water.

We get larger and smaller units of capacity in the metric system by using the same prefixes on *liter* that we used for *meter* to describe length. For example, 1 kilometer (km) = 1000 meters (m). Likewise, 1 kiloliter (kL) = 1000 liters (L). *Kilo* means 1000 no matter what measurement unit we are using.

Metric Units of Capacity

1 kiloliter (kL) = 1000 L
1 hectoliter (hL) = 100 L
1 dekaliter (daL) = 10 L
1 liter (L) = 1 L
1 deciliter (dL) = $\frac{1}{10}$ L or 0.1 L
1 centiliter (cL) = $\frac{1}{100}$ L or 0.01 L
1 milliliter (mL) = $\frac{1}{1000}$ L or 0.001 L

Note
The units used most often are liter and milliliter.

Here are some examples that will give you an idea of the size of a milliliter.

A box measuring 1 cm on each side has a volume of
1 cm · 1 cm · 1 cm = 1 cubic centimeter, and
1 cubic cm = 1 milliliter. **This box holds exactly 1 mL.**

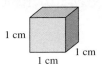

A teaspoon is about 5 mL.

A can of soda contains about 355 mL.

Let's practice deciding which metric unit is appropriate for measuring a particular capacity.

EXAMPLE 1

Fill in the blank with the appropriate metric unit. Use mL or L.

a) A dishwasher uses about 34 _____ of water.

b) A can contains 480 _____ of soup.

Solution

a) A dishwasher uses about 34 L of water. (This is about 9 gal.) We do not use mL because that would be a *very* small quantity.

b) A can contains 480 mL of soup (480 mL ≈ 2 c). Liters would not make sense because 480 L ≈ 127 gal.

[YOU TRY 1]

Fill in the blank with the appropriate metric unit. Use mL or L.

a) The juice box holds 180 _____.

b) The car's tank holds 53 _____ of gas.

---

2 Convert Between Metric Units of Capacity

We learned two methods for converting between metric units of length: unit fractions and the metric conversion line. We can use those same methods to convert between metric units of capacity. Here is the metric conversion chart.

Metric Conversion Chart

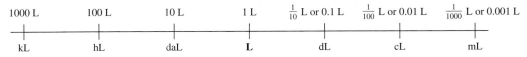

1000 L	100 L	10 L	1 L	$\frac{1}{10}$ L or 0.1 L	$\frac{1}{100}$ L or 0.01 L	$\frac{1}{1000}$ L or 0.001 L
kL	hL	daL	L	dL	cL	mL

Because we use milliliters often, you should remember that 1000 mL = 1 L.

EXAMPLE 2

Change to the indicated unit using unit fractions and using the metric conversion chart.

a)   1.4 L to milliliters   b)   250 mL to liters

Solution

a)   **Unit fractions:** We are given *liters* and want to convert to *milliliters*. Use the relationship **1000 mL = 1 L.**

$$1.4 \text{ L} \cdot \frac{\textbf{1000 mL}}{\textbf{1 L}} = \frac{1.4 \cancel{\text{ L}}}{1} \cdot \frac{1000 \text{ mL}}{1 \cancel{\text{ L}}} \qquad \text{Divide out the unit of liters.}$$

$$= \frac{1400 \text{ mL}}{1} \qquad \text{Multiply.}$$

$$= 1400 \text{ mL} \qquad \text{Simplify.}$$

So, 1.4 L = 1400 mL.

Metric Conversion Chart:

Start at L on the metric conversion chart. To change from L to mL, we move *three places to the right*. Therefore, move the decimal point *three places to the right*.

$$1.4 \text{ L} = 1400. \text{ ml} \qquad \text{Put in zeros as placeholders.}$$

Move the decimal point three places to the right.

Therefore, 1.4 L = 1400 mL.

b)   **Unit fractions:** We are given *milliliters* and want to convert to *liters*. Use the relationship **1000 mL = 1 L.**

$$250 \text{ mL} \cdot \frac{\textbf{1 L}}{\textbf{1000 mL}} = \frac{250 \cancel{\text{ mL}}}{1} \cdot \frac{1 \text{ L}}{1000 \cancel{\text{ mL}}} \qquad \text{Divide out the unit of milliliters.}$$

$$= \frac{250 \text{ L}}{1000} \qquad \text{Multiply.}$$

$$= 0.250 \text{ L} \qquad \text{Simplify.}$$

So, 250 mL = 0.250 L or 0.25 L.

Metric Conversion Chart:

Start at mL on the metric conversion chart. Move *three places to the left* to get from mL to L. Move the decimal point *three places to the left*.

$$250. \text{ mL} = .250 \text{ L}$$

Put the decimal point at   Move the decimal point
the end of the number.   three places to the left.

250 mL = 0.250 L or 0.25 L

Ⓦ Hint

You can use the same procedures that you have already learned in Section 7.2!

[YOU TRY 2]

Change to the indicated unit using unit fractions and using the metric conversion chart.

a)   120 mL to liters   b)   3.7 L to milliliters

3 Learn the Basic Units of Weight (Mass)

In the metric system, the basic unit of *mass* is the **gram.** Mass is commonly referred to as *weight,* but they are two different things. The **mass** of an object is the amount of matter in the object. The **weight** of an object is a measure of the force of gravity on an object. The farther we get from the center of the Earth, the less the pull of gravity. Therefore, an object on the surface of the Earth weighs more than if the object is far from the Earth. An astronaut, for example, is weightless in space but his or her mass is the same whether on Earth or in space.

In science classes, especially physics, it is important to understand the difference between mass and weight. If an object is on the Earth, the weight and mass of that object are the same. Therefore, we will use the word *weight.*

The abbreviation for gram is g. The gram is related to the metric length.

If each side of a box has a length of 1 cm, and we fill this tiny box with water, the weight of the water in the box is 1 gram.

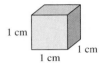

The weight of the water in this box is exactly 1 gram.

Here are some examples that will help you understand the measure of grams.

A large paper clip weighs about 1 g.

A nickel weighs 5 g.

The iPhone 7 weighs 138 g. (www.apple.com)

To describe larger and smaller units of weight (mass), use the same prefixes that are used for length and capacity. For example, 1 kilometer (km) = 1000 meters (m), 1 kiloliter (kL) = 1000 liters (L), and 1 kilogram (kg) = 1000 grams (g). *Kilo* means 1000. Here are other units of weight.

Metric Units of Weight

1 kilogram (kg) = 1000 g
1 hectogram (hg) = 100 g
1 dekagram (dag) = 10 g
1 gram (g) = 1 g
1 decigram (dg) = $\frac{1}{10}$ g or 0.1 g
1 centigram (cg) = $\frac{1}{100}$ g or 0.01 g
1 milligram (mg) = $\frac{1}{1000}$ g or 0.001 g

Note
The units used most often are kilogram, gram, and milligram.

In the metric system, kilograms are used instead of pounds. 1 kg ≈ 2.2 lb

Tom bought 1 kg of potatoes.

This newborn baby weighs 3.2 kg.

This man weighs 84 kg.

1 kg ≈ 2.2 lb

3.2 kg ≈ 7 lb

84 kg ≈ 185 lb

Milligrams are used to measure very small weights. For example, a certain multivitamin contains 60 mg of vitamin C.

Vitamin

Which metric unit of weight (mass) would we use in each case in Example 3?

EXAMPLE 3

Fill in the blank with the appropriate metric unit. Use milligrams, grams, or kilograms.

a) A full-size pickup truck weighs about 4600 _____.

b) Tim bought an apple that weighed 170 _____.

c) A frozen diet dinner contains 650 _____ of sodium.

Solution

a) A full-size pickup truck weighs about 4600 kg. (This is about 10,141 lb.) The metric system uses kilograms instead of pounds.

b) Tim bought an apple that weighed 170 g (170 g ≈ 6 oz). The unit of kg would be too large for an apple, and mg would be too small.

c) A frozen diet dinner contains 650 mg of sodium (650 mg ≈ 0.02 oz). Milligrams are used to measure *very* small amounts.

[YOU TRY 3] Fill in the blank with the appropriate metric unit. Use milligrams, grams, or kilograms.

a) One packet of sugar substitute contains 1 _____.

b) One serving of canned spaghetti contains 450 _____ of potassium.

c) Melinda's laptop weighs 1.8 _____.

4 Convert Between Metric Units of Weight (Mass)

We can use unit fractions or the metric conversion chart to convert from one unit of weight (mass) to another.

Metric Conversion Chart

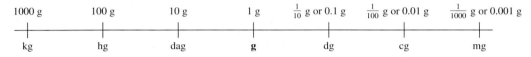

1000 g	100 g	10 g	1 g	$\frac{1}{10}$ g or 0.1 g	$\frac{1}{100}$ g or 0.01 g	$\frac{1}{1000}$ g or 0.001 g
kg	hg	dag	**g**	dg	cg	mg

Since we use grams, kilograms, and milligrams most often, remember the following relationships:

$$1 \text{ kg} = 1000 \text{ g} \qquad 1000 \text{ mg} = 1 \text{ g}$$

EXAMPLE 4

Change to the indicated unit using unit fractions and using the metric conversion chart.

a) Charise weighs 54 kg. Change kilograms to grams.

b) The chemist mixed 175 mg of sodium chloride with water. Change milligrams to grams.

Solution

a) **Unit fractions:** To change *kilograms* to *grams,* use the relationship **1 kg = 1000 g.**

$$54 \text{ kg} \cdot \frac{\mathbf{1000 \text{ g}}}{\mathbf{1 \text{ kg}}} = \frac{54 \cancel{\text{ kg}}}{1} \cdot \frac{1000 \text{ g}}{1 \cancel{\text{ kg}}} \qquad \text{Divide out the unit of kilograms.}$$

$$= \frac{54{,}000 \text{ g}}{1} \qquad \text{Multiply.}$$

$$= 54{,}000 \text{ g} \qquad \text{Simplify.}$$

So, 54 kg = 54,000 g. (This is about 119 lb.)

Metric Conversion Chart:

Start at kg on the metric conversion chart and move *three places to the right* to get to g. Now, move the decimal point *three places to the right.*

$$54. \text{ kg} = 54000. \text{ g}$$

Put the decimal point at the end of the number.   Move the decimal point three places to the right.

So, 54 kg = 54,000 g.

b) **Unit fractions:** To change 175 mg to grams, use the relationship **1000 mg = 1 g.**

$$175 \text{ mg} \cdot \frac{1 \text{ g}}{1000 \text{ mg}} = \frac{175 \text{ mg}}{1} \cdot \frac{1 \text{ g}}{1000 \text{ mg}} \qquad \text{Divide out the unit of milligrams.}$$

$$= \frac{175 \text{ g}}{1000} \qquad \text{Multiply.}$$

$$= 0.175 \text{ g} \qquad \text{Simplify.}$$

So, 175 mg = 0.175 g. (This is approximately 0.006 oz.)

Metric Conversion Chart:

Start at mg on the metric conversion chart and move *three places to the left* to get to g. Move the decimal point *three places to the left.*

$$175. \text{ mg} = .175 \text{ g}$$

Put the decimal point at the end of the number.   Move the decimal point three places to the left.

Therefore, 175 mg = 0.175 g.

[YOU TRY 4]

Change to the indicated unit using unit fractions and using the metric conversion chart.

a)   The tablet contains 200 mg of ibuprofen. Change milligrams to grams.

b)   The refrigerator weighs 91 kg. Change kilograms to grams.

---

5 Distinguish Between the Different Metric Units

Now that we have learned the metric units for length, capacity, and weight (mass), let's learn how to determine which metric unit should be used to measure certain quantities. Remember that the basic unit of measurement for length is *meter,* the basic unit of measurement for capacity is *liter,* and the basic unit of measurement for weight (mass) is *gram.*

EXAMPLE 5

Fill in the blank with the appropriate metric unit. Use m, mm, cm, km, L, mL, g, mg, or kg.

a)   Trent bought a 1.5 -_____ carton of orange juice.

b)   The CD case is 14 _____ wide.

c)   The dog weighs 21 _____.

Solution

a)   Trent bought a 1.5-<u>L</u> carton of orange juice. (This is about 0.4 gal.) To measure an amount of a liquid, we use a unit of capacity. Use liters instead of milliliters because milliliters would be too small: 1.5 mL ≈ 0.0004 gal.

b)   The CD case is 14 <u>cm</u> wide. (14 cm ≈ 5.5 in.) Use centimeters because it is the appropriate unit of length. 14 km is too large (14 km ≈ 8.7 mi) and 14 mm is too small (14 mm ≈ 0.0055 in.).

c) The dog weighs 21 kg. (21 kg ≈ 46 lb) The word *weighs* tells us to use a measurement for weight. Choose kilograms because both grams and milligrams are too small: 21 g ≈ 0.05 lb and 21 mg ≈ 0.00005 lb.

[YOU TRY 5]

Fill in the blank with the appropriate metric unit. Use m, mm, cm, km, L, mL, g, mg, or kg.

a) The jar holds 35 _____ of basil leaves.

b) The bottle contains 240 _____ of cough syrup.

c) The distance from Minneapolis to Orlando is about 2526 _____.

ANSWERS TO [YOU TRY] EXERCISES

1) a) mL (180 mL ≈ 0.75 b) L (53 L ≈ 14 gal)   2) a) 0.12 L (0.12 L ≈ 0.5 c)
b) 3700 mL (3700 mL ≈ 0.98 gal)   3) a) g (1 g ≈ 0.035 oz) b) mg (450 mg ≈ 0.016 oz)
c) kg (1.8 kg ≈ 4 lb)   4) a) 0.2 g b) 91,000 g   5) a) mg b) mL c) km

---

E Evaluate   7.3   Exercises   Do the exercises, and check your work.

Objective 1: Learn the Basic Units of Capacity
Fill in the blank with the appropriate metric unit. Use milliliters or liters.

1) A teaspoon holds approximately 5 _____ of liquid.

2) A tablespoon holds approximately 14 _____ of liquid.

3) A can of soda pop contains 0.355 _____ of soda pop.

4) A small cup of coffee contains 0.236 _____ of coffee.

 5) An eye dropper delivers 0.05 _____ per drop.

6) A raindrop has a volume of 1 _____.

Determine whether the indicated quantity seems reasonable or unreasonable for each application. If it is unreasonable, specify whether it is too much or too little.

7) To create a bubble bath requires adding 10 L of bubble bath soap to a tub of warm water.

8) Houng is going to make 6 servings of Vietnamese chicken noodle soup (Pho Ga). She needs to use $4\frac{1}{2}$ mL of water to make this popular Vietnamese cuisine.

9) An antacid tablet is dissolved in 120 mL of water.

10) A patient was given 15 mL of cough syrup every four hours.

©Pixtal/Age Fotostock

11) When Bret washes his car, he uses approximately 450 mL of water.

12) To help tired eyes, one should use 100 mL of saline solution eye drops twice a day.

Objective 2: Convert Between Metric Units of Capacity
Change to the indicated unit using unit fractions and using the metric conversion chart.

13) 325 mL to liters   14) 206 mL to liters

15) 4000 mL to liters   16) 5000 mL to liters

17) 7.1 L to milliliters   18) 3.4 L to milliliters

 19) 0.06 L to milliliters   20) 0.01 L to milliliters

21) 35 mL to liters   22) 85 mL to liters

Objective 3: Learn the Basic Units of Weight (Mass)

Fill in the blank with the appropriate metric unit. Use milligrams, grams, or kilograms.

23) One grain of rice weighs 25 _____.

24) One grain of salt weighs 2 _____.

25) The large bag of wild bird seed weighs 9 _____.

26) David's laptop weighs 3125 _____.

27) The weight of a computer mouse is 90 _____.

28) A quarter weighs 5.67 _____.

29) A dime weighs 2268 _____.

30) A penny weighs 2500 _____.

31) The bag of groceries weighs 3.7 _____.

32) The sign in an elevator says that it can carry at most 600 _____.

Determine whether the indicated quantity seems reasonable or unreasonable for each application. If it is unreasonable, specify whether it is *too much* or *too little*.

33) The average banana contains 465 mg of potassium.

©Ivan Danik/Getty Images

34) An orange contains approximately 70 mg of vitamin C.

35) A recipe for four servings of chicken soup requires 1 kg of salt.

36) To make 6 servings of sweet lemonade requires 200 mg of sugar.

37) A recipe to make 40 chocolate chip cookies requires 200 g of chocolate chips.

38) Sahar is planning to serve pita bread for her dinner guests. To make 16 pitas, she needs 750 g of flour.

Objective 4: Convert Between Metric Units of Weight (Mass)

Change to the indicated unit using unit fractions and using the metric conversion chart.

39) 72 g to milligrams   40) 38 g to milligrams

41) 0.82 kg to grams   42) 0.79 kg to grams

43) 194 g to kilograms   44) 611 g to kilograms

45) 2.01 g to milligrams   46) 5.002 g to milligrams

47) $\frac{3}{4}$ kg to grams   48) $\frac{1}{4}$ kg to grams

Change to the indicated unit using unit fractions and using the metric conversion chart.

49) The patient receives 300 mg of medication every six hours. Change milligrams to grams.

50) The bag of dog food weighs 8 kg. Change kilograms to grams.

51) A large egg weighs approximately 50 g. Change grams to milligrams.

52) A lightweight running shoe weighs 361 g. Change grams to kilograms.

53) Isaac's briefcase weighs 4.1 kg. Change kilograms to grams.

54) A recipe for garam marsala uses 15 g of ginger powder. Change grams to milligrams.

55) Juanita worked out with a kettlebell that weighed 3000 g. Change grams to kilograms.

56) The kitten weighed 97,000 mg at birth. Change milligrams to grams.

Objective 5: Distinguish Between the Different Metric Units

Fill in the blank with the appropriate metric unit. Use m, mm, cm, km, L, mL, g, mg, or kg.

57) The height of a basketball hoop is approximately 3 _____.

58) The car needed 1 _____ of oil.

59) A hot dog weighs approximately 60 _____.

60) A hamburger contains approximately 100 _____ of cholesterol.

61) Some mechanical pencils use lead that has a diameter of 0.7 _____.

62) A handheld eraser has a length of approximately 6 _____.

63) One gallon is approximately 3.785 _____.

64) One tablespoon of liquid is approximately 14 _____.

65) Joseph drives his car to work a distance of 23 ____ every day.

66) The track around the football field measures 400 ____.

©Aaron Roeth Photography

Mixed Exercises

Determine whether the indicated quantity seems reasonable or unreasonable for each application. If it is unreasonable, specify whether it is *too much* or *too little*.

67) The diameter of a regulation table tennis ball is approximately 44 mm.

68) Martin takes 200 mg of vitamin C on a daily basis.

69) Agneza fills her car's gas tank with 16 mL of gasoline.

70) The driveway is 3.7 km wide.

Change to the indicated unit.

71) 340 mL to liters

72) 2 km to meters

73) 9.1 g to milligrams

74) 6000 g to kilograms

75) $15\frac{1}{2}$ m to kilometers

76) $\frac{1}{2}$ L to milliliters

77) A communications satellite weighs about 315 kg. Change kilograms to grams.

78) A dining room table is 160 cm long. Change centimeters to meters.

R1) Describe steps you can take to check your answers for the exercises.

R2) Can you explain to a friend the quickest way to convert metric units? What would you tell her?

7.4 Solve Applied Problems Involving Metric Units

What are your objectives for Section 7.4?	How can you accomplish each objective?
1 Perform Operations with Metric Units	• Create a procedure for performing operations on numbers with different metric units. • Complete the given examples on your own. • Complete You Trys 1 and 2.
2 Solve Applied Problems Involving Metric Units	• Use the Five Steps for Solving Applied Problems, and add any additional steps needed. • Complete the given examples on your own. • Complete You Trys 3–5.

W Work   **Read the explanations, follow the examples, take notes, and complete the You Trys.**

1 Perform Operations with Metric Units

To solve some applications, we have to add, subtract, multiply, or divide numbers with metric units. For example, 4.6 kg + 2.1 kg = 6.7 kg. Because the units are the same, we add the numbers and keep the units of kg. However, if these units are not the same, we may have to express them with the same units before performing the operations.

EXAMPLE 1

Add 2 m + 31 cm.

Solution

The units are not the same, so we have to convert one of the units so that it is the same as the other. We can convert to either meters or centimeters using the metric conversion chart or unit fractions. Let's use the relationship between meters and centimeters to move the decimal point.

Express each measurement in *meters*.	Express each measurement in *centimeters*.
31 cm = 0.31 m Change centimeters to meters.	2 m = 200 cm Change meters to centimeters.
Now, add the measurements.	Now, add the measurements.
2 m + 31 cm = 2 m + 0.31 m = 2.31 m	2 m + 31 cm = 200 cm + 31 cm = 231 cm

W Hint

Is it easier for you to understand what 2.31 m represents or what 231 cm represents?

The answer can be expressed either way: 2 m + 31 cm = 2.31 m or 231 cm.

[YOU TRY 1]   Add 4 m + 742 mm.

EXAMPLE 2

Divide 36 L by 4.

Solution

Only one number is a unit of measurement. This problem means *divide 36 L into four equal parts.* We just divide 36 by 4, and keep the units of L.

$$36 \text{ L} \div 4 = 9 \text{ L}$$

[YOU TRY 2]   Multiply 7 g by 8.

2 Solve Applied Problems Involving Metric Units

Let's solve some applied problems involving metric units.

EXAMPLE 3

Latrice has a piece of fabric that is 1 m 20 cm long. She cuts off a piece that is 85 cm long. Find the length of fabric remaining, in centimeters.

Solution

Because Latrice cuts off an 85-cm piece of fabric, we have to *subtract* that from the total length of 1 m 20 cm. The units are not the same, so convert 1 m 20 cm to centimeters since we are asked to express the answer in centimeters.

Change 1 m 20 cm to centimeters.	Subtract 85 cm from 120 cm.
1 m 20 cm = 100 cm + 20 cm = 120 cm	11 10 1 2 0 − 8 5 3 5 120 cm − 85 cm = 35 cm

©Jupiterimages/Getty Images

35 cm of fabric remains.

Check the answer by working backward.

Amount of fabric remaining + Amount of fabric cut off = Original length of fabric

$$35 \text{ cm} \quad + \quad 85 \text{ cm} \quad = \quad 120 \text{ cm}$$

We can think of 120 cm as 100 cm + 20 cm = 1 m + 20 cm. This is the same as the original length of fabric stated in the problem.

YOU TRY 3

A carton of orange juice contains 1 L 450 mL. If Zach uses 600 mL, how many milliliters of orange juice is left in the carton?

When we are working with problems involving money, round answers to the nearest cent.

EXAMPLE 4

At the deli, prosciutto costs $48.99/kg. Giovanni needs 120 g for a chicken saltimbocca recipe. How much will the prosciutto cost?

Solution

The $48.99 *per* kilogram is the *unit cost* of the prosciutto. To find the cost of 120 g of prosciutto, we have to multiply the unit cost by the amount of prosciutto Giovanni is buying. *First*, however, we have to change 120 g to kilograms so that the units are the same.

To change 120 g to kilograms, use either unit fractions or the metric conversion chart. We will use unit fractions. Then, use unit fractions to find the cost.

Change 120 g to kilograms.	Find the cost of the prosciutto.
$\dfrac{120 \text{ g}}{1} \cdot \dfrac{1 \text{ kg}}{1000 \text{ g}} = \dfrac{120 \text{ kg}}{1000} = 0.120 \text{ kg} = 0.12 \text{ kg}$	Multiply the amount of prosciutto by the unit cost. $\dfrac{0.12 \text{ kg}}{1} \cdot \dfrac{\$48.99}{\text{kg}} = \$5.8788 \approx \$5.88$

Giovanni spent $5.88 for 120 g of prosciutto.

Check the answer by working backward. Divide $5.88 by 0.12 kg to get the unit cost. $5.88 ÷ 0.12 kg = $49.00/kg. This is close to the given unit cost of $48.99/kg. The difference is due to the rounding of the cost of the prosciutto.

[YOU TRY 4]

In the bulk foods section of the grocery store, lentils cost $3.49/kg. If Devyani buys 640 g, how much does she pay?

EXAMPLE 5

On a production line, 280 L of shampoo is divided evenly into 800 bottles. How many milliliters of shampoo is in each bottle?

Solution

First, change 280 L to milliliters since the answer should be in milliliters. Then, divide the total amount of shampoo produced by 800.

$$280 \text{ L} = 280,000 \text{ mL}$$

Divide 280,000 mL by 800 to determine the number of milliliters of shampoo in each bottle:

$$800 \overline{)280,000}^{\,350}$$

Each bottle will contain 350 mL of shampoo.

Check the answer by working backward.

If 800 bottles each contain 350 mL of shampoo, then the total amount of shampoo is 800 · 350 mL = 280,000 mL. Changing from milliliters to liters, we get 280 L.

[YOU TRY 5]

Twenty kilograms of granola is packaged evenly into 50 bags. How many grams of granola are in each bag?

ANSWERS TO [YOU TRY] **EXERCISES**

1) 4.742 m or 4742 mm   2) 56 g   3) 850 mL   4) $2.23   5) 400 g

E **Evaluate**   **7.4**   Exercises   Do the exercises, and check your work.

Objective 1: Perform Operations with Metric Units
Perform the indicated operation.

1) 3 m + 55 cm

2) 9 m + 23 cm

3) 2.4 kg + 160 g

4) 1.3 kg + 420 g

5) 0.8 L + 619 mL

6) 0.7 L + 525 mL

7) 1.75 m + 32 cm

8) 4.64 m + 72 cm

9) 0.27 m + 57 cm

10) 0.38 m + 49 cm

11) 2.25 kg − 455 g

12) 0.68 kg − 325 g

13) 0.75 L − 75 mL

14) 0.13 L − 13 mL

15) 2.5 g + 300 mg

16) 3.5 cm + 50 mm

17) Divide 24 L by 8.

18) Divide 30 g by 5.

19) Divide 750 mL by 3.

20) Divide 900 cm by 6.

21) Divide 4.8 m by 12.

22) Divide 9.6 m by 16.

23) Multiply 14 cm by 8.

24) Multiply 32 cm by 7.

25) Multiply 1.77 g by 9.

26) Multiply 5.18 g by 4.

27) Divide 9 kg by 2.

28) Divide 14 km by 4.

29) Divide 0.612 kg by 18.

30) Divide 0.552 kg by 24.

31) Multiply 0.05 mL by 60.

32) Multiply 0.05 mL by 80.

33) Multiply 355 cm by 6.

34) Multiply 720 mm by 4.

35) Multiply 0.048 km by 5.

36) Multiply 0.025 kg by 6.

Objective 2: Solve Applied Problems Involving Metric Units

Solve each problem.

37) Tanhoa measured a length of wood and found it to be 1 m 4 cm long. If he cuts off 32 cm, what is the length of the remaining board, in centimeters?

38) If Corina cuts off 70 cm from a piece of fabric measuring 1 m 50 cm in length, how long is the remaining piece, in centimeters?

39) White cheddar cheese costs $34.75/kg. Angelique needs 350 g for an au gratin potato recipe. Find the cost of the cheese.

40) Luke bought 600 g of chocolates for his girlfriend for Valentine's Day. If the chocolates cost $39.99/kg, how much did he pay?

©duckycards/Getty Images

41) A home soda maker can make a total of 60 L of soda. If Yuka plans to make 200 bottles containing the same amount of soda, how many milliliters will be in each bottle?

42) Ninety kilograms of rice is divided evenly into 120 bags. How many grams of rice are in each bag?

43) If the weight of toothpaste applied to a toothbrush is 430 mg, approximately how many brushes can you get from a 170-g tube of toothpaste?

44) A can of soda contains approximately 355 mL. The volume of four 2-L bottles of soda has the same volume as how many cans? Round the answer to the nearest tenth.

45) If the average rate of human hair growth is 12 mm per month, approximately how many years will it take to grow hair that is 1 meter in length? Round the answer to the nearest year.

©arekmalang/123RF

(http://emedicine.medscape.com/article/837994-overview)

46) How many drops of solution are in a 0.5-L container if each drop is 0.05 mL?

47) A patient is prescribed a total of 560 mg of medication to be taken twice a day over a 4-day period. How much should each dosage be?

48) A patient is prescribed a total of 315 mg of medication to be taken three times a day over a 1-week period. How much should each dosage be?

49) An eye drop bottle contains 15 mL of saline solution. If you put one drop of solution in each eye every day, how many days will it take to empty the bottle? Assume each drop is 0.05 mL.

50) Abigail's new lipstick container weighed 18.5 g (including the cap). After putting on lipstick 30 times, the lipstick container weighed 18.2 g (including the cap). What is the weight of one application, in mg?

51) A drug prescription of 45 pills costs $427.95. If each pill contains 3 mg of medication, what is the cost per milligram?

©j4m3z/Getty Images

52) A drug prescription of 60 pills costs $326.40. If each pill contains 4 mg of medication, what is the cost per milligram?

Mixed Exercises: Objectives 1 and 2

Perform the indicated operation.

53) Multiply 118 mL by 15.

54) 1.85 L + 216 mL

55) 1.65 kg − 980 g

56) Divide 6.3 m by 7.

57) 0.5 cm + 50 mm

58) 0.35 kg − 35 g

59) Divide 1.2 m by 4.

60) 4.7 cm + 47 mm

Solve each problem.

61) Charlotte's lip gloss costs $5.10 for 10.2 mL. If each application uses 0.06 mL, what is the cost per application?

62) A 25.5-mL bottle of Jerry's favorite cologne costs $27.20. Each stroke of the spray applicator delivers 0.15 mL of cologne. If he paid $27.20 for the bottle, what is the cost of each spray?

©Ingram Publishing

Use the table for Exercises 63–68.

12-oz Beverage or as Noted	Caffeine Amount (milligrams)
Red Bull (8.2 oz)	80.0
Jolt	71.2
Pepsi One	38.0
Mountain Dew	54.0
Dr. Pepper	42.0

(National Soft Drink Association)

63) Which has more caffeine, a four-pack of Red Bull or a six-pack of Pepsi?

64) Which has more caffeine, two Jolts or four Dr. Peppers?

65) If Rene drinks two cans of Red Bull per day, how many grams of caffeine does he receive each week?

66) If Maurice drinks two cans of Mountain Dew per day, how many grams of caffeine does he receive each week?

67) How many grams of caffeine are in a six-pack of Dr. Pepper cans?

68) How many grams of caffeine are in a four-pack of 8.2-oz Red Bull cans?

R Rethink

R1) Think back on the main objectives involving metric units. Write a paragraph describing what makes the metric system easy to use.

R2) Which objectives do you still need to master?

7.5 Metric–U.S. Customary Conversions and Temperature

P Prepare

O Organize

What are your objectives for Section 7.5?	How can you accomplish each objective?
1 Convert Between U.S. Customary and Metric Units	• Understand the **Relationships Between U.S. Customary Units and Metric Units.** • Use unit fractions to convert between units. • Complete the given examples on your own. • Complete You Trys 1 and 2.
2 Understand the Relationship Between the Fahrenheit and Celsius Scales	• Learn the *freezing and boiling points of water.* • Use your basic knowledge relating Celsius and Fahrenheit to answer questions relating to temperature. • Complete the given example on your own. • Complete You Try 3.
3 Convert Between Fahrenheit and Celsius Units	• Learn the **Procedure for Converting Between Fahrenheit and Celsius Temperatures.** • Complete the given examples on your own. • Complete You Trys 4 and 5.

W Work   Read the explanations, follow the examples, take notes, and complete the You Trys.

1 Convert Between U.S. Customary and Metric Units

Even though we use the U.S. customary system of measurement, we still come across metric units in the United States. In sports, for example, a track-and-field athlete might compete in a 400-m race. How long is this in yards? A pharmacist might mix 50 mg of antibiotic in a solution for a patient in a hospital. How many ounces is this?

We can use the following relationships to convert between U.S. customary units and metric units.

 Hint

This section contains many relationships that may be new to you. Try using flash cards to learn them.

Relationships Between U.S. Customary Units and Metric Units

Length	Capacity	Weight (Mass)
1 inch = 2.54 centimeters	1 cup ≈ 236.59 milliliters	1 ounce ≈ 28.35 grams
1 foot ≈ 0.30 meter	1 quart ≈ 0.95 liter	1 pound ≈ 0.45 kilogram
1 yard ≈ 0.91 meter	1 gallon ≈ 3.79 liters	
1 mile ≈ 1.61 kilometers		

Note

All of the relationships are approximations *except* for the relationship between inches and centimeters. Since 1 inch *exactly equals* 2.54 centimeters, we can use the equal sign: = . For all other conversions, we use the symbol for *is approximately equal to:* ≈ .

We will use unit fractions to convert between the two measurement systems. This method is one that is used often in nursing and science courses as well.

EXAMPLE 1

As of June 2017, Wayde Van Niekirk of South Africa was the world record holder in the 400-m race. Convert 400 m to yards. Round the answer to the nearest tenth. (www.iaaf.org)

Solution

First, identify the units we are given and the units we want to get. Since we are given *meters* and we want to convert to *yards,* write down the relationship between meters and yards: **1 yd ≈ 0.91 m**.

Next, multiply the measurement we are given, 400 m, by the unit fraction relating yards and meters so that the meters divide out and we will be left with yards.

©Chris Ryan/age fotostock

$$400 \text{ m} \cdot \frac{\textbf{1 yd}}{\textbf{0.91 m}} = \frac{400 \text{ m}}{1} \cdot \frac{1 \text{ yd}}{0.91 \text{ m}} \qquad \text{Divide out the unit of meters.}$$

$$= \frac{400 \text{ yd}}{0.91} \qquad \text{Multiply.}$$

$$\approx 439.6 \text{ yd} \qquad \text{Simplify.}$$

400 m ≈ 439.6 yd. Therefore, a 400-meter race is approximately 439.6 yd.

[YOU TRY 1] A marathon is, officially, 42.195 km. Convert this number to miles. Round the answer to the nearest tenth.

BE CAREFUL Look at Example 1. If we had tried to multiply 400 m by the unit fraction relating meters and yards as $\dfrac{400 \text{ m}}{1} \cdot \dfrac{0.91 \text{ m}}{1 \text{ yd}}$, the units of meters would not divide out. That's how we know this is the wrong way to do this problem! If this happens the first time you try to convert between units, don't panic. Write the unit fraction the other way to see whether the correct units will divide out.

Let's convert between other units.

EXAMPLE 2

Convert to the indicated unit. Round the answer to the nearest tenth, if necessary.

a) A recipe calls for 2 kg of beef stew meat. Convert this number to pounds.

b) Huong wants to buy a 50-in. plasma television. Convert this number to centimeters.

Solution

a) We are given *kilograms* and want to convert to *pounds*. The relationship is

$$1 \text{ lb} \approx 0.45 \text{ kg}$$

Multiply 2 kg by the unit fraction relating kilograms and pounds so that kilograms will divide out and leave us with pounds.

$$2 \text{ kg} \cdot \frac{1 \text{ lb}}{0.45 \text{ kg}} = \frac{2 \cancel{\text{ kg}}}{1} \cdot \frac{1 \text{ lb}}{0.45 \cancel{\text{ kg}}} \qquad \text{Divide out the unit of kilograms.}$$

$$= \frac{2 \text{ lb}}{0.45} \qquad \text{Multiply.}$$

$$\approx 4.4 \text{ lb} \qquad \text{Divide.}$$

Therefore, 2 kg of beef $\approx$ 4.4 lb.

 Hint
Write out the example as you are reading it.

b) We are given *inches* and want to convert to *centimeters*. The relationship is

$$1 \text{ in.} = 2.54 \text{ cm}$$

Because this is an *exact* conversion, we use = instead of $\approx$.

Multiply 50 in. by the unit fraction relating inches and centimeters so that inches will divide out and leave us with centimeters.

$$50 \text{ in.} \cdot \frac{2.54 \text{ cm}}{1 \text{ in.}} = \frac{50 \cancel{\text{ in.}}}{1} \cdot \frac{2.54 \text{ cm}}{1 \cancel{\text{ in.}}} \qquad \text{Divide out the unit of inches.}$$

$$= \frac{127 \text{ cm}}{1} \qquad \text{Multiply.}$$

$$= 127 \text{ cm} \qquad \text{Simplify.}$$

Therefore, 50 in. = 127 cm.

[YOU TRY 2]

Convert to the indicated unit. Round the answer to the nearest tenth, if necessary.

a) An art teacher mixes 3.5 c of water with some flour to make plaster of Paris. Convert this number to milliliters.

b) Anastasia is 152.4 cm tall. Convert this number to inches.

c) A box contains 9 oz of fruit snacks. Convert this number to grams.

---

www.mhhe.com/messersmith

2 Understand the Relationship Between the Fahrenheit and Celsius Scales

In the United States, we use the Fahrenheit scale to measure temperature. In most of the rest of the world, the Celsius scale is used. (Science courses, even in the United States, usually use the Celsius scale.) To get a feel for how these two measures of temperature are related, let's look at some widely known Fahrenheit temperatures and give their Celsius equivalents.

Relationships Between Fahrenheit and Celsius

Freezing Point of Water		Boiling Point of Water	
In Degrees Fahrenheit	In Degrees Celsius	In Degrees Fahrenheit	In Degrees Celsius
32°F	0°C	212°F	100°C

Note
The symbol ° represents degrees. We read 32°F as "32 degrees Fahrenheit," and we read 100°C as "100 degrees Celsius."

Here are some other temperatures that might help you understand how the Fahrenheit and Celsius scales are related.

The average high temperature in Miami in July is 91°F. This is approximately 33°C. (www.weather.com)

©Glow Images

The temperature inside a refrigerator should be at most 40°F. This is approximately 4°C. (www.usda.gov)

Source: U.S. Department of Agriculture

This thermometer compares Fahrenheit and Celsius temperatures.

EXAMPLE 3   Determine which Celsius temperature is appropriate for the situation.

a) The average low temperature in Fairbanks, Alaska, in February is _____. Choose from −27°C, 10°C, and 27°C. (www.weather.com)

b) A comfortable household temperature is _____. Choose from 21°C, 50°C, and 68°C.

Solution

a) The average low temperature in Fairbanks, Alaska, in February is $-27°C$. This is the only reasonable answer because $-27°C \approx -17°F$. The other two choices do not make sense because $10°C = 50°F$ and $27°C \approx 81°F$.

b) A comfortable household temperature is $21°C$. This is approximately 70°F. The other temperatures are too hot: $50°C = 122°F$ and $68°C \approx 154°F$.

[YOU TRY 3] Determine which Celsius temperature is appropriate for the situation.

a) The oven temperature to bake a cake might be _____. Choose from 30°C, 180°C, and 350°C.

b) The average low temperature in Denver, Colorado, in January is about _____. Choose from −9°C, 15°C, and 22°C. (www.weather.com)

3 Convert Between Fahrenheit and Celsius Units

To convert between Fahrenheit temperature and Celsius temperature, we can use these formulas.

Procedure How to Convert Between Fahrenheit and Celsius Temperatures

Changing from °F to °C	Changing from °C to °F
$C = \dfrac{5(F - 32)}{9}$	$F = \dfrac{9C}{5} + 32$
To change a Fahrenheit temperature (F) to Celsius (C):	To change a Celsius temperature (C) to Fahrenheit (F):
1) **Substitute** the Fahrenheit temperature for F in the formula.	1) **Substitute** the Celsius temperature for C in the formula.
2) **Evaluate** the expression using the order of operations.	2) **Evaluate** the expression using the order of operations.
3) **Round** the answer to the nearest degree, if necessary.	3) **Round** the answer to the nearest degree, if necessary.

EXAMPLE 4   Convert 77°F to Celsius.

Solution

To change from degrees Fahrenheit to Celsius, use $C = \dfrac{5(F - 32)}{9}$. Substitute 77 for F, and simplify.

$$C = \frac{5(F - 32)}{9} = \frac{5(77 - 32)}{9} = \frac{5(\overset{5}{45})}{\underset{1}{9}} = 25$$

Therefore, $77°F = 25°C$.

[**YOU TRY 4**] Convert 41°F to Celsius.

EXAMPLE 5   Convert 7°C to Fahrenheit.

Solution

To change from degrees Celsius to Fahrenheit, use $F = \dfrac{9C}{5} + 32$. Substitute 7 for C, and simplify.

$$F = \frac{9C}{5} + 32 = \frac{9(7)}{5} + 32 = \frac{63}{5} + 32 = 12.6 + 32 = 44.6$$

Round the answer to the nearest degree: $44.6°F \approx 45°F$. Therefore, $7°C \approx 45°F$.

[**YOU TRY 5**] Convert 17°C to Fahrenheit.

ANSWERS TO [**YOU TRY**] **EXERCISES**

1) 26.2 mi   2) a) 3.5 c $\approx$ 828.1 mL   b) 152.4 cm = 60 in.   c) 9 oz $\approx$ 255.2 g
3) a) 180°C (180°C = 356°F) b) −9°C(−9°C $\approx$ 16°F)   4) 5°C   5) 17°C $\approx$ 63°F

Using Technology

Fahrenheit to Celsius

When converting 91°F to Celsius, we use the formula $C = \dfrac{5(F - 32)}{9}$ and replace F with 91. This gives us the equation $C = \dfrac{5(91 - 32)}{9}$. Using a calculator to find the value of the right-hand side of the equation, we must follow the order of operations. In this case, we first simplify the expression in the parentheses, then multiply by 5, and then divide by 9. Enter $(\boxed{9}\,\boxed{1}\,\boxed{-}\,\boxed{3}\,\boxed{2})\,\boxed{\times}\,\boxed{5}\,\boxed{\div}\,\boxed{9}\,\boxed{=}$ into the calculator.

The display will likely show 32.77777777. Rounding to the nearest tenth, we get 32.8. This means $91°F \approx 32.8°C$.

Celsius to Fahrenheit

When converting 18°C to Fahrenheit, we use the formula $F = \dfrac{9C}{5} + 32$ and replace C with 18. This gives us the equation $F = \dfrac{9(18)}{5} + 32$. Using a calculator to find the value of the right-hand side of the equation, we must follow the order of operations. In this case, we first multiply 9 by 18, then divide by 5, and then add 32. We now enter $\boxed{9}\,\boxed{\times}\,\boxed{1}\,\boxed{8}\,\boxed{\div}\,\boxed{5}\,\boxed{+}\,\boxed{3}\,\boxed{2}\,\boxed{=}$ into the calculator.

The display will show 64.4. This means $18°C = 64.4°F$.

Objective 1: Convert Between U.S. Customary and Metric Units

Convert to the indicated unit. Round the answer to the nearest tenth, if necessary. Use the table before Example 1.

1) 50 ft to meters

2) 100 yd to meters

3) 200 yd to meters

4) 100 ft to meters

5) 2 c to milliliters

6) 3 gal to liters

7) 4 kg to pounds

8) 150 g to ounces

9) 70 cm to inches

10) 120 km to miles

11) $6\frac{1}{2}$ qt to liters

12) 2.4 gal to liters

13) 40 mi to kilometers

14) 7 in. to centimeters

15) 18 cm to inches

16) 130 km to miles

17) 63 kg to pounds

18) 470 g to ounces

19) 12.6 oz to grams

20) $23\frac{3}{4}$ lb to kilograms

21) 11 L to gallons

22) 7 L to quarts

23) 39 m to yards

24) 15 m to feet

For Exercises 25–28, convert to the indicated metric unit:

  a) using the values in the table before Example 1 to get an approximation of the equivalent metric unit.

  b) using the exact relationships of 1 ft = 12 in., 1 yd = 36 in., 1 in. = 2.54 cm, and 100 cm = 1 m.

25) 10 ft to meters

26) 18 ft to meters

27) 50 yd to meters

28) 1 yd to meters

Solve each problem.

29) The average length of goldfish ranges from 2 to 18 in., depending on the variety. Find the length, in centimeters, of an 18-in. goldfish. (www.petco.com)

©Anthony Bradshaw/Getty Images

30) Marcia measures the length of her walking stride and finds it to be 26 in. How many centimeters is this?

31) A 9-kg bag of cat litter weighs approximately how many pounds?

32) A bag of dry dog food is available in a 15.5-lb bag. Approximately how many kilograms is this?

33) A dishwasher earns an Energy Star rating if it uses less than 5.8 gal of water per cycle. Does a dishwasher that uses 21 L of water per cycle qualify? (www.energystar.gov)

34) Mehdi's luggage weighs 23 kg. If the airline charges an additional $25 to check a piece of luggage that weighs over 50 lb, will he have to pay the extra money for his bag?

35) To conserve water, Mark installs a new, ultralow-flow toilet in his home that uses 6 L of water per flush. The old toilet uses 3.5 gal of water per flush. How much water, in gallons, does the new toilet save per flush? Round the answer to the nearest tenth.

36) Janet's top-loading washing machine uses 28 gal of water to wash a medium-sized load while Trisha's front-loading machine uses 50 L. Which machine uses less water, and by how much, in gallons? Round the answer to the nearest tenth.

37) Jeremy searches a field with a metal detector and finds a pure gold nugget weighing 18.7 g. If the price of gold is $925.00 per ounce, how much is the nugget worth?

38) Ysela's wedding band contains 8.4 g of platinum. If the price of platinum is $1713.00 per ounce, how much is the platinum content worth?

Objective 2: Understand the Relationship Between the Fahrenheit and Celsius Scales

Determine which Celsius temperature is appropriate for the situation.

39) On a snowy day, the temperature outdoors is _____. Choose from −4°C, 8°C, and 30°C.

40) Hot cocoa is best served at _____. Choose from 10°C, 46°C, and 100°C.

©Radius Images/Alamy

41) To bake a casserole, the oven temperature might be _____. Choose from 30°C, 200°C, and 375°C.

42) A hot tub might have a water temperature of _____. Choose from 10°C, 35°C, and 80°C.

43) If the outdoor temperature is 29°C, choose clothing best suited for this temperature. Choose from a T-shirt and shorts, jeans and sweater, heavy coat and gloves.

44) Is a temperature change of 10°C equal to a temperature change of 38°F? Explain your answer.

Objective 3: Convert Between Fahrenheit and Celsius Units

45) Explain how to change a temperature from °F to °C.

46) Explain how to change a temperature from °C to °F.

Perform the following temperature conversions. Round to the nearest degree.

47) 41°F to Celsius   48) 95°F to Celsius

49) 82°F to Celsius   50) 34°F to Celsius

51) 104°F to Celsius   52) 45°F to Celsius

53) 72°F to Celsius   54) 61°F to Celsius

55) 20°C to Fahrenheit   56) 85°C to Fahrenheit

57) 73°C to Fahrenheit   58) 6°C to Fahrenheit

59) 25°C to Fahrenheit   60) 55°C to Fahrenheit

61) 9°C to Fahrenheit   62) 2°C to Fahrenheit

Mixed Exercises: Objectives 1–3

Convert to the indicated unit. Round the answer to the nearest tenth, if necessary.

63) 75 ft to meters   64) 12 gal to liters

65) $14\frac{1}{5}$ oz to grams   66) 6.5 in. to centimeters

67) 9 L to quarts   68) 58 kg to pounds

69) 500 mL to cups   70) 200 km to miles

71) Silvia rode her bike 15 mi on the boardwalk at the beach. How many kilometers did she ride? Round to the nearest tenth.

72) The length of Gabriella's walking stride is 26 in. How many centimeters is this?

©Lars A. Niki

73) A cool air-conditioned room might have a temperature of _____. Choose from 20°C, 40°C, and 66°C.

74) A person with a high fever might have a temperature of _____. Choose from 39°C, 99°C, and 102°C.

75) Convert 52°F to Celsius. Round to the nearest degree.

76) Convert 26°C to Fahrenheit. Round to the nearest degree.

77) The highest temperature ever recorded on Earth was 56.7°C in Death Valley, California, in 1913. What is this temperature in Fahrenheit? Round to the nearest degree. (www.wmo.int)

©Frank Lukasseck/Getty Images

78) The highest temperature ever recorded on the continent of Antarctica was 63.5°F in 2015. What is this temperature in Celsius? Round to the nearest degree. (www.wmo.int)

R Rethink

R1) If you had to use one relationship between U.S. customary units and metric units for each of length, capacity, and weight, which would you pick and why?

R2) If you memorized only one of the formulas for converting between Fahrenheit and Celsius, how could you derive the other formula?

Group Activity – Measurement and Conversion

Activity 1: Measurement

A. Match each object listed on the left with the most appropriate measurement from the list on the right. [Hint: Think about whether the object is measuring a length, capacity, or weight (mass).]

1) The thickness of a dime   a. 1827.35 km

2) The weight of a peanut   b. 136 g

3) The amount of water in a swimming pool   c. 6.4 cm

4) The amount of water in a raindrop   d. 0.98 m

5) The length of your little finger   e. 30,000 mg

6) The width of a twin bed   f. 0.5 mL

7) The weight of an apple   g. 2 L

8) The distance between Los Angeles and Seattle   h. 68.22 kL

9) The weight of a compact car   i. 1700 kg

10) The amount of soda in a soda bottle   j. 1.35 mm

B. Use the measurements in A. to answer the following questions.

1) Convert the measurement for the weight of a compact car to grams. What do you notice about the numerical part of your answer?

2) Convert the measurement for the thickness of a dime to centimeters. What do you notice about the numerical part of your answer?

3) Convert the measurement for the amount of water in a raindrop to deciliters. What do you notice about the numerical part of your answer?

4) Convert the measurement for the width of a twin bed to centimeters. What do you notice about the numerical part of your answer?

5) Convert the measurement for the amount of water in a swimming pool to liters. What do you notice about the numerical part of your answer?

6) Describe any patterns that you notice after answering the five questions above.

Activity 2: Conversions

- Work in groups of two or three.
- Name something that could be measured by each metric measurement listed below (make sure you state whether it is a length, capacity, or weight).
- Convert each metric measurement to an appropriate U.S. customary measurement.
- Determine whether your original answer was reasonable.
- If your first answer was not reasonable, name something that is more appropriate.
- The first one has been done for you.

1) 6.75 kg: The weight of a computer monitor

2) 90.96 L

3) 28.65 m

4) 350 g

5) 2.8 kg

Every school—whether it's a high school, community college, college, or university—
is different and has its own rules and procedures. Understanding how *your* school works
and where to go for help are essential parts of being successful in college. It's important to
understand how your school works so that, for example, you know where and when to turn
in your financial aid application and you know where to get help if you have questions
about choosing the classes you need for graduation. Take this survey to learn how well
you know your school. Check all boxes that apply.

- ☐ I know the address of my school's website.
- ☐ I can navigate the school's website to find most information that I need.
- ☐ I am aware of whether my school has a handbook containing useful information.
- ☐ I have signed up to receive emergency campus messages by email, text, or automated phone call.
- ☐ I am aware of important dates such as when to register for classes, when tuition is due, and when financial aid forms are due.
- ☐ I know where to register for classes on campus.
- ☐ On campus, I know where to ask questions about financial aid.
- ☐ I can locate the bookstore.
- ☐ I know the location of the library.
- ☐ I know the difference between an adviser and a counselor.
- ☐ I know the location of the campus health center.
- ☐ I know the location of student services offices that might be of interest to me. Some examples are veterans' support services, the office to help students with disabilities, and child care.
- ☐ I know the locations of all of my instructors' offices as well as their office hours.
- ☐ I know the location of the tutoring center/math lab, and I know their procedures for getting help when I need it.
- ☐ I can locate the Testing Center and know its rules and hours of operation.
- ☐ I am aware of clubs, organizations, and activities on campus, and I know where to go to become involved in those that interest me.
- ☐ I know the location of the office where I can go if I have questions about or want help finding a job.

Think about the items that you have, and have *not*, checked in this survey. Which apply to
you and might contribute to your success in college? In the Study Strategies at the
beginning of this chapter, you will learn how to get to know your school.

Chapter 7:   Summary

Definition/Procedure	Example

7.1 Using U.S. Customary Measurements

The relationships between U.S. Customary units can be found in this section.

12 in. = 1 ft   4 qt = 1 gal   16 oz = 1 lb

Using Multiplication or Division to Convert Between Units

1) When converting from a *larger unit* to a *smaller unit,* use *multiplication.*

2) When converting from a *smaller unit* to a *larger unit,* use *division.*

Convert 1.5 gal to quarts.

Since we are going from a larger unit to a smaller unit, we will multiply.

$$1.5 \text{ gal} \cdot 4 = 6 \text{ qt} \qquad 4 \text{ qt} = 1 \text{ gal}$$

So, 1.5 gal = 6 qt.

Unit Fractions

A **unit fraction** is a fraction that is equivalent to 1. We can write measurement equivalencies as unit fractions.

We can write 12 in. = 1 ft as the unit fraction $\dfrac{12 \text{ in.}}{1 \text{ ft}}$ or $\dfrac{1 \text{ ft}}{12 \text{ in.}}$.

Using Unit Analysis to Convert Between Units

To convert from one unit of measurement to another, we multiply the measurement we are given by the appropriate unit fraction.

Use unit analysis to convert 54 in. to feet.

$$54 \text{ in.} \cdot \frac{1 \text{ ft}}{12 \text{ in.}} = \frac{\overset{9}{\cancel{54} \text{ in.}}}{1} \cdot \frac{1 \text{ ft}}{\underset{2}{\cancel{12} \text{ in.}}}$$

Divide 54 and 12 by 6; divide out the unit of inches.

$$= \frac{9 \text{ ft}}{2}$$

Multiply.

$$= \frac{9}{2} \text{ ft}$$

Simplify.

$$= 4\frac{1}{2} \text{ ft or } 4.5 \text{ ft}$$

Write the final answer as a mixed number or decimal.

7.2 The Metric System: Length

The Metric System

The **metric system** is based on multiples of 10. The basic unit of length is the **meter.** A meter is about 39.37 in. long, so it is a little longer than a yard.

Meterstick

Yardstick

Since the metric units of length are related to the meter, the units are described by using a prefix on the word *meter.*

The measurements on the left allow us to state these additional relationships:

Metric Units of Length

1 kilometer (km) = 1000 m

1 hectometer (hm) = 100 m

1 dekameter (dam) = 10 m

1 meter (m) = 1 m

$1 \text{ decimeter (dm)} = \dfrac{1}{10} \text{ m or } 0.1 \text{ m}$

$1 \text{ centimeter (cm)} = \dfrac{1}{100} \text{ m or } 0.01 \text{ m}$

$1 \text{ millimeter (mm)} = \dfrac{1}{1000} \text{ m or } 0.001 \text{ m}$

The units used most often are kilometer, meter, centimeter, and millimeter.

Additional Relationships Between Metric Units of Length

10 decimeters = 1 meter

100 centimeters = 1 meter

1000 millimeters = 1 meter

10 millimeters = 1 centimeter

10 centimeters = 1 decimeter

Definition/Procedure	Example
Use Unit Fractions to Convert Between Units We can use unit fractions and the relationships between metric units to convert from one unit to another.	Use unit fractions to convert 27 km to meters. $27 \text{ km} \cdot \dfrac{1000 \text{ m}}{1 \text{ km}} = \dfrac{27 \text{ km}}{1} \cdot \dfrac{1000 \text{ m}}{1 \text{ km}}$ Divide out the unit of kilometers. $= \dfrac{27{,}000 \text{ m}}{1}$ Multiply. $= 27{,}000 \text{ m}$ Simplify. Therefore, 27 km = 27,000 m.
Use the Metric Conversion Chart to Convert Between Units This is how to use the metric conversion chart to convert from one unit to another: 1) Find the unit that you are given. 2) Count the number of places you must move to go from the unit you are given to the unit you want to get. 3) Move the decimal point the same number of places and in the same direction that you did on the metric conversion chart.	Use the metric conversion chart to change 45 cm to meters. Find cm on the chart. Since we move *two places to the left* to reach m, we move the decimal point in 45 cm *two places to the left* to change to meters. Remember that the whole number 45 can be written with a decimal point to the right of the 5. $45. \text{ cm} = .45 \text{ m}$ Put the decimal point at the end of the number. Move the decimal point two places to the left. So, 45 cm = 0.45 m.

7.3 The Metric System: Capacity and Weight (Mass)

Units of Capacity

Units of capacity measure quantities of liquids. In the United States, we use units such as quarts and gallons. In the metric system, the basic unit of capacity is the **liter,** abbreviated as L.

A liter is a little more than a quart.

To describe larger and smaller units of capacity in the metric system, we use the same prefixes on *liter* that we used for *meter* to describe length.

The units used most often are liter and milliliter.

1 kiloliter (kL) = 1000 liters (L)

1000 milliliters (mL) = 1 liter (L)

Definition/Procedure	Example

Converting Between Metric Units of Capacity

We can use unit fractions or the metric conversion chart to convert between metric units of capacity.

Change 350 mL to liters using a) a unit fraction and b) the metric conversion chart.

a) $350 \text{ mL} \cdot \dfrac{1 \text{ L}}{1000 \text{ mL}} = \dfrac{350 \text{ mL}}{1} \cdot \dfrac{1 \text{ L}}{1000 \text{ mL}}$ Divide out the unit of milliliters.

$= \dfrac{350 \text{ L}}{1000}$ Multiply.

$= 0.350 \text{ L}$ Simplify.

So, 350 mL = 0.350 L or 0.35 L.

b) Start at mL on the metric conversion chart. Move *three places to the left* to get from mL to L. Move the decimal point *three places to the left*.

350. mL = .350 L

Put the decimal point at the end of the number. Move the decimal point three places to the left.

So, 350 mL = 0.350 L or 0.35 L.

Units of Weight (Mass)

In the metric system, the basic unit of mass is the **gram,** abbreviated as g. If an object is on the Earth, the weight and mass of that object are the same. So, we will use the word *weight*.

A penny weighs approximately 2.5 g. (www.usmint.gov)

To describe larger and smaller units of weight (mass) in the metric system, we use the same prefixes on *grams* that we used for *meter* to describe length.

The units used most often are kilogram, gram, and milligram.

1 kilogram (kg) = 1000 grams (g)

1000 milligrams (mg) = 1 gram (g)

Converting Between Metric Units of Weight (Mass)

We can use unit fractions or the metric conversion chart to convert between metric units of weight (mass).

Change 4.8 kg to grams using a) a unit fraction and b) the metric conversion chart.

a) $4.8 \text{ kg} \cdot \dfrac{1000 \text{ g}}{1 \text{ kg}} = \dfrac{4.8 \text{ kg}}{1} \cdot \dfrac{1000 \text{ g}}{1 \text{ kg}}$ Divide out the unit of kilograms.

$= \dfrac{4800 \text{ g}}{1}$ Multiply.

$= 4800 \text{ g}$ Simplify.

So, 4.8 kg = 4800 g.

b) Start at kg on the metric conversion chart. Move *three places to the right* to get from kg to g. Move the decimal point *three places to the right*.

4.8 kg = 4800. g

Move the decimal point three places to the right.

4.8 kg = 4800 g

Definition/Procedure	Example

7.4 Solve Applied Problems Involving Metric Units

Perform Operations with Metric Units

We can add, subtract, multiply, and divide numbers with metric units. Sometimes, it is necessary to express the numbers with the same units before performing the operations.

Add 3 m + 215 mm.

The units are not the same, so we will convert one of the units so that it is the same as the other.

Express each measurement in *meters*.	Express each measurement in *millimeters*.
215 mm = 0.215 m	3 m = 3000 mm
Now, add the measurements.	Now, add the measurements.
3 m + 215 mm = 3 m + 0.215 m = 3.215 m	3 m + 215 mm = 3000 mm + 215 mm = 3215 mm

Solve an Applied Problem Involving Metric Units

We can solve applications involving metric units.

The instructions on a 140-mL bottle of prescription cough medicine say to take 10 mL of the medicine two times per day until it is gone. How many days will the medicine last?

First, determine how much medicine is taken per day. Then, divide the total amount of medicine, 140 mL, by the amount of medicine taken each day.

Since the dosage is 10 mL twice a day, the total amount of medicine taken each day is 10 mL · 2 = 20 mL.

Divide 140 mL by 20 mL to determine the number of days the medicine will last: $\dfrac{140 \, \text{mL}}{20 \, \text{mL}} = 7$

The medicine will last 7 days. The check is left to the student.

7.5 Metric–U.S. Customary Conversions and Temperature

We can convert between U.S. customary and metric units.

Relationships Between U.S. Customary Units and Metric Units

Length

1 in. = 2.54 cm	1 yd ≈ 0.91 m
1 ft ≈ 0.30 m	1 mi ≈ 1.61 km

Capacity

1 cup ≈ 236.59 mL	1 gal ≈ 3.79 L
1 qt ≈ 0.95 L	

Weight (Mass)

1 oz ≈ 28.35 g	1 lb ≈ 0.45 kg

Definition/Procedure	Example
We can use unit fractions to convert between the two measurement systems.	Change 20 in. to cm. Multiply the given measurement of 20 in. by the unit fraction relating inches and centimeters so that *inches* will divide out and leave us with centimeters. $$20 \text{ in.} \cdot \frac{\textbf{2.54 cm}}{\textbf{1 in.}} = \frac{20 \text{ in.}}{1} \cdot \frac{2.54 \text{ cm}}{1 \text{ in.}} \qquad \text{Divide out the unit of inches.}$$ $$= \frac{50.8 \text{ cm}}{1} \qquad \text{Multiply.}$$ $$= 50.8 \text{ cm} \qquad \text{Simplify.}$$ Therefore, 20 in. = 50.8 cm.
Changing from °F to °C Use this formula to change a Fahrenheit temperature (F) to Celsius (C). $$C = \frac{5(F - 32)}{9}$$	Convert 95°F to Celsius. $$C = \frac{5(F - 32)}{9} = \frac{5(95 - 32)}{9} = \frac{5(63)^7}{9^1} = 35$$ Therefore, 95°F = 35°C.
Changing from °C to °F Use this formula to change a Celsius temperature (C) to Fahrenheit (F). $$F = \frac{9C}{5} + 32$$	Convert 11°C to Fahrenheit. $$F = \frac{9C}{5} + 32 = \frac{9(11)}{5} + 32 = \frac{99}{5} + 32 = 19.8 + 32 = 51.8$$ Round the answer to the nearest degree: 51.8°F ≈ 52°F. Therefore, 11°C ≈ 52°F.

Chapter 7: Review Exercises

(7.1) Fill in the blank.

1) _____ ft = 1 yd

2) 1 pt = _____ c

3) 1 lb = _____ oz

4) _____ min = 1 hr

5) To convert from seconds to minutes, do we use multiplication or division? Explain your answer.

6) To convert from feet to inches, do we use multiplication or division? Explain your answer.

Use multiplication or division to convert to the indicated unit.

7) 10 qt to gallons

8) 52 oz to pounds

9) $1\frac{1}{4}$ days to hours

10) 3.5 ft to inches

Use unit analysis to convert to the indicated unit.

11) 7 min to seconds

12) 81 in. to feet

13) 5 pt to quarts

14) 4.6 hr to minutes

15) $6\frac{2}{3}$ yd to feet

16) 630 in. to yards

Solve each problem.

17) Vlad is comparing two brands of hot dogs. The label for Brand A says that its unit cost is $3.49/lb. The unit cost for Brand B is $0.24/oz. Which is the better buy?

©Ingram Publishing/Alamy

18) It takes Helen 20 min to crochet each square for an afghan. How many hours will it take for her to make the 96 squares she needs for the blanket?

(7.2) Fill in the blank.

19) 1 m = _____ cm

20) 1 km = _____ m

21) _____ mm = 1 m

22) _____ mm = 1 cm

For Exercises 23 and 24, determine which unit of metric length would be most appropriate for each situation. Choose from m, mm, cm, and km.

23) The distance between New York City and Los Angeles

24) The length of a key

Use unit fractions to change to the indicated unit.

25) 631 mm to meters

26) 892 m to kilometers

27) 1.2 km to meters

28) 5.75 m to centimeters

Use the metric conversion chart to change to the indicated unit.

29) 7 m to centimeters

30) 4 m to millimeters

31) 96.5 mm to meters

32) 28.3 m to kilometers

33) 0.8 km to centimeters

34) 0.7 km to millimeters

(7.3) For Exercises 35–38, determine which metric unit of capacity would be more appropriate for each situation. Choose from mL and L.

35) The amount of gas it takes to fill a car's tank

36) The amount of water used to do dishes in the kitchen sink

37) The amount of milk in a baby bottle

38) The amount of soup in a can

Use unit fractions to convert to the indicated unit.

39) 3.25 L to milliliters

40) 4.75 L to milliliters

41) 140 mL to liters

42) 290 mL to liters

Use the metric conversion chart to change to the indicated unit.

43) 630 mL to liters

44) 850 mL to liters

45) 2 L to milliliters

46) 9 mL to milliliters

Fill in the blank.

47) _____ g = 1 kg

48) 1000 mg = _____ g

For Exercises 49–52, fill in the blank with mg, g, or kg.

49) The tablet contains 200 _____ of ibuprofen.

50) A female elephant can weigh about 3500 _____.

51) Jada's backpack weighs 4 _____.

52) The strawberry weighed 23 _____.

Use unit fractions to convert to the indicated unit.

53) 82 mg to grams

54) 61 mg to grams

55) 13 kg to grams

56) 22 kg to grams

57) $4\frac{1}{2}$ g to milligrams

58) 7.9 g to milligrams

Use the metric conversion chart to change to the indicated unit.

59) 9750 g to kilograms

60) 8140 g to kilograms

61) 0.02 kg to milligrams

62) 0.06 kg to milligrams

(7.4) Perform the indicated operation.

63) Divide 48 g by 8.

64) Multiply 0.05 mL by 4.

65) 7 m + 47 cm

66) 0.09 g − 37 mg

Solve each problem.

67) If Ruby cuts off 85 cm from a piece of fabric measuring 2 m 50 cm in length, how long is the remaining piece, in centimeters?

68) Danny cuts a metal pipe measuring 1 m 80 cm into three equal lengths. Find the length of each piece, in centimeters.

69) Masa harina costs about $3.98/kg. Lupita's tortilla recipe uses 250 g of the flour and makes about 15 tortillas. If she wants to make 120 tortillas so that she can share them with friends and family, find the cost of the masa harina.

©Glow Images

70) Davida's 50-mL pump spray perfume bottle delivers 0.1 mL of her favorite perfume per stroke. If she applies the perfume twice a day, how many days will the bottle last?

(7.5) For Exercises 71–78, convert to the indicated unit.

71) During the 2016-2017 season, the year the New England Patriots won Super Bowl LI, quarterback Tom Brady threw for 4691 yd. Convert this number to meters. Round to the nearest meter. (www.espn.com)

72) According to the *Guinness World Records,* the world's tallest man was about 2.7 m tall. Convert this number to feet. (www.guinnessworldrecords.com)

73) Ralph caught a lake trout that weighed 11 kg. Convert this number to pounds. Round to the nearest tenth.

©lynx/iconotec.com/Glow Images

74) A 5-min shower uses about 12.5 gal of water. Convert this number to liters. Round to the nearest tenth.

75) The builder installed a 190-L water heater in a new home. Convert this number to gallons. Round to the nearest gallon.

76) A bakery ordered a 22.5-kg sack of Swiss pastry flour. Convert this number to pounds.

77) Crystal's cell phone is 4 in. long. Convert this number to centimeters. Round to the nearest tenth.

78) The driving distance between Little Rock, AR, and Nashville, TN, is about 350 mi. Convert this number to kilometers. Round to the nearest kilometer.

Determine which Celsius temperature is appropriate for the situation.

79) Water freezes at _____. Choose from 0°C, 10°C, and 32°C.

80) The high temperature on a summer day in Hawaii might be _____. Choose from 5°C, 84°C, and 31°C.

Perform the following temperature conversions. Round to the nearest degree, if necessary.

81) Convert 50°F to Celsius.

82) Convert 86°F to Celsius.

83) Convert 79°F to Celsius.

84) Convert 37°F to Celsius.

85) Convert 5°C to Fahrenheit.

86) Convert 65°C to Fahrenheit.

87) Convert 132°C to Fahrenheit.

88) Convert 19°C to Fahrenheit.

Mixed Exercises

Fill in the blank with the appropriate metric unit. Choose from mm, cm, m, km, mL, L, mg, g, and kg.

89) The diameter of a DVD is 120 _____.

90) The glass contains 350 _____ of orange juice.

91) Heinrich's car weighs 1620 _____.

92) Tina's bedroom is 4.5 _____ wide.

93) The aquarium in Celina's house holds 75 _____ of water.

94) A kitten weighs about 100 _____ at birth.
(http://veterinarymedicine.dvm360.com)

Convert to the indicated unit.

95) 2.25 L to milliliters

96) 0.6 kg to grams

97) 48 cm to meters

98) 1294 mm to meters

99) $5\frac{1}{3}$ min to seconds

100) $2\frac{3}{4}$ days to hours

101) 5800 mg to kilograms

102) 1.7 km to millimeters

Perform the indicated operation.

103) 89 cm − 52 mm

104) Divide 20 L by 4.

105) Multiply 0.6 kg by 7.

106) 2 g + 85 mg

Convert to the indicated unit. Round the answer to the nearest tenth, if necessary.

107) 160 km to miles

108) 80 g to ounces

109) 6 qt to liters

110) 2 c to milliliters

111) 9.8 kg to pounds

112) $6\frac{1}{2}$ in. to centimeters

Perform the following temperature conversions. Round to the nearest degree, if necessary.

113) Convert 22°C to Fahrenheit.

114) Convert 59°F to Celsius.

Solve each problem.

115) Marilyn needs to buy 110 g of pepperoni for a pizza, and the pepperoni costs $22.00/kg. How much will Marilyn pay for the pepperoni?

©a9photo/Shutterstock

116) Neda has a length of ribbon that is 1 m 60 cm long. If she cuts off 75 cm, how much ribbon is left, in centimeters?

Chapter 7: Test

Fill in the blank.

1) _____ qt = 1 gal

2) 1 min = _____ sec

Convert to the indicated unit.

3) 21 ft to yards

4) 6 pt to quarts

5) 24 oz to pounds

6) $4\frac{2}{3}$ ft to inches

7) 3.5 days to minutes

Fill in the blank with the appropriate metric unit. Choose from mm, cm, m, km, mL, L, mg, g, and kg.

8) The mechanic put 1.5 _____ of oil in the car's engine.

9) Sonya is 7 years old and 112 _____ tall.

10) Christopher weighs 79 _____.

11) The diameter of a quarter is 24 _____.

12) The bottle contains 475 _____ of salad dressing.

13) The cookie weighs about 8 _____.

14) The traffic light is 8.5 _____ tall.

Convert to the indicated unit.

15) 250 mg to grams

16) 7.3 m to centimeters

17) 62 mL to liters

18) $2\frac{1}{2}$ kg to grams

19) 19.35 cm to millimeters

20) 800,000 mm to kilometers

21) Add 3 kg + 45 g.

Determine which Celsius temperature is appropriate for the situation.

22) The temperature when Sven shoveled snow could have been _____. Choose from 23°C, 11°C, and −7°C.

23) The temperature when Monique drove her convertible with the top down could have been _____. Choose from 26°C, 68°C, and 80°C.

Convert to the indicated unit. Round the answer to the nearest tenth, if necessary.

24) 50 in. to centimeters

25) 6 kg to pounds

©Ingram Publishing

26) 320 km to miles

27) 14.8 gal to liters

28) $1\frac{1}{2}$ cups to milliliters

Perform the following temperature conversions. Round to the nearest degree, if necessary.

29) 77°F to Celsius

30) 4°C to Fahrenheit

Solve each problem.

31) PVC pipe costs $2.48 per foot. Find the cost of 54 in. of this pipe.

32) If your hair grows about 12 mm per month, how long would it take to grow 6 cm?

33) At an Asian market, dried shrimp costs $33.80/kg. Oki needs 1 lb of the shrimp for a stir fry recipe. How much will she pay for the shrimp?

©Glow Images

34) Parvesh has a piece of plastic landscape edging that is 5 m long. If he cuts off 72 cm, find the length of edging remaining, in meters.

35) A patient is prescribed 2100 mg of diclofenac for arthritis. The instructions say to take one tablet twice a day for two weeks. How much diclofenac is in each tablet?

Chapter 7: Cumulative Review for Chapters 1–7

1) Explain the difference between evaluating $(-2)^4$ and -2^4, and evaluate each.

2) Write 14,209.37 in words.

Perform the indicated operations. Write all answers in lowest terms.

3) $-1323 \div 27$

4) $59 - 2085 - (-647) - 18{,}996$

5) 0.0882×1000

6) $3\frac{5}{9} \div 3\frac{1}{3}$

7) $\frac{2}{3} - \frac{7}{8}$

8) $413.6 - 257.92$

9) $\frac{14}{27} \cdot \frac{15}{28}$

10) $-1.44 \div (-32)$

11) $\frac{-87}{0}$

12) $\left(\frac{1}{3}\right)^4 \cdot \left(-\frac{3}{2}\right)^3$

13) $\sqrt{49} - 120 \div 8 - 9^2$

14) $\frac{54}{6} - 4[12^2 - 3(1+5)^2]$

Find the area and perimeter of each figure. Be sure to include the correct units.

15)

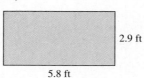

2.9 ft

5.8 ft

16)

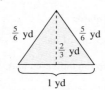

$\frac{5}{6}$ yd $\frac{5}{6}$ yd

$\frac{2}{3}$ yd

1 yd

Look at Exercises 17–20 and identify each as an expression or an equation. If it is an expression, simplify it. If it is an equation, solve it.

17) $6k - 17 = 2(4k - 3) + 7$

18) $6k - 17 + 2(4k - 3) + 7$

19) $-\dfrac{2}{3}y + 11 + \dfrac{1}{2}y - 3$ 20) $-\dfrac{2}{3}y + 11 = 3$

21) What is the difference between a ratio and a rate?

22) Convert 63°F to Celsius. Round to the nearest degree.

Solve each problem.

23) A cookie recipe uses $2\dfrac{1}{2}$ c flour and makes 36 cookies. If Lisa wants to make 54 cookies, how much flour will she need?

©McGraw-Hill Education

24) Adenuga ran 400 m in 48 sec. What was his speed in meters per minute?

25) Lorena's new washer cost $50 less than the dryer. Together, they cost $846. Find the cost of each item.

8 Percents

©Paul Bradbury/Getty Images

Math at Work:
Environmental Inspector

Vikki Chun grew up loving the outdoors. Now, she makes her living protecting it. Vikki works as an environmental inspector for her town, ensuring that local businesses and residents comply with environmental regulations.

"My job involves many different tasks," Vikki says. "I visit offices, factories, even private residences, and perform inspections to confirm that everyone is doing their part to keep our environment clean. Then, I offer suggestions for how they can make their spaces even more environmentally friendly."

Behind the scenes, however, Vikki uses online resources to stay up-to-date on environmental regulations, to learn what other communities are doing, and to learn how to make homes and businesses as healthy for the environment as they can be. "Using the Internet and understanding percentages are a big part of my job," explains Vikki. "For example, I learned that recycling just 5% more waste can lead to a 20% or 30% reduction in the growth of our town's landfill. Not only do I need to understand what that means, I need to know how to help our businesses and individuals recycle more so that less waste ends up in the landfill."

For Vikki, using percentages *and* the Internet are essential parts of her job. In this chapter, you will learn how to use both, too.

Study Strategies

The Internet provides access to millions of resources to supplement your learning. Using these resources can help you master difficult math concepts. But just as you may have a particular learning style in the classroom, it is the same for online resources. It is important for you to find websites that fit your style of learning, and it is *equally* important that you learn how to use those resources in a way that will be most helpful to you.

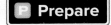

- **I will investigate and use online resources to help me learn math.**

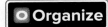

- Complete the emPOWERme survey that appears before the Chapter Summary to help you understand how much you know about online learning resources.
- Gather the materials you will need such as a computer, your book, paper, and pencils.
- Find a workspace that is quiet and that has enough room to work on the computer *and* to take notes.
- Decide which math skills you want to learn or review. The more carefully you define the skills, the more easily you will be able to identify an appropriate online resource.
- Be aware of any online resources that are available with your textbook or other course materials. Familiarize yourself with the types of educational materials contained within this resource. If you are using an interactive e-book, you should realize that videos are embedded in the book so that you can watch them as you read.
- Ask your instructor for recommendations, or search the Internet to find appropriate learning materials.
- Think about your own learning style and decide which type of resource is best for you. Do you want a website that will solve an equation for you, step by step, after you type in a specific equation? Or do you prefer to watch a video where a teacher explains a topic? Would you like a discussion board where you can pose questions to other students?
- Be sure to consider the source of the resource before using it. Some well-known sources are more likely to be accurate and of high quality, such as those from the publisher of this textbook, major organizations, and major research universities.
- Some well-known resources are: the many online resources that accompany this book (see the front of the book or ask your instructor for information), www.youtube.com, and www.khanacademy.org.

- Explore the online resource. Learn which types of educational tools it contains such as explanations that you read, problem-solvers, and videos. Does this resource contain the type of mathematics that you are studying?
- Work with the resource thoroughly. Watch one or two complete videos, or work through a couple of problems step by step.
- **Be an active user!** Take notes as you go along. Don't just watch or read passively!
- Keep a list of the websites or resources that are most helpful to you. Bookmark them so that you can find them easily in the future.
- Be sure that the methods used in the resources are consistent with what your instructor has taught you and what is shown in your book. If not, ask your instructor about the differences. Sometimes, it can be confusing if instructors solve a problem one way but an outside resource uses a different method.

- Did you find any websites or online resources that were helpful to you? Are you aware of the different types of content available on those resources? Did you understand the material better after using that resource for learning?

- If you identified some helpful resources, be sure to note what they were so that you can use them in the future. Next time, explore it more; you may find other types of learning resources that are even better!
- If you did not find anything that was helpful, think about why. Did you use the wrong content within that resource? Should you have watched a video instead of just having a website solve problems for you? Or, should you try a different website or resource next time?
- If you are still struggling with a concept, plan a time to get extra help from your instructor or from your school's tutoring center.

Chapter 8 **POWER** Plan

P Prepare	**O** Organize
What are your goals for Chapter 8?	**How can you accomplish each goal? (Write in the steps you will take to succeed.)**
1 Be prepared before and during class.	• _____ • _____ • _____ • _____
2 Understand the homework to the point where you could do it without needing any help or hints.	• _____ • _____ • _____
3 Use the P.O.W.E.R. framework to learn how to use the Internet to supplement your learning.	• Complete the emPOWERme that appears before the Chapter Summary to determine how much you know about using the Internet for learning. • Read the Study Strategies that explain how to use the P.O.W.E.R. framework to help you find and use resources to help you learn math. • _____
4 Write your own goal. _____ _____	• _____
What are your objectives for Chapter 8?	**How can you accomplish each objective?**
1 Understand percents, and be able to convert between percents, fractions, and decimals.	• Understand what a percent represents and how it compares to a fraction or decimal. • Understand the procedures that allow you to convert between percents and decimals. • Determine standard procedures to convert more complex percents to decimals or fractions. • Know three different methods to write a fraction as a percent.
2 Know how to compute percents of a number, and solve applied problems involving percents by using an equation.	• Be able to, mentally, find percents that are multiples of 5, 10, or 100. • Understand what you are being asked to find, and use the correct formula. • Use the Five Steps for Solving Applied Problems to solve application problems that contain percents.
3 Know how to compute common percents by using formulas.	• Understand the formulas and procedures for *sales tax*, *commission*, *sale price*, and *percent increase/decrease*. • Know how to use the correct formula with the Five Steps for Solving Applied Problems.
4 Learn how to find simple interest and compound interest.	• Understand the definitions of *interest*, *simple interest*, *principal*, and *compound interest*. • Learn the formulas to compute simple interest and compound interest.
5 Write your own goal. _____ _____	• _____

E Evaluate Complete the Chapter Review
and Chapter Test. How did
you do?

R Rethink
- Can you mentally compute percents well enough
  to estimate sale prices and determine the amount
  of a tip?
- Do you feel like you have a thorough understanding of percents?
- Did you use any online resources while studying this chapter? If so, which
  ones? Which were most helpful?

8.1 Percents, Fractions, and Decimals

P Prepare

O Organize

What are your objectives for Section 8.1?	How can you accomplish each objective?
1 Understand the Meaning of Percent	<ul><li>Know the definition of a *percent*.</li><li>Be able to write a sentence that describes a percentage.</li><li>Complete the given examples on your own.</li><li>Complete You Trys 1 and 2.</li></ul>
2 Relate Percents to Fractions and Decimals	<ul><li>Recognize how to write a percent as a fraction and/or a decimal.</li><li>Complete the given example on your own.</li><li>Complete You Try 3.</li></ul>
3 Change Percents to Decimals	<ul><li>Write the procedure for **How to Change a Percent to a Decimal** in your own words.</li><li>Complete the given example on your own.</li><li>Complete You Try 4.</li></ul>
4 Write Percents as Fractions in Lowest Terms	<ul><li>Create a step-by-step procedure for writing percents as fractions in lowest terms.</li><li>Know the two different techniques for writing a percent that contains a fraction as a fraction.</li><li>Write a statement that explains how to change a percent that contains a fraction to a fraction.</li><li>Complete the given examples on your own.</li><li>Complete You Trys 5–8.</li></ul>
5 Change Decimals to Percents	<ul><li>Write the procedure for **How to Change a Decimal to a Percent** in your own words.</li><li>Complete the given example on your own.</li><li>Complete You Try 9.</li></ul>
6 Write Fractions as Percents	<ul><li>Know how to write a fraction as a percent by finding an equivalent fraction with a denominator of 100, using long division, and using a proportion.</li><li>Complete the given examples on your own.</li><li>Complete You Trys 10–13.</li></ul>

1 Understand the Meaning of Percent

Percents are everywhere: 68% of adults in the United States use Facebook, or the unemployment rate in the United States is 5%. These are just a couple of ways percents are used. But what does percent mean?

Definition

Percent means *out of 100*.

Let's think about the meaning of each percent in the first paragraph.

> *68% of adults in the United States use Facebook* means that in the United States, 68 out of 100 adults use Facebook. *The unemployment rate in the United States is 5%* means that 5 out of every 100 people in the United States are unemployed.

We can think about percentages (percents) with a picture.

The figure to the right is divided into 100 squares of equal size, and 23 of them are shaded. Therefore, 23 *out of* 100 squares are shaded, or 23% of the squares are shaded.

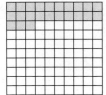

EXAMPLE 1

Explain the meaning of the following statement: *Approximately 73% of full-time undergraduates received financial aid during the 2011–2012 academic year.* (nces.ed.gov)

Solution

73% means 73 *out of* 100, so the statement means that approximately 73 out of 100 undergraduates received financial aid in the 2011–2012 academic year.

[YOU TRY 1]

Explain the meaning of the following statement: *In 2016, approximately 70% of married women with children under 18 were working outside the home.* (www.census.gov)

EXAMPLE 2

Rewrite the statement using a percent: *In 2015, approximately 24 out of every 100 employers in the United States allowed employees to work some of their hours at home.* (www.bls.gov)

Solution

Percent means *out of 100*, so 24 *out of* 100 is the same as 24%. We can rewrite the statement as follows: *In 2015, approximately 24% of employers in the United States allowed employees to work some of their hours at home.*

©LWA/Sharie Kennedy/Blend Images LLC

Rewrite the statement using a percent: *In 2015, approximately 7 out of 100 students played on their high school's football team.* (www.cnsnews.com)

2 Relate Percents to Fractions and Decimals

In Chapter 6, we learned that we could write an expression like *3 out of 8* as the fraction $\frac{3}{8}$. Because percent means *out of 100*, we can write a percent as a fraction. In Example 1, because 79% means *79 out of 100*, we can write 79% as $\frac{79}{100}$. And, because fractions can be written as decimals, we can write percents as decimals as well:

$$79\% = \frac{79}{100} = 0.79$$

EXAMPLE 3

Write each percent as a fraction and then as a decimal.

a) 83%   b) 7%

Solution

a) Because 83% means 83 out of 100, we write

$$83\% = \frac{83}{100} = 0.83$$

b) Because 7% means 7 out of 100, write

$$7\% = \frac{7}{100} = 0.07$$

[YOU TRY 3]

Write each percent as a fraction and then as a decimal.

a) 61%   b) 9%

3 Change Percents to Decimals

We can change percents directly to decimals by remembering that when we divide a number by 100, we move the decimal point two places to the left. For example,

$$\frac{31}{100} = 31 \div 100 = 0.31$$

Let's use this method to change a percent to a decimal.

Procedure How to Change a Percent to a Decimal

1) Remove the percent symbol.
2) Move the decimal point two places to the left.

EXAMPLE 4

Write each percent as a decimal.

a) 25%   b) 4%   c) 0.7%   d) 100%   e) 316%

Solution

a) 25% = 25.%   Put the decimal point at the end of the number.
        = 0.25   Remove the percent symbol, and move the decimal point two places to the left.

b) 4% = 4.%   Put the decimal point at the end of the number.
        = 0.04   Remove the percent symbol, and move the decimal point two places to the left.

c) 0.7% = 0.007   Remove the percent symbol, and move the decimal point two places to the left.

d) 100% = 100.%   Put the decimal point at the end of the number.
        = 1.00   Remove the percent symbol, and move the decimal point two places to the left.
        = 1

e) 316% = 316.%   Put the decimal point at the end of the number.
        = 3.16   Remove the percent symbol, and move the decimal point two places to the left.

Note

1) We see in Example 4d that 100% = 1.

2) Example 4e shows that 316% = 3.16, a number that is greater than 1.

In general, we can say that a percent greater than 100 is equivalent to a decimal number greater than 1.

[YOU TRY 4]

Write each percent as a decimal.

a) 48%   b) 6%   c) 0.5%   d) 120%   e) 500%

---

4 Write Percents as Fractions in Lowest Terms

When we change a percent to a fraction, we usually write it in lowest terms.

EXAMPLE 5

Write each percent as a fraction or mixed number in lowest terms.

a) 25%   b) 50%   c) 75%   d) 64%   e) 120%

Solution

a) $25\% = \dfrac{25}{100} = \dfrac{25 \div 25}{100 \div 25} = \dfrac{1}{4}$ Divide numerator and denominator by 25.

b) $50\% = \dfrac{50}{100} = \dfrac{50 \div 50}{100 \div 50} = \dfrac{1}{2}$ Divide numerator and denominator by 50.

c) $75\% = \dfrac{75}{100} = \dfrac{75 \div 25}{100 \div 25} = \dfrac{3}{4}$ Divide numerator and denominator by 25.

---

d) $64\% = \dfrac{64}{100} = \dfrac{64 \div 4}{100 \div 4} = \dfrac{16}{25}$ Divide numerator and denominator by 4.

e) $120\% = \dfrac{120}{100} = \dfrac{120 \div 20}{100 \div 20}$ Divide numerator and denominator by 20.

         $= \dfrac{6}{5}$ Perform the division.

         $= 1\dfrac{1}{5}$ Write as a mixed number.

So, $120\% = \dfrac{6}{5}$ or $1\dfrac{1}{5}$.

W Hint

Use this note to help develop a procedure for writing percents as fractions in lowest terms.

Note

1) A percent *less than* 100% will be a fraction *less than* 1.
2) A percent *greater than* 100% will be a fraction (or mixed number) *greater than* 1.
3) 100% = 1

[YOU TRY 5]

Write each percent as a fraction in lowest terms.

a) 30%   b) 55%   c) 81%   d) 88%   e) 190%

Some percents contain decimals. We will learn two ways to change a percent like 16.5% to a fraction in lowest terms. One method is to first change the percent to a decimal, and then change the decimal to a fraction.

EXAMPLE 6

Write 16.5% as a fraction in lowest terms by first writing 16.5% as a decimal.

Solution

First, change 16.5% to a decimal: $16.5\% = 0.165$

Next, change the decimal to a fraction: $0.165 = \dfrac{165}{1000}$ ← denominator of 1000

             ↑
       thousandths place

Now, simplify $\dfrac{165}{1000}$: $\dfrac{165}{1000} = \dfrac{165 \div 5}{1000 \div 5} = \dfrac{33}{200}$

Therefore, $16.5\% = \dfrac{33}{200}$.

Another way to change 16.5% to a fraction in lowest terms is to first write it as a fraction with a denominator of 100.

EXAMPLE 7 Write 16.5% as a fraction in lowest terms by first writing it as a fraction with a denominator of 100.

Solution

$$16.5\% = \frac{16.5}{100}$$

A fraction is *not* in lowest terms if it contains a decimal. Therefore, the next step is to eliminate the decimal from the fraction.

Because the digit farthest to the right in 16.5 is in the tenths place, multiplying 16.5 by 10 will move the decimal point to the end of the number.

$$16.5 \cdot 10 = 165$$

W Hint

Compare the methods in Examples 6 and 7. Which do you prefer, and why?

If we multiply the numerator of $\frac{16.5}{100}$ by 10, we must multiply the denominator by 10 so that we do not change the value of the fraction:

$$\frac{16.5}{100} \cdot \frac{10}{10} = \frac{165}{1000}$$

Now, simplify $\frac{165}{1000}$: $\frac{165}{1000} = \frac{165 \div 5}{1000 \div 5} = \frac{33}{200}$

Therefore, $16.5\% = \frac{33}{200}$. The result is the same no matter which method is used.

If a percent contains a fraction, like $33\frac{1}{3}\%$, we can change it to a fraction by first writing $33\frac{1}{3}$ over 100.

EXAMPLE 8

Write $33\frac{1}{3}\%$ as a fraction in lowest terms.

Solution

Begin by writing $33\frac{1}{3}\%$ as a fraction with a denominator of 100: $33\frac{1}{3}\% = \dfrac{33\frac{1}{3}}{100}$.

This fraction may look complicated, but remember that a fraction is another way to represent a division problem. So, we can write

$$\frac{33\frac{1}{3}}{100} = 33\frac{1}{3} \div 100 = \frac{100}{3} \div 100 = \frac{\overset{1}{\cancel{100}}}{3} \cdot \frac{1}{\underset{1}{\cancel{100}}} = \frac{1}{3}$$

Change the mixed number to an improper fraction.

Therefore, $33\frac{1}{3}\% = \dfrac{1}{3}$.

> **W Hint**
>
> How would you describe the quickest way to do Example 8?

[YOU TRY 8]

Write $66\frac{2}{3}\%$ as a fraction in lowest terms.

---

5 Change Decimals to Percents

If we remember the relationship between a decimal and a fraction, we can then write the decimal as a percent. For example, $0.74 = 74\%$ because

$$0.74 = \frac{74}{100} = 74\%$$

So, to change 0.74 to 74%, we moved the decimal point two places to the right and put the percent symbol at the end of the number.

> **Procedure How to Change a Decimal to a Percent**
>
> 1) Move the decimal point two places to the right.
> 2) Put the percent symbol at the end of the number.

EXAMPLE 9

Write each decimal as a percent.

a) 0.99   b) 0.0085   c) 4.13   d) 1.5   e) 1

Solution

a) $0.99 = 99.$ Move the decimal point two places to the right.

   $= 99\%$ Put the % symbol at the end of the number.
                 Remove the decimal point from the end of the number.

b) $0.0085 = 0.85$ Move the decimal point two places to the right.

   $= 0.85\%$ Put the % symbol at the end of the number.

c) $4.13 = 413.$ Move the decimal point two places to the right.

        $= 413\%$ Put the % symbol at the end of the number.
                          Remove the decimal point from the end of the number.

d) $1.5 = 150.$ Move the decimal point two places to the right.

        $= 150\%$ Put the % symbol at the end of the number.
                          Remove the decimal point from the end of the number.

e) $1 = 1.$ Put the decimal point at the end of the number.

       $= 100.$ Move the decimal point two places to the right.

       $= 100\%$ Put the % symbol at the end of the number.
                         Remove the decimal point from the end of the number.

Note

1) Examples 9a and b show that a number less than 1 is a percent less than 100%.

2) Examples 9c and d show that a number greater than 1 is a percent greater than 100%.

3) Example 9e shows that 1 = 100%.

[YOU TRY 9]   Write each decimal as a percent.

a) 0.62   b) 0.0047   c) 8.95   d) 2.7   e) 4

6 Write Fractions as Percents

We can use several different methods to write a fraction as a percent. Because percent means *out of 100,* one way is to start by writing the given fraction as a fraction with a denominator of 100.

EXAMPLE 10

Write $\dfrac{4}{5}$ as a percent by first writing it as a fraction with a denominator of 100.

Solution

To write $\dfrac{4}{5}$ with a denominator of 100, multiply the fraction by $\dfrac{20}{20}$.

$$\frac{4}{5} \cdot \frac{20}{20} = \frac{80}{100} = 80\%$$

This means *80 out of 100* or *80%.*

W Hint

When do you think this would be the quickest method to use?

So, $\dfrac{4}{5} = 80\%$.

[YOU TRY 10]

Write $\dfrac{19}{50}$ as a percent by first writing it as a fraction with a denominator of 100.

We can also change a fraction to a percent using long division.

EXAMPLE 11

Write $\dfrac{7}{8}$ as a percent using long division.

Solution

Use long division to change $\dfrac{7}{8}$ to a decimal, then change the decimal to a percent.

$$\begin{array}{r} .875 \\ 8\overline{)7.000} \end{array}$$

Therefore, $\dfrac{7}{8} = 0.875 = 87.5\%$.

YOU TRY 11

Write $\dfrac{9}{16}$ as a percent using long division.

A third way to write a fraction as a percent is to use a proportion.

EXAMPLE 12

Use a proportion to change $\dfrac{3}{8}$ to a percent.

Solution

Remember, percent means *out of 100*. To change $\dfrac{3}{8}$ to a percent, we can think of this problem as the proportion "3 is to 8 as what number is to 100?" Write this statement as the proportion $\dfrac{3}{8} = \dfrac{x}{100}$, where x is the percent.

Solve the equation for x by setting the cross products equal to each other.

W Hint

Would it be helpful to create a procedure in your notes to outline these steps?

$$\dfrac{3}{8} = \dfrac{x}{100}$$

$8 \cdot x = 8x$

$3 \cdot 100 = 300$

$$8x = 300$$

$$\dfrac{8x}{8} = \dfrac{300}{8} \qquad \text{Divide each side by 8.}$$

$$x = 37.5 \qquad \text{Simplify.}$$

Therefore, $\dfrac{3}{8} = \dfrac{37.5}{100}$. So, $\dfrac{3}{8} = 37.5\%$. We can also think of 37.5% as $37\frac{1}{2}\%$.

YOU TRY 12

Use a proportion to change $\dfrac{7}{8}$ to a percent.

Sometimes, we might be asked to round a percent.

EXAMPLE 13

Use a proportion to change $\dfrac{5}{6}$ to a percent. Give the exact answer, and also round the answer to the nearest tenth of a percent.

Solution

Set up the proportion $\dfrac{5}{6} = \dfrac{x}{100}$, where x is the percent. Solve the equation.

$6 \cdot x = 6x$

$5 \cdot 100 = 500$

$6x = 500$

$\dfrac{6x}{6} = \dfrac{500}{6}$ Divide each side by 6.

$x = 83.\overline{3}$

To find the *exact* answer, we write $\dfrac{5}{6} = \dfrac{83.\overline{3}}{100} = 83.\overline{3}\%$.

We can *round to the nearest tenth of a percent* by first rounding $83.\overline{3}$ to 83.3. Then,

$\dfrac{5}{6} \approx \dfrac{83.3}{100} = 83.3\%$.

Note

When we solve the equation in Example 13, if we change $\dfrac{500}{6}$ to the mixed number $83\dfrac{1}{3}$, then we can also say that $\dfrac{5}{6} = 83\dfrac{1}{3}\%$.

[YOU TRY 13]

Use a proportion to change $\dfrac{4}{9}$ to a percent. Give the exact answer, and also round the answer to the nearest tenth of a percent.

Summary Writing a Fraction as a Percent

We can write a fraction as a percent using one of these methods:

1) Write the fraction with a denominator of 100.
2) Use long division to change the fraction to a decimal, then change the decimal to a percent.
3) Use a proportion.

Using Technology

When using a calculator to change a decimal to a percent, we multiply the decimal by 100 and put the % symbol at the end of the number. To convert 0.38 to a percent, we enter [.][3][8][×][1][0][0][=] into the calculator. The display screen will likely show 38. as the result. Finally, we remove the decimal and write 38% for our final answer.

We can use a calculator to change a fraction to a percent. For example, to convert $\frac{7}{9}$ to a percent, we enter [7][÷][9][×][1][0][0][=] into the calculator.

(Multiplying by 100 is the same as moving the decimal point two places to the right.) The display screen will likely show 77.77778 for the result. This means that the exact answer is 77.$\overline{7}$%. If we are asked to round our answer to the *nearest tenth of a percent*, we would write 77.8% for our final answer.

[E] Evaluate **8.1** Exercises

Do the exercises, and check your work.

Objective 1: Understand the Meaning of Percent
Explain the meaning of each statement.

1) 23% of students who consume media approximately 3 hr per day earn mostly C grades or lower. (www.zdnet.com)

©Ingram Publishing

2) As of 2015, 33% of American women aged 25 and older obtained a bachelor's degree or more. (www.census.gov)

3) 81% of all Americans have a profile on one or more social networking websites. (www.statista.com)

4) In 2015, 24% of employees did some or all of their work at home. (www.bls.gov)

Rewrite each statement using a percent.

5) 43 out of 100 teens have been victims of cyberbullying. (www.ncpc.org)

6) In 2014, 27 out of 100 veterans were aged 25 and older with at least a bachelor's degree. (factfinder.census.gov)

7) As of June 2017, 15 out of 100 members of the armed forces were women. (www.census.gov)

8) Thirty-eight out of 100 Americans read the news online. (pewresearch.org)

Source: U.S. Air Force, photo by Staff Sgt. Bryan Bouchard

Objective 2: Relate Percents to Fractions and Decimals

Write each percent as a fraction and a decimal.

9) 67%

10) 43%

11) 59%

12) 97%

13) 1%

14) 9%

Objective 3: Change Percents to Decimals

15) Explain how to change a percent to a decimal by moving the decimal point.

16) Is this statement true or false? 3.4% = 340

Write each percent as a decimal by moving the decimal point.

17) 72%

18) 68%

19) 3%

20) 5%

21) 45.2%

22) 13.5%

23) 0.4%

24) 0.1%

25) 200%

26) 400%

27) 150%

28) 209%

29) 0.25%

30) 0.91%

31) 0.01%

32) 0.07%

33) Any percent greater than 100% is greater than what decimal number?

34) Any percent less than 100% is less than what decimal number?

Objective 4: Write Percents as Fractions in Lowest Terms

Write each percent as a fraction or mixed number in lowest terms.

35) 20%

36) 90%

37) 85%

38) 16%

39) 67%

40) 51%

41) 2%

42) 4%

43) 275%

44) 225%

Write each percent as a fraction in lowest terms by first writing the percent as a decimal.

45) 52%

46) 64%

47) 22.5%

48) 17.5%

49) 8.4%

50) 7.6%

51) 0.3%

52) 0.9%

53) 93.75%

54) 56.25%

Write each percent as a fraction in lowest terms by first writing it as a fraction with a denominator of 100.

55) 80.5%

56) 51.2%

57) 9.9%

58) 4.3%

59) 1.58%

60) 2.94%

61) 0.4%

62) 0.6%

63) Do you prefer to use the method in Exercises 45–54 or the method in Exercises 55–62 to change a percent to a decimal? Why?

64) When George writes 11.9% as a fraction, he gets a final answer of $\dfrac{11.9}{100}$. Is this correct? Explain.

Write each percent as a fraction in lowest terms.

65) $62\dfrac{1}{2}\%$

66) $87\dfrac{1}{2}\%$

67) $6\dfrac{1}{4}\%$

68) $81\dfrac{1}{4}\%$

69) $83\dfrac{1}{3}\%$

70) $6\dfrac{2}{3}\%$

71) $25\dfrac{4}{5}\%$

72) $51\dfrac{2}{5}\%$

Objective 5: Change Decimals to Percents

73) Explain how to change a decimal to a percent.

74) Is this statement true or false? If a decimal number is greater than 1, then when it is changed to a percent, it will be greater than 100%.

Write each decimal as a percent.

75) 0.82

76) 0.59

77) 0.08

78) 0.09

79) 0.7

80) 0.4

81) 0.0125

82) 0.0672

83) 8.6

84) 1.7

85) 9

86) 10

Objective 6: Write Fractions as Percents

Write the fraction as a percent by first writing it as a fraction with a denominator of 100.

87) $\frac{7}{10}$

88) $\frac{1}{10}$

89) $\frac{1}{20}$

90) $\frac{1}{25}$

91) $\frac{49}{50}$

92) $\frac{17}{20}$

93) $\frac{31}{10}$

94) $\frac{23}{10}$

Write the fraction as a percent using long division.

95) $\frac{5}{8}$

96) $\frac{3}{8}$

97) $\frac{12}{25}$

98) $\frac{3}{20}$

99) $\frac{9}{5}$

100) $\frac{27}{25}$

101) $\frac{1}{16}$

102) $\frac{13}{16}$

Use a proportion to change the fraction to a percent.

103) $\frac{13}{20}$

104) $\frac{24}{25}$

105) $\frac{7}{40}$

106) $\frac{3}{40}$

107) $\frac{13}{8}$

108) $\frac{9}{8}$

Use a proportion to change the fraction to a percent. Give the exact answer, and also round the answer to the nearest tenth of a percent.

109) $\frac{2}{3}$

110) $\frac{1}{3}$

111) $\frac{1}{9}$

112) $\frac{8}{9}$

113) $\frac{5}{12}$

114) $\frac{7}{12}$

115) $\frac{11}{6}$

116) $\frac{7}{6}$

117) Name three methods for writing a fraction as a percent.

118) Which method do you prefer for writing a fraction as a percent? Why?

Solve each problem. Write your answer as a fraction in lowest terms, as a decimal, and as a percent.

119) A calculus class started the semester with 50 students and ended with 42 students. What portion of the students who started the class finished it?

120) The current polar bear population is about 25,000, and approximately 15,000 of them live in Canada. What portion of the polar bear population lives in Canada? (www.worldwildlifefund.org)

©Ingram Publishing/Age Fotostock

121) A Banana Nut Bread Clif Bar contains 9 g of protein. The recommended daily allowance of protein is 50 g. What portion of the recommended daily allowance of protein is in the Clif bar? (www.clifbar.com)

122) A soccer team won 9 of its 16 soccer games. What portion of the games did the team win?

R Rethink

R1) Explain how your knowledge of decimals and fractions helped you complete the exercises.

R2) Why is it important to be able to write a percent as a fraction or decimal?

R3) Look at Exercise 91. Change it to a percent using the three different methods. Then, compare and contrast the methods. What do you see that is similar?

8.2 Compute Basic Percents Mentally

P Prepare

What are your objectives for Section 8.2?	How can you accomplish each objective?
1 Find 10% of a Number	<ul><li>Write the procedure for **Finding 10% of a Number** in your own words.</li><li>Complete the given examples on your own.</li><li>Complete You Trys 1 and 2.</li></ul>
2 Compute Percents That Are Multiples of 10	<ul><li>Write the procedure for **Finding Percents That Are Multiples of 10** in your own words.</li><li>Complete the given examples on your own.</li><li>Complete You Trys 3–5.</li></ul>
3 Find and Use 5% of a Number	<ul><li>Write the procedure for **Finding 5% of a Number** in your own words.</li><li>Understand how to combine Objectives 2 and 3.</li><li>Complete the given examples on your own.</li><li>Complete You Trys 6–8.</li></ul>
4 Perform Computations with Multiples of 100%	<ul><li>Write the procedure for **Performing Computations with Multiples of 100%** in your own words.</li><li>Complete the given example on your own.</li><li>Complete You Try 9.</li></ul>
5 Solve Applications Containing Basic	<ul><li>Keep in mind the meaning of the word *of* in the context of percent problems.</li><li>Complete the given examples on your own.</li><li>Complete You Trys 10 and 11.</li></ul>

O Organize

W Work

Read the explanations, follow the examples, take notes, and complete the You Trys.

We use percents every day. We may want to buy a purse that is marked down 20% or leave a 15% tip at a restaurant. These computations can be done "in our heads." Let's begin by finding 10% of a number mentally.

1 Find 10% of a Number

EXAMPLE 1

Find:

a) $\dfrac{1}{10}$ of 40   b)   10% of 40

Solution

a)   To find $\dfrac{1}{10}$ *of* 40, we multiply: $\dfrac{1}{10} \cdot 40 = \dfrac{1}{\cancel{10}} \cdot \dfrac{\cancel{40}}{1} = \dfrac{4}{1} = 4$

We learned in earlier chapters that, in a problem like this, *of* means multiply.

b) As in part a), the *of* in 10% *of* 40 means multiply. First, we can change 10% to a decimal or a fraction. Let's change 10% to a decimal.

$$10\% = \frac{10}{100} = \frac{1}{10} = 0.1$$

So, to find 10% of 40, we multiply $0.1 \cdot 40. = 4$.

To multiply a number by 0.1, move the decimal point one place to the left.

Notice that this is the same as the result in a) because $\frac{1}{10}$ and 10% are equivalent.

Note

Recall from our study of decimals that to multiply a number by 0.1, we move the decimal point one place to the left as in part b) of Example 1: $0.1 \cdot 40 = 4$. This fact is very important and will allow us to mentally compute percents that are multiples of 10.

[YOU TRY 1] Find:

a) $\frac{1}{10}$ of 80   b) 10% of 80

Procedure How to Find 10% of a Number

To find 10% of a number, locate the decimal point in the number and move it one place to the left.

Example: 10% of 50. = 5

EXAMPLE 2 Find:

a) 10% of 90   b) 10% of 3500   c) 10% of 78

Solution

a) To find 10% of 90, locate the decimal point in 90 and move it one place to the left:
10% of 90. = 9

b) To find 10% of 3500, locate the decimal point in 3500 and move it one place to the left: 10% of 3500. = 350

c) To find 10% of 78, locate the decimal point in 78 and move it one place to the left:
10% of 78. = 7.8

[YOU TRY 2] Find:

a) 10% of 70   b) 10% of 5100   c) 10% of 16

2 Compute Percents That Are Multiples of 10

Now that we know how to find 10% of a number, let's use that information to find percents that are multiples of 10, like 20%, 30%, 40%, etc.

EXAMPLE 3

Find:

a) 20% of 40   b) 30% of 40

Solution

a) 20% *of* 40 means 0.20 · 40 or 0.2 · 40.

$$\begin{array}{r} 4\,0 \\ \times\ 0.2 \\ \hline 8.0 \end{array}$$

20% of 40 = 8

b) 30% *of* 40 means 0.30 · 40 or 0.3 · 40.

$$\begin{array}{r} 4\,0 \\ \times\ 0.3 \\ \hline 1\,2.0 \end{array}$$

30% of 40 = 12

[YOU TRY 3] Find:

a) 20% of 80   b) 30% of 80

Do you see a pattern? Let's compare both parts of Example 3 to 10% of 40 in Example 1b.

10% of 40 = 4   10% of 40 = 4
20% of 40 = 8   30% of 40 = 12

> **Procedure** How to Find Percents That Are Multiples of 10
>
> If we know the value of 10% of a number,
>
> 1) find 20% of the number by multiplying the value by 2;
> 2) find 30% of the number by multiplying the value by 3;
> 3) find 40% of the number by multiplying the value by 4;
>
> and so on.

If we learn these rules, we can do many percentage calculations in our heads. For example, if we want to know how much money we will save if an item is on sale or if we want to compute the amount of a tip, we can use what we learn here.

EXAMPLE 4

Find 70% of 40.

Solution

In Example 1, we found that 10% of 40 = 4. Therefore, 70% of 40 = 7 · 4 = 28.

[YOU TRY 4]   Find 70% of 80.

EXAMPLE 5   Find:

   a) 60% of 80   b) 90% of 500   c) 20% of 35

Solution

   a) To find 60% of 80, first find 10% of 80. Then, multiply the result by 6.

$$10\% \text{ of } 80. = 8$$
$$60\% \text{ of } 80 = 6 \cdot 8 = 48$$

 Hint

Make up a percent problem on your own, and explain to someone how to compute it mentally.

---

   b) To find 90% of 500, first find 10% of 500. Then, multiply the result by 9.

$$10\% \text{ of } 500. = 50$$
$$90\% \text{ of } 500 = 9 \cdot 50 = 450$$

   c) To find 20% of 35, first find 10% of 35. Then, multiply the result by 2.

$$10\% \text{ of } 35. = 3.5$$
$$20\% \text{ of } 35 = 2 \cdot 3.5 = 7$$

[YOU TRY 5]   Find:

   a) 80% of 50   b) 40% of 700   c) 20% of 41

3   Find and Use 5% of a Number

If we know what 10% of a number equals, then how can we find 5% of the number? We will use the same reasoning that we have used throughout this section. For example, we know that 10% of 40 = 4. What is 5% of 40?

$$10\% \text{ of } 40 = 4$$

5% is *half* of 10%

$$5\% \text{ of } 40 = \frac{1}{2} \cdot 4 = 2$$

Multiply 4 by $\frac{1}{2}$.

Therefore, 5% of 40 = 2. Remember that multiplying a number by $\frac{1}{2}$ is the same as dividing the number by 2.

> **Procedure** How to Find 5% of a Number
>
> If we know the value of 10% of a number, then 5% of the number is *half* of that value.
>
>   *Example:*   Since 10% of 140 = 14, 5% of 140 = $\frac{1}{2} \cdot 14 = 7$.

EXAMPLE 6

Find:

a) 5% of 160

b) 5% of 34

Solution

a) First, find 10% of 160. Then, find 5% of 160 by finding $\frac{1}{2}$ of that result.

$$10\% \text{ of } 160 = 16$$

5% is half of 10%

$$5\% \text{ of } 160 = \frac{1}{2} \cdot 16 = 8$$

Multiply 16 by $\frac{1}{2}$.

W Hint

In your notes, summarize the procedures for computing percents mentally.

b) First, find 10% of 34. Then, find 5% of 34 by finding $\frac{1}{2}$ of that result.

$$10\% \text{ of } 34 = 3.4$$

5% is half of 10%

$$5\% \text{ of } 34 = \frac{1}{2} \cdot 3.4 = 3.4 \div 2 = 1.7$$

Multiply 3.4 by $\frac{1}{2}$.

[YOU TRY 6] Find:

a) 5% of 220

b) 5% of 62

Using what we have learned, how can we find 15% of a number?

EXAMPLE 7

Find 15% of 60.

Solution

To find 15% of 60, find 10% of 60 and 5% of 60, and then add the results.

$$10\% \text{ of } 60 = 6 \qquad 5\% \text{ of } 60 = 3$$

$$15\% \text{ of } 60 = 10\% \text{ of } 60 + 5\% \text{ of } 60$$

$$= \quad 6 \quad + \quad 3 \qquad 5\% \text{ of } 60 \text{ is } 3 \text{ since } \frac{1}{2} \text{ of } 6 \text{ is } 3.$$

$$= 9$$

Therefore, 15% of 60 = 9.

[YOU TRY 7] Find 15% of 40.

We can use the same reasoning to find other percentages.

EXAMPLE 8

EXAMPLE 8   Find 35% of 120.

Solution

We want to think of 35% of 120 as 30% of 120 + 5% of 120. We need to know 10% of 120 to find these values.

$$10\% \text{ of } 120 = 12, \text{ so } 30\% \text{ of } 120 = 36 \text{ and } 5\% \text{ of } 120 = 6$$

$$35\% \text{ of } 120 = 30\% \text{ of } 120 + 5\% \text{ of } 120$$

$$= \quad 36 \quad + \quad 6$$

$$= 42$$

So, 35% of 120 = 42.

> **W Hint**
> Write a new procedure to help you complete exercises similar to Example 8.

[YOU TRY 8]   Find 45% of 20.

Note

In Example 7, we could have found 15% of 60 another way. Since 5% of 60 = 3, then 15% of 60 = 3 · 5% of 60 = 3 · 3 = 9. Likewise, in Example 8, we could have found 35% of 120 another way. Since 5% of 120 = 6, then 35% of 120 = 7 · 5% of 120 = 7 · 6 = 42.

4   Perform Computations with Multiples of 100%

Earlier in this chapter, we learned that 100% = 1. Additionally, we can change other percents to decimal numbers by moving the decimal point in the percent: 200% = 2, 300% = 3, and so on.

   If we remember that 100% = 1, 200% = 2, 300% = 3, etc., we can often perform calculations with multiples of 100% "in our heads."

> **Procedure Perform Computations with Multiples of 100%**
>
> 100% *of* a number equals the number. To find 200% *of* a number, multiply the number by 2, to find 300% *of* a number, multiply the number by 3, and so on.

EXAMPLE 9   Find:

a)   100% of 79   b)   200% of 300   c)   400% of 8

Solution

a)   100% of 79 means 1 · 79 = 79.

b)   200% of 300 means 2 · 300 = 600.

c)   400% of 8 means 4 · 8 = 32.

Find:

a) 100% of 53   b) 300% of 400   c) 600% of 7

5 Solve Applications Containing Basic Percentages

One way we use percentages is to compute a tip.

EXAMPLE 10

A restaurant bill is $28.00. If Charlie wants to leave a 15% tip, find the amount of the tip and the total amount of money Charlie will pay.

Solution

To find the amount of the tip, compute 15% *of* $28.00. Then, add the tip to the $28.00 to determine the total amount Charlie will pay.

So, let's find the amount of the tip, 15% *of* $28.00.

©Don Mason/Blend Images

$$15\% \text{ of } \$28.00 = 10\% \text{ of } \$28.00 + 5\% \text{ of } \$28.00$$
$$= \quad \$2.80 \quad + \quad \$1.40 \qquad 5\% \text{ of } \$28.00 = \$1.40 \text{ since } \frac{1}{2} \text{ of } \$2.80 = \$1.40.$$
$$= \$4.20$$

The tip is $4.20.

Now, find the total amount Charlie will pay.

$$\text{Total amount Charlie will pay} = \text{Amount of the bill} + \text{Amount of the tip}$$
$$= \quad \$28.00 \quad + \quad \$4.20$$
$$= \$32.20$$

The amount of the tip is $4.20, and the total amount Charlie will pay is $32.20.

[YOU TRY 10]

Siddhani wants to leave a 15% tip on the $24.00 restaurant bill. Find the amount of the tip and the final amount she pays.

When we are shopping, we often see signs like "Save 20%!" or "Everything is 30% off!" How do we know whether we are paying the correct amount when we get to the cash registers? (Cash registers DO make mistakes sometimes!) This is when it is very helpful for us to know how to perform mental calculations with percentages. And whether we are figuring out how much money we will save in a store or computing a tip, *rounding* the numbers we are working with can give us a ballpark figure of how much we are saving or how much we should leave as a tip.

EXAMPLE 11

A sign in a store says that all purses are 20% off. Ting likes a purse with a regular price of $49.95. Approximately how much money will she save if she buys this purse, and what is the sale price?

Solution

First, round $49.95 to $50.00 to make the number easier to work with. Find the approximate amount of the discount: 20% *of* $50.00. Then, subtract that amount from $50.00 to determine the approximate sale price of the purse.

Approximate amount of the discount: 20% *of* $50.00 = 2 · $5.00 = $10.00

Now, find the approximate sale price of the purse.

$$\text{Sale price} = \text{Original price} - \text{Amount of discount}$$
$$= \quad \$50.00 \quad - \quad \$10.00$$
$$= \$40.00$$

Ting will save approximately $10.00, and the sale price will be about $40.00.

[YOU TRY 11]

Damien wants to buy a GPS for his car. The regular price is $129.89, and it is on sale for 30% off. Approximately how much money will he save, and what will be the approximate sale price?

ANSWERS TO [YOU TRY] EXERCISES

1) a) 8 b) 8   2) a) 7 b) 510 c) 1.6   3) a) 16 b) 24   4) 56   5) a) 40 b) 280 c) 8.2
6) a) 11 b) 3.1   7) 6   8) 9   9) a) 53 b) 1200 c) 42
10) amount of tip: $3.60; final amount: $27.60
11) approximate amount of the discount: $39.00; approximate sale price: $91.00

E Evaluate **8.2** Exercises   Do the exercises, and check your work.

Objective 1: Find 10% of a Number

 1) Find:

a) $\dfrac{1}{10}$ of 60

b) 10% of 60

3) Find:

a) $\dfrac{1}{10}$ of 320

b) 10% of 320

5) Find:

a) $\dfrac{1}{10}$ of 3650

b) 10% of 3650

2) Find:

a) $\dfrac{1}{10}$ of 50

b) 10% of 50

4) Find:

a) $\dfrac{1}{10}$ of 470

b) 10% of 470

6) Find:

a) $\dfrac{1}{10}$ of 2370

b) 10% of 2370

7) Explain how to find 10% of a number "in your head."

8) Why do $\dfrac{1}{10}$ of a number and 10% of the same number give the same result?

Find 10% of each of the following numbers "in your head."

9) 70

 11) 410

13) 600

15) 1000

17) 3870

19) 23

21) 5

10) 90

12) 850

14) 200

16) 4000

18) 5440

20) 76

22) 8

Objective 2: Compute Percents That Are Multiples of 10

23) Explain how to find 20% of a number "in your head."

24) Explain how to find 30% of a number "in your head."

Find each of the following percentages "in your head."

25) a) 10% of 60
    b) 20% of 60
    c) 70% of 60

26) a) 10% of 20
    b) 20% of 20
    c) 90% of 20

27) a) 10% of 70
    b) 30% of 70
    c) 80% of 70

28) a) 10% of 10
    b) 30% of 10
    c) 40% of 10

29) a) 30% of 200
    b) 70% of 200

30) a) 20% of 100
    b) 80% of 100

31) a) 20% of 800
    b) 50% of 800

32) a) 40% of 600
    b) 90% of 600

33) a) 60% of 110
    b) 40% of 110

34) a) 70% of 120
    b) 30% of 120

35) a) 10% of 31
    b) 20% of 31

36) a) 10% of 42
    b) 20% of 42

Objective 3: Find and Use 5% of a Number

37) Explain how to find 5% of a number "in your head."

38) Explain how to find 15% of a number "in your head."

Find 5% of each of the following numbers "in your head."

39) 80

40) 20

41) 120

42) 140

43) 500

44) 300

45) 400

46) 600

47) 68

48) 84

49) 22

50) 46

Find each of the following percentages "in your head."

51) a) 10% of 60
    b) 5% of 60
    c) 15% of 60

52) a) 10% of 80
    b) 5% of 80
    c) 15% of 80

53) a) 10% of 200
    b) 5% of 200
    c) 15% of 200

54) a) 10% of 400
    b) 5% of 400
    c) 15% of 400

55) a) 10% of 180
    b) 5% of 180
    c) 15% of 180

56) a) 10% of 120
    b) 5% of 120
    c) 15% of 120

Find 15% of each of the following numbers.

57) 20

58) 40

59) 800

60) 600

61) 440

62) 320

63) 500

64) 900

65) 4000

66) 6000

67) 88

68) 66

Find each of the following percentages "in your head."

69) 45% of 80

70) 35% of 40

71) 35% of 60

72) 45% of 20

73) 25% of 200

74) 25% of 600

75) 75% of 120

76) 55% of 120

Objective 4: Perform Computations with Multiples of 100%

Find each of the following percentages "in your head."

77) a) 100% of 35
    b) 200% of 35

78) a) 100% of 23
    b) 200% of 23

79) a) 300% of 5
    b) 500% of 5

80) a) 400% of 7
    b) 900% of 7

81) a) 400% of 30
    b) 600% of 30

82) a) 300% of 80
    b) 700% of 80

83) a) 700% of 200
    b) 200% of 200

84) a) 600% of 500
    b) 500% of 200

Objective 5: Solve Applications Containing Basic Percentages

Solve each problem.

85) A restaurant bill is $24.00. If Joan wants to leave a 15% tip, find the amount of the tip and the total amount of money she will pay.

86) Antonio wants to leave a 15% tip on his $8.00 restaurant bill. Find the amount of the tip and the final amount he pays.

87) Habib wants to leave a 15% tip on his $18.00 restaurant bill. Find the amount of the tip and the final amount he pays.

88) A restaurant bill is $32.00. If Georgette wants to leave a 15% tip, find the amount of the tip and the total amount of money Georgette will pay.

89) Dawn gets a manicure and a pedicure for $60. If she wants to leave a 15% tip, find the amount of the tip and the total amount of money she will pay.

90) Yesla gets a haircut, color treatment, and a deep conditioning hair treatment. She wants to leave a 15% tip. If the total cost for these services is $80.00, find the amount of the tip and the final amount she pays.

©Ingram Publishing

91) Between flights, Russ gets a shoe shine in the airport terminal for $8.00. If Russ wants to leave a 20% tip, find the amount of the tip and the total amount of money he will pay.

92) Kimberly's gardener charges her $35.00 a month for weekly landscape service. If she wants to include a 20% tip with the monthly payment, find the amount of the tip and her final monthly payment.

93) Rene's weekly pool service costs $54.00 per month. If he wants to include a 5% tip with the monthly payment, find the amount of the tip and his final monthly payment.

94) Philip hires a tree trimming service to trim and shape a tree on his property. The service costs $180.00. If he wants to include a 5% tip with the cost, find the amount of the tip and the final cost.

©Denise McCullough

Solve each problem by first rounding the regular price of the item to the nearest dollar.

95) Kelly finds a dress marked 20% off its regular price of $69.95. Approximately how much money will she save if she buys the dress, and what is the approximate sale price?

96) Stella finds a pair of leather boots marked 30% off their regular price of $119.95. Approximately how much money will she save if she buys the boots, and what is the approximate sale price?

97) Amir and his wife Anna like a refrigerator for their new home that is marked 25% off its regular price of $679.95. Approximately how much money will they save if they buy the refrigerator, and what is the approximate sale price?

98) Duyen sees that her favorite jeans are marked 40% off their regular price of $59.95. Approximately how much money will she save if she buys the jeans, and what is the approximate sale price?

99) Daryl finds a wallet marked 20% off its regular price of $29.95. Approximately how much money will he save if he buys the wallet, and what is the approximate sale price?

100) Ely's favorite running shoes are marked 30% off their regular $89.95 price. Approximately how much money will he save if he buys the shoes, and what is the approximate sale price?

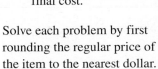
©McGraw-Hill Education

R Rethink

R1) How could you use the techniques you learned about percents to help you with multiplication and division of other numbers?

R2) The next time you are buying something on sale or you are computing a tip, do you feel confident that you could use what you have learned here to estimate how much money you will save or how much tip you should leave?

R3) In your own words, write out an explanation of the methods you learned in this section. Include examples.

8.3 Use an Equation to Solve Percent Problems

What are your objectives for Section 8.3?	How can you accomplish each objective?
1 Find the Percent of a Number Using Multiplication	• Write your own procedure for **Finding a Percent of a Number Using Multiplication.** • Complete the given examples on your own. • Complete You Trys 1 and 2.
2 Use an Equation to Find the Number	• Use the first paragraph of this section to write a procedure. • Complete the given examples on your own. • Complete You Trys 3 and 4.
3 Use an Equation to Find the Percentage	• Follow Example 5 to create a procedure for **Using an Equation to Find the Percentage.** • Complete You Try 5.

Read the explanations, follow the examples, take notes, and complete the You Trys.

At the beginning of Section 8.2, we learned that a statement like *10% of 40* means we multiply $0.1 \cdot 40 = 4$. We also did this calculation mentally: 10% of 40 = 4.

We computed percents like 10%, 15%, 20%, 25%, 100%, and 200% mentally because it is not too complicated to work with multiples of 5% and 10%. But what if we have to find 63% of 84? How should we find this?

1 Find the Percent of a Number Using Multiplication

EXAMPLE 1

Find:

a)   63% of 84   b)   9.5% of 700

Solution

a)   The *of* in 63% *of* 84 means multiply. Change 63% to a decimal, then multiply.

$$63\% \text{ of } 84 = 0.63 \cdot 84 = 52.92$$

W Hint

What would be a quick way to determine whether you were close to the correct answer?

b)   To find 9.5% *of* 700, change 9.5% to a decimal, then multiply.

$$9.5\% \text{ of } 700 = 0.095 \cdot 700 = 66.5$$

[YOU TRY 1]

Find:

a)   47% of 92   b)   3.2% of 180

Many real-world problems involve computations containing percents.

EXAMPLE 2

In 2013, a female pharmacist earned 86.1% of what a male earned for doing the same job. If a male pharmacist earned $108,000 in 2013, how much did his female colleague earn? (www.bls.gov)

Solution

We can think of the problem statement as *a female pharmacist earned 86.1% of $108,000*. What does *of* tell us to do? Multiply.

$$86.1\% \text{ of } \$108,000 = 0.861 \cdot \$108,000 = \$92,988$$

The female pharmacist earned $92,988.

©Plush Studios/Blend Images LLC

[YOU TRY 2]

The number of people who watched the Oscars telecast in 2017 was about 95.6% of the number who watched in 2016. If approximately 34,400,000 people watched the Oscars in 2016, how many watched in 2017? (www.time.com)

2 Use an Equation to Find the Number

We can also use an equation to solve problems like those in Examples 1 and 2 as well as to solve other problems involving percents. To do this, we will first write a statement or question in English and use that to write a mathematical equation. Read the statement slowly and carefully, and think about what the words mean in terms of math.

EXAMPLE 3

Use an equation to find:

a)   29% of 400   b)   130% of 92

Solution

W Hint

Sometimes it is helpful to have an estimate of what the answer should be close to. For instance, what would 30% of 400 be?

a)   Let's think of finding 29% of 400 as the question, "What is 29% of 400?" Remember that we can use a variable, x, to represent the unknown quantity. Therefore,

$$x = \text{the unknown quantity, 29\% of 400}$$

Write the question in English, understand its meaning in terms of math, and write an equation.

English:	What	is	29%	of	400?
Meaning:	What number	equals	29%	times	400?
	↓	↓	↓	↓	↓
Equation:	x	=	0.29	·	400

The equation is $x = 0.29 \cdot 400$. Solve the equation.

$$x = 116 \qquad \text{Multiply.}$$

Therefore, 116 is 29% of 400.

b) Think of 130% of 92 as "What is 130% of 92?" Let x represent the unknown quantity.

$$x = \text{the unknown quantity, 130\% of 92}$$

Write the question in English, understand its meaning in terms of math, and write an equation.

English:	What	is	130%	of	92?
Meaning:	What number	equals	130%	times	92?

| *Equation:* | x | $=$ | 1.30 | $\cdot$ | 92 |

The equation is $x = 1.3 \cdot 92$. Solve the equation.

$$x = 119.6 \qquad \text{Multiply.}$$

Therefore, 119.6 is 130% of 92. Notice that because 130% is greater than 100%, the answer is greater than 92.

[YOU TRY 3] Use an equation to find:

a) 72% of 600   b) 160% of 34

We can use an equation to find other quantities involving percentages.

EXAMPLE 4 Use an equation to solve each problem.

a) 7 is 10% of what number?   b) $3\frac{1}{2}\%$ of what number is 56?

Solution

a) Let x represent the unknown value.

$$x = \text{the number}$$

Hint

Don't just try to memorize procedures. If you begin by writing an equation in English, it will be easier for you to write the correct mathematical equation.

Write the question in English, understand its meaning in terms of math, and write an equation.

English:	7	is	10%	of	what number?
Meaning:	7	equals	10%	times	what number?

| *Equation:* | 7 | $=$ | 0.1 | $\cdot$ | x |

The equation is $7 = 0.1 \cdot x$. Solve it.

$$7 = 0.1x \qquad \text{Write the equation without the multiplication symbol.}$$

$$\frac{7}{0.1} = \frac{0.1x}{0.1} \qquad \text{Divide both sides by 0.1.}$$

$$70 = x \qquad \text{Perform the division.}$$

Therefore, 7 is 10% of 70. We can check the answer by finding 10% of 70.

$$10\% \text{ of } 70 = 7 \quad \checkmark$$

The answer is correct.

b) Let x represent the unknown value.

$$x = \text{the number}$$

Write the question in English, understand its meaning in terms of math, and write an equation.

English:	$3\frac{1}{2}\%$	of	what number	is	56?
Meaning:	3.5%	times	what number	equals	56?
	↓	↓	↓	↓	↓
Equation:	0.035	·	x	=	56

The equation is $0.035 \cdot x = 56$. Solve it.

$$0.035x = 56 \qquad \text{Write the equation without the multiplication symbol.}$$

$$\frac{0.035x}{0.035} = \frac{56}{0.035} \qquad \text{Divide both sides by 0.035.}$$

$$x = 1600 \qquad \text{Perform the division.}$$

Therefore, $3\frac{1}{2}\%$ of 1600 is 56. Check the answer by multiplying.

$$3\frac{1}{2}\% \text{ of } 1600 = 3.5\% \text{ of } 1600 = 0.035 \cdot 1600 = 56 \quad ✓$$

The answer is correct.

[YOU TRY 4] Use an equation to solve each problem.

a)   12 is 20% of what number?   b) $7\frac{1}{2}\%$ of what number is 105?

3 Use an Equation to Find the Percentage

Sometimes we are asked to find a percentage. We can use an equation to do this, too.

EXAMPLE 5   Use an equation to solve each problem.

a)   20 is what percent of 80?   b)   What percent of 3500 is 7?

c)   What percent of 40 is 64?

Solution

a)   We will solve these problems like we solved the problems in Examples 3 and 4. What is the unknown quantity? It is a percent. Let x represent the percent.

$$x = \text{the percent}$$

Write the question in English, understand its meaning in terms of math, and write an equation.

English:	20	is	what percent	of	80?
Meaning:	20	equals	what percent	times	80?
	↓	↓	↓	↓	↓
Equation:	20	=	x	·	80

The equation is $20 = x \cdot 80$. The commutative property says we can also write $x \cdot 80$ as $80 \cdot x$. So, we can think of the equation as $20 = 80 \cdot x$ or $20 = 80x$. Solve this equation.

$$20 = 80x$$

$$\frac{20}{80} = \frac{80x}{80} \qquad \text{Divide both sides by 80.}$$

$$0.25 = x \qquad \text{Perform the division.}$$

The final answer is *not* 0.25 because x represents a percent. The last step is to change 0.25 to a percent.

$$0.25 = 25\% \qquad \text{Change 0.25 to a percent.}$$

Therefore, 20 is 25% of 80. We can check the answer using multiplication.

$$25\% \text{ of } 80 = 0.25 \cdot 80 = 20$$

The answer is correct.

 BE CAREFUL When you are asked to find a percent, the value that you get for x is *not* the final answer. You must change that number to a percent.

 Note
We could have solved $20 = 80x$ by simplifying the fraction $\frac{20}{80}$ like this:

$$20 = 80x$$

$$\frac{20}{80} = \frac{80x}{80} \qquad \text{Divide both sides by 80.}$$

$$\frac{1}{4} = x \qquad \text{Simplify the fraction to solve for } x.$$

Then, rewrite $\frac{1}{4}$ as a percent: $\frac{1}{4} = 25\%$. Either method will give us the same result.

b) To determine *what percent of 3500 is 7,* first let x represent the unknown quantity, the percent.

$$x = \text{the percent}$$

Write the question in English, understand its meaning in terms of math, and write an equation.

English:	What percent	of	3500	is	7?
Meaning:	What percent	times	3500	equals	7?
	↓	↓	↓	↓	↓
Equation:	x	$\cdot$	3500	$=$	7

The equation is $x \cdot 3500 = 7$, or $3500x = 7$. Solve this equation.

$$3500x = 7$$

$$\frac{3500x}{3500} = \frac{7}{3500} \qquad \text{Divide both sides by 3500.}$$

$$x = 0.002 \qquad \text{Perform the division.}$$

Change 0.002 to a percent: $0.002 = 0.2\%$ $\left(\text{Because } 0.2 = \dfrac{2}{10} = \dfrac{1}{5}, \text{ the answer may also be written as } \dfrac{1}{5}\%.\right)$

Therefore, 7 is 0.2% of 3500.

c) In *what percent of 40 is 64,* the unknown quantity is the percent. Let the variable represent this unknown.

$$x = \text{the percent}$$

Write the question in English, understand its meaning in terms of math, and write an equation.

English:	What percent	of	40	is	64?
Meaning:	What percent	times	40	equals	64?
	↓	↓	↓	↓	↓
Equation:	x	$\cdot$	40	$=$	64

The equation is $x \cdot 40 = 64$. The commutative property lets us think of the equation as $40x = 64$. Solve $40x = 64$.

$$40x = 64$$

$$\frac{40x}{40} = \frac{64}{40} \qquad \text{Divide both sides by 40.}$$

$$x = 1.6 \qquad \text{Perform the division.}$$

Change 1.6 to a percent: $1.6 = 160\%$

Therefore, 64 is 160% of 40. Since 64 is greater than 40, it makes sense that the percentage is greater than 100.

[YOU TRY 5] Use an equation to solve each problem.

a) 40 is what percent of 5000? b) What percent of 1800 is 153?

c) What percent of 90 is 117?

Using Technology

We can use a calculator to solve problems involving percent. But first we must learn how to construct and solve equations for these problems.

Suppose we wanted to know, for example, what percent of 2500 is 75? First, set up an equation using x to represent the unknown quantity, and do the calculation by hand. Be sure to write down all the steps so you can understand the order in which the numbers and operations are entered into the calculator. Once you have completed the calculation by hand, enter $\boxed{7}\boxed{5}\boxed{\div}\boxed{2}\boxed{5}\boxed{0}\boxed{0}\boxed{\times}\boxed{1}\boxed{0}\boxed{0}\boxed{=}$ into the calculator. The display will likely show 3. as the result. Because we calculated a percent, we remove the decimal point and write our final answer as 3%. Therefore 3% of 2500 is 75.

Did you get the right answer by hand? Do you understand the order in which the numbers and operations were entered into the calculator?

E Evaluate ## 8.3 Exercises   Do the exercises, and check your work.

Objective 1: Find the Percent of a Number Using Multiplication

Find the percent using multiplication.

1) 57% of 63
2) 29% of 39
3) 76% of 161
4) 93% of 157
5) 6.1% of 300
6) 8.2% of 800
7) 1.9% of 420
8) 4.5% of 540
9) 52.4% of 3600
10) 81.7% of 2200
11) 126% of 95
12) 165% of 72

Solve each problem.

13) In 2015, women working full-time in Indiana earned 76% of what men earned working similar jobs in that state. If a man earned $55,000, what did a woman earn doing a similar job in Indiana? (www.aauw.org)

14) A 2015 study found that 10.8% of high school students in the United States smoke cigarettes. In a high school with 1650 students, how many would be expected to be smokers? Round the answer to the nearest whole number. (www.cdc.gov)

15) The percent of state hospital employees who had nonfatal workplace injuries in 2015 was 7.6%. If a state hospital has 250 workers, how many employees are expected to have a nonfatal workplace injury? (www.bls.gov)

16) Thirty-two percent of adults who do not use the Internet say they do not use it because it is too difficult. In a group of 50 adults who do not use the Internet, how many would be expected to cite this as their reason? (www.pewresearch.org)

17) A bank loan officer recommends that a young couple put 15% down on a new home costing $275,000. What is the amount of the recommended down payment?

©Ingram Publishing

18) Sharon's weekly salary is $2490.00, and 27% of this amount is withheld for taxes. How much of her weekly salary is withheld for taxes?

19) At a community college, approximately 57% of the students are female. If the college has 10,400 students, how many students are female?

20) The Tampa Bay Rays won approximately 42% of their games during the 2016 season. The team played a total of 162 games. How many games did they win? (www.baseball-reference.com)

21) Approximately 81% of households in the United States have at least one high-definition television. Out of 450,000 U.S. households, approximately how many have a high-definition television? (www.leichtmanresearch.com)

©Onoky/SuperStock

22) In 2016, only the wife was employed in 7.1% of married couples in the United States. Out of 2000 married couples, how many women would have been expected to be the only working spouse? (www.bls.gov)

23) In the first quarter of 2017, 78.6% of American home-owners were 65 and older. Out of 250,000 American homeowners, how many are expected to have been 65 and older in the first quarter of 2017? (www.census.gov)

24) Between 2006 and 2010, 11% of American children aged 11–17 repeated a grade in school. In a school district with 7500 students aged 11–17, how many are expected to have repeated a grade between 2006 and 2010? (www.census.gov)

Objective 2: Use an Equation to Find the Number
Use an equation to find each number.

25) 20% of 70

26) 30% of 60

27) 64% of 800

28) 43% of 900

29) 58% of 200

30) 79% of 500

31) 0.6% of 150

32) 0.4% of 140

33) 170% of 13

34) 190% of 18

35) 250% of 57

36) 230% of 96

Use an equation to solve each problem.

37) Find 90% of 150 problems.

38) Find 30% of 40 cars.

39) Find 55% of 700 computers.

40) Find 85% of 20 dogs.

41) Find 0.5% of 1200 coupons.

42) Find 0.3% of 9000 cats.

43) Find 120% of 620 textbooks.

44) Find 110% of 470 in.

Use an equation to solve each problem.

45) 6 is 30% of what number?

46) 12 is 40% of what number?

47) 78 is 65% of what number?

48) 24 is 15% of what number?

49) 75% of what number is 300?

50) 62% of what number is 186?

51) 160% of what number is 72?

52) 180% of what number is 63?

53) $5\frac{1}{2}$% of what number is 110?

54) $6\frac{1}{2}$% of what number is 143?

55) 555 is $92\frac{1}{2}$% of what number?

56) 322 is $80\frac{1}{2}$% of what number?

Objective 3: Use an Equation to Find the Percentage
Use an equation to solve each problem.

57) 24 is what percent of 40?

58) 18 is what percent of 60?

59) 27 is what percent of 60?

60) 28 is what percent of 80?

61) What percent of 400 is 108?

62) What percent of 300 is 222?

63) 170 is what percent of 5000?

64) 688 is what percent of 8000?

65) 323 is what percent of 3400?

66) 90 is what percent of 7500?

67) What percent of 135 is 162?

68) What percent of 102 is 153?

69) What percent of 700 is 4.9?

70) What percent of 900 is 3.6?

Solve each problem.

71) What is 17% of 90?

72) What percent of 2500 is 130?

 73) $7\frac{1}{2}$% of what number is 225?

74) What is 83.5% of 4800 flowers?

75) 40 is what percent of 250?

76) 85% of what number is 221?

77) Find 53.1% of $8610.

78) What percent of 135 is 162?

79) According to a 2014 report, approximately 7.7% of the 700,000 U.S. service members sent to the 1991 Gulf War filed disability claims related to their service. Find the number of veterans who filed disability claims. (www.militarytimes.com)

80) In December 2016, insurance benefit costs were 8% of a private-industry employee's total earnings. If a private-industry employee earned $4500 in December 2016, what was the employee's cost for insurance benefits? (www.bls.gov)

R Rethink

R1) Write a procedure that can help you check your answers using estimation.

R2) Which topics in this section do you know well, and which do you need to practice more?

8.4 Solve Applications Involving Percents

P Prepare **O Organize**

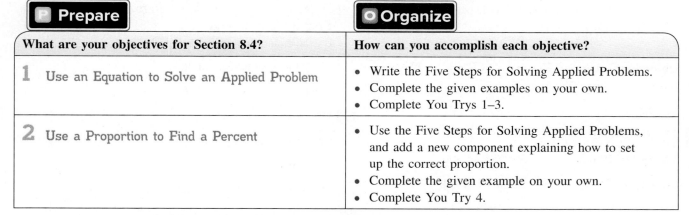

What are your objectives for Section 8.4?	How can you accomplish each objective?
1 Use an Equation to Solve an Applied Problem	• Write the Five Steps for Solving Applied Problems. • Complete the given examples on your own. • Complete You Trys 1–3.
2 Use a Proportion to Find a Percent	• Use the Five Steps for Solving Applied Problems, and add a new component explaining how to set up the correct proportion. • Complete the given example on your own. • Complete You Try 4.

W Work   **Read the explanations, follow the examples, take notes, and complete the You Trys.**

1 Use an Equation to Solve an Applied Problem

In the previous section, we practiced using equations to solve basic percent problems. Now, we will use equations to solve applications involving percents. We will use the five-step problem-solving process we have been using throughout the book.

EXAMPLE 1

Use an equation to solve the problem.

When Monika gets her paycheck every two weeks, she deposits 5% of her take-home pay into her savings account. If she puts $60 into her savings account every two weeks, how much is her take-home pay?

Solution

Step 1: **Read** the problem carefully, and identify what we are being asked to find.

We must find Monika's take-home pay.

Step 2: **Choose a variable** to represent the unknown.

$x =$ Monika's take-home pay

W Hint

Be sure to **write down** what the variable represents!

Step 3: **Translate** the information that appears in English into an algebraic equation.

We can think of the problem as *$60 is 5% of Monika's take-home pay.*
Then, write an equation.

English:	$60	is	5%	of	Monika's take-home pay.
Meaning:	$60	equals	5%	times	Monika's take-home pay.
	↓	↓	↓	↓	↓
Equation:	60	=	0.05	·	x

The equation is $60 = 0.05x$.

Step 4: **Solve** the equation.

$$60 = 0.05x$$
$$\frac{60}{0.05} = \frac{0.05x}{0.05} \qquad \text{Divide both sides by 0.05.}$$
$$x = 1200 \qquad \text{Perform the division.}$$

Step 5: **Check** the answer, and **interpret** the solution as it relates to the problem.

Monika's take-home pay is $1200.

Check: 5% of $1200 = 0.05 · $1200 = $60 ✓ The answer is correct.

[YOU TRY 1] Use an equation to solve the problem.

Daryl's car payment is 7% of his monthly salary. Determine how much Daryl earns each month if his car payment is $210 per month.

EXAMPLE 2

Use an equation to solve the problem.

During the 2015–2016 academic year, the University of Texas at Austin had 50,950 students. Approximately 19.5% of them were Hispanic. How many students were Hispanic? (www.utexas.edu)

©Ariel Skelley/Blend Images LLC

Solution

Step 1: **Read** the problem carefully, and identify what we are being asked to find.

We must determine how many students were Hispanic.

Step 2: **Choose** a variable to represent the unknown.

$$x = \text{the number of Hispanic students}$$

Step 3: **Translate** the information that appears in English into an algebraic equation.

We can think of the problem as *19.5% of students were Hispanic. Find the number of Hispanic students.*

Let's write this as an equation.

English:	19.5%	of	50,950	were	Hispanic
Meaning:	19.5%	times	50,950	equals	number of Hispanics.
	↓	↓	↓	↓	↓
Equation:	0.195	·	50,950	=	x

The equation is $0.195 \cdot 50{,}950 = x$.

Step 4: **Solve** the equation.

$$0.195 \cdot 50{,}950 = x$$
$$9935.25 = x \qquad \text{Multiply.}$$

Step 5: **Check** the answer, and **interpret** the solution as it relates to the problem.

Because we are finding the number of students, we should round to the nearest whole number: $9935.25 \approx 9935$.

In 2015–2016, approximately 9935 students were Hispanic.

Verify that $9935 \approx 19.5\%$ of 50,950.

[YOU TRY 2] Use an equation to solve the problem.

At the 2014 Winter Olympics in Sochi, Russia, the United States won 28 medals. Approximately 32.1% of them were gold medals. How many gold medals did American athletes win? (www.olympic.org)

We can use an equation to find a percentage.

EXAMPLE 3

Use an equation to solve the problem.

In 2010, 14 out of 35 automakers installed oil life monitoring systems in their cars to let drivers know when an oil change is needed. What percent of the automakers used this system? What percent did not use this system? (www.edmunds.com)

Solution

Step 1: **Read** the problem carefully, and identify what we are being asked to find.

We must determine the percent of automakers that installed oil life monitoring systems. Then, determine the percent that did not.

Step 2: **Choose** a variable to represent the unknown.

x = the percent of automakers that installed oil life monitoring systems

(We will find the percent that did *not* install the system later.)

Step 3: **Translate** the information that appears in English into an algebraic equation.

Because 14 out of 35 automakers installed oil life monitoring systems in their cars, we can think of the first part of the problem as *14 automakers is what percent of 35 automakers?*

Write the statement in English; then write an equation.

English:	14 automakers	is	what percent	of	35 automakers?
Meaning:	14 automakers	equals	what percent	times	35 automakers?
	↓	↓	↓	↓	↓
Equation:	14	=	x	·	35

The equation is $14 = x \cdot 35$. We can rewrite it as $14 = 35x$.

Step 4: **Solve** the equation.

$$14 = 35x$$
$$\frac{14}{35} = \frac{35x}{35} \qquad \text{Divide both sides by 35.}$$
$$0.4 = x \qquad \text{Perform the division.}$$

Step 5: **Check** the answer, and **interpret** the solution as it relates to the problem.

Because x represents a percent, change 0.4 to a percent: $0.4 = 40\%$.

So, 40% of the automakers installed the oil life monitoring system. To determine the percent that did not, subtract 40% from 100% since 100% equals all the automakers: $100\% - 40\% = 60\%$.

In 2010, 40% of automakers installed oil life monitoring systems and 60% did not.

Check: To determine the number that installed the system, find 40% of 35: 40% of $35 = 0.4 \cdot 35 = 14$. The number that did not install the system is 60% of $35 = 0.6 \cdot 35 = 21$. Add $14 + 21 = 35$, the total number of automakers. The answer is correct.

[YOU TRY 3]   Use an equation to solve the problem.

A college cheerleading squad has 20 members, and 17 of them were gymnasts before they went to college. What percent of the cheerleaders were gymnasts? What percent were not gymnasts?

2 Use a Proportion to Find a Percent

Recall that percents can be written as fractions. For example, $31\% = \dfrac{31}{100}$, which can be thought of as a ratio, 31 out of 100. Because a proportion is a statement that two ratios or rates are equal, we can also use proportions to find a percent.

EXAMPLE 4   Use a proportion to solve the problem.

Statistics show that about 1 out of 5 Las Vegas tourists are from foreign countries. What percent of the tourists are from outside the United States? (www.lvcva.com)

Solution

Step 1: **Read** the problem carefully, and identify what we are being asked to find.

> We must find the percent of tourists from outside the United States.

Step 2: **Choose** a variable to represent the unknown.

©Brand X Pictures

> We can write *1 foreign tourist out of 5 total tourists* as the rate, or fraction, $\dfrac{1}{5}$.
>
> If we write this fraction with a denominator of 100, **the numerator will be the percent** since percent means *out of 100*. We can do this by writing a proportion. Think of the stated problem as
>
> 1 foreign tourist *is to* 5 total tourists *as* how many foreign tourists *are to* 100 total tourists?
>
> x = the number of foreign tourists out of 100 tourists, or **the percent**

Step 3: **Translate** the information that appears in English into an algebraic equation.

> Write a proportion using x for the percent.
>
> foreign tourists $\rightarrow$ $\dfrac{1 \text{ foreign tourist}}{5 \text{ total tourists}} = \dfrac{x}{100 \text{ total tourists}}$ $\leftarrow$ foreign tourists
> total tourists $\rightarrow$ $\leftarrow$ total tourists
>
> 1 foreign tourist *is to* 5 total tourists *as* x foreign tourists *are to* 100 total tourists.

Step 4: **Solve** the equation. We do not need the units when solving for x.

> $$\frac{1}{5} = \frac{x}{100}$$ Write the equation without the units.
>
> $$5x = 1 \cdot 100$$ Find the cross products.
>
> $$5x = 100$$ Multiply.
>
> $$\frac{5x}{5} = \frac{100}{5}$$ Divide by 5.
>
> $$x = 20$$ Perform the division.

Step 5: **Check** the answer, and **interpret** the solution as it relates to the problem.

Finding that $x = 20$ means that $\dfrac{1}{5} = \dfrac{20}{100}$, and $\dfrac{20}{100} = 20\%$.

Therefore, 20% of Las Vegas tourists are from foreign countries.

Use the cross products to check $x = 20$ in the proportion $\dfrac{1}{5} = \dfrac{x}{100}$.

$$5 \cdot 20 = 100$$

$$\dfrac{1}{5} = \dfrac{20}{100} \quad \text{Substitute 20 for } x.$$

$$1 \cdot 100 = 100$$

The cross products are equal, so the answer is correct.

[**YOU TRY 4**] Use a proportion to solve the problem.

A study done by the Centers for Disease Control and Prevention found that 8 out of 25 births in the United States in 2015 were by Cesarean section. What percent of U.S. births in 2015 were by C-section? (www.cdc.gov)

Note
We could have solved Example 3 using a proportion. Let $x =$ the number of automakers out of 100 that have installed oil life monitoring systems in their cars. This is the same as the **percent**. Then, we can write the proportion $\dfrac{14}{35} = \dfrac{x}{100}$, and solve for x. We get $x = 40$. Therefore, $\dfrac{14}{35} = \dfrac{40}{100}$,

and $\dfrac{40}{100} = 40\%$.

ANSWERS TO [**YOU TRY**] **EXERCISES**

1) $3000   2) 9   3) 85% were gymnasts; 15% were not gymnasts   4) 32%

[E] **Evaluate**   **8.4** Exercises   Do the exercises, and check your work.

Objective 1: Use an Equation to Solve an Applied Problem

Use an equation to solve each problem.

(24) 1) Kathy owns a small business and sets aside 7% of her weekly sales to pay her monthly utility bill. If she set aside $84 for utilities, what was her weekly sales amount?

2) A small liberal arts college has 2300 students enrolled. If 55% of the students are female, how many male students are enrolled in the college?

©Kristy-Anne Glubish/Design Pics

3) In 2012, there were 1,700,000 new cases of diagnosed diabetes among people aged 20 years or older. Approximately 21.8% of those were people aged 20 to 44. How many 20- to 44-year-olds were diagnosed with diabetes in 2012? (www.cdc.gov)

4) Of the approximately 53,000 workers in Berkeley, California, 8.9% of them ride their bikes to work. How many people ride their bikes to work? (www.governing.com)

5) In 2016, 103,100,000 people paid for streaming music services. Approximately 37.8% of those people used Spotify. Determine the number of people who paid for Spotify. (www.digitalmusicnews.com)

6) Watermelon is 91% water. How many pounds of an 8-lb watermelon are not water?

In 2015, there were 28,500,000 nonelderly, uninsured Americans. The pie chart shows the percent of these Americans without health insurance by race/ethnicity. Use this information for Exercises 7–10. (www.kff.org)

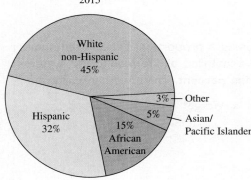

Nonelderly, Uninsured Americans
2015

7) How many of the uninsured were of Hispanic descent?

8) How many of the uninsured were African American?

9) How many of the uninsured were white?

10) How many of the uninsured were Asian/Pacific Islanders?

The table below compares the attitudes of adults 18 and older to teenagers' attitudes with regard to cell phone usage. Use these data for Exercises 11–16.

Adult vs. Teen Attitudes Toward Cell Phone Usage

	% Who Agree	
	Adults	Teens
I feel safer because I can always use my cell phone to get help.	91	93
I like that my cell phone makes it easy to arrange plans with other people.	88	84
I think it's rude when someone repeatedly interrupts a conversation or meeting to check their cell phone.	86	n/a
I get irritated when a call or text on my cell phone interrupts me.	42	48
When I am bored, I use my cell phone to entertain myself.	39	69

(Pew Research Center's Internet & American Life Project, April 29–May 30, 2010)

11) When comparing two equally sized groups of 1500 adults and teens, how many more teens are expected to use a cell phone to entertain themselves when they are bored?

12) When comparing two equally sized groups of 1200 adults and teens, how many more adults are expected to like that their cell phone makes it easy to arrange plans with other people?

13) In a group of 750 adults, how many are expected to think that it's rude when someone repeatedly interrupts a conversation or meeting to check their cell phone?

14) In a large lecture hall containing 300 adult students, how many of these students are expected to use their cell phone for entertainment if they become bored?

15) In a group of 200 teens, how many are expected *not* to feel safer because they can always use their cell phone to get help?

16) In a classroom of 25 teens, how many do *not* like that their cell phone makes it easy to arrange plans with other people?

Use an equation to solve each problem.

17) Hector's parents sent 84 text messages last month. This is 20% of the total recorded on the bill. How many total text messages did the family send last month?

18) Janice bought a coupon book and has already used 33 coupons. If this is 60% of all the coupons, how many coupons were in the book?

19) During a weeklong clearance sale, a bike store sold 30% of its inventory, which amounted to 75 bicycles. How many bikes did the store have in stock?

©DreamPictures/Pam Ostrow/ Blend Images

20) A salesperson earned a commission of $144 one week. If his commission rate is 12%, find the value of the merchandise he sold.

21) Five percent of the units in an apartment building are vacant. If there are six empty apartments, find the total number of units in the apartment building.

22) Alex got 12 questions correct on his math quiz, for a grade of 80%. How many questions were on the quiz?

23) A high school basketball varsity team includes three sophomores. If there are 15 players on the team, what percent of the team were sophomores? What percent were not sophomores?

24) Steve decides to take a mountain bike ride on a 36-mile-loop trail. When his mountain bike mileage gauge indicates that he has traveled 27 miles, what percent of the trail has he traveled? What percent of the trail remains?

25) Susan has $240 deducted from her weekly $1200 salary to pay for child care. What percent of Susan's salary is deducted for child care? What percent is not?

26) Lorena has $54 deducted from her $450 weekly salary to pay for car pool transportation to work. What percent is deducted for the car pool? What percent is not?

©Emely/Cultura RF/Getty Images

27) In a math class of 40 students, five students received an A for the letter grade at the end of the semester. What percent of students received an A for their letter grade? What percent did not?

28) In a large auditorium with 500 seats, 36 seats were reserved for VIP guests. What percent of the seating was reserved for VIP guests? What percent was not?

29) Approximately 2 out of 5 American workers check their email while in the bathroom. What percent is this? (www.techcrunch.com)

30) In 2014, 21 out of 50 American adult workers checked work email while on vacation. What percent is this? (www.forbes.com)

Objective 2: Use a Proportion to Find a Percent
Use a proportion to solve each problem.

31) 70 freshmen is what percent of 280 freshmen?

32) 94 births is what percent of 470 births?

33) 92 mL is what percent of 400 mL?

34) 42 kg is what percent of 120 kg?

35) 2 weeks is what percent of 14 days?

36) 1 day is what percent of 48 hr?

37) 350 mL is what percent of 1.75 L?

38) 1 L is what percent of 1250 mL?

39) In April 2017, 19 out of 25 Allegiant Airlines flights arrived on time. What percent of flights were on time? (www.flightstats.com)

©Ingram Publishing

40) During the 2014–2015 school year, approximately 9 out of 20 high school athletes in Connecticut were female. What percent is this? (www.nfhs.org)

41) In 2016, approximately 27 out of 50 American workers did not use all of their vacation days. What percent is this? (www.marketwatch.com)

42) When American families travel on vacation within the United States, they spend, on average, 13 out of every 50 dollars on lodging. What percent of the family vacation budget is spent on lodging? (www.bls.gov)

43) In 2014, approximately 3 out of 25 American elementary school teachers were male. What percent were male? (www.data.uis.unesco.org)

44) The U.S. Postal Service is one of the leading employers of minorities. In 2017, approximately 2 out of 5 of its employees were from a minority group. What percent of U.S. Postal employees were minorities in 2017? (about.usps.com)

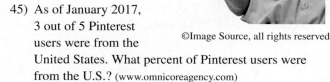

©Image Source, all rights reserved

45) As of January 2017, 3 out of 5 Pinterest users were from the United States. What percent of Pinterest users were from the U.S.? (www.omnicoreagency.com)

46) A human brain weighing 1400 g contains approximately 980 g of water. What percent of the human brain weight is water? (ga.water.usgs.gov)

47) In 2015, there were approximately 97,200 ATV-related injuries in the United States, and 27,216 of those injuries were to children under 16. What percent of all injuries were to children? (www.cpsc.gov)

48) Doctors recommend that adults get 8 hr of sleep per day. What percent of one day is 8 hr? Round the answer to the nearest tenth of a percent. (www.mayoclinic.com)

49) In a classroom of 40 students, six students were absent. What percentage of the students were present?

50) On the first day of classes, a math professor has two seats open in her class of 40 students. With two seats open, what percent of the class is filled?

51) A coed softball team of 12 players has three females. What percentage of the team is male?

52) The Department of Health and Human Services recommends that adults get at least 30 min of exercise per day. What percent of

©Don Mason/Blend Images

one day is 30 min? Round your answer to the nearest tenth of a percent. (www.hhs.gov)

Mixed Exercises: Objectives 1 and 2
Solve each problem.

53) At a college basketball game, 16% of the tickets sold were to the visiting team. If 14,200 tickets were sold, how many were sold to the visitors?

54) A school computer lab has 55 laptop computers for students to check out. If 44 computers are currently checked out, what percentage of the 55 laptop computers is still available for checkout?

55) After a recent flood, 144 people spent the night at a Red Cross shelter. If 25% of the total number of cots were still available, how many cots were in the shelter?

56) Popular belief says that adults should drink eight 8-oz glasses of water per day. According to the Mayo Clinic, this is actually about 61.5% of the recommended daily amount for men. How much water should men drink each day? Round to the nearest whole number. (www.mayoclinic.com)

57) 50 cm is what percent of 2 m?

58) 205 days is what percent of 250 days?

59) If 28 out of 32 chemistry students plan to take a physics course, what percent do not think they will take physics?

60) A survey of 2500 people revealed that 63% of them would take a train to work if they had commuter rail service in their town. How many people said they would like to ride a train to work?

©Mark Williamson/Getty Images

R Rethink

R1) Think of a situation where you needed to determine a percent or a percent of a number. Write a problem similar to the exercises presented, and solve it.

R2) How could you use the Five Steps for Solving Applied Problems to help you in other courses?

Putting It All Together

P Prepare

O Organize

What are your objectives?	How can you accomplish each objective?
1 Review the Concepts of Sections 8.1–8.4	• Be sure that you can apply the objectives you have learned in the previous sections. • If you are not confident about a certain example, go back to the section that gives more explanation. • Be sure to note the summaries and the procedures to refresh your memory on topics already covered. • Complete the given examples on your own. • Complete all the You Trys.

W Work

Read the explanations, follow the examples, take notes, and complete the You Trys.

1 Review the Concepts of Sections 8.1–8.4

In Section 8.1, we said that percent means *out of 100*. Therefore, we can write a percent like 58% as $58\% = \dfrac{58}{100} = 0.58$ or $58\% = \dfrac{58}{100} = \dfrac{29}{50}$. We can change a percent to a fraction with a denominator of 100, and then write it as a decimal or as a fraction in lowest terms.

We can also change a percent to a decimal by moving the decimal point two places to the left and removing the percent symbol.

EXAMPLE 1

Write each percent as a decimal.

a) 16%   b) 0.3%   c) 225%

Solution

a) 16% = 16.% = 0.16   b) 0.3% = 0.003

c) 225% = 225.% = 2.25

Note

A percent greater than 100 is equivalent to a decimal number greater than 1.

[YOU TRY 1] Write each percent as a decimal.

a) 47%   b) 130%   c) 6.8%

To change a decimal to a percent, we move the decimal point two places to the right and then put the percent symbol at the end of the number.

EXAMPLE 2

Write each decimal as a percent.

a) 0.73   b) 0.09   c) 1

Solution

a) 0.73 = 73%   b) 0.09 = 9%   c) 1 = 1. = 100%
└── Put a decimal point
    at the end of the number.

YOU TRY 2

Write each decimal as a percent.

a) 0.008   b) 4   c) 0.51

There are three different ways to change a fraction to a percent.

EXAMPLE 3

Change each fraction to a percent.

a) $\dfrac{5}{8}$ b) $\dfrac{13}{20}$ c) $\dfrac{7}{6}$

Solution

a) Since the denominator, 8, does not divide evenly into 100, we will *not* write this fraction with a denominator of 100. To change $\dfrac{5}{8}$ to a percent, we should use either long division or a proportion. Let's use **long division.**

$$\begin{array}{r} .625 \\ 8\overline{)5.000} \end{array}$$

Therefore, $\dfrac{5}{8} = 0.625 = 62.5\%$ or $62\dfrac{1}{2}\%$.

W Hint

Write down, for yourself, when one method is easier to use than another.

b) Look at the denominator of $\dfrac{13}{20}$. If we multiply 20 by 5, we will get a denominator of 100. Therefore, we will change $\dfrac{13}{20}$ to a percent by **writing the fraction with a denominator of 100.**

$$\dfrac{13}{20} = \dfrac{13}{20} \cdot \dfrac{5}{5} = \dfrac{65}{100} = 65\% \qquad \text{So, } \dfrac{13}{20} = 65\%.$$

c) The denominator of 6 does not divide evenly into 100, so we will not try to write $\dfrac{7}{6}$ with a denominator of 100. We could use either long division or a proportion.

Let's **use a proportion.** Set up the proportion $\dfrac{7}{6} = \dfrac{x}{100}$, where x is the percent, and solve for x.

$$\dfrac{7}{6} \diagup\!\!\!\!\times\!\!\!\!\diagdown \dfrac{x}{100} \quad \begin{array}{l} 6 \cdot x = 6x \\ 7 \cdot 100 = 700 \end{array}$$

$$6x = 700$$

$$\dfrac{6x}{6} = \dfrac{700}{6} \qquad \text{Divide each side by 6.}$$

$$x = 116.\overline{6} \qquad \text{Simplify.}$$

So, $\dfrac{7}{6} = \dfrac{116.\overline{6}}{100} = 116.\overline{6}\%$. If we change $\dfrac{700}{6}$ to the mixed number $116\dfrac{2}{3}$ when solving the equation, then we can also see that $\dfrac{7}{6} = 116\dfrac{2}{3}\%$.

[YOU TRY 3] Change each fraction to a percent.

a) $\dfrac{22}{25}$ b) $\dfrac{17}{12}$ c) $\dfrac{9}{16}$

In Section 8.2, we learned how to compute percentages of some numbers mentally. Recall that to find 10% of a number, we move the decimal point in that number one place to the left. We can also find percents that are multiples of 10.

EXAMPLE 4 Perform each of the calculations mentally.

a) 10% of 120 b) 40% of 120 c) 5% of 120 d) 45% of 120

Solution

a) 10% of 120 = 12   Move the decimal point one place to the left.

b) In part a), we found that 10% of 120 = 12. To find 40% of 120, multiply that result by 4.

$$10\% \text{ of } 120 = 12$$
$$40\% \text{ of } 120 = 4 \cdot 12 = 48$$

40% of 120 = 48

W Hint

In your own words, summarize how to find percents that are multiples of 10%.

c) Because 10% of 120 is 12, 5% of 120 is $\dfrac{1}{2}$ of that result.

5% is *half* of 10%
$$10\% \text{ of } 120 = 12$$
$$5\% \text{ of } 120 = \dfrac{1}{2} \cdot 12 = 6$$

5% of 120 = 6

d) We want to think of 45% of 120 as 40% of 120 + 5% of 120.

$$45\% \text{ of } 120 = 40\% \text{ of } 120 + 5\% \text{ of } 120$$
$$= \quad\;\; 48 \quad\;\; + \quad 6$$
$$= 54$$

So, 45% of 120 = 54.

[YOU TRY 4] Find:

a) 10% of 60 b) 60% of 60 c) 5% of 60 d) 65% of 60

Putting It All Together

We can use these types of mental calculations when we leave a tip at a restaurant or when we buy something on sale.

To solve a problem such as *find 87% of 195*, we multiply because the *of* indicates multiplication:

$$87\% \text{ of } 195 = 0.87 \cdot 195 = 169.65$$

Sometimes, however, it is helpful to use an equation to solve a problem involving percents.

EXAMPLE 5   Use an equation to solve each problem.

a)   160% of what number is 384?   b)   27 is what percent of 36?

Solution

a)   Let x represent the unknown value: $x = $ the number

Write the question in English, understand its meaning in terms of math, and write an equation.

Hint
Are you writing out the example as you are reading it?

English:	160%	of	what number	is	384?
Meaning:	160%	times	what number	equals	384?
	↓	↓	↓	↓	↓
Equation:	1.6	·	x	=	384

The equation is $1.6 \cdot x = 384$. Solve this equation.

$$1.6x = 384 \qquad \text{Write the equation without the multiplication symbol.}$$

$$\frac{1.6x}{1.6} = \frac{384}{1.6} \qquad \text{Divide both sides by 1.6.}$$

$$x = 240 \qquad \text{Perform the division.}$$

Therefore, 160% of 240 is 384.

b)   Begin by letting x represent the unknown quantity: $x = $ the percent

Write the question in English, understand its meaning in terms of math, and write an equation.

English:	27	is	what percent	of	36?
Meaning:	27	equals	what number	times	36?
	↓	↓	↓	↓	↓
Equation:	27	=	x	·	36

The equation is $27 = x \cdot 36$, which we can also write as $27 = 36x$.

$$27 = 36x$$

$$\frac{27}{36} = \frac{36x}{36} \qquad \text{Divide both sides by 36.}$$

$$0.75 = x \qquad \text{Perform the division.}$$

Change 0.75 to a percent: 0.75 = 75%. Therefore, 27 is 75% of 36. We can check the answer using multiplication.

$$75\% \text{ of } 36 = 0.75 \cdot 36 = 27 \qquad ✓$$

    a) 153 is what percent of 180?   b) 8% of what number is 120?

ANSWERS TO [YOU TRY] EXERCISES

1) a) 0.47 b) 1.3 c) 0.068   2) a) 0.8% b) 400% c) 51%

3) a) 88% b) $141.\overline{6}\%$ or $141\frac{2}{3}\%$ c) 56.25% or $56\frac{1}{4}\%$ 4) a) 6 b) 36 c) 3 d) 39

5) a) 85% b) 1500

Putting It All Together Exercises

 **Evaluate** Do the exercises, and check your work.

Objective 1: Review the Concepts of Sections 8.1–8.4

Write each percent as a decimal.

1) a) 96%   b) 0.02%

  c) $4\frac{1}{2}\%$

2) a) 172%   b) $\frac{3}{5}\%$

  c) 23.8%

Write each percent as a fraction or mixed number in lowest terms.

 3) a) 8%   b) $\frac{3}{4}\%$

  c) 250%

4) a) 31%   b) $18\frac{1}{3}\%$

  c) 50.2%

Write each number as a percent.

 5) a) 0.413   b) $\frac{17}{25}$

  c) $\frac{2}{3}$ d) $3\frac{1}{4}$

6) a) 2   b) $\frac{9}{8}$

  c) 0.003   d) $\frac{9}{20}$

In Exercises 7–14, perform the calculations mentally.

 7) Find:

  a) 10% of 20   b) 5% of 20

  c) 30% of 20   d) 35% of 20

8) Find:

  a) 10% of 40   b) 5% of 40

  c) 70% of 40   d) 75% of 40

9) Find 80% of 70.   10) Find 60% of 90.

11) Find 45% of 60.   12) Find 95% of 30.

13) Find 400% of 9.   14) Find 300% of 50.

15) Find 16% of 89.   16) Find 3.8% of 5600.

17) Find $72\frac{1}{4}\%$ of 9000. 18) Find 0.5% of 700.

In Exercises 19–22, solve each problem by doing the calculations in your head.

19) A video game has a regular price of $59.89, and it is on sale for 15% off. Find the approximate amount of the discount and the sale price.

20) A box of diapers has a regular price of $23.99, and it is on sale for 10% off. Find the approximate amount of the discount and the sale price.

21) Leah leaves a 20% tip on an $18.00 restaurant bill. Find the amount of the tip and the total amount Leah pays the server.

22) Marcus' restaurant bill is $40.17, and he wants to leave a 15% tip. Round the bill to the nearest dollar, and compute the tip based upon that amount. How much tip will he leave, and how much does he owe if he adds this tip to the original amount he owes?

Use an equation to solve each problem.

23) 7% of what number is 10.5?

24) What is 38% of 4000?

25) 108 is what percent of 90?

26) 664.3 is 91% of what number?

27) What is $23\frac{1}{2}$% of 8200?

28) What percent of 175 is 28?

Use an equation to solve each problem.

29) A study shows that 3.3% of American adults are vegetarians. If the American adult population is about 245,000,000, how many adults are vegetarians? (www.vrg.org)

©Ingram Publishing

30) In 2014, approximately 2 out of 3 Americans aged 2 years and older visited a dentist. What percent is this? Round the answer to the nearest tenth of a percent. (www.cdc.gov)

31) At the end of 2016, 47% of Netflix subscribers, or 44,086,000 people, were from outside the United States. Determine the *total* number of people who subscribed to Netflix. (www.businessinsider.com)

32) A bag of candy-covered chocolates contained 9 yellow candies. If this is approximately 17.3% of the total, how many candies were in the bag? Round to the nearest whole number.

33) Through 2016, the New York Yankees had won 27 World Series titles, more than any other team. Find the percentage of titles won by the Yankees given that there had been 112 World Series until that time. Round to the nearest tenth of a percent. (mlb.mlb.com)

34) A survey done by the Pew Research Center showed that about 22% of adults used Twitter or a social networking site to get information about the 2010 midterm elections. Given that 2257 adults were surveyed, how many used a social networking site to learn about the elections in 2010? (pewresearch.org)

©takayuki/Shutterstock

R Rethink

R1) Were you able to easily distinguish between the different types of percent problems in this section? If not, what could you do to improve?

R2) Explain how the information you've learned about percents can help you to better understand what you hear in the news or how to be a wiser consumer.

8.5 More Applications with Percents

What are your objectives for Section 8.5?	How can you accomplish each objective?
1 Compute Sales Tax	• Learn the formula to compute **Sales Tax,** and add it to the Five Steps for Solving Applied Problems. • Complete the given examples on your own. • Complete You Trys 1 and 2.
2 Compute Commissions	• Learn the formula to compute **Amount of Commission,** and add it to the Five Steps for Solving Applied Problems. • Complete the given examples on your own. • Complete You Trys 3 and 4.
3 Find a Sale Price	• Learn the formula to compute **Amount of Discount and Sale Price,** and add it to the Five Steps for Solving Applied Problems. • Complete the given example on your own. • Complete You Try 5.
4 Find Percent Increase and Decrease	• Write the procedure for **Finding Percent Increase or Decrease** in your own words, and add it to the Five Steps for Solving Applied Problems. • Complete the given examples on your own. • Complete You Trys 6 and 7.

W Work   **Read the explanations, follow the examples, take notes, and complete the You Trys.**

In this section, we will learn how to solve more everyday problems involving percents: sales tax, commissions, sale price, and percent increase and decrease.

1 Compute Sales Tax

When we buy something, we often pay *sales tax*. **Sales tax** is a percent of the cost of a purchase. Cities, states, and even countries charge sales tax to help fund their governments. We can use this formula to find the amount of sales tax on an item.

> **Formula** Sales Tax
>
> Amount of sales tax = Rate of tax · Cost of item
>
> The rate of tax is usually given as a percent. To use it in the formula, *change the percent to a decimal.*

We will use the five-step problem solving process to solve these applications.

EXAMPLE 1

A pair of jeans costs $39.99. If the sales tax rate is 7%, find the amount of the tax and the total cost of the jeans.

Solution

Step 1: **Read** the problem carefully, and identify what we are being asked to find.

We must determine how much is paid in sales tax and also find the total cost of the jeans.

Step 2: **Choose** a variable to represent the unknown.

We will let x = the amount of sales tax. After finding this value, we will find the total cost of the jeans.

Step 3: **Translate** the information that appears in English into an algebraic equation.

First, use the formula to find the amount of sales tax. Change the sales tax rate from a percent to a decimal: $7\% = 0.07$

$$\text{Amount of sales tax} = \text{Rate of tax} \cdot \text{Cost of item}$$
$$x \quad\quad = \quad 0.07 \quad \cdot \quad \$39.99$$

The equation is $x = 0.07 \cdot \$39.99$.

Step 4: **Solve** the equation.

$$x = 0.07 \cdot \$39.99$$
$$x = \$2.7993 \approx \$2.80 \quad\quad \text{Round to the nearest cent.}$$

Find the total cost of the jeans.

$$\text{Total cost of the jeans} = \text{Price of the jeans} + \text{Amount of sales tax}$$
$$\text{Total cost of the jeans} = \quad \$39.99 \quad + \quad \$2.80$$
$$= \$42.79$$

Step 5: **Check** the answer, and **interpret** the solution as it relates to the problem.

The amount of sales tax is $2.80. The total cost of the jeans is $42.79.

Double-check the calculations to be sure all the arithmetic is correct.

Amount of tax = $0.07 \cdot \$39.99 = \$2.7993 \approx \$2.80$. Then the total cost of the jeans is $\$39.99 + \$2.80 = \$42.79$. ✓

The answer is correct.

[YOU TRY 1] A digital camera costs $189.99. If the sales tax rate is 8%, find the amount of the sales tax and the total cost of the camera.

We can use the sales tax formula to find any quantity in the formula that we do not know.

EXAMPLE 2

A price tag shows that a laptop computer costs $1200. The sales tax on the computer is $102. What is the sales tax rate?

Solution

Step 1: **Read** the problem carefully, and identify what we are being asked to find.

We must find the rate of the sales tax.

Step 2: **Choose** a variable to represent the unknown.

$$x = \text{rate of tax}$$

W Hint

Write out the example as you are reading it.

Step 3: **Translate** the information that appears in English into an algebraic equation.

We will use the formula

Amount of sales tax = Rate of tax · Cost of item

Substitute $102 for the amount of sales tax, x for the rate of tax, and $1200 for the cost of the laptop.

©Creative Crop/Getty Images

$$\text{Amount of sales tax} = \text{Rate of tax} \cdot \text{Cost of item}$$
$$\$102 \quad = \quad x \quad \cdot \quad \$1200$$

The equation is $\$102 = x \cdot \$1200$ or $102 = 1200x$.

Step 4: **Solve** the equation.

$$102 = 1200x$$
$$\frac{102}{1200} = \frac{1200x}{1200} \qquad \text{Divide both sides by 1200.}$$
$$0.085 = x \qquad \text{Perform the division.}$$

Step 5: **Check** the answer, and **interpret** the solution as it relates to the problem.

Because tax rates are usually given as a percent, change the decimal to a percent: $0.085 = 8.5\%$.

The sales tax rate is 8.5%.

To check the answer, find 8.5% of $1200 to see whether the sales tax is $102.

$$0.085 \cdot \$1200 = \$102 \quad \checkmark$$

The answer is correct.

[YOU TRY 2] An engagement ring costs $3600 before tax. If the sales tax is $189, what is the rate of the sales tax?

---

2 Compute Commissions

Salespeople and real estate agents are often paid on *commission*. That is, the amount of money they earn, their **commission,** is a percent *of* total sales. (As in other percent problems we have studied so far, the word *of* indicates multiplication.) The formula for computing commission is similar to the formula used to find the amount of sales tax.

> **Formula** Amount of Commission
>
> $$\text{Amount of commission} = \text{Rate of commission} \cdot \text{Amount of sales}$$
>
> The rate of commission is usually given as a percent. To use it in the formula, *change the percent to a decimal.*

EXAMPLE 3

Vinny sells advertising space on websites, and his commission rate is 7.5%. This month his advertising sales total $35,800. Find the amount of his commission.

Solution

Step 1: **Read** the problem carefully, and identify what we are being asked to find.

We must find the amount of Vinny's commission.

Step 2: **Choose** a variable to represent the unknown.

$$x = \text{the amount of commission}$$

Step 3: **Translate** the information that appears in English into an algebraic equation.

First, use the formula to find the amount of commission. Change the rate of commission from a percent to a decimal: 7.5% = 0.075.

$$\text{Amount of commission} = \text{Rate of commission} \cdot \text{Amount of sales}$$
$$x \qquad = \qquad 0.075 \qquad \cdot \qquad \$35{,}800$$

The equation is $x = 0.075 \cdot \$35{,}800$.

Step 4: **Solve** the equation.

$$x = 0.075 \cdot \$35.800$$
$$x = \$2685$$

Step 5: **Check** the answer, and **interpret** the solution as it relates to the problem.

Vinny's commission is $2685.

Double-check the calculations to be sure all the arithmetic is correct.

Amount of commission = $0.075 \cdot \$35{,}800 = \$2685$ ✓

The answer is correct.

We can use the formula to find the *rate* of commission.

EXAMPLE 4

Last month, Vanitra earned a commission of $520 for selling $13,000 worth of appliances. What is her rate of commission?

Solution

Step 1: **Read** the problem carefully, and identify what we are being asked to find.

We must find Vanitra's rate of commission.

Step 2: **Choose** a variable to represent the unknown.

$$x = \text{rate of commission}$$

W Hint

Remember to **write down** what the variable represents!

Step 3: **Translate** the information that appears in English into an algebraic equation.

We will use the formula

Amount of commission = Rate of commission · Amount of sales

Substitute $520 for the amount of commission, x for the rate of commission, and $13,000 for the amount of sales.

Amount of commission	=	Rate of commission	·	Amount of sales
$520	=	x	·	$13,000

The equation is $520 = x \cdot \$13{,}000$ or $520 = 13{,}000x$.

Step 4: **Solve** the equation.

$$520 = 13{,}000x$$

$$\frac{520}{13{,}000} = \frac{13{,}000x}{13{,}000} \qquad \text{Divide both sides by 13,000.}$$

$$0.04 = x \qquad \text{Perform the division.}$$

Step 5: **Check** the answer, and **interpret** the solution as it relates to the problem.

Because a *rate* of commission is usually given as a percent, change the decimal to a percent: $0.04 = 4\%$

The rate of commission is 4%.

To check the answer, find 4% of $13,000 to see whether the commission is $520.

$$0.04 \cdot \$13{,}000 = \$520 \quad \checkmark$$

The answer is correct.

[YOU TRY 4] One month, a salesman sold $19,800 worth of furniture. If his commission was $990, what was his rate of commission?

3 Find a Sale Price

In Section 8.2, we learned how to find or approximate the sale price of an item when the numbers were easy to work with. We will look at more problems here. Let's review the process we used to find the amount of the discount and then the sale price.

> **Formulas** Amount of Discount and Sale Price
>
> Amount of discount = Rate of discount · Original price
>
> Sale price = Original price – Amount of discount
>
> The rate of discount is usually given as a percent. To use it in the formula, change the percent to a decimal.

EXAMPLE 5   The original price of a winter coat is $120. At the end of the winter season, the coat is on sale for 33% off. What is the sale price of the coat?

Solution

Step 1: **Read** the problem carefully, and identify what we are being asked to find.

   We must find the sale price of the coat. But, first, we must find the amount of the discount. This will be x, the unknown.

Step 2: **Choose** a variable to represent the unknown.

   We will let x = the amount of the discount.
   Then, we will find the sale price.

Step 3: **Translate** the information that appears in English into an algebraic equation.

©Ingram Publishing

   First, use the formula to find the amount of the discount.

   Change the percent to a decimal: 33% = 0.33

$$\underset{x}{\text{Amount of discount}} = \underset{0.33}{\text{Rate of discount}} \cdot \underset{\$120}{\text{Original price}}$$

   The equation is $x = 0.33 \cdot \$120$.

Step 4: **Solve** the equation.

$$x = 0.33 \cdot \$120$$
$$x = \$39.60$$

   The amount of the discount is $39.60.

   Now, find the sale price of the coat.

$$\text{Sale price} = \text{Original price} - \text{Amount of discount}$$
$$\text{Sale price} = \quad \$120 \quad - \quad \$39.60$$
$$= \quad \$80.40$$

Step 5: **Check** the answer, and **interpret** the solution as it relates to the problem.

The sale price of the coat is $80.40.

Double-check the calculations to be sure all the arithmetic is correct.

Amount of discount = 0.33 · $120 = $39.60

Sale price = $120 − $39.60 = $80.40 ✓

The answer is correct.

[YOU TRY 5] The original price of a gas grill is $549. At the end of the summer, it is 35% off. Find the sale price of the grill.

---

4 Find Percent Increase and Decrease

One way to describe an increase or decrease in a number is with a percentage. We call this a **percent change.** Specifically, if a quantity increases, we can find the **percent increase** in the quantity. If a quantity decreases, we can find the **percent decrease** in the quantity.

> **W Hint**
> You will subtract to find the amount of the increase *or* the amount of the decrease.

Procedure How to Find Percent Increase or Decrease

1) First, **find the amount of the increase or decrease** by subtracting the smaller number from the larger number.

2) Then, **use a proportion** to find the percent increase or decrease.

$$\frac{\text{Amount of increase or decrease}}{\text{Original amount}} = \frac{\text{Percent increase or decrease}}{100}$$

EXAMPLE 6 In 1995, a town had a population of 3000. In 2015, its population had grown to 6600. Find the percent increase in the population from 1995 to 2015.

Solution

Step 1: **Read** the problem carefully, and identify what we are being asked to find.

We must find the percent increase in the population from 1995 to 2015.

This is a *percent increase* problem because the population increased from 1995 to 2015.

Step 2: **Choose** a variable to represent the unknown.

$$x = \text{the percent increase}$$

Step 3: **Translate** the information that appears in English into an algebraic equation.

We will use the formula

$$\frac{\text{Amount of increase}}{\text{Original amount}} = \frac{\text{Percent increase}}{100}$$

First, find the *amount* of the increase: $6600 - 3000 = 3600$.

The original amount in the formula is 3000, the population in the earlier year. Set up a proportion.

$$\text{Amount of increase} \rightarrow \frac{3600}{3000} = \frac{x}{100} \leftarrow \text{Percent increase}$$
$$\text{Original amount} \rightarrow$$

Step 4: **Solve** the equation.

Solve the proportion using cross products.

$$\frac{3600}{3000} \gtrless \frac{x}{100}$$
$$\begin{array}{l} 3000 \cdot x = 3000x \\ 3600 \cdot 100 = 360{,}000 \end{array}$$

$3000x = 360{,}000$ Set the cross products equal to each other.

$$\frac{3000x}{3000} = \frac{360{,}000}{3000}$$ Divide both sides by 3000.

$x = 120$ Perform the division.

The percent increase is 120%.

Step 5: **Check** the answer, and **interpret** the solution as it relates to the problem.

The population increased by 120% from 1995 to 2015. This answer makes sense because the *amount* of the increase, 3600, is more than the original population of 3000.

Check: The amount of the increase was 120% of $3000 = 1.2 \cdot 3000 = 3600$.

Then, the population in 2015 was $3000 + 3600 = 6600$. ✓

The answer is correct.

[YOU TRY 6] Five years ago, a start-up company had 15 employees. Now it has 39. Find the percent increase in the number of employees.

In a *percent decrease* application, we will see a *decrease* in a quantity.

EXAMPLE 7

In 2000, Americans smoked a total of 430 billion cigarettes. In 2011, that number was approximately 290 billion. Find the percent decrease in the number of cigarettes smoked. Round the answer to the nearest tenth of a percent. (www.cdc.gov)

Solution

Step 1: **Read** the problem carefully, and identify what we are being asked to find.

We must find the percent decrease in the number of cigarettes smoked.

This is a *percent decrease* problem because the number of cigarettes *decreased* from 2000 to 2011.

Step 2: **Choose** a variable to represent the unknown.

$$x = \text{the percent decrease}$$

Step 3: **Translate** the information that appears in English into an algebraic equation.

We will use the formula

©Kristy-Anne Glubish/Design Pics

$$\frac{\text{Amount of decrease}}{\text{Original amount}} = \frac{\text{Percent decrease}}{100}$$

First, find the *amount* of the decrease: 430 billion − 290 billion = 140 billion.

The original amount in the formula is 430 billion, the number of cigarettes smoked in the earlier year. **Set up a proportion.** We can leave out the unit of billions since they are the same in the number of cigarettes in both years.

$$\text{Amount of decrease} \rightarrow \quad \frac{140}{430} = \frac{x}{100} \quad \leftarrow \text{Percent decrease}$$
$$\text{Original amount} \rightarrow$$

Step 4: **Solve** the equation.

Solve the proportion using cross products.

$$\frac{140}{430} \diagup \frac{x}{100} \quad \begin{matrix} 430 \cdot x = 430x \\ 140 \cdot 100 = 14{,}000 \end{matrix}$$

$$430x = 14{,}000 \qquad \text{Set the cross products equal to each other.}$$

$$\frac{430x}{430} = \frac{14{,}000}{430} \qquad \text{Divide both sides by 430.}$$

$$x \approx 32.6 \qquad \text{Round the quotient to the nearest tenth.}$$

The percent decrease is approximately 32.6%.

Step 5: **Check** the answer, and **interpret** the solution as it relates to the problem.

The number of cigarettes smoked by Americans decreased by approximately 32.6% from 2000 to 2011.

Check: The amount of the decrease was approximately 32.6% of 430 billion: 0.326 · 430 billion = 140.18 billion.

Then, the number of cigarettes smoked in 2011 was approximately 430 billion − 140.18 billion = 289.82 billion. This is close to the number given in the problem, 290 billion. The difference is due to rounding to the tenth of a percent.

The answer is correct.

[YOU TRY 7]

In 1990, 17,695,000 people had manufacturing jobs in the United States. That number fell to approximately 12,398,000 in 2017. Find the percent decrease in the number of Americans working in manufacturing. (www.bls.gov)

ANSWERS TO [YOU TRY] EXERCISES

1) amount of tax: $15.20; total cost: $205.19   2) 5.25%   3) $10,440   4) 5%   5) $356.85
6) 160%   7) 29.9%

Using Technology

Simple arithmetic calculators often have a percent key that can be used to quickly calculate the sale price of an item. The percent key is generally labeled %. Suppose the original price of a cookware set is $150. During a clearance sale, the cookware is on sale for 35% off. To find the sale price of the cookware, we enter [1][5][0][−][3][5][%] into the calculator. The display will likely show 97.5 as the result. This means that the sale price of the cookware is $97.50.

E Evaluate **8.5** Exercises   Do the exercises, and check your work.

Objective 1: Compute Sales Tax
Solve each problem.

 1) A pair of tennis shoes costs $55. If the sales tax rate is 6%, find the amount of the tax and the total cost of the shoes.

2) An office supply store sells software to make personal business cards for $32. If the sales tax rate is 5%, find the amount of the tax and the total cost of the software.

3) A price tag shows that a wireless telephone system costs $78. If the sales tax rate is 8%, find the amount of the tax and the total cost of the phone system.

4) A price tag shows that a portable DVD player costs $56. If the sales tax rate is 7%, find the amount of the tax and the total cost of the DVD player.

5) A 4-GB memory stick costs $4. If the sales tax rate is 7.75%, find the amount of the tax and the total cost of the memory stick.

6) A gel ink ballpoint pen and pencil set costs $8. If the sales tax rate is 8.25%, find the amount of the tax and the total cost of the set.

7) The price of a pay-as-you-go cell phone is $75. If the sales tax rate is 7.8%, find the amount of the tax and the total cost of the set.

8) The sales tax rate for an online purchase of a four-person tableware set was 8.5%. If the purchase price was $32, find the amount of the tax and the total cost of the set.

9) A microwave oven is on sale for $54. Find the amount of the tax and the total cost of the microwave if the sales tax rate is 7%.

©Masterchief_Productions/ Shutterstock

10) A package of four avocados costs $7.50. Find the amount of the tax and the total cost of the avocados if the sales tax rate is 6%.

Find each sales tax rate.

11) A picture frame sold for $12, and the amount of the sales tax was $0.96. What was the sales tax rate?

12) A computer monitor sells for $110, and the sales tax is $9.90. What is the sales tax rate?

13) A six-pack of soft pastels costs $26. Hailey also pays $1.17 for sales tax. What is the sales tax rate?

14) A two-day sale is offering a cookware set for $140. If the sales tax amount for the purchase is $11.62, what is the sales tax rate?

15) A sweater costs $79.99 plus $6.76 in sales tax. What is the sales tax rate? Round the answer to the nearest hundredth of a percent.

16) Suppose the regular price of headphones is $39.99. If the sales tax amount is $3.54, what is the sales tax rate? Round the answer to the nearest hundredth of a percent.

17) Janessa's total bill for a dress was $91.77, which included $7.77 for sales tax. What was the sales tax rate?

18) A coffeemaker costs $64.71, including the sales tax. If the tax was $4.71, what was the sales tax rate?

19) Charles paid $1363.20 for a road bike, including tax. Find the sales tax rate if the amount of the tax was $83.20.

©Eric Misko/Elite Images/ McGraw-Hill Education

20) The cost of a new desktop computer, including tax, is $1446.24. If $66.24 of this is sales tax, what is the sales tax rate?

Objective 2: Compute Commissions
Solve each problem.

 21) How much commission will a real estate agent earn on a sale of a $425,000 home if the agent is paid 3.5% commission on the selling price?

©Ariel Skelley/Blend Images LLC

22) A cosmetics salesperson is paid a 12.5% rate on beauty products she sells from her home. Find her commission if she sold $1650 worth of products.

23) Edina receives 4% commission for finding customers who want the exterior of their homes painted. If she finds two customers who will each pay $3500 for the job, what is Edina's total commission?

24) Randy is paid 3% commission to find homeowners who want to have their roofs replaced. If he arranges two jobs, each costing $8500, how much commission does Randy earn for the two jobs?

25) A Caribbean vacation property sales agent receives 5% commission for each property she sells. If she sells three $150,000 condos, how much commission does she earn?

26) Suzanne earns 6% commission at an electronics warehouse store. How much commission will Suzanne earn on the sale of three HD televisions each worth $1250?

©Juice Images/Alamy

27) A shoe salesman sold 14 pairs of shoes, giving him total sales of $950 for the day. If he received a commission of $114, what is his commission rate?

28) Junji received a commission check for $1020 for his monthly sales of $6800. What is his commission rate?

29) A medical equipment salesperson sold six refurbished ultrasound machines to local hospitals for $3000 each, earning a commission of $2880. What is the salesperson's commission rate?

30) A dental X-ray machine salesperson sold five $1800 machines. If the salesperson received a $1260 commission check for the sales, what is her commission rate?

31) Joanna sells cleaning supplies as a door-to-door salesperson working on a commission. If she received a commission check for $293.75 for selling $2350 worth of cleaning supplies, what is her commission rate?

32) A pawn shop salesperson made $343.65 commission for selling $2370 worth of pawned merchandise. What is his commission rate?

Objective 3: Find a Sale Price
Solve each problem.

33) A pair of leather gloves that regularly sells for $54 is marked 30% off. What is the sale price of the gloves?

34) A handbag that regularly sells for $180 is marked 40% off. What is the sale price of the handbag?

35) An outlet store is selling a designer dress that regularly sells for $235 at 35% off. What is the sale price of the dress?

36) A ski outlet store is selling its ski boots at 45% off. If the boots regularly sell for $240, what is the sale price?

37) A pet groomer has a 25%-off weekend special on a pet grooming package for large dogs, which includes a flea-killing shampoo and a hand-cut trim. What is the sale price if the package normally costs $90?

©Ingram Publishing/Alamy

38) The regular price of a graphite fishing pole is $86. What is the sale price if the fishing pole is marked 35% off?

39) A caterer offers a 25% discount per guest to wedding parties serving over 200 guests. The regular price per guest is $12. If Sandra hires the caterer to serve 250 guests at her wedding reception, what is the total cost to serve all 250 guests?

40) A 20% group discount rate is given to groups larger than 20 people attending a play. If the tickets regularly sell for $45 per person, what is the discounted price for a group of 30 people to attend the play?

41) The regular price of a raincoat is $79.99, and it is on sale for 25% off. Find the total cost of the raincoat if the sales tax rate is 7.5%.

42) During an end-of-the-year clearance sale, a department store is selling its most popular pair of jeans at 60% off. If the regular price of the jeans is $124.99, find the total cost if the sales tax rate is 8.5%.

43) During a Memorial Day sale, a retailer sells bedroom sets at 35% off. If a bedroom set regularly sells for $1259.99, find the total cost if the sales tax rate is 9%.

44) An office supply store sells a computer desk for $349.99. If the desk is marked 40% off, find the total cost if the sales tax rate is 9.5%.

Objective 4: Find Percent Increase and Decrease
Solve each problem.

45) It took $40 for Stephanie to fill her tank with gas four months ago. Today, it takes $50 to fill her tank. Find the percent increase in price.

©ImageSource/Age fotostock

46) Anthony scored 60% on his first math exam. He was determined to do better on his second math exam by studying more. His hard work paid off, and he scored 96%. Find the percent increase in the exam score.

47) Approximately 50 years ago, a U.S. postage stamp cost 4 cents. In 2017, it cost 49 cents. Find the percent increase in price. (www.usps.com)

48) When Tatiana first bought her Japanese maple tree, it was 2 ft high. It is now 9 ft high. Find the percent increase in height.

49) Sebastian bought a 4-GB flash drive two years ago for $50. Today he bought one for $8. Find the percent decrease in price.

50) Cinta bought her first car four years ago for $15,000. Today she purchased a new car, and the dealership gave her $1500 for her old car. Find the percent decrease in the value of Cinta's first car.

51) The number of daily active Snapchat users in the fourth quarter of 2016 was 158 million. In the first quarter of 2017, that number rose to 166 million. Find the percent increase in the number of daily active Snapchat users. Round the answer to the nearest percent. (techcrunch.com)

52) In 1896, the first modern Olympics in Athens had 14 teams. In 2016, the Summer Olympics in Rio de Janeiro had 207 teams. Find the percent increase in the number of teams participating in the Olympics. Round the answer to the nearest tenth of a percent. (www.olympic.org)

53) The median sale price of new homes in the United States in April 2016 was $321,300. In April 2017, it was $309,200. Find the percent decrease in price. Round the answer to the nearest tenth of a percent. (www.census.gov)

54) The average sale price of new American homes in April 2016 was $380,000. In April 2017, it was $368,300. Find the percent decrease in price. Round the answer to the nearest tenth of a percent. (www.census.gov)

55) In 2007, Christina was 3 ft 4 in. tall. Ten years later, she was 5 ft 10 in. tall. Find the percent increase in height.

56) Anthony discovered the shoes he wore as a toddler in his parents' attic. As a toddler, his feet were approximately 56 mm long. His feet are now 19.6 cm long. Find the percent increase in length.

©IC Squared Studios/Getty Images

Mixed Exercises: Objectives 1–4
Solve each problem.

57) Sophie is a jewelry sales representative who is paid 12.5% commission on her weekly sales. If her weekly sales totaled $3800, what was the amount of her commission?

58) Tadas sometimes rides his bike to work, adding 40 min to his commute time. It takes Tadas 10 min to get to work if he travels by car. Find the percent decrease in commute time when Tadas drives his car to work.

59) A table saw with a price tag of $180 is marked 33% off. What is the sale price for the table saw?

60) Adam is a fitness machine sales representative who is paid on commission. If he received a commission of $238 on an elliptical machine he sold for $2800, what was Adam's commission rate?

61) The sales tax amount for a new wireless router is $3.90. If the router costs $60, what is the sales tax rate?

62) A price tag shows that a skateboard costs $80. Find the amount of the sales tax and the total cost of the skateboard if the sales tax rate is 5.25%.

©Ingram Publishing /SuperStock

63) In 2011, Subway had 35,920 stores worldwide. In 2016, they had 44,702 stores. Find the percent increase in the number of Subway stores. Round the answer to the nearest tenth of a percent. (www.statista.com)

64) A washer-dryer pair regularly sells for $1250. During a Memorial Day sales event, the pair is marked down by 40% off. What is the sale price for the washer-dryer pair?

R Rethink

R1) Are computing sales tax and finding the amount of commission similar? If so, how?

R2) How can you use the objectives you just mastered every day?

8.6 Simple and Compound Interest

P Prepare   **O Organize**

What are your objectives for Section 8.6?	How can you accomplish each objective?
1 Find Simple Interest	• Know the definitions of *interest, simple interest,* and *principal.* • Learn the formula for finding **Simple Interest.** • Write a procedure for computing the original amount plus interest. • Complete the given examples on your own. • Complete You Trys 1–4.
2 Understand Compound Interest	• Know the definitions of *compound interest* and *compounded annually.* • Understand the process of finding compound interest using $I = P \cdot R \cdot T$. • Complete the given example on your own. • Complete You Try 5.
3 Solve Compound Interest Problems	• Learn the formula for computing **Compound Interest.** • Complete the given examples on your own. • Complete You Trys 6 and 7.

 **W Work**   Read the explanations, follow the examples, take notes, and complete the You Trys.

1 Find Simple Interest

Why do banks pay us interest when we deposit money in a bank account? Why do *we* pay interest when we borrow money? What *is* interest? Let's look at two examples.

> Kelly has a savings account containing $4700. The bank pays Kelly *interest*, or a *fee*, to use her money to loan to other people and businesses.

> Omar wants to borrow $12,000 from his credit union to buy a car. The credit union will loan him the money, but it will charge him *interest*, or a *fee*, to borrow this money. In the end, Omar will owe the credit union more than the $12,000 that he borrowed.

Interest is a fee paid for borrowing money. Banks pay interest to customers who deposit money in savings accounts for using their money to loan to people and businesses. This is why people earn interest on bank accounts. Likewise, when we borrow money from banks, credit unions, and credit card companies, we pay them interest. Often, interest is computed as a percentage of the money borrowed for a given length of time.

There are different ways to compute interest. First, we will learn about *simple interest*. **Simple interest** is calculated based on the original amount of money borrowed. The original amount of money borrowed is called the **principal.**

Here is the formula for computing simple interest.

Formula Simple Interest

$$\text{Interest} = \text{Principal} \cdot \text{Rate} \cdot \text{Time}$$
$$I = P \cdot R \cdot T$$

P = the principal. This is the original amount of money borrowed or the original amount of money in a bank account.

R = the interest rate. It is usually given as a percent. **We must change it to a decimal when we use it in the equation.**

T = the time, *in years*, that the money will be borrowed or the length of time an amount of money is in an account.

 When using the formula $I = P \cdot R \cdot T$:

1) the interest rate *must* be in decimal form.
2) the time *must* be in years.

EXAMPLE 1   Saori borrows $1200 for 1 year at an interest rate of 7%. Find the interest.

Solution

Use the formula $I = P \cdot R \cdot T$.

The principal, P, is the amount of money borrowed: $P = \$1200$.

The interest rate, R, is 7%. Change this to a decimal: $R = 0.07$.

The time, T, is the length of the loan, 1 year: $T = 1$.

$$I = \quad P \cdot R \cdot T$$
$$I = (1200)(0.07)(1) \quad \text{Substitute the values.}$$
$$I = 84 \quad\quad\quad\quad\quad \text{Multiply.}$$

The interest is \$84.

$\left[\text{YOU TRY 1}\right]$ Gao borrows \$1400 for 1 year at an interest rate of 5%. Find the interest.

EXAMPLE 2 Arkady borrows \$960 for 3 months at $6\frac{1}{2}\%$ interest. How much interest will he owe?

Solution

Change the interest rate to a decimal: $R = 6\frac{1}{2}\% = 6.5\% = 0.065$

Rewrite the time, T, in terms of years: 3 months $= \dfrac{3}{12}$ yr $= \dfrac{1}{4}$ yr

Identify P, R, and T:

$$P = \$960 \qquad R = 0.065 \qquad T = \frac{1}{4}$$

> ### W Hint
> If the time is not given in years, you must rewrite it in terms of years.

$$I = \quad P \cdot R \cdot T$$
$$I = (960)(0.065)\left(\frac{1}{4}\right) \quad \text{Substitute the values.}$$
$$I = (62.4)\left(\frac{1}{4}\right) \quad\quad \text{Multiply the first two numbers.}$$
$$I = 15.60$$

Arkady will owe \$15.60 in interest.

Using Technology

We can use a calculator to quickly calculate I where $I = (960)(0.065)\left(\frac{1}{4}\right)$. To calculate $(960)(0.065)\left(\frac{1}{4}\right)$, we enter $\boxed{9}\,\boxed{6}\,\boxed{0}\,\boxed{\times}\,\boxed{.}\,\boxed{0}\,\boxed{6}\,\boxed{5}\,\boxed{\div}\,\boxed{4}\,\boxed{=}$ into the calculator. The display will likely show 15.6 for the result. Since we are dealing with money, we will round to the nearest hundredth and write our final answer as \$15.60 for the value of I.

$\left[\text{YOU TRY 2}\right]$ Celia borrows \$450 at $7\frac{1}{2}\%$ interest for 4 months. How much interest will she owe?

To determine the *total* amount of money that must be paid back on a loan or to determine the *total* amount of money in a bank account after being paid interest, we must add the interest to the original amount borrowed or deposited in an account.

EXAMPLE 3

Freija deposited $2850 into an account at 5% interest for 18 months. How much will be in her account after this time?

©Keith Brofsky/Getty Images

Solution

First, find the amount of interest Freija will earn.

Change 18 months to years: 18 months $= \dfrac{18}{12}$ yr $= \dfrac{3}{2}$ yr or 1.5 yr

Use $I = P \cdot R \cdot T$.

$$P = \$2850 \qquad R = 0.05 \qquad T = 1.5$$
$$\uparrow$$
Change the percent to a decimal.

$$I = \quad P \quad \cdot \quad R \cdot T$$
$$I = (2850)(0.05)(1.5) \qquad \text{Substitute the values.}$$
$$I = 213.75 \qquad \text{Multiply.}$$

Freija will earn $213.75 in interest.

Add $213.75 to the amount deposited to determine the total amount in the account after 18 months.

$$\text{Total amount} = \text{Amount deposited} + \text{Amount of interest}$$
$$= \qquad \$2850 \qquad + \qquad \$213.75$$
$$= \$3063.75$$

After 18 months, Freija will have $3063.75 in her account.

[YOU TRY 3]

Javier deposited $5720 into an account for 30 months at 6% interest. How much money will be in his account after this time?

Many people (maybe even you) borrow money to buy a car. If that's the case, when you figure in the interest, how much does the car *really* cost?

In the beginning of the section, we said that if Omar borrows $12,000 to buy a car, he will pay interest on that money. Let's determine the total amount he will owe for borrowing that money.

Tara deposits $2000 into an account for 3 years at 6% interest.

a)   Determine the amount of interest if she earns *simple interest* for 3 years. Then, find the total amount in her account after 3 years.

b)   Determine the amount of interest she earns if the interest is *compounded annually* for 3 years. Then, find the total amount in her account after 3 years.

c)   Will the amount of money in her account after 3 years be different if the interest is computed differently? Explain your answer.

3 Solve Compound Interest Problems

We can use a single formula to find the amount in an account when interest is compounded in different ways.

Formula Compound Interest

Let A = the total amount in an account after T years. Use this formula to find the value of A when interest is compounded more than one time per year:

$$A = P\left(1 + \frac{R}{n}\right)^{n \cdot T}$$

P = the principal, R = the interest rate in decimal form, and n = the number of times interest is compounded each year.

EXAMPLE 6   Steve invests $6000 at 3% interest compounded annually for 5 years. Find the amount of money in his account after 5 years and the amount of interest earned.

Solution

Use $A = P\left(1 + \frac{R}{n}\right)^{n \cdot T}$. We want to find A, the total amount after 5 years.

Identify all of the values that we will use in the formula, substitute them into the formula, and use a calculator to evaluate.

$$P = 6000 \qquad R = 3\% = 0.03 \qquad T = 5 \qquad n = 1$$

W Hint

Be sure to write down the value of each variable on your paper.

$$A = P\left(1 + \frac{R}{n}\right)^{n \cdot T}$$

$$A = 6000\left(1 + \frac{0.03}{1}\right)^{1 \cdot 5} \qquad \text{Substitute the values.}$$

$$A = 6000(1 + 0.03)^5 \qquad \text{Simplify } \frac{0.03}{1} \text{ and multiply } 1 \cdot 5.$$

$$A = 6000(1.03)^5 \qquad \text{Add inside the parentheses.}$$

$$A = 6955.64 \qquad \text{Evaluate. Round the answer to the nearest hundredth.}$$

After 6 years, the account will contain $6955.64.

Interest earned = $6955.64 − $6000 = $955.64.

To calculate A, where $A = 6000(1.03)^5$, using a calculator, we will need to use the exponent function key. In most cases, the exponent function key is labeled $\boxed{y^x}$ or $\boxed{\wedge}$. To calculate $6000(1.03)^5$ following the order of operations, we enter $\boxed{1}\ \boxed{.}\ \boxed{0}\ \boxed{3}\ \boxed{y^x}$ $\boxed{5}\ \boxed{=}\ \boxed{\times}\ \boxed{6}\ \boxed{0}\ \boxed{0}\ \boxed{0}\ \boxed{=}$ into the calculator. The display will likely show 6955.644446 for the result. Because we are dealing with money, we will round to the nearest hundredth and write our final answer as $6955.64 for the value of A.

[YOU TRY 6]

Dimos invests $10,000 at 5% interest compounded annually for 7 years. Find the amount of money in the account after this time and the amount of interest earned.

Interest can be compounded more often than every year. It can be compounded

W Hint

Learn these relationships!

Semiannually	Quarterly	Monthly	Weekly	Daily
2 times per year	4 times per year	12 times per year	52 times per year	365 times per year (some banks use 360)

EXAMPLE 7

Divya invests $8300 at $5\frac{1}{2}\%$ interest for 6 years compounded quarterly.

a) Find the amount of money in her account after 6 years.

b) How much interest did Divya earn?

Solution

a) Use $A = P\left(1 + \dfrac{R}{n}\right)^{n \cdot T}$. We have to find A, the total amount in Divya's account

after 6 years. Because interest is **compounded quarterly**, or **4** times per year, $n = 4$. Identify all the values that we will use in the formula, substitute them into the formula, and use a calculator to evaluate.

$$P = 8300 \qquad R = 5\frac{1}{2}\% = 5.5\% = 0.055 \qquad T = 6 \qquad n = 4$$

$$\uparrow$$

Compounded quarterly means 4 times per year.

$$A = P\left(1 + \frac{R}{n}\right)^{n \cdot T}$$

$$A = 8300\left(1 + \frac{0.055}{4}\right)^{4 \cdot 6} \qquad \text{Substitute the values.}$$

$$A = 8300(1 + 0.01375)^{24} \qquad \text{Divide } \frac{0.055}{4} \text{ and multiply } 4 \cdot 6.$$

$$A = 8300(1.01375)^{24} \qquad \text{Add inside the parentheses.}$$

$$A = 11{,}519.11 \qquad \text{Evaluate } (1.01375)^{24} \text{ first; then multiply by 8300.}$$

After 6 years, the account will contain $11,519.11.

b) Interest earned = $11,519.11 − $8300 = $3219.11

YOU TRY 7

Matt invests $7800 at $6\frac{1}{2}$% interest for 5 years compounded monthly. Find the amount of money in the account after 5 years, and determine the amount of interest Matt earned.

ANSWERS TO **YOU TRY** **EXERCISES**

1) $70   2) $11.25   3) $6578   4) $18,410

5) a) $360; $2360   b) $382.03; $2382.03   c) If the interest is compounded yearly, Tara will have $22.03 more in her account than if interest were computed as simple interest.

6) total amount: $14,071.00; amount of interest: $4071.00

7) total amount: $10,785.98; amount of interest: $2985.98

E Evaluate **8.6** Exercises   Do the exercises, and check your work.

Objective 1: Find Simple Interest

Solve each problem.

1) Marshall borrows $1600 for 1 year at an interest rate of 6%. Find the interest.

2) Donnell borrows $900 for 1 year at an interest rate of 5%. Find the interest.

3) Bret borrows $1400 for 2 years at an interest rate of 4%. Find the interest.

4) Carrie borrows $800 for 2 years at an interest rate of 8%. Find the interest.

5) If Gustavo borrowed $1580 from a lender who charges 4.25% interest, how much interest does Gustavo owe after 4 years?

6) If a person borrows $1750 from a lender who charges 3.75% interest, how much interest does the person owe after 4 years?

7) Zachary deposits $6000 into an account earning 4% interest for $3\frac{1}{2}$ years. How much interest will he earn?

8) Kashif deposits $3000 into an account earning 6% interest for $2\frac{1}{4}$ years. How much interest will he earn?

Write each time in terms of years.

 9) 2 months

10) 9 months

11) 8 months

12) 6 months

13) 26 weeks

14) 4 weeks

15) 65 weeks

16) 78 weeks

17) Noor borrows $6000 for 10 months at 4% interest. How much interest will she owe?

18) Raymond borrows $2000 for 3 months at a 5% interest rate. How much interest will he owe?

19) If Tyler borrowed $1950 for 40 weeks at $5\frac{3}{4}$% interest, how much interest did he owe?

20) If Caroline borrowed $1200 for 13 weeks at a $4\frac{1}{4}$% interest rate, how much interest will she owe?

21) If Yasmeen borrows $15,000 for 48 months at $3\frac{1}{2}$% interest, how much interest will she owe?

22) If Aleksos borrows $18,000 for 60 months at a $5\frac{1}{2}$% interest rate, how much interest will he owe?

Solve each simple interest problem.

23) A bank loaned Baharah $1500 at an interest rate of 8%. If she wants to pay back the loan in 6 months, how much does Baharah pay back to the bank including interest?

24) Wade took out an emergency student loan for $900 at an interest rate of 4.5%. If Wade wants to pay back the loan in 3 months, how much does he pay back to the lender, including interest?

25) A bank loaned Roger $2000 at an interest rate of 6.5%. If he wants to pay back the loan in 15 months, how much does he pay back to the bank, including interest?

26) Sonia took out a personal loan for $700 at an interest rate of 5.5%. If Sonia wants to pay back the loan in 18 months, how much does she pay back to the lender, including interest?

27) Rene puts $3000 into a savings account at a 4% interest rate. How much is in the account after two years?

28) Sukhon decides to invest $5000 in an account that pays 3.5% interest. If she does not withdraw any money, how much is in the account after 5 years?

29) A bank loaned Tessa $4500 at an interest rate of $3\frac{3}{4}$%. If her payment plan is over 3 years, how much does she owe?

30) Cleveland borrowed $1500 from a lender charging $5\frac{1}{4}$% interest. How much does Cleveland owe if he plans to pay the loan back in 2 years?

 31) If Ryan takes out a loan for $5200 and plans to pay it back in 15 weeks, how much does he owe if the interest rate is $4\frac{3}{4}$%?

32) Naoto borrows money for some unexpected car repairs. He will take out an emergency loan for $624 at $6\frac{1}{4}$% interest and plans to pay off the loan in 10 weeks. How much will he owe?

33) Fredericka is going to put a $2000 down payment on a new car costing $14,000. If the finance company charges 4.5% interest, how much does Fredericka owe, and what is her monthly payment if she chooses one of the following three payment plans?

  a) a 3-yr loan   b) a 4-yr loan

  c) a 5-yr loan

  d) How much does Fredericka save if she pays off the loan in 3 years instead of 5 years?

34) Jeremy wants to buy a new truck costing $24,000 and makes a $4000 down payment. If the finance company charges 3.5% interest, how much does Jeremy owe, and what is his monthly payment if he chooses one of the following three payment plans?

  a) a 3-yr loan

  b) a 4-yr loan

  c) a 5-yr loan

  d) How much does Jeremy save if he pays off the loan in 3 years instead of 5 years?

35) Paavo takes out a home improvement loan for $30,000 at an interest rate of 5.5%. How much does he owe, and what is his monthly payment if he chooses one of the following three payment plans?

©Terry Vine/Blend Images LLC

  a) a 5-yr loan

  b) a 6-yr loan

  c) a 7-yr loan

  d) How much does Paavo save if he pays off the loan in 5 years instead of 7 years?

36) Clara needs to take out a student loan for $20,000 at an interest rate of 4.5%. Once she graduates, how much does Clara owe, and what is her monthly payment if she chooses one of the following three payment plans?

  a) a 4-yr loan

  b) a 5-yr loan

  c) a 6-yr loan

  d) How much does Clara save if she pays off the loan in 4 years instead of 6 years?

Objective 2: Understand Compound Interest

Answer the following questions. Find the amount in b) using the method in Example 5.

37) Romero deposits $1000 into an account for 2 years at 5% interest.

  a) Determine the amount of interest if he earns *simple interest* for 2 years. Then, find the total amount in his account after 2 years.

  b) Determine the amount of interest he earns if the interest is *compounded annually* for 2 years. Then, find the total amount in his account after 2 years.

  c) Will the amount of money in his account after 2 years be different if the interest is computed differently? Explain your answer.

38) Alnira deposits $6000 into an account for 2 years at 6% interest.

  a) Determine the amount of interest if she earns *simple interest* for 2 years. Then, find the total amount in her account after 2 years.

  b) Determine the amount of interest she earns if the interest is *compounded annually* for 2 years. Then, find the total amount in her account after 2 years.

  c) Will the amount of money in her account after 2 years be different if the interest is computed differently? Explain your answer.

39) Chau deposits $5000 into an account for 3 years at 3% interest.

   a) Determine the amount of interest if he earns *simple interest* for 3 years. Then, find the total amount in his account after 3 years.

   b) Determine the amount of interest he earns if the interest is *compounded annually* for 3 years. Then, find the total amount in his account after 3 years.

   c) Will the amount of money in his account after 3 years be different if the interest is computed differently? Explain your answer.

40) Grace deposits $3000 into an account for 3 years at 4% interest.

   a) Determine the amount of interest if she earns *simple interest* for 3 years. Then, find the total amount in her account after 3 years.

   b) Determine the amount of interest she earns if the interest is *compounded annually* for 3 years. Then, find the total amount in her account after 3 years.

   c) Will the amount of money in her account after 3 years be different if the interest is computed differently? Explain your answer.

Objective 3: Solve Compound Interest Problems
Solve each problem.

41) Thao invests $7000 at 4% interest compounded annually for 8 years. Find the amount of money in the account after this time.

42) Felipe invests $12,000 at 6% interest compounded annually for 6 years. Find the amount of money in the account after this time.

43) Alison invests $11,500 at 5.25% interest compounded annually for 9 years. Find the amount of money in the account after this time, and determine how much interest she earned.

44) Tiana invests $13,500 at 6.25% interest compounded annually for 11 years. Find the amount of money in the account after this time, and determine how much interest she earned.

45) Mike invests $22,000 at 6.75% interest compounded annually for 20 years. Find the amount of money in the account after this time, and determine how much interest he earned.

46) Roger invests $31,000 at 7.25% interest compounded annually for 15 years. Find the amount of money in the account after this time, and determine the amount of interest he earned.

47) Herman invests $6400 at 5.5% interest for 7 years compounded quarterly. Find the amount of money in the account after 7 years, and determine the amount of interest Herman earned.

48) Raffi invests $7200 at 4.5% interest for 9 years compounded quarterly. Find the amount of money in the account after 9 years, and determine the amount of interest he earned.

49) Peggy invests $8400 at $6\frac{1}{2}$% interest for 10 years compounded semiannually. Find the amount of money in the account after 10 years, and determine the amount of interest she earned.

50) Jesse invests $5900 at $7\frac{1}{2}$% interest for 12 years compounded semiannually. Find the amount of money in the account after 12 years, and determine the amount of interest Jesse earned.

51) Alejandro invests $12,000 at 5.75% interest for 6 years compounded monthly. Find the amount of money in the account after 6 years, and determine the amount of interest Alejandro earned.

52) Vivian invests $8200 at 6.75% interest for 8 years compounded monthly. Find the amount of money in the account after 5 years, and determine the amount of interest she earned.

53) Lauren invests $14,000 at 3% interest for 5 years compounded weekly. Find the amount of money in the account after 5 years, and determine the amount of interest Lauren earned.

54) Olaf invests $5700 at 4% interest for 6 years compounded weekly. Find the amount of money in the account after 6 years, and determine the amount of interest he earned.

55) Parwana deposits her bonus of $21,450 into a savings account that pays $4\frac{3}{4}$% interest compounded daily. Find the amount of money in the account after 7 years, and determine the amount of interest she will earn.

56) Joseph and Kayla sell their home and make a profit of $28,750. They decide to invest it in a savings account that pays $4\frac{1}{4}$% interest compounded monthly. Find the amount of money in the account after 4 years, and determine the amount of interest Joseph and Kayla will earn.

57) Alex finances $16,500 to buy a work truck. If the lender provides him with a 5-year loan at 3.75% interest compounded monthly, how much does Alex owe? Also find the amount of interest he will pay the lender.

58) Dagmar borrows $19,800 from her bank to buy a new car. If she gets a 6-year loan at 4.75% interest compounded monthly, how much does Dagmar owe? How much interest will she pay the bank?

R **Rethink**

R1) How will you think differently about the way you borrow money? (Think about your student loans or credit cards.)

R2) How will you think differently about the way you save money? (Think about your savings account.)

Group Activity – Percents

Students should work in groups of two or three to complete this activity.

At Ernie's Electronics, employees earn a base salary of $500 per week, plus a 5% commission rate on all sales. Suppose you work at Ernie's Electronics. Answer the following questions about your earnings.

1) Last week, you sold $5800 worth of merchandise. How much did you earn in commission last week (try to calculate this mentally)?

2) How much was your *gross pay* for last week? Hint: *Gross pay* refers to your weekly pay before taxes are deducted.

3) Approximately 7% of your weekly pay is deducted for government taxes. What was your *net pay* for last week? Hint: *Net pay* refers to the amount of money that you receive after taxes are deducted. This is also known as your *take-home pay*.

4) Last week, you bought an iPod, which cost $245.99. The sales tax rate was 6.6%. What was the total cost of the iPod (round to the nearest cent)?

5) What percentage of last week's *net pay* was the total cost of the iPod (round to the nearest tenth of a percent)?

6) After working at Ernie's Electronics for several months, you were able to save $1500. You want to put your savings into an interest-bearing account. Your bank has three options: **Option A:** $3\frac{1}{4}$% interest rate, simple interest; **Option B:** 3.15% interest rate, compounded annually; **Option C:** 3% interest rate, compounded monthly. Predict which account will generate the greatest account balance after 5 years. Then, calculate the total account balance after 5 years for each option. Was your prediction correct?

How much do *you* know about Internet resources and how to use them? Read each statement, and check T for *true* or F for *false*.

		T	F
1.	I am aware of and know how to access all of the online resources available with this textbook.	☐	☐
2.	I have used some of the online resources that accompany this textbook.	☐	☐
3.	I have asked my instructor whether he or she can recommend any websites that are a good fit for the text we are using.	☐	☐
4.	I frequently search the Internet for supplemental learning resources.	☐	☐
5.	I have a good understanding of which Web sources are trustworthy, or I am willing to put in the time to determine which sources can be trusted.	☐	☐
6.	I have decided which type of Internet resource best fits my learning style.	☐	☐
7.	I understand that Internet resources are presented at different difficulty levels, and I am willing to determine whether a given level of a lesson is appropriate for my class.	☐	☐
8.	I feel I learn more when I am watching a video because I can rewind it when I do not understand the material.	☐	☐
9.	I understand that some video formats will not fit my particular learning style.	☐	☐
10.	I understand that I may learn techniques that are different from what my instructor used in class. In these cases, I will talk with my instructor before I use the technique on the exam.	☐	☐
11.	I understand how to take into account user statistics regarding the number of viewers who have watched a video.	☐	☐
12.	I know that there is value in taking into account the user comments relating to videos posted on the Internet.	☐	☐

Scoring:

Scoring of this emPOWERme is simple: The more statements you agree are true about yourself, the greater your ability and readiness to use online resources as a way to supplement your instruction. Each of these statements involves important considerations in the use of instructional resources you find on the Web.

Chapter 8:   Summary

Definition/Procedure	Example

8.1 Percents, Fractions, and Decimals

Percent means *out of 100*.

Explain the meaning of the statement. *In 2015, approximately 19% of men in the United States smoked cigarettes.* (www.kff.org)

19% means 19 *out of* 100, so the statement means that approximately 19 out of 100 men in the United States smoked cigarettes in 2015.

Because percent means *out of 100,* we can write a percent as a fraction, and because fractions can be written as decimals, we can also write percents as decimals.

Write 57% as a fraction and then as a decimal.

Since 57% means 57 out of 100, we write

$$57\% = \frac{57}{100} = 0.57$$

How to Change a Percent to a Decimal

1) Remove the percent symbol.

2) Move the decimal point two places to the left.

Write 21% as a decimal.

$21\% = 21.\%$ Put the decimal point at the end of the number.

$= 0.21$ Remove the percent symbol, and move the decimal point two places to the left.

When we change a percent to a fraction, we usually write it in lowest terms.

Write 40% as a fraction in lowest terms.

$$40\% = \frac{40}{100}$$

$$= \frac{40 \div 20}{100 \div 20} = \frac{2}{5}$$ Divide numerator and denominator by 20.

Write Percents Containing Decimals or Fractions as Fractions in Lowest Terms

One method for changing a percent to a fraction is to first change the percent to a decimal, and then change the decimal to a fraction.

Write 22.5% as a fraction in lowest terms by first writing 22.5% as a decimal.

First, change 22.5% to a decimal: $22.5\% = 0.225$

Next, change the decimal to a fraction and simplify.

$$0.225 = \frac{225}{1000} = \frac{225 \div 25}{1000 \div 25} = \frac{9}{40}$$

Therefore, $22.5\% = \frac{9}{40}$.

Another way to change a percent to a fraction in lowest terms is to first write it as a fraction with a denominator of 100.

Write 22.5% as a fraction in lowest terms by first writing it as a fraction with a denominator of 100.

$$22.5\% = \frac{22.5}{100}$$

The next step is to eliminate the decimal from the fraction.

$$\frac{22.5}{100} \cdot \frac{10}{10} = \frac{225}{1000}$$

Now, simplify $\frac{225}{1000}$: $\frac{225}{1000} = \frac{225 \div 25}{1000 \div 25} = \frac{9}{40}$

Therefore, $22.5\% = \frac{9}{40}$.

Definition/Procedure	Example

How to Change a Decimal to a Percent

1) Move the decimal point two places to the right.

2) Put the percent symbol at the end of the number.

Write 0.43 as a percent.

$0.43 = 43.$ Move the decimal point two places to the right.

$= 43\%$ Put the % symbol at the end of the number. Remove the decimal point from the end of the number.

Write Fractions as Percents

We can use several different methods to write a fraction as a percent.

1) Since percent means *out of 100,* one way is to start by writing the given fraction as a fraction with a denominator of 100.

2) We can also change a fraction to a percent using long division.

3) A third way to write a fraction as a percent is to use a proportion.

1) Write $\dfrac{3}{5}$ as a percent by first writing it as a fraction with a denominator of 100.

Multiply the fraction by $\dfrac{20}{20}$.

$$\dfrac{3}{5} \cdot \dfrac{20}{20} = \dfrac{60}{100} = 60\% \qquad \text{So, } \dfrac{3}{5} = 60\%.$$

2) Write $\dfrac{1}{8}$ as a percent using long division.

$$8)\overline{1.000}^{\;0.125} \qquad \text{Therefore, } \dfrac{1}{8} = 0.125 = 12.5\%.$$

3) Use a proportion to change $\dfrac{3}{25}$ to a percent.

We want to write $\dfrac{3}{25}$ with a denominator of 100. Set up the proportion $\dfrac{3}{25} = \dfrac{x}{100}$, where x is the percent, and solve for x.

$$\dfrac{3}{25} = \dfrac{x}{100} \qquad \begin{array}{l} 25 \cdot x = 25x \\ 3 \cdot 100 = 300 \end{array}$$

$$25x = 300$$

$$\dfrac{25x}{25} = \dfrac{300}{25} \qquad \text{Divide each side by 25.}$$

$$\dfrac{\overset{1}{\cancel{25}}x}{\underset{1}{\cancel{25}}} = 12 \qquad \text{Divide out the common factor.}$$

$$x = 12 \qquad \text{Simplify.}$$

Therefore, $\dfrac{3}{25} = \dfrac{12}{100}$. So, $\dfrac{3}{25} = 12\%$.

Definition/Procedure	Example

8.2 Compute Basic Percents Mentally

Definition/Procedure	Example
How to Find 10% of a Number To find 10% of a number, locate the decimal point in the number and move it one place to the left.	Find 10% of 875. To find 10% of 875, locate the decimal point in 875 and move it one place to the left. 10% of 875. = 87.5
How to Find Percents That Are Multiples of 10 If we know the value of 10% of a number, 1) find 20% of the number by multiplying the value by 2 2) find 30% of the number by multiplying the value by 3 3) find 40% of the number by multiplying the value by 4 and so on.	Find 30% of 70. To find 30% of 70, first find 10% of 70. Then, multiply the result by 3. $$10\% \text{ of } 70 = 7$$ $$30\% \text{ of } 70 = 3 \cdot 7 = 21$$
How to Find 5% of a Number Find the value of 10% of a number, and then multiply that value by $\frac{1}{2}$.	Find 5% of 64. First, find 10% of 64. Then, find 5% of 64 by finding $\frac{1}{2}$ of that result. $$10\% \text{ of } 64 = 6.4$$ $$5\% \text{ of } 64 = \frac{1}{2} \cdot 6.4 = 3.2$$
How to Find Percents That Are Multiples of 5, but Not Multiples of 10 1) Split the percentage into the greatest multiple of 10% and the remaining 5%, and find those percentages. 2) Add the results.	Find 25% of 240. To find 25% of 240, find 20% of 240 and 5% of 240 and then add the results. 10% of 240 = 24 so 20% of 240 = 48 and 5% of 240 = 12 25% of 240 = 20% of 240 + 5% of 240 $$= \quad 48 \quad + \quad 12$$ $$= 60$$
Perform Computations with Multiples of 100% 100% *of* a number equals the number. To find 200% *of* a number, multiply the number by 2, to find 300% *of* a number, multiply the number by 3, and so on.	Find 500% of 43. 500% of 43 means $5 \cdot 43 = 215$.
Solve Applications Containing Basic Percentages Many everyday situations involve percentages.	A restaurant bill at Karma is $34.00. If Deena wants to leave a 20% tip, find the amount of the tip and the total amount of money Deena will pay. Find the amount of the tip, 20% *of* $34.00. 20% of $34.00 = 2 \cdot \$3.40 = \$6.80 The tip is $6.80. Now, find the total amount Deena will pay. Total paid = Amount of the bill + Amount of the tip $$= \quad \$34.00 \quad + \quad \$6.80$$ $$= \$40.80$$ The amount of the tip is $6.80, and the total amount Deena will pay is $40.80.

Definition/Procedure	Example

8.3 Use an Equation to Solve Percent Problems

Find the Percent of a Number Using Multiplication by first changing the percent to a decimal and then multiplying.

Find 33% of 55.

The *of* in 33% of 55 means multiply. Change 33% to a decimal, then multiply.

$$33\% \text{ of } 55 = 0.33 \cdot 55 = 18.15$$

Use an Equation to Find the Number

We will first write a statement or question in English and use that to write a mathematical equation. Read the statement slowly and carefully, and think about what the words mean in terms of math.

Use an equation to find 33% of 55.

Let's think of finding 33% of 55 as the question, "What is 33% of 55?"

Let x = the unknown quantity, 33% of 55.

Write the question in English, understand its meaning in terms of math, and write an equation.

English:	What	is	33%	of	55?
Meaning:	What	equals	33%	times	55?
	↓	↓	↓	↓	↓
Equation:	x	=	0.33	·	55

The equation is $x = 0.33 \cdot 55$. Solve the equation.

$$x = 18.15 \qquad \text{Multiply.}$$

Therefore, 18.15 is 33% of 55.

We can also use an equation to find other quantities involving percentages.

Use an equation to solve the following problem.

4 is 20% of what number?

Let x represent the unknown value.

$$x = \text{the number}$$

Write the question in English, understand its meaning in terms of math, and write an equation.

English:	4	is	20%	of	what number?
Meaning:	4	equals	20%	times	what number?
	↓	↓	↓	↓	↓
Equation:	4	=	0.2	·	x

The equation is $4 = 0.2 \cdot x$. Solve it.

$4 = 0.2x$ Write the equation without the multiplication symbol.

$\dfrac{4}{0.2} = \dfrac{0.2x}{0.2}$ Divide both sides by 0.2 to get x by itself.

$20 = x$ Perform the division.

Therefore, 4 is 20% of 20. We can check the answer by finding 20% of 20:   20% of 20 = 4 ✓

The answer is correct.

Definition/Procedure	Example

Use an Equation to Find the Percentage

We will first write a statement or question in English and use that to write a mathematical equation. Read the statement slowly and carefully, and think about what the words mean in terms of math.

33 is what percent of 150?

What is the unknown quantity? It is a percent.

Let x = the percent.

Write the question in English, understand its meaning in terms of math, and write an equation.

English:	33	is	what percent	of	150?
Meaning:	33	equals	what percent	times	150?
	↓	↓	↓	↓	↓
Equation:	33	=	x	·	150

The equation is $30 = x \cdot 150$. The commutative property says we can also write $x \cdot 150$ as $150 \cdot x$. So, we can think of the equation as $33 = 150 \cdot x$ or $33 = 150x$. Solve this equation.

$$33 = 150x$$
$$\frac{33}{150} = \frac{150x}{150} \quad \text{Divide both sides by 150 to get x by itself.}$$
$$0.22 = x \quad \text{Perform the division.}$$

The final answer is not 0.22 because x represents a percent. The last step is to change 0.22 to a percent.

$$0.22 = 22\% \quad \text{Change 0.22 to a percent.}$$

Therefore, 33 is 22% of 150. We can check the answer using multiplication.

$$22\% \text{ of } 150 = 0.22 \cdot 150 = 33 \quad ✓$$

The answer is correct.

BE CAREFUL When you are asked to find a percent, the value that you get for x is *not* the final answer. You must change that number to a percent.

8.4 Solve Applications Involving Percents

Use an Equation to Solve an Applied Problem

Step 1: **Read** the problem carefully, and identify what we are being asked to find.

Step 2: **Choose** a variable to represent the unknown.

Step 3: **Translate** the information that appears in English into an algebraic equation.

Step 4: **Solve** the equation.

Step 5: **Check** the answer, and **interpret** the solution as it relates to the problem.

Use an equation to solve the problem.

Mitchell spends 8% of his yearly salary on food every year. If he spent $7000 on food last year, how much money did he make that year?

Solution

Step 1: **Read** the problem carefully, and identify what we are being asked to find.

We must find Mitchell's salary for the year.

Step 2: **Choose** a variable to represent the unknown.

x = Mitchell's annual salary

Definition/Procedure	Example

Step 3: **Translate** the information that appears in English into an algebraic equation.

We can also think of the situation in the problem as *$7000 is 8% of his annual salary.* Find his annual salary.

The equation will come from the statement *$7000 is 8% of his annual salary.* Write out the statement in English, then write an equation and solve it.

Step 4: **Solve** the equation.

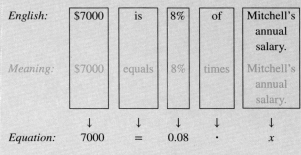

English: | $7000 | is | 8% | of | Mitchell's annual salary.

Meaning: $7000 equals 8% times Mitchell's annual salary.

↓ ↓ ↓ ↓ ↓

Equation: 7000 = 0.08 · x

The equation is $7000 = 0.08x$.

$$7000 = 0.08x$$

$$\frac{7000}{0.08} = \frac{0.08x}{0.08}$$ Divide both sides by 0.08.

$$87,500 = x$$ Perform the division.

Step 5: **Check** the answer, and **interpret** the solution as it relates to the problem.

Mitchell's annual salary is $87,500.

Check by finding 8% of $87,500.

$$8\% \text{ of } \$87,500 = 0.08 \cdot \$87,500 = \$7000 \quad ✓$$

The answer is correct.

Use a Proportion to Find a Percent

Recall that percents can be written as fractions, and since a proportion is a statement that two ratios or rates are equal, we can also use proportions to find a percent.

Use a proportion to solve the problem.

At a local coffee shop, 6 out of every 40 drinks served are decaffeinated. What percent of the drinks served are decaffeinated?

Solution

Step 1: **Read** the problem carefully, and identify what we are being asked to find.

We must find the percent of the drinks that are decaffeinated.

Definition/Procedure	Example

Step 2: **Choose** a variable to represent the unknown.

We can write *6 decaffeinated drinks out of 40 total drinks* as the rate, or fraction, $\dfrac{6}{40}$. If we write this fraction with a denominator of 100, **the numerator will be the percent** since percent means *out of 100*. We can do this by writing a proportion. Think of the stated problem as *6 decaffeinated drinks <u>are to</u> 40 total drinks <u>as</u> how many decaffeinated drinks <u>are to</u> 100 total drinks?*

x = the number of decaffeinated drinks out of 100 drinks, or **the percent**

Step 3: **Translate** the information that appears in English into an algebraic equation.

Write a proportion using x for the percent.

$$\frac{6 \text{ decaffeinated drinks}}{40 \text{ total drinks}} = \frac{x}{100 \text{ total drinks}}$$

Step 4: **Solve** the equation. We do not need the units when solving for x.

$$40 \cdot x = 40x$$

$$\frac{6}{40} = \frac{x}{100}$$

$$6 \cdot 100 = 600$$

$40x = 6 \cdot 100$	Find the cross products.
$40x = 600$	Multiply.
$\dfrac{40x}{40} = \dfrac{600}{40}$	Divide by 40.
$x = 15$	Perform the division.

Finding that $x = 15$ means that $\dfrac{6}{40} = \dfrac{15}{100}$, and $\dfrac{15}{100} = 15\%$.

Step 5: **Check** the answer, and **interpret** the solution as it relates to the problem.

Therefore, 15% of drinks served are decaffeinated.

Use the cross products to check $x = 15$ in the proportion $\dfrac{6}{40} = \dfrac{x}{100}$.

$$\frac{6}{40} = \frac{15}{100} \qquad \text{Substitute 15 for } x.$$

The cross products are equal, so the answer is correct.

Definition/Procedure	Example

8.5 More Applications with Percents

Compute Sales Tax

Amount of sales tax = Rate of tax · Cost of item

Use the Five Steps for Solving Applied Problems.

A pair of sandals costs $29.99. If the sales tax rate is 9%, find the amount of tax and the total cost of the sandals.

Solution

Step 1: **Read** the problem carefully, and identify what we are being asked to find.

We must determine how much is paid in sales tax and also find the total cost of the sandals.

Step 2: **Choose** a variable to represent the unknown.

Let x = the amount of sales tax. After finding this value, we will find the total cost of the sandals.

Step 3: **Translate** the information that appears in English into an algebraic equation.

Change the sales tax rate from a percent to a decimal: 9% = 0.09

Amount of sales tax = Rate of tax · Cost of item
$$x = 0.09 \cdot \$29.99$$

The equation is $x = 0.09 \cdot \$29.99$.

Step 4: **Solve** the equation.

$x = 0.09 \cdot \$29.99$
$x = \$2.6991 \approx \$2.70$ Round to the nearest cent.

The amount of sales tax is $2.70.

Find the total cost of the sandals.

$$\frac{\text{Total cost of}}{\text{the sandals}} = \frac{\text{Price of}}{\text{the sandals}} + \frac{\text{Amount of}}{\text{sales tax}}$$

$$\frac{\text{Total cost of}}{\text{the sandals}} = \$29.99 + \$2.70$$

$$= \$32.69$$

Step 5: **Check** the answer, and **interpret** the solution as it relates to the problem. **The amount of sales tax is $2.70. The total cost of the sandals is $32.69.** Double-check the calculations to be sure all the arithmetic is correct.

Definition/Procedure	Example

Compute Commissions

$$\frac{\text{Amount of}}{\text{commission}} = \text{Rate of commission} \cdot \text{Amount of sales}$$

Use the Five Steps for Solving Applied Problems.

Everett is a salesman at a furniture store, and his commission rate is 9.5%. Today he sold $4800 worth of furniture. Find the amount of his commission.

Solution

Step 1: **Read** the problem carefully, and identify what we are being asked to find.

We must find the amount of Everett's commission.

Step 2: **Choose** a variable to represent the unknown.

$$x = \text{the amount of commission}$$

Step 3: **Translate** the information that appears in English into an algebraic equation.

Change the rate of commission from a percent to a decimal: $9.5\% = 0.095$

$$\text{Commission} = \frac{\text{Rate of}}{\text{commission}} \cdot \text{Amount of sales}$$
$$x \quad = \quad 0.095 \quad \cdot \quad \$4800$$
$$= \$456$$

The equation is $x = 0.095 \cdot \$4800$.

Step 4: **Solve** the equation.

$$x = 0.095 \cdot \$4800$$
$$x = \$456$$

Step 5: **Check** the answer, and **interpret** the solution as it relates to the problem. **Everett's commission is $456.** Double-check the calculations to be sure all the arithmetic is correct.

Amount of commission $= 0.095 \cdot \$4800 = \$456.$ ✓

The answer is correct.

8.6 Simple and Compound Interest

Interest is a fee paid for borrowing money. **Simple interest** is calculated on the original amount of money borrowed. The original amount of money borrowed is called the **principal.**

To find the simple interest earned, use this formula.

$$\text{Interest} = \text{Principal} \cdot \text{Rate} \cdot \text{Time}$$
$$I \quad = \quad P \quad \cdot \quad R \quad \cdot \quad T$$

$P =$ the principal. This is the original amount of money borrowed or the original amount of money in a bank account.

$R =$ the interest rate. It is usually given as a percent. **We must change it to a decimal when we use it in the equation.**

$T =$ the time, *in years,* that the money will be borrowed or the amount of time an amount of money is in an account.

Zoe borrows $2800 for 2 years at 7% simple interest. How much interest will she owe?

$$P = \$2800 \qquad R = 7\% = 0.07 \qquad T = 2$$
$$I = \quad P \cdot R \cdot T$$
$$I = (2800)(0.07)(2) \qquad \text{Substitute the values.}$$
$$I = 392 \qquad \text{Multiply.}$$

Zoe will owe $392 in interest.

Definition/Procedure	Example
Compound Interest **Compound interest** is interest that is paid on the original principal *plus* the interest that has already been paid. When interest is computed at the end of each year, we say that the interest is **compounded annually.** Interest can be compounded in other intervals, too. Let A = the total amount in an account after T years. Use this formula to find the value of A when interest is compounded more than one time per year: $$A = P\left(1 + \frac{R}{n}\right)^{n \cdot T}$$ P = the principal, R = the interest rate in decimal form, and n = the number of times interest is compounded each year.	Norah invests \$7000 at 6% interest compounded monthly for 3 years. Find the amount of interest in her account after 3 years and the amount of interest Norah will earn. We will use a calculator. $P = \$7000 \qquad R = 6\% = 0.06 \qquad T = 3 \qquad n = 12$ $$A = P\left(1 + \frac{R}{n}\right)^{n \cdot T}$$ Compounded monthly means 12 times per year. $A = 7000\left(1 + \dfrac{0.06}{12}\right)^{12 \cdot 3}$ Substitute the values. $A = 7000(1 + 0.005)^{36}$ Divide $\dfrac{0.06}{12}$ and multiply $12 \cdot 3$. $A \approx 8376.76$ Evaluate. Round to the nearest cent. Norah will have \$8376.76 in her account. The amount of interest she will earn is \$8376.76 − \$7000 = \$1376.76.

Chapter 8: Review Exercises

(8.1) For Exercises 1 – 3, rewrite each statement using a percent.

1) Approximately 37 out of 100 American households have at least one dog. (www.avma.org)

©Sherri Messersmith

2) Approximately 30 out of 100 Americans with associate's degrees earn more than those with bachelor's degrees. (www.theatlantic.com)

3) Approximately 14 out of 100 jobs in San Diego are in the education or health services fields. (www.hemispheresmagazine.com)

4) Explain the meaning of this statement: When the Cubs won Game 7 of the 2016 World Series, 71% of the people watching television in the Chicago area were watching the game. (www.foxbusiness.com)

Write each percent as a fraction and a decimal.

5) 14%

6) 18%

7) 34%

8) 31%

Write each percent as a decimal.

9) 0.35%

10) 0.47%

11) 201%

12) 305%

Write each decimal as a percent and each percent as a decimal.

13) During the 2015–2016 academic year, residents of Georgia were 0.87 of the entire student body at the University of Georgia. (www.admissions.uga.edu)

14) The sales tax in Chicago, Illinois, is 0.1025

15) The state unemployment rate in Nevada for April 2017 was 4.7%. (www.bls.gov)

16) The interest rate for a car loan is 3.25%.

Write each percent as a fraction or mixed number in lowest terms.

17) 23%

18) 77%

19) 210%

20) 370%

Write each percent as a fraction in lowest terms by first writing the percent as a decimal.

21) 42.5%

22) 57.5%

23) 43.75%

24) 68.75%

Write each percent as a fraction in lowest terms.

25) $26\frac{2}{3}\%$

26) $53\frac{1}{3}\%$

Write the fraction as a percent by first writing it as a fraction with a denominator of 100.

27) $\dfrac{11}{20}$

28) $\dfrac{11}{50}$

29) $\dfrac{49}{50}$

30) $\dfrac{11}{25}$

Write the fraction as a percent using long division.

31) $\dfrac{5}{8}$

32) $\dfrac{1}{16}$

33) $\frac{12}{5}$

34) $\frac{7}{4}$

(8.2)

35) First find 10% of 70, then find the following.

   a) 60% of 70

   b) 90% of 70

   c) 20% of 70

36) First find 10% of 30, then find the following.

   a) 60% of 30

   b) 90% of 30

   c) 30% of 30

37) Find:

   a) 90% of 140

   b) 70% of 140

38) Find:

   a) 60% of 60

   b) 20% of 60

39) Find:

   a) 15% of 130

   b) 15% of 72

40) Find:

   a) 15% of 150

   b) 15% of 24

41) Find:

   a) 100% of 49

   b) 200% of 38

42) Find:

   a) 200% of 157

   b) 400% of 34

43) An electronics warehouse store is having a one-day 35% off sale on its big-screen 3D televisions that regularly sell for $1799. If a buyer purchases the television on the day of the sale, approximately how much money will the buyer save, and what is the approximate sale price?

44) A home improvement store is having a 35% off clearance sale on its large-capacity steam washer-dryer sets that regularly sell for $1899. Approximately how much money will a buyer save during the clearance sale, and what is the approximate clearance price?

©Pixtal/AGE Fotostock

(8.3) Find the percent using multiplication.

45) Find:

   a) 72% of 44

   b) 6.7% of 500

46) Find:

   a) 28% of 36

   b) 3.5% of 700

47) Find:

   a) 24% of 132

   b) 1.8% of 240

48) Find:

   a) 36% of 152

   b) 3.1% of 360

49) A bank loan officer recommends that a young couple put 20% down on a new home costing $325,000. What is the amount of the recommended down payment?

50) At a local community college, approximately 43% of the students are male. If the college has 9500 students, how many students are female?

51) Use an equation to solve each problem.

   a) 16 is 8% of what number?

   b) 15% of what number is 75?

   c) $4\frac{1}{2}$% of what number is 90?

52) Use an equation to solve each problem.

   a) 28 is 14% of what number?

   b) 37% of what number is 148?

   c) $4\frac{1}{2}$% of what number is 135?

53) Use an equation to solve each problem.

   a) 70 is 20% of what number?

   b) 13% of what number is 117?

   c) $8\frac{1}{2}$% of what number is 204?

54) Use an equation to solve each problem.

   a) 48 is 6% of what number?

   b) 63% of what number is 504?

   c) $8\frac{1}{2}$% of what number is 136?

(8.4) Solve each problem.

55) In a class of 40 math students, 24 students claimed that they did not spend enough time over the weekend preparing for the exam. What percent of students did not spend enough time over the weekend preparing for the exam? What percent did?

56) Maria races in a triathlon that requires her to swim 0.5 mile, bike 7 miles, and then run 2.5 miles. What percent of the race is swimming? What percent of the race is biking? What percent of the race is running?

57) In 2016, approximately 3 out of 5 new vehicle sales were light trucks. What percent is this? (www.trucks.com)

©Fuse/Getty Images

58) Sylvia decides to give $450 from her $3000 savings account to the Red Cross to help people in need. What percent of her savings did she give to the Red Cross? What percent of her savings did she keep?

Use a proportion to solve the problem.

59) A coed soccer team of 20 players has eight females on the team. What is the percentage of males on the team?

60) During basketball practice, Joseph makes 52 out of 80 free throws. What percent of free throw attempts does Joseph miss?

Find the percent using a proportion.

61) 80 cm is what percent of 2 m?

62) 0.3 m is what percent of 120 cm?

63) 6 ft is what percent of 1 yd?

64) 36 in. is what percent of 1 yd?

65) 2 qt is what percent of 1 gal?

66) 1 c is what percent of 1 qt?

67) 500 mL is what percent of 0.25 L?

68) 2 L is what percent of 1250 mL?

(8.5) Solve each problem.

69) A price tag shows that a computer printer costs $86. If the sales tax rate is 8%, find the amount of tax and the total cost of the computer printer.

70) Carly finds a paper shredder for her home business that costs $57. If the sales tax rate is 7%, find the amount of tax and the total cost of the paper shredder.

71) If the sales tax amount was $11.40 for a purchase of $150, what was the sales tax rate?

72) Jeremy paid $1.62 in sales tax on a package of blank compact disks with a list price of $18. What is the tax rate?

73) A salesperson for an office furniture store is paid a commission rate of 6%. If he sells $12,500 worth of office furniture, how much commission does he earn?

74) Shelly sells $6560 worth of jewelry in one week. What is the amount of her commission check for her weekly sales if she is paid 8% commission?

©Robert Kneschke/Shutterstock

75) A desktop computer that regularly sells for $1350 is marked 40% off. What is the sale price of the desktop computer?

76) During a holiday sale, the first 50 customers get a 60% discount on one item in the department store. If Tanomo is one of the first 50 customers and she finds a dress costing $120, what is the sale price for her dress?

77) Randy bought a Les Paul custom guitar in 1987 for $900. Today it is worth $3150. Find the percent increase in the price.

78) In Fall 2016, a professor taught 300 students. In Fall 2017, she had 348 students. Find the percent increase in the number of students.

79) Ramona sometimes rides the bus to work to help save the environment. It takes 1 hour 30 minutes when she rides the bus. When she takes her car, it takes 45 minutes to get to work. Find the percent decrease in time for Ramona's commute to work when she uses the car.

©Doug Sherman/Geofile

80) In the past, Mike's heating bill during the winter months averaged $150. This year he replaced his windows with dual-pane, energy-saving windows, and his average heating bill during the winter dropped to $90. Find the percent decrease in Mike's heating bill.

(8.6) In Exercises 81–92, solve each simple interest problem.

81) Julia borrowed $1200 at a high interest rate of 16%. Find the amount of interest after 1 year.

82) Jamal borrowed $2500 at an interest rate of 6%. Find the amount of interest after 2 years.

83) If Andy borrowed $1600 from a lender who charges 3.75% interest, how much interest does Andy owe after 2 years?

84) If a person borrows $2500 from a lender who charges 4.25% interest, how much interest does the person owe after 3 years?

85) Write each in terms of years.

   a) 13 months   b) 39 weeks

86) Write each in terms of years.

   a) 5 months   b) 3 weeks

87) How much interest will Josephine owe if she borrows $960 for 10 weeks at a $2\frac{3}{4}\%$ interest rate?

88) How much interest will Russell owe if he borrows $2200 for 30 months at a $5\frac{1}{4}\%$ interest rate?

89) Glenda took out a personal loan for $900 at an interest rate of 4.2%. If Glenda wants to pay back the loan in 18 months, how much does she pay back to the lender, including interest?

90) Brad puts $4000 into a savings account at a 5.25% interest rate. How much is in the account after 3 years?

91) Capri is going to put a $3000 down payment on a new car costing $18,000. If the finance company charges 3.5% interest, how much does Capri owe and what is her monthly payment if she chooses one of the following three payment plans?

   a) a 3-yr loan

   b) a 4-yr loan

   c) a 5-yr loan

   d) How much does Capri save if she pays off the loan in 3 years instead of 5 years?

92) April is going to put a $3500 down payment on a new car costing $19,500. If the finance company charges 2.5% interest, how much does April owe and what is her monthly payment if she chooses one of the following three payment plans?

a) a 3-yr loan

b) a 4-yr loan

©Blend Images/Getty Images

c) a 5-yr loan

d) How much does April save if she pays off the loan in 3 years instead of 5 years?

93) Ahmed deposits $6000 into an account for 3 years at 4% interest.

a) Determine the amount of interest if he earns *simple interest* for 3 years. Then, find the total amount in his account after 3 years.

b) Determine the amount of interest he earns if the interest is *compounded annually* for 3 years. Then, find the total amount in his account after 3 years.

c) Will the amount of money in his account after 3 years be different if the interest is computed differently? Explain your answer.

94) Tina deposits $5000 into an account for 4 years at 5% interest.

a) Determine the amount of interest if she earns *simple interest* for 4 years. Then, find the total amount in her account after 3 years.

b) Determine the amount of interest she earns if the interest is *compounded annually* for 4 years. Then, find the total amount in her account after 4 years.

c) Will the amount of money in her account after 4 years be different if the interest is computed differently? Explain your answer.

95) Shufan invests $6700 at $4\frac{3}{4}$% for 8 years compounded weekly. Find the amount of money in the account after 8 years, and also determine the amount of interest Shufan earned.

96) Wade and Carol deposit their inheritance of $325,000 into a savings account that pays 5.25% compounded monthly. If they leave the money in the account for 15 years, how much money will they have in the account? How much interest will they earn?

Chapter 8: Test

1) What does *percent* mean?

Write each percent as a decimal and as a fraction in lowest terms.

2) 45%

3) 0.7%

4) 6%

5) 14.8%

Write each number as a percent.

6) 0.6

7) 0.0924

8) $\frac{4}{5}$

9) $\frac{7}{8}$

10) $3\frac{1}{2}$

11) 2

12) a) Explain how to compute 10% of 40 "in your head." Then, find 10% of 40.

b) Find 80% of 40.

c) Find 5% of 40.

d) Find 85% of 40.

13) Find 300% of 15.

14) A restaurant bill is $38. If Liz wants to leave a 20% tip, determine the amount of the tip and the total amount of money she would pay.

15) The original price of a pair of boots is $60. Now, they are on sale for 30% off. What is the sale price of the boots?

16) Find 27.3% of 480.

Solve each problem using an equation.

17) 35% of what number is 280?

18) 360 is what percent of 4000?

Solve each problem.

19) Vernon's mortgage payment is 20% of his monthly income. If his mortgage payment is $1280, how much does he make each month?

20) A dining room set costs $2478 plus a sales tax of $7\frac{1}{2}$%. What is the total cost of the dining room set, including sales tax?

21) A 2015 study revealed that 16 out of 25 American adults drink coffee daily. What percent of American adults drink coffee every day? (www.gallup.com)

22) When Soldier Field in Chicago opened in 1924, its seating capacity was 74,000. When the new Soldier Field opened in 2003, its new seating capacity was 61,500. Find the percent decrease in the seating capacity. Round the answer to the nearest tenth of a percent. (www.stadiumsofprofootball.com)

©Sam Edwards/Glow Images

23) Keiko borrows $9000 for 4 years at 4.5% simple interest. How much interest will she owe on this loan?

24) Paul deposits $5000 for 9 months into an account paying 3% simple interest. How much money will be in the account after 9 months?

25) Trisha deposits $8000 at 7% interest compounded annually in a college fund for her son. The money will remain in the account for 10 years.

a) How much will be in the college fund after 10 years?

b) How much interest did Trisha earn?

Chapter 8:   Cumulative Review for Chapters 1–8

1) Rewrite $30 + 17$ using the commutative property.

2) Round 19,483,350 to the nearest hundred thousand.

3) The following pictograph shows the preferred sport in a high school classroom. Use the pictograph to answer the following exercises.

Favorite Sports

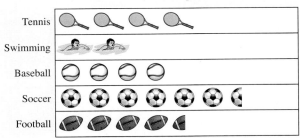

Each picture represents 2 students.

a) Find the number of students who prefer baseball.

b) How many students are in the class?

c) Which sport is most popular among these students? How many students prefer this sport?

d) Which sport is least popular?

e) How many fewer students prefer swimming than football?

f) How many more students prefer football than baseball?

4) Write $4 \cdot 4 \cdot 4 \cdot 4 \cdot 4$ using an exponent.

Perform the indicated operations. Write all answers in lowest terms.

5) 4100×0.065

6) $-\dfrac{8}{9} \div 10$

7) $\$32 - \$17.79$

8) $\dfrac{3}{10} + \dfrac{1}{4} + \dfrac{5}{8}$

9) $-72 - 18 \div 6 + 5^3 \cdot \sqrt{121}$

10) $7.8 \div 0.03$

Solve.

11) $3(a + 4) = 5a + 12$

12) $\dfrac{1}{6}p - 1 = \dfrac{2}{9}p + \dfrac{7}{18}$

Combine like terms.

13) $-8y^2 + 5y - 14 + 2y^2 - y + 3$

14) Write the ratio of 8 ft to 4 yd.

15) Change 500 mL to liters.

16) Convert 2 wk to hours.

17) When he played with the Houston Rockets, Yao Ming's height was listed as 7 ft 6 in., and his weight was 310 lb. Find his height in centimeters and his weight in kilograms. Use 1 in. = 2.54 cm and 1 lb ≈ 0.45 kg. (www.nba.com)

18) Write 70.5% as a fraction in lowest terms.

19) Write 1.8 as a percent.   20) Find 90% of 60.

21) Solve using an equation: 29.4 is 30% of what number?

Solve each problem.

22) An airline survey revealed that 11 out of 20 passengers do not check their luggage. On a flight with 148 passengers, how many carried all their luggage onto the plane? Round to the nearest person.

©ColorBlind Images/Getty Images

23) Manuela spent $263.24 on food for Thanksgiving dinner for her whole family. This was $145.89 more than her grocery bill the week before. How much did Manuela spend on groceries the week before Thanksgiving?

24) Manuela's stuffing recipe uses $3\frac{1}{2}$ c of chicken broth and feeds 10 people. If she will be feeding 25 people and wants to make enough stuffing for everyone, how much chicken broth will she need?

25) Concertgoers were asked to bring canned goods to donate to a food pantry. If 15,400 people attended the concert and 62% of them brought at least one item, how many people donated canned goods for the food pantry?

©Ingram Publishing

Graphs and the Rectangular Coordinate System

©Getty Images/Comstock Images

Math at Work:

Loan Officer

If you want to own a home, Mariano Marquez is a good person to talk to. As a loan officer for a bank, his job is to help people obtain a mortgage loan to buy a house. "I think we all share the dream of owning the roof over our head," says Mariano. "I try to make that happen for people."

In assisting his customers, Mariano gathers all their financial data in a table—including their savings, their income, their credit card debt, and any outstanding loans they might hold—to create a picture of where they stand in terms of money. He uses their information, along with financial tables and graphs, to determine what sort of mortgage loan they can afford. "There's a lot of sophisticated math that goes into my work," Mariano explains. "But on a basic level, it's a matter of adding up all the projected income people have and subtracting their projected expenses. If the resulting number is positive, we can move forward; if the number is negative, we have a problem."

Mariano has a piece of advice for any future homebuyers: Get a handle on your finances early. "Even people in college should keep track of how they spend," Mariano says. "You'd be surprised at how easy it is to save a little bit of money each month." Of course, the more you have saved up, the easier it is to get a mortgage, or any other kind of loan.

We explore tables and graphs (along with linear equations in two variables) in this chapter. We also introduce some financial literacy skills that you can start using right away.

It is very easy to lose track of how you are spending your money on a daily, weekly, and monthly basis. Unfortunately, spending this way can lead to credit card debt, overdrawn accounts, and other financial problems. Luckily, we can use the P.O.W.E.R. framework to create a budget that will help put you in the driver's seat of your financial life.

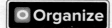

- **I will create and use a budget that will allow me to achieve my long-term and short-term goals.**

- Complete the emPOWERme survey that appears before the Chapter Summary to learn about your financial philosophy. Keep that information in mind as you create a budget.
- Write down your financial goals. These might include paying tuition, paying off a car loan, saving money for a large purchase, or putting money in a savings account every month.
- Figure out how you spend your money by making a list of expenses. For a week, write down or use a phone app to keep track of *every* purchase you make, large and small. This will help you understand how you spend your money on a daily basis. It may be helpful to organize them into categories such as living expenses and entertainment.
- Make a list of everything you will need to spend money on in the coming year. This list should include things like tuition, rent or mortgage, food, gas money, insurance payments, entertainment, and so forth.
- Make a list of long-term goals as well, such as saving a certain amount of money every year.
- Make a list of all of your income for the coming year such as the amount of money you earn at your job and any financial aid you might receive.
- Add up all of your sources of income as well as everything you spend money on, including small, daily purchases and your larger expenses.

- Create your budget. It could be a weekly budget or a monthly budget, whichever works best for you. Look at your income and list of expenses.
- Prioritize your spending. Assign money in your budget to the most important things first such as rent, food, and transportation to and from work and/or school. Give lower priority to things like entertainment. If possible, try to save a little bit of money every week or month.
- If you find that your spending is greater than your income, look at your app or what you wrote down when you were keeping track of how you spend your money. Think about where you could cut some costs, and adjust your budget accordingly.
- If possible, look for ways to increase your income such as by finding a part-time job, if your schedule allows.
- When you feel like you have created a reasonable budget, start using it right away!

- After using your budget for a couple of weeks or months, ask yourself, "Was I able to stick to my budget and pay all expenses? Did I spend less money on things I really didn't need compared to *before* I made a budget? Was I able to save some money every month?"

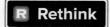

- If you were able to stick to your budget and have enough money to pay all of your expenses as well as have some money for fun, continue using the budget. No matter what, however, think about ways you may want to or need to be flexible. In the future, for example, you may need to adjust the budget if your rent increases.
- Were you able to save a little bit of money while on your budget? Look at your income and expenses and, if possible, try to save some money every week or month.
- If you did not stick to the budget, think about the reason why. Was the budget unrealistic? Did you spend more on extra things that you didn't really need that took money away from more important expenses? Think about what you could do differently the next month or week. Adjust your budget accordingly.
- Be on the lookout for issues that might create budget nightmares such as credit cards with interest rates that spike after a certain amount of time.

Chapter 9 ⟦POWER⟧ Plan

What are your goals for Chapter 9?	How can you accomplish each goal? (Write in the steps you will take to succeed.)
1 Be prepared before and during class.	• _____ • _____ • _____
2 Understand the homework to the point where you could do it without needing any help or hints.	• _____ • _____
3 Use the P.O.W.E.R. framework to develop financial literacy.	• Complete the emPOWERme that appears before the Chapter Summary to learn about your attitudes toward money. • Read the Study Strategies to learn how to use the P.O.W.E.R. framework to create a budget. • _____
4 Write your own goal. _____ _____	• _____

What are your objectives for Chapter 9?	How can you accomplish each objective?
1 Learn how to read and make different types of graphs and perform operations on the data.	• Know how to recognize, use, and make tables and different graphs: the pictograph, bar graph, line graph, histogram, and circle graph.
2 Understand the basic concepts of linear equations.	• Know how to identify a linear equation in two variables, and use the Cartesian coordinate system. • Be able to work with ordered pairs: determine whether they are a solution, complete an ordered pair, and plot them.
3 Know how to graph a linear equation in two variables.	• Know the properties for solutions of linear equations in two variables and the graphs of linear equations in two variables. • Graph by plotting points. • Understand how to recognize the equations of horizontal and vertical lines and how to graph them.
4 Write your own goal. _____ _____	• _____ _____

| **W** Work | Read Sections 9.1–9.5, and complete the exercises. |

| **E** Evaluate | Complete the Chapter Review and Chapter Test. How did you do? | **R** Rethink | • Thinking about the different types of graphs in Sections 9.1–9.3, which are easy for you to make, and which are more difficult? Why do you think some give you more trouble than others? |

• What do you understand well about the graphing in Sections 9.4 and 9.5, and what do you need to practice more?

• How are mathematics and creating a budget related? Do you think the P.O.W.E.R. framework can help you create a budget and save for the future? Explain.

9.1 Reading Tables, Pictographs, Bar Graphs, and Line Graphs

P Prepare   **O** Organize

What are your objectives for Section 9.1?	How can you accomplish each objective?
1 Read a Table	• Review how to use a table. • Complete the given example on your own. • Complete You Try 1.
2 Read a Pictograph	• Write the definition of a *pictograph* in your own words. • Complete the given example on your own. • Complete You Try 2.
3 Read and Make a Bar Graph	• Write the definition of a *bar graph* in your own words. • Complete the given examples on your own. • Complete You Trys 3–5.
4 Read a Line Graph	• Write the definition of a *line graph* in your own words. • Complete the given examples on your own. • Complete You Trys 6 and 7.

W Work   **Read the explanations, follow the examples, take notes, and complete the You Trys.**

1 Read a Table

Throughout the book, we have read and interpreted information from tables in order to answer questions.

EXAMPLE 1

The table shows the amount of money won by certain golfers on the LPGA tour in 2016. Use the information for parts a) and b).

Golfer	Tournament Winnings
Ariya Jutanugarn	$2,550,947
Cristie Kerr	$456,215
Lydia Ko	$2,493,059
Anna Nordqvist	$1,424,945
Lexi Thompson	$888,571

(www.lpga.com)

a) How much more did Ariya Jutanugarn earn than Cristie Kerr?

b) Find the total winnings of the top two golfers.

Solution

 Hint

Remember, graph paper helps us line up the numbers correctly.

a)

$$
\begin{array}{r}
2,550,947 \\
-\ \ \ 456,215 \\
\hline
2,094,732
\end{array}
$$

← Ariya Jutanugarn's winnings
← Cristie Kerr's winnings
← Difference in their winnings

Ariya Jutanugarn earned $2,094,732 more than Cristie Kerr.

b) The top two golfers were Ariya Jutanugarn and Lydia Ko. Find the sum of their earnings.

$$
\begin{array}{r}
2,550,947 \\
+\ 2,493,059 \\
\hline
5,044,006
\end{array}
$$

← Ariya Jutanugarn's winnings
← Lydia Ko's winnings
← Their total winnings

Together, the top two golfers earned $5,044,006.

[YOU TRY 1]

Use the table in Example 1 to answer these questions.

a) How much less did Lexi Thompson earn than Anna Nordqvist?

b) Find the total tournament winnings of all the players.

We can also use *graphs* to display information. **Graphs** are visual ways to represent information. We will study three types of graphs in this section: pictographs, bar graphs, and line graphs.

2 Read a Pictograph

A **pictograph** is a graph that uses pictures or symbols to represent information. While this type of graph is visually appealing, it may be difficult to read fractional parts of a whole.

EXAMPLE 2

Use the given pictograph to answer the following questions.

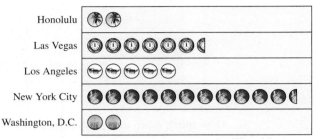

Approximate Number of Foreign Visitors to
Selected United States Cities in 2012

Each symbol represents 1 million visitors.
(www.euromonitor.com)

a) Which city had the most foreign visitors, and approximately how many visitors did it have?

b) Which two cities had the same number of visitors?

c) How many fewer people visited Los Angeles than Las Vegas?

Solution

a) New York City had the most foreign visitors because it has the greatest number of symbols. Since there are 11.5 pictures of the Statue of Liberty and each of them represents 1 million people, there were 11.5 · 1 million = 11.5 million foreign visitors to New York City.

b) Because Honolulu and Washington, D.C., have the same number of pictures, they had the same number of visitors.

c) Las Vegas had 6.5 million visitors and Los Angeles had 5 million visitors, so to determine how many fewer people visited Los Angeles, we subtract:
6.5 million − 5 million = 1.5 million fewer visitors to Los Angeles.

[YOU TRY 2]

Use the pictograph in Example 2 to answer the following questions.

a) How many people visited Honolulu?

b) How many more people visited New York City than Las Vegas?

3 Read and Make a Bar Graph

A **bar graph** is another way to display and read information. In Example 3, the graph displays the number of farms in certain states in 2016.

EXAMPLE 3   Use the given bar graph to answer the following questions.

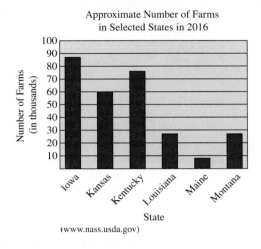

Approximate Number of Farms
in Selected States in 2016

(www.nass.usda.gov)

a)   How many farms did Kansas have in 2016?

b)   Which state had the fewest farms, and approximately how many did it have?

c)   Which state had approximately 76,000 farms?

Solution

a)   Go to the top of the bar for Kansas. It touches the line that is a guide for reading the graph. On the left, that line is labeled 60. Notice that the number of farms is in thousands, so the "60" represents 60,000. **Therefore, Kansas had 60,000 farms.**

b)   The state with the lowest bar had the fewest farms. **That state is Maine.** To determine how many farms were in Maine, go to the top of the bar for Maine and move in a straight line to the left where you see the scale for the numbers. The top of the bar is a little lower than 10, let's say at 8, so **we will estimate it as 8000 farms.**

c)   The number 76,000 is a little more than halfway between 70,000 and 80,000. Since the units on the left are in thousands, go to the left of the graph and locate the region between 70 and 80. Move to the right to see that the bar that reaches a little above halfway between 70 and 80 is for Kentucky. **Therefore, Kentucky had approximately 76,000 farms.**

[YOU TRY 3]   Use the bar graph in Example 3 to answer the following questions.

a)   Which states had the same number of farms, and how many did they have?

b)   How many fewer farms were in Louisiana than in Iowa?

Next, let's learn how to *make* a bar graph.

EXAMPLE 4

The table gives the approximate heights of some of the tallest waterslides in the world. Make a bar graph to display this information. (www.popularmechanics.com, www.yahoo.com)

Name of Waterslide (location)	Height (in ft)	Name of Waterslide (location)	Height (in ft)
The Cliffhanger (Texas)	81	Scorpion's Tail (Wisconsin)	100
Insano (Brazil)	134	Summit Plummet (Florida)	120
Leap of Faith (The Bahamas)	60	The Wildebeest (Indiana)	64

Solution

The bars on a bar graph can be either vertical or horizontal. Let's make them vertical. Each bar will represent the height of each waterslide. The names of the waterslides will go on the *horizontal* axis, at the bottom of the graph, and the heights will go on the sides, on the *vertical* axis. We need to decide how to number the vertical axis.

Look at the heights in the table and list them from smallest to largest: 60, 64, 81, 100, 120, and 134. If we label the axis from 0 to 140, all of the numbers will be included and the number 140 is not *too* much larger than the biggest number, 134. Let's label the axis in multiples of 20 because we must include a large range of numbers. Also, draw light, horizontal lines at each multiple of 20 to make it easier to read the height of each bar. Then, for each waterslide on the horizontal axis, draw a bar up to its height on the vertical axis.

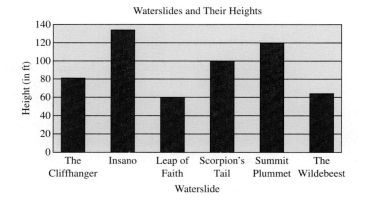

Hint

List the numbers from smallest to largest to help you determine the range and scale of numbers on the vertical axis.

YOU TRY 4

The table lists the five best-selling movie soundtracks of all time and the number of albums sold. Make a bar graph to display this information. (www.ppcorn.com)

Top-Selling Movie Soundtracks

Movie	Number of Albums (in millions)
The Bodyguard	47
Dirty Dancing	42
Grease	28
Saturday Night Fever	45
Titanic	30

A **double-bar graph** can be used to compare two sets of data. Use the double-bar graph to answer the questions in Example 5.

EXAMPLE 5 The double-bar graph shows the approximate number of new Netflix subscribers during each quarter of 2015 and 2016. Use the graph to answer the questions. (https://ir.netflix.com)

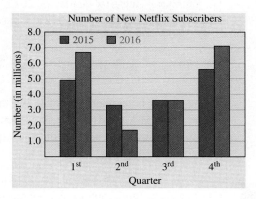

Number of New Netflix Subscribers

a) Find the number of new subscribers in the second quarter of 2015.

b) Find the number of new subscribers in the second quarter of 2016.

c) Find the ratio of the number of new subscribers in the fourth quarter of 2015 to the number of new subscribers in the third quarter of 2015.

Solution

a) First, notice that the graph tells us that the blue bars represent the numbers for 2015, and the red bars represent the numbers for 2016. Along the bottom of the graph, locate the 2nd quarter. The blue bar represents the number of new subscribers in the second quarter of 2015, and it reaches up to about 3.3 million. **In 2015, Netflix gained about 3.3 million new subscribers.**

b) The red bar above the 2nd quarter on the graph represents the number of new subscribers in the second quarter of 2016, and it reaches up to about 1.7 million. **In 2016, Netflix gained about 1.7 million new subscribers.**

c) Locate the number of new subscribers in the fourth quarter of 2015. Go to the 4th quarter column. The blue bar reaches up to 5.6 million. Now, go to the 3rd quarter column. The blue bar reaches up to approximately 3.6 million. Find the ratio.

$$\frac{\text{number in } 4^{th} \text{ quarter of } 2015}{\text{number in } 3^{rd} \text{ quarter of } 2015} = \frac{5.6 \text{ million}}{3.6 \text{ million}} = \frac{5.6}{3.6} = \frac{5.6 \cdot 10}{3.6 \cdot 10} = \frac{56}{36} = \frac{14}{9}$$

The ratio of the number of new subscribers in the fourth quarter of 2015 to the number in the third quarter of 2015 is $\frac{14}{9}$. This means that for every 14 new subscribers in the fourth quarter of 2015, there were 9 new subscribers in the third quarter of 2015.

[YOU TRY 5] Use the graph in Example 5.

a) How many more new subscribers did Netflix have in the first quarter of 2016 than in the same quarter of 2015?

b) During which quarter did Netflix gain the same number of subscribers both years?

c) Find the ratio of the number of new subscribers in the last quarter of 2015 to the number in the first quarter of 2015. Explain the meaning of the answer.

4 Read a Line Graph

A **line graph** is another type of graph that is often used to represent information. A line graph is a good way to show trends in information over time. The graph in Example 6 shows the number of hours spent playing video games per person per year for various years.

EXAMPLE 6 Use the given line graph to answer the following questions.

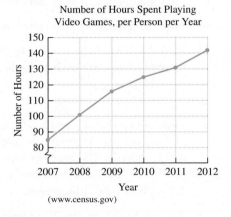

Number of Hours Spent Playing
Video Games, per Person per Year

(www.census.gov)

a) How many hours per person per year were spent playing video games in 2007?

b) In what year did the number of hours equal 125?

c) What is the general trend in the number of hours spent playing video games over the time period shown?

[W] Hint
Be sure to read the questions very carefully.

Solution

a) Locate the year 2007 at the bottom of the graph. Go up to the point on the graph. The dot on the line graph is halfway between the 80 and 90, so we will estimate the number at 85. **In 2007, the amount of time spent playing video games was about 85 hr per person per year.**

b) On the left side of the graph, 125 is halfway between 120 and 130. Locate that place on the left, then move straight over to the right until you reach the line graph. Then, read down; **the year is 2010.**

c) As we move along the graph from 2007 to 2012, the numbers increase. **Therefore, the number of hours people spend playing video games is increasing each year.**

[YOU TRY 6]

Use the line graph in Example 6 to answer the following questions.

a) In what year did a person spend about 101 hr per year playing video games?

b) How many more hours did a person spend playing video games in 2012 than in 2007?

We can also use a **comparison line graph** to compare two sets of data.

EXAMPLE 7

The comparison line graph shows the corn crop yield (the amount of corn produced per acre), in bushels per acre, for Illinois and Minnesota from 2008 to 2015. Use the graph to answer the questions. (www.nass.usda.gov)

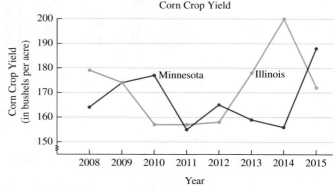

Corn Crop Yield

a) In which year was Minnesota's yield 20 bushels more per acre than Illinois'? What were their yields during that year?

b) During which year were their yields the same? What was the yield?

Solution

a) First, notice that the blue line represents the crop yield for Illinois. The red line represents the crop yield for Minnesota. To determine when Minnesota had a greater corn crop yield than Illinois, locate the points where the red graph is above the blue graph. Those years are 2010, 2012, and 2015. Locate the heights of the red points for those years as well as the heights of the blue points for those years by looking at the scale on the left of the graph. Which red and blue points for the same years look like they are about 20 units apart? That is in 2010. **In 2010, the corn crop yield for Minnesota was 20 more than that of Illinois. Minnesota's yield was 177 bushels per acre, and Illinois' yield was 157 bushels per acre.** Now, locate the height of the red point for 2010 by looking at the scale to the left. It matches up to about 177. **In 2010, the corn crop yield for Minnesota was 177 bushels per acre.**

b) How do we know when their corn crop yields were the same? It is where their graphs *intersect*. (The place where two lines **intersect** is the place where they meet.) Locate this point. Read the year and yield. The year is 2009, and the yield is about 174 bushels per acre. **In 2009, the corn crop yield for both Illinois and Minnesota was 174 bushels per acre.**

[YOU TRY 7]

Use the graph in Example 7.

a) In which year was Illinois' yield about 2 bushels per acre more than Minnesota's?

b) Which state had the highest yield, and in which year was this achieved? What was that yield?

c) Find the percent change in Illinois' yield from 2009 to 2010. Round the answer to the nearest tenth of a percent.

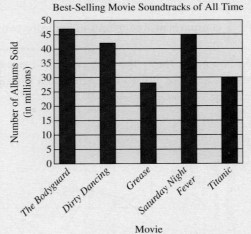
E Evaluate **9.1** Exercises   Do the exercises, and check your work.

Objective 1: Read a Table

The table shows the number of female prisoners in the
United States from 2011 to 2015. Use the information
in the table for Exercises 1–6. (www.bjs.gov)

Year	Number of Female Prisoners
2011	111,407
2012	108,772
2013	111,358
2014	113,028
2015	111,495

1) How many female prisoners were there in
   2013?

2) How many female prisoners were there in
   2011?

3) In which year were there the fewest female
   prisoners?

4) In which year were there the most female
   prisoners?

5) Find the percent change in the number of female
   prisoners from 2014 to 2015. Round the answer to
   the nearest tenth of a percent.

6) Find the percent change in the number of female
   prisoners from 2012 to 2013. Round the answer to
   the nearest tenth of a percent.

Objective 2: Read a Pictograph

The pictograph shows the amounts of food items sold
at a community fair. Use the pictograph to answer the
questions in Exercises 7–12.

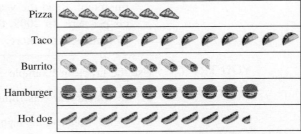

Amount of Food Sold at a Fair

Each symbol = 100 units

7) How many hamburgers were sold at this event?

8) Approximately how many hot dogs were sold?

9) Which food item was the most popular at the fair? How many of these food items were sold?

10) Which food item had the least sales?

11) Approximately how many more hamburgers than burritos were sold?

12) Approximately how many hot dogs and hamburgers were sold at this event?

Objective 3: Read and Make a Bar Graph

The bar graph shows the number of miles of interstate highways for six selected states. Use the graph to answer the questions in Exercises 13–20.

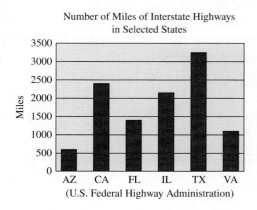

Number of Miles of Interstate Highways in Selected States

(U.S. Federal Highway Administration)

13) Which state has the greatest number of miles of interstate highways?

14) Which state has the least interstate mileage?

15) Which state has approximately 2400 mi of interstate highways?

16) Which states have less than 1500 mi of interstate highways?

17) How many more miles of interstate highway does Texas have than Illinois? Approximate your answer.

18) How many more miles of interstate highway does California have than Arizona? Approximate your answer.

19) How many miles of interstate highway do Florida and Virginia have combined? Approximate your answer.

20) How many miles of interstate highway do California and Texas have combined? Approximate your answer.

The bar graph shows the average total student loan amounts for graduates with bachelor's degrees at selected schools

in 2016. Use the graph to answer the questions in Exercises 21–26. (www.collegefactual.com)

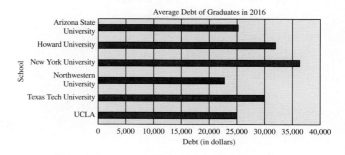

Average Debt of Graduates in 2016

21) Which school's students graduate with the most debt, on average? Approximately how much debt does each student have?

22) Which school's students graduate with the least debt, on average? Approximately how much debt does each student have?

23) Which schools' students graduate with between $25,000 and $30,000 of debt?

24) Which schools' students graduate with more than $30,000 of debt?

25) Approximate the difference in the average debt of a New York University graduate and a graduate of Northwestern.

26) What is the average debt of a graduate of UCLA?

Use the information in the table to make a vertical bar graph for Exercises 27–30.

27) **Patents Issued in the United States**

Year	Number
2011	247,713
2012	276,788
2013	302,948
2014	326,032
2015	325,979

(www.uspto.gov)

28) **Median Salary of Zoologists and Wildlife Biologists in Alaska**

Year	Median Annual Salary
2013	$67,910
2014	$69,660
2015	$72,500
2016	$73,730

(www.usawage.com)

29) **Magnitude of Some Major Earthquakes in April 2012**

Location	Magnitude
Argentina	6.7
Bulgaria	5.6
Mexico	6.5
Northern Sumatra	8.6
Oregon	5.9
Tonga	6.6
Vanuatu	6.5

(earthquake.usgs.gov)

30) **Number of Medals Won at the 2014 Winter Olympics in Sochi**

Country	Number
Canada	25
Germany	19
Norway	26
Russia	33
Switzerland	11
United States	28

(www.espn.com)

Punkin chunkin is the sport of *chunking,* or hurling, a pumpkin using human power or a mechanical device. The World Championship Punkin Chunkin competition is held the first weekend after Halloween in Sussex County, Delaware. The double-bar graph shows the longest distance chunked by three teams in the 2015 and 2016 World Championship in the Adult Catapult division. Use the graph for Exercises 31–38. (www.punkinchunkin.com)

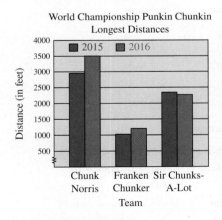

31) How far did team Chunk Norris chunk a pumpkin in 2016?

32) How far did team Franken Chunker chunk a pumpkin in 2015?

33) How far did team Chunk Norris chunk a pumpkin in 2015?

34) How far did team Franken Chunker chunk a pumpkin in 2016?

35) How much farther did Chunk Norris chunk a pumpkin in 2016 than in 2015?

36) How much farther did Franken Chunker chunk a pumpkin in 2016 than in 2015?

37) Which team chunked the pumpkin farther in 2015 than in 2016?

38) See Exercise 37. Use a signed number to represent the change from 2015 to 2016.

Objective 4: Read a Line Graph

The line graph shows the cost of a U.S. first-class postage stamp for various years. Use the graph to answer the questions in Exercises 39–46.

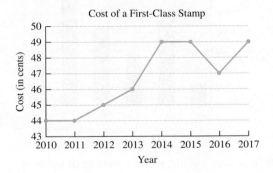

39) What was the cost of a stamp in 2010?

40) What was the cost of a stamp in 2015?

41) In what year was the cost of a stamp 46¢?

42) In what year was the cost of a stamp 45¢?

43) During which years did the price of a stamp decrease? By how much did it decrease?

44) During which years did the price increase the most? By how much did it increase?

45) How much less did a first-class stamp cost in 2010 than in 2017?

46) During which consecutive years did the cost of a stamp stay the same? What was the price at that time?

The comparison line graph displays the average number of viewers each night during a given week in 2017 for *The Tonight Show* Starring Jimmy Fallon and *Jimmy Kimmel Live*. Use the graph for Exercises 47–52.

(tvbythenumbers.zap2it.com)

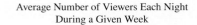

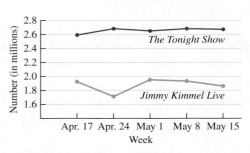

Average Number of Viewers Each Night During a Given Week

47) Which show had more viewers each week?

48) How many more people watched *The Tonight Show* each night during the week of April 17? Write your answer in thousands.

49) How many fewer people watched *Jimmy Kimmel Live* each night during the week of May 15? Write your answer in thousands.

50) During which week did *Jimmy Kimmel Live* average about 1 million fewer viewers per night than *The Tonight Show*?

51) Find the percent change in the average number of nightly viewers of *Jimmy Kimmel Live* from the week of April 17 to April 24. Round the answer to the nearest tenth of a percent.

52) Find the percent change in the average number of nightly viewers of *The Tonight Show* from the week of April 17 to April 24. Round the answer to the nearest tenth of a percent.

R Rethink

R1) What did you find most interesting about these types of problems?

R2) Which of the graphs did you find easiest to work with, and why?

9.2 Frequency Distributions and Histograms

P Prepare

O Organize

What are your objectives for Section 9.2?	How can you accomplish each objective?
1 Make a Frequency Distribution Table	• Explain, in your own words, how to make a frequency distribution table. • Complete the given example on your own. • Complete You Try 1.
2 Make a Table Using Class Intervals	• Write the definition of a *class interval* in your own words. • Explain how a table with class intervals is related to a frequency distribution table. • Complete the given example on your own. • Complete You Try 2.
3 Use a Histogram	• Write the definition of a *histogram* in your own words. • Complete the given example on your own. • Complete You Try 3.

Read the explanations, follow the examples, take notes, and complete the You Trys.

Data are facts or pieces of information. **Frequency** is defined as the number of times something occurs. If we have a lot of data, we can organize it in a *frequency distribution table*.

1 Make a Frequency Distribution Table

Joe is a construction worker. Here is a list of the number of hours he worked each week in 2017.

8	26	26	26	28	28	28	27	30	34	0	37	39
37	38	40	39	40	40	39	32	44	40	40	44	42
32	0	16	40	40	39	24	0	40	40	26	40	40
39	40	32	39	40	40	30	24	30	27	26	24	16

It seems that there is no organization in this table. Let's organize the data in a **frequency distribution table** so that they are easier to read.

EXAMPLE 1

Make a frequency distribution table to organize the data. Then, answer the questions.

a) How many times did Joe work 26 hr in a week?

b) The only time Joe did not work in a given week was when he took vacation time. How many times did this happen?

Solution

Begin by making a column listing the number of hours worked each week. Then, go through the list and put a tally mark, | , next to the number of hours worked each time that number appears. Finally, count the tally marks in each row and put that total in a third column labeled *Frequency*. This tells us the number of times that each number of hours appears in the list.

Number of Hours Worked	Tally	Frequency	Number of Hours Worked	Tally	Frequency
0					3	32					3
8			1	34			1
16				2	37				2
24					3	38			1
26	ⅢⅠ	5	39	ⅢⅠ	6
27	||	2	40	Ⅲ Ⅲ ||||	14
28					3	42			1
30					3	44				2

(Notice that the sum of the numbers in the frequency columns is 52, the total number of weeks in the first table.)

a) Now that we have organized the information in this table, it is easier for us to see that Joe worked 26 hr in a week 5 times.

b) Joe took vacation time 3 times during 2017.

[YOU TRY 1]

Use the table in Example 1.

a) How many times did Joe work 16 hr in one week?

b) Which number of hours did he work 5 times?

2 Make a Table Using Class Intervals

In Example 1, notice that there are 16 different categories for the number of hours worked. We can simplify the table by organizing the number of hours worked into *class intervals*. A **class interval** is a range of values into which data are organized.

EXAMPLE 2

Make a new table from the data in Example 1 by organizing the data into class intervals. Then, answer the questions.

a) How many times did Joe work between 20 and 29 hr per week?

b) Which range of hours did he work the least?

Solution

Let's organize the number of hours worked into intervals 0–9, 10–19, 20–29, etc. (These are the class intervals.) Add the frequencies in those intervals in the table in Example 1 to get the numbers for the Class Frequency column in Example 2.

Number of Hours Worked (Class Intervals)	Class Frequency
0–9	4
10–19	2
20–29	13
30–39	16
40–49	17

 Hint

Do you understand how to use a frequency distribution table to make a table using class intervals?

a) Looking at the class interval 20–29, we see that the corresponding class frequency is 13. **Joe worked 20–29 hr in a week 13 times.**

b) Look for the smallest class frequency. That is 2. The corresponding class interval (number of hours worked) is 10–19. **The range of hours that Joe worked the least was 10–19 hr.**

 Note

We can use any range of numbers in a class interval. Look at the numbers in the table to decide which range would make sense.

3 Use a Histogram

From the table containing the class intervals in Example 2, we can make a special type of bar graph called a *histogram*. In a **histogram,** the width of each bar represents the class intervals, and the height represents the class frequency. Also, the width of each bar, and therefore the size of each class interval, is the same in a histogram.

EXAMPLE 3 The histogram displays the information in the table in Example 2. Use the histogram to answer the questions.

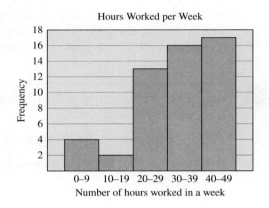

W Hint
Use a ruler (or even your driver's license) as a straightedge to help you make your graphs.

a) Which range of hours did Joe work most often?

b) How many weeks did Joe work less than 30 hr?

Solution

a) The highest bar is for the 40–49 hr class interval. **Therefore, Joe worked 40–49 hr most often.**

b) To determine the number of weeks he worked less than 30 hr, add the number of weeks in the categories 0–9 hr, 10–19 hr, and 20–29 hr.

$$4 \quad + \quad 2 \quad + \quad 13 \quad = \quad \textbf{19 weeks}$$

| number of times worked 0–9 hr | number of times worked 10–19 hr | number of times worked 20–29 hr |

[YOU TRY 3] Use the histogram in Example 3.

How many times did Joe work 30 hr per week or more?

ANSWERS TO [YOU TRY] EXERCISES

1) a) 2 times b) 26 hr   2) 16 times   3) 33 times

Objective 3: Use a Histogram

Seasonal snowfall is computed from July 1 of one year through June 30 of the next year. (For example, the 2010 seasonal snowfall amount is computed from July 1, 2010 through June 30, 2016.) The histogram shows the seasonal snowfall amounts in Syracuse, New York, from 1970 to 2016. Use the histogram for Exercises 1–8. (www.tsforecast.com)

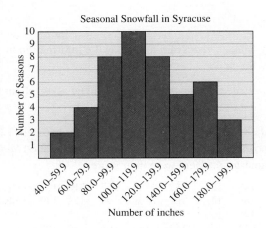

Seasonal Snowfall in Syracuse

1) During how many seasons did Syracuse have 140.0–159.9 in. of snow?

2) During how many seasons did Syracuse have 60.0–79.9 in. of snow?

3) Which snowfall amount occurred least often? How many seasons had this amount?

4) Which snowfall amount occurred most often? How many seasons did this amount occur?

5) Which snowfall amount occurred six times?

6) Which snowfall amount occurred eight times?

7) How many seasons did Syracuse have at least 120.0 in. of snow?

8) How many seasons did Syracuse have less than 100.0 in. of snow?

A manufacturing company has 80 employees. The histogram shows the number of years its employees have worked at the company. Use the histogram for Exercises 9–20.

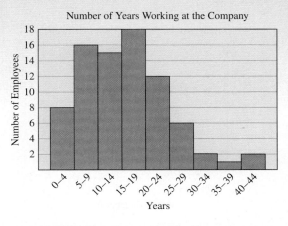

Number of Years Working at the Company

9) How many people have worked at the company for 10–14 yr?

10) How many people have worked at the company for 25–29 yr?

11) The greatest number of employees are in which range of years worked?

12) The least number of employees are in which range of years worked?

13) Find the ratio of the number of employees who have worked at the company 0–4 years to the number who have worked 40–44 yr. Explain the meaning of this ratio.

14) Find the ratio of the number of employees who have worked at the company 20–24 yr to the number who have worked 15–19 yr. Explain the meaning of this ratio.

15) How many people have worked at the company for at least 20 yr?

16) How many people have worked at the company less than 10 yr?

17) What percent of the employees have worked at the company for at least 20 yr?

18) What percent of the employees have worked at the company less than 10 yr?

19) What percent of the employees have been working the shortest time?

20) What percent of the employees have been working the longest time?

The histogram shows the heights, in inches, of members of a professional basketball team. Use the histogram for Exercises 21–30.

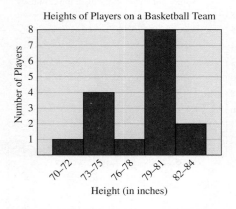

Heights of Players on a Basketball Team

21) The least number of players are in which height category? How many are in this category?

22) The greatest number of players are in which height category? How many are in this category?

23) a) How many team members are 82–84 in. tall?

b) We can also think of height in terms of feet and inches. For example, 63 in. = 5 ft 3 in. Rewrite the height range 82–84 in. as feet and inches.

24) a) How many team members are 73–75 in. tall?

b) See Exercise 23b. Rewrite the height range 73–75 in. as feet and inches.

25) How many players are less than 79 in. (6 ft 7 in.) tall?

26) How many players are 79 in. (6 ft 7 in.) or taller?

 27) What percent of the team is less than 79 in. tall?

28) What percent of the team is 79 in. or taller?

29) How many more players are in the tallest group than in the shortest group?

30) How many fewer players are 76–78 in. tall compared with 79–81 in. tall?

Mixed Exercises: Objectives 1–3

 31) Here are the test scores, in percent, for the 32 students in Mr. Garcia's Botany class.

73	82	51	94	76	71	68	73
85	90	72	63	76	83	79	73
81	59	66	77	91	60	70	46
75	54	78	43	88	73	80	74

a) Make a table like the one here to organize the information. Make a tally of the frequency of the scores in the class intervals, then count the tally marks and fill in the class frequency in the last column.

Test Scores (Class Intervals)	Tally	Number of Scores (Class Frequency)
40–49		
50–59		
60–69		
70–79		
80–89		
90–99		

b) Make a histogram to display the information in the table.

c) Write three questions that could be asked about the data. Answer the questions.

32) At a wellness fair, 40 college employees had their LDL cholesterol levels checked. (This is the so-called *bad* cholesterol.) Here were their scores.

93	148	127	82	103	165	210	131	119	74
158	186	143	107	191	99	172	149	154	126
166	140	159	170	100	216	152	138	125	172
98	121	147	163	130	195	85	97	116	184

a) The medical community classifies the LDL scores as shown here. Make a table like this one to organize the information. Make a tally of the frequency of the scores in the class intervals, then count the tally marks and fill in the class frequency in the last column. (www.heart.org)

LDL Scores (Class Intervals)	Tally	Number of Scores (Class Frequency)
70–99 (Good)		
100–129 (Near optimal)		
130–159 (Borderline high)		
160–189 (High)		
190–219 (Very high)		

b) Make a histogram to display the information in the table.

c) Write three questions that could be asked about the data. Answer the questions.

R Rethink

R1) When is a histogram a good way to display information?

R2) Compare histograms to the bar graphs we learned about in Section 9.1. How are they similar, and how are they different?

9.3 Using and Making Circle Graphs

P Prepare   ## O Organize

What are your objectives for Section 9.3?	How can you accomplish each objective?
1 Review How to Use a Circle Graph to Solve a Problem	• Complete the given example on your own. • Complete You Try 1.
2 Make a Circle Graph	• Understand how to compute the number of degrees needed in a sector of a circle graph. • Learn how to use a protractor. • Explain, in your own words, how to make a circle graph, and write it in your notes. • Complete the given examples on your own. • Complete You Trys 2 and 3.

W Work

Read the explanations, follow the examples, take notes, and complete the You Trys.

Throughout this book, we have analyzed data given in the form of a circle graph or pie chart. Circle graphs can be used to organize information about parts of a total amount. Let's review how to use a circle graph to solve a problem.

1 Review How to Use a Circle Graph to Solve a Problem

EXAMPLE 1

The large, deep-dish sausage pizza at Mario's Pizzeria is portioned by weight according to the pie chart. The pizza weighs 5 lb.

a) How many pounds of crust are used to make the pizza?

b) How many pounds of sausage are used to make the pizza?

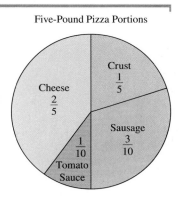

Five-Pound Pizza Portions

Solution

a) Find the fraction in the *crust* section of the pie chart. The $\frac{1}{5}$ means that $\frac{1}{5}$ *of* the total weight of the pizza is the crust. The *of* indicates multiplication. To determine the number of pounds of crust in the pizza, find $\frac{1}{5}$ *of* 5 lb or $\frac{1}{5} \cdot 5$.

$$\text{Number of pounds of crust} = \frac{1}{5} \cdot 5 = \frac{1}{5} \cdot \frac{5}{1} = \frac{1}{\cancel{5}} \cdot \frac{\overset{1}{\cancel{5}}}{1} = 1$$

One pound of crust is used to make the pizza.

b) Find the fraction in the *sausage* section of the pie chart. The $\frac{3}{10}$ means that $\frac{3}{10}$ *of* the total weight of the pizza is the sausage. The number of pounds of sausage in the pizza is $\frac{3}{10}$ *of* 5 lb or $\frac{3}{10} \cdot 5$

$$\text{Number of pounds of sausage} = \frac{3}{10} \cdot 5 = \frac{3}{10} \cdot \frac{5}{1} = \frac{3}{\underset{2}{\cancel{10}}} \cdot \frac{\overset{1}{\cancel{5}}}{1} = \frac{3}{2} \text{ or } 1\frac{1}{2}$$

$1\frac{1}{2}$ lb of sausage are used to make the pizza.

[YOU TRY 1] Use the pie chart in Example 1 to answer these questions.

a) How many pounds of sauce are used to make the pizza?

b) How many pounds of cheese are used to make the pizza?

---

2 Make a Circle Graph

Circle graphs can be used to organize information about parts of a total amount. Throughout this book, we have analyzed data given in the form of a circle graph. Now, we will learn how to *make* a circle graph.

Recall that a circle has 360°. If we split a circle in half, each *sector* has 180°. (A **sector** of the circle is a pie-shaped portion of the circle.)

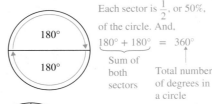

Each sector is $\frac{1}{2}$, or 50%, of the circle. And,

$\underbrace{180° + 180°}_{\substack{\text{Sum of} \\ \text{both} \\ \text{sectors}}} = \underset{\substack{\uparrow \\ \text{Total number} \\ \text{of degrees in} \\ \text{a circle}}}{360°}$

If we divide the circle in half again, how many degrees are in each sector? 90°

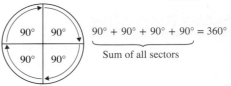

$\underbrace{90° + 90° + 90° + 90°}_{\text{Sum of all sectors}} = 360°$

What *fraction* of the circle does each 90° sector represent?

$$\frac{\text{number of degrees in the sector}}{\text{total number of degrees in a circle}} = \frac{90°}{360°} = \frac{1}{4}$$

Each 90° sector is $\frac{1}{4}$ of the circle.

What *percent* of the circle does each 90° sector represent?

$$\frac{90°}{360°} = \frac{1}{4} = 0.25 = 25\%$$

Each 90° sector is 25% of the circle.

We use a tool called a **protractor** to draw a circle graph. A protractor measures the number of degrees in an angle, and it allows us to draw angles of certain degrees. To determine the size of each sector in a circle graph, we calculate the number of degrees in each sector using percents.

EXAMPLE 2 Mrs. Szymanski's English Composition class has 20 students. The table gives us information about the number of hours worked each week by the students in her class. Make a circle graph to display this information.

Number of Hours Worked Each Week	Percent of the Students in the Class
Do not work at all	10%
Less than 20 hr	35%
20–40 hr	50%
More than 40 hr	5%
Total	100%

Solution

Each category for number of hours worked will be represented by a sector of the circle. We use the percentages to determine the size of each sector.

$$\begin{array}{cc} \text{Number of degrees} \\ \text{in a sector} \end{array} = \begin{array}{c} \text{Percent of} \\ \text{students} \end{array} \cdot \begin{array}{c} \text{Total degrees in} \\ \text{a circle} \end{array}$$

$$\begin{array}{cc} \text{Number of degrees} \\ \text{in a sector} \end{array} = \begin{array}{c} \text{Percent of} \\ \text{students} \end{array} \cdot \quad 360°$$

Before we make the circle graph, let's figure out the number of degrees each sector will have.

Hours Worked	Percent	Number of Degrees in the Sector
None	10%	10% of 360° = 0.10 · 360° = 36°
Less than 20	35%	35% of 360° = 0.35 · 360° = 126°
20–40	50%	50% of 360° = 0.50 · 360° = 180°
More than 40	5%	5% of 360° = 0.05 · 360° = 18°
		Total number of degrees = 360°

Note

If the degrees in the table do not add up to 360°, then there is a mistake somewhere. Double-check your work.

Now let's draw the circle graph. We'll begin with the sector containing 36° and work our way down the table.

W Hint

Practice drawing angles using a protractor before drawing them in a circle graph.

Draw a circle. Use a straightedge to draw a line from the center to the left edge of the circle. Place the protractor on the circle so that the hole in the protractor is over the center of the circle and so that the 0 on the left lines up with the line on the left. Read 36° from the left and make a mark at this point. Draw a line from the center of the circle to the mark at 36°. Label the sector.

36°
Do not work: 10%

Next, let's make the sector that represents the percent of students who work less than 20 hr per week. According to the table, this sector will measure 126°.

Line up the protractor on the second line you drew, putting the hole in the protractor over the center of the circle. Starting at the left, move over 126°, and make a mark. Draw a line from the center of the circle to the mark. Label this sector.

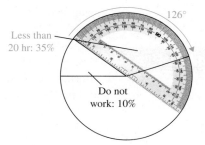

Now we will make the sector that represents the 50% of students who work 20–40 hr per week. Our calculations tell us that this sector will contain 180°. (To draw this sector, it may be helpful if you turn your paper so that you are not working upside down.)

Line up the protractor on the last line drawn with the hole in the protractor over the center of the circle. Make a mark at 180°, and label the sector.

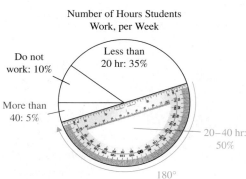

Number of Hours Students Work, per Week

The remaining, fourth sector represents the students who work more than 40 hr per week. (You can double-check that this sector has 18°.)

This is the circle graph that represents the data given in the table.

[YOU TRY 2]

During the month of May, 340 people had their hair dyed at a certain salon. The table shows the percent of people who chose each color for dyeing their hair. Make a circle graph to display this information.

Hair Dyeing in May

Color	Percent of People Who Chose the Color
Blonde	50%
Brown	25%
Black	15%
Red	10%

We can use the circle graph to answer questions like we did in other sections of the book.

EXAMPLE 3

Use the circle graph in Example 2 to answer the questions.

a) How many students work 20–40 hr per week?

b) How many students do not work at all?

Solution

a) There are 20 students in the class, and the circle graph shows that 50% of the students work 20–40 hr per week. Therefore, the number of students who work 20–40 hr per week is 50% *of* 20.

$$50\% \text{ of } 20 = 0.50 \cdot 20 = 10$$

Ten students work 20–40 hr per week.

b) 10% of the students in the class do not work at all. Therefore, the number of students who do not work is 10% *of* 20.

$$10\% \text{ of } 20 = 0.10 \cdot 20 = 2$$

Two students do not work.

[YOU TRY 3] Use the circle graph in You Try 2 to answer the questions.

a) How many people dyed their hair blonde?

b) How many people dyed their hair red?

c) How many more people dyed their hair blonde compared with red?

ANSWERS TO [YOU TRY] EXERCISES

1) a) $\frac{1}{2}$ lb   b) 2 lb

2) Use the following number of degrees for each sector.
blonde: 180°, brown: 90°, black: 54°, red: 36°

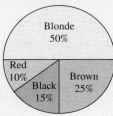

Hair Dyeing in May

Blonde 50%
Brown 25%
Black 15%
Red 10%

3) a) 170 people   b) 34 people   c) 136 people

9.3 Exercises

Do the exercises, and check your work.

Get Ready

Solve each problem.

1) A class of 24 students has 15 boys. Find the percentage of boys in the class.

2) A bike-racing team has 20 members, and 7 of them are from the United States. What percentage of the team consists of Americans?

3) A county has 600 miles of roads, and 120 miles of them need to be repaired. What percentage of the roads need repair?

4) Priyanka has read 306 pages of a 720-page book. What percentage of the book has she read?

Find each percentage.

5) 40% of 360   6) 30% of 360

7) 25% of 360   8) 75% of 360

9) 5% of 360   10) 55% of 360

Objective 1: Review How to Use a Circle Graph to Solve a Problem

A small manufacturing company recently moved into a new facility that has 18,000 square feet of floor space. The circle graph below shows how much floor space will be given to each department. Use this information for Exercises 11–16.

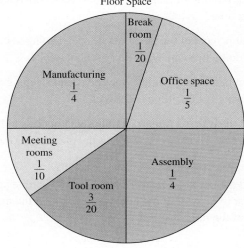

Floor Space

Break room $\frac{1}{20}$
Office space $\frac{1}{5}$
Manufacturing $\frac{1}{4}$
Assembly $\frac{1}{4}$
Meeting rooms $\frac{1}{10}$
Tool room $\frac{3}{20}$

11) How much floor space is given to the break room?

12) How much floor space is given to the tool room?

 13) How much floor space is provided for office space?

14) What is the combined area for manufacturing and assembly?

15) Verify that the sum of the floor space for breaks, offices, and assembly equals half of the 18,000-square-foot facility. What does this tell you about the sum of $\frac{1}{20}$, $\frac{1}{5}$, and $\frac{1}{4}$?

16) Verify that the sum of the floor space for meetings and tools equals the floor space of the assembly area. What does this tell you about the sum of $\frac{1}{10}$ and $\frac{3}{20}$?

Angarra is a full-time community college student. The circle graph shows how he budgets his time during his 120-hour school week. Use this information for Exercises 17–22.

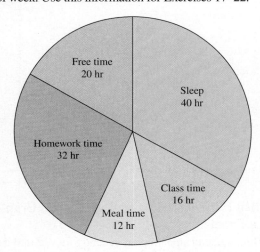

Free time
20 hr

Sleep
40 hr

Homework time
32 hr

Class time
16 hr

Meal time
12 hr

 17) What fraction of the school week is Angarra in class?

18) What fraction of Angarra's school week consists of meal time?

19) What percent of the time does he spend going to class and doing his homework every school week?

20) What percent of the time does he spend sleeping and enjoying his free time each week?

21) Find the ratio of Angarra's time spent in class to his time doing homework. Explain the meaning of the ratio.

22) Find the ratio of Angarra's free time to his time spent either in class or doing homework. Explain the meaning of the ratio.

Objective 2: Make a Circle Graph

23) How many degrees are in a circle?

24) What is a sector of a circle?

25) What is a protractor?

26) A table gives information in terms of the percent of the whole represented by each category. How do you determine the number of degrees in a sector of a circle graph that displays this information?

Each percent in Exercises 27–30 represents a share of the total in a table. If you are making a circle graph, determine the number of degrees in the sector.

27) 45%   28) 65%

29) 30%   30) 50%

31) After calculating the number of degrees in every sector of a circle graph, the degrees should add up to what number?

32) In your own words, explain how to make a circle graph.

The table shows how Scott classifies his 320 Facebook friends.

Scott's Facebook Friends

Scott's Facebook Friends	Percent of the Total Number of Friends
Work-related	40%
Family	10%
Close friends	5%
Casual acquaintances	20%
Barely know them	25%

Follow the steps to make a circle graph. Then, answer the questions in Exercises 33–40.

33) a) Determine the number of degrees needed for each sector in the circle graph.

   b) Draw the circle graph.

34) How many of Scott's Facebook friends are family members?

35) How many of Scott's Facebook friends are close friends?

36) How many of Scott's Facebook friends does he know through work?

37) How many of Scott's Facebook friends does he barely know?

38) Find the ratio of family members to work-related friends. Explain what the ratio means.

39) Find the ratio of close friends to the people he barely knows. Explain what the ratio means.

40) How many more casual acquaintances does he have compared with those he calls close friends?

Holly earns $2500 per month. The table displays how she spends her money each month.

Holly's Monthly Expenses

Expenses	Percent of Monthly Income
Rent	30%
Child care	20%
Food	15%
Utilities	10%
Other	25%

Follow the steps to make a circle graph. Then, answer the questions in Exercises 41–48.

41) a) Determine the number of degrees needed for each sector in the circle graph.

b) Draw the circle graph.

42) How much does Holly spend on rent?

43) How much does Holly pay for child care every month?

44) How much more does Holly spend on food than on utilities?

45) Find the ratio of the amount spent on rent to the amount spent on child care. Explain what the ratio means.

46) Find the ratio of the amount spent on food to the amount spent on utilities. Explain what the ratio means.

47) What are some expenses that might be included in the *Other* category?

48) After taking care of the first four expenses in the table, how much does Holly have left for *Other* expenses?

49) A total of 1200 students are taking Beginning Algebra at a community college. The course is offered in four different formats: in a *traditional classroom,* in a *math lab,* as a *hybrid* of lecture and online, and completely *online.* The table shows the number of students taking Beginning Algebra in each format.

Students Enrolled in Beginning Algebra

Format	Number of Students	Percent of Total	Degrees in the Sector
Traditional	540		
Math lab	300		
Hybrid	240		
Online	120		

a) Determine the percent of the total number of students taking Beginning Algebra in each format.

b) Determine the number of degrees needed to represent each category as a sector in a circle graph.

c) Make a circle graph to display the number of students taking Beginning Algebra in each format.

50) A total of 800 people were asked how many nights their family eats dinner together each week. The table displays the results of the survey.

Number of Nights Families Eat Dinner Together Each Week

Number of Nights Families Eat Dinner Together Each Week	Number of Families	Percent of Total	Degrees in the Sector
0	40		
1–2	80		
3–4	200		
5–6	280		
7	200		

a) Determine the percent of the total for each category.

b) Determine the number of degrees needed to represent each category as a sector in a circle graph.

c) Make a circle graph to display the information labeled with the percent of the families who eat dinner together a given number of times each week.

R Rethink

R1) What do you understand well about making circle graphs, and what, if anything, do you find confusing?

R2) Of all the graphs we have learned how to make, which do you prefer to use, and why?

9.4 Introduction to Linear Equations in Two Variables

What are your objectives for Section 9.4?	How can you accomplish each objective?
1 Define a Linear Equation in Two Variables	• Write the definition of *linear equation in two variables* in your own words, and write an example. • Understand the *Cartesian coordinate system* or *rectangular coordinate system*.
2 Decide Whether an Ordered Pair Is a Solution of an Equation	• Know how to substitute values into an equation. • Complete the given example on your own. • Complete You Try 1.
3 Complete Ordered Pairs for a Given Equation	• Be able to substitute either the *x*-value or the *y*-value, and solve for the missing variable. • Complete the given examples on your own. • Complete You Trys 2 and 3.
4 Plot Ordered Pairs	• Draw the *Cartesian coordinate system* in your notes with the quadrants, *x*-axis, *y*-axis, and origin labeled. • Follow the examples to create a procedure for plotting points. Write the procedure in your notes. • Complete You Try 4.

W Work

Read the explanations, follow the examples, take notes, and complete the You Trys.

In Section 9.1, we learned that different types of graphs can be used to communicate information. For example, the information about mobile phones given in the table can also be displayed on a line graph. (www.emarketer.com)

Number of Mobile Phone Users Worldwide from 2015 to 2020 (*projected)

Year	Number (in billions)
2015	4.15
2016	4.30
2017	4.43
2018	4.57*
2019	4.68*
2020	4.78*

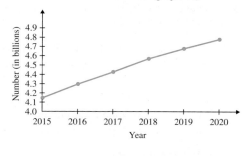

Number of Mobile Phone Users Worldwide from 2015–2020 (projected)

The table and the graph each display information as **paired data.** That is, they show the relationship between two quantities, *year* and the *number of mobile phone users.* These paired data can be represented with ordered pairs, data written in parentheses, separated by commas, with the year followed by the number: (year, number). For example, the ordered pair (2015, 4.15) tells us that *in the year* 2015, *the number of mobile phone users was* 4.15 billion. On the graph, the years are on the **horizontal axis** and the number of users is on the **vertical axis.** Usually, the first value of an ordered pair is represented on the horizontal axis and the second value is represented on the vertical axis.

We can apply these ideas of paired data and ordered pairs to extend work we have already done with equations.

1 Define a Linear Equation in Two Variables

Later in this section, we will see that graphs like the one just described are based on the *Cartesian coordinate system,* also known as the *rectangular coordinate system,* which gives us a way to graphically represent the relationship between two quantities. First, we will learn about *linear equations in two variables.* Let's begin with a definition.

Definition

A **linear equation in two variables** can be written in the form $Ax + By = C$, where A, B, and C are numbers and where both A and B do not equal zero.

Some examples of linear equations in two variables are

$$4x - 3y = 10 \qquad y = \frac{1}{2}x + 1 \qquad y = x$$

2 Decide Whether an Ordered Pair Is a Solution of an Equation

A solution of a linear equation in two variables is written as an *ordered pair* so that when the values are substituted for the appropriate variables, we obtain a true statement.

EXAMPLE 1

Determine whether each ordered pair is a solution of $4x - 3y = 10$.

a) (1, −2)   b) $\left(\dfrac{1}{4}, 5\right)$

Solution

a) Solutions to the equation $4x - 3y = 10$ are written in the form (x, y), where (x, y) is called an **ordered pair.** Therefore, the ordered pair (1, −2) means that $x = 1$ and $y = -2$.

$$(1, -2)$$
$$\nearrow \qquad \nwarrow$$
$$x\text{-coordinate} \qquad y\text{-coordinate}$$

To determine whether $(1, -2)$ is a solution of $4x - 3y = 10$, we substitute 1 for x and -2 for y. Remember to put these values in parentheses.

$$4x - 3y = 10$$
$$4(1) - 3(-2) = 10 \qquad \text{Substitute 1 for } x \text{ and } -2 \text{ for } y.$$
$$4 + 6 = 10 \qquad \text{Multiply.}$$
$$10 = 10 \qquad \text{True}$$

Because substituting $x = 1$ and $y = -2$ into the equation gives the true statement $10 = 10$, $(1, -2)$ *is a solution* of $4x - 3y = 10$. We say that $(1, -2)$ *satisfies* $4x - 3y = 10$.

b)   The ordered pair $\left(\dfrac{1}{4}, 5\right)$ tells us that $x = \dfrac{1}{4}$ and $y = 5$.

$$4x - 3y = 10$$
$$4\left(\dfrac{1}{4}\right) - 3(5) = 10 \qquad \text{Substitute } \dfrac{1}{4} \text{ for } x \text{ and 5 for } y.$$
$$1 - 15 = 10 \qquad \text{Multiply.}$$
$$-14 = 10 \qquad \text{False}$$

Because substituting $\left(\dfrac{1}{4}, 5\right)$ into the equation gives the false statement $-14 = 10$, the ordered pair is *not* a solution of the equation.

[YOU TRY 1]

Determine whether each ordered pair is a solution of the equation $2x + 9y = 15$.

a) $(3, 1)$ b) $(-4, -2)$ c) $\left(0, \dfrac{5}{3}\right)$

---

3   Complete Ordered Pairs for a Given Equation

If we are given the value of one variable in an equation, we can find the value of the other variable that makes the equation true.

EXAMPLE 2

Complete the ordered pair $(-2, \quad)$ for $y = 3x + 11$.

Solution

To complete the ordered pair $(-2, \quad)$, we must find the value of y from $y = 3x + 11$ when $x = -2$.

$$y = 3x + 11$$
$$y = 3(-2) + 11 \qquad \text{Substitute } -2 \text{ for } x.$$
$$y = -6 + 11$$
$$y = 5$$

When $x = -2$, $y = 5$. The ordered pair is $(-2, 5)$.

[YOU TRY 2]

Complete the ordered pair $(2, \quad)$ for $y = 6x - 8$.

If we want to complete more than one ordered pair for a particular equation, we can organize the information in a **table of values.**

EXAMPLE 3

Complete the table of values for the equation, and write the information as ordered pairs.

$-x + 2y = 9$

x	y
1	
	-6
	$\dfrac{3}{2}$

Solution

$-x + 2y = 9$

x	y
1	
	-6
	$\dfrac{3}{2}$

The first ordered pair is $(1,\ \ )$, and we must find y.

$$-x + 2y = 9$$
$$-(1) + 2y = 9 \qquad \text{Substitute 1 for } x.$$
$$-1 + 2y = 9$$
$$2y = 10 \qquad \text{Add 1 to each side.}$$
$$y = 5 \qquad \text{Divide by 2.}$$

The ordered pair is $(1, 5)$.

The second ordered pair is $(\ \ , -6)$, and we must find x.

$$-x + 2y = 9$$
$$-x + 2(-6) = 9 \qquad \text{Substitute } -6 \text{ for } y.$$
$$-x + (-12) = 9 \qquad \text{Multiply.}$$
$$-x = 21 \qquad \text{Add 12 to each side.}$$
$$x = -21 \qquad \text{Divide by } -1.$$

The ordered pair is $(-21, -6)$.

The third ordered pair is $\left(\ \ , \dfrac{3}{2}\right)$, and we must find x.

$$-x + 2y = 9$$
$$-x + 2\left(\dfrac{3}{2}\right) = 9 \qquad \text{Substitute } \dfrac{3}{2} \text{ for } y.$$
$$-x + 3 = 9 \qquad \text{Multiply.}$$
$$-x = 6 \qquad \text{Subtract 3 from each side.}$$
$$x = -6 \qquad \text{Divide by } -1.$$

The ordered pair is $\left(-6, \dfrac{3}{2}\right)$.

As you complete each ordered pair, fill in the table of values. The completed table will look like this:

x	y
1	5
-21	-6
-6	$\dfrac{3}{2}$

[YOU TRY 3]

Complete the table of values for the equation, and write the information as ordered pairs.

$x - 6y = -3$

x	y
9	
	-1
	$\dfrac{1}{2}$

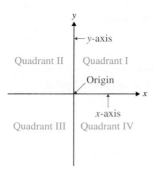

Quadrant II   Quadrant I

Origin

Quadrant III   Quadrant IV

4 Plot Ordered Pairs

When we completed the table of values for the equation in Example 3, we were finding solutions of the linear equation in two variables.

How can we represent the solutions on a graph? We will use the **Cartesian coordinate system,** also known as the **rectangular coordinate system,** to graph the ordered pairs (x, y).

In the Cartesian coordinate system, we have a horizontal number line, called the **x-axis,** and a vertical number line, called the **y-axis.**

The x-axis and y-axis in the Cartesian coordinate system determine a flat surface called a **plane.** The axes divide this plane into four **quadrants,** as shown in the figure. The point at which the x-axis and y-axis intersect is called the **origin.** The arrow at one end of the x-axis and one end of the y-axis indicates the positive direction on each axis.

Ordered pairs can be represented by **points** in the plane. Therefore, to graph the ordered pair (4, 2) we *plot the point* (4, 2). We will do this in Example 4.

W Hint

You move counterclockwise from Quadrant I to move through the quadrants in order.

EXAMPLE 4

Plot the point (4, 2).

Solution

Since $x = 4$, we say that the *x-coordinate* of the point is 4. Likewise, the *y-coordinate* is 2.

The *origin* has coordinates (0, 0). The **coordinates** of a point tell us how far from the origin, in the x-direction and y-direction, the point is located. So, the coordinates of the point (4, 2) tell us that to locate the point we do the following:

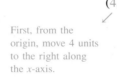

First, from the origin, move 4 units to the right along the x-axis.

Then, from the current position, move 2 units up, parallel to the y-axis.

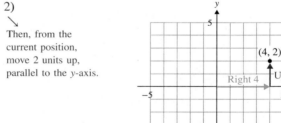

W Hint

Always start at the origin to first move horizontally and then move vertically.

EXAMPLE 5

Plot the points.

a)   (−3, 4)   b)   (1, −4)   c) $\left(\dfrac{5}{2}, 3\right)$

d)   (−5, −2)   e)   (0, 1)   f)   (−4, 0)

Solution

The points are plotted on the graph below.

a) (−3, 4)

<u>First</u>
From the origin, move left 3 units on the *x*-axis.

<u>Then</u>
From the current position, move 4 units up, parallel to the *y*-axis.

b) (1, −4)

<u>First</u>
From the origin, move right 1 unit on the *x*-axis.

<u>Then</u>
From the current position, move 4 units down, parallel to the *y*-axis.

c) $\left(\dfrac{5}{2}, 3\right)$

W Hint

Negative *x*-values make you move left, and negative *y*-values make you move down.

Think of $\dfrac{5}{2}$ as $2\dfrac{1}{2}$. From the origin, move right $2\dfrac{1}{2}$ units, then up 3 units.

d) (−5, −2) From the origin, move left 5 units, then down 2 units.

e) (0, 1) The *x*-coordinate of 0 means that we don't move in the *x*-direction (horizontally). From the origin, move up 1 on the *y*-axis.

f) (−4, 0) From the origin, move left 4 units. Since the *y*-coordinate is zero, we do not move in the *y*-direction (vertically).

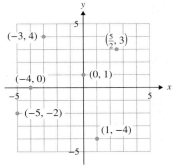

[YOU TRY 4]

Plot the points.

a) (3, 1)   b) (−2, 4)   c) (0, −5)   d) (2, 0)   e) (−4, −3)   f) $\left(1, \dfrac{7}{2}\right)$

ANSWERS TO [YOU TRY] EXERCISES

1) a) yes   b) no   c) yes   2) (2, 4)   3) (9, 2), (−9, −1), $\left(0, \dfrac{1}{2}\right)$

4)

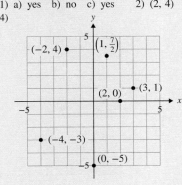

9.4 Exercises

Do the exercises, and check your work.

Mixed Exercises: Objectives 1 and 2

Determine whether each statement is *true* or *false*.

1) The ordered pair for a linear equation of the form $Ax + By = C$ is written in the form (x, y).

2) The first value of an ordered pair (x, y) is represented on the vertical axis.

3) The y-coordinate of the ordered pair $(-7, 3)$ is -7.

4) The equation $y = 3x + 8$ is an example of a linear equation in two variables.

Determine whether each ordered pair is a solution of the given equation.

5) $x - y = 7$

   a) $(10, 3)$

   b) $(1, -6)$

   c) $(-2, 5)$

6) $x + y = -5$

   a) $(-9, 4)$

   b) $(2, 3)$

   c) $(6, -11)$

7) $-4x + 3y = 18$

   a) $(-5, -1)$

   b) $(-3, 2)$

   c) $(0, 6)$

8) $2x - 7y = 4$

   a) $(5, -1)$

   b) $(2, 0)$

   c) $(9, 2)$

(24) 9) $y = -6x - 11$

   a) $(0, 11)$

   b) $(-5, 0)$

   c) $\left(-\dfrac{1}{3}, -9\right)$

10) $y = -8x + 3$

   a) $(0, -5)$

   b) $(3, 0)$

   c) $\left(-\dfrac{1}{4}, 5\right)$

11) $y = \dfrac{1}{2}x - 9$

   a) $(-10, -14)$

   b) $(18, 0)$

   c) $(2, -8)$

12) $y = \dfrac{1}{3}x - 10$

   a) $(12, -6)$

   b) $(0, -10)$

   c) $(-27, -19)$

Objective 3: Complete Ordered Pairs for a Given Equation

Complete the ordered pair for each equation.

13) $y = 5x + 2$; $(1, \ )$

14) $y = 4x + 1$; $(3, \ )$

15) $x + 8y = -6$; $(\ , -3)$

16) $x + 3y = -8$; $(\ , -5)$

17) $-3x + 10y = 15$; $(9, \ )$

18) $-2x + 9y = 16$; $(4, \ )$

19) $4x - 3y = -10$; $\left(\ , \dfrac{2}{3}\right)$

20) $5x - 4y = -3$; $\left(\ , \dfrac{3}{4}\right)$

Complete the table of values for each equation.

21) $y = x - 6$

x	y
0	
9	
-2	

22) $y = x - 9$

x	y
0	
10	
-4	

23) $y = -3x + 2$

x	y
0	
5	
-1	

24) $y = -8x + 7$

x	y
0	
2	
-3	

(24) 25) $-x + 4y = 12$

x	y
0	
	0
	4

26) $-x + 2y = 8$

x	y
0	
	0
	5

27) $2x - 5y = -6$

x	y
	0
$-\dfrac{1}{2}$	
	-2

28) $3x - 4y = -9$

x	y
	0
$-\dfrac{1}{3}$	
	-6

29) $y = x$

x	y
0	
	5
-1	

30) $y = -x$

x	y
0	
	3
-2	

31) $x = 1$ (Hint: Think of the equation as $x + 0y = 1$.)

x	y
	-6
	2
	0

32) $y = 2$ (Hint: Think of the equation as $0x + y = 2$.)

x	y
-5	
1	
0	

Objective 4: Plot Ordered Pairs

Name each point with an ordered pair, and identify the quadrant in which each point lies.

33)

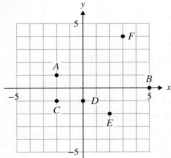

34)

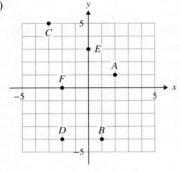

In Exercises 35–40, graph the four ordered pairs on the same axes. Clearly label the points.

35) $(4, 1), (-2, 5), (0, 3), (1, 0)$

36) $(2, 4), (-5, 3), (0, 2), (3, 0)$

37) $(-2, -3), (5, -4), (0, 0), \left(-\frac{1}{2}, 2\right)$

38) $(-3, -5), (4, 2), (0, 0), \left(-\frac{1}{2}, 5\right)$

39) $\left(0, \frac{7}{3}\right), (-1, -6), \left(\frac{1}{2}, \frac{3}{2}\right), (4, -4)$

40) $\left(0, \frac{11}{3}\right), (-1, -4), \left(\frac{9}{2}, \frac{1}{2}\right), (3, -3)$

In Exercises 41–46, fill in the blank with *positive, negative,* or *zero.*

41) The y-coordinate of every point in quadrant II is _____.

42) The x-coordinate of every point in quadrant III is _____.

43) The x-coordinate of every point in quadrant II is _____.

44) The y-coordinate of every point in quadrant I is _____.

45) The y-coordinate of every point on the x-axis is _____.

46) The x-coordinate of every point on the y-axis is _____.

R Rethink

R1) Does a table of values help you organize your ordered pairs? Explain.

R2) How does your knowledge of plotting fractions on a number line help you to plot points?

9.5 Graphing Linear Equations in Two Variables

P Prepare

What are your objectives for Section 9.5?	
1 Graph a Linear Equation by Plotting Points	
2 Graph Horizontal and Vertical Lines	

O Organize

How can you accomplish each objective?
• Learn the properties for **Solutions of Linear Equations in Two Variables** and **The Graph of a Linear Equation in Two Variables.**   • After following the examples, create a procedure for graphing linear equations by plotting points.   • Complete the given examples on your own.   • Complete You Trys 1–3.
• Write the properties **The Graph of $x = a$** and **The Graph of $y = b$** in your own words and memorize.   • Complete the given examples on your own.   • Complete You Trys 4 and 5.

W Work

Read the explanations, follow the examples, take notes, and complete the You Trys.

In Example 3 of Section 9.4, we found that the ordered pairs (1, 5), (−21, −6), and $\left(-6, \frac{3}{2}\right)$ are three solutions of the equation $-x + 2y = 9$. But how many solutions does the equation have? *It has an infinite number of solutions.* Every linear equation in two variables has an infinite number of solutions because we can choose any number for one of the variables and we will get another number for the other variable.

Property Solutions of Linear Equations in Two Variables

Every linear equation in two variables has an infinite number of solutions, and the solutions are ordered pairs.

How can we represent all of the solutions to a linear equation in two variables? We can represent them with a graph, and that graph is a *line*.

Property The Graph of a Linear Equation in Two Variables

The graph of a linear equation in two variables, $Ax + By = C$, is a straight **line.** Each point on the line is a solution of the equation.

---

1 Graph a Linear Equation by Plotting Points

EXAMPLE 1

Complete the table of values and graph $3x - y = 6$.

x	y
1	−3
0	−6
3	
	0

Solution

When $x = 3$, we get

$$3x - y = 6$$
$$3(3) - y = 6 \quad \text{Substitute 3 for } x.$$
$$9 - y = 6$$
$$-y = -3 \quad \text{Divide by } -1.$$
$$y = 3 \quad \text{Solve for } y.$$

When $y = 0$, we get

$$3x - y = 6$$
$$3x - (0) = 6 \quad \text{Substitute 0 for } y.$$
$$3x = 6 \quad \text{Divide by 3.}$$
$$x = 2 \quad \text{Solve for } x.$$

The completed table of values is

x	y
1	−3
0	−6
3	3
2	0

This gives us the ordered pairs (1, −3), (0, −6), (3, 3), and (2, 0). Each is a solution of the equation $3x - y = 6$.

Plot the points. They lie on a straight line. We draw the line through these points to get the graph.

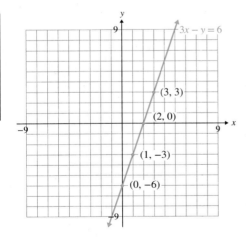

The line represents all solutions of the equation $3x - y = 6$. Every point on the line is a solution of the equation. The arrows on the ends of the line indicate that the line extends indefinitely in each direction.

Complete the table of values and graph $x - 2y = 5$.

x	y
3	−1
5	0
0	
	−2

Note

Although it is true that we need to find only two points to graph a line, it is best to plot at least three as a check.

EXAMPLE 2

Graph $-x + 4y = 8$.

Solution

We can use any values for x and y to find ordered pairs that satisfy the equation. Let's complete a table of values for $x = 0$, $x = 4$, and $x = -4$.

$x = 0$:
$$-x + 4y = 8$$
$$-(0) + 4y = 8$$
$$4y = 8$$
$$y = 2$$

$x = 4$:
$$-x + 4y = 8$$
$$-(4) + 4y = 8$$
$$-4 + 4y = 8$$
$$4y = 12$$
$$y = 3$$

$x = -4$:
$$-x + 4y = 8$$
$$-(-4) + 4y = 8$$
$$4 + 4y = 8$$
$$4y = 4$$
$$y = 1$$

We get the table of values

x	y
0	2
4	3
−4	1

W Hint

Use a ruler (or your driver's license) as a straightedge to help you draw lines.

Plot the points (0, 2), (4, 3), and (−4, 1), and draw the line through them.

Graph $2x - 3y = 4$.

EXAMPLE 3

Graph $y = -\dfrac{1}{3}x$.

Solution

Because the equation gives y in terms of x, $y = -\dfrac{1}{3}x$, we will choose values for x and calculate the corresponding values of y. We can choose any value we want for x, but let's think about it carefully.

The number 0 is usually a good value to substitute for the variable, so we will choose that. But, what would be some other good values to substitute for x? Again, look at the equation $y = -\dfrac{1}{3}x$. *If we choose a value for x that is a multiple of 3, then when we multiply x by* $-\dfrac{1}{3}$*, the value of y will not be a fraction.* Therefore, we will choose two values for x that are multiples of 3: Let's choose 3 and -3.

$$x = 0: \quad y = -\frac{1}{3}x \qquad\qquad x = 3: \quad y = -\frac{1}{3}x \qquad\qquad x = -3: \quad y = -\frac{1}{3}x$$

$$y = -\frac{1}{3}(0) \qquad\qquad\qquad y = -\frac{1}{3}(3) \qquad\qquad\qquad y = -\frac{1}{3}(-3)$$

$$y = 0 \qquad\qquad\qquad\qquad y = -1 \qquad\qquad\qquad\qquad y = 1$$

We get the table of values

x	y
0	0
3	−1
−3	1

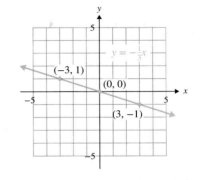

Plot the points $(0, 0)$, $(3, -1)$, and $(-3, 1)$, and graph the line.

[**YOU TRY 3**] Graph $y = -\dfrac{4}{3}x$.

Note

The **slope** of a line describes its steepness and tells us about the slant of the line.

1) If a line slants upward as you move from left to right, the line has a *positive* slope. It is also true that if a line has a positive slope, then as the values of x increase, the values of y increase as well.

   The lines in Examples 1 and 2 have positive slopes. They slant upward as you move from left to right, and as the values of x get larger, the values of y increase, too.

2) If a line slants downward as you move from left to right, the line has a *negative* slope. If a line has a negative slope, then as the values of x increase, the values of y decrease.

   Look at the line in Example 3. It has a negative slope. The line slants downward as you move from left to right, and as the values of x increase, the values of y decrease.

2 Graph Horizontal and Vertical Lines

An equation like $y = -2$ is a linear equation in two variables because it can be written as $0x + y = -2$. Likewise, an equation such as $x = 3$ can be written as $x + 0y = 3$. Let's see what the graphs of these equations look like.

EXAMPLE 4

Graph $y = -2$.

Solution

Because $y = -2$ is the same as $0x + y = -2$, any value we substitute for x will result in $0x = 0$ so that y always equals -2. The equation $y = -2$ means that *no matter the value of x, y always equals* -2. Make a table of values where we choose any value for x, but y is always -2.

x	y
0	-2
2	-2
-2	-2

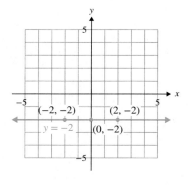

W Hint

If y is constant, then the graphed line is horizontal, parallel to the x-axis.

Plot the points, and draw a line through them. The graph of $y = -2$ is a **horizontal line.**

We can generalize the result as follows:

Property The Graph of $y = b$

If b is a constant (a number), then the graph of $y = b$ is a **horizontal line** going through the point $(0, b)$.

[**YOU TRY 4**]

Graph $y = 2$.

EXAMPLE 5

Graph $x = 3$.

Solution

The equation $x = 3$ means that *no matter the value of y, x always equals* 3. We can make a table of values where we choose any value for y, but x is always 3.

Plot the points, and draw a line through them. The graph of $x = 3$ is a **vertical line.**

x	y
3	0
3	1
3	-2

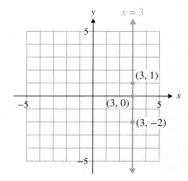

We can generalize the result as follows:

Property The Graph of $x = a$

If a is a constant (a number), then the graph of $x = a$ is a **vertical line** going through the point $(a, 0)$.

[YOU TRY 5]   Graph $x = -4$.

ANSWERS TO [YOU TRY] **EXERCISES**

1) $\left(0, -\dfrac{5}{2}\right)$, $(1, -2)$

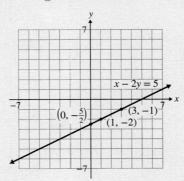

2)
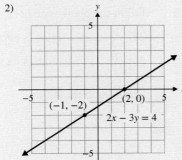

3)
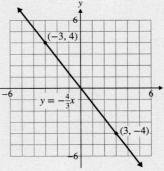

4)
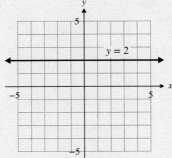

5)
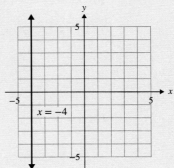

Objective 1: Graph a Linear Equation by Plotting Points

1) The graph of a linear equation in two variables is a _____.

2) Every linear equation in two variables has how many solutions?

Complete the table of values and graph each equation.

3) $y = x - 4$

x	y
0	
2	
4	

4) $y = x + 2$

x	y
0	
1	
-2	

5) $y = -2x + 3$

x	y
0	
3	
-1	

6) $y = -2x - 1$

x	y
0	
2	
-2	

7) $x + 2y = -2$

x	y
0	
	0
	-3

8) $x + 3y = 3$

x	y
0	
	0
	3

9) $4x - 3y = 15$

x	y
0	
	3
2	

10) $3x - 2y = -4$

x	y
0	
	8
-1	

11) $y = -4$

x	y
0	
-5	
1	

12) $x = 5$

x	y
	0
	-2
	4

13) What is the minimum number of points you should plot when you are graphing a line? Why?

14) Suppose that you are asked to graph $y = \frac{1}{6}x$. In addition to choosing $x = 0$, what other types of numbers are good choices for x, and why?

Graph each equation. Plot at least three points.

15) $y = -4x + 3$

16) $y = -3x + 5$

17) $-x + 2y = 4$

18) $-x + 4y = -8$

19) $2y + 3x = -1$

20) $3y + x = -4$

21) $4x - y = 3$

22) $2x - y = 5$

23) $y = x$

24) $y = -x$

25) $y = \frac{1}{4}x$

26) $y = \frac{1}{2}x$

27) $y = -\frac{2}{3}x - 1$

28) $y = -\frac{3}{4}x + 5$

Objective 2: Graph Horizontal and Vertical Lines

Fill in the blank with *horizontal* or *vertical*.

29) A _____ line is parallel to the y-axis.

30) A _____ line is parallel to the x-axis.

Graph each equation.

31) $y = 3$

32) $y = 1$

33) $x = -2$

34) $x = -4$

35) $x = \frac{7}{2}$

36) $x = \frac{5}{2}$

37) $y = -4.5$

38) $y = -0.5$

39) $x = 0$

40) $y = 0$

Mixed Exercises: Objectives 1 and 2

Graph each equation.

41) $y = \frac{1}{2}x - 3$

42) $x = 2$

43) $5x + 2y = 6$

44) $y = \frac{5}{3}x - 4$

45) $x = 3y$

46) $-2x - 3y = 9$

47) $y = -\frac{8}{3}$

48) $x = -4y$

Read the Note that follows You Try 3, then determine whether the line in the indicated exercise has a *positive* or *negative* slope.

49) Exercise 43

50) Exercise 44

51) Exercise 41

52) Exercise 46

53) Exercise 21

54) Exercise 20

 55) Exercise 27

56) Exercise 26

**R Rethink**

R1) In which other courses, besides math, have you had to graph a line?

R2) Did you use graph paper to graph these lines? If not, do you think that using graph paper could make your work easier?

Group Activity – Graphs and the Rectangular Coordinate System

You are going to be moving into a new apartment and you have contacted two local moving companies, U-Haul-It and Mighty Movers, to obtain prices for renting a moving truck. The total cost for renting a moving truck from U-Haul-It can be represented by the equation $y = 0.10x + 60$, where x is the amount of miles driven and y is the total cost. The total cost for renting a moving truck from Mighty Movers can be represented by the equation $y = 0.60x + 35$, where x is the amount of miles driven and y is the total cost.

1) Create two tables and find seven solutions for each equation. Use 0, 10, 20, 30, 40, 50, and 60 as your values for x.

2) Describe any patterns that you notice in the tables. Your description should relate to the situation involving miles driven and cost.

3) Graph both equations on the following graph. Label each axis.

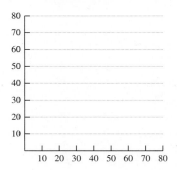

4) Describe what the point (50, 65) means in practical terms.

5) Explain how you will decide which company to use. Your answer should address the cost and the miles driven.

6) Pick two ordered pairs from the U-Haul-It table. Calculate the slope of the line between these two points. What does the slope represent in the context of this situation?

7) Pick two ordered pairs from the Mighty Movers table. Calculate the slope of the line between these two points. What does the slope represent in the context of this situation?

8) Look at your graph. Which company has a steeper graph? Why is this so?

9) The y-intercept for the U-Haul-It graph is _____ and the y-intercept for the Mighty Movers graph is _____. Explain what the y-intercepts represent in the context of this situation.

10) Suppose there were a third company that charged a flat fee of $80 to rent a truck, no matter how many miles you drive. Write an equation that would model the cost for this company. What would the graph of your equation look like?

In order to manage your financial life the way you want, it is important to understand just how much money matters to you. Complete the following exercise to learn about your personal financial philosophy. (The table continues on the next page.)

A. Attitudes Toward Money

	Strongly Disagree	Disagree	Neutral	Agree	Strongly Agree
1. Money is essential for happiness.					
2. Having money guarantees happiness.					
3. Money makes no difference to one's happiness.					
4. More money equals more happiness.					
5. Beyond having enough to live on modestly, money doesn't make much of a difference.					
6. I frequently worry about money.					
7. I frequently daydream about having a lot of money.					
8. If I suddenly had to live on very little money, I could adjust easily.					
9. If I suddenly won a lot of money, I would go on a spending spree.					
10. If I suddenly won a lot of money, I would share it with my relatives.					
11. If I suddenly won a lot of money, I would give a large percentage to charity.					
12. If I found a substantial amount of cash in a bag, I would try hard to find its rightful owner.					
13. If I could carry only a briefcase full of $100 bills out of a burning building or my pet dog, I would take the dog.					
14. I plan to make a lot of money in my career.					
15. I plan to make only enough money to live in reasonable comfort.					
16. It's great to have money.					
17. Money is a necessary evil.					
18. Money is the root of all evil.					

B. Sources of Satisfaction

1. Which activities that you engaged in over the last five years have given you the greatest satisfaction?

2. How much money did those activities cost?

3. How would you spend your time if you could do anything you chose?

4. How much money would this cost each year?

C. Personal Financial Philosophy

Consider your attitudes toward money and the sources of your satisfaction. Use this thinking to help shape your budget, as well as your long-term financial and career goals.

Chapter 9:   Summary

Definition/Procedure	Example

9.1 Reading Tables, Pictographs, Bar Graphs, and Line Graphs

Graphs are visual ways to represent information. In this section, we learned how to read and interpret **tables, pictographs, bar graphs,** and **line graphs.**

Use the given bar graph to answer the following questions.

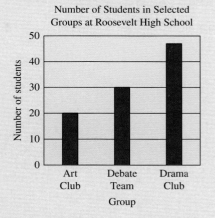

Number of Students in Selected Groups at Roosevelt High School

a) Which group is the largest, and approximately how many students are in it? The Drama Club is the largest with about 47 students.

b) How many more people are on the Debate Team than in the Art Club? The Debate Team has 10 more members since $30 - 20 = 10$.

9.2 Frequency Distributions and Histograms

Data are facts or pieces of information.

Frequency is the number of times something occurs.

A **class interval** is a range of values into which data are organized.

A **histogram** is a special type of bar graph in which the width of each bar represents the class intervals, and the height represents the class frequency. The width of each bar is the same.

A total of 50 high school math students were asked how many hours they spent on their homework each week. The histogram shows the results of the survey. Use the histogram to answer the questions.

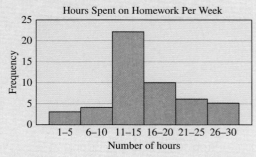

Hours Spent on Homework Per Week

1) How many students spent 6–10 hr on their homework per week? Four students spent 6–10 hr per week on their homework.

2) How many students spent the greatest amount of time on their homework? How many hours did they spend on their homework? Twenty-two students spent the greatest amount of time on homework. They did homework for 11–15 hr per week.

3) Which range of hours did exactly five students spend on their homework? Exactly five students spent 26–30 hr on their homework.

| Definition/Procedure | Example |

9.3 Using and Making Circle Graphs

A **circle graph** (also called a **pie chart**) can be used to organize information about parts of a total amount.

A circle has 360°, so we use percentages to determine the size of a sector in a circle graph.

We use a **protractor** to draw a circle graph.

An ice cream shop surveys 200 customers to determine their favorite toppings. The circle graph displays the results. Use the graph to answer the questions.

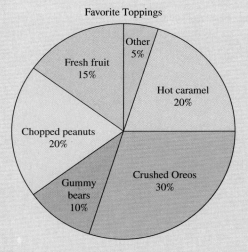

Favorite Toppings

1) How many customers chose fresh fruit?

Find 15% of 200 customers: 0.15(200) = 30 customers

2) How many customers chose hot caramel?

Find 20% of 200 customers: 0.20(200) = 40 customers

3) Find the ratio of customers who chose fresh fruit to those who chose hot caramel. Explain what the ratio means.

$$\text{ratio} = \frac{\text{number who chose fruit}}{\text{number who chose caramel}}$$

$$= \frac{30}{40}$$

$$= \frac{3}{4}$$

For every three customers who chose fresh fruit, four customers chose hot caramel.

9.4 Introduction to Linear Equations in Two Variables

A **linear equation in two variables** can be written in the form $Ax + By = C$, where A, B, and C are numbers and where both A and B do not equal zero.

To determine whether an ordered pair is a solution of an equation, substitute the values for the variables.

Is $(4, -1)$ a solution of $2x - 3y = 11$?

$$2x - 3y = 11$$
$$2(4) - 3(-1) = 11 \quad \text{Substitute 4 for } x \text{ and } -1 \text{ for } y.$$
$$8 + 3 = 11$$
$$11 = 11 \checkmark$$

Yes, $(4, -1)$ is a solution.

Definition/Procedure	Example

9.5 Graphing Linear Equations in Two Variables

The graph of a linear equation in two variables, $Ax + By = C$, is a straight line. Each point on the line is a solution to the equation.

We can graph the line by plotting the points and drawing the line through them.

Graph $y = \frac{1}{3}x - 2$ by plotting points.

Make a table of values. Plot the points, and draw a line through them.

x	y
0	−2
3	−1
−3	−3

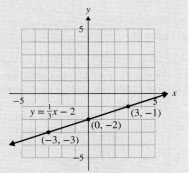

If a is a constant, then the graph of $x = a$ is a *vertical line* going through the point $(a, 0)$.

If b is a constant, then the graph of $y = b$ is a *horizontal line* going through the point $(0, b)$.

Graph $x = 2$.

Graph $y = -3$.

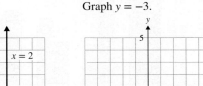

Chapter 9:   Review Exercises

(9.1) **The Daytona 500 is a 500-mile-long NASCAR race held each year at the Daytona International Speedway in Daytona Beach, Florida. Listed here are four winners and their winning times. Use this information for Exercises 1–4. (Note: The average speed will be in miles per hour, mph.)** (www.daytonainternationalspeedway.com)

Year	Winner	Time
1961	Marvin Panch	3.34 hours
1981	Richard Petty	2.95 hours
1998	Dale Earnhardt	2.90 hours
2016	Denny Hamlin	3.17 hours

1) Who had the fastest time? What was his average speed, rounded to the nearest hundredth?

2) Who had the slowest time? What was his average speed, rounded to the nearest hundredth?

3) How much less time did it take for the fastest driver to complete the course compared to the slowest driver?

4) How many minutes separated the fastest and slowest winning times?

In a survey, middle school students were asked to write down their favorite type of summer vacation. The following pictograph shows the results of the survey. Use the pictograph to answer Exercises 5–10.

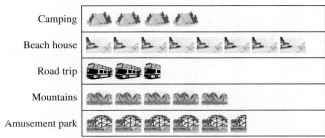

Each picture = 20 students

5) How many students chose the road trip?

5) How many students chose the amusement park?

7) How many more students chose the beach house than the mountains?

8) How many more students chose the amusement park than camping?

9) What percent of the total number of students chose the mountains?

10) What percent of the total number of students chose the beach house?

The double-bar graph displays the earnings of some of the world's highest paid hip-hop artists in 2015 and 2016. Use the graph for Exercises 11–15. (www.forbes.com)

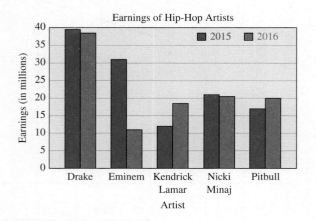

11) Which artists earned about the same amount of money in 2016? Approximately how much did they earn?

12) Which artists earned more in 2015 than in 2016?

13) How much more did Kendrick Lamar earn in 2016 than in 2015? What was the percent increase in his earnings from 2015 to 2016?

14) Find the ratio of Nicki Minaj's earnings in 2015 to Kendrick Lamar's earnings in 2015. Explain the meaning of this ratio.

15) Who earned the most and who earned the least in 2016? What is the difference in their earnings?

16) The table displays the number of times different celebrities have hosted *Saturday Night Live* through the end of Season 42 in May 2017. Make a bar graph to display the information. (www.nbc.com)

Number of Times Hosting
Saturday Night Live

Host	Number
Alec Baldwin	17
Tom Hanks	9
Dwayne Johnson	5
Melissa McCarthy	5
Betty White	1

The following line graph shows the number of ticket sales, in thousands, over a seven-month period. Use the graph to answer the questions in Exercises 17–20.

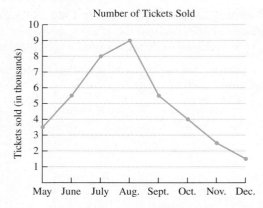

17) What month had the greatest number of ticket sales and how many tickets were sold?

18) What month had the lowest number of ticket sales and how many tickets were sold?

19) Between what two consecutive months did the greatest decrease in ticket sales occur?

20) Approximately how much did the ticket sales increase between May and August?

(9.2) A large department store surveyed its employees to determine how many miles each employee must drive to get to work. The histogram shows the results of the survey. Use the histogram for Exercises 21–24.

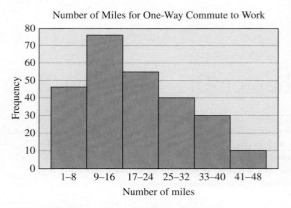

21) How many employees drive 25–32 miles to get to work?

22) How many employees drive the furthest to get to work? How many miles do they drive to work?

23) Exactly 80 employees live within what range of miles from the department store?

24) How many employees drive at least 17 miles to get to the department store?

25) Here are the ages of United States presidents at the time of Inauguration. (www.whitehouse.gov)

57	61	57	57	58	57	61	54	68	51
49	64	50	48	65	52	56	46	54	49
50	47	55	55	54	42	51	56	55	54
51	60	62	43	55	56	61	52	69	64
46	54	47	70

a) Make a table like the one here to organize the information. Make a tally of the frequency of United States presidents in the class intervals, then count the tally marks and fill in the class frequency in the last column.

Age at Inauguration (Class Intervals)	Tally	Number of Presidents (Class Frequency)
42–45		
46–49		
50–53		
54–57		
58–61		
62–65		
66–69		
70+		

b) Make a histogram to display the information in the table.

c) Write three questions that could be asked about the data. Answer the questions.

(9.3) Miss Sorenson teaches a math class that meets 140 min per day, two days per week. The circle graph shows how Miss Sorenson manages her class time. Use this information for Exercises 26–29.

Daily Use of Class Time

26) How much break time does Miss Sorenson give her students?

27) How much time is allowed for the Quiz?

28) What is the combined Q&A and Quiz time?

29) Based on the pie chart, what is the sum of all six fractions:
$$\frac{3}{28} + \frac{1}{14} + \frac{1}{7} + \frac{3}{14} + \frac{3}{28} + \frac{5}{14} = ?$$

30) A total of 900 students at a local community college were asked to specify where they purchased their textbooks. The table shows the results of the survey.

Where Students Get Their Textbooks

Location	Number of Students	Percent of Total	Degrees in the Sector
Online	270		
From a friend	90		
College bookstore	405		
Off-campus bookstore	135		

a) Determine the percent of the total number of students who purchased their textbooks in each category.

b) Determine the number of degrees needed to represent each category as a sector in a circle graph.

c) Make a circle graph to display the information. Label the percent of the students who purchased their textbooks from each location.

(9.4) Determine whether each ordered pair is a solution of the given equation.

31) $6x - y = 17; (2, -5)$

32) $y = -2x + 1; (-1, -1)$

33) $y = \frac{1}{4}x + 5; (8, 7)$

34) $y = -\frac{2}{5}x; (0, 0)$

Complete the ordered pair for each equation.

35) $y = x + 4; (6, \quad)$

36) $y = \frac{3}{2}x - 9; (-12, \quad)$

37) $x + 5y = -11; (\quad, -4)$

38) $-3x + 8y = 2; (\quad, 0)$

Complete the table of values for each equation.

39) $-2x + 7y = -14$

x	y
0	
	0
−7	
	4

40) $y = 6x + 5$

x	y
0	
	0
	1
$\frac{1}{2}$	

Plot the ordered pairs on the same axes. Clearly label the points.

41) $(2, 4), (-5, 0), (-1, -3), (0, 3)$

42) $(-4, 1), (0, -2), (5, 0), (3, 3)$

Fill in the blank with *positive* or *negative*.

43) The *y*-coordinate of every point in quadrant II is
_____.

44) The *x*-coordinate of every point in quadrant II is
_____.

(9.5) Complete the table of values and graph each equation.

45) $y = 3x - 4$

x	y
0	
1	
2	

46) $y = -\frac{1}{2}x$

x	y
0	
4	
-2	

47) $x + 2y = 5$

x	y
0	
	0
	4

48) $-x + y = 1$

x	y
0	
	0
	-3

Graph each equation.

49) $y = -x + 3$

50) $y = \frac{2}{3}x - 4$

51) $-5x + 2y = -2$

52) $y = -x$

53) $x = -3$

54) $y = 2.5$

Chapter 9: Test

The table shows the number of games played by selected Major League Baseball players from 2014 to 2016. Use the information to solve the problems in Exercises 1–4. (www.espn.com)

Player	2014	2015	2016
Ryan Howard	153	129	112
Albert Pujols	159	157	152
Jose Ramirez	68	97	152
Anthony Rizzo	140	160	155

1) Find the total number of games played by the four players in 2016. Then, find the average number of games played by each player.

2) Find the percent increase in the number of games played by Jose Ramirez from 2015 to 2016. Round the answer to the nearest tenth of a percent.

3) In 2014, Jose Ramirez played in how many fewer games than Anthony Rizzo?

4) Find the total number of games played by Ryan Howard during these three years.

The pictograph shows the number of vehicles sold in one month. Use the pictograph to solve the problems in Exercises 5–8.

Each picture = 10 vehicles

5) How many hybrid vehicles were sold this month?

6) How many minivans were sold this month?

7) How many more trucks were sold when compared to minivans?

8) What percent of the total number of vehicles sold were hybrid cars?

9) A survey asked 150 college students to reveal their favorite television show. The four most popular answers and the number of students who preferred each show are listed in the table. Make a bar graph to display this information.

Favorite Television Show	Number of Students
Game of Thrones	46
The Walking Dead	31
Empire	28
House of Cards	20

The comparison line graph shows the approximate number of visitors to the Statue of Liberty and Yellowstone National Park during the years 2012–2016. Use the graph to answer the questions in Exercises 10–13. (http://irma.nps.gov)

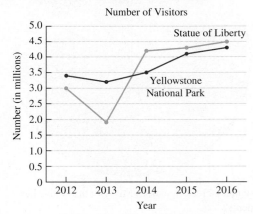

10) How many people visited Yellowstone in 2012?

11) During which year was the difference between their number of visitors the greatest? Which had more visitors, and by how much?

12) During which years did the Statue of Liberty and Yellowstone have the same number of visitors, and how many did they have?

13) During which years did the Statue of Liberty have more visitors than Yellowstone National Park?

14) Here are the sodium levels in milligrams (mg) for 30 breakfast cereals.

15	65	280	150	240	0	290	115	85	55
25	85	160	110	205	95	120	200	60	75
95	75	210	205	180	70	210	100	45	50

a) Make a table like the one here to organize the information. Make a tally of the frequency of the sodium levels in the class intervals, then count the tally marks and fill in the class frequency in the last column.

Sodium Levels (mg) (Class Intervals)	Tally	Number of Cereals (Class Frequency)
0–49		
50–99		
100–149		
150–199		
200–249		
250–299		

b) Make a histogram to display the information in the table.

c) Write three questions that could be asked about the data. Answer the questions.

The circle graph represents the end-of-semester grade distribution for a 180-student biology lecture course. Use the graph for Exercises 15–18.

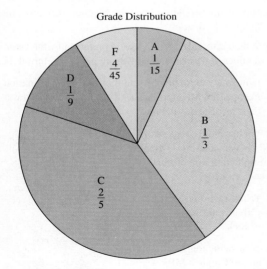

Grade Distribution

15) What fractional amount of students passed the course with a C or better?

16) How many students passed the course with a C or better?

17) How many students did not earn an F?

18) What is the sum of all the fractional parts?

19) A total of 1800 members of a local health club were asked to specify the exercise activity on which they spend most of their workout time. The table shows the results of the survey.

Preferred Exercise Activity

Activity	Number of Members	Percent of Total	Degrees in the Sector
Jogging	270		
Swimming	90		
Weight lifting	450		
Aerobics class	630		
Exercise machines	360		

a) Determine the percent of the total members for each exercise activity.

b) Determine the number of degrees needed to represent each exercise activity as a sector in a circle graph.

c) Make a circle graph to display the information labeled with the percent of the members who prefer each exercise activity.

Determine whether each ordered pair is a solution of the given equation.

20) $y = 6x - 2$; $(1, 8)$

21) $-4x + 3y = -7$; $(-2, -5)$

22) Complete the ordered pair: $x - 7y = 10$; $(\ \ , -3)$

Plot the ordered pairs on the same axes. Clearly label the points.

23) $(4, 0)$, $(-1, -5)$, $(2, 3)$, $(2, -3)$, $(0, 1)$

24) Complete the table of values and graph $2x + y = 3$.

x	y
0	
	0
	-1

Graph each equation.

25) $y = x - 2$

26) $y = -\dfrac{1}{3}x$

27) $-3x + 4y = 4$

28) $y = -1$

29) $x = 2$

Fill in the blank with positive, negative, or zero.

30) The x-coordinate of every point in quadrant IV is _____.

31) The y-coordinate of every point on the x-axis is _____.

32) The y-coordinate of every point in quadrant III is _____.

Perform the indicated operations.

1) $-\dfrac{8}{15} \cdot \dfrac{5}{12}$

2) $23 - 9.74$

3) $-612 - (-809)$

4) $14 + 56 \div (-7) + \sqrt{9} - (1+2)^2$

5) $\dfrac{4}{9a} + \dfrac{1}{6a}$

6) Simplify $5x^2 - 9x - 2 + 2x^2 + x + 10$.

7) Explain the difference between evaluating $(-7)^2$ and -7^2, and evaluate each.

Solve each equation.

8) $4p - 7 = 25$

9) $\dfrac{4}{9}c + \dfrac{5}{9} = \dfrac{2}{3}(c+2)$

10) $\dfrac{n}{18} = \dfrac{40}{12}$

11) Find the area of the triangle.

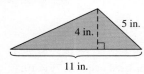

12) Find the exact area and circumference of the circle as well as an approximation of each using 3.14 for π.

For Exercises 13 and 14, write an equation and solve.

13) A company is putting a fence around its rectangular parking lot that is twice as long as it is wide. If they use 390 ft of fencing, what are the dimensions of the parking lot?

14) In Spring 2018, Carlos transferred to a state university and his tuition and fees were $5465. This was $335 more than three times the amount he paid at the community college in the Fall 2017 semester. Find the amount of his tuition and fees at the community college.

15) Change 5 min to seconds.

16) Change 2400 m to km.

17) Change $\dfrac{13}{20}$ to a percent.

18) Write 3.2% as a fraction in lowest terms.

Compute each percent mentally.

19) 40% of 90

20) 15% of 400

21) Christian is a licensed pilot, and these are the number of miles he has flown each month over the last six months: 807, 714, 1092, 388, 824, and 675. Find the average number of miles he flew each month.

22) In 2011, Sayuri paid $480 per month for child care. In 2012, the cost rose to $504 per month. Find the percent increase in the cost of child care.

23) One hundred people who have taken more than one cruise were asked to reveal their favorite destination. The results are displayed in the bar graph. Use the graph to answer the questions.

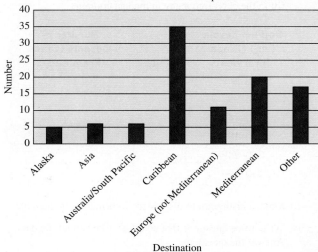

a) Which destination was the favorite among the people surveyed, and how many people declared this their favorite?

b) How many people said that the Mediterranean was their favorite destination?

c) Which cruise destinations were preferred by the same number of people, and how many people preferred them?

d) How many more people said their favorite destination was Europe (not Mediterranean) than said that Alaska was their favorite destination?

24) Calvin has 2400 songs on his MP3 player. The circle graph shows the fraction of music in each category.

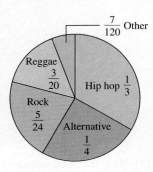

a) What fraction of Calvin's music is hip hop or rock?

b) How many reggae songs are on Calvin's MP3 player?

c) What fraction of music is *not* classified as alternative?

d) How many songs are *not* classified as alternative?

25) Graph $2x + 3y = -9$.

Appendix A Whole-Number Arithmetic

A.1 Adding Whole Numbers

P Prepare

O Organize

What are your objectives for Section A.1?	How can you accomplish each objective?
1 Add Numbers Using a Number Line	• Write the definition of a *number line* in your own words. • Complete the given example on your own. • Complete You Try 1.
2 Use the Commutative Property	• Write the **Commutative Property of Addition** in your own words. • Complete the given examples on your own. • Complete You Trys 2 and 3.
3 Use the Associative Property	• Write the **Associative Property of Addition** in your own words. • Complete the given examples on your own. • Complete You Trys 4–8.
4 Add Numbers with No Regrouping (Carrying)	• Write the procedure for **Adding Numbers with More Than One Digit** in your own words. • In this objective, the sum of the digits in each column is a single-digit number. • Complete the given examples on your own. • Complete You Trys 9 and 10.
5 Add Numbers with Regrouping (Carrying)	• In this objective, the sum of the digits in each column might be a two-digit number. • Complete the given examples on your own. • Complete You Trys 11 and 12.
6 Solve Applied Problems Using Addition	• Read the applied problem twice and be sure to understand what is being asked. • Write the definition of *perimeter* in your notes. • Complete the given examples on your own. • Complete You Trys 13 and 14.

W Work

Read the explanations, follow the examples, take notes, and complete the You Trys.

Recall that **whole numbers** are 0, 1, 2, 3, 4, 5, . . . , where the three dots mean that the list continues in the same way forever. **Even numbers,** like 2, 4, 6, 8, 10, 714, etc., end in 0, 2, 4, 6, or 8. **Odd numbers,** like 1, 3, 5, 7, 9, 11, 267, etc., end in 1, 3, 5, 7, or 9.

We can strengthen our addition, subtraction, multiplication, and division skills with whole numbers if we begin with the basics and understand how these operations are related. To do this, we will use a number line.

1 Add Numbers Using a Number Line

Definition

A **number line** is a line used to represent numbers, in order, on evenly spaced, marked points on the line.

Here is a number line marked with the numbers 0 through 10:

Note

1) It is important that the spacing between the tick marks be the same.

2) As you move from left to right on the number line, the numbers get larger.

We can represent a number on the number line with a dot. For example, the number 4 can be placed on the number line like this:

We can use a number line to add numbers.

EXAMPLE 1

Use a number line to add 2 + 5.

Solution

Start at 0 and move 2 spaces to the right to reach 2. To add 5, move 5 more spaces to the right. We finish at 7.

$$2 + 5 = 7$$

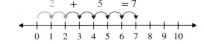

W Hint

Does this feel like you are counting?

[YOU TRY 1] Use a number line to add 7 + 3.

In an addition problem, the numbers being added together are called the **addends,** and the answer is called the **sum.** In Example 1, the addends are 2 and 5. The sum is 7.

2 Use the Commutative Property

If we change the order in which we add numbers, does it change the sum?

EXAMPLE 2 Use a number line to add $5 + 2$.

Solution

Start at 0 and move 5 spaces to the right to reach 5. To add 2, move 2 more spaces to the right. We finish at 7.

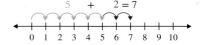

$$5 + 2 = 7$$

Notice that this is the same result we obtained in Example 1 when we found that $2 + 5 = 7$.

YOU TRY 2 Use a number line to add $3 + 7$.

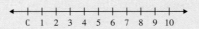

We can add numbers in any order and the result, or sum, will be the same. The *commutative property of addition* tells us this is true.

W Hint
You can add left to right or right to left!

Property The Commutative Property of Addition

The **commutative property of addition** says that changing the order in which we add numbers does not change the sum.

Note

It may be helpful to remember the commutative property this way: To *commute* to work means that each day we travel from home to our place of business and then back home again. Therefore, commuting refers to changing location. When we *commute* numbers, we are changing the locations of those numbers.

EXAMPLE 3 Add $8 + 1$, then rewrite the addition problem using the commutative property.

Solution

$8 + 1 = 9$. Using the commutative property, we get $1 + 8 = 9$.

YOU TRY 3 Add $4 + 7$, then rewrite the addition problem using the commutative property.

3 Use the Associative Property

Let's add three numbers on a number line. Then we will learn another property of addition.

EXAMPLE 4

Use a number line to add $3 + 6 + 4$.

Solution

We will extend the number line to 15.

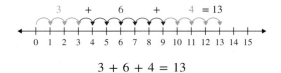

$$3 + 6 + 4 = 13$$

$\left[\text{YOU TRY 4}\right]$ Use a number line to add $5 + 2 + 8$.

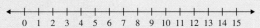

We can also use *parentheses* to add numbers. **Parentheses** are grouping symbols that tell us the order in which we should perform arithmetic operations. Usually, we do the operations in parentheses before other operations in a problem.

EXAMPLE 5

Use a number line to add $(3 + 6) + 4$.

Solution

Let's add the numbers in parentheses first; then we will add on the number line.

 Hint
Notice that the same numbers are being added in Examples 4, 5, and 6.

$(3 + 6) + 4 =$
$\quad 9 \quad + 4 =$
$\quad\quad 13$

$$(3 + 6) + 4 = 13$$

$\left[\text{YOU TRY 5}\right]$ Use a number line to add $(5 + 2) + 8$.

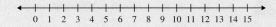

When we add several numbers, can we change the placement of the parentheses without changing the sum?

EXAMPLE 6   Use a number line to add 3 + (6 + 4).

Solution

Add the numbers in parentheses first.

3 + (6 + 4) =
3 +   10   =
         13

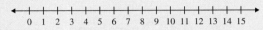

3 + (6 + 4) = 13

This is the same as the result in Example 5 when the parentheses were in a different place: (3 + 6) + 4 = 13.

[YOU TRY 6]   Use a number line to add 5 + (2 + 8).

Examples 5 and 6 show that (3 + 6) + 4 = 3 + (6 + 4) = 13. Changing the position of the parentheses in an addition problem does *not* change the sum. This is an example of the *associative property of addition*.

> **Property The Associative Property of Addition**
>
> The **associative property of addition** says that we can change the position of grouping symbols when adding numbers and the sum remains the same.

EXAMPLE 7   Add 9 + (3 + 8), then rewrite the addition problem using the associative property.

Solution

Original Sum	**Using the Associative Property**
9 + (3 + 8) =	(9 + 3) + 8 =
9 + 11 = 20	12 + 8 = 20

Notice that we changed the position of the parentheses. We did not change the position of the numbers.

[YOU TRY 7]   Add (5 + 2) + 8, then rewrite the addition problem using the associative property.

Sometimes, using the commutative and associative properties together makes it easier to find the sum of numbers.

EXAMPLE 8

Add $4 + 7 + 6 + 3$.

Solution

The commutative property says that we can add numbers in any order and the sum does not change. Let's rearrange the numbers.

$$4 + 7 + 6 + 3 = 4 + 6 + 7 + 3 \qquad \text{Commutative property}$$

The associative property says that we can *group* the addends in any way.

$$= (4 + 6) + (7 + 3)$$
$$= \quad 10 \quad + \quad 10 \quad = 20$$

W Hint

If possible, rearrange the numbers so that groups add up to 10.

[YOU TRY 8]

Add $9 + 2 + 7 + 8$.

4 Add Numbers with No Regrouping (Carrying)

We have added more than two numbers using the associative and commutative properties. In this first example, we will add several numbers by aligning them vertically.

EXAMPLE 9

Add $5 + 3 + 7 + 2$.

Solution

Line up the numbers vertically, one under the other.

Add the first two numbers, then add that sum to the third number.

We continue until we have added all the numbers.

$$
\begin{array}{l}
5 \\
3 \\
7 \\
+2 \\
\hline
17
\end{array}
$$

$5 + 3 = 8$

$8 + 7 = 15$

$15 + 2 = 17$

$\leftarrow$ Write the final sum under the horizontal line.

The sum of the numbers is 17.

[YOU TRY 9]

Add $8 + 2 + 6 + 3$.

Next, we will add numbers containing more than one digit.

Procedure Adding Numbers with More Than One Digit

1) Line up the numbers vertically so that the ones digits are in the same column, the tens digits are in the same column, the hundreds digits are in the same column, and so on.

2) Begin the addition process in the column farthest to the right, and then move to the left, adding numbers column by column. That is, add the numbers in the ones place, then add the numbers in the tens place, then add the numbers in the hundreds place, and so on.

Note

Using graph paper to perform operations with numbers can help us line up the numbers in their proper places.

EXAMPLE 10

Add 702 + 41 + 33 + 120.

Solution

Line up the numbers in columns starting with the ones at the right.

> Begin adding in the ones column, the column farthest to the right.
>
> Next, add the numbers in the tens column.
>
> Then, add the numbers in the hundreds column.

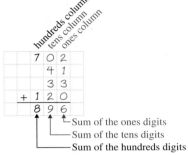

Sum of the ones digits
Sum of the tens digits
Sum of the hundreds digits

The sum is 896.

YOU TRY 10

Add 221 + 12 + 453 + 101.

5 Add Numbers with Regrouping (Carrying)

Let's extend what we have learned about digits and place value.

For example, we can think of the number 18 in different ways:

1) 18 = 18 ones   or   2) 18 = 1 ten + 8 ones

We can think of the number 250 like this:

1) 250 = 250 ones   or   2) 250 = 2 hundreds + 5 tens + 0 ones

                 or   3) 250 = 25 tens

These other ways of describing numbers are what we use when we add numbers that require **regrouping** or **carrying.**

EXAMPLE 11

Add 38 + 27.

Solution

Line up the numbers in their correct columns.

> Add the 8 and 7 in the ones column: 8 + 7 = 15.
>
> Write the 5 in the ones column and "carry the one" to the tens column. We can do this because we are regrouping 15 as 15 = 1 ten + 5 ones.
>
> Add all the digits in the tens column, including the 1.

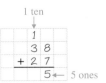

The sum is 65.

EXAMPLE 12   Add 256 + 14 + 5,418 + 5 + 743.

Solution

Line up the numbers in their correct columns, and add the digits in the ones column.

W Hint

As you are reading the example, write it out on your paper too!

```
          2
      2 5 6
        1 4
    5, 4 1 8
          5
  +   7 4 3
          6 ↑
```

The sum of the digits in the ones column is 26. Regroup this as 2 tens + 6 ones. Carry the 2 to the tens column.

Add the digits in the tens column, including the carried 2.

```
      1 2
      2 5 6
        1 4
    5, 4 1 8
          5
  +   7 4 3
        3 6 ↑
```

The sum of the digits in the tens column is 13 tens. Regroup this as 1 hundred and 3 tens. Carry the 1 to the hundreds column.

Add the digits in the hundreds column, including the carried 1.

```
      1 2
      2 5 6
    1   1 4
    5, 4 1 8
          5
  +   7 4 3
      4 3 6 ↑
```

The sum of the digits in the hundreds column is 14 hundreds. Regroup this as 1 one-thousand and 4 hundreds. Carry the 1 to the thousands column.

Add the digits in the thousands column, including the carried 1.

```
      1 2
      2 5 6
    1   1 4
    5, 4 1 8
          5
  +   7 4 3
    6, 4 3 6  ← The final sum is 6,436.
         ↑
```

The sum in the thousands column is 6.

The answer is 6,436.

[YOU TRY 12]   Add 46 + 7,518 + 947 + 1,136.

6 Solve Applied Problems Using Addition

We use addition in many ways to solve applied problems.

| EXAMPLE 13 | Phil lives in Los Angeles and picked up friends to go to Las Vegas. First, he drove to Bakersfield to pick up Stu, then they drove to Barstow to get Doug and Alan, and then they went on to Las Vegas. Find the total number of miles Phil drove to get to Las Vegas. |

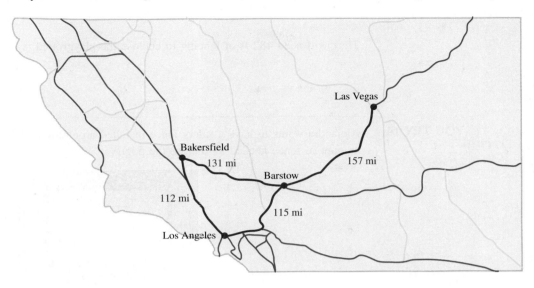

Solution

W Hint

Read the problem carefully, maybe more than once.

W Hint

Do we need the 115-mile distance from LA to Barstow?

The distance from Los Angeles to Bakersfield is 112 mi, the distance from Bakersfield to Barstow is 131 mi, and the distance from Barstow to Las Vegas is 157 mi. Add these numbers to find the total distance that Phil drove.

$$
\begin{array}{r}
1\ 1 \\
1\ 1\ 2 \leftarrow \text{Distance from Los Angeles to Bakersfield} \\
1\ 3\ 1 \leftarrow \text{Distance from Bakersfield to Barstow} \\
+\ 1\ 5\ 7 \leftarrow \text{Distance from Barstow to Las Vegas} \\
\hline
4\ 0\ 0 \leftarrow \text{Total number of miles Phil drove}
\end{array}
$$

Phil drove a total of 400 mi.

[YOU TRY 13] Use the map in Example 13. When their vacation was over, Phil, Stu, Doug, and Alan took the shortest route from Las Vegas to Phil's house in Los Angeles. Find the total number of miles they drove back from Las Vegas.

The **perimeter** of a figure is the distance around a figure. We use perimeter to solve many everyday problems.

| EXAMPLE 14 | A city parks department wants to put a fence around a rectangular playground. How many feet of fencing will they need? |

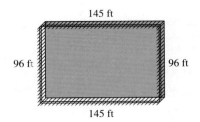

Solution

We must find the *perimeter* of the playground, the total distance around the playground. Add the lengths of all of the sides.

They will need 482 ft of fencing to enclose the playground.

		2	2	
	1	4	5	
		9	6	
	1	4	5	
+		9	6	
	4	8	2	

[YOU TRY 14] Alejandra wants to have a safety fence installed around her backyard pool. Find the amount of fence needed to enclose the pool.

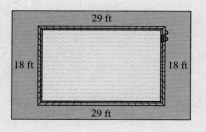

ANSWERS TO [YOU TRY] EXERCISES

1)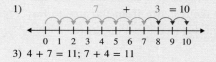

2)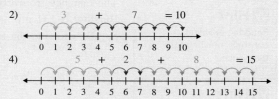

3) $4 + 7 = 11$; $7 + 4 = 11$

4)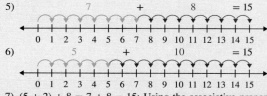

5)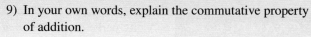

6)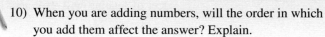

7) $(5 + 2) + 8 = 7 + 8 = 15$; Using the associative property: $5 + (2 + 8) = 5 + 10 = 15$
8) 26   9) 19   10) 787   11) 74   12) 9,647   13) 272 mi   14) 94 ft

---

E Evaluate **A.1** Exercises   Do the exercises, and check your work.

Objective 1: Add Numbers Using a Number Line

Identify the addends and sum in each addition problem.

1) $8 + 7 = 15$

2) $1 + 9 = 10$

Use a number line to add the numbers.

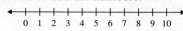

3) $5 + 4$ 4) $4 + 5$

5) $6 + 4$ 6) $7 + 1$

7) $2 + 3 + 4$ 8) $5 + 2 + 1$

Objective 2: Use the Commutative Property

9) In your own words, explain the commutative property of addition.

10) When you are adding numbers, will the order in which you add them affect the answer? Explain.

Add, then rewrite the problem using the commutative property.

11) $8 + 7$ 12) $2 + 11$

13) $1 + 16$ 14) $7 + 0$

15) $2 + 4 + 8$

16) $9 + 7 + 1$

 17) $7 + 6 + 3$

18) $5 + 0 + 8$

Objective 3: Use the Associative Property

 19) In your own words, explain the associative property of addition.

20) Answer true or false. $7 + (2 + 3) = (7 + 2) + 3$
Which property helped you get the answer?

Add, then rewrite the problem using the associative property.

21) $3 + (7 + 5)$

22) $(6 + 3) + 2$

 23) $4 + (7 + 3)$

24) $2 + (4 + 6)$

25) $(3 + 5) + 5$

26) $(6 + 2) + 8$

Find the missing length labeled with ?.

27)

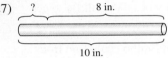
?   8 in.
10 in.

28)

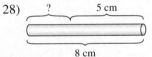
?   5 cm
8 cm

29)

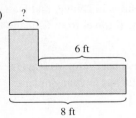
?
6 ft
8 ft

30)

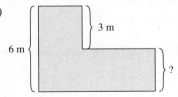
3 m
6 m
?

Objective 4: Add Numbers with No Regrouping (Carrying)
Add.

31) $4 + 1 + 7 + 6$

32) $8 + 5 + 6 + 3$

33)
```
  56
+ 23
```

34)
```
  67
+ 31
```

 35)
```
  432
+ 125
```

36)
```
  608
+ 231
```

 37)
```
  4,205
+ 3,581
```

38)
```
  1,613
+ 4,152
```

39)
```
  2,014
     33
+   921
```

40)
```
      16
     542
+  8,431
```

41) $452 + 5,103 + 34$

42) $61 + 8,200 + 137$

43) $60,142 + 11 + 5,200 + 516$

44) $2,431 + 210 + 71,002 + 45$

Objective 5: Add Numbers with Regrouping (Carrying)
Add.

45) $47 + 39$

46) $75 + 18$

47) $84 + 36$

48) $93 + 57$

49) $375 + 486$

50) $594 + 187$

51)
```
  6,594
+ 1,822
```

52)
```
  2,917
+ 1,846
```

53)
```
  5,475
+ 4,925
```

54)
```
  7,692
+ 2,318
```

55)
```
  89
  26
+ 17
```

56)
```
  79
  18
+ 32
```

57)
```
     571
  19,680
      85
   2,712
     803
+  5,957
```

58)
```
  37,092
     863
   4,216
      28
     511
+  8,003
```

59) $268 + 564 + 17$

60) $365 + 52 + 456$

61) $8,256 + 936 + 36,589$

62) $1,001 + 89,201 + 199$

63) $12 + 36,987 + 185 + 4 + 2,066$

64) $85,645 + 8 + 3,214 + 198 + 70,301$

65) $23,584 + 1,965 + 354 + 42,000 + 26 + 3,750$

66) $189 + 45,256 + 3,658 + 7,000 + 87 + 810$

67) $9,400 + 78,228 + 55 + 546 + 41,617 + 154$

68) $300 + 20 + 68,000 + 9,500 + 4 + 25,000$

Objective 6: Solve Applied Problems Using Addition

Solve each problem.

69) A builder installs the window pictured here but still needs to install the trim that goes around it. How many inches of trim will he need?

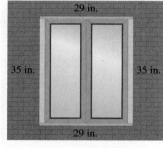

70) Farzan wants to put a fence around his garden to keep out the rabbits. How many feet of fencing will he need?

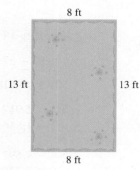

71) Sheng-Li paid $1,199 for a new television and $449 for a home theater system. How much did he pay for these electronics?

72) For the fall semester, Latrice paid $1,624 for tuition and $582 for books. How much did she pay for tuition and books?

73) Pauly, Ronnie, and Vinny spend 75 min at the gym, then they go to the tanning salon for 25 min, and finally they spend 110 min doing laundry. How long did they spend on these activities?

74) Mike did 120 crunches on Monday, 150 crunches on Wednesday, 120 crunches on Friday, and 100 crunches on Sunday. How many crunches did he do all together?

75) A community college has 2,687 students taking math courses on campus, 1,385 at an off-campus facility, and 887 students in online sections. How many total students are enrolled in math courses?

76) An appliance manufacturer has 1,235 employees at its plant in Ohio, 672 employees at its Michigan plant, and 834 people working at its factory in Pennsylvania. Find the total number of employees at these three plants.

Use the map for Exercises 77–82.

Flying Distances Between Cities

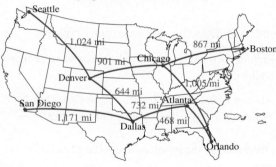

77) Lara is meeting her friends in Colorado to go skiing over Christmas break. She will fly from San Diego to Dallas and then on to Denver. How many miles will she fly to get to Denver?

78) The Anderson family is flying from Boston to go to Disney World, and they have to change planes in Chicago. Find the total miles they will fly to get from Boston to Orlando.

79) The cheapest flight that Emilio could find from Seattle to Boston makes stops in Denver and Chicago. How many miles will he travel if he purchases this itinerary?

80) Corinne gets a cheap flight from her home in Chicago to San Diego if she makes stops in Denver and Dallas. Find the total distance she will travel from Chicago to San Diego.

81) Find the total, round-trip distance Veronica travels from Dallas to Orlando if she has to change planes in Atlanta.

82) Using the cities given on the map, find the shortest flying distance from Denver to Orlando.

Find the perimeter of each figure in Exercises 83 and 84.

83)
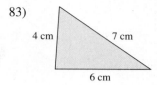

84)
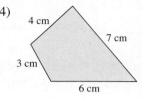

Solve each problem involving perimeter.

85) How many inches of wood does James need to make a frame for this picture?

86) Mr. Rizzo wants to put a chain-link fence around his property. How many feet of fencing will he need?

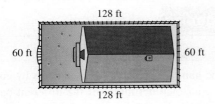

128 ft

60 ft   60 ft

128 ft

87) A community garden is on a corner lot and has the dimensions shown here. If the director wants to put a fence around it, how many feet will she need?

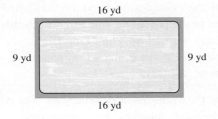

30 ft   50 ft

40 ft

88) Find the length of the border around the backyard ice skating rink.

16 yd

9 yd   9 yd

16 yd

89) A construction site is enclosed by yellow tape to keep people out. How many feet of tape were used?

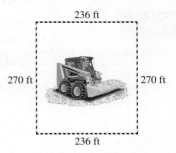

236 ft

270 ft   270 ft

236 ft

90) Haruko is sewing a border around a scarf for her daughter. How many centimeters of the border will she need?

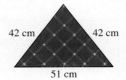

42 cm   42 cm

51 cm

Find the length of each missing side, then find the perimeter of the figure.

91)

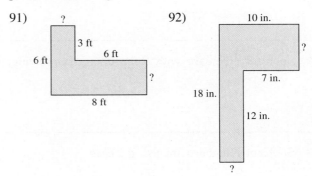

?

3 ft

6 ft   6 ft

?

8 ft

92)

10 in.

?

7 in.

18 in.

12 in.

?

R1) Did you use graph paper to help line up the numbers correctly?

R2) Could you explain how to regroup (or carry) to a friend?

R3) Would you be able to complete similar exercises without needing any help?

R4) Think of a situation where you needed to find the sum of two or more "large" numbers. How did you find the sum? How would you find the sum now?

A.2 Subtracting Whole Numbers

What are your objectives for Section A.2?	How can you accomplish each objective?
1 Subtract Numbers Using a Number Line	• Use the number line to subtract numbers by counting. • Write your own definition of the different parts of a subtraction problem: *minuend*, *subtrahend*, and *difference*. • Complete the given example on your own. • Complete You Try 1.
2 Relate Subtraction and Addition	• Check a subtraction problem by using addition. • Check an addition problem by using subtraction. • Complete the given examples on your own. • Complete You Trys 2 and 3.
3 Subtract Without Regrouping (Borrowing)	• Write the procedure for **Subtracting Numbers with More Than One Digit** in your own words. • Complete the given examples on your own. • Complete You Trys 4 and 5.
4 Subtract Numbers with Regrouping (Borrowing)	• Follow the examples, and add any additional steps needed to the procedure you wrote for **Subtracting Numbers with More Than One Digit.** • Complete the given examples on your own. • Complete You Trys 6 and 7.
5 Subtract Numbers Involving Zeros	• Follow the examples, and write your own procedure for regrouping numbers that contain zeros. • Complete the given examples on your own. • Complete You Trys 8 and 9.
6 Solve Applied Problems Using Subtraction	• Read the applied problem twice and be sure to understand what is being asked. • Complete the given example on your own. • Complete You Try 10.

W Work   **Read the explanations, follow the examples, take notes, and complete the You Trys.**

What does $8 - 3$ mean? Let's look at a number line to answer this question.

1 Subtract Numbers Using a Number Line

EXAMPLE 1   Use a number line to subtract $8 - 3$.

Solution

Start at 0 and move 8 units to the right to reach 8.
To subtract 3, move 3 units to the left. We finish at 5.

$$8 - 3 = 5$$

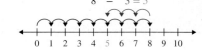

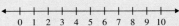

[YOU TRY 1] Use a number line to subtract 9 − 5.

0 1 2 3 4 5 6 7 8 9 10

Note

Notice, also, that if 8 − 3 = 5, then 8 − 5 = 3.

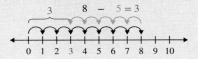

$$8 - 5 = 3$$

0 1 2 3 4 5 6 7 8 9 10

W Hint

Why do you think each number has a different name?

In Example 1, 8 − 3 = 5, the number 8 is called the **minuend,** the number 3 is called the **subtrahend,** and the number 5 is called the **difference.** You should know the names for the different parts of a subtraction problem.

2 Relate Subtraction and Addition

How is the subtraction problem 8 − 3 = 5 related to the addition problem 5 + 3 = 8? Let's look at these problems when they are written vertically.

$$\begin{array}{r} 8 \\ -\ 3 \\ \hline 5 \end{array} \uparrow 5 + 3 = 8 \qquad \begin{array}{r} 5 \\ +\ 3 \\ \hline 8 \end{array}$$

This means that we can check the answer to a subtraction problem with an addition problem, and we can check the answer to an addition problem with a subtraction problem.

EXAMPLE 2 Find 9 − 8, then check the answer using addition.

Solution

$$\begin{array}{r} 9 \\ -\ 8 \\ \hline 1 \end{array} \uparrow \begin{array}{l} \text{Check:} \\ 1 + 8 = 9 \end{array} \qquad \text{Check: } \begin{array}{r} 1 \\ +\ 8 \\ \hline 9 \end{array}$$

[YOU TRY 2] Find 7 − 2, then check the answer using addition.

EXAMPLE 3

Find 2 + 4, then check the answer using subtraction.

Solution

$$\begin{array}{r} 2 \\ + 4 \\ \hline 6 \end{array} \quad \text{Check: } 6 - 4 = 2 \qquad \text{Check: } \begin{array}{r} 6 \\ - 4 \\ \hline 2 \end{array}$$

[YOU TRY 3]

Find 6 + 1, then check the answer using subtraction.

3 Subtract Without Regrouping (Borrowing)

Next, we will subtract numbers containing more than one digit.

> **Procedure** Subtracting Numbers with More Than One Digit
>
> 1) Line up the numbers vertically so that the ones digits are in the same column, the tens digits are in the same column, the hundreds are in the same column, and so on.
>
> 2) Begin the subtraction process in the column farthest to the right, and then move to the left, subtracting numbers column by column. That is, subtract the numbers in the ones place, then subtract the numbers in the tens place, then subtract the numbers in the hundreds place, and so on.

Once again, we will use graph paper to help us line up our numbers correctly.

EXAMPLE 4

Subtract 38 − 13, then check the answer using addition.

Solution

Line up the numbers in columns starting with the ones at the right so that the ones are in the same column and the tens are in the same column.

W Hint

Have you tried using graph paper to help you add and subtract?

Begin subtracting in the ones column, the column farthest to the right.

Next, subtract the numbers in the tens column.

$$\begin{array}{r} 3\ 8 \\ -\ 1\ 3 \\ \hline 2\ 5 \end{array}$$
└─ Difference of the ones digits
└───── Difference of the tens digits

The difference is 25. Let's check the answer:

Check:
$$\begin{array}{r} 2\ 5 \\ +\ 1\ 3 \\ \hline 3\ 8 \end{array}$$

[YOU TRY 4]

Find 75 − 42, then check the answer using addition.

EXAMPLE 5

Subtract 964 − 250, then check the answer using addition.

Solution

Line up the numbers in columns starting with the ones at the right so that the ones are in the same column, the tens are in the same column, and the hundreds are in the same column.

Hint

What are the similarities between this procedure and the addition procedure?

Begin subtracting in the ones column, the column farthest to the right.

Next, subtract the numbers in the tens column.

Finally, subtract the numbers in the hundreds column.

$$\begin{array}{r} 9\ 6\ 4 \\ -\ 2\ 5\ 0 \\ \hline 7\ 1\ 4 \end{array}$$

4 ones − 0 ones = 4 ones
6 tens − 5 tens = 1 ten
9 hundreds − 2 hundreds = 7 hundreds

The difference is 714. Check:

$$\begin{array}{r} 7\ 1\ 4 \\ +\ 2\ 5\ 0 \\ \hline 9\ 6\ 4 \end{array}$$

[YOU TRY 5]

Find 498 − 192, then check the answer using addition.

Next, let's look at some subtraction problems where we must regroup or *borrow*.

4 Subtract Numbers with Regrouping (Borrowing)

Sometimes, we have to regroup when we subtract. For example, we can write

8 tens = 7 tens + 1 ten = 7 tens + 10 ones
5 hundreds = 4 hundreds + 1 hundred = 4 hundreds + 10 tens

EXAMPLE 6

Subtract $\begin{array}{r} 83 \\ -\ 16 \end{array}$

Solution

Normally, we would begin by subtracting 3 − 6 in the ones column, but because 6 is greater than 3 we must regroup the 8 tens in 83.

Look at the number 83. We will "borrow" one ten from the tens column and add it to the ones column as 10 ones.

Borrow 1 ten from the 8 tens: → $\begin{array}{r} 7\ 13 \\ \not{8}\ \not{3} \\ -\ 1\ 6 \end{array}$ ← Add the 1 ten to the 3 ones:

8 tens − 1 ten = 7 tens   1 ten + 3 ones = 10 ones + 3 ones
= 13 ones

Next, subtract the numbers in the ones column. Then, subtract the numbers in the tens column.

$$
\begin{array}{r}
^{7}\!\!\!\not 8 \ ^{13}\!\!\!\not 3 \\
-\ 1 \ 6 \\
\hline
6 \ 7
\end{array}
$$

7 tens − 1 ten = 6 tens   13 ones − 6 ones = 7 ones

The difference is 67.

Check using addition:

$$
\begin{array}{r}
1 \\
6 \ 7 \\
+\ 1 \ 6 \\
\hline
8 \ 3
\end{array}
$$

[YOU TRY 6]

Subtract $\begin{array}{r} 72 \\ -\ 38 \\ \hline \end{array}$

EXAMPLE 7

Subtract.

a) $\begin{array}{r} 572 \\ -\ 249 \\ \hline \end{array}$ b) $\begin{array}{r} 341 \\ -\ 263 \\ \hline \end{array}$ c) $\begin{array}{r} 1{,}254 \\ -\ 178 \\ \hline \end{array}$

Solution

a) We cannot just subtract the numbers in the ones column because the 9 is bigger than the 2. So, we have to borrow a ten from the tens column:

7 tens − 1 ten = 6 tens $\begin{array}{r} ^{6}\!\!\!\not 5 \ ^{12}\!\!\!\not 7 \not 2 \\ -\ 2 \ 4 \ 9 \\ \hline \end{array}$ 1 ten + 2 ones =
10 ones + 2 ones = 12 ones

> **W Hint**
> Notice how each part of this example is different from the others.

Subtract the numbers in the ones column, then the numbers in the tens column, then the numbers in the hundreds column.

$\begin{array}{r} ^{6}\!\!\!\not 5 \ ^{12}\!\!\!\not 7 \not 2 \\ -\ 2 \ 4 \ 9 \\ \hline 3 \ 2 \ 3 \end{array}$ ← 12 ones − 9 ones = 3 ones
6 tens − 4 tens = 2 tens
5 hundreds − 2 hundreds = 3 hundreds

The difference is 323.

b) First, regroup in the tens column.

4 tens − 1 ten = 3 tens
$$\begin{array}{r} {}^{3}\;{}^{11} \\ 3\;\not{4}\;\not{1} \\ -\;2\;6\;3 \end{array}$$
1 ten + 1 one =
10 ones + 1 one = 11 ones

The 6 in the tens column is bigger than the 3 in the tens column. So, we have to regroup again. Then, subtract.

3 hundreds − 1 hundred = 2 hundreds
$$\begin{array}{r} {}^{13} \\ {}^{2}\;\not{3}\;{}^{11} \\ \not{3}\;\not{4}\;\not{1} \\ -\;2\;6\;3 \\ \hline 7\;8 \end{array}$$
1 hundred + 3 tens =
10 tens + 3 tens = 13 tens
Subtract.

The difference is 78.

c) How do we start? Regroup in the tens column because the 8 in the ones column is larger than the 4.

5 tens − 1 ten = 4 tens
$$\begin{array}{r} {}^{4}\;{}^{14} \\ 1,\;2\;\not{5}\;\not{4} \\ -\;\;\;1\;7\;8 \end{array}$$
1 ten + 4 ones =
10 ones + 4 ones = 14 ones

The 7 in the tens column is bigger than the 4 above it. Regroup again, then subtract.

2 hundreds − 1 hundred = 1 hundred
$$\begin{array}{r} {}^{14} \\ {}^{1}\;\not{4}\;{}^{14} \\ 1,\;\not{2}\;\not{5}\;\not{4} \\ -\;\;\;1\;7\;8 \\ \hline 1,\;0\;7\;6 \end{array}$$
1 hundred + 4 tens =
10 tens + 4 tens = 14 tens
Subtract.

The difference is 1,076.

[**YOU TRY 7**] Subtract.

a) $\begin{array}{r} 985 \\ -\,317 \end{array}$

b) $\begin{array}{r} 634 \\ -\,159 \end{array}$

c) $\begin{array}{r} 1,623 \\ -\,596 \end{array}$

5 Subtract Numbers Involving Zeros

EXAMPLE 8

Subtract $\begin{array}{r} 9,502 \\ -\,4,356 \end{array}$

Solution

In the ones column, the 6 is greater than the 2 so we must regroup and borrow from the tens column. However, there are no tens in the tens column. Therefore, we must regroup 1 hundred as 10 tens.

W Hint

If you have to regroup during subtraction, you will need to regroup when checking your answer.

5 hundreds − 1 hundred = 4 hundreds
$$\begin{array}{r} {}^{4}\;{}^{10} \\ 9,\;\not{5}\;\not{0}\;2 \\ -\;4,\;3\;5\;6 \end{array}$$
1 hundred + 0 tens =
10 tens + 0 tens = 10 tens

Now we can borrow from the tens column.

$$
\begin{array}{r}
\overset{9}{4}\ \overset{}{\cancel{10}}\ \overset{}{12} \\
9,\ \cancel{5}\ \cancel{0}\ \cancel{2} \\
-\ 4,\ 3\ 5\ 6 \\
\hline
5,\ 1\ 4\ 6
\end{array}
$$

10 tens − 1 ten = 9 tens   1 ten + 2 ones = 10 ones + 2 ones = 12 ones   Subtract.

The difference is 5,146. Verify that $5{,}146 + 4{,}356 = 9{,}502$.

[**YOU TRY 8**]

Subtract
$$
\begin{array}{r}
5{,}804 \\
-\ 1{,}287 \\
\end{array}
$$

EXAMPLE 9   Subtract.

a)
$$
\begin{array}{r}
890 \\
-\ 514 \\
\end{array}
$$

b)
$$
\begin{array}{r}
601 \\
-\ 273 \\
\end{array}
$$

c)
$$
\begin{array}{r}
4{,}000 \\
-\ 1{,}699 \\
\end{array}
$$

Solution

a)   Regroup 1 ten as 10 ones. Then, subtract.

$$
\begin{array}{r}
\overset{8}{\cancel{8}}\ \overset{10}{\cancel{9}}\ \cancel{0} \\
-\ 5\ 1\ 4 \\
\hline
3\ 7\ 6
\end{array}
$$

9 tens − 1 ten = 8 tens   1 ten + 0 ones = 10 ones + 0 ones = 10 ones   Subtract.

The difference is 376.

b)   To subtract in the ones column, we have to "borrow" a ten from the tens column. But, as the problem is written now, there are zero tens. Therefore, regroup 1 hundred as 10 tens. Then, borrow.

Regroup 1 hundred as 10 tens.   Regroup 1 ten as 10 ones.

$$
\begin{array}{r}
\overset{5}{\cancel{6}}\ \overset{10}{\cancel{0}}\ 1 \\
-\ 2\ 7\ 3 \\
\end{array}
\qquad
\begin{array}{r}
\overset{}{}\ \overset{9}{}\ \\
\overset{5}{\cancel{6}}\ \overset{\cancel{10}}{\cancel{0}}\ \overset{11}{\cancel{1}} \\
-\ 2\ 7\ 3 \\
\hline
3\ 2\ 8
\end{array}
$$

Subtract.

The difference is 328.

c)   We have to regroup 1 thousand, 1 hundred, *and* 1 ten in order to subtract.

$$
\begin{array}{r}
\overset{}{}\ \overset{9}{}\ \overset{9}{}\ \\
\overset{3}{\cancel{4}}\ \overset{\cancel{10}}{\cancel{0}}\ \overset{\cancel{10}}{\cancel{0}}\ \overset{10}{\cancel{0}} \\
-\ 1,\ 6\ 9\ 9 \\
\hline
2,\ 3\ 0\ 1
\end{array}
$$

The difference is 2,301.

[**YOU TRY 9**]

Subtract.

a)
$$
\begin{array}{r}
470 \\
-\ 238 \\
\end{array}
$$

b)
$$
\begin{array}{r}
705 \\
-\ 389 \\
\end{array}
$$

c)
$$
\begin{array}{r}
8{,}000 \\
-\ 4{,}899 \\
\end{array}
$$

6 Solve Applied Problems Using Subtraction

We use subtraction in many ways to solve applied problems.

EXAMPLE 10

The table shows the amount of money won by certain bowlers on the PBA tour in 2016. Who won more money, Tommy Jones or Tom Smallwood? How much more money did he win?

Bowler	Tournament Winnings
Dom Barrett	$116,890
Tommy Jones	$99,267
Anthony Simonsen	$143,278
Tom Smallwood	$57,157
EJ Tackett	$168,290

(www.pba.com)

Solution

Tommy Jones won more money because $99,267 is more than the $57,157 won by Tom Smallwood. To determine how much *more* money Tommy won, we subtract the smaller number from the larger one.

$$
\begin{array}{r}
9\,9{,}2\,6\,7 \\
-\;5\,7{,}1\,5\,7 \\
\hline
4\,2{,}1\,1\,0
\end{array}
$$ Subtract.

Tommy Jones won $42,110 more than Tom Smallwood.

YOU TRY 10

Use the table in Example 10. Who won more money, Anthony Simonsen or EJ Tackett? How much more did he win?

ANSWERS TO YOU TRY **EXERCISES**

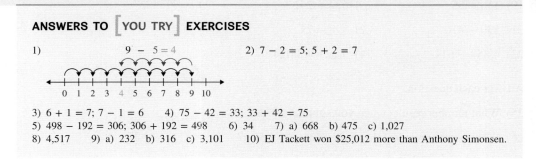

1)   9 − 5 = 4   2) 7 − 2 = 5; 5 + 2 = 7

3) 6 + 1 = 7; 7 − 1 = 6   4) 75 − 42 = 33; 33 + 42 = 75
5) 498 − 192 = 306; 306 + 192 = 498   6) 34   7) a) 668   b) 475   c) 1,027
8) 4,517   9) a) 232   b) 316   c) 3,101   10) EJ Tackett won $25,012 more than Anthony Simonsen.

Objective 1: Subtract Numbers Using a Number Line

Identify the minuend, subtrahend, and difference in each subtraction problem.

1) $5 - 4 = 1$ 2) $6 - 2 = 4$

Use a number line to subtract the numbers.

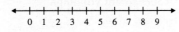

3) $9 - 4$ 4) $10 - 6$

5) $8 - 5$ 6) $9 - 6$

Objective 2: Relate Subtraction and Addition

Subtract. Then, check the answer using addition.

7) $9 - 7$ 8) $7 - 3$

9) $5 - 1$ 10) $9 - 6$

Add. Then, check the answer using subtraction.

11) $1 + 5$ 12) $4 + 3$

13) $7 + 2$ 14) $2 + 6$

Objective 3: Subtract Without Regrouping (Borrowing)

Subtract. Then, check the answer using addition.

15) $94 - 52$ 16) $69 - 13$

17) $528 - 301$ 18) $495 - 260$

19) $156 - 55$ 20) $674 - 251$

21) $530 - 410$ 22) $398 - 71$

23) $4,859 - 614$ 24) $9,227 - 3,024$

Answer each question.

25) What number results when you subtract 2 from 9?

26) What number results when you subtract 1 from 6?

27) What number results when you subtract 518 from 948?

28) What number results when you subtract 122 from 726?

Find the missing length labeled with ?.

29)
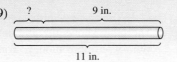

30)
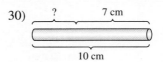

31)
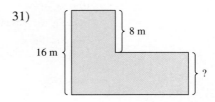

32)
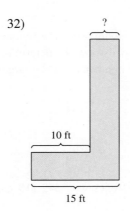

Objective 4: Subtract Numbers with Regrouping (Borrowing)

Subtract.

33) $\begin{array}{r} 21 \\ -\ 7 \end{array}$ 34) $\begin{array}{r} 72 \\ -\ 9 \end{array}$

35) $\begin{array}{r} 62 \\ -48 \end{array}$ 36) $\begin{array}{r} 54 \\ -26 \end{array}$

37) $\begin{array}{r} 95 \\ -37 \end{array}$ 38) $\begin{array}{r} 71 \\ -52 \end{array}$

39) $\begin{array}{r} 678 \\ -159 \end{array}$ 40) $\begin{array}{r} 491 \\ -183 \end{array}$

41) $\begin{array}{r} 833 \\ -747 \end{array}$ 42) $\begin{array}{r} 672 \\ -593 \end{array}$

43) 2,513
    − 364

44) 4,922
    − 755

45) 7,436
    − 4,058

46) 9,227
    − 5,139

47) 2,549
    − 1,985

48) 5,794
    − 4,979

Objective 5: Subtract Numbers Involving Zeros
Subtract.

49)   70
    − 7

50)   60
    − 8

51) 920
    − 207

52) 580
    − 151

53) 6,074
    − 2,519

54) 9,063
    − 7,825

55) 3,008
    − 1,478

56) 5,003
    − 2,611

57) 19,021
    − 16,998

58) 75,106
    − 56,197

59) 800
    − 164

60) 600
    − 358

61) 7,000
    − 2,564

62) 2,000
    − 1,316

63) 66,005
    − 37,894

64) 81,009
    − 22,654

Mixed Exercises: Objectives 4 and 5
Subtract.

65) 947 − 215

66) 4,693 − 1,277

67) 5,000
    − 1,799

68) 7,000
    − 2,999

69) 8,945
    − 1,898

70) 87,020
    − 26,415

71) Subtract 62,078 from 283,120.

72) Subtract 15,907 from 485,428.

Objective 6: Solve Applied Problems Using Subtraction
Solve each problem.

73) Aris needs to complete 42 semester units to meet his general education requirement to earn his degree. If he has already completed 24 units in this area, how many more units does he need to complete his general education requirement?

©Kelly Redinger/Design Pics

74) Sumaya received $2,378 in financial aid last year. This year she received $2,501. How much more money did she receive this year?

75) A Burger King Double Whopper with cheese has 990 calories, while the BK Big Fish Sandwich has 640 calories. How many fewer calories does the fish sandwich have? (www.bk.com)

76) The Applegate subdivision contains 365 homes while the Tallgrass subdivision contains 473 homes. How many more homes are in the Tallgrass subdivision?

©Greg K__ca/Shutterstock

77) The average annual salary of a registered nurse in Montana in May 2016 was $64,300. In Texas, the average annual salary was $72,180. How much more was the average annual salary in Texas? (www.bls.gov)

78) The sticker price on a car was $24,145, but Jignesh paid $21,300. How much money did he save off the sticker price?

79) In 2017, a university had 5,366 students. This is 371 more than in 2015. How many students attended in 2015?

80) At Seneca High School, 694 students are taking Spanish. This is 257 more than the number of students studying French. How many are taking French?

81) In 2006, approximately 6,071,000 people participated in a book club at least once. The estimate for 2008 was 6,720,000. In which year did more people participate in a book club? How many more? (www.census.gov)

82) In 2015, the population of Vermont was approximately 626,000, and the population of Delaware was about 945,900. Which state had the larger population? By how much? (www.census.gov)

83) The driving distance from Washington, D.C., to Des Moines, IA, is about 1,060 mi. The distance from Washington, D.C., to Oklahoma City, OK, is about 1,330 mi. How much farther is it from Washington D.C. to Oklahoma City?

84) In 2016, 74,340 people attended the championship game of the NCAA Final Four tournament. In 2017, that number was 76,168. How many fewer people attended the championship game in 2016? (www.ncaa.com)

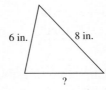 85) If the perimeter of the figure is 21 in., find the missing side length.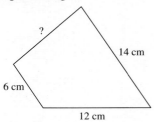

86) If the perimeter of the figure is 40 cm, find the missing side length.

R **Rethink**

R1) Do you understand how to subtract when you have to borrow?

R2) Which types of problems could you do easily, and which problems were difficult for you? How could you get more practice?

A.3 Multiplying Whole Numbers

P Prepare

O Organize

What are your objectives for Section A.3?	How can you accomplish each objective?
1 Understand the Meaning of Multiplication	• Write the definition of *multiplication* in your own words. • Understand the terms *multiplicand, multiplier, factor,* and *product.* • Memorize the multiplication facts from 1 through 12 by reviewing the table provided. • Complete the given examples on your own. • Complete You Trys 1 and 2.
2 Use the Commutative Property of Multiplication	• Learn the commutative property of multiplication. • Complete the given example on your own. • Complete You Try 3.
3 Use the Associative Property of Multiplication	• Learn the associative property of multiplication. • Complete the given example on your own. • Complete You Try 4.
4 Multiply a Number by a One-Digit Number	• Follow the examples, and write a procedure for multiplying a number by a one-digit number. • Complete the given examples on your own. • Complete You Trys 5 and 6.
5 Multiply a Number by a Number with More Than One Digit	• Be sure to follow each example carefully, as each one will point out different scenarios you might encounter while multiplying. • Follow the examples, and write a procedure for multiplying a number by a number with more than one digit. • Complete the given examples on your own. • Complete You Trys 7 and 8.
6 Multiply a Number by a Number Ending in Zero	• Write the procedure for **Multiplying a Number by 10, 100, 1,000, 10,000, etc.** in your own words. • Complete the given example on your own. • Complete You Try 9.
7 Solve Applied Problems Using Multiplication	• Be sure to carefully read the application problem and understand what you are being asked to find. • Complete the given example on your own. • Complete You Try 10.

Read the explanations, follow the examples, take notes, and complete the You Trys.

1 Understand the Meaning of Multiplication

What is multiplication? **Multiplication** is a shorthand way to represent repeated addition of the same number.

Let's add $3 + 3 + 3 + 3$ on a number line, then see how we can write it as a multiplication problem.

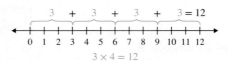

$$3 \times 4 = 12$$

We found that $3 + 3 + 3 + 3 = 12$. Since we are adding the 3 *four times,* we can say that 3 *times* 4 *equals* 12, which can be written as $3 \times 4 = 12$. The 3 is called the **multiplicand,** and 4 is called the **multiplier.** The numbers being multiplied together are also called **factors,** so 3 and 4 are factors of 12. The answer to a multiplication problem is called the **product.** So, 12 is the product of 3 and 4.

We can also write the multiplication problem vertically:

$$
\begin{array}{r}
3 \leftarrow \text{Multiplicand} \\
\times\ 4 \leftarrow \text{Multiplier} \\
\hline
1\ 2 \leftarrow \text{Product}
\end{array}
$$

3 and 4 are *factors.*

EXAMPLE 1

Use a number line to add $2 + 2 + 2$. Then,

a)   write the addition problem as a multiplication problem.

b)   identify the multiplicand and the multiplier.

c)   identify the factors and the product.

Solution

We can add $2 + 2 + 2$ on the number line like this:

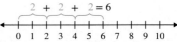

$$2 + 2 + 2 = 6$$

a)   Since we are adding 2 *three times,* we can write $2 + 2 + 2 = 6$ as the multiplication problem $2 \times 3 = 6$.

b)   The multiplicand is 2, and the multiplier is 3.

c)   The factors are 2 and 3. The product is 6.

W Hint

Pay attention to the relationship between addition and multiplication.

[YOU TRY 1]

Use a number line to add $5 + 5$. Then,

a)   write the addition problem as a multiplication problem.

b)   identify the multiplicand and the multiplier.

c)   identify the factors and the product.

Multiplication can be written in several different ways. For example, other ways to write $2 \times 3 = 6$ are

$$
\begin{array}{r}
2 \\
\times\ 3 \\
\hline
6
\end{array}
\qquad \text{or} \qquad 2 \cdot 3 = 6 \qquad \text{or} \qquad 2(3) = 6 \qquad \text{or} \qquad (2)(3) = 6
$$

Multiplication dot

No operation symbol means multiplication.

So, multiplication may also be represented by a multiplication dot or with no symbol at all before or between parentheses.

EXAMPLE 2

Find each product.

a) 4 · 5   b) 9(3)   c) (0)(8)

Solution

a) Here, the multiplication dot is used.   4 · 5 = 20

b) In this problem, there is no operation symbol between the 9 and the parenthesis right next to it. So, the operation is multiplication.   9(3) = 27

c) There is no operation symbol between the parentheses, so the operation is multiplication. (0)(8) = 0   (Zero times any number equals zero.)

[YOU TRY 2] Find each product.

a) 3 · 7   b) 5(0)   c) (9)(6)

If you do not remember the multiplication facts from 1 through 12, you can review them using this table.

Multiplication Table

×	1	2	3	4	5	6	7	8	9	10	11	12
1	1	2	3	4	5	6	7	8	9	10	11	12
2	2	4	6	8	10	12	14	16	18	20	22	24
3	3	6	9	12	15	18	21	24	27	30	33	36
4	4	8	12	16	20	24	28	32	36	40	44	48
5	5	10	15	20	25	30	35	40	45	50	55	60
6	6	12	18	24	30	36	42	48	54	60	66	72
7	7	14	21	28	35	42	49	56	63	70	77	84
8	8	16	24	32	40	48	56	64	72	80	88	96
9	9	18	27	36	45	54	63	72	81	90	99	108
10	10	20	30	40	50	60	70	80	90	100	110	120
11	11	22	33	44	55	66	77	88	99	110	121	132
12	12	24	36	48	60	72	84	96	108	120	132	144

W Hint

Make yourself some flash cards to review the multiplication facts.

2 Use the Commutative Property of Multiplication

Earlier, we learned that addition is commutative. Is multiplication commutative? In Example 1, we found that 2 × 3 = 6. Does 3 × 2 = 6? Yes, 2 × 3 = 6 and 3 × 2 = 6. This is just one example that shows that multiplication is commutative.

Property The Commutative Property of Multiplication

The **commutative property of multiplication** says that changing the order in which we multiply numbers does not change the product. For example,

$$2 \times 3 = 6 \quad \text{and} \quad 3 \times 2 = 6$$

EXAMPLE 3   Multiply 8 · 9, then rewrite the problem using the commutative property of multiplication.

Solution

8 · 9 = 72. Using the commutative property of multiplication, we get 9 · 8 = 72.

[YOU TRY 3]   Multiply 7 × 6, then rewrite the problem using the commutative property of multiplication.

3 Use the Associative Property of Multiplication

In Section A.1, we learned that addition is associative. The associative property also applies to multiplication.

> **Property** The Associative Property of Multiplication
>
> The **associative property of multiplication** says that we can group factors in any order and the product will remain the same.

EXAMPLE 4   Multiply 2 × 4 × 3.

Solution

When we are multiplying more than two numbers, we can group them any way we like. Let's do this two ways.

1) 2 × 4 × 3 = (2 × 4) × 3   Use the associative property of multiplication to group the factors.
         =   8   × 3   Perform the operation inside the parentheses first.
         = 24   Multiply.

Or,

2) 2 × 4 × 3 = 2 × (4 × 3)   Use the associative property of multiplication to group the factors.
         = 2 ×   12   Perform the operation inside the parentheses first.
         = 24   Multiply.

No matter which way we group the factors, the product is the same.

[YOU TRY 4]   Multiply 7 × 1 × 6.

So far, the multiplication problems we have seen all use the basic multiplication facts from 1 through 12. Next we will learn how to multiply larger numbers by one-digit numbers.

4 Multiply a Number by a One-Digit Number

EXAMPLE 5

Multiply.

a) 21
   × 3

b) 102
   × 4

Solution

a) Begin with the multiplier, 3. We will multiply each place in the number 21 by 3, starting with the ones column of 21. So, first multiply 3 × 1 one = 3 ones. Then, multiply 3 × 2 tens = 6 tens.

	2	1
×		3
		3

3 × 1 one = 3 ones

	2	1
×		3
	6	3

3 × 2 tens = 6 tens

The product is 63.

b) Begin with the multiplier, 4. Multiply each place in the number 102 by 4, starting with the ones column of 102. So, first multiply 4 × 2 ones = 8 ones, then multiply 4 × 0 tens = 0 tens, and finally multiply 4 × 1 hundred = 4 hundreds.

	1	0	2
×			4
	4	0	8

← 4 × 2 ones = 8 ones
  4 × 0 tens = 0 tens
  4 × 1 hundred = 4 hundreds

The product is 408.

[YOU TRY 5]

Multiply.

a) 14
   × 2

b) 204
   × 2

Sometimes we need to use regrouping to multiply numbers.

EXAMPLE 6

Multiply.

a) 473
   × 2

b) 738
   × 5

Solution

a) Begin with the multiplier, 2. Multiply 2 × 3 ones = 6 ones.

	4	7	3
×			2
			6

2 × 3 ones = 6 ones

W Hint

Can you write a procedure to generalize this objective?

Now, multiply 2 × 7 tens = 14 tens. Regroup this as 1 hundred + 4 tens. Write the 4 in the tens column of the product and write the 1 above the 4 in the hundreds column.

2 × 7 tens = 14 tens;
14 tens = 1 hundred + 4 tens
Write the 4 tens in the product;
write the 1 above the 4 in the hundreds column.

Multiply 2 × 4 hundreds = 8 hundreds and add the 1 hundred above the 4 to get 9 hundreds in the product.

2 × 4 hundreds = 8 hundreds
Add the 8 hundreds and regrouped 1 hundred:
8 hundreds + 1 hundred = 9 hundreds

The product is 946.

b)   Multiply each digit in 738 by 5 as shown here.

$$
\begin{array}{r}
1\ 4 \\
7\ 3\ 8 \\
\times\ 5 \\
\hline
3\ 6\ 9\ 0
\end{array}
$$

5 × 8 = 40 ones; regroup as 4 tens + 0 ones; write 0 in the ones column and 4 above the 3.

5 × 3 = 15 tens; add 15 tens + 4 tens = 19 tens; regroup as 1 hundred + 9 tens; write 9 in the tens column and 1 above the 7.

5 × 7 = 35 hundreds; add 35 hundreds + 1 hundred = 36 hundreds

The product is 3,690.

[**YOU TRY 6**]   Multiply.

a) $\begin{array}{r} 291 \\ \times\ \ 3 \\ \hline \end{array}$ b) $\begin{array}{r} 845 \\ \times\ \ 6 \\ \hline \end{array}$

5 Multiply a Number by a Number with More Than One Digit

How do we multiply 34 × 16?
Since 16 = 1 ten + 6 ones, one way to find 34 × 16 is like this:

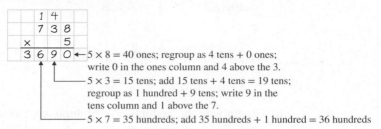

Add the results to get

$$
\begin{array}{r}
3\ 4 \\
\times\ 1\ 6 \\
\hline
2\ 0\ 4 \\
+\ 3\ 4\ 0 \\
\hline
5\ 4\ 4
\end{array}
$$

34 × 6
34 × 10
Add to get the product.

204 and 340 are called **partial products.** We can also find 34 × 16 by multiplying in a single step using partial products. Usually, we do not write the 0 in 340.

$$
\begin{array}{r}
3\ 4 \\
\times\ 1\ 6 \\
\hline
2\ 0\ 4 \\
3\ 4 \\
\hline
5\ 4\ 4
\end{array}
$$

Leave off the 0. Line up the 4 in the tens column.

The product is 544.

Let's look at another example.

EXAMPLE 7 Multiply.

a) $\begin{array}{r} 322 \\ \times\, 213 \end{array}$ b) $\begin{array}{r} 268 \\ \times\, 49 \end{array}$

Solution

a) We will use partial products to write all multiplication steps in a single problem.

		3	2	2	
	×	2	1	3	
		9	6	6	Multiply 3 ones × 322. Line up the rightmost digit in the ones column.
	3	2	2		Multiply 1 ten × 322. Line up on the right in the tens column.
6	4	4			Multiply 2 hundreds × 322. Line up on the right in the hundreds place.
6	8,	5	8	6	Add to get the product.

The product is 68,586.

b) Begin by multiplying by 9. Notice that we must regroup.

		6	7	
		2	6	8
	×		4	9
	2,	4	1	2	Multiply 268 by 9.

W Hint
Remember, work out the example on your paper as you are reading it!

Now, multiply by 4. Again, we must regroup.

		2	3	← This is the regrouping from multiplying by 4.
		6	7	← This is the regrouping from multiplying by 9.
		2	6	8
	×		4	9
	2	4	1	2	Multiply 268 by 9; line up in the ones column.
1	0	7	2		Multiply 268 by 4; line up in the tens column.
1	3,	1	3	2	Add to get the product.

The product is 13,132.

YOU TRY 7 Multiply.

a) $\begin{array}{r} 413 \\ \times\, 122 \end{array}$ b) $\begin{array}{r} 276 \\ \times\, 83 \end{array}$

Be careful when the multiplier contains zeros. We still have to account for their positions in the multiplication problem.

EXAMPLE 8 Multiply.

a) $\begin{array}{r} 254 \\ \times\, 301 \end{array}$ b) $\begin{array}{r} 6{,}132 \\ \times\, 4{,}003 \end{array}$

Solution

a)

		2	5	4	
	×	3	0	1	
		2	5	4	Multiply 254 by 1. Line up in the ones column.
	0	0	0		Multiply 254 by 0. Line up in the tens column.
7	6	2			Multiply 254 by 3. Line up in the hundreds column.
7	6,	4	5	4	Add to get the product.

The product is 76,454.

b) Let's look at two methods for working with zeros.

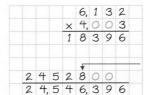

```
              6, 1 3 2                              6, 1 3 2
          ×   4, 0 0 3                          ×   4, 0 0 3
          1 8 3 9 6                              1 8 3 9 6
          0 0 0 0            or
        0 0 0 0
      2 4 5 2 8                               2 4 5 2 8 0 0
      2 4, 5 4 6, 3 9 6                       2 4, 5 4 6, 3 9 6
```

Start this partial
product in the
thousands column.

The product is 24,546,396. The method on the right is a shorthand way to account for the two zeros in the multiplier.

[**YOU TRY 8**]   Multiply.

a)   186
    × 204

b)   5,283
    × 3,007

Multiplying a whole number by 10 or 100 or 1,000 can be simple if we notice a pattern.

6 Multiply a Number by a Number Ending in Zero

Let's multiply 25 by 10, 100, 1,000, and 10,000 and see what happens:

$$25 \times 10 = 250$$
$$25 \times 100 = 2,500$$
$$25 \times 1,000 = 25,000$$
$$25 \times 10,000 = 250,000$$

Do you notice the pattern?

Procedure Multiplying a Whole Number by 10, 100, 1,000, 10,000, etc.

When you multiply a whole number by 10, 100, 1,000, 10,000, and so on, the result is the number followed by the number of zeros in the multiplier. For example, $382 \times 100 = 38,200$.

We can use this property to multiply a whole number by other *multiples* of 10 as well. For example, 20 is a multiple of 10 since $2 \cdot 10 = 20$. (A product is a **multiple** of each of its factors.) So if we use what we know about multiplying a whole number by 10 or 100 or 1,000, we can also use a shortcut to multiply by other multiples of 10.

EXAMPLE 9   Multiply.

a) $9,722 \times 1,000$ b) $83 \cdot 20$ c) 240×300

Solution

a) $9,722 \times 1,000 = 9,722,000$ Add three zeros to the end of 9,722.

b) 20 is a multiple of 10. So, to find $83 \cdot 20$ we can multiply 83 by 2, then add one zero to the end of that result.

$$83 \cdot 2 = 166, \text{then add one zero to } 166:1,660$$
$$83 \cdot 20 = 1,660$$

c) In 240×300, 240 is a multiple of 10 and 300 is a multiple of 100. To find 240×300, we can multiply 24 by 3 then add three zeros to the end of that result since there are a total of three zeros in the multiples of 10 and 100.

$$24 \times 3 = 72, \text{ then add three zeros to } 72:72,000$$
$$240 \times 300 = 72,000$$

[YOU TRY 9]

Multiply.

a) $68 \cdot 100$ b) 32×40 c) $530 \cdot 600$

7 Solve Applied Problems Using Multiplication

Multiplication is often used to solve real-world problems.

EXAMPLE 10

Adelita bought 17 gift cards worth $25 each for her employees. How much did she spend?

Solution

To determine the total amount of money Adelita spent, we multiply the value of each gift card, $25, by the number she bought, 17.

Adelita spent $425 on the gift cards.

$$
\begin{array}{r}
2\ 5 \leftarrow \text{Value of each gift card}\\
\times\ 1\ 7 \leftarrow \text{Number of cards}\\
\hline
1\ 7\ 5\\
2\ 5\quad\\
\hline
4\ 2\ 5
\end{array}
$$

[YOU TRY 10]

Terrance bought 87 matching t-shirts for all of his relatives attending a family picnic. Each shirt cost $12. Find the total cost of the shirts.

ANSWERS TO [YOU TRY] EXERCISES

1)   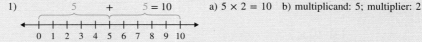   a) $5 \times 2 = 10$ b) multiplicand: 5; multiplier: 2

c) factors: 5 and 2; product: 10   2) a) 21   b) 0   c) 54   3) $7 \times 6 = 42$; $6 \times 7 = 42$
4) 42   5) a) 28   b) 408   6) a) 373   b) 5,070   7) a) 50,386   b) 22,908
8) a) 37,944   b) 15,885,981   9) a) 6,800   b) 1,280   c) 318,000   10) $1,044

Objective 1: Understand the Meaning of Multiplication

Use a number line to add the following numbers. Then,

a) write the addition problem as a multiplication problem.

b) identify the multiplicand and the multiplier.

c) identify the factors and the product.

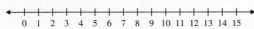

0 1 2 3 4 5 6 7 8 9 10 11 12 13 14 15

1) $4 + 4 + 4$ 2) $5 + 5 + 5$

3) $3 + 3 + 3 + 3 + 3$ 4) $2 + 2 + 2 + 2$

Find each product.

5) a) 9×3 6) a) 6×5

   b) $4(7)$ b) $12(3)$

   c) $5 \cdot 11$ c) $7 \cdot 9$

7) a) $(10)(6)$ 8) a) $4 \cdot 3$

   b) 2×9 b) $(5)(9)$

   c) $12 \cdot 4$ c) 7×8

9) a) 6×6 10) a) 4×0

   b) $1 \cdot 7$ b) $(9)(9)$

   c) $\begin{array}{r} 12 \\ \times\ 8 \end{array}$ c) $\begin{array}{r} 6 \\ \times 7 \end{array}$

Objective 2: Use the Commutative Property of Multiplication

11) Explain, in your own words, the commutative property of multiplication.

12) Rewrite $8(3)$ in three other ways.

Find each product, then rewrite the problem using the commutative property.

13) 3×5 14) 10×7

15) $12 \cdot 9$ 16) $9 \cdot 6$

17) $8(4)$ 18) $3(11)$

19) $0 \cdot 2$ 20) $7 \cdot 0$

21) 10×1 22) $1 \cdot 5$

Answer the following questions.

23) What number do you multiply by 4 to get 48?

24) What number do you multiply by 9 to get 36?

25) 7 times what number equals 56?

26) 6 times what number equals 54?

Objective 3: Use the Associative Property of Multiplication

27) In your own words, explain the associative property of multiplication.

28) How can using the associative property of multiplication make multiplying three numbers easier? Give an example.

Multiply in two different ways using the associative property of multiplication.

29) $3 \times 2 \times 4$ 30) $6 \times 1 \times 9$

31) $9 \cdot 0 \cdot 7$ 32) $2 \cdot 4 \cdot 0$

Objective 4: Multiply a Number by a One-Digit Number

Multiply.

33) $\begin{array}{r} 31 \\ \times\ 2 \end{array}$ 34) $\begin{array}{r} 22 \\ \times\ 3 \end{array}$

35) $\begin{array}{r} 413 \\ \times\ \ 3 \end{array}$ 36) $\begin{array}{r} 701 \\ \times\ \ 4 \end{array}$

37) $\begin{array}{r} 804 \\ \times\ \ 2 \end{array}$ 38) $\begin{array}{r} 633 \\ \times\ \ 2 \end{array}$

39) $\begin{array}{r} 94 \\ \times\ 7 \end{array}$ 40) $\begin{array}{r} 56 \\ \times\ 8 \end{array}$

41) $\begin{array}{r} 635 \\ \times\ \ 9 \end{array}$ 42) $\begin{array}{r} 423 \\ \times\ \ 6 \end{array}$

43) $185 \cdot 8$ 44) $872 \cdot 5$

45) $(6)(4,809)$ 46) $(7)(6,044)$

Objective 5: Multiply a Number by a Number with More Than One Digit

47) $\begin{array}{r} 14 \\ \times 12 \end{array}$ 48) $\begin{array}{r} 32 \\ \times 21 \end{array}$

49) 312
    × 23

50) 879
    × 11

51) 38
    × 25

52) 55
    × 34

53) 613
    × 64

54) 228
    × 93

55) 1,773
    × 48

56) 9,161
    × 29

57) 599
    ×781

58) 806
    ×322

59) 8,104
    × 216

60) 4,043
    × 597

61) 403 · 517

62) 602 · 773

63) 2,533(5,004)

64) 1,882(7,006)

Objective 6: Multiply a Number by a Number Ending in Zero

65) In your own words, explain how to multiply a number by 100. Then, give an example.

66) Fill in the blank: 38 × _____ = 38,000.

Multiply.

67) a) 94 × 10
    b) 55 × 100
    c) 67 × 1,000

68) a) 58 × 10
    b) 86 × 100
    c) 23 × 1,000

69) a) 817 · 10,000
    b) 261 · 100
    c) 150 · 100

70) a) 995 · 10,000
    b) 384 · 100
    c) 670 · 100

71) a) 41 · 20
    b) 9 · 50
    c) 256 · 30

72) a) 32 · 20
    b) 8 · 40
    c) 198 · 50

73) a) 150 × 300
    b) 70 × 8,000
    c) 490 × 6,000

74) a) 120 × 700
    b) 40 × 3,000
    c) 580 × 5,000

75) What is the product of 420 and 650?

76) What is the product of 700 and 340?

Objective 7: Solve Applied Problems Using Multiplication

©fotog/Getty Images

77) A classroom contains 5 rows of desks. If each row contains 9 desks, how many desks are in the classroom?

78) Kim stores her collection of horror movies on shelves. If she has 5 shelves of movies and there are 10 movies on each shelf, how many horror movies does Kim have?

79) A box contains 24 packages of paper cups, and each package holds 30 cups. Find the total number of cups in the box.

80) If Pete did 80 push-ups every day for 30 days in a row, find the total number of push-ups he did.

81) Cora bought 7 packages of coffee costing $6 each. How much did she spend?

82) Johnny sold 54 bales of hay for $23 each. How much money did he get for the hay?

83) Haishin pays $129 per month for his gym membership. How much does he pay in a year?

84) Lily's rent is $635 per month. How much rent does she pay in a year?

85) Valeria has to wait 8 weeks for her new sofa to be delivered. How many days does she have to wait?

©Image Source/Blend Images

86) Rajnish's birthday is 6 weeks away. How many days is it until his birthday?

R Rethink

R1) Were you able to complete these exercises without looking back at the multiplication table?

R2) Check some of your answers for Exercises 47 to 56 by rounding the numbers before multiplying.

Was this an effective way to spot-check your answers?

R3) Write an application problem similar to those you just solved.

A.4 Introduction to Division and Short Division

P Prepare

O Organize

What are your objectives for Section A.4?	How can you accomplish each objective?
1 Understand the Meaning of Division	• Understand how to divide by using a number line. • Know the terms *dividend, divisor,* and *quotient.* • Recognize that division is the opposite of multiplication. • Complete the given example on your own. • Complete You Try 1.
2 Perform Division Involving Zero	• Learn the properties for **Dividing a Number by Zero** and **Dividing Zero by a Nonzero Number.** • Complete the given examples on your own. • Complete You Trys 2 and 3.
3 Understand Remainders, and Check a Division Problem Using Multiplication	• Write a statement that summarizes two different results when dividing numbers, and include the word *remainder.* • Write the procedure for **Checking a Division Problem Using Multiplication** in your own words. • Complete the given examples on your own. • Complete You Trys 4 and 5.
4 Use the Divisibility Rules	• Write the definition of *divisible* in your own words. • Spend some time to understand and learn the divisibility rules. • Complete the given examples on your own. • Complete You Trys 6 and 7.
5 Use Short Division	• Understand that *short division* involves division by a one-digit number. • Write a procedure that outlines how to perform short division. • Complete the given examples on your own. • Complete You Trys 8–10.
6 Solve Applied Problems Using Division	• Carefully read the application problem, and understand what you are being asked to find. • Complete the given example on your own. • Complete You Try 11.

W Work   **Read the explanations, follow the examples, take notes, and complete the You Trys.**

1 Understand the Meaning of Division

What does $12 \div 3$ mean? The expression $12 \div 3$ is read as "12 divided by 3," and it means that we have to figure out how many threes it takes to make 12. Let's look at this on a number line.

How many threes does it take to make 12?   4

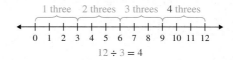

$$12 \div 3 = 4$$

It takes 4 threes to make 12, so $12 \div 3 = 4$. In a division problem, the number being divided is called the **dividend,** the number you are dividing by is the **divisor,** and the result is the **quotient.**

$$\overset{\text{Divisor}}{\underset{\downarrow}{}}$$
$$\text{Dividend} \rightarrow 12 \div 3 = 4 \leftarrow \text{Quotient}$$

There are several different ways to write a division problem. $12 \div 3 = 4$ can also be written as

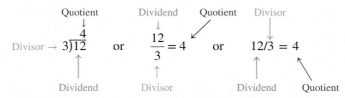

Division and multiplication are opposite operations. Let's see how the division problem $12 \div 3 = 4$ is related to the multiplication problem $3 \times 4 = 12$. (Remember that in the multiplication problem $3 \times 4 = 12$, the multiplier is 4.) We can represent $3 \times 4 = 12$ on a number line as

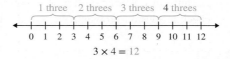

$$3 \times 4 = 12$$

Notice that the number line for $12 \div 3 = 4$ and the number line for $3 \times 4 = 12$ look the same. We can use this relationship to help us perform division by thinking of it in terms of multiplication.

EXAMPLE 1   Divide. Then identify the dividend, the divisor, and the quotient.

a) $10 \div 2$ b) $1\overline{)3}$ c) $\dfrac{4}{4}$

Solution

a)   To find $10 \div 2$, ask yourself, "*How many twos does it take to make* 10?" or "*Two times what number equals* 10?" That number is 5; $2 \times 5 = 10$. We can verify this on the number line.

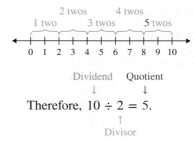

Therefore, $10 \div 2 = 5$.

Hint

Think in terms of multiplying or finding the missing factor to help you divide.

b) To find $1\overline{)3}$, ask yourself, "*One times what number equals 3?*" That number is 3 since $1 \cdot 3 = 3$.

$$\text{Divisor} \rightarrow 1\overset{3}{\overline{)3}} \;\leftarrow \text{Quotient}$$
$$\underset{\text{Dividend}}{\uparrow}$$

Note

Any number divided by 1 equals itself.

c) To find $\dfrac{4}{4}$, ask yourself, "*Four times what number equals 4?*" That number is 1.

$$\underset{\text{Divisor} \rightarrow}{\overset{\text{Dividend} \rightarrow}{}}\dfrac{4}{4} = \overset{\text{Quotient}}{\underset{\downarrow}{1}} \quad \text{since} \quad 4 \cdot 1 = 4$$

Note

Any nonzero number divided by itself equals 1.

Because of this relationship between division and multiplication, we can use multiplication to check a division problem. We will do this in Example 4.

[YOU TRY 1] Divide. Then identify the dividend, the divisor, and the quotient.

a) $\dfrac{35}{7}$ b) $3\overline{)24}$ c) $10 \div 10$ d) $\dfrac{7}{1}$

2 Perform Division Involving Zero

Next we will look at division problems involving zero.

EXAMPLE 2 Divide $\dfrac{8}{0}$.

Solution

Ask yourself, "*Zero times what number equals 8?*" **There is no such number!** We say that $\dfrac{8}{0}$ is *undefined*. This means there is no answer to $\dfrac{8}{0}$.

Property Dividing a Number by Zero

Any number divided by zero is **undefined.** That is, there is no answer to a number divided by zero. For example, $\dfrac{8}{0}$ is undefined.

[YOU TRY 2]

Divide $6 \div 0$.

EXAMPLE 3

Divide $0 \div 5$.

Solution

Ask yourself, "*Five times what number equals* 0?" That number is 0. Therefore,

$$0 \div 5 = 0 \quad \text{since} \quad 5 \cdot 0 = 0$$

Because any number multiplied by 0 equals 0, we have the following property.

Property Dividing Zero by a Nonzero Number

Zero divided by any nonzero number equals zero.   *Example:* $0 \div 5 = 0$.

[YOU TRY 3]

Divide $\dfrac{0}{2}$.

Note

$0 \div 0$ is undefined.

3 Understand Remainders, and Check a Division Problem Using Multiplication

All of the division problems we have seen so far divide exactly. For example, $12 \div 3 = 4$. But what happens if the division does *not* work out to be exact? Then, we get a *remainder*.

Let's look at $9 \div 2$. On a number line we can think of this as, "*How many twos does it take to make* 9?"

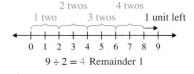

$9 \div 2 = 4$ Remainder 1

$9 \div 2$ does *not* divide exactly. After reaching 8 on the number line, there is no more room for another 2. It takes 4 twos to get to 8, then it takes 1 more unit to get to 9. We say that

$$9 \div 2 = 4 \quad \text{Remainder 1} \quad \text{or} \quad 9 \div 2 = 4 \text{ R1}$$

The quotient is 4 and the *remainder* is 1. The **remainder** is how many units are left when you divide the dividend by the divisor. The remainder is always less than the divisor. If the remainder equals 0, we say that the dividend **divides evenly** by the divisor. One such example is $12 \div 3 = 4$. We say that 12 divides evenly by 3. If the remainder is not equal to 0, we say that the dividend **does not divide evenly** by the divisor, as in $9 \div 2 = 4$ R1. In this case, 9 does *not* divide evenly by 2.

EXAMPLE 4

Find

a) $42 \div 7$ b) $11 \div 4$

Solution

a) $42 \div 7 = 6$ Because there is no remainder, 42 divides evenly by 7. To check our answer, we can write the related multiplication problem $7 \times 6 = 42$. This is how we can check to be sure that our quotient is correct.

b) The number line shows that there are 2 fours in 11 and then there are 3 units left over. So, $11 \div 4 = 2$ R3.

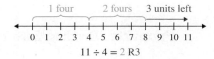

$$11 \div 4 = 2 \text{ R3}$$

The number line also shows how we can check our result to this division problem with multiplication and division:

Divisor   Remainder
    ↓   ↓
$(4 \times 2) + 3 = 11$ ← Dividend
        ↑
    Quotient

W Hint

After seeing this number line (for part b), do you think you could solve a similar problem without a number line?

[YOU TRY 4]

Find

a) $32 \div 8$ b) $37 \div 5$

Let's generalize this procedure for checking a division problem with multiplication.

Procedure Checking a Division Problem Using Multiplication

To check a division problem using multiplication, multiply the divisor by the quotient, then add the remainder. The result should be the dividend.

$$(\text{Divisor} \times \text{Quotient}) + \text{Remainder} = \text{Dividend}$$

Perform the multiplication in parentheses first, then add the remainder.

EXAMPLE 5

Check each division problem to determine whether the statement is true or false.

a) $78 \div 6 = 13$ b) $\dfrac{52}{9} = 5 \text{ R7}$ c) $7\overline{)47}\overset{6 \text{ R3}}{}$

Solution

a)   The statement $78 \div 6 = 13$ is true because $6 \times 13 = 78$.

b)   To determine whether $\dfrac{52}{9} = 5 \text{ R7}$ is true, multiply the divisor by the quotient, then add the remainder. The result should be the dividend.

$$(9 \times 5) + 7 = 45 + 7 = 52 \quad \checkmark$$

The statement $\dfrac{52}{9} = 5 \text{ R7}$ is true.

c)   To determine whether $7\overline{)47}\overset{6 \text{ R3}}{}$ is true, multiply the divisor by the quotient, then add the remainder. The result should be the dividend.

$$(7 \times 6) + 3 = 42 + 3 = 45$$

Because $45 \neq 47$, the statement $7\overline{)47}\overset{6 \text{ R3}}{}$ is false.

[YOU TRY 5]

Check each division problem to determine whether the statement is true or false.

a) $98 \div 7 = 14$ b) $\dfrac{67}{4} = 16 \text{ R2}$ c) $6\overline{)53}\overset{8 \text{ R5}}{}$

This characteristic of one number dividing evenly by another number leads us to the important topic of divisibility and the divisibility rules.

4   Use the Divisibility Rules

Definition

A whole number is **divisible** by another whole number if the remainder equals 0.

In Example 5a), we can say that 78 *is divisible by* 6 since $78 \div 6 = 13$. But in Example 5b), 52 is not divisible by 9 since there is a remainder when dividing 52 by 9.

EXAMPLE 6

a)   Is 66 divisible by 11?   Yes, because $66 \div 11 = 6$.

b)   Is 37 divisible by 9?   No, because $37 \div 9 = 4 \text{ R1}$.

[YOU TRY 6]

a)   Is 25 divisible by 3?   b)   Is 54 divisible by 6?

When we are working with larger numbers, it is helpful to have rules to determine whether one number is divisible by another number. These are called the **divisibility rules.** We have not included a divisibility rule for 7 because it is difficult to use.

Divisibility Rules

A number is divisible by	Example
…2 if it ends in 0, 2, 4, 6, or 8. If a number is divisible by 2, it is an **even number.**	7,394 is divisible by 2 because it ends in 4. It is an even number.
…3 if the sum of its digits is divisible by 3.	837—Add its digits: $8 + 3 + 7 = 18$. Since 18 is divisible by 3, the number 837 is divisible by 3.
…4 if its last two digits form a number that is divisible by 4.	5,932—The last two digits form the number 32. Since 32 is divisible by 4, the number 5,932 is divisible by 4.
…5 if the number ends in 0 or 5.	645 is divisible by 5 since it ends in 5.
…6 if it is divisible by 2 and by 3.	1,248—The number is divisible by 2 since it is an even number. The number is divisible by 3 since the sum of its digits is divisible by 3: $1 + 2 + 4 + 8 = 15$. Therefore, the number 1,248 is divisible by 6.
…8 if its last three digits form a number that is divisible by 8.	5,800—The last three digits form the number 800. Since 800 is divisible by 8, the number 5,800 is divisible by 8.
…9 if the sum of its digits is divisible by 9.	79,542—Add its digits: $7 + 9 + 5 + 4 + 2 = 27$. Since 27 is divisible by 9, the number 79,542 is divisible by 9.
…10 if it ends in a zero.	490 is divisible by 10 because it ends in 0.

> **⚠ Hint**
> The fastest way to memorize this table is to practice!

The tests we use most often are those for 2, 3, 5, and 10.

EXAMPLE 7

Determine whether each number is divisible by 2, 3, 5, and/or 10 or none of these.

a)   516   b)   7,425   c)   2,890

Solution

a)

Is 516 divisible by	
2?	**Yes.** 516 is an even number.
3?	**Yes.** Add the digits: $5 + 1 + 6 = 12$. Because 12 is divisible by 3, the number 516 is divisible by 3.
5?	**No.** 516 does *not* end in 0 or 5.
10?	**No.** 516 does *not* end in 0.

b)

Is 7,425 divisible by	
2?	**No.** 7,425 is *not* an even number.
3?	**Yes.** Add the digits: $7 + 4 + 2 + 5 = 18$. Because 18 is divisible by 3, the number 7,425 is divisible by 3.
5?	**Yes.** 7,425 ends in 5.
10?	**No.** 7,425 does *not* end in 0.

c)

Is 2,890 divisible by	
2?	**Yes.** 2,890 is an even number.
3?	**No.** Add the digits: $2 + 8 + 9 + 0 = 19$. Because 19 is *not* divisible by 3, the number 2,890 is *not* divisible by 3.
5?	**Yes.** 2,890 ends in 5.
10?	**Yes.** 2,890 ends in 0.

[**YOU TRY 7**] Determine whether each number is divisible by 2, 3, 5, and/or 10 or none of these.

a)   165   b)   4,170   c)   20,423

Next, we will learn how to divide larger numbers by one-digit numbers using a method called *short division*.

5 Use Short Division

Short division is a method for dividing a number by a one-digit divisor.

EXAMPLE 8 Use short division to find $2\overline{)68}$.

Solution

First, divide 6 by 2: $2\overline{)68}^{\,3}$ $6 \div 2 = 3$

Next, divide 8 by 2: $2\overline{)68}^{\,34}$ $8 \div 2 = 4$

Therefore, $2\overline{)68}^{\,34}$. Notice that there is no remainder. We can check the answer using multiplication: $2 \times 34 = 68$. ✓

[**YOU TRY 8**] Use short division to find $3\overline{)93}$.

EXAMPLE 9 Use short division to find $\dfrac{958}{3}$.

Solution

Rewrite the problem as $3\overline{)958}$.

Divide 9 by 3: $3\overline{)958}^{\,3}$ $9 \div 3 = 3$

Next, divide 5 by 3:

$3\overline{)95^{2}8}^{\,31}$ $5 \div 3 = 1\,\text{R}2$

Write the remainder of 2 in front of the 8. The remainder together with the 8 forms the number 28.

Now, divide 28 by 3:

$$3\overline{)95^28}\quad{}^{\,319\,R1}\qquad 28 \div 3 = 9\,R1$$

Since we have reached the end of the number 958, we have finished the division process. Therefore, $\dfrac{958}{3} = 319\,R1$.

Check: $(3 \times 319) + 1 = 957 + 1 = 958.$ ✓

[YOU TRY 9] Use short division to find $\dfrac{475}{4}$.

EXAMPLE 10 Use short division to divide 1,967 by 6.

Solution

Rewrite the problem as $6\overline{)1,967}$. Because 1 does not divide by 6, look at the first two digits of 1,967 together. Divide 19 by 6: $\quad 6\overline{)1,9^167}\;{}^{\,3}\qquad 19 \div 6 = 3\,R1$

Next, divide 16 by 6: $\quad 6\overline{)1,9^16^47}\;{}^{\,3\,2}\qquad 16 \div 6 = 2\,R4$

Finally, divide 47 by 6:

$$6\overline{)1,9^16^47}\;{}^{\,3\,2\,7\,R5}\qquad 47 \div 6 = 7\,R5$$

W Hint

Can you make a general statement that outlines how to perform short division?

So 1,967 divided by 6 is 327 R5.

Check: $(6 \times 327) + 5 = 1,962 + 5 = 1,967.$ ✓

[YOU TRY 10] Use short division to divide 2,053 by 7.

6 Solve Applied Problems Using Division

EXAMPLE 11 Jessica buys a package of 48 cookies for her daughter's playgroup of 9 children. If she gives the same number of cookies to each child, how many cookies can each child have? Will there be any left over?

Solution

We need to determine how many times 9 divides into 48.

$$48 \div 9 = 5\,R3$$

Each child can have 5 cookies, and there will be 3 left over.

©John Lund/Sam Diephuis/Blend Images LLC

E Evaluate **A.4** Exercises   Do the exercises, and check your work.

Objective 1: Understand the Meaning of Division
Use a number line to divide the numbers. Then,
  a) identify the dividend, divisor, and quotient.
  b) rewrite the problem using two other notations.

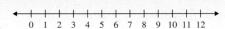
0 1 2 3 4 5 6 7 8 9 10 11 12

1) $6 \div 2$ 2) $10 \div 5$

3) $6\overline{)12}$ 4) $4\overline{)12}$

5) $\dfrac{7}{1}$ 6) $\dfrac{5}{5}$

Find the quotient, and write a related multiplication problem.

7) $\dfrac{36}{9}$ 8) $\dfrac{42}{6}$

9) $7\overline{)84}$ 10) $5\overline{)40}$

11) $3 \div 1$ 12) $8 \div 8$

Objective 2: Perform Division Involving Zero

13) Is it possible to divide 0 by a nonzero number? Explain your answer. Yes it's 0

14) Is it possible to divide a number by 0? Explain your answer.

Divide, if possible.

15) $6 \div 0$ 16) $11 \div 0$

17) $\dfrac{0}{7}$ 18) $\dfrac{0}{6}$

19) $0\overline{)8}$ 20) $0\overline{)5}$

21) $0 \div 10$ 22) $0 \div 9$

Fill in the missing number.

23) $20 \div \underline{\quad} = 4$ 24) $21 \div \underline{\quad} = 3$

25) $\underline{\quad} \div 9 = 7$ 26) $\underline{\quad} \div 7 = 8$

27) $5 \div \underline{\quad} = 1$ 28) $4 \div \underline{\quad} = 4$

Objective 3: Understand Remainders, and Check a Division Problem Using Multiplication

29) What does it mean to say that a dividend divides evenly by a divisor?

30) What does it mean to say that a dividend does not divide evenly by a divisor?

31) Find
    a) $18 \div 6$ b) $20 \div 6$

32) Find
    a) $30 \div 5$ b) $34 \div 5$

33) Find
    a) $\dfrac{36}{4}$ b) $\dfrac{39}{4}$

34) Find
    a) $9\overline{)45}$ b) $9\overline{)51}$

35) How do you check the result of a division problem?

36) Nivaj checks the division problem $43 \div 4 = 10$ R3 like this: $(10 \times 3) + 4 = 34$. Is the division result wrong, or did he make a mistake when checking the answer? Explain.

Check each division problem to determine whether the statement is true or false.

37) $95 \div 9 = 10$ R5

38) $46 \div 8 = 5$ R6

39) $\dfrac{58}{6} = 9$ R2

40) $\dfrac{16}{7} = 2$ R3

41) True or False: $31 \div 6 = 4$ R7. Explain your answer.

42) Does 23 divide evenly by 3? Explain your answer.

Divide. Then, check your answer.

43) $4\overline{)29}$

44) $9\overline{)53}$

45) $\dfrac{62}{7}$

46) $\dfrac{23}{8}$

47) $3\overline{)15}$

48) $6\overline{)24}$

49) $48 \div 5$

50) $37 \div 7$

Objective 4: Use the Divisibility Rules

51) What is an even number?

52) Every even number is divisible by what number?

53) How do you know whether a number is divisible by 2?

54) How do you know whether a number is divisible by 10?

55) How do you know whether a number is divisible by 3?

56) Make up an example of a three-digit number that is divisible by 3.

Determine whether each number is divisible by 2, 3, 5, and/or 10, or none of these.

57) 52 no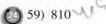

58) 75

59) 810 yes

60) 708

61) 4,863

62) 1,740

63) 92,725

64) 73,660

65) 100,176

66) 872,619

67) 54,623

68) 6,089

If a number is divisible by both 2 *and* 3, then the number is also divisible by 6. Determine whether each of the following numbers is divisible by 6.

69) 84

70) 78

71) 429

72) 513

73) 2,520

74) 4,110

Objective 5: Use Short Division

75) Can short division be used to divide 748 by 12? Explain your answer.

76) Can short division be used to divide 953 by 7? Explain your answer.

Use short division to divide.

77) $2\overline{)46}$

78) $3\overline{)39}$

79) $\dfrac{492}{4}$

80) $\dfrac{684}{6}$

81) $959 \div 3$

82) $871 \div 2$

83) $6\overline{)7,526}$

84) $3\overline{)5,350}$

85) Divide 2,496 by 8.

86) Divide 3,598 by 7.

87) $\dfrac{3,380}{6}$

88) $\dfrac{7,954}{8}$

89) $4\overline{)81,920}$

90) $5\overline{)53,450}$

Objective 6: Solve Applied Problems Using Division

91) At the end of business Thursday, a tip jar contains $54. The six people who worked that day will split the tips evenly. How much tip money will each employee receive?

92) A banquet room can hold a total of 216 people. If each table seats eight people, how many tables are in the room?

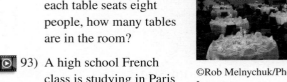

©Rob Melnychuk/Photodisc/Getty Images

93) A high school French class is studying in Paris for the summer. Each student checked two pieces of luggage, and a total of 78 bags were checked. How many students went to Paris?

94) Tyra runs a modeling agency, and each day, Monday through Friday, she interviews the same number of potential models. If she interviews 30 people per week, how many does she see each day?

95) A lecture hall holds 204 people. If six people can sit at each table, how many tables are in the lecture hall?

96) Manoli is cleaning out his office and is stacking his books in piles of eight. If he has a total of 112 books, how many stacks are there?

97) Tickets to a college play cost $9 each. If the ticket revenue was $2,016, how many people attended the play?

98) Tickets to a college rugby game cost $7 each. If the ticket revenue was $441, how many people watched the game?

©Don Hammond/Design Pics

99) Min did the same number of sit-ups every day for a week. If she did a total of 420 sit-ups, how many did she do each day?

100) A hotel is being remodeled, and every room will have three lamps. The supplier delivers 138 lamps. How many rooms are in the hotel?

R Rethink

R1) Explain the procedure used to check an answer that contains a remainder. Why must you always multiply inside the parentheses first?

R2) After completing the exercises, which divisibility rules do you still need to review?

R3) Where have you encountered a division problem in the last week?

A.5 Long Division of Whole Numbers

P Prepare

O Organize

What are your objectives for Section A.5?	How can you accomplish each objective?
1 Use Long Division	• Understand the difference between *long division* and short division. • Complete the given examples on your own. • Complete You Trys 1–3.
2 Divide Numbers Ending in Zeros	• Write the procedure for **Dividing Whole Numbers Ending in Zeros by 10, 100, 1,000, etc.** in your own words. • Complete the given examples on your own. • Complete You Trys 4 and 5.
3 Solve Applied Problems Using Long Division	• Read the applied problem twice, and be sure to understand what is being asked. • Complete the given example on your own. • Complete You Try 6.

W Work   **Read the explanations, follow the examples, take notes, and complete the You Trys.**

If the total cost for 23 art history students to visit a museum is $322, what is the cost per student? We will answer this question using *long division* in Example 6. **Long division** is a method for dividing a number by a divisor with any number of digits. (Recall from the previous section that short division can be used to divide a number by a *one-digit* divisor.)

1 Use Long Division

EXAMPLE 1

Use long division to find $8\overline{)2{,}520}$.

Solution

We begin as we do with short division. Since 8 does not divide into the first 2 in the dividend, ask yourself, "*How many times does* 25 *divide evenly by* 8?" 3. Write 3 above the 5, then perform the following steps.

Multiply 3×8 to get 24. Write the 24 under the 25 and subtract.

$$
\begin{array}{r}
3 \\
8\overline{)2{,}5\,2\,0} \\
-2\,4 \quad \longleftarrow \quad 3 \times 8 = 24 \\
\hline
1 \quad \longleftarrow \quad 25 - 24 = 1
\end{array}
$$

Bring down the 2.

$$
\begin{array}{r}
3 \\
8\overline{)2{,}5\,2\,0} \\
-2\,4\downarrow \\
\hline
1\,2
\end{array}
$$

W Hint
Work out the example on your paper as you are reading it.

Ask yourself, "*How many times does* 12 *divide evenly by* 8?" 1. Write 1 above the second 2 in the dividend, then

Multiply 1×8 to get 8. Write 8 under the 2 in 12, and subtract.

$$
\begin{array}{r}
3\ \ 1 \\
8\overline{)2{,}5\,2\,0} \\
-2\,4 \\
\hline
1\,2 \\
-\ \ 8 \quad \longleftarrow \quad 1 \times 8 = 8 \\
\hline
4 \quad \longleftarrow \quad 12 - 8 = 4
\end{array}
$$

Bring down the 0.

$$
\begin{array}{r}
3\ \ 1 \\
8\overline{)2{,}5\,2\,0} \\
-2\,4 \\
\hline
1\,2 \\
-\ \ 8\ \ \downarrow \\
\hline
4\,0
\end{array}
$$

Ask yourself, "*How many times does* 40 *divide evenly by* 8?"   5

Multiply 5×8 to get 40. Write this 40 under the 40, and subtract. The remainder is 0.

$$
\text{Therefore, } 8\overline{)2{,}520}. \quad \overset{315}{\phantom{8)2{,}520}}
$$

Check using multiplication: $8 \times 315 = 2{,}520$. ✓

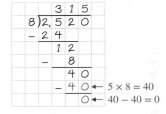

YOU TRY 1

Use long division to find $7\overline{)4{,}151}$.

If the divisor has more than one digit, we cannot use short division. We will use long division.

EXAMPLE 2

Use long division to find $19\overline{)6{,}217}$.

Solution

Ask yourself, "*How many times does* 62 *divide evenly by* 19?" We can estimate this by rounding 19 to 20: $\dfrac{62}{20} = 3$ R2. So, we will try 3 as the first digit in the quotient. Write 3 above the 2, then

Multiply 3×19 to get 57. Write the 57 under the 62 and subtract.

```
        3
19)6,2 1 7
  − 5 7  ←——— 3 × 19 = 57
      5  ←——— 62 − 57 = 5
```

Bring down the 1.

```
        3
19)6,2 1 7
  − 5 7↓
      5 1
```

Ask yourself, "*How many times does* 51 *divide evenly by* 19?" Again, round 19 to 20 and divide to estimate: $\dfrac{51}{20} = 2$ R11. So, we will try 2 as the second digit in the quotient. Write 2 above the 1 in the dividend, then

Multiply 2×19 to get 38. Write 38 under the 51, and subtract.

```
        3 2
19)6,2 1 7
  − 5 7
      5 1
    − 3 8  ←——— 2 × 19 = 38
      1 3  ←——— 51 − 38 = 13
```

Bring down the 7.

```
        3 2
19)6,2 1 7
  − 5 7
      5 1
    − 3 8↓
      1 3 7
```

Ask yourself, "*How many times does* 137 *divide evenly by* 19?" Round 19 to 20 and divide: $\dfrac{137}{20} = 6$ R17. Write 6 above the 7, then

Multiply 6×19 to get 114. Write the 114 under the 137, and subtract. We get 23. Since the remainder is greater than the divisor, increase the 6 in the quotient to 7.

Change the 6 in the quotient to 7. Multiply 7×19 to get 133. Write the 133 under the 137, and subtract. The remainder is 4.

W Hint

Notice that, like in this example, you will not always guess the correct number in the quotient the first time around!

```
            3 2 6 ← Change the 6 to 7.
  → 19)6,2 1 7
      − 5 7
The remainder,   5 1
23, is greater  − 3 8
than the divisor,  1 3 7
19.           − 1 1 4 ← 6 × 19 = 114
            → 2 3 ← 137 − 114 = 23
```
The remainder cannot be greater than the divisor.

```
            3 2 7
  19)6,2 1 7
    − 5 7
        5 1
      − 3 8
        1 3 7
      − 1 3 3 ← 7 × 19 = 133
        → 4 ← 137 − 133 = 4
```
The remainder is less than the divisor.

Therefore, $19\overline{)6{,}217}$ = $\overset{327 \text{ R4}}{\phantom{)6{,}217}}$

Check: $(19 \times 327) + 4 = 6{,}213 + 4 = 6{,}217$. ✓

[YOU TRY 2] Use long division to find $17\overline{)8{,}372}$.

In Example 3, we will see that, sometimes, we must write a 0 in the quotient.

EXAMPLE 3

Find 12,782 ÷ 42.

Solution

Begin by setting up the problem as 42)12,782.

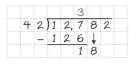

1) How many times does 127 divide evenly by 42?   3

2) Multiply 3 × 42 = 126.

3) Subtract 127 − 126 = 1.

4) Bring down the 8.

Start the process again. Ask yourself, "*How many times does* 18 *divide evenly by 42?*"   0

 Write 0 in the quotient as a placeholder.

Bring down the 2, and do the division process again.

 Hint

Get in the habit of asking yourself the questions that are asked in the solutions of the examples!

1) How many times does 182 divide evenly by 42?   4

2) Multiply 4 × 42 = 168.

3) Subtract 182 − 168 = 14.

4) The remainder is 14.

Therefore, 12,782 ÷ 42 = 304 R14.

Check: (42 × 304) + 14 = 12,768 + 14 = 12,782. ✓

[YOU TRY 3] Find 21,557 ÷ 53.

2 Divide Numbers Ending in Zeros

In Section A.3, we saw a pattern for multiplying a number by a number ending in 0. For example, 36 × 1,000 = 36,000. That is, to multiply a number by 1,000, we add three zeros to the end of the number.

Similarly, there is a pattern for dividing a number ending in zeros by another number ending in zeros. Do you see the pattern below?

$$15,000 \div 1 = 15,000$$
$$15,000 \div 10 = 1,500$$
$$15,000 \div 100 = 150$$
$$15,000 \div 1,000 = 15$$

Procedure Dividing Whole Numbers Ending in Zeros by 10, 100, 1,000, etc.

When you divide a whole number ending in zeros by

1) 10, you get the quotient by dropping one zero from the whole number.

   Example: $2,700 \div 10 = 270$

2) 100, you get the quotient by dropping two zeros from the whole number.

   Example: $630,000 \div 100 = 6,300$

3) 1,000, you get the quotient by dropping three zeros from the whole number.

   Example: $78,000 \div 1,000 = 78$

This pattern continues for divisors 10,000, 100,000, and so on.

EXAMPLE 4

Divide.

a) $940 \div 10$ b) $810,000 \div 1,000$ c) $8,000 \div 200$

Solution

a) Since we are dividing by 10, we get the quotient by dropping one zero from 940.
    $940 \div 10 = 94$

b) Dividing by 1,000 means that we get the quotient by dropping three zeros from 810,000. $810,000 \div 1,000 = 810$

c) Here we are dividing 8,000 by a *multiple* of 100. To find $8,000 \div 200$, first drop two zeros from each number to get $80 \div 2$. (This is the same as dividing each number by 100.) Now, finish the division: $80 \div 2 = 4$. So, $8,000 \div 200 = 4$ since $80 \div 2 = 4$.

[YOU TRY 4]

Divide.

a) $420 \div 10$ b) $39,000 \div 1,000$ c) $6,000 \div 300$

Example 4c) shows that we can use this pattern of dropping zeros to divide by multiples of 10.

EXAMPLE 5

Find $1,200 \overline{)18,000}$.

Solution

Because the divisor, 1,200, is a multiple of 100, we can drop two zeros from each number.

$1,200 \overline{)18,000}$ Drop two zeros to get $12 \overline{)180}$

Divide.

$$\begin{array}{r} 1\ 5 \\ 1\ 2\overline{)1\ 8\ 0} \\ -1\ 2\downarrow \\ \overline{6\ 0} \\ -6\ 0 \\ \overline{0} \end{array}$$

Since $180 \div 12 = 15$, it follows that $18{,}000 \div 1{,}200 = 15$.

Check using multiplication: $1{,}200 \times 15 = 18{,}000$. ✓

[YOU TRY 5] Find $1{,}500\overline{)45{,}000}$.

3 Solve Applied Problems Using Long Division

In the next example, we will answer the question that was asked at the beginning of this section.

EXAMPLE 6 If the total cost for 23 art history students to visit a museum is $322, what is the cost per student?

Solution

We will use long division to find the answer.

$$\begin{array}{r} 1\ 4 \\ 2\ 3\overline{)3\ 2\ 2} \\ -2\ 3 \\ \overline{9\ 2} \\ -9\ 2 \\ \overline{0} \end{array}$$

©John Lund/Marc Romanelli/
Getty Images

The cost is $14 per student.

[YOU TRY 6] Clarissa pays a caterer $2,698 to feed her 142 guests. What is the cost per person?

ANSWERS TO [YOU TRY] EXERCISES

1) 593   2) 492 R8   3) 406 R39   4) a) 42 b) 39 c) 20   5) 30   6) $19 per person

E Evaluate **A.5** Exercises   Do the exercises, and check your work.

Objective 1: Use Long Division

Use long division to find each quotient. Check your answer.

1) $3\overline{)192}$

2) $4\overline{)312}$

3) $6\overline{)3{,}486}$

4) $7\overline{)3{,}031}$

5) $4\overline{)1{,}071}$

6) $9\overline{)6{,}478}$

7) $4{,}272 \div 13$

8) $8{,}849 \div 18$

9) $4{,}758 \div 61$

10) $5{,}394 \div 93$

11) $52\overline{)4{,}940}$

12) $75\overline{)1{,}650}$

13) $\dfrac{2{,}834}{39}$

14) $\dfrac{1{,}139}{42}$

15) $47\overline{)65{,}153}$

16) $86\overline{)97{,}822}$

17) $7{,}923 \div 4$

18) $16{,}204 \div 5$

19) $71\overline{)61,118}$ 20) $63\overline{)29,639}$

21) $4,015 \div 5$ 22) $2,472 \div 8$

23) $\dfrac{15,042}{25}$ 24) $\dfrac{12,692}{31}$

25) $68\overline{)170,279}$ 26) $49\overline{)279,696}$

27) a) Find $5,396 \div 8$ using short division.

  b) Find $5,396 \div 8$ using long division.

  c) Which method do you prefer? Why?

28) Can we use short division to find $32,714 \div 27$? Explain your answer.

Objective 2: Divide Numbers Ending in Zeros

29) If you divide a whole number ending in 0 by 10, how do you get the quotient?

30) If you divide a whole number ending in zeros by 100, how do you get the quotient?

31) Divide.

  a) $30 \div 10$

  b) $4,800 \div 100$

  c) $312,000 \div 1,000$

32) Divide.

  a) $70 \div 10$

  b) $6,100 \div 100$

  c) $177,000 \div 1,000$

33) Divide.

  a) $9,260,000 \div 10$

  b) $21,100,000 \div 1,000$

  c) $78,000 \div 10$

34) Divide.

  a) $835,000 \div 10$

  b) $670,000 \div 100$

  c) $554,000,000 \div 1,000$

Find each quotient.

35) $50\overline{)850}$ 36) $30\overline{)390}$

37) $237,900 \div 3,900$ 38) $275,600 \div 5,300$

39) $4,600\overline{)3,358,000}$ 40) $2,100\overline{)1,764,000}$

41) $\dfrac{7,000,000}{28,000}$ 42) $\dfrac{11,160,000}{36,000}$

Objective 3: Solve Applied Problems Using Long Division

Solve each problem.

43) Landon is the captain of his intramural soccer team and orders uniforms for each of the 18 players. If the total cost of the uniforms is $756, how much does each player owe?

44) A company spends $6,600 to send 24 employees to a conference. What is the cost per employee?

45) Each of the 230 guests at a fundraising dinner paid the same amount to attend. If $34,500 was raised, what was the cost per person?

46) On opening day, the revenue for single-day adult lift tickets at a ski resort was $34,776. If each of these lift tickets cost $92, how many were sold?

©McGraw-Hill Education/John Flournoy, photographer

47) In 2016, the average person spent approximately 730 hr on social media. How many hours per day does the average person spend on social media? (Hint: 1 yr = 365 days) (www.socialmediatoday.com)

48) In 2016, approximately 15,137,500,000 bushels of corn were harvested from 86,500,000 acres in the United States. How many bushels were harvested per acre? (www.nass.usda.gov)

49) In 2009, approximately 46,000,000 turkeys weighing a total of 736,000,000 lb were eaten on Thanksgiving day. Find the average weight of a turkey cooked on Thanksgiving day. (National Turkey Federation)

©Ariel Skelley/Blend Images LLC

50) Five hundred full-grown turkeys have about 1,750,000 feathers. How many feathers does one turkey have?

51) A whale migrated 3,000 miles in about 60 days. Find the average number of miles the whale traveled each day.

52) A company will give each of its 132 employees the same end-of-the-year bonus. How much will each person receive if the total set aside for the bonuses is $330,000?

Mixed Exercises: Objectives 1–3

Divide.

53) 40,693 ÷ 67

54) 413,000 ÷ 1,000

55) $\dfrac{1,280}{40}$

56) 9)$\overline{2,214}$

57) 8,600 ÷ 10

58) $\dfrac{9,180}{16}$

Solve each problem.

59) A group of 16 friends went out to dinner and split the bill evenly. With tax and tip, the total bill was $336. How much did each person owe?

60) In 2014, Americans ate fresh citrus fruits at a rate of approximately 22 lb per person per year. Find the population of a town if it was expected that the people in that town ate 50,600 lb of citrus fruits in 2014. (www.statista.com)

©evdayan/Getty Images

R Rethink

R1) Are you using graph paper to help you line up numbers correctly?

R2) Where do you need more help in this section?

Appendix A: Review Exercises

Determine whether each number is divisible by 2, 3, 5, and/or 10 or none of these.

1) 6,105

2) 730

3) 283

4) 594

5) 19,950

6) 30,439

Perform the indicated operation.

7) 476
   × 59

8) 7,534 − 621

9) 21,308 + 72 + 9,665 + 147

10) 5,000 ÷ 23

11) 700 − 286

12) 39 · 10,000

13) $\dfrac{424}{8}$

14) 26 × 74

15) 780,000 ÷ 1,000

16) 6,413 + 9 + 18,007 + 521

17) (100)(152)

18) 9)$\overline{648}$

19) 349,914
    + 680,276

20) 40,512 − 16,708

21) 5,729
    − 493

22) $\dfrac{18,063}{9}$

23) 15,045 ÷ 5

24) 92,503
    + 48,726

25) 34 · 96

26) (8,509)(708)

27) 60,772 − 23,059

28) 4,250,000 ÷ 100

29) 26)$\overline{4,500}$

30) 500
    − 449

Answers to Selected Exercises

Chapter 1

Section 1.1

1) A digit is a single character in a numbering system, but a number is what you get when you write digits together in a certain order.

3) 5 hundreds   5) 2 ten-billions

7) a) 8 b) 3 c) 6 d) 5   9) a) 0 b) 3 c) 3

11) a) 9 b) 5 c) 6   13) 72 thousands, 544 ones

15) 803 millions, 1 thousands, 216 ones

17) 41 billions, 906 millions, 553 thousands, 213 ones

19) six hundred one

21) five million, four hundred forty-nine

23) four hundred fifty million, six hundred twenty-nine thousand, eight hundred seventy-five

25) 7,283   27) 48,902,020   29) 11,002,000,004

31) 1,940,000,000   33) 891,827   35) 39,250,017

37) Answers may vary.   39) 40   41) 200

43) 77,000   45) 4,600   47) 146,000   49) 0

51) 40,000   53) 500,000   55) 3,500,000

57) 400,000,000   59) 300,000   61) 1,000

63) 80; 100; 0   65) 1,400; 1,400; 1,000

67) 619,760; 619,800; 620,000   69) 1,000,000

71) Answers may vary. One such number is 7,231.

73) Answers may vary. One such number is 33,852.

Section 1.2

1) A negative number is a number that is less than zero.

3) −$23   5) −3 lb   7) 14,494 ft

9)
$$-5\;-4\;-3\;-2\;-1\;\;0\;\;1\;\;2\;\;3\;\;4\;\;5$$

11)
$$-5\;-4\;-3\;-2\;-1\;\;0\;\;1\;\;2\;\;3\;\;4\;\;5$$

13) smaller   15) <   17) <

19) <   21) <   23) >   25) >

27) <   29) <   31) >   33) >

35) It is the distance of the number from zero.

37) 8   39) 19   41) 17   43) 0

45) −35   47) −24

49) They are both a distance of 5 from zero on the number line.

51) −1   53) −47   55) 23   57) 147

59) false   61) true   63) true   65) >   67) <

Section 1.3

1) 1,049   3) 433,154

5) 3;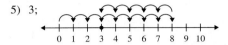

7) −4;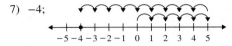

9) −2;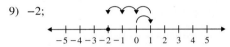

11) −3;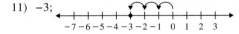

13) −6;

15) Answers may vary.   17) always   19) −11

21) −24   23) −98   25) −590   27) −32   29) −125

31) Answers may vary. One expression is −40 + (−5).

33) Answers may vary.   35) −3   37) 8   39) 6

41) −11   43) −15   45) 193   47) 92   49) −249

51) Answers may vary. One expression is −30 + 2.

53) Answers may vary. One expression is 55 + (−4).

55) $736 + (−$258); $478   57) −12 yd + 9 yd; −3 yd

59) 159 GB   61) $157   63) Answers may vary.

65) 15; 7 + 8   67) −7; 9 + (−16)   69) 7; 0 + 7

71) 14; Answers may vary.   73) Answers may vary.

75) 15; (3 + 7) + 5   77) 14; (4 + 7) + 3

79) −4; −11 + (5 + 2)   81) 0; (−7 + (−1)) + 8

83) 11   85) 30   87) 0   89) 9   91) −37

93) −929   95) −165   97) −325 ft   99) 5°F

Section 1.4

1) 631   3) 4,722   5) 5,443

7) They are the same distance from 0 on the number line, but they are on opposite sides of 0; their sum is 0.

9) −6   11) 0   13) 8   15) 23

17) 6 + (−5); 1;

19) 3 + (−7); −4;

21) −2 + (−3); −5;

23) −2 + 2; 0;

25) Answers may vary.   27) 15 + (−6) = 9

29) 5 + (−14) = −9   31) 125 + (−183) = −58

33) −8 + (−3) = −11   35) −7 + (−12) = −19

37) −134 + (−925) = −1,059   39) −10 + 18 = 8

41) −23 + 27 = 4   43) −29 + 15 = −14

45) −791 + (−683) = −1,474   47) 3,508 + 2,917 = 6,425

49) −54 + (−54) = −108   51) −21 + 21 = 0

53) Answers may vary. One expression is 7 − 20.

55) Answers may vary. One expression is −1 − (−10).

57) 20,547 ft   59) 127 m   61) −223 ft

63) −12   65) 8   67) 22   69) −539

71) −56   73) 10   75) −1,003   77) 53

79) −11   81) −78   83) 51   85) −61

Section 1.5

1) 79;   60
        + 20
         80

3) 143;   60
          60
        + 20
         140

5) 23;   70
        − 40
         30

7) 61;   180
        − 120
         60

9) −172; −330
        + 160
        −170

11) 903;   700
          + 200
           900

13) 1,379;   200
             600
           + 600
           1,400

15) 338;   700
          − 400
           300

17) 1,644;   4,100
            − 2,400
             1,700

19) −1,284;   −800
            + (−500)
             −1,300

21) 6,663;   −1,600
            + 8,300
             6,700

23) estimate: approximately 9,100 miles; exact: 9,183 miles

25) estimate: In 2022, there will be approximately 208,000,000 more smart phone users than in 2010; exact: In 2022, there will be approximately 208,060,000 more smart phone users than in 2010.

27) In front-end rounding, a number is rounded to the place furthest to the left.

29) estimate: 120; exact: 121   31) estimate: 1,000; exact: 937

33) estimate: 26,000; exact: 28,465

35) estimate: −80; exact: −78   37) estimate: 200; exact: 213

39) estimate: 50; exact: 57   41) estimate: 200; exact: 173

43) estimate: 30,000; exact: 34,293

45) estimate: −40; exact: −45

47) estimate: 11,000; exact: 10,761

49) estimate: 90; exact: 94   51) estimate: −90; exact: −81

53) estimate: 10 more boxes; exact: 16 more boxes

55) estimate: $2,900; exact: $2,717

57) estimate: 30; exact: 27

59) estimate: $600,000,000; exact: $500,228,170

Section 1.6

1) 32,798   3) 12,721,020   5) 28,242,440

7) positive   9) −15   11) −56   13) −32

15) −96   17) 0   19) −38   21) −918

23) −9   25) 7   27) negative   29) 6

31) 45   33) 984   35) −8   37) −1

39) 24   41) −280   43) −84   45) 30

47) −120   49) 504   51) 30   53) −7

55) 210   57) −88   59) commutative property

61) distributive property   63) identity property

65) 70; 5 · (2 · 7) = 5 · 14 = 70

67) −72; 8 × (−9) = −72

69) −32; −4 · 5 + (−4) · 3 = −20 + (−12) = −32

71) −24; −3 · 9 − (−3) · 1 = −27 − (−3) = −27 + 3 = −24

73) estimate: 45,000; exact: 44,496

75) estimate: −15,000,000; exact: −17,930,570

77) estimate: 80,000,000; exact: 72,936,344

79) estimate: −3,500 customers; exact: −3,536 customers

81) estimate: $1,000; exact: $1,152

83) estimate: −$10,000; exact: −$14,196

85) −2,688   87) 627,587   89) −30

91) 576   93) 210   95) 88

Section 1.7

1) 1,874   3) 76   5) 2,056   7) 90,000

9) 67,941 Remainder 8   11) negative   13) 8

15) −5   17) −8   19) undefined   21) −63

23) 14   25) 5   27) 0   29) 1   31) −1

33) 29   35) −11   37) −42

39) Answers may vary. One expression is $-56 \div 7 = -8$.

41) -6 43) 40   45) -15 47) 4

49) -30 51) 4   53) 60   55) $-2,000$

57) estimate: 40; exact: 34   59) estimate: -500; exact: -461

61) estimate: 600; exact: 591   63) -63 employees

65) 16,800 people

67) estimate: 100 points per game; exact: 95 points per game

69) estimate: 1,500 downloads per hour; exact: 1,370 downloads per hour

71) 7 teachers   73) 17 trips   75) 26 tenders

Putting It All Together Exercises

1) -155 2) -54 3) -181 4) 0

5) 19   6) $-4,391$ 7) -120 8) 61

9) undefined   10) -35 11) -30 12) 0

13) 56   14) -203 15) 108   16) 6

17) 0   18) 152   19) -108 20) -84

21) 0   22) 5   23) 240   24) -197

25) -8 26) 140   27) $-2,451$ 28) undefined

29) -12 30) 275   31) 347   32) -8

33) 2,700,000   34) 526,000   35) $-17,000$

36) $-84,000,000$ 37) estimate: 278,000; exact: 278,616

38) estimate: $-70,000$; exact: $-69,896$

39) estimate: $-150,000$; exact: $-152,859$

40) estimate: $-15,000$; exact: $-14,321$

41) estimate: 6,900; exact: 7,029

42) estimate: 500; exact: 417

43) -18; $9 \cdot 4 - 9 \cdot 6 = 36 - 54 = -18$

44) 30; $-10(-3) = 30$

45) 17; $(-5 + 15) + 7 = 10 + 7 = 17$

46) -66; $-6 \cdot 9 + (-6) \cdot 2 = -54 + (-12) = -66$

47) -5 yd   48) $-\$12$

49) estimate: $600,000; exact: $736,332

50) estimate: 10 buses; exact: 11 buses

Section 1.8

1) base: 10; exponent: 2; 100   3) base: 5; exponent: 2; 25

5) base: 0; exponent: 2; 0   7) base: 2; exponent: 3; 8

9) base: 3; exponent: 4; 81   11) base: 20; exponent: 2; 400

13) 9^4 15) 4^6 17) always   19) $(-10)^3$

21) $(-13)^2$ 23) 49   25) 64   27) -32

29) $-1,000$ 31) 1   33) Answers may vary.

35) 4   37) 9   39) 1   41) 7   43) 3

45) 20   47) Answers may vary.   49) 9   51) -8

53) 2   55) 18   57) -17 59) 12

61) a) 81   b) -81 63) a) 25   b) -25

65) The base of -7^2 is 7, and $-7^2 = -49$.
The base of $(-7)^2$ is -7, and $(-7)^2 = 49$.

67) 16   69) -16 71) $-1,000$ 73) 144

75) -1 77) 65   79) -56 81) -3

83) 62   85) 0   87) 2   89) -12

91) 124   93) -34 95) -491

97) It is incorrect to do the subtraction first. Multiplication should be done before subtraction.

$$48 - 4 \cdot (-2) = 48 - (-8) = 48 + 8 = 56$$

99) $-10^2 = -100$, not 100.
$$\begin{aligned} -10^2 \div |12 - 37| &= -100 \div |12 - 37| \\ &= -100 \div |-25| \\ &= -100 \div 25 \\ &= -4 \end{aligned}$$

100) The expression $(8 - 2)^2$ was evaluated incorrectly. First, subtract inside the parentheses, then square the value.
$$\begin{aligned} \sqrt{49} + (8 - 2)^2 &= 7 + (6)^2 \\ &= 7 + 36 \\ &= 43 \end{aligned}$$

Chapter 1: Group Activity

Activity #1 Answers

1) negative   2) negative   3) positive   4) negative

5) negative   6) positive   7) negative   8) positive

9) negative   10) positive

Activity #2 Answers

1) -96 2) -10 3) 50

Chapter 1: Review Exercises

1) 7 billions   3) 138 millions, 952 thousands, 600 ones

5) four hundred ninety billion, six hundred seventeen million, five thousand, nine hundred fifteen

7) 6,200   9) 10,000,000

11) It is a number that is less than 0.   13) $-13,000$ ft

15)

17) $>$ 19) $>$ 21) $<$ 23) $>$ 25) $>$

27) It is the distance of a number from 0.

29) 15   31) -26 33) 35

35) Answers may vary.   37) -5 39) -11 41) 7

43) 250   45) $16 + (-7)$; 9 yd   47) $-\$56$

49) 11; $8 + 3$ 51) 1; answers may vary

53) $15; (1 + 9) + 5$ 55) It is the opposite of the number.

57) -28 59) always   61) $8 + (-20) = -12$

63) $-52 + 89 = 37$ 65) $-6,142 + (-1,087) = -7,229$

67) 0 69) -81 71) $215°F$

73) estimate: 3,500; exact: 3,473

75) estimate: $-12,000$; exact: $-9,144$

77) estimate: -70; exact: -73

79) negative   81) -60 83) 48

85) $-27; -3 \cdot 7 + (-3) \cdot 2 = -21 + (-6) = -27$

87) $54; 9(6) = 54$ 89) estimate: 30,000; exact: 28,853

91) $-3,375$ ft   93) 4 95) -45 97) undefined

99) 15 101) -12 103) estimate: -100; exact: -97

105) estimate: 25 boxes; exact: 29 boxes   107) 8^5

109) 36 111) 16 113) 49 115) -49 117) -8

119) 10 121) 13 123) -18 125) -16

127) The expression $(7 - 4)^2$ was evaluated incorrectly. First, subtract inside the parentheses, then square the value.

$$\sqrt{81} + (7 - 4)^2 = 9 + (3)^2$$
$$= 9 + 9$$
$$= 18$$

129) $-3,654$ 131) 264 133) -466

135) -6 137) 6

139) Answers may vary. One expression is $-27 \div 3 = -9$.

141) Answers may vary. One expression is $30 + (-4) = 26$.

143) Answers may vary. One expression is $-42 - (-5) = -37$.

145) 9,518 ft

Chapter 1: Test

1) a) thousands b) ten-millions

2) a) three thousand ninety-four b) 8,115,622

3) a) 7,430,000 b) 7,000,000 c) 7,426,000

4) a) $>$ b) $<$ c) $<$ d) $>$

5)
$$
\begin{array}{ccccccc}
-5 & & -2 & 0 & & 3 & 4 \\
\end{array}
$$
$-5\,-4\,-3\,-2\,-1\ \ 0\ \ 1\ \ 2\ \ 3\ \ 4\ \ 5$

6) Answers may vary. The absolute value of a number is its distance from zero.

7) a) 8 b) 32 c) 0 d) -5 8) a) 13 b) -6

9) -10 10) 28 11) -8 12) -61 13) -4

14) -635 15) undefined   16) 0 17) 0

18) 59 19) -40 20) -8 21) 64

22) -25 23) 32 24) 6 25) 9

26) $(-7)^6$ 27) -5 28) 44 29) 0

30) a) $7; (-8 + 12) + 3 = 4 + 3 = 7$
   b) $-8; 4 \cdot 5 - 4 \cdot 7 = 20 - 28 = -8$ c) $15; 6 + 9 = 15$

31) estimate: 40,000; exact: 41,272

32) estimate: $-8,400$; exact: $-8,292$

33) estimate: -250; exact: -275

34) Answers may vary. One expression is $-7 \cdot 5 = -35$.

35) Answers may vary. One expression is $-3 - (-75) = 72$.

36) -454 ft   37) -53 members   38) $375

39) 62 boxes   40) -1

Chapter 2

Section 2.1

1) When Rebecca works 4 hr, she earns $56. When Rebecca works 18 hr, she earns $252.

3) When David works 5 days, he earns $425. When David works 8 days, he earns $680.

5) When Julian hikes 3 hr, he travels 9 mi. When Julian hikes 8 hr, he travels 24 mi.

7) After 15 min, 195 drops of fluid have fallen. After 45 min, 585 drops of fluid have fallen.

9) a) 40 b) 15 c) 5 d) 125

11) a) 5 b) -19 c) -3 d) -48

13) a) 14 b) 4 c) -7 15) a) -19 b) -51 c) -49

17) The expression increases by 2.   19) 29 21) -82

23) -31 25) 134 27) 5^2 29) x^2 31) r^8

33) a^3b^2 35) k^2t^5 37) $3q^4$ 39) $-8d^3f$

41) $-15q^2r^4s$ 43) $12b^3c^2d^3$ 45) a) 16 b) 16

47) 18 49) -48 51) -54 53) -216

55) never   57) sometimes   59) Answers may vary.

61) terms: $3x^3, -y^2, 5z, -6$; coefficients: 3, -1, 5; constant: -6

63) terms: $-7s^3, 6t^2, -12u, 16$; coefficients: -7, 6, -12; constant: 16

65) terms: $-7a^2b, 36b^3c^2, -51c, 32$; coefficients: -7, 36, -51; constant: 32

67) terms: $r^2, 5r, 7$; coefficients: 1, 5; constant: 7; value of the expression: 73

69) terms: $3x^2, -4xy, 7$; coefficients: 3, -4; constant: 7; value of the expression: 58

71) terms: $-4g^2, 8gh, -14$; coefficients: -4, 8; constant: -14; value of the expression: 6

73) terms: $9m^3, 2n^2, mn, 3n$; coefficients: 9, 2, 1, 3; constant: 0; value of the expression: -17

75) terms: $3d^2f, -4df^2, -126$; coefficients: 3, -4; constant: -126; value of the expression: 0

Section 2.2

1) $2 + 9 = 9 + 2; 11 = 11$

3) $-4 \cdot 7 = 7 \cdot (-4); -28 = -28$

5) Numbers can be multiplied in any order and the result is the same; answers may vary, but one example is $3 \cdot 5 = 5 \cdot 3$.

7) $w + 3$ 9) $2 \cdot c$ or $2c$ 11) $8y + (-7)$ or $8y - 7$

13) $(-9 + 4) + 3 = -5 + 3 = -2;$
$-9 + (4 + 3) = -9 + 7 = -2$

15) $3 \cdot (2 \cdot 8) = 3 \cdot 16 = 48; (3 \cdot 2) \cdot 8 = 6 \cdot 8 = 48$

17) $h + (6 + 7) = h + 13$

19) $(-10 + 2) + 5k = -8 + 5k$ (Using the commutative property, we can write $5k - 8$.)

21) $(3 \cdot 4)u = 12u$ 23) $(-12 \cdot 5)p^2 = -60p^2$

25) $(-6 \cdot (-8))ab = 48ab$

27) $9(-n) = 9(-1n) = (9 \cdot (-1))n = -9n$

29) $7 \cdot 2 + 7 \cdot 9 = 14 + 63 = 77$

31) $-1 \cdot 6 - (-1) \cdot 10 = -6 - (-10) = -6 + 10 = 4$

33) $3m + 15$ 35) $2c - 18$ 37) $-4t - 4$

39) $16v - 24$ 41) $-18 + 21r$ or $21r - 18$

43) $-g - 4$ 45) $-b + 23$

47) They have the same variables with the same exponents.

49) yes   51) yes   53) no   55) no   57) $8g$

59) $10w$ 61) $-7p$ 63) 0 65) $-2d$

67) $9k + 15$ 69) $8b^2 - 2$ 71) $-5xy$

73) $8mn + 1$ 75) $-6p + 10q$ 77) $5v^2 - 4v + 19$

79) $-3a + 10ab + 5b$ 81) $6c^2d + 14cd^2 - 7cd$

83) a) associative; $30m$ b) distributive; $30 + 5m$

85) a) distributive; $-36 + 9c$ or $9c - 36$ b) associative; $-36c$

87) Bashir is right because he used the order of operations correctly. Multiplication is performed before addition. Kelly did the problem wrong because she added before multiplying.

89) $5x + 47$ 91) $9c - 22$ 93) $-21h + 1$

95) $21r + 5$ 97) 0

99) Answers may vary. One expression is $2p - 11p$.

Section 2.3

1) expression   3) equation   5) No, it is an expression.

7) An equation contains an equal sign but an expression does not.

9) no   11) yes   13) yes   15) no

17) yes   19) $b = 14$ 21) $y = -8$ 23) $m = -10$

25) $k = 21$ 27) $x = 0$ 29) $t = 23$ 31) $w = -24$

33) $y = 8$ 35) Answers may vary. One equation is $w + 7 = 3$.

37) $a = 5$ 39) $u = -8$ 41) $d = 2$ 43) $r = 6$

45) $k = -4$ 47) $b = 18$ 49) $c = -1$ 51) $w = 10$

53) Answers may vary. One equation is $3n = 18$.

55) a) Use the subtraction property of equality because 4 is being added to k, so to solve for k, we subtract 4 from each side; $k = 16$
b) Use the division property of equality because k is being multiplied by 4, so to solve for k, we divide each side by 4; $k = 5$

57) a) Use the division property of equality because y is being multiplied by -3, so to solve for y, we divide each side by -3; $y = -6$
b) Use the addition property of equality because 3 is being subtracted from y, so to solve for y, we add 3 to each side; $y = 21$

59) $c = -9$ 61) $w = 9$ 63) $k = -47$

65) $m = -2$ 67) $x = 0$ 69) $p = 1$

Section 2.4

1) $k = 3$ 3) $a = -4$ 5) $m = 2$ 7) $r = -6$

9) $k = 12$ 11) $w = 0$ 13) $x = 1$ 15) $b = 9$

17) $n = -5$ 19) $z = 8$ 21) $x = -9$

23) Answers may vary.

25) Combine like terms; $8x + 14 - 14 = 30 - 14;$
Combine like terms; $\dfrac{8x}{8} = \dfrac{16}{8}; x = 2$

27) $h = 4$ 29) $d = -2$ 31) $p = 10$

33) $x = -15$ 35) $c = 1$

37) Distribute the 3.   39) $z = -4$ 41) $h = 1$

43) $w = -11$ 45) $n = -7$ 47) $k = 0$

49) $a = 3$ 51) $m = -1$ 53) $w = 11$ 55) $x = 5$

Section 2.5

1) Answers may vary.   3) $x = -5$ 5) $a = 1$

7) $k = 7$ 9) $p = -7$ 11) $g = 0$ 13) $r = 1$

15) $z = -4$ 17) $y = -5$ 19) $w = 16$

21) $v = -5$ 23) $n = 0$ 25) $x = -18$

27) $r = 1$ 29) multiplication   31) addition

33) subtraction   35) division

37) No. Subtraction is not commutative. For example, $9 - 5 = 4$ but $5 - 9 = -4$.

39) a) $1 + 8$ or $8 + 1$ b) $x + 8$ or $8 + x$

41) a) $9 \cdot 4$ b) $9x$ 43) a) $10 - 3$ b) $10 - x$

45) a) $\dfrac{-42}{-7}$ b) $\dfrac{-42}{x}$ 47) a) $2 \cdot 15$ b) $2x$

49) $x + 9$ or $9 + x$ 51) $x - 4$ 53) $\dfrac{x}{-35}$ 55) $-8x$

57) $2x$ 59) $x - 17$ 61) $10x$ 63) $2x + 11$

65) $2x - 6$ 67) $3x + 1$ 69) $8x - 10$ 71) $\dfrac{x}{5} - 3$

73) Answers may vary. One phrase is *the quotient of a number and 9*.

75) Answers may vary. One phrase is *the negative of a number*.

77) Answers may vary. One phrase is *15 less than a number*.

79) Answers may vary. One phrase is *11 more than twice a number*.

Section 2.6

1) Answers may vary.   3) $x + 9 = 24$; 15

5) $x - 16 = 11$; 27   7) $2x + 8 = 6$; -1

9) $3x - 10 = -4$; 2   11) $-12 + x = 5x$; -3

13) $2x - 7 = x + 8$; 15   15) $5x = 6x + 4$; -4

17) $2x + 6 = 4x - 8$; 7

19) Answers may vary. One statement is *1 less than a number is 9*.

21) Answers may vary. One statement is *4 more than three times a number is −11*.

23) Answers may vary. One statement is *4 times a number is the same as 18 less than 7 times the number*.

25) An age cannot be negative.   27) $396   29) $233

31) $-$59   33) 27,435 points   35) 411 mi

37) 10 points   39) $75 million   41) $47 per month

43) 13 students   45) 414 sq mi   47) 28 mpg

49) Answers may vary. One problem is *A certain puppy video had 21,043 more views than a certain kitten video. If the puppy video had 93,688 views, how many views did the kitten video have?*

Section 2.7

1) $g - 4$ 3) $m + 13$ 5) $2v$

7) Sig: 194, Keith: 157   9) Bill: 64, Sherri: 35

11) ice cream: 260 calories, frozen yogurt: 200 calories

13) Justin Bieber: 83 million, Rihanna: 61 million

15) Nicole: 578, Nicole's mom: 201

17) 1993: 375,000; 2013: 752,000

19) student: $43, adult: $86   21) 14 in., 22 in.

23) bracelet: 12 in., necklace: 21 in.   25) 8 ft, 12 ft

27) 7 ft, 21 ft   29) 5 ft, 11 ft

Chapter 2: Group Activity

1) Answers may vary. A possible response is shown below.

Operations	Equations
Begin with	$x = 8$
Divide both sides by 2.	$\dfrac{x}{2} = 4$
Add 3 to both sides.	$\dfrac{x}{2} + 3 = 7$
Write the translation problem.	**"The sum of 3 and the quotient of a number and 2 yields 7. Find the number."**

2) Answers may vary. A possible response is shown below.

Operations	Equations
Begin with	$x = -3$
Multiply both sides by -4.	$-4x = 12$
Subtract 8 from both sides.	$-4x - 8 = 4$
Write the translation problem:	**"Eight less than the product of −4 and a number results in 4. Find the number."**

3) Answers may vary. A possible response is shown below.

Operations	Equations
Begin with	$x = -12$
Multiply both sides by 7.	$7x = -84$
Subtract 1 from both sides.	$7x - 1 = -85$
Write the translation problem:	**"The difference of seven times a number and 1 is −85. Find the number."**

4) Answers may vary. A possible response is shown below.

Operations	Equations
Begin with	$x = 42$
Divide both sides by 6.	$\dfrac{x}{6} = 7$
Add -1 to both sides.	$-1 + \dfrac{x}{6} = 6$; *or* $\dfrac{x}{6} - 1 = 6$
Write the translation problem:	**"One less than the quotient of a number and 6 results in 6. Find the number."**

5) Answers may vary. A possible response is shown below.

Operations	Equations
Begin with	$x = -4$
Add 7 to both sides.	$x + 7 = 3$
Multiply both sides by 2.	$2(x + 7) = 6$
Write the translation problem:	**"Twice the sum of 7 and a number is 6. Find the number."**

6) Answers may vary. A possible response is shown below.

Operations	Equations
Begin with	$x = 17$
Add x to both sides.	$2x = 17 + x$
Subtract 3 from both sides.	$2x - 3 = 14 + x$
Write the translation problem:	**"Three less than twice a number results in 14 more than the number. Find the number."**

Chapter 2: Review Exercises

1) -23 3) -5

5) terms: $t^2, 8t, -3$; coefficients: 1, 8; constant: -3; value of the expression: 81

7) $k + 12$ 9) $(-8 \cdot 7)p = -56p$

11) $7 \cdot k + 7 \cdot 3 = 7k + 21$

13) $-1 \cdot v - (-1) \cdot 6 = -v + 6$

15) No. The exponent on the variable, x, is different in each term.

17) $-7m - 2n + 1$ 19) $9h^2k + 8hk^2 + 8h$

21) An equation contains an equal sign, but an expression does not.

23) yes   25) $y = 7$ 27) $w = 6$ 29) $x = 1$

31) $z = 8$ 33) $m = 9$ 35) $h = 3$

37) $t = 0$ 39) $q = -4$ 41) $x = 12$

43) $a = 6$ 45) $x - 6$ 47) $\dfrac{x}{8}$

49) $2x + 11 = 5$; -3 51) Answers may vary.

53) 25 min   55) Hayle: $150,000; Desisa: $75,000

57) Maria: 53,500 mi; Maria's dad: 21,500 mi

59) equation; $t = -1$ 61) expression; $12h - 9k - 15$

63) equation; $a = -6$ 65) equation; $c = 2$

67) always   69) Answers may vary. One expression is $2n - 7n$.

71) Answers may vary. One equation is $2k = 6$.

73) Answers may vary. One statement is *4 less than a number is 10.*

75) 20 ft, 28 ft

Chapter 2: Test

1) An equation contains an equal sign, but an expression does not.

2) 19   3) -10 4) $-6n^3p^2$

5) terms: $4x^2, 7y^2, -5x, y, -8$; coefficients: 4, 7, -5, 1; constant: -8

6) $4 \cdot 9 + 4 \cdot n = 36 + 4n$

7) $-1 \cdot y - (-1) \cdot (3) = -y + 3$

8) $6c - 6d - 11$ 9) $11k + 18$

10) yes   11) $x = -9$ 12) $w = 6$ 13) $p = 4$

14) $y = 1$ 15) $m = 0$ 16) $h = -2$ 17) $k = -13$

18) $10x$ 19) $x - 7$ 20) $15 + x$ or $x + 15$

21) $\dfrac{x}{-6}$ 22) $2x + 8$ or $8 + 2x$ 23) $x + 5 = 16$; 11

24) $3x - 4 = x - 8$; -2 25) $65

26) 9 A.M.: 28 students, 11 A.M.: 34 students

27) 12 in., 24 in.   28) sometimes   29) always

30) Answers may vary. One phrase is *5 less than six times a number.*

31) Answers may vary. One statement is *the quotient of a number and 4 is -3.*

32) Answers may vary. One expression is $-20a + a$.

Chapter 2: Cumulative Review for Chapters 1 and 2

1) one thousand, eight hundred, seventy-three

2) 45,002,009

3) a) 9,981,000   b) 10,000,000   c) 10,000,000

4) a) It is the distance of a number from 0.
   b) 15   c) 8   d) -3

5) -6708 6) 17   7) -346 8) -178

9) 6   10) 120   11) a) 64   b) -9 c) 32   d) 7

12) Answers may vary.   13) 2   14) -26

15) -15 16) 0   17) $567

18) a) terms: $x^2, 3y^2, -7x, 2y, -4$; coefficients: 1, 3, -7, 2; constant: -4 b) -6

19) $w + (9 + 5)$; $w + 14$

20) Like terms are terms that have the same variables with the same exponents.

21) $14p - 13$ 22) $h = -12$ 23) $n = 5$

24) a) $\dfrac{x}{8}$ b) $x - 5$ c) $-4x$ d) $x + 9$ or $9 + x$

25) 15 years old

Chapter 3

Section 3.1

1) The numerator is 4, and the denominator is 7.

3)

5)

7)

9)
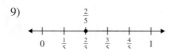

11)

13)
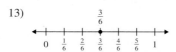

15)

17)

19)

21)

23)

25) Answers may vary. One figure is

27) $\dfrac{1}{2} = \dfrac{2}{4}$ 29) $\dfrac{1}{3} = \dfrac{2}{6}$ 31) $\dfrac{4}{6} = \dfrac{2}{3}$

33) a) improper fraction   b) proper fraction
c) proper fraction

35) a) improper fraction   b) proper fraction
c) improper fraction

37) The numerator of an improper fraction is greater than or equal to the denominator. The numerator of a proper fraction is less than the denominator.

39) Answers may vary. One fraction is $\dfrac{13}{10}$.

41)

43)

45)

47)

49) Answers may vary. One group of figures is
 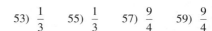

51) to the left of 0

53) $\dfrac{1}{3}$ 55) $\dfrac{1}{3}$ 57) $\dfrac{9}{4}$ 59) $\dfrac{9}{4}$

61)

63)

65)

67)

69)

71)

Answers may vary.

73)

75)

Section 3.2

1) 3 and 5   3) 2, 4, 5, and 10   5) 3 and 5

7) 1, 2, 4, and 8   9) 1, 3, 5, and 15   11) 1 and 17

13) 1, 3, 7, 9, 21, and 63   15) Answers may vary.

17) 2   19) prime   21) composite

23) prime   25) composite

27) Writing the prime factorization of a number means writing the number as a product of its prime factors.

29) $15 = 3 \cdot 5$ 31) $24 = 2 \cdot 2 \cdot 2 \cdot 3$ or $2^3 \cdot 3$

33) $78 = 2 \cdot 3 \cdot 13$ 35) $270 = 2 \cdot 3 \cdot 3 \cdot 3 \cdot 5$ or $2 \cdot 3^3 \cdot 5$

37) $1300 = 2 \cdot 2 \cdot 5 \cdot 5 \cdot 13$ or $2^2 \cdot 5^2 \cdot 13$

39) $8 = 2 \cdot 2 \cdot 2$ or 2^3 41) $99 = 3 \cdot 3 \cdot 11$ or $3^2 \cdot 11$

43) $330 = 2 \cdot 3 \cdot 5 \cdot 11$

45) $495 = 3 \cdot 3 \cdot 5 \cdot 11$ or $3^2 \cdot 5 \cdot 11$

47) a) 1, 2, 3, 4, 6, 8, 12, and 24   b) $24 = 2 \cdot 2 \cdot 2 \cdot 3$ or $2^3 \cdot 3$

49) A fraction is in lowest terms if the numerator and denominator have no common factors other than 1.

51) yes   53) no   55) no

57) Answers may vary.   59) $\dfrac{5}{8}$ 61) $\dfrac{3}{7}$

63) $-\dfrac{1}{2}$ 65) $\dfrac{11}{7}$ 67) 5   69) $-\dfrac{3}{7}$

71) $\dfrac{2m}{3}$ 73) $\dfrac{12}{17t^2}$ 75) $\dfrac{2}{5}$ 77) $-\dfrac{4}{9}$

79) $\dfrac{1}{12}$ 81) $\dfrac{1}{4t}$ 83) $\dfrac{4x^4}{9}$ 85) $-\dfrac{17h^2}{13}$

87) yes   89) no   91) yes   93) yes   95) no

97) yes   99) true   101) false   103) false

Section 3.3

1) multiplication   3) 72   5) 4

7) $\dfrac{15}{32}$ 9) $-\dfrac{7}{26}$ 11) $\dfrac{1}{10}$

13) Answers may vary. Two fractions are $\dfrac{11}{8}\cdot\dfrac{5}{6}$.

15) $\dfrac{28}{45}$ 17) $\dfrac{1}{8}$ 19) $\dfrac{3}{35}$ 21) $-\dfrac{5}{84}$ 23) $\dfrac{y}{x^2}$

25) $6v^4$ 27) $-\dfrac{1}{2}$ 29) $\dfrac{9}{5}$

31) Write the whole number with a denominator of 1, then multiply.

33) 12   35) -49 37) $-\dfrac{60x}{7}$ 39) $-\dfrac{2}{3h}$

41) 1   43) Answers may vary.   45) $\dfrac{5}{4}$ 47) $\dfrac{2}{9k}$

49) -7 51) $\dfrac{1}{2}$ 53) Answers may vary.

55) $\dfrac{3}{4}$ 57) $-\dfrac{63}{80}$ 59) $\dfrac{21y}{4}$ 61) $\dfrac{3d}{7}$

63) $-\dfrac{1}{12}$ 65) $18x$

67) It is a fraction that contains a fraction in its numerator, its denominator, or both.

69) $\dfrac{6}{7}$ 71) $\dfrac{7}{10}$ 73) $-\dfrac{55u}{6t^2}$ 75) $\dfrac{1}{100}$

77) There are 24 female students and 16 male students.

79) 17,020 were female; 19,980 were male

81) $\dfrac{1}{12}$ of dinner plates are vegan; 3 out of 36 dinner plates are vegan.

83) \$354   85) \$826   87) 48 cupcakes

89) $\dfrac{1}{8}$ of a quart   91) 24 mi   93) 36 gifts

95) $-\dfrac{22}{9w}$ 97) a^2b^3 99) $-\dfrac{4}{5}$ 101) $\dfrac{1}{24}$

103) Parents deposited \$2400; total amount is \$6000.

105) 150 songs

Section 3.4

1)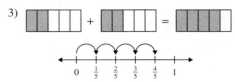

$\dfrac{5}{6}$; Shaded answer may vary.

3)

$\dfrac{4}{5}$; Shaded answer may vary.

5) *Like* fractions have the same denominator, but *unlike* fractions have different denominators.

7) Is the answer in lowest terms?

9) $\dfrac{8}{11}$ 11) $\dfrac{1}{2}$ 13) $\dfrac{4}{13}$ 15) $\dfrac{3}{5}$

17) $-\dfrac{5}{7}$ 19) -3 21) $-\dfrac{1}{3}$ 23) 1

25) $\dfrac{5}{9y}$ 27) $\dfrac{1}{5c}$ 29) $\dfrac{1}{2z}$ 31) $\dfrac{a+3}{4}$

33) $\dfrac{2}{7}$ 35) $\dfrac{6}{3}$ or 2   37) $\dfrac{2}{6}$ 39) $\dfrac{3}{6},\dfrac{2}{4},\dfrac{1}{2}$

41) $\dfrac{4}{32}$ 43) $\dfrac{21}{27}$ 45) $-\dfrac{24}{36}$ 47) $-\dfrac{33}{45}$

49) $\dfrac{90}{96}$ 51) $\dfrac{4p}{7p}$ 53) $-\dfrac{20k}{8k}$ 55) $\dfrac{7c}{35}$

57) Answers may vary.   59) 12   61) 20   63) 21

65) 60   67) 75   69) 36   71) 40

73) 108   75) 180   77) 360   79) 168

81) It is the least common multiple of the denominators.

83) LCD $= 14;\ \dfrac{4}{7}=\dfrac{8}{14};\ \dfrac{3}{14}$ already has the LCD

85) LCD $= 24;\ \dfrac{3}{8}=\dfrac{9}{24},\ \dfrac{5}{12}=\dfrac{10}{24}$

87) LCD $= 24;\ -\dfrac{7}{8}=-\dfrac{21}{24},\ -\dfrac{5}{6}=-\dfrac{20}{24}$

89) LCD $= 54;\ \dfrac{2}{9}=\dfrac{12}{54},\ \dfrac{31}{54}$ already has the LCD

91) LCD $= 60;\ \dfrac{7}{12}=\dfrac{35}{60},\ \dfrac{3}{20}=\dfrac{9}{60}$

93) LCD $= 252;\ -\dfrac{19}{28}=-\dfrac{171}{252},\ -\dfrac{7}{18}=-\dfrac{98}{252}$

95) $LCD = 260; \dfrac{1}{52} = \dfrac{5}{260}, \dfrac{42}{65} = \dfrac{168}{260}$

97) $LCD = 10; \dfrac{w}{10}$ already has the LCD, $\dfrac{3}{5} = \dfrac{6}{10}$

99) $LCD = 18; \dfrac{1}{6} = \dfrac{3}{18}, \dfrac{x}{9} = \dfrac{2x}{18}$

101) $LCD = 8c; \dfrac{3}{8} = \dfrac{3c}{8c}, \dfrac{5}{8c}$ already has the LCD

103) $LCD = 30w; \dfrac{9}{10w} = \dfrac{27}{30w}, \dfrac{1}{6} = \dfrac{5w}{30w}$

105) $LCD = 36; \dfrac{5}{12} = \dfrac{15}{36}, \dfrac{5}{6} = \dfrac{30}{36}, \dfrac{4}{9} = \dfrac{16}{36}$

107) $LCD = 20; \dfrac{7}{10} = \dfrac{14}{20}, \dfrac{3}{4} = \dfrac{15}{20}, \dfrac{1}{5} = \dfrac{4}{20}$

109) Answers may vary. Two fractions are $\dfrac{3}{4}$ and $\dfrac{2}{5}$.

Section 3.5

1) $\dfrac{6}{7}$ 3) $\dfrac{2}{5}$ 5) $-\dfrac{1}{5}$

7) $LCD = 6; \dfrac{5}{6}$ 9) $LCD = 12; \dfrac{2}{3}$

11) Answers may vary.   13) $\dfrac{3}{4}$ 15) $\dfrac{7}{6}$ 17) 1

19) $-\dfrac{1}{24}$ 21) $\dfrac{23}{60}$ 23) $\dfrac{7}{8}$ 25) $\dfrac{35 + 4x}{5x}$

27) $\dfrac{22k + 3}{6k}$ 29) $\dfrac{3a + 32}{36}$ 31) $\dfrac{6 + 7c}{c^2}$

33) $\dfrac{5}{21}$ 35) $\dfrac{1}{3}$ 37) $-\dfrac{22}{45}$ 39) $\dfrac{13}{30}$

41) $-\dfrac{91}{132}$ 43) $\dfrac{9m - 10}{10m}$ 45) $\dfrac{4h - 15}{24}$

47) when the denominators are large   49) $\dfrac{79}{90}$

51) $-\dfrac{1}{168}$ 53) $\dfrac{23}{168}$ 55) $\dfrac{21}{4}$ 57) $\dfrac{28 + a}{7}$

59) $\dfrac{7}{9}$ 61) $\dfrac{z - 15}{5}$ 63) $\dfrac{-16 + w}{8}$ or $\dfrac{w - 16}{8}$

65) $\dfrac{7}{18}$ yd   67) $\dfrac{5}{12}$ ft

69) The lifeguard applicants travel a total distance of $\dfrac{7}{10}$ mi.

71) The remaining fractional amount of the scholarship is $\dfrac{7}{24}$.

73) Fiona's car used $\dfrac{8}{15}$ of its battery's electric capacity on her way to work.

75) $\dfrac{3}{20}$ of the seniors chose math or science as an area of study.

77) 93 seniors chose math or science as an area of study.

79) $\dfrac{3}{5}$ of the seniors did not choose business as an area of study.

81) Exactly one-half the senior class chose business or science as their area of study.

83) $-\dfrac{3}{4}$ 85) $\dfrac{23}{40}$ 87) $\dfrac{33}{10}$

89) $\dfrac{k + 21}{7}$ 91) $\dfrac{3a + 28}{12a}$

93) $\dfrac{4}{5}$ of the students passed the course with a C or better.

95) 144 students passed the course with a C or better.

Section 3.6

1)

3)

5)

7)

9) Answers may vary.   11) $\dfrac{4}{3}$ 13) $\dfrac{13}{5}$ 15) $\dfrac{58}{9}$

17) $\dfrac{65}{6}$ 19) proper   21) proper   23) $1\dfrac{2}{3}$

25) $2\dfrac{1}{4}$ 27) $6\dfrac{4}{9}$ 29) $12\dfrac{4}{5}$ 31) 3

33) Answers may vary.   35) $3\dfrac{4}{7}$ 37) $9\dfrac{4}{5}$

39) -15 41) $17\dfrac{1}{3}$ 43) $\dfrac{4}{5}$

45) Answers may vary.   47) $1\dfrac{3}{5}$ 49) $-\dfrac{2}{3}$

51) $\dfrac{27}{32}$ 53) $1\dfrac{5}{8}$ 55) $3\dfrac{1}{3}$ 57) $-1\dfrac{1}{4}$

59) To add mixed numbers, add the whole-number parts and add the fractional parts. To subtract, subtract the whole-number parts and subtract the fractional parts. In both cases, write the final answer in lowest terms.

61) $2\dfrac{5}{9}$ 63) $4\dfrac{9}{14}$ 65) $11\dfrac{3}{4}$ 67) $2\dfrac{7}{9}$ 69) $5\dfrac{13}{21}$

71) $-1\dfrac{7}{24}$ 73) $8\dfrac{17}{30}$ 75) $-14\dfrac{43}{72}$

77) No. It is not in simplest form because the fractional part is an improper fraction.

79) $4\dfrac{2}{5}$ 81) $3\dfrac{1}{7}$ 83) $7\dfrac{1}{3}$ 85) $13\dfrac{1}{3}$ 87) $28\dfrac{3}{40}$

89) $7\dfrac{3}{2}$ 91) $8\dfrac{29}{16}$ 93) $3\dfrac{1}{2}$ 95) $7\dfrac{11}{30}$ 97) $\dfrac{47}{84}$

99) $\dfrac{3}{5}$ 101) $2\dfrac{4}{7}$ 103) $5\dfrac{3}{4}$ 105) $\dfrac{11}{15}$

107) $-4\dfrac{1}{12}$ 109) $-3\dfrac{23}{24}$ 111) Answers may vary.

113) $2\dfrac{1}{4}$ in.   115) $2\dfrac{5}{6}$ lb   117) gain; $\dfrac{3}{4}$ lb

119) 16 squares   121) $37\dfrac{1}{2}$ hr per week

123) a) $18\dfrac{1}{4}$ hr   b) $2\dfrac{3}{4}$ hr more

125) $1\dfrac{1}{8}$ cups   127) 72 megabytes   129) 35 patties

Putting It All Together Exercises

1) False. The denominator cannot equal 0.

2) False. In lowest terms, $\dfrac{7}{1} = 7$.

3) False. 35 has factors other than 1 and 35. Other factors are 5 and 7.

4) true   5) $\dfrac{8}{15}$ 6) $\dfrac{1}{4}$ 7) $5t$ 8) $\dfrac{14}{55m}$

9) yes   10) no   11) $\dfrac{138}{11}$ 12) $\dfrac{25}{8}$

13) $14\dfrac{11}{12}$ 14) $1\dfrac{3}{7}$

15) False. A common denominator is needed only for adding and subtracting fractions.

16) true   17) true

18) False. The fractional part of a mixed number must be a proper fraction. In simplest form, $3\dfrac{11}{6} = 4\dfrac{5}{6}$.

19) False. $\dfrac{1}{2}$ is less than 1 whole, and $\dfrac{3}{2}$ is greater than 1 whole, so $\dfrac{1}{2}$ is less than $\dfrac{3}{2}$.

20) true   21) $-\dfrac{5}{24}$ 22) $\dfrac{2}{5}$ 23) $9\dfrac{15}{28}$

24) $-\dfrac{44}{27}$ or $-1\dfrac{17}{27}$ 25) $\dfrac{11-3z}{18}$ 26) $15\dfrac{17}{30}$

27) $\dfrac{1}{6}$ 28) $\dfrac{93}{8}$ or $11\dfrac{5}{8}$ 29) $1\dfrac{17}{36}$ 30) $\dfrac{37}{55}$

31) $\dfrac{25}{2}$ or $12\dfrac{1}{2}$ 32) $-\dfrac{103}{72}$ or $-1\dfrac{31}{72}$ 33) 80

34) $\dfrac{4}{9}$ 35) $6\dfrac{3}{4}$ 36) $\dfrac{25}{72}$ 37) $\dfrac{72}{11}$ or $6\dfrac{6}{11}$

38) $\dfrac{633}{560}$ or $1\dfrac{73}{560}$ 39) $-\dfrac{31}{44}$ 40) $\dfrac{5}{8}$

41) $\dfrac{53}{48}$ or $1\dfrac{5}{48}$ 42) $-2\dfrac{23}{42}$ 43) $-\dfrac{20}{7}$ or $-2\dfrac{6}{7}$

44) $12\dfrac{17}{45}$ 45) $8\dfrac{5}{12}$ 46) $\dfrac{5d^2}{39}$ 47) $\dfrac{3}{22}$

48) $8\dfrac{13}{16}$ 49) $\dfrac{28r+15}{40r}$ 50) $-\dfrac{1}{x^3y^2}$

51) $11\dfrac{5}{24}$ 52) -24 53) a) 8   b) 24

54) a) 13   b) $\dfrac{1}{6}$ 55) a) 56   b) $\dfrac{8}{9}$

56) a) 12   b) $\dfrac{2}{15}$

57) Answers may vary. Two fractions are $\dfrac{10}{29} + \dfrac{8}{29}$.

58) Answers may vary. Two fractions are $\dfrac{5}{7} - \dfrac{3}{7}$.

59) Answers may vary. Two fractions are $\dfrac{3}{10} \div \dfrac{5}{7}$.

60) Answers may vary. Two fractions are $-\dfrac{5}{11} \cdot \dfrac{2}{3}$.

61) Answers may vary. Two mixed numbers are $6\dfrac{10}{21} - 2\dfrac{5}{21}$.

62) Answers may vary. Two mixed numbers are $5\dfrac{4}{27} + 4\dfrac{7}{27}$.

63)
$$-3\tfrac{1}{2} \quad -1\tfrac{5}{6} \quad -\tfrac{5}{8} \quad \tfrac{3}{4} \quad 2\tfrac{1}{5} \quad\quad 4\tfrac{2}{3}$$
$$-5\;-4\;-3\;-2\;-1\;\;0\;\;1\;\;2\;\;3\;\;4\;\;5$$

64)
$$-4\tfrac{1}{2} \quad -2\tfrac{3}{4} \quad\quad -\tfrac{2}{3}\;\;\tfrac{1}{4} \quad 1\tfrac{4}{5} \quad 3\tfrac{3}{8}$$
$$-5\;-4\;-3\;-2\;-1\;\;0\;\;1\;\;2\;\;3\;\;4\;\;5$$

65) 3 bottles   66) 16 strips   67) $1\dfrac{3}{4}$ mi

68) $3\dfrac{11}{12}$ hr   69) 20 min   70) 372 free throws

Section 3.7

1) <   3) >   5) <   7) >   9) <

11) <   13) >   15) <   17) <

19) >   21) Answers may vary.   23) <

25) >   27) >   29) <   31) >   33) >

35) $<$ 37) $<$ 39) $\left(\dfrac{1}{5}\right)^4$ 41) $\left(-\dfrac{3}{4}\right)^6$

43) $\dfrac{1}{64}$ 45) $\dfrac{25}{36}$ 47) $\dfrac{100}{81}$ 49) $\dfrac{8}{27}$

51) $-\dfrac{1}{64}$ 53) $\dfrac{1}{81}$ 55) $\dfrac{1000}{27}$

57) $\dfrac{16}{81}$ 59) $<$ 61) $>$ 63) $=$

65) Answers may vary.   67) a) $\dfrac{1}{64}$ b) $-\dfrac{1}{64}$

69) a) $-\dfrac{8}{27}$ b) $-\dfrac{8}{27}$ 71) a) $-\dfrac{1}{16}$ b) $\dfrac{1}{16}$

73) $\dfrac{100}{49}$ or $2\dfrac{2}{49}$ 75) $\dfrac{17}{30}$ 77) $\dfrac{35}{36}$ 79) -1

81) $-\dfrac{11}{8}$ or $-1\dfrac{3}{8}$ 83) $\dfrac{10}{3}$ or $3\dfrac{1}{3}$ 85) $-\dfrac{5}{96}$

87) $\dfrac{20}{81}$ 89) $\dfrac{121}{144}$ 91) $\dfrac{9}{4}$ or $2\dfrac{1}{4}$ 93) $\dfrac{17}{32}$

Section 3.8

1) $n = -4$ 3) $p = 18$ 5) $z = 9$

7) $w = \dfrac{1}{10}$ 9) $b = -\dfrac{1}{4}$ 11) $a = \dfrac{11}{18}$

13) $d = \dfrac{11}{35}$ 15) $m = \dfrac{19}{8}$ 17) $-\dfrac{3}{2}$ 19) $a = 5$

21) $d = 28$ 23) $x = -45$ 25) $h = \dfrac{16}{3}$

27) $q = 24$ 29) $w = \dfrac{7}{2}$ 31) $w = -\dfrac{1}{3}$

33) $x = -14$ 35) $r = 32$ 37) $p = -\dfrac{2}{21}$

39) $a = \dfrac{1}{30}$ 41) $d = \dfrac{11}{18}$ 43) $g = \dfrac{1}{2}$ 45) $x = -\dfrac{5}{3}$

47) $r = -10$ 49) $q = \dfrac{5}{2}$ 51) $w = -\dfrac{6}{11}$

53) $z = 12$ 55) $t = \dfrac{3}{2}$ 57) Answers may vary.

59) $t = -\dfrac{3}{20}$ 61) $x = \dfrac{1}{2}$ 63) $n = -\dfrac{1}{5}$

65) $y = 4$ 67) $k = -12$ 69) $h = \dfrac{10}{3}$

71) An expression does not contain an equal sign. An equation contains an equal sign.

73) a) equation; $r = 8$ b) expression; $\dfrac{r+4}{8}$

75) a) expression; $\dfrac{5x}{12}$ b) equation; $x = 2$ 77) $w = 6$

79) $k = -\dfrac{13}{20}$ 81) $d = 60$ 83) $p = \dfrac{4}{3}$

85) $\dfrac{1}{2}x$ 87) $\dfrac{2}{3}x + 7$ or $7 + \dfrac{2}{3}x$ 89) $\dfrac{1}{3}x - 4$

91) $\dfrac{1}{2}x + 1 = 8$; 14   93) $\dfrac{3}{4}x = -63$; -84

95) Answers may vary. One statement is *2 less than one-third of a number is 9.*

97) decaf: 37 cups, regular: 74 cups   99) 20 ft, 30 ft

Chapter 3: Group Activity

Equation Set #1	Equation Set #2
A: $x = \dfrac{5}{8}$	A: $x = 1$
B: $x = -\dfrac{1}{3}$	B: $x = 0$
C: $x = -\dfrac{1}{2}$	C: $x = -\dfrac{1}{5}$
D: $x = -2$	D: $x = -2$
E: $x = \dfrac{3}{2}$	E: $x = 4$

Smallest to largest: Smallest to largest:
D, C, B, A, E   D, C, B, A, E

Chapter 3: Review Exercises

1) $\dfrac{3}{4}$;

3) It cannot equal 0. If the denominator of a fraction is 0, we say that it is undefined.

5) proper   7) $\dfrac{8}{11}$

9)

11) The sum of the digits is divisible by 3. One such example is 4152. Examples may vary.

13) composite   15) $2 \cdot 3 \cdot 7$ 17) $2^3 \cdot 3^2 \cdot 5^2$

19) no   21) $\dfrac{2}{5}$ 23) $-\dfrac{11}{14}$ 25) $-\dfrac{7z}{11}$

27) $\dfrac{1}{3a^3}$ 29) no

31) Multiply the first fraction by the reciprocal of the second fraction.

33) $-\dfrac{9}{14}$ 35) $39x$ 37) $\dfrac{9}{22}$ 39) $\dfrac{b^2}{2a^2}$

41) $-\dfrac{27}{28m}$ 43) $\dfrac{1}{9}$ 45) $\dfrac{1}{8}$ of a liter

47) *Like* fractions have the same denominator, but *unlike* fractions have different denominators.

49) $\dfrac{8}{9}$ 51) 1   53) $-\dfrac{2}{3y}$ 55) $\dfrac{15}{50}$ 57) 144

59) LCD = 40; $\dfrac{7}{10} = \dfrac{28}{40}, \dfrac{5}{8} = \dfrac{25}{40}$

61) Answers may vary.

63) $\dfrac{3}{8}$ 65) $-\dfrac{1}{20}$ 67) $\dfrac{5h+36}{45}$ 69) $\dfrac{15x-2}{20x}$

71) $-\dfrac{51}{140}$ 73) $\dfrac{7}{11}$ 75) $-\dfrac{1}{15}$

77) The total amount of rain during May, June, and July was $\dfrac{23}{80}$ in.

79) $2\dfrac{1}{3}$

81)

83) $\dfrac{22}{5}$ 85) $8\dfrac{2}{3}$ 87) addition and subtraction

89) $1\dfrac{17}{26}$ 91) $-6\dfrac{1}{2}$ 93) $9\dfrac{13}{24}$ 95) 39

97) $14\dfrac{1}{8}$ lb   99) <   101) >   103) $\dfrac{4}{9}$

105) $-\dfrac{4}{9}$ 107) $-\dfrac{1}{30}$ 109) $\dfrac{77}{75}$ or $1\dfrac{2}{75}$

111) $-\dfrac{9}{8}$ or $-1\dfrac{1}{8}$ 113) a) $t = -\dfrac{7}{20}$ b) $t = -\dfrac{7}{20}$

115) $p = -21$ 117) $k = 20$ 119) $x = \dfrac{2}{33}$

121) $n = -6$ 123) a) expression; $\dfrac{d}{10}$ b) equation; $d = 2$

125) $-\dfrac{1}{6}$ 127) $-\dfrac{4k^3}{9}$ 129) $\dfrac{45}{88}$

Chapter 3: Test

1) never   2) always   3) sometimes   4) always

5) always   6) sometimes   7) never   8) sometimes

9) $\dfrac{5}{8}$ 10) $3\dfrac{1}{4}$

11)

12) a) 1, 2, 3, 4, 6, 8, 12, 24   b) $2^3 \cdot 3$ or $2 \cdot 2 \cdot 2 \cdot 3$

13) $-\dfrac{2}{3}$ 14) $\dfrac{17w^3}{35}$

15) Yes. Each can be written as $\dfrac{3}{4}$ in lowest terms.

16) a) $\dfrac{36}{63}$ b) $-\dfrac{15z}{18z}$

17) No. It is not in lowest terms because the fractional part of the mixed number is improper. In lowest terms, $6\dfrac{5}{3} = 7\dfrac{2}{3}$.

18) a) no   b) Multiply the first fraction by the reciprocal of the second fraction.

19) $\dfrac{1}{3}$ 20) $\dfrac{3}{10}$ 21) $7\dfrac{37}{56}$ 22) $\dfrac{2}{3x}$

23) $\dfrac{5c}{27d^2}$ 24) $\dfrac{16}{15}$ or $1\dfrac{1}{15}$ 25) $-\dfrac{2}{9}$ 26) $17\dfrac{11}{40}$

27) $-\dfrac{1}{18}$ 28) $-12r$ 29) $\dfrac{3}{32}$ 30) $-1\dfrac{1}{3}$

31) $-\dfrac{41}{30}$ or $-1\dfrac{11}{30}$ 32) $\dfrac{a-12}{6}$ 33) $\dfrac{8m-3}{2m^2}$

34) $\dfrac{24z+5}{40z}$ 35) >   36) <

37) a) $\dfrac{8}{125}$ b) $\dfrac{9}{64}$ c) $-\dfrac{9}{64}$

38) $-\dfrac{5}{18}$ 39) $\dfrac{5}{12}$ 40) $-\dfrac{1}{216}$ 41) 8 oz

42) 200   43) a) $1\dfrac{7}{12}$ tsp   b) $\dfrac{5}{12}$ tsp   44) $p = 24$

45) $n = -\dfrac{7}{30}$ 46) $x = -9$ 47) $y = \dfrac{2}{11}$

48) $h = -25$ 49) $c = 8$ 50) $\dfrac{1}{2}x - 9$

Chapter 3: Cumulative Review for Chapters 1–3

1) 2636   2) 1726   3) −319,652   4) 601 R17

5) 680; 700; 1000   6) 58,400; 58,400; 58,000

7) $445,234   8) 13   9) 60   10) $\dfrac{19}{62}$

11) $\dfrac{35}{12}$ or $2\dfrac{11}{12}$ 12) $\dfrac{21x-5}{30x}$ 13) $12\dfrac{5}{8}$

14) $-\dfrac{5}{12c}$ 15) $-\dfrac{10}{33}$ 16) $\dfrac{29}{40}$ 17) $\dfrac{4k-15}{36}$

18) true   19) true   20) $3\dfrac{7}{24}$ in.   21) $y = -6$

22) $v = -\dfrac{3}{2}$ 23) $w = \dfrac{1}{4}$ 24) $a = 4$

25) first game: 19,000 points; second game: 25,000 points

Chapter 4

Section 4.1

1) Answers may vary.   3) line; $\overleftrightarrow{TU}$ or $\overleftrightarrow{UT}$

5) line segment; $\overline{KM}$ or $\overline{MK}$ 7) ray; $\overrightarrow{UV}$

9) ray; $\overrightarrow{KJ}$ 11) Answers may vary.

13) An obtuse angle has a measure greater than 90° and less than 180°.

15) A right angle has a measure of exactly 90°, and it is indicated by a small square in the angle.

17) ∠1: ∠TZU or ∠UZT
    ∠2: ∠UZN or ∠NZU
    ∠3: ∠NZP or ∠PZN
    ∠4: ∠PZT or ∠TZP

19) ∠1: ∠DTW or ∠WTD
    ∠2: ∠WTP or ∠PTW
    ∠3: ∠PTV or ∠VTP
    ∠4: ∠VTD or ∠DTV

21) ∠1: ∠EBA or ∠ABE
    ∠2: ∠ABC or ∠CBA
    ∠3: ∠CBD or ∠DBC
    ∠4: ∠DBE or ∠EBD

23) acute   25) obtuse   27) straight

29) right   31) Answers may vary.   33) 120°

35) parallel; $\overleftrightarrow{GK} \parallel \overleftrightarrow{LP}$ 37) neither

39) perpendicular; $c \perp d$ 41) obtuse

43) right   45) $\overrightarrow{EF}$ 47) true   49) false

51) true   53) false

Section 4.2

1) 414   3) $2\frac{3}{4}$ 5) $6\frac{7}{8}$

7) It is the distance around the rectangle. $P = 2l + 2w$

9) 18 in.; 18 in^2 11) $1\frac{3}{4}$ yd; $\frac{3}{32}$ yd^2 13) 20 in.; $22\frac{3}{4}$ in^2

15) 1523 in.; 29,880 in^2 17) 2 ft   19) $2\frac{1}{2}$ ft

21) 48 cm; 144 cm^2 23) $2\frac{2}{5}$ m; $\frac{9}{25}$ m^2

25) $5\frac{1}{2}$ m; $1\frac{57}{64}$ m^2 27) 5 ft

29) $346\frac{2}{3}$ yd; 6400 yd^2 31) 8100 ft^2; 360 ft

33)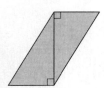

35) 40 ft; 78 ft^2 37) 30 cm; 24 cm^2

39) $9\frac{3}{4}$ ft; $3\frac{3}{16}$ ft^2 41) always   43) 1620 in^2

45) $Area = \frac{1}{2} \cdot height \cdot (short\ base + long\ base)$

or $A = \frac{1}{2}h(b + B)$

47) 48 cm; 117 cm^2 49) 75 yd; 342 yd^2

51) $19\frac{1}{2}$ ft; $26\frac{1}{8}$ ft^2 53) never   55) 50 in.

57) 20 cm; 12 cm^2 59) 42 m; 104 m^2 61) $6800

63) $2\frac{2}{3}$ mi; $\frac{4}{9}$ mi^2 65) 53 ft; $74\frac{1}{4}$ ft^2

67) 50 cm; 78 cm^2 69) $108   71) $1836

Section 4.3

1) $A = \frac{1}{2} \cdot base \cdot height$ or $A = \frac{1}{2}bh$ 3) 16 yd

5) 35 in.   7) $5\frac{5}{8}$ m^2 9) $9\frac{3}{4}$ cm^2 11) 60 ft^2

13) 342 in^2 15) 30 m; 42 m^2 17) 60 yd; 165 yd^2

19) 20 in^2; 20 in^2; 40 in^2; The sum of the areas of the two triangles equals the area of the rectangle.

21) 180°   23) m∠x = 59°   25) m∠x = 36°

27) m∠x = 13°   29) m∠x = 62°

31) a triangle with three acute angles

33) a triangle with three sides of different lengths

35) acute   37) obtuse   39) right   41) isosceles

43) scalene   45) isosceles   47) equilateral

49) false   51) true   53) 72 ft^2 55) $26,880

57) Answers may vary.   59) Answers may vary.

Section 4.4

1) 98   3) $45\frac{5}{6}$ 5) $39\frac{38}{125}$

7) $V = lwh$ 9) 72 cm^3 11) 216 cm^3 13) $36\frac{24}{25}$ in^3

15) 25 mm^3 17) 28 in^3 19) $\frac{112}{3}$ m^3 or $37\frac{1}{3}$ m^3

21) $\frac{15}{4}$ yd^3 or $3\frac{3}{4}$ yd^3 23) 2,592,100 m^3 25) $1397\frac{1}{2}$ in^3

27) 3 cm   29) 62 ft^2 31) 434 cm^2 33) 394 in^2

35) $V = 420$ m^3; $SA = 368$ m^2

37) $V = \frac{8}{27}$ yd^3; $SA = 2\frac{2}{3}$ yd^2

Putting It All Together Exercises

1) ray; $\overrightarrow{RT}$ 2) line segment; $\overline{AB}$ or $\overline{BA}$

3) line; $\overleftrightarrow{PN}$ or $\overleftrightarrow{NP}$ 4) Answers may vary.

5) obtuse; $\angle ABC$, $\angle CBA$ or $\angle B$ 6) right; $\angle H$

7) acute; $\angle V$ 8) straight; $\angle XYZ$, $\angle ZYX$ or $\angle Y$

9) Answers may vary.   10) Answers may vary.

11) $A = lw$; $P = 2l + 2w$ 12) $A = \frac{1}{2}bh$

13) area: 30 cm^2; perimeter: 30 cm

14) area: $\frac{15}{16}$ yd^2; perimeter: $4\frac{1}{4}$ yd

15) area: $\frac{1}{16}$ ft^2; perimeter: 1 ft

16) area: 104 m^2; perimeter: 46 m

17) area: 34 in^2; perimeter: 26 in.

18) area: 18 cm^2; perimeter: $21\frac{1}{5}$ cm

19) area: 76 m^2; perimeter: 44 m

20) area: 126 ft^2; perimeter: 54 ft

21) 82 in^2 22) 325 cm^2 23) 180°

24) a triangle with an obtuse angle

25) a triangle with all of the sides of the same length

26) a triangle with two sides of the same length

27) never   28) never   29) always   30) never

31) $m\angle x = 26°$ 32) $m\angle x = 135°$

33) $m\angle x = 72°$ 34) $m\angle x = 23°$

35) right   36) obtuse   37) acute

38) right   39) scalene   40) equilateral

41) isosceles   42) isosceles

43) $V = \frac{3}{32}$ m^3; $SA = 1\frac{3}{8}$ m^2

44) $V = 3\frac{3}{8}$ ft^3; $SA = 13\frac{1}{2}$ ft^2

45) 108 in^3 46) 80 cm^3 47) $7800

48) 48 in^2 49) $2864   50) 604,200 mm^3

Section 4.5

1) $P = 2l + 2w$ 3) $A = \frac{1}{2}bh$ 5) $a + b + c = 180°$

7) feet   9) 9 ft   11) 12 cm   13) 17 in.

15) $\frac{3}{4}$ m   17) 3 ft   19) 11 mm

21) Answers may vary.   23) 13 cm   25) 4 m

27) $3\frac{1}{2}$ in.   29) 53 cm   31) 36 in.   33) 6 cm

35) width: 5 ft, length: 9 ft   37) 20 ft by 26 ft

39) 85 ft by 170 ft   41) 10 in., 18 in., and 18 in.

43) 180°   45) $m\angle C = 25°$

47) $m\angle L = 71°$, $m\angle N = 28°$

49) $m\angle U = 68°$, $m\angle V = 68°$

51) $m\angle B = 65°$, $m\angle V = 25°$

53) $m\angle F = 40°$, $m\angle G = 120°$, $m\angle H = 20°$

Chapter 4: Group Activity

1) Mrs. Buyer is correct. She calculated the perimeter of the room and Mr. Buyer calculated the area of the room. The floor molding covers the distance around the room so the formula for perimeter should be used.

2) Answers may vary.

3)

Room	Length	Width	Calculations: $2l + 2w = P$	Perimeter
Living	24 ft	15 ft	$2(24) + 2(15) = 48 + 30 = 78$	78 ft
Dining	16 ft	12 ft	$2(16) + 2(12) = 32 + 24 = 56$	56 ft
Family	30 ft	16 ft	$2(30) + 2(16) = 60 + 32 = 92$	92 ft
Bedroom 1	18 ft	15 ft	$2(18) + 2(15) = 36 + 30 = 66$	66 ft
Bedroom 2	14 ft	14 ft	$2(14) + 2(14) = 28 + 28 = 56$	56 ft
Bedroom 3	13 ft	13 ft	$2(13) + 2(13) = 26 + 26 = 52$	52 ft

4) 363 ft

5) 31 lengths; $93

Chapter 4: Review Exercises

1) $\angle 1$: $\angle AGX$ or $\angle XGA$
$\angle 2$: $\angle AGZ$ or $\angle ZGA$
$\angle 3$: $\angle ZGB$ or $\angle BGZ$
$\angle 4$: $\angle BGX$ or $\angle XGB$

3) straight   5) acute   7) neither

9) perpendicular; $\overleftrightarrow{d} \perp \overleftrightarrow{v}$ 11) 36 cm; 80 cm^2

13) 3 ft; $\frac{9}{16}$ ft^2 15) 56 cm; 96 cm^2 17) 4600 yd^2

19) $25\frac{3}{5}$ in.; $24\frac{3}{4}$ in^2 21) 32 m; 49 m^2

23) 50 m   25) $3\frac{1}{2}$ in^2 27) 379 ft^2

29) $336   31) $m\angle x = 38°$ 33) acute

35) scalene   37) $V = 108$ mm^3; $SA = 150$ mm^2

39) $V = \frac{8}{125}$ ft^3; $SA = \frac{24}{25}$ ft^2 41) 288 in^3

43) 5 cm   45) 9 ft   47) 10 ft

49) 11 ft by 20 ft   51) 180°   53) 26°

55) $m\angle R = 35°$, $m\angle S = 55°$ 57) scalene

59) $\overrightarrow{ML}$ 61) false   63) $1400

65) 35,000 ft^3 67) 228 ft^2 69) never

71) always   73) always   75) $30\frac{3}{8}$ in^2

Chapter 4: Test

1) a) ray; $\overrightarrow{NK}$ b) line; $\overleftrightarrow{FG}$ or $\overleftrightarrow{GF}$ c) line segment; $\overline{AR}$ or $\overline{RA}$

2) a) They intersect at 90° angles.   b) They never intersect.

3) An acute angle has a measure less than 90°, but the measure of an obtuse angle is greater than 90° and less than 180°.

4) a) obtuse   b) acute   c) right

5) a) triangle   b) 150 cm^2 c) 60 cm

6) a) trapezoid   b) 138 in^2 c) 48 in.

7) a) rectangle   b) $\frac{253}{6}$ ft^2 or $42\frac{1}{6}$ ft^2 c) $26\frac{1}{3}$ ft

8) a) parallelogram   b) 252 m^2 c) 86 m

9) a) 78 ft^2 b) 44 ft   10) 180°

11) a) acute   b) isosceles   12) a) right   b) scalene

13) 96 m^3 14) $V = 27$ ft^3; $SA = 57$ ft^2 15) $576

16) 54 in^2 17) 6 mm   18) 4 in.   19) 18 in.

20) $5\frac{1}{2}$ ft by $7\frac{1}{2}$ ft   21) $m\angle A = 33°$, $m\angle C = 82°$

22) never   23) always   24) always   25) never

Chapter 4: Cumulative Review for Chapters 1–4

1) 19,646   2) −3568   3) $-\frac{8}{27}$ 4) $\frac{2}{3}$

5) $\frac{5}{8}$ 6) −49   7) $-\frac{1}{18}$ 8) −33   9) 45

10) estimate: 900; exact: 684   11) $-12x - 1$

12) $\frac{5}{24}n - \frac{1}{4}$ 13) $p = -4$ 14) $a = 14$

15) $k = \frac{1}{4}$ 16) $y = -\frac{5}{3}$

17) a) line; $\overleftrightarrow{RT}$ or $\overleftrightarrow{TR}$ b) ray; $\overrightarrow{PN}$

18) 52 mm^2 19) $A = 1$ yd^2; $P = 4\frac{1}{3}$ yd

20) $V = 9$ ft^3; $SA = 27$ ft^2 21) 18

22) 2015: $83 billion; 2016: $92 billion

23) 34 in. and 17 in.

24) $m\angle P = 130°$; $m\angle Q = 36°$; $m\angle R = 14°$ 25) 240 ft

Chapter 5
Section 5.1

1) Decimals represent fractions with denominators that are powers of 10.

3)
```
        9/10
<--+--+--+--+--+--+--+--+--•-+-->
 0  0.1 0.2 0.3 0.4 0.5 0.6 0.7 0.8 0.9 1
```

5)
```
              6/10
<--+--+--+--+--+--•--+--+--+--+-->
 0  0.1 0.2 0.3 0.4 0.5 0.6 0.7 0.8 0.9 1
```

7) $\frac{73}{100} = 0.73$ 9) $\frac{40}{100} = \frac{2}{5} = 0.40 = 0.4$

11)
```
                    0.58
<--+--+--+--+--+--•+--+--+--+--+-->
 0  0.1 0.2 0.3 0.4 0.5 0.6 0.7 0.8 0.9 1
```

13)
```
  0.06
<-+•-+--+--+--+--+--+--+--+--+-->
 0  0.1 0.2 0.3 0.4 0.5 0.6 0.7 0.8 0.9 1
```

15)
```
        0.25
<--+--+--•+--+--+--+--+--+--+--+-->
 0  0.1 0.2 0.3 0.4 0.5 0.6 0.7 0.8 0.9 1
```

17) 0.03   19) 0.81   21) −0.141   23) 0.067

25) 0.0893   27) 0.2051   29) 6.7   31) −38.001

33) 10.533   35) 8.1   37) 4.09   39) −2.443

41) 0—ones, 3—tenths, 5—hundredths, 7—thousandths, 2—ten-thousandths

43) 2—hundreds, 3—tens, 7—ones, 8—tenths, 0—hundredths, 4—thousandths

45) 4—tens, 0—ones, 1—tenths, 6—hundredths, 2—thousandths, 5—ten-thousandths, 9—hundred-thousandths

47) four tenths   49) thirty-six hundredths

51) seven thousandths

53) seven thousand four hundred fifteen ten-thousandths

55) fifty-seven and three tenths

57) negative eight hundred nine and fifty-six hundredths

59) three and five hundred seventy-six ten-thousandths

61) negative six and seventeen hundred-thousandths

63) 0.15   65) 96.7   67) 0.032   69) −8.0004

71) 5005.00005   73) Answers may vary.   75) $\frac{73}{100}$

77) $\frac{207}{10,000}$ 79) $4\frac{9}{10}$ 81) $-\frac{3}{5}$ 83) $-\frac{3}{5}$

85) $\frac{17}{25}$ 87) $9\frac{4}{5}$ 89) $-5\frac{18}{125}$ 91) $1\frac{3}{20,000}$

93) Answers may vary.   95) 8.0014

97) negative two hundred four and eight tenths

99) $\frac{3}{4} = 0.75$ 101) 0.037   103) $-2\frac{21}{25}$

Section 5.2

1)

0.4

0 0.1 0.2 0.3 0.4 0.5 0.6 0.7 0.8 0.9 1

3)

0.0

0 0.1 0.2 0.3 0.4 0.5 0.6 0.7 0.8 0.9 1

5)

0.7

0 0.1 0.2 0.3 0.4 0.5 0.6 0.7 0.8 0.9 1

7) a) 0.7   b) 0.68   9) a) 94.35   b) 90

11) a) 5628.492   b) 6000   13) a) 7600   b) 7620.1838

15) a) 0   b) 0.10   17) a) 5.3   b) 5.27

19) a) 43.91800   b) 43.9180   21) Answers may vary.

23) hundred-thousandths; 0.00059

25) >   27) <   29) =   31) >   33) $5.23

35) $722.12   37) $298.23   39) $1.10   41) $0.67

43) $18   45) $39   47) $43   49) $682

51) $1600   53) $46   55) $0   57) $1

59) $268   61) 1.9   63) 0.0072   65) 7000

67) a) $375.85   b) $376   69) a) $0.36   b) $0

Section 5.3

1) 14.5   3) 0.95   5) 12.964   7) 6.55

9) 21.301   11) 52.312   13) 460.7

15) $379 + \dfrac{9}{10}$ cents or $379 + 0.9$ cents; 3.799 dollars

17) Answers may vary.   19) 5.3   21) 21.16

23) 230.899   25) 61.34   27) 25.16   29) 81.739

31) 38.417   33) 18.793   35) 36.8

37) 25.47   39) 3.4   41) 5.93   43) 121.5048

45) −8.41   47) −0.652   49) 7.3

51) −68.705   53) −0.06   55) −4.53

57) 38.02   59) 882.179   61) −106.125

63) −5.9946   65) 3.97   67) −2.98

69) −107.542   71) $4.33   73) $10.14

75) $3.39   77) $5.00   79) $1.50   81) $2.62

83) −571.12   85) 185.107   87) 11.75

89) 0.51   91) 33.403   93) 26.38

95) $31.37

Section 5.4

1) Answers may vary.   3) 25.83   5) −68.58

7) 312.916   9) 116.754   11) 530.19

13) −2.0093   15) 651.63   17) −0.00072

19) 164,131.2   21) 603   23) 0.125 × 5; 0.625

25) 0.16   27) 1.44   29) 0.0144

31) 0.0025   33) 0.000049

35) Move the decimal point in the number 3 places to the right.

37) 25.87   39) −36.6   41) 0.608   43) 43,670

45) −6400   47) 3.801   49) 19.258

51) −0.00702   53) 9.4   55) −320

57) $657.00   59) $11,160   61) 314.4 lb

63) $68.20   65) $22.71   67) 297.46 mi

69) area: 36 in^2; perimeter: 24.6 in.   71) $14.74

73) a) 38.25 hr   b) $913.79   c) $82.24   d) $776.51

75) 0.0038   77) −1.6974   79) 106.552

81) 1.9881   83) 25,900   85) 0.00175

87) 30   89) $211.20

Section 5.5

1) Answers may vary.   3) 2.94   5) 5.12

7) −0.00528   9) −9.2   11) 23.09   13) 0.3865

15) 3.035   17) 0.4325   19) −1.25   21) 1.157

23) −3.267   25) 0.013   27) −79.771

29) exact: 1.4333… or $1.4\overline{3}$; approximation: 1.433

31) exact: −9.44666… or $-9.44\overline{6}$; approximation: −9.447

33) exact: 0.10222… or $0.10\overline{2}$; approximation: 0.102

35) exact: 0.4272727… or $0.4\overline{27}$; approximation: 0.427

37) Answers may vary.   39) 11.2   41) 6.85

43) −1.5   45) 0.037   47) −5210   49) 490

51) 15.2   53) 13,050   55) −2.81   57) 0.53

59) 36.76   61) −1.73   63) 913.63   65) 4.05

67) −3.7   69) 8.34   71) 3.96   73) −38

75) 0   77) $0.32   79) 108.7 yd per game

81) $2.56 per gal   83) 1300   85) 5.9   87) −200

89) exact: 17.4333… or $17.4\overline{3}$; approximation: 17.43

91) −3.810   93) 0   95) $1437.28

Putting It All Together Exercises

1) 7.9   2) 0.023   3) thirty-one hundredths; $\dfrac{31}{100}$

4) one hundred twenty-five thousandths; $\dfrac{1}{8}$

5) one and six tenths; $1\dfrac{3}{5}$ 6) 2.08   7) 831.6

8) 74.070   9) $34

10) Answers may vary. Write the numbers vertically, lining up the decimal points. Add like you would add whole numbers. Put the decimal point in the answer directly below the decimal in the problem.

11) 445.54   12) 0.0067   13) −0.027

14) −9.03   15) 4100   16) −23.64

17) 66.09931   18) 350.48   19) 421.676

20) 1.96   21) 7.91   22) 64   23) 13.051

24) 24.91   25) 55.38   26) −605.472

27) exact: 47.$\overline{7}$; approximation: 47.778

28) exact: 0.85$\overline{5}$; approximation: 0.856

29) exact: −0.0$\overline{72}$; approximation: −0.073

30) exact: 81.$\overline{81}$; approximation: 81.818

31) 0.23   32) −269.57   33) $38.93

34) 0.194 billion or 194 million   35) 261 tickets

36) $392.37   37) 7.6 in.   38) 37.1 in.

Section 5.6

1) 0.5   3) 0.6   5) −0.125

7) 0.4375   9) −2.4   11) 4.0625

13)

Fraction	Decimal Equivalent	Fraction	Decimal Equivalent
$\frac{1}{16}$	0.0625	$\frac{1}{8}$	0.125
$\frac{2}{16} = \frac{1}{8}$	0.1250	$\frac{2}{8} = \frac{1}{4}$	0.250
$\frac{3}{16}$	0.1875	$\frac{3}{8}$	0.375
$\frac{4}{16} = \frac{2}{8} = \frac{1}{4}$	0.2500	$\frac{4}{8} = \frac{1}{2}$	0.500
$\frac{5}{16}$	0.3125	$\frac{5}{8}$	0.625
$\frac{6}{16} = \frac{3}{8}$	0.3750	$\frac{6}{8} = \frac{3}{4}$	0.750
$\frac{7}{16}$	0.4375	$\frac{7}{8}$	0.875
$\frac{8}{16} = \frac{4}{8} = \frac{1}{2}$	0.5000	$\frac{8}{8} = 1$	1.000
$\frac{9}{16}$	0.5625		
$\frac{10}{16} = \frac{5}{8}$	0.6250		
$\frac{11}{16}$	0.6875		
$\frac{12}{16} = \frac{6}{8} = \frac{3}{4}$	0.7500		
$\frac{13}{16}$	0.8125		
$\frac{14}{16} = \frac{7}{8}$	0.8750		
$\frac{15}{16}$	0.9375		
$\frac{16}{16} = 1$	1.0000		

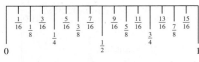

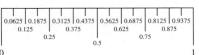

15) 1.125 in.   17) 4.9375 in.

19) exact: 0.$\overline{3}$; approximation: 0.333

21) exact: −0.$\overline{7}$; approximation: −0.778

23) exact: 0.8$\overline{3}$; approximation: 0.833

25) exact: 0.9$\overline{6}$; approximation: 0.967

27) exact: −0.$\overline{09}$; approximation: −0.091

29) 1) Divide 17 by 20.   2) Write $\frac{17}{20}$ as a fraction with a denominator of 100, then write it as a decimal.

31) 0.6   33) −0.75   35) −0.45

37) 0.64   39) 0.945   41) −4.5

43) <   45) <   47) =   49) >   51) >

53) =   55) <   57) <   59) >   61) >

63) =   65) >   67) <   69) =

71) $4\frac{1}{5}$, 4.259, 4.26, $4\frac{3}{10}$ 73) $2\frac{1}{16}$, $2\frac{1}{8}$, 2.7, 2.75

75) $-\frac{5}{16}$, −0.3, $\frac{7}{8}$, 0.97   77) 1.8   79) $-0.\overline{18}$

81) 0.006   83) −2.5625

85) exact: 0.4$\overline{6}$; approximation: 0.467

87) =   89) <   91) >

93) $-\frac{2}{3}$, −0.6, $\frac{59}{100}$, 0.7

Section 5.7

1) average   3) 29   5) 42 laps

7) 11,740   9) 26 ppg   11) $545.50

13) $65.6 million   15) 71.8%   17) 66.3%

19) 4.4 people   21) 15.2 cars   23) $9.07

25) 3.08   27) 1.19   29) 2.93   31) 2.67

33) To find the *mean*, add all of the numbers in the list and divide by the number of numbers. The *median* is the middle number in the list.

35) Arrange the numbers from lowest to highest. The number in the middle is the median.

37) $24   39) 22   41) $14.50   43) 75,710.5

45) 999.5   47) $94.50

49) It is the number that appears most frequently in the list.

51) $5.00   53) no mode

55) The list is bimodal. The modes are 71 and 78.

57) 59 in.   59) $389.99   61) no mode

63) The list is bimodal. The modes are 102 and 114.

65) The list is bimodal. The modes are 80% and 85%.

67) no mode

69) mean: 43.25 medals; median: 39.5 medals; mode: 36 medals and 46 medals

71) mean: 73.2%; median: 73.5%; mode: 85%

73) 39 mg   75) 3.71

Section 5.8

1) $b = 1.4$ 3) $y = 3.5$ 5) $k = -0.05$

7) $c = -0.2$ 9) $m = 2.4$ 11) $a = 0.9$

13) $u = -1.3$ 15) $r = 60$ 17) $b = 18$

19) $n = 2.2$ 21) Answers may vary.   23) $g = 4$

25) $c = 0$ 27) $n = 0.4$ 29) $v = 25$

31) $b = -12.5$ 33) $p = 5.75$ 35) Answers may vary.

37) $m = 18$ 39) $n = 0.75$ 41) $b = -2$

43) $k = -2.2$ 45) 146 hours   47) 3 ft

49) 7.2 in., 12.3 in.   51) equation; $p = -2$

53) expression; $-4m$ 55) equation; $a = 60$

57) expression; $-6.1h - 3.2$ 59) equation; $c = -4.6$

Section 5.9

1) perfect squares: 1, 4, 9, 16; their square roots: 1, 2, 3, 4

3) $\sqrt{2} \approx 1.4$

5) $\sqrt{8} \approx 2.8$

7) $\sqrt{20} \approx 4.5$

9) $\sqrt{10} \approx 3.2$

11) $\sqrt{45} \approx 6.7$

13) 4.472   15) 6.708   17) 12.530

19) unknown leg $= \sqrt{(\text{hypotenuse})^2 - (\text{known leg})^2}$

21) 4 km   23) $\sqrt{11}$ ft ≈ 3.3 ft   25) 10 in.

27) $\sqrt{34}$ mi ≈ 5.8 mi   29) $\sqrt{84}$ cm ≈ 9.2 cm

31) 12 cm   33) 2.831 km   35) 3.113 m

37) No, it applies only to right triangles.

39) $\sqrt{175}$ m ≈ 13.2 m   41) $\sqrt{16,200}$ ft ≈ 127.3 ft

43) $\sqrt{65}$ yd ≈ 8.1 yd

45) $\sqrt{15} \approx 3.9$

47) 4.123   49) $\sqrt{45}$ km ≈ 6.7 km

51) $\sqrt{21}$ ft ≈ 4.6 ft   53) 1.204 ft   55) 12 ft

Section 5.10

1) The radius is the distance from the center to a point on the circle; the diameter is the distance across the circle passing through the center.

3) 14 km   5) 3.2 ft   7) 9 cm   9) $\frac{1}{12}$ cm

11) 7 in.   13) the circumference   15) $C = \pi \cdot d$

17) exact: 10π in.; approximation: 31.4 in.

19) exact: 16π cm; approximation: 50.24 cm

21) exact: 6.8π in.; approximation: 21.352 in.

23) exact: 3π ft; approximation: 9.42 ft

25) exact: $4 + 2\pi$ m; approximation: 10.28 m

27) exact: 7π m; approximation: 22 m

29) exact: $\frac{1}{4}\pi$ mi; approximation: $\frac{11}{14}$ mi   31) $A = \pi \cdot r^2$

33) exact: 9π in^2; approximation: 28.26 in^2

35) exact: 81π cm^2; approximation: 254.34 cm^2

37) exact: 5.76π ft^2; approximation: 18.0864 ft^2

39) exact: 2500π ft^2; approximation: 7850 ft^2

41) exact: 0.32π m^2; approximation: 1.0048 m^2

43) exact: 32π yd^2; approximation: 100.48 yd^2

45) 114.61 ft; approximately 21 humans   47) $226.08

49) $7284   51) $V = \frac{4}{3}\pi r^3$; $SA = 4\pi r^2$

53) Volume: exact: $\frac{256}{3}\pi$ in^3; approximation: 267.95 in^3;

SA: exact: 64π in^2; approximation: 200.96 in^2

55) Volume: exact: $\frac{4}{375}\pi$ mm^3; approximation:

$\frac{88}{2625}$ mm$^3 \approx 0.03$ mm^3; SA: exact: $\frac{4}{25}\pi$ mm^2;

approximation: $\frac{88}{175}$ mm$^2 \approx 0.50$ mm^2

57) Volume: exact: $\frac{9}{16}\pi$ m^3; approximation: 1.77 m^3;

SA: exact: $\frac{9}{4}\pi$ mi^2; approximation: 7.07 m^2

59) exact: 144π mm^3; approximation: 452.16 mm^3

61) $V = \pi r^2 h$; $SA = 2\pi rh + 2\pi r^2$

63) Volume: exact: 1.25π m^3; approximation: 3.925 m^3;
SA: exact: 13.5π m^2; approximation: 42.39 m^2

65) Volume: exact: 250π mm^3; approximation:

$\frac{5500}{7}$ mm$^3 \approx 785.71$ mm^3; SA: exact: 150π mm^2;

approximation: $\frac{3300}{7}$ mm$^2 \approx 471.43$ mm^2

67) Volume: exact: 360π ft^3; approximation: 1130.4 ft^3;
    SA: exact: 192π ft^2; approximation: 602.88 ft^2

69) $V = \dfrac{1}{3}\pi r^2 h$

71) exact: 50π in^3; approximation: 157 in^3

73) exact: 12π cm^3; approximation: 37.68 cm^3

75) 25,120 mi   77) 343.36 mm^3 79) 1186.92 m^3

81) Area = 136 cm^2; Perimeter = 50 cm

83) Area: exact: 0.0064π mm^2; approximation: 0.020096 mm^2
    Circumference: exact: 0.16π mm; approximation: 0.5024 mm

85) 20.52 cm^2 87) exact: 600π in^3; approximation: 1884 in^3

89) $\dfrac{15}{4}$ yd^3 or $3\dfrac{3}{4}$ yd^3 or 3.75 yd^3 91) 1948.83 cm^3

Chapter 5: Group Activity

Evaluating Expressions

1) 0.295   2) -26.75 3) -3.7 4) 6.315

Solving Equations

1) $x = -1.3$ 2) $x = -0.2$ 3) $x = 0.85$ 4) $x = 4.05$

Chapter 5: Review Exercises

1) $\dfrac{3}{10} = 0.3$;

3)

5) 0.097   7) -3.867

9) 5—tens, 2—ones, 4—tenths, 0—hundredths,
   6—thousandths, 7—ten-thousandths,
   9—hundred-thousandths, 8—millionths

11) 50,072.36   13) $-\dfrac{49}{50}$ 15) $1\dfrac{1}{2}$

17) a) 40,000 b) 40,000 c) 39,605.0 d) 39,605.00

19) $585.74

21) Write the numbers vertically so that the decimal points are
    lined up. If any numbers are missing digits to the right of the
    decimal point, insert zeros. Then, add the same way we add
    whole numbers. Place the decimal point in the answer *directly
    below* where the decimal point appears in the problem.

23) 7.24   25) -317.743 27) 16.416

29) 16,623.456   31) $3.75   33) -47.0556

35) 0.00192   37) 85.72284   39) 22.09

41) -0.02005 43) 12,064.5 sq yd   45) 7.58

47) 0.0058   49) exact: $65.\overline{7}$; approximation: 65.778

51) -29.44 53) $242.80   55) 0.8

57) -5.0625 59) exact: $0.\overline{18}$; approximation: 0.18

61) 0.036   63) $>$ 65) $=$

67) $-0.76, -0.7, \dfrac{1}{2}, \dfrac{3}{4}$ 69) 54,102 mi   71) 3.33

73) 20 mi   75) 11.6   77) no mode

79) 7.0658   81) 3005.48   83) -56.0306

85) a) $74 b) $73.90   87) $7\dfrac{66}{125}$

89) a) 34 hours b) $1.80 per hour c) Vernon: $13.50, Jane:
    $7.20, Aliyah: $6.30, Domingo: $14.40, Nayana: $12.60,
    Steve: $7.20

91) $z = -0.77$ 93) $x = 4$ 95) $n = 1.15$

97) 38 hr   99)

$\sqrt{32} \approx 5.7$

101) 8 in.   103) $\sqrt{80}$ m ≈ 8.9 m   105) 30 ft^2

107) 4.8 km   109) 6.5 in.

111) Area: exact: 42.25π mm^2; approximation: 132.67 mm^2;
     Circumference: exact: 13π mm; approximation: 40.82 mm

113) 125,600 mi^2

115) exact: $\dfrac{32}{3}\pi$ in^3; approximation: 33.49 in^3

117) exact: 1200π mm^3; approximation: 3768.00 mm^3

119) exact: 16π in^2; approximation: 50.24 in^2

Chapter 5: Test

1) 0.6;

2) a) $\dfrac{73}{100}$ b) $-\dfrac{4}{5}$ c) $2\dfrac{3}{40}$

3) a) four and nine hundredths

   b) six hundred fourteen ten-thousandths

4) 16.573   5) 9.47   6) 311.0

7) $2375   8) $62.14

9) Answers may vary. You do not need to line up the decimals
   when you multiply decimals. Write the problem vertically,
   line up the numbers on the right side, and multiply as if they
   were whole numbers. The number of decimal places in the
   answer equals the *total* number of decimal places in the
   factors.

10) 22.774   11) 7250   12) 227.8672

13) -45.34 14) -0.0251 15) 3.91   16) 0.829

17) 272.594   18) exact: $6.\overline{6}$; approximation: 6.667

19) 9.375   20) $>$ 21) $=$ 22) $<$

23) $-1.207, -1.2, 0.73, 1.073, 1.25$

24) $1057.81   25) $0.95

26) mean: 33.9 customers per hour; median: 30 customers per hour; mode: 26 customers per hour

27) 3.21   28) a) $n = -7.7$ b) $k = 3.5$ c) $w = 6$

29) 1.8 ft   30) No. The Pythagorean theorem can be used only with right triangles.

31) $\sqrt{53}$ in. ≈ 7.3 in.   32) 12 ft   33) circumference

34) a) exact: 100π in^2; approximation: 314 in^2
    b) exact: 20π in.; approximation: 62.8 in.

35) a) sphere   b) exact: $\dfrac{500}{3}\pi$ ft^3; approximation: 523.33 ft^3
    c) exact: 100π ft^2; approximation: 314 ft^2

36) 1648.5 mm^3

Chapter 5: Cumulative Review for Chapters 1–5

1) 83,210   2) 42,401   3) 189,054   4) 730

5) $-\dfrac{4}{45}$ 6) $-\dfrac{11}{24}$ 7) 12 cm

8) 3 in.; 6 in.; 56 in.   9) $\dfrac{3}{20} = 0.15$

10)

11) 83.0023   12) 6,400,000

13) 0.5499   14) $11\dfrac{4}{5}$ 15) $\dfrac{1}{625}$ 16) 38

17) $\dfrac{21}{8}$ or $2\dfrac{5}{8}$ 18) -0.0096 19) 59   20) $t = -\dfrac{7}{2}$

21) $n = 29$ 22) $1\dfrac{3}{4}$ cups   23) \$12.67

24) \$93.6 million   25) Alec: 201,037; Kullen: 183,204

Chapter 6
Section 6.1

1) 52   3) 2

5) A ratio is a comparison of two quantities.

7) $\dfrac{11}{16}$ 9) $\dfrac{1}{2}$ 11) $\dfrac{3}{2}$ 13) $\dfrac{4}{1}$

15) $\dfrac{8}{3}$; for every 8 part-time employees, there are 3 full-time employees.

17) $\dfrac{31}{250}$ 19) $\dfrac{219}{31}$ 21) $\dfrac{1}{2}$

23) For every 219 digital albums sold, there were 31 vinyl albums sold.

25) $\dfrac{8}{3}$ 27) $\dfrac{3}{22}$ 29) $\dfrac{4}{5}$ 31) $\dfrac{8}{13}$ 33) $\dfrac{34}{21}$

35) $\dfrac{4}{1}$ 37) $\dfrac{19}{11}$ 39) $\dfrac{7}{8}$ 41) $\dfrac{3}{1}$ 43) $\dfrac{16}{23}$ 45) $\dfrac{1}{4}$

47) $\dfrac{19}{900}$ 49) $\dfrac{16}{17}$ 51) $\dfrac{7}{12}$ 53) $\dfrac{2}{3}$ 55) $\dfrac{1}{10}$

57) $\dfrac{3}{16}$ 59) $\dfrac{5}{1}$ 61) $\dfrac{7}{1}$ 63) $\dfrac{9}{4}$ 65) $\dfrac{3}{2}$

67) $\dfrac{5}{9}$ 69) \$3250   71) $\dfrac{5}{1}$ 73) $\dfrac{9}{1}$ 75) $\dfrac{1}{20}$

77) $\dfrac{3}{2}$ 79) $\dfrac{7}{1}$

Section 6.2

1) A rate compares quantities with different units, and a ratio compares quantities with the same units.

3) $\dfrac{8\text{ ft}}{3\text{ sec}}$ 5) $\dfrac{45\text{ mi}}{2\text{ gal}}$ 7) $\dfrac{6\text{ cups}}{5\text{ servings}}$ 9) $\dfrac{140\text{ mi}}{3\text{ hr}}$

11) $\dfrac{\$1}{4\text{ daisies}}$ 13) $\dfrac{2\text{ oz}}{\$15}$ 15) $\dfrac{\$199}{1\text{ wk}}$

17) It is a rate with a denominator of 1.   19) 79 mi/hr

21) 26 mpg   23) \$40/hr   25) \$55.50/day

27) 23.4 mpg   29) \$1.22/oz   31) \$0.213/oz

33) \$0.375/can   35) 18.2 mph   37) 18.7 mpg

39) 76 mpg   41) 19.6 hr/wk; 2.8 hr/day

43) \$40.95   45) \$79.00   47) 48.7 yd/return

49) \$13.50/hr   51) 160 mph   53) 36 oz; \$0.178/oz

55) 20 oz; \$0.113/oz   57) Brand A; \$1.23/lightbulb

59)

Dimensions in feet	Monthly Cost	Floor Area in square feet	Cost/ft^2
15×15	\$281.25	225	\$1.25
10×10	\$150.00	100	\$1.50
8×8	\$112.00	64	\$1.75
8×6	\$88.80	48	\$1.85
5×5	\$48.75	25	\$1.95

Section 6.3

1) A proportion is a statement that two ratios or two rates are equal.

3) $\dfrac{8}{11} = \dfrac{24}{33}$ 5) $\dfrac{14}{8} = \dfrac{35}{20}$ 7) $\dfrac{50}{18} = \dfrac{25}{9}$

9) $\dfrac{2}{5} = \dfrac{8}{20}$ 11) $\dfrac{480\text{ mi}}{8\text{ hr}} = \dfrac{180\text{ mi}}{3\text{ hr}}$

13) $\dfrac{2\text{ in.}}{5\text{ ft}} = \dfrac{10\text{ in.}}{25\text{ ft}}$ 15) $\dfrac{\$8}{12\text{ songs}} = \dfrac{\$6}{24\text{ songs}}$

17) 1) Write each fraction in lowest terms.
    2) Find the cross products.

19) $\frac{1}{3} = \frac{1}{3}$; true   21) $\frac{8}{7} \neq \frac{9}{7}$; false   23) $\frac{2}{3} \neq \frac{5}{9}$; false

25) $\frac{31}{9} = \frac{31}{9}$; true   27) $\frac{9}{2} = \frac{9}{2}$; true

29) $16 = 16$; true   31) $42 \neq 40$; false

33) $380 = 380$; true   35) $192 \neq 216$; false

37) $103.5 \neq 73.1$; false   39) $19.2 = 19.2$; true

41) $\frac{731}{2} = \frac{731}{2}$; true   43) $10 \neq 14\frac{27}{32}$; false

45) $32 = 32$; true

47) No. To determine whether a proportion is true, you find the cross products or write each fraction in lowest terms. To multiply fractions, you multiply the numerators and multiply the denominators.

49) a) $441 = 441$; true   b) $\frac{9}{49}$

51) a) $\frac{6}{7}$ b) $336 \neq 288$; false

53) a) $\frac{33}{91}$ b) $84 \neq 143$; false

55) $\frac{15}{18} = \frac{35}{42}$; $630 = 630$; the claim is correct.

Section 6.4

1) 42   3) 39.2   5) $2\frac{4}{9}$ 7) $6\frac{1}{2}$

9) To solve a proportion means to find the missing number that will make the proportion true.

11) First, find the cross products. Then, set the cross products equal to each other, and solve the equation. Finally, check the solution by substituting it into the original proportion and finding the cross products.

13) $x = 2$ 15) $x = 14$ 17) $x = 18$ 19) $x = 10$

21) $x = \frac{45}{2}$ 23) $x = \frac{16}{27}$ 25) $x = 2$

27) $x = 3\frac{1}{7}$ 29) $x = \frac{13}{15}$ 31) $x = 4\frac{5}{18}$

33) $x = 3.36$ 35) $x = 1.68$ 37) $x = 104$

39) $x = 4\frac{3}{8}$ 41) $x = 6\frac{2}{9}$ 43) $x = 3\frac{1}{3}$

45) $x = 3$ 47) $x = 5$ 49) $x = 24$

51) $x = 6$ 53) $x = 2\frac{2}{3}$ 55) $x = 1.2$

Section 6.5

1) 595 mi   3) 20 g   5) 640 mi   7) $261.45

9) 320 mi   11) 274 cal   13) 110 citizens

15) length = 9 ft; width = 20 ft

17) length = 14 ft; width = 10 ft

19) length = 28 ft; width = 20 ft   21) 560 sq ft

23) 48 drops   25) 80 hr   27) 21 ft   29) 18

31) 64   33) 3384 votes   35) 0.25 gal; 32 oz

37) 105 teachers   39) 1126 students

Section 6.6

1) Supplementary angles have measures that sum to 180°.

3) 29°   5) 67°   7) 83°   9) 77°

11) 126°   13) 63°   15) 38°   17) 138°

19) 2°   21) 129°   23) 56°   25) always

27) Their angle measures are the same.

29) Their angle measures are the same.

31) $\angle B$ and $\angle D$ are vertical angles; therefore, $\angle B \cong \angle D$. $\angle C$ and $\angle E$ are vertical angles; therefore, $\angle C \cong \angle E$.

33) $\angle G$ and $\angle L$ are vertical angles; therefore, $\angle G \cong \angle L$. $\angle M$ and $\angle H$ are vertical angles; therefore, $\angle M \cong \angle H$.

35) m$\angle TJQ = 94°$; m$\angle DJQ = 86°$

37) m$\angle ADF = 124°$; m$\angle FDB = 56°$

39) $\angle T$ and $\angle U$; $\angle U$ and $\angle V$; $\angle V$ and $\angle W$; $\angle W$ and $\angle T$

41) Answers may vary.

43) m$\angle A = 101°$; m$\angle C = 79°$; m$\angle F = 101°$

45) m$\angle T = 55°$; m$\angle U = 125°$; m$\angle V = 55°$

47) m$\angle FBH = 71°$; m$\angle PBK = 91°$; m$\angle PBQ = 71°$; m$\angle QBD = 18°$

49) m$\angle HKT = 120°$; m$\angle VKF = 120°$; m$\angle FKG = 19°$; m$\angle GKH = 41°$

51) 10 angles; 360°   53) transversal

55) alternate interior

57) Corresponding angles are congruent.

59) corresponding angles: $\angle 1$ and $\angle 3$, $\angle 2$ and $\angle 4$, $\angle 5$ and $\angle 7$, $\angle 6$ and $\angle 8$; alternate interior angles: $\angle 2$ and $\angle 7$, $\angle 3$ and $\angle 6$

61) m$\angle 1 = 112°$; m$\angle 2 = 68°$; m$\angle 3 = 68°$; m$\angle 4 = 112°$; m$\angle 5 = 68°$; m$\angle 6 = 68°$; m$\angle 7 = 112°$

63) m$\angle 1 = 83°$; m$\angle 2 = 97°$; m$\angle 3 = 97°$; m$\angle 4 = 97°$; m$\angle 5 = 83°$; m$\angle 6 = 97°$; m$\angle 7 = 83°$

65) $\angle K$ and $\angle M$ are vertical angles; therefore, $\angle K \cong \angle M$. $\angle N$ and $\angle L$ are vertical angles; therefore, $\angle N \cong \angle L$.

67) 148°

69) m$\angle AGF = 80°$; m$\angle EGD = 47°$; m$\angle DGC = 80°$; m$\angle CGB = 53°$

71) m$\angle 1 = 53°$; m$\angle 2 = 127°$; m$\angle 3 = 53°$; m$\angle 4 = 53°$; m$\angle 5 = 127°$; m$\angle 6 = 127°$; m$\angle 7 = 53°$

73) always   75) never   77) sometimes

Section 6.7

1) They are exactly the same shape and size.

3) corresponding angles:
$\angle M$ and $\angle D$, $\angle N$ and $\angle E$, $\angle P$ and $\angle F$;
corresponding sides:
$\overline{MN}$ and $\overline{DE}$, $\overline{NP}$ and $\overline{EF}$, $\overline{MP}$ and $\overline{DF}$

5) corresponding angles:
$\angle T$ and $\angle K$, $\angle U$ and $\angle J$, $\angle V$ and $\angle H$;
corresponding sides:
$\overline{TV}$ and $\overline{KH}$, $\overline{TU}$ and $\overline{KJ}$, $\overline{UV}$ and $\overline{JH}$

7) SAS   9) SSS   11) ASA

13) Similar triangles have the same shape, the measures of their corresponding angles are the same, and the lengths of their corresponding sides are proportional.

15) yes; answers may vary

17) corresponding angles:
$\angle A$ and $\angle F$, $\angle C$ and $\angle D$, $\angle B$ and $\angle E$;
corresponding sides:
$\overline{AC}$ and $\overline{FD}$, $\overline{CB}$ and $\overline{DE}$, $\overline{BA}$ and $\overline{EF}$

19) corresponding angles:
$\angle K$ and $\angle H$, $\angle L$ and $\angle F$, $\angle M$ and $\angle G$;
corresponding sides:
$\overline{KL}$ and $\overline{HF}$, $\overline{LM}$ and $\overline{FG}$, $\overline{MK}$ and $\overline{GH}$

21) corresponding angles:
$\angle M$ and $\angle B$, $\angle N$ and $\angle A$, $\angle P$ and $\angle C$;
corresponding sides:
$\overline{MN}$ and $\overline{BA}$, $\overline{NP}$ and $\overline{AC}$, $\overline{PM}$ and $\overline{CB}$

23) 15 m; 36 m   25) 27 dm; 57 dm

27) 72 ft; 351 ft   29) 96 km; 234 km

31) 24 km; 57 km   33) 12 in.; 36 in.

35) 13.5 in.; 31.5 in.   37) 26 m; 48 m

39) proportional   41) congruent

43) 45.5 ft   45) 21 ft   47) 3.75 m

Chapter 6: Group Activity

1) No; yes; explanations will vary.

2) b; Answers may vary.

3) a; Answers may vary.

4) Possible examples are $\frac{3}{8}=\frac{90}{x}$ and $\frac{3}{90}=\frac{8}{x}$. He will sign into his account 240 times in 90 days.

5) Possible examples are $\frac{16}{25}=\frac{800}{x}$ and $\frac{16}{800}=\frac{25}{x}$. There were 1250 people in the group.

6) Possible examples are $\frac{110}{\frac{3}{4}}=\frac{x}{2}$ and $\frac{\frac{3}{4}}{110}=\frac{2}{x}$. There are approximately 293 calories in 2 cups of Lucky Charms.

Chapter 6: Review Exercises

1) A ratio compares two quantities with the same units.

3) $\frac{7}{12}$ 5) $\frac{5}{1}$ 7) $\frac{20}{21}$ 9) $\frac{5}{16}$ 11) $\frac{47}{23}$ 13) $\frac{49}{50}$

15) *Jurassic World* to *Beauty and the Beast*   17) $\frac{3}{7}$

19) A ratio compares two quantities with the same units, and a rate compares quantities with different units.

21) $\frac{1 \text{ cup}}{3 \text{ servings}}$ 23) It is a rate with a denominator of 1.

25) 60 text messages/hr   27) $1.50/day

29) 400-count; $0.012/napkin   31) 1.2 mph

33) $\frac{16}{14}=\frac{24}{21}$ 35) $\frac{350 \text{ cell phones}}{450 \text{ adults}}=\frac{7 \text{ cell phones}}{9 \text{ adults}}$

37) 1) Write each ratio in lowest terms.
2) Find the cross products.

39) $\frac{14}{3}=\frac{14}{3}$; true   41) $\frac{6}{17}\neq\frac{10}{23}$; false

43) $126 \neq 96$; false   45) $\frac{49}{2}=\frac{49}{2}$; true

47) a) $\frac{79}{45}$ or $1\frac{34}{45}$ b) $120 \neq 117$; false

49) $x=12$ 51) $x=\frac{15}{4}$ 53) $x=\frac{3}{8}$

55) $x=3.8$ 57) 450 ft   59) 50 drops

61) $43°$ 63) $74°$ 65) $18°$ 67) $44°$

69) $m\angle T=146°$; $m\angle N=34°$; $m\angle Z=34°$

71) corresponding angles:
$\angle 1$ and $\angle 8$, $\angle 2$ and $\angle 4$, $\angle 3$ and $\angle 6$, $\angle 5$ and $\angle 7$;
alternate interior angles: $\angle 2$ and $\angle 5$, $\angle 6$ and $\angle 8$

73) corresponding angles:
$\angle A$ and $\angle X$, $\angle B$ and $\angle Y$, $\angle C$ and $\angle Z$;
corresponding sides:
$\overline{AB}$ and $\overline{XY}$, $\overline{BC}$ and $\overline{YZ}$, $\overline{AC}$ and $\overline{XY}$

75) SAS   77) 40 ft   79) 24.5 yd   81) 42 m

83) always   85) never

87) $\frac{3}{5}$ 89) 64 mi/hr   91) $44 \neq 42$; false

93) $x=16$ 95) $\frac{2}{3}$ cup   97) $24°$ 99) 27 in.

Chapter 6: Test

1) A ratio compares two quantities with the same units, but a rate compares quantities with different units.

2) $\frac{9}{5}$ 3) $\frac{2}{3}$

4) $\frac{5}{14}$; for every 5 g of cholesterol in the yogurt, there are 14 g of protein.

5) $2.49 per lb or $2.49/lb   6) 42.5 mi/hr or 42.5 mph

7) 110.6 mi/wk   8) 50 drops per minute

9) 17 oz; $0.482/oz   10) $\frac{\$12}{9 \text{ lb}} = \frac{\$20}{15 \text{ lb}}$

11) a) true b) false

12) Elsa did not multiply correctly. Cross products are used to determine whether a proportion is true or false. We do not use cross products to multiply fractions. Correct:

$$\frac{3}{10} \cdot \frac{4}{9} = \frac{\overset{1}{\cancel{3}}}{\underset{5}{\cancel{10}}} \cdot \frac{\overset{2}{\cancel{4}}}{\underset{3}{\cancel{9}}} = \frac{2}{15}$$

13) $x = 15$ 14) $x = 4\frac{1}{5}$ 15) 4.5 in.

16) 9.45 mg   17) 6 crashes   18) a) 159° b) 69°

19) m∠RAT = 29°; m∠WAX = 87°; m∠TAW = 64°; m∠YAZ = 64°

20) m∠1 = 72°; m∠2 = 108°; m∠3 = 72°; m∠4 = 72°; m∠5 = 108°; m∠6 = 72°; m∠7 = 108°

21) In both triangles, the corresponding sides and the angle between them have the same measure. Therefore, SAS (Side-Angle-Side) proves they are congruent.

22) 10 km   23) false   24) true   25) false

Chapter 6: Cumulative Review for Chapters 1–6

1) $-\frac{6}{5}$ or $-1\frac{1}{5}$ 2) 8.39   3) 3,295,057   4) $7\frac{11}{12}$

5) $-\frac{23}{24}$ 6) 1.971   7) 5.6   8) $-\frac{1}{30}$

9) 4,720,000   10) −5600   11) 29   12) −16

13) 174.82   14) 170   15) $\frac{16}{25}$ 16) $\frac{5}{12}$

17) $n = -1$ 18) $c = 2.75$ 19) $x = 5$

20)

Date	Odometer Start	Odometer End	Miles Traveled	Gallons Purchased	Miles per Gallon
8/24	15,352.8	15,625.2	272.4	12.0	22.7
9/2	15,625.2	15,958.9	333.7	14.2	23.5
9/13	15,958.9	16,248.7	289.8	13.8	21.0
9/22	16,248.7	16,457.8	209.1	8.5	24.6

21) 121°   22) a) 130° b) 50° c) 4 d) 180°

23) No. The price should be $29.99.

24) $19.08   25) 1323

Chapter 7

Section 7.1

1) Answers may vary.   3) 12 in. = 1 ft

5) 60; 60   7) 3; 5280   9) 2; 8

11) when you are converting from a smaller unit to a larger unit

13) 180 min   15) 3 ft   17) $1\frac{1}{2}$ gal

19) $\frac{3}{4}$ ton or 0.75 ton   21) 150 min   23) 21 in.

25) 5 qt   27) 1.75 qt or $1\frac{3}{4}$ qt

29) 15 yd   31) $\frac{60 \text{ min}}{1 \text{ hr}}$ or $\frac{1 \text{ hr}}{60 \text{ min}}$

33) Step 1: Identify the units given and the units we want to get. Write down the relationship between those units. Step 2: Multiply the given measurement by the unit fraction relating the unit given and the unit we want to get so that the given unit will divide out and leave us with the unit we want.

35) 84 in.   37) 20 pt   39) $3\frac{1}{4}$ lb or 3.25 lb

41) 84 hr   43) 15   45) $4\frac{1}{2}$ or 4.5   47) 105

49) 7   51) 2500   53) $\frac{1}{2}$ or 0.5

55) a) 2880 min b) 2880 min   57) 86,400

59) $4\frac{1}{2}$ 61) 3   63) 40   65) $3\frac{3}{4}$

67) 5 pt   69) 900 min   71) $6.08

73) Brand B is the better buy because its unit cost is $2.88/lb.

75) Brand B is the better buy because the unit cost of Brand A is $3.04/lb.

77) 25,500 times   79) $114.75

Section 7.2

1) $\frac{1}{1000}; \frac{1}{1000}$ 3) 1000; 1000   5) $\frac{1}{100}; \frac{1}{100}$

7) a) 15.6 cm b) 156 mm   9) a) 24 mm b) 2.4 cm

11) Answers may vary.   13) Answers may vary.

15) 3 cm   40 mm   0.5 dm   0.06 m   17) cm   19) m

21) cm   23) km   25) m   27) 600 cm

29) 3800 mm   31) 7.3 km   33) 0.25 m

35) 0.0182 km   37) 130 m   39) 2.5 km

41) 34.28 m   43) 7.5 mm   45) 1500 cm   47) 75 mm

49) 0.97 km   51) 0.82 m   53) 0.016 km   55) 86 mm

57) 3490 m   59) 23.4 dm   61) 4.3 hm

63) 9,440,000 mm   65) 100 cm   1.01 m   12 dm   1400 mm

67) 96,000,000 m   69) 12 mm   71) cm   73) m

75) 3.4 m   77) 1500 m   79) 2,000,000 cm

81) 32,100 cm

Section 7.3

1) mL   3) L   5) mL   7) unreasonable; too much

9) reasonable   11) unreasonable; too little

13) 0.325 L   15) 4 L   17) 7100 mL   19) 60 mL

21) 0.035 L   23) mg   25) kg   27) g   29) mg

31) kg   33) reasonable   35) unreasonable; too much

37) reasonable   39) 72,000 g   41) 820 g

43) 0.194 kg   45) 2010 mg   47) 750 g   49) 0.3 g

51) 50,000 mg   53) 4100 g   55) 3 kg   57) m

59) g   61) mm   63) L   65) km   67) reasonable

69) unreasonable; too little   71) 0.34 L   73) 9100 mg

75) 0.0155 km   77) 315,000 g

Section 7.4

1) 3.55 m or 355 cm   3) 2.56 kg or 2560 g

5) 1.419 L or 1419 mL   7) 2.07 m or 207 cm

9) 0.84 m or 84 cm   11) 1.795 kg or 1795 g

13) 0.675 L or 675 mL   15) 2.8 g or 2800 mg

17) 3 L   19) 250 mL   21) 0.4 m   23) 112 cm

25) 15.93 g   27) 4.5 kg   29) 0.034 kg   31) 3 mL

33) 2130 cm   35) 0.24 km   37) 72 cm   39) $12.16

41) 300 mL   43) 395 brushes   45) 7 yr   47) 70 mg

49) 150 days   51) $3.17/mg   53) 1770 mL or 1.77 L

55) 0.67 kg or 670 g   57) 5.5 cm or 55 mm

59) 0.3 m or 30 cm   61) $0.03 per application

63) a four-pack of Red Bull   65) 1.12 g   67) 0.252 g

Section 7.5

1) 15 m   3) 182 m   5) 473.2 mL   7) 8.9 lb

9) 27.6 in.   11) 6.2 L   13) 64.4 km   15) 7.1 in.

17) 140 lb   19) 357.2 g   21) 2.9 gal   23) 42.9 yd

25) a) 10 ft ≈ 3 m   b) 10 ft = 3.048 m

27) a) 50 yd ≈ 45.50 m   b) 50 yd = 45.72 m

29) 45.72 cm   31) 20 lb   33) yes

35) 1.9 gal per flush   37) $610.14   39) −4°C

41) 200°C   43) T-shirt and shorts

45) Use the formula $C = \dfrac{5(F - 32)}{9}$.   47) 5°C

49) 28°C   51) 40°C   53) 22°C   55) 68°F

57) 163°F   59) 77°F   61) 48°F   63) 22.5 m

65) 402.6 g   67) 9.5 qt   69) 2.1 c   71) 24.2 km

73) 20°C   75) 11°C   77) 134°F

Chapter 7: Group Activity Answers

Activity 1 Answers

A. 1) j   2) e   3) h   4) f   5) c

   6) d   7) b   8) a   9) i   10) g

B. 1) 1700 kg = 1,700,000 g. The numerical part of the answer increased.

   2) 1.35 mm = 0.135 cm. The numerical part of the answer decreased.

   3) 0.5 mL = 0.005 dL. The numerical part of the answer decreased.

   4) 0.98 m = 98 cm. The numerical part of the answer increased.

   5) 68.22 kL = 68,220 L. The numerical part of the answer increased.

   6) Answers will vary. Possible answer: When we convert a larger unit to a smaller unit, the numerical part of the answer gets larger. When we convert a smaller unit to a larger unit, the numerical part of the answer gets smaller. When we convert from a larger unit to a smaller unit, we multiply to get our answer. When we convert from a smaller unit to a larger unit, we divide to get our answer.

Activity 2 Answers

1) The weight of a computer monitor (6.75 kg is about 15 pounds)

2) Possible answer: The amount of water in a bathtub: 90.96 L ≈ 24 gal

3) Possible answer: The length of a basketball court: 28.65 m ≈ 95.5 ft

4) Possible answer: The weight of a cell phone: 350 g ≈ 12.35 oz

5) Possible answer: The weight of a textbook: 2.8 kg ≈ 6.2 lb

Compare your answers with the answers of another group. Were your answers similar in length, capacity, and weight?

Chapter 7: Review Exercises

1) 3   3) 16

5) Use division because we are converting from a smaller unit to a larger unit.

7) $2\frac{1}{2}$ gal or 2.5 gal   9) 30 hr   11) 420 sec

13) $2\frac{1}{2}$ qt or 2.5 qt   15) 20 ft

17) Brand A; the unit cost of Brand B in dollars/pound is $3.84/lb.

19) 100   21) 1000   23) km   25) 0.631 m

27) 1200 m   29) 700 cm   31) 0.0965 m

33) 80,000 cm   35) L   37) mL   39) 3250 mL

41) 0.14 L   43) 0.63 L   45) 2000 mL   47) 1000

49) mg   51) kg   53) 0.082 g   55) 13,000 g

57) 4500 mg   59) 9.75 kg   61) 20,000 mg   63) 6 g

65) 7.47 m or 747 cm   67) 165 cm   69) $7.96

71) 4269 m   73) 24.4 lb   75) 50 gal   77) 10.2 cm

79) 0°C   81) 10°C   83) 26°C   85) 41°F

87) 270°F   89) mm   91) kg   93) L

95) 2250 mL   97) 0.48 m   99) 320 sec

101) 0.0058 kg   103) 83.8 cm or 838 mm

105) 4.2 kg   107) 99.4 mi   109) 5.7 L

111) 21.8 lb   113) 72°F   115) $2.42

Chapter 7: Test

1) 4   2) 60   3) 7 yd   4) 3 qt   5) $1\frac{1}{2}$ lb or 1.5 lb

6) 56 in.   7) 5040 min   8) L   9) cm   10) kg

11) mm   12) mL   13) g   14) m   15) 0.25 g

16) 730 cm   17) 0.062 L   18) 2500 g

19) 193.5 mm   20) 0.8 km   21) 3.045 kg or 3045 g

22) −7°C   23) 26°C   24) 127 cm   25) 13.3 lb

26) 198.8 mi   27) 56.1 L   28) 354.9 mL

29) 25°C   30) 39°F   31) $11.16   32) 5 months

33) $15.21   34) 4.28 m   35) 75 mg

Chapter 7: Cumulative Review for Chapters 1–7

1) The base of $(-2)^4$ is −2, so $(-2)^4$ means
   $-2 \cdot (-2) \cdot (-2) \cdot (-2) = 16$. The base of -2^4 is 2,
   and -2^4 means $-1 \cdot 2 \cdot 2 \cdot 2 \cdot 2 = -16$.

2) fourteen thousand, two hundred nine and thirty-seven
   hundredths

3) −49   4) −20,375   5) 88.2   6) $1\frac{1}{15}$

7) $-\frac{5}{24}$ 8) 155.68   9) $\frac{5}{18}$ 10) 0.045

11) undefined   12) $-\frac{1}{24}$ 13) −89   14) −135

15) area = 16.82 ft²; perimeter = 17.4 ft

16) area = $\frac{1}{3}$ yd²; perimeter = $\frac{8}{3}$ yd

17) equation; $k = -9$ 18) expression; $14k - 16$

19) expression; $-\frac{1}{6}y + 8$ 20) equation; $y = 12$

21) A ratio compares two quantities with the same units,
    but a rate compares quantities with different units.

22) 17°C   23) $3\frac{3}{4}$ c   24) 500 m/min

25) washer: $398; dryer: $448

Chapter 8

Section 8.1

1) 23% means 23 out of 100, so the statement means that
   23 out of 100 students who consume media approximately
   3 hours per day earn mostly C grades or lower.

3) 81% means 81 out of 100, so the statement means 81 out of
   100 Americans have a profile on one or more social
   networking websites.

5) 43% of teens have been victims of cyberbullying.

7) As of June 2017, 15% of all members of the armed forces
   were women.

9) $\frac{67}{100} = 0.67$ 11) $\frac{59}{100} = 0.59$ 13) $\frac{1}{100} = 0.01$

15) Remove the percent symbol, then move the decimal point
    two places to the left.

17) 0.72   19) 0.03   21) 0.452   23) 0.004

25) 2   27) 1.5   29) 0.0025   31) 0.0001

33) 1   35) $\frac{1}{5}$ 37) $\frac{17}{20}$ 39) $\frac{67}{100}$ 41) $\frac{1}{50}$

43) $2\frac{3}{4}$ 45) $0.52 = \frac{13}{25}$ 47) $0.225 = \frac{9}{40}$

49) $0.084 = \frac{21}{250}$ 51) $0.003 = \frac{3}{1000}$

53) $0.9375 = \frac{15}{16}$ 55) $\frac{80.5}{100} = \frac{161}{200}$ 57) $\frac{9.9}{100} = \frac{99}{1000}$

59) $\frac{1.58}{100} = \frac{79}{5000}$ 61) $\frac{0.4}{100} = \frac{1}{250}$

63) Answers may vary.

65) $\frac{5}{8}$ 67) $\frac{1}{16}$ 69) $\frac{5}{6}$ 71) $\frac{129}{500}$

73) Move the decimal point two places to the right, and put the
    percent symbol at the end of the number.

75) 82%   77) 8%   79) 70%   81) 1.25%

83) 860%   85) 900%   87) $\frac{70}{100} = 70\%$

89) $\frac{5}{100} = 5\%$ 91) $\frac{98}{100} = 98\%$ 93) $\frac{310}{100} = 310\%$

95) 62.5%   97) 48%   99) 180%   101) 6.25%

103) 65%   105) 17.5%   107) 162.5%

109) exact: $66.\overline{6}\%$; rounded: 66.7%

111) exact: $11.\overline{1}\%$; rounded: 11.1%

113) exact: $41.\overline{6}\%$; rounded: 41.7%

115) exact: $183.\overline{3}\%$; rounded: 183.3%

117) 1) Write the fraction with a denominator of 100. 2) Use long
     division to change the fraction to a decimal, then change the
     decimal to a percent. 3) Use a proportion.

119) $\frac{21}{25}$; 0.84; 84%   121) $\frac{9}{50}$; 0.18; 18%

Section 8.2

1) a) 6 b) 6   3) a) 32 b) 32   5) a) 365 b) 365

7) Locate the decimal point in the number, and move it one place to the left.

9) 7   11) 41   13) 60   15) 100   17) 387

19) 2.3   21) 0.5

23) Find 10% of the number, then multiply by 2.

25) a) 6 b) 12 c) 42   27) a) 7 b) 21 c) 56

29) a) 60 b) 140   31) a) 160 b) 400

33) a) 66 b) 44   35) a) 3.1 b) 6.2

37) Find 10% of the number, then multiply that result by $\frac{1}{2}$, or divide the result by 2.

39) 4   41) 6   43) 25   45) 20   47) 3.4

49) 1.1   51) a) 6 b) 3 c) 9   53) a) 20 b) 10 c) 30

55) a) 18 b) 9 c) 27   57) 3   59) 120   61) 66

63) 75   65) 600   67) 13.2   69) 36   71) 21

73) 50   75) 90   77) a) 35 b) 70   79) a) 15 b) 25

81) a) 120 b) 180   83) a) 1400 b) 400

85) $3.60; $27.60   87) $2.70; $20.70

89) $9.00; $69.00   91) $1.60; $9.60

93) $2.70; $56.70   95) $14.00; $56.00

97) $170.00; $510.00   99) $6.00; $24.00

Section 8.3

1) 35.91   3) 122.36   5) 18.3   7) 7.98

9) 1886.4   11) 119.7   13) $41,800   15) 19 employees

17) $41,250   19) 5928   21) 364,500 households

23) 196,500   25) 14   27) 512   29) 116

31) 0.9   33) 22.1   35) 142.5   37) 135 problems

39) 385 computers   41) 6 coupons   43) 744 textbooks

45) 20   47) 120   49) 400   51) 45

53) 2000   55) 600   57) 60%   59) 45%

61) 27%   63) 3.4%   65) 9.5%   67) 120%

69) 0.7%   71) 15.3   73) 3000   75) 16%

77) $4571.91   79) 53,900 veterans

Section 8.4

1) $1200   3) 370,600   5) 38,971,800   7) 9,120,000

9) 12,825,000   11) 450   13) 645   15) 14

17) 420   19) 250   21) 120   23) 20%; 80%

25) 20%; 80%   27) 12.5%; 87.5%   29) 40%

31) 25%   33) 23%   35) 100%   37) 20%

39) 76%   41) 54%   43) 12%   45) 60%

47) 28%   49) 85%   51) 75%   53) 2272

55) 192   57) 25%   59) 12.5%

Putting It All Together Exercises

1) a) 0.96 b) 0.0002 c) 0.045

2) a) 1.72 b) 0.006 c) 0.238

3) a) $\frac{2}{25}$ b) $\frac{3}{400}$ c) $\frac{5}{2}$ or $2\frac{1}{2}$

4) a) $\frac{31}{100}$ b) $\frac{11}{60}$ c) $\frac{251}{500}$

5) a) 41.3% b) 68% c) $66.\overline{6}$% or $66\frac{2}{3}$% d) 325%

6) a) 200% b) 112.5% c) 0.3% d) 45%

7) a) 2 b) 1 c) 6 d) 7   8) a) 4 b) 2 c) 28 d) 30

9) 56   10) 54   11) 27   12) 28.5   13) 36

14) 150   15) 14.24   16) 212.8   17) 6502.5

18) 3.5   19) $9.00; $51.00   20) $2.40; $21.60

21) $3.60; $21.60   22) $6.00; $46.17   23) 150

24) 1520   25) 120%   26) 730   27) 1927

28) 16%   29) 8,085,000   30) 66.7%

31) 93,800,000   32) 52   33) 24.1%   34) 497

Section 8.5

1) $3.30; $58.30   3) $6.24; $84.24   5) $0.31; $4.31

7) $5.85; $80.85   9) $3.78; $57.78   11) 8%

13) 4.5%   15) 8.45%   17) 9.25%   19) 6.5%

21) $14,875   23) $280   25) $22,500   27) 12%

29) 16%   31) 12.5%   33) $37.80   35) $152.75

37) $67.50   39) $2250   41) $64.49   43) $892.70

45) 25%   47) 1125%   49) 84%   51) 5%

53) 3.8%   55) 75%   57) $475   59) $120.60

61) 6.5%   63) 24.4%

Section 8.6

1) $96   3) $112   5) $268.60   7) $840   9) $\frac{1}{6}$ yr

11) $\frac{2}{3}$ yr   13) $\frac{1}{2}$ yr   15) $\frac{5}{4}$ yr or 1.25 yr   17) $200

19) $86.25   21) $2100   23) $1560   25) $2162.50

27) $3240   29) $5006.25   31) $5271.25

33) a) $13,620; $378.33 b) $14,160; $295.00
c) $14,700; $245.00 d) $1080

35) a) $38,250; $637.50 b) $39,900; $554.17
    c) $41,550; $494.64 d) $3300.00

37) a) interest: $100; total: $1100
    b) interest: $102.50; total: $1102.50
    c) If interest is compounded annually, Romero will earn
    $2.50 more than if it is computed as simple interest.

39) a) interest: $450; total: $5450
    b) interest: $463.64; total: $5463.64
    c) If interest is compounded annually, Chau will earn
    $13.64 more than if it is computed as simple interest.

41) $9579.98   43) total: $18,226.23; interest: $6726.23

45) total: $81,241.95; interest: $59,241.95

47) total: $9380.89; interest: $2980.89

49) total: $15,925.04; interest: $7525.04

51) total: $16,929.92; interest: $4929.92

53) total: $16,264.98; interest: $2264.98

55) total: $29,910.30; interest: $8460.30

57) owe: $19,896.98; interest: $3396.98

Chapter 8: Group Activity

1) $290   2) $790   3) $734.70

4) $262.23   5) 35.7%

6) Student predictions will vary. Option A: $1743.75; Option B:
$1751.61; Option C: $1742.43

Chapter 8: Review Exercises

1) Approximately 37% of American households have at least
one dog.

3) Approximately 14% of jobs in San Diego are in the education
or health services fields.

5) $\frac{14}{100} = 0.14$ 7) $\frac{34}{100} = 0.34$ 9) 0.0035

11) 2.01   13) 87%   15) 0.047   17) $\frac{23}{100}$

19) $2\frac{1}{10}$ 21) $0.425 = \frac{17}{40}$ 23) $0.4375 = \frac{7}{16}$

25) $\frac{4}{15}$ 27) $\frac{55}{100} = 55\%$ 29) $\frac{98}{100} = 98\%$

31) 62.5%   33) 240%   35) a) 42 b) 63 c) 14

37) a) 126 b) 98   39) a) 19.5 b) 10.8

41) a) 49 b) 76   43) $630; $1170

45) a) 31.68 b) 33.5   47) a) 31.68 b) 4.32

49) $65,000   51) a) 200 b) 500 c) 2000

53) a) 350 b) 900 c) 2400   55) 60%; 40%

57) 60%   59) 60%   61) 40%   63) 200%

65) 50%   67) 200%   69) $6.88; $92.88

71) 7.6%   73) $750   75) $810   77) 250%

79) 50%   81) $192   83) $120

85) a) $\frac{13}{12}$ yr b) $\frac{3}{4}$ yr   87) $5.08   89) $956.70

91) a) $16,575; $460.42 b) $17,100; $356.25
    c) $17,625; $293.75 d) $1050

93) a) interest: $720; total: $6720
    b) interest: $749.18; total: $6749.18
    c) If interest is compounded annually, Ahmed will earn
    $29.18 more than if it is computed as simple interest.

95) total: $9795.61; interest: $3095.61

Chapter 8: Test

1) out of 100   2) 0.45; $\frac{9}{20}$ 3) 0.007; $\frac{7}{1000}$

4) 0.06; $\frac{3}{50}$ 5) 0.148; $\frac{37}{250}$ 6) 60%   7) 9.24%

8) 80%   9) 87.5%   10) 350%   11) 200%

12) a) The decimal point in the number 40 is after the zero. To
find 10% of 40, move the decimal point one place to the left.
10% of 40 = 4. b) 32 c) 2 d) 34

13) 45   14) $7.60; $45.60   15) $42   16) 131.04

17) 800   18) 9%   19) $6400   20) $2663.85

21) 64%   22) 16.9%   23) $1620   24) $5112.50

25) a) $15,737.21 b) $7737.21

Chapter 8: Cumulative Review for Chapters 1–8

1) 17 + 30   2) 19,500,000

3) a) 8 b) 42 c) soccer; 13 d) swimming e) 5 f) 1

4) 4^5 5) 266.5   6) $-\frac{4}{45}$ 7) $14.21

8) $\frac{47}{40}$ or $1\frac{7}{40}$ 9) 1300   10) 260   11) $a = 0$

12) $p = -25$ 13) $-6y^2 + 4y - 11$ 14) $\frac{2}{3}$

15) 0.5 L   16) 336 hr

17) height: 228.6 cm; weight: 139.5 kg   18) $\frac{141}{200}$

19) 180%   20) 54   21) 98   22) 81 people

23) $117.35   24) $8\frac{3}{4}$ c   25) 9548 people

Chapter 9
Section 9.1

1) 111,358   3) 2012   5) −1.4%   7) 1000

9) tacos; 1200   11) 250   13) Texas

15) California   17) 1100 mi   19) 2500 mi

21) New York University; $36,500

23) Arizona State University and Texas Tech University

25) $13,500

27)

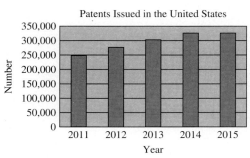

Patents Issued in the United States

29)

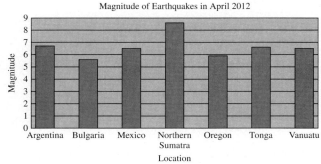

Magnitude of Earthquakes in April 2012

31) 3508 ft   33) 2948 ft   35) 560 ft

37) Sir Chunks-A-Lot   39) 44¢   41) 2013

43) From 2015 to 2016, the price decreased by 2¢.   45) 5¢ less

47) *The Tonight Show*   49) 800,000   51) −10.9%

Section 9.2

1) 5   3) 40.0–59.9 in.; 2   5) 160.0–179.9 in.

7) 22 seasons   9) 15   11) 15–19 yr

13) ratio $= \dfrac{4}{1}$. For every four employees who have been at the company 0–4 yr, there is one employee who has been at the company 40–44 yr.

15) 23   17) 28.75%   19) 10%

21) Two height ranges tie for having the least number of players: 70–72 in. and 76–78 in. There is one player in each of these categories.

23) a) 2 players   b) 6 ft 10 in. −7 ft   25) 6

27) 37.5%   29) 1

31) a)

Test Scores (Class Intervals)	Tally	Number of Scores (Class Frequency)
40–49	||	2
50–59	|||	3
60–69	||||	4
70–79	ⅢⅢ ||||	14
80–89	Ⅲ |	6
90–99	|||	3

b)

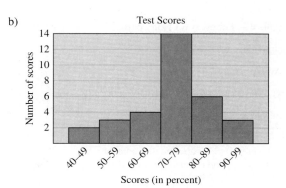

Test Scores

c) Answers may vary.

Section 9.3

1) 62.5%   3) 20%   5) 144   7) 90   9) 18

11) 900 ft²   13) 3600 ft²   15) Their sum must equal $\dfrac{1}{2}$

17) $\dfrac{2}{15}$ 19) 40%

21) The ratio is $\dfrac{1}{2}$. For every hour Angarra is in class, he spends 2 hr doing homework.

23) 360°

25) It is a tool that measures the number of degrees in an angle and that allows us to draw angles of certain degrees.

27) 162°   29) 108°   31) 360°

33) a) work-related: 144°, family: 36°, close friends: 18°, casual acquaintances: 72°, barely know them: 90°

b) Scott's Facebook Friends

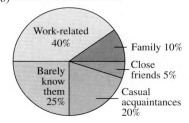

35) 16   37) 80

39) ratio $= \dfrac{1}{5}$. For every Facebook friend who is a close friend, he has five Facebook friends he barely knows.

41) a) rent: 108°, child care: 72°, food: 54°, utilities: 36°, other: 90°

b) Holly's Monthly Expenses

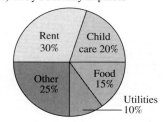

43) $500

45) ratio $= \dfrac{3}{2}$. For every \$3 Holly pays in rent, she pays \$2 for child care.

47) Answers may vary.

49) a) 45%; 25%; 20%; 10% b) 162°; 90°; 72°; 36°

c) Number of Students Enrolled in
Beginning Algebra

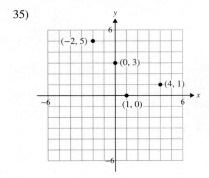

Section 9.4

1) true   3) false   5) a) yes b) yes c) no

7) a) no b) yes c) yes   9) a) no b) no c) yes

11) a) yes b) yes c) yes   13) 7   15) 18

17) $\dfrac{21}{5}$ 19) -2

21)
x	y
0	−6
9	3
−2	−8

23)
x	y
0	2
5	−13
−1	5

25)
x	y
0	3
−12	0
4	4

27)
x	y
−3	0
$-\dfrac{1}{2}$	1
−8	−2

29)
x	y
0	0
5	5
−1	−1

31)
x	y
1	−6
1	2
1	0

33) A: (−2, 1); quadrant II; B: (5, 0); no quadrant;
C: (−2, −1); quadrant III; D: (0, −1); no quadrant;
E: (2, −2); quadrant IV; F: (3, 4); quadrant I

35)

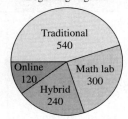

37)

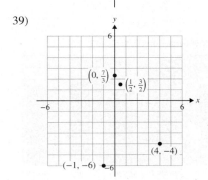

39)

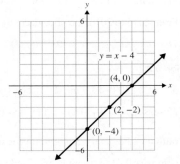

41) positive   43) negative   45) zero

Section 9.5

1) line

3)
x	y
0	−4
2	−2
4	0

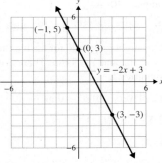

5)
x	y
0	3
3	−3
−1	5

7)
x	y
0	−1
−2	0
4	−3

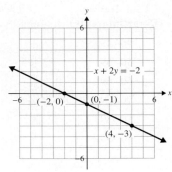

9)

x	y
0	−5
6	3
2	$-\frac{7}{3}$

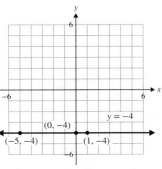

11)

x	y
0	−4
−5	−4
1	−4

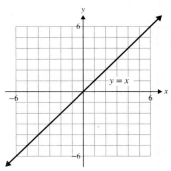

13) You should use at least three. Any two points will give you a line, but the third point can act as a "check." If all three points do not lie on the same line, then you know you made a mistake somewhere.

15)

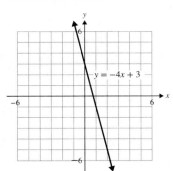

17)

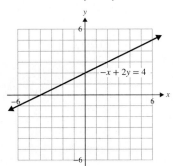

19)

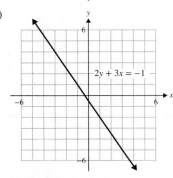

21)

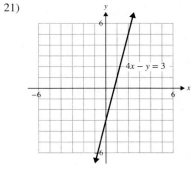

23)

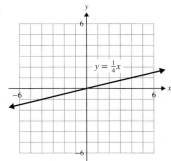

25)

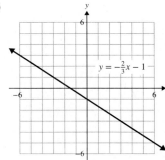

27)

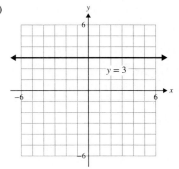

29) vertical

31)

33)
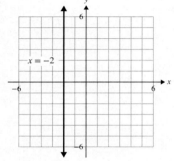
$x = -2$

35)
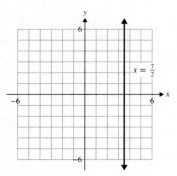
$x = \frac{7}{2}$

37)
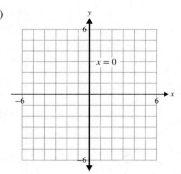
$y = -4.5$

39)
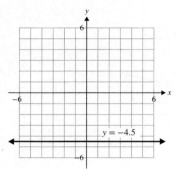
$x = 0$

41)
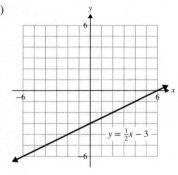
$y = \frac{1}{2}x - 3$

43)
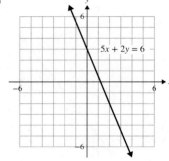
$5x + 2y = 6$

45)
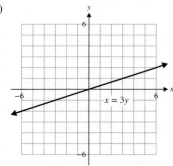
$x = 3y$

47)
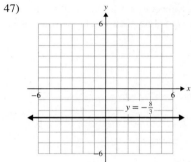
$y = -\frac{8}{3}$

49) negative   51) positive

53) positive   55) negative

Chapter 9: Group Activity

1) U-Haul-It

x	y
0	60
10	61
20	62
30	63
40	64
50	65
60	66

Mighty Movers

x	y
0	35
10	41
20	47
30	53
40	59
50	65
60	71

2) For U-Haul-It, the cost increases $1 for every 10 miles driven, or 10 cents per mile. For Mighty Movers, the cost increases $6 for every 10 miles driven, or 60 cents per mile.

3)

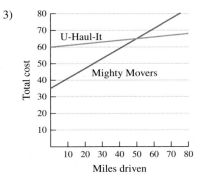

4) It will cost $65 to drive 50 miles at U-Haul-It and it will cost $65 to drive 50 miles at Mighty Movers. The cost is the same at both companies for 50 miles driven.

5) Answers may vary. Sample response: If I am going to drive less than 50 miles,
I would choose Mighty Movers because the cost for Mighty Movers is cheaper than for U-Haul-It before you reach 50 miles. If I am going to drive more than 50 miles, I would choose U-Haul-It.

6) The slope is 1/10, or 0.10. I notice that this is the coefficient of x in the equation and that this is the cost per mile.

7) The slope is 6/10, or 0.60. I notice that this is the coefficient of x in the equation and that this is the cost per mile.

8) Mighty Movers has a steeper graph because they have a higher cost per mile.

9) 60; 35; In addition to the cost per mile, each rental company charges a flat fee, or an up-front fee, for renting the truck. The flat fee at U-Haul-It is $60 and the flat fee at Mighty Movers is $35.

10) $y = 80$; The graph will be a horizontal line, crossing the y-axis at 80.

Chapter 9: Review Exercises

1) Dale Earnhardt; 172.41 mph

3) It took Dale Earnhardt 0.44 hr less than Marvin Panch.

5) 50   **7)** 60   **9)** 20%

11) Nicki Minaj and Pitbull; Nicki Minaj earned approximately $20.5 million, and Pitbull earned about $20 million.

13) He earned approximately $6.5 million more in 2016 than in 2015. This is a percent increase of approximately 54%.

15) Drake earned the most and Eminem earned the least. The difference in their earnings was $27.5 million.

17) August; 9000   **19)** August and September

21) 41   **23)** 9–16 mi

25) a)

Age at Inauguration (Class Intervals)	Tally	Number of Presidents (Class Frequency)
42–45	||	2
46–49	Ж||	7
50–53	Ж||	7
54–57	Ж Ж Ж |	16
58–61	Ж	5
62–65	||||	4
66–69	||	2
70+	|	1

b)

Age of Presidents at Inauguration

c) Answers may vary.

27) 10 min   **29)** The sum must equal 1.   **31)** yes

33) yes   **35)** 10   **37)** 9

39)

x	y
0	−2
7	0
−7	−4
21	4

41)

43) positive

45)

x	y
0	-4
1	-1
2	2

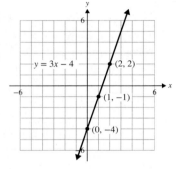

47)

x	y
0	$\frac{5}{2}$
5	0
-3	4

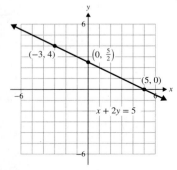

49)

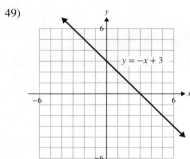

51)

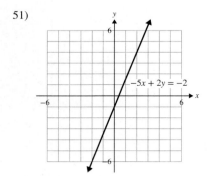

53)

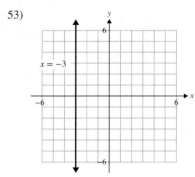

Chapter 9: Test

1) total = 571; average = 142.75 games per player

2) 56.7%   3) 72   4) 394   5) 30

6) 45   7) 30   8) 10%

9)

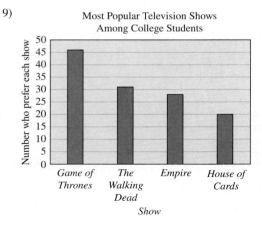

10) 3.4 million

11) 2013; Yellowstone National Park had approximately 1.3 million more visitors than the Statue of Liberty.

12) In 2015, the Statue of Liberty had 4.3 million visitors, and Yellowstone had the same number of visitors in 2016.

13) 2014 – 2016

14) a)

Sodium Levels (mg) (Class Intervals)	Tally	Number of Cereals (Class Frequency)
0–49	IIII	4
50–99	IIII IIII I	11
100–149	IIII	4
150–199	III	3
200–249	IIII I	6
250–299	II	2

b)

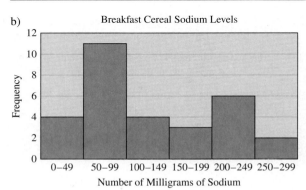

c) Answers may vary.

15) $\frac{4}{5}$ of the students   16) 144 students

17) 164 students   18) 1

19) a) 15%; 5%; 25%; 35%; 20%

b) 54°; 18°; 90°; 126°; 72°

c)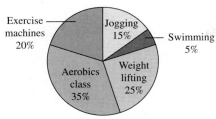

Preferred Exercise Activity

20) no   21) yes   22) −11

23)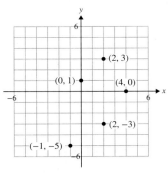

24)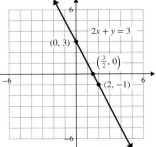

x	y
0	3
$\frac{3}{2}$	0
2	−1

25)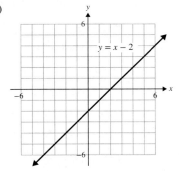

26)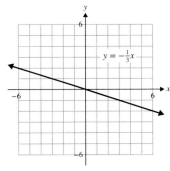

27)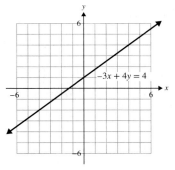

28)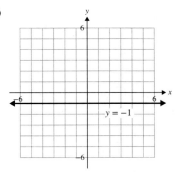

29)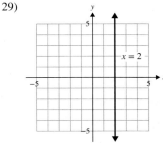

30) positive   31) zero   32) negative

Chapter 9: Cumulative Review for Chapters 1–9

1) $-\frac{2}{9}$ 2) 13.26   3) 197   4) 0   5) $\frac{11}{18a}$

6) $7x^2 - 8x + 8$

7) The base of $(-7)^2$ is −7, so $(-7)^2$ means $-7 \cdot (-7) = 49$.
The base of -7^2 is 7, and -7^2 means $-1 \cdot 7 \cdot 7 = -49$.

8) $p = 8$ 9) $c = -\frac{7}{2}$ 10) $n = 60$ 11) 22 in²

12) exact area: 25π cm²; approximate area: 78.5 cm²
exact circumference: 10π cm;
approximate circumference: 31.4 cm

13) 65 ft by 130 ft   14) $1710   15) 300 sec

16) 2.4 km   17) 65%   18) $\frac{4}{125}$ 19) 36

20) 60   21) 750 miles per month   22) 5%

23) a) Caribbean; 35 b) 20 c) Asia and Australia/South
Pacific were each preferred by six people.   d) 6

24) a) $\frac{13}{24}$ b) 360   c) $\frac{3}{4}$ d) 1800

25)

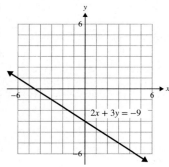

Appendix A
Section A.1

1) addends: 8, 7; sum: 15

3) 9;

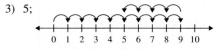

5) 10;

7) 9;

9) Answers may vary.   11) 15; 7 + 8

13) 17; 16 + 1   15) 14; Answers may vary.

17) 16; Answers may vary.   19) Answers may vary.

21) 15; (3 + 7) + 5   23) 14; (4 + 7) + 3

25) 13; 3 + (5 + 5)   27) 2 in.   29) 2 ft   31) 18

33) 79   35) 557   37) 7,786   39) 2,968

41) 5,589   43) 65,869   45) 86   47) 120

49) 861   51) 8,416   53) 10,400   55) 132

57) 29,808   59) 849   61) 45,781   63) 39,254

65) 71,679   67) 130,000   69) 128 in.   71) $1,648

73) 210 min   75) 4,959 students   77) 1,815 mi

79) 2,792 mi   81) 2,400 mi   83) 17 cm

85) 64 in.   87) 120 ft   89) 1,012 ft

91) 2 ft; 3 ft; 28 ft

Section A.2

1) minuend: 5; subtrahend: 4; difference: 1

3) 5;

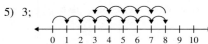

5) 3;

7) 2; 2 + 7 = 9   9) 4; 4 + 1 = 5   11) 6; 6 − 5 = 1

13) 9; 9 − 2 = 7   15) 42; 42 + 52 = 94

17) 227; 227 + 301 = 528   19) 101; 101 + 55 = 156

21) 120; 120 + 410 = 530   23) 4,245; 4,245 + 614 = 4,859

25) 7   27) 430   29) 2 in.   31) 8 m   33) 14

35) 14   37) 58   39) 519   41) 86   43) 2,149

45) 3,378   47) 564   49) 63   51) 713

53) 3,555   55) 1,530   57) 2,023   59) 636

61) 4,436   63) 28,111   65) 732   67) 3,201

69) 7,047   71) 221,042   73) 18   75) 350

77) $7,880   79) 4,995   81) 2008; 649,000

83) 270 mi   85) 7 in.

Section A.3

1)

a) $4 \times 3 = 12$ b) multiplicand: 4; multiplier: 3

c) factors: 4 and 3; product: 12

3)

a) $3 \times 5 = 15$ b) multiplicand: 3; multiplier: 5

c) factors: 3 and 5; product: 15

5) a) 27   b) 28   c) 55   7) a) 60   b) 18   c) 48

9) a) 36   b) 7   c) 96   11) Answers may vary.

13) 15; $5 \times 3 = 15$ 15) 108; $9 \cdot 12 = 108$

17) 32; 4(8) = 32   19) 0; $2 \cdot 0 = 0$ 21) 10; $1 \times 10 = 10$

23) 12   25) 8   27) Answers may vary.

29) $(3 \times 2) \times 4 = 6 \times 4 = 24; 3 \times (2 \times 4) = 3 \times 8 = 24$

31) $(9 \cdot 0) \cdot 7 = 0 \cdot 7 = 0; 9 \cdot (0 \cdot 7) = 9 \cdot 0 = 0$ 33) 62

35) 1,239   37) 1,608   39) 658   41) 5,715

43) 1,480   45) 28,854   47) 168   49) 7,176

51) 950   53) 39,232   55) 85,104   57) 467,819

59) 1,750,464   61) 208,351   63) 12,675,132

65) The product is the number followed by two zeros.
Examples may vary.

67) a) 940   b) 5,500   c) 67,000

69) a) 8,170,000   b) 26,100   c) 15,000

71) a) 820   b) 450   c) 7,680

73) a) 45,000   b) 560,000   c) 2,940,000

75) 273,000   77) 45   79) 720   81) $42

83) $1,548   85) 56

Section A.4

1) 3;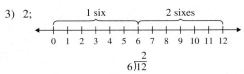

$6 \div 2 = 3$

   a) dividend: 6; divisor: 2; quotient: 3

   b) $\dfrac{6}{2} = 3$, $2\overline{)6}^{\,3}$, or $6/2 = 3$

3) 2;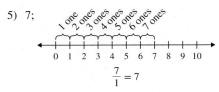

$6\overline{)12}^{\,2}$

   a) dividend: 12; divisor: 6; quotient: 2

   b) $\dfrac{12}{6} = 2$, $12 \div 6 = 2$, or $12/6 = 2$

5) 7;

a) dividend: 7; divisor: 1; quotient: 7

   $\dfrac{7}{1} = 7$

   b) $7 \div 1 = 7$, $1\overline{)7}^{\,7}$, or $7/1 = 7$

7) 4; $9 \cdot 4 = 36$ 9) 12; $7 \cdot 12 = 84$ 11) 3; $1 \cdot 3 = 3$

13) Yes. 0 divided by a nonzero number equals 0.

15) undefined   17) 0   19) undefined   21) 0

23) 5   25) 63   27) 5   29) The remainder equals 0.

31) a) 3   b) 3 R2   33) a) 9   b) 9 R3

35) Multiply the divisor by the quotient, then add the remainder. The result should be the dividend.

37) $(9 \times 10) + 5 = 95$; true   39) $(6 \times 9) + 2 = 56$; false

41) False. The remainder is larger than the divisor.

43) 7 R1; $(4 \times 7) + 1 = 29$ 45) 8 R6; $(7 \times 8) + 6 = 62$

47) 5; $3 \times 5 = 15$ 49) 9 R3; $(5 \times 9) + 3 = 48$

51) a number that is divisible by 0, 2, 4, 6, or 8

53) The number ends in 0, 2, 4, 6, or 8.

55) Add the digits in the number. If that sum is divisible by 3, then the number is divisible by 3.

57) 2   59) 2, 3, 5, 10   61) 3   63) 5   65) 2, 3

67) none of these   69) yes   71) no   73) yes

75) No, we cannot use short division because the divisor has two digits.

77) 23   79) 123   81) 319 R2   83) 1,254 R2

85) 312   87) 563 R2   89) 20,480   91) $9

93) 39   95) 34   97) 224   99) 60

Section A.5

1) 64   3) 581   5) 267 R3   7) 328 R8

9) 78   11) 95   13) 72 R26   15) 1,386 R11

17) 1,980 R3   19) 860 R58   21) 803

23) 601 R17   25) 2,504 R7

27) a) 674 R4   b) 674 R4   c) Answers may vary.

29) Drop one zero from the whole number.

31) a) 3   b) 48   c) 312

33) a) 926,000   b) 21,100   c) 7,800

35) 17   37) 61   39) 730   41) 250   43) $42

45) $150   47) 2 hr per day   49) 16 lb   51) 50 mi

53) 607 R24   55) 32   57) 860   59) $21

Appendix A: Review Exercises

1) 3 and 5   3) none of these   5) 2, 3, 5, and 10

7) 28,084   9) 31,192   11) 414   13) 53

15) 780   17) 15,200   19) 1,030,190   21) 5,236

23) 3,009   25) 3,264   27) 37,713   29) 173 R2

Index

Endpoints, 285, 286

English expressions, written as algebraic expressions, 123–126, 129–131

Equality
addition and subtraction properties of, 109–111, 252–253, 433–434
division property of, 111–113, 253–254, 434–435
multiplication property of, 253–254
solving equations using properties of, 433–435

Equations. *See also* Linear equations; Linear equations in two variables
combining like terms to solve, 116–119
containing decimals, 433–438
containing fractions, 252–259
containing variables on both sides of equal sign, 121–123
determining if number is solution of, 108–109
eliminating fractions to solve, 257–259
equivalent, 109
explanation of, 108, 259
expressions vs., 259–260
finding numbers with, 651–652
solutions to, 108–109
to solve percent problems, 651–655, 658–661

Equilateral triangles, 314, 330

Equivalent equations, 109

Equivalent fractions
determining if two fractions are, 176
explanation of, 170–171
method to write, 198–200, 202
writing fraction as decimal using, 416–417

Estimation, rounding for, 37–39

Even numbers
divisibility rules and, 166
explanation of, A–1
of negative factors, 46, 62

Exponential expressions, simplifying, 69

Exponents
explanation of, 64–65
integers used with, 66
used with fractions, 247
with variables, 93–95

Expressions. *See also* Algebraic expressions
equations vs., 259–260
evaluation of, 91–93, 144
explanation of, 90–91, 259
identifying parts of, 95–96
method to simplify, 102–106, 144–145

F

Factorization, prime, 167–170, 174–175, 201–203, 240

Factors
common, 171–173
dividing out common, 182–185
explanation of, 44, 165, A–26
greatest common, 172–173
prime, 182

Factor tree, 168, 169

Fahrenheit scale, 607–609

Fraction bar, 54, 157

Fractions. *See also* Mixed numbers
addition of like, 194–204
addition of unlike, 207–215
in applied problems, 188–189, 214–215
comparing decimals and, 417–418
complex, 188
division of, 186–190
equivalent, 170–171, 176, 198–200, 202, 416–417
explanation of, 156–159, 239
exponents used with, 247
improper, 159, 160, 222–224, 232, 239
with large denominators, 212
like, 195–204
linear equations containing, 257–259
in lowest terms, 171–175, 180, 181, 240, 368–369, 505, 630–632
method to compare, 245–247
multiplication of, 179–186, 188–189
negative, 160–161
on number line, 157, 158, 161, 198, 245, 246
order of operations with, 248–249
percent written as, 629–633
proper, 159, 239
rates written as, 498
ratios that compare, 489–491

reciprocals of, 185–187, 254–256
review of, 239–242, 265–276
rewritten with least common denominator, 202–204, 208–213, 241–242
simplifying, 171
solving equations containing, 252–257
subtraction of, 194–204
unit, 570–573, 583–584
unlike, 195, 207–215
writing decimals as, 367–369
written as decimals, 414–417
written as percent, 634–636

Frequency, 423, 728

Frequency distribution tables, 728–729

Front-end rounding
in applications, 57–58
to estimate product, 49–50
to estimate quotient, 56–57
explanation of, 40

G

Geometry. *See also specific geometric shapes*
angle measures in triangles, 312
in applied problems, 333–337
explanation of, 285
lines, line segments, and rays, 285–290
perimeter and area of irregular figures, 302–303
perimeter and area of parallelograms, 299–300
perimeter and area of rectangles, 294–297
perimeter and area of squares, 297–298
perimeter and area of trapezoids, 300–302
perimeter and area of triangles, 308–311
review of, 326–330, 343–349
surface area of rectangular solids, 322–323
triangle classification, 313–314
volume of rectangular pyramids, 321–322
volume of rectangular solids, 319–321

Gram, 593

Graphing calculators. *See* Calculators

U.S. customary units
  in applied problems, 574–575
  capacity in, 568
  explanation of, 568
  length in, 568
  time in, 568
  use of multiplication or division to
      convert between, 569
  use of unit analysis to convert
      between, 570–574
  weight in, 568

V

Variables
  on both sides of equal sign in
      equations, 121–123
  explanation of, 91, 144
  exponents with, 93–95
  isolating the, 110, 111
Vertex, of angles, 286
Vertical angles, 523–524, 526–527
Vertical axis, 741
Vertical lines, 752, 753

Volume
  explanation of, 319, 327
  of rectangular pyramids,
      321–322
  of rectangular solids, 319–320

W

Weight, 425
  conversion between U.S. customary
      units and metric units of,
      605, 606
  in metric system, 593–596
Weighted mean, 423–425
Whole numbers
  addition of, A–2 – A–10
  in applied problems, A–8 – A–10
  dividing decimals by, 400–403
  division of, A–36 – A–45,
      A–48 – A–52
  explanation of, 6, A–1
  with exponents, 64–65
  long division of, A–47 – A–50
  multiplication of, A–25 – A–33

number line to add, A–2
powers of, 65
subtraction of, A–14 – A–21
Words, writing numbers in, 8–9

X

x-axis, 744

Y

y-axis, 744

Z

Zero
  in decimals, 382, 383, 386,
      393, 402
  division involving, A–38 – A–39,
      A–50 – A–52
  multiplication by number ending
      in, A–32 – A–33
  multiplication property of,
      46–47
  subtracting numbers involving,
      A–19 – A–20